Main trends of research in the social and human sciences

Part two/Volume two

New Babylon

Studies in the social sciences

21/2

Mouton Publishers/Unesco
The Hague · Paris · New York
mcmlxxviii

Main trends of research in the social and human sciences

Part two/Volume two:
Legal science/Philosophy

Under the editorship of
Jàcques Havet

Mouton Publishers/Unesco
The Hague · Paris · New York
mcmlxxviii

ISBN: 90-279-7693-7 (Mouton, The Hague)
2-7193-0801-3 (Mouton, Paris)
92-3-101013-1 (Unesco, Paris)

Printed in Great Britain

General Table of Contents

Table of contents: Volume II

3. Legal science

CHAPTER VI

Legal science

VIKTOR KNAPP

Professor at Charles University, Prague
Member of the Czechoslovak Academy of Sciences

CONTENTS

FOREWORD

The origin of this study on the trends of research in legal science in its broad concept dates from a first discussion held by a committee of experts meeting at Unesco in Paris from 7 to 13 July 1967. On the strength of this discussion the international rapporteur appointed for the purpose drew up a draft questionnaire. This draft, reviewed in the light of suggestions received from the associate international rapporteurs,[1] was circulated to a number of 'national correspondents' who were invited to submit, under each heading of the questionnaire, a synopsis of the main trends of research in the field of law, primarily within their own countries but also where appropriate in neighbouring countries, and to provide a selective list of bibliographical examples. Replies were received from 17 'national correspondents'.[2]

In addition to these 17 replies seven 'special contributions' were drawn up[3] at Unesco's request. Five National Commissions for Unesco, those of

1. Dr. Allah Buksh K. Brohi, former Minister, barrister (Karachi, Pakistan);
Professor Hilding Eek (University of Stockholm, Sweden);
Professor F. H. Lawson (University of Lancaster, United Kingdom);
Professor Max Rheinstein (University of Chicago, United States of America);
Professor André Tunc (University of Paris, France).

2. Dr. Allah Buksh K. Brohi (Pakistan);
Professor René David (France);
Professor Hilding Eek (Sweden), assisted by Professors T. Opsahl (Norway) and H. Jokela (Finland), for the Nordic countries as a whole (with the co-operation of the Unesco National Commissions concerned);
Professor Hector Fix Zamudio (Mexico);
Dr. Ahmad bin Mohd. Ibrahim (Singapore), for Malaysia and Singapore;
Institute of State and Law of the USSR Academy of Sciences;
Inter-academic Committee of the Council of Academies of the Socialist Federal Republic of Yugoslavia (Chairman, Professor Lado Vavpetić);
Mr. Justice Adeyinka Morgan (Nigeria);
National Research Council of Italy (Rapporteur, Professor Massimo Severo Giannini);
Professor Zoltan Péteri (Hungary);
Mr. Edilbert Razafindralambo (Madagascar);
Professor Max Rheinstein (U.S.A.);
Professor G. S. Sharma (India);
Professor Kenneth R. Simmonds (United Kingdom);
Professor Seydou Madani Sy (Senegal);
Professor Ukai Nobushige (Japan);
Professor Reinhold Zippelius (Federal Republic of Germany).

3. Professor Norberto Bobbio (Italy): 'L'objet de la science juridique';
Professor René David (France): 'Géographie juridique';
Professor Hilding Eek (Sweden): 'Coercion versus voluntary submission to law';
Professor Victor Knapp and Dr. Vladimir Vrecion (Czechoslovakia): L'applicabilité des méthodes de la cybernétique au droit et à l'administration';
Mr. Y. Linant de Bellefonds (France): 'Les tendances actuelles du droit comparé en pays d'Islam';
Professor Radomir Lukić (Yugoslavia): 'Rapport entre l'histoire du droit et l'histoire des théories juridiques et politiques';
Professor Jean Poirier (France): 'Situation et programme de travail de l'ethnologie

Belgium, India, Japan, the Arab Republic of Egypt and the Ukrainian Soviet Socialist Republic, kindly provided the international rapporteur with written contributions prepared by the Belgian Royal Academy of Science, Letters and Fine Arts, by Professors B. S. MURTY and T. S. RAMA RAO, by Professor AOMI Junichi, by Professors Sarawat ANIS, Soufi Abou TALEB, Salah Eldin ABDEL-WAHAB, Fouad EL-ATTAR and Abdel Moneim EL-BADRAWI and by the Ukrainian Academy of Sciences, respectively. The USSR National Commission forwarded the text of an essay by Professor Zinovi ČERNILOVSKIJ (CHERNILOVSKY) on the theme 'Contemporary views in the USSR on the origin and rôle of the State'. Lastly, a number of spontaneous contributions and miscellaneous comments were also of great help to the Rapporteur in adding to his records and to the range of opinions expressed; special mention may here be made of the contributions of Professor F. H. LAWSON, already named among the associate rapporteurs, and at that time Secretary-General of the International Association of Legal Science, who was good enough to draft detailed replies to the questionnaire, particularly for the Common Law countries, of Judge J. MALHERBE (Lyons, France) and of Father François RUSSO (France).[4]

The present article, as drawn up by the international rapporteur, is based on the replies to the questionnaire, on the 'special contributions', on the contributions of the National Commissions and on the further sources mentioned. It is, however, not confined to a compilation of these texts: the author has been selective, he has recast the structure originally proposed and has in addition mentioned certain developments, on the strength of his own knowledge of current trends, which were not covered in the documentation received.

If it has seemed necessary to recast the structure of this study as originally conceived, the reason lies chiefly in the fact that the replies to the questionnaire showed that the range of questions first proposed was too broad, and that for many of them there were no trends of research in contemporary legal science of sufficient significance to deserve mention. Accordingly a process of selection was applied to the headings with a view to orientating the study towards the more notable trends.

The manuscript was submitted in draft to the asociate rapporteurs for scrutiny; the author has, in addition, invited comments from his friends Professor V. M. ČHIKVADZE (CHKHIKVADZE), corresponding member of the Academy of Sciences of the USSR, and Academician Imre SZABÓ, Director of the Institute of State and Law of the Hungarian Academy of

juridique'.

4. The replies of Professors FIX ZAMUDIO, GIANNINI, PÉTERI and RHEINSTEIN, the contributions of Professors AOMI, KNAPP and VRECION, POIRIER and ČERNILOVSKIJ (CHERNILOVSKY), and a compilation of the contributions from Egyptian authors made by Professor ABDEL-HAMID and Mr. TAGELDIN, have been published in the *International Social Science Journal* (Unesco), XXII (3), 1970 (special issue: *Trends in Legal Learning*), on the basis of a selection made by the editorial staff of this journal and one for which it alone assumes responsibility.

Sciences. The author has, in completing the final draft of the manuscript, taken into account all the comments and suggestions obtained in this way.

These principles governing the preparation of this survey give it a specific character, the more so in that, although the work of its author, it is the fruit of close co-operation with the associate rapporteurs. They have given their views on all the successive drafts and have recommended a number of changes and additions. This is particularly important since the author – as the reader will soon realize – is a Marxist and speaks as one, whereas the associate reporters do not share this theoretical concept. Hence it was often necessary for the author to strive, in the spirit of the recommendations formulated for the entire undertaking at the two meetings convened by Unesco in July 1967 (see above) and in June 1969, to find forms of expression which, without, of course, relinquishing his own philosophical convictions, might still be acceptable to his five colleagues. In accordance with the recommendations of the above-mentioned meetings, the author has been careful to refrain from adopting any controversial attitude with regard to the opinions of his associate rapporteurs (whom, moreover, he holds in particular respect and esteem). Without dissimulating the differences that divide his conceptions from theirs, he has above all sought to offer the reader, in line with Unesco's intention, a faithful description of the main trends which are in fact discernible at present in legal science – having due regard to the fact that since law is determined by the social and economic system of the country in which it is in force, a survey even of the main trends in research in legal science is influenced in turn (as the replies to the questionnaire show) by the division of the contemporary world into major socio-economic systems, separated from one another by deep-rooted differences. It follows that the aim of this study is to identify, describe and classify current developments in legal science throughout the world, rather than itself to solve scientific problems. It is for the same reason that the author, in drawing up a survey of trends in legal research, has refrained as a general rule from expressing an opinion as to their value and that, apart from what has just been said by way of avowing his ideological point of departure, he has not expressed – apart from a few exceptions which the reader will easily recognize – his own scientific views on the problems connected with the trends described.

Since this survey, although, we repeat, the work of its author, is also based upon the comments of the associate rapporteurs and the replies of the national correspondents, etc., the references given naturally take on a somewhat unusual character: the author was unable to follow the normal practice of referring only to published works, and has had very often to cite unpublished documents: comments from the associate rapporteurs, replies from national rapporteurs, and so on. In view of the fact that the replies and contributions published in the *International Social Science Journal* [5] were abridged and recast, or even rewritten for the purpose of publication in this

5. See above, p. 926, n. 4.

journal, the quotations given here are, as a general rule, taken from the manuscripts deposited in Unesco's archives; however, in cases where the passage referred to appears in the published version, we have also given the exact reference for the convenience of readers who wish to consult it.[6]

With respect to the bibliographical references, the bibliography covering the subject-matter of this study is immense. It was therefore impossible to quote even a representative selection. Such a selection would have been all the more difficult to make inasmuch as the author cannot claim that he is not overlooking certain works, even perhaps very noteworthy ones, which may have been published in languages not accessible to him and in countries in regard to which no bibliographical information was furnished in the national reports. He regrets this especially, being himself from a country where it was customary to quote, not without resentment, the dictum: *bohemica non leguntur*, and craves indulgence if he has omitted to cite a particular work which the reader would have liked to see among the bibliographical references.

To ensure maximum objectivity, the author has culled these references first of all from the lists of works submitted to him by the national rapporteurs. Since these lists, however, were drawn up in very heterogeneous fashion, it was necessary to make a selection from them, generally of a very restrictive kind. In certain instances, particularly when no bibliographical references – or only very brief details – were given in the national reports, and also in the case of works on very specialized subjects (such as space law, the legal implications of organ transplants, the applicability of exact methods in legal research, etc.), the author has drawn on his own knowledge of existing documentation or, when necessary, has consulted specialists. Furthermore, since the present chapter represents more than five years' work during which time the production of books on legal science has by no means stood still, the author has been obliged to follow the development of literature in this field to the best of his ability and to add to the bibliographical annex a certain number of works of his own choice. Lastly, the author has included in the bibliographical annex and the footnotes certain works which the associate rapporteurs wished to have appear there.

It must be recalled once again that the object of the present study is not to describe the present world status of legal science or to draw up a balance-sheet of the problems that this science faces; in accordance with the intended nature of the work of which this study forms part, the object here is solely to survey the main *trends* at present visible in legal science.

Nor does this study claim or aim at an encyclopaedic character. Research trends are revealed more by questions than by answers. Hence it is proposed in the first place to draw the reader's attention to those questions and problems likely to throw light on current trends in research in the field of legal

6. In the footnotes that follow, only the names of authors whose *published* works or studies are referred to (these also appear, with detailed references, in the bibliographical annex to the chapter) are printed in small capitals.

science; but an attempt has not been made in every case to add an encyclopaedic view of the solutions applied.

In closing this foreword the author wishes to express his warmest thanks and sincere gratitude to the general rapporteur, Mr. Jacques Havet, whose help has been invaluable, and to all those who have helped him in his work through their advice, suggestions and criticism, their replies, data and judgements – to the associate international rapporteurs, the national correspondents, the authors of special contributions and the various people who have come forward spontaneously, as well as to those Unesco National Commissions which were good enough to contribute to the inquiry, and also to Professor Čhikvadze and Academician Szabó for their critical scrutiny of the first draft of the manuscript. In view of the very wide range of problems forming the subject-matter of the present work, the author has also on numerous occasions sought help and advice from his colleagues in Prague, particularly Professors Vladimír KOPAL (for Parts VII,1 and VIII,1) and Pavel KALENSKÝ (for Part VII,2); to them also he expresses his deep appreciation.

SECTION ONE: GENERAL CONSIDERATIONS

I. THE NOTION OF TRENDS OF RESEARCH IN LEGAL SCIENCE

1. The very notion of a *trend* has different meanings, while the trends visible at any given time in a specific science may be and are of highly varied types. Thus we can detect in the first place, at any stated period and in respect of a particular discipline, tendencies towards stagnation, immobility, a certain preference for pursuing research within the limits of familiar horizons and using well-tested methods. Elsewhere, on the contrary, we may discover dynamic tendencies, efforts to alter prevailing opinions, to broaden the field of vision of a scientific discipline, to launch novel undertakings and approaches, and to perfect methods as yet untried.

We propose here to study precisely those dynamic trends as they reveal themselves in the evolving legal science of today. We will seek to show in what direction research is moving and especially to identify any new elements emerging in legal science even if those elements are not yet dominant or even fully established, even if they are only sketchy and of recent birth – in so far, at least, as it is possible to detect the first signs of an emerging innovation in research.

Nevertheless, in attempting to discover the new trends of research in the field of legal science, it is important to keep in mind certain specific conditions of the development of our science. This is so, first of all, because legal science, in general, operates by means of notions which reflect a high degree of abstraction (even, in our view, when its subject is case law), and which, while their outlines may frequently alter, nevertheless retain the abstract form which has been theirs throughout the ages and has to some extent survived social transformations.[7]

We must also make allowances for the fact that the conditions in which legal science develops are quite different from those involved in the development of the natural sciences and we cannot expect, in our discipline, discoveries of the kind made in the natural sciences. On the other hand, as we shall see later, the striking advances in the exact and natural sciences, which affect the development of social relations, are beginning to have a marked influence on legal science and promote new trends within it.

We should also note a further feature making for conservatism in legal science, viz. the marked weight of tradition in the 'legal way of thinking'. This 'professional bias' of lawyers is of considerable importance: what we call the 'legal way of thinking' – whether based on the traditions of Roman law or on those of Common Law – is a particularly efficacious weapon in legal practice (something like the delicacy of touch required of a surgeon); but for the research worker in law, traditional 'legal thought' acts as a retarding brake inhibiting him from venturing on a quest for new horizons beyond the classic frontiers of 'legal thought' (limits within which, we venture to say, it is unlikely that any important new horizons may be discovered).

However, all the specific characteristics of the development of legal science have in no way kept it from undergoing considerable development since the end of the nineteenth century and this is what forms the historical 'background' of the present trends which we are investigating. It cannot be denied that, as we shall see, the legal science which we know today scarcely resembles that of a hundred years ago and that it is primarily at the philosophical and methodological levels that our discipline has undergone the most significant and far-reaching changes since the end of the nineteenth century. Yet even as regards research in the realm of positive law – where, for the reasons given above, it is relatively conservative – legal science does not live only on the inheritance of the past. We have to recognize certain new trends – often daring, even truly revolutionary ones – that distinguish the present era. These trends, however, are not uniformly noticeable over the whole field (which moreover, as we shall see, is defined in a somewhat vague manner), but only manifest themselves in the solution of certain problems or within the limits of specific branches or sectors of legal science. For example, the powerful thrust of research in the field of comparative law, which we shall examine in greater detail below, constitutes one of the most marked trends in contemporary legal science; in continental Europe, however, this trend is noticeably stronger in the sector of private law than in that of public law,[8] whereas in the Common Law countries, comparative law has

7. ENGELS, *Ludwig Feuerbach*, Chap. IV.

8. This in no way implies that comparative public law is neglected in continental Europe; this branch of study also exists there although it has not the same vitality as comparative studies in private law. For comparative public law see, for example, the collection *Vergleichung im öffentlichen Recht* (1964); ROZMARYN, 'Quelques questions de la théorie des constitutions socialistes' (1965); JOVIČIĆ, *Savremeni federalizam. Uporednopravna studija* (1973); etc. We may also mention

focused its attention to a greater extent on administrative law.[9]

We find another illustration of this fragmentary character of certain new orientations in countries in which up to the present a positivist conception of the knowledge of law has prevailed and in which, nevertheless, a sociological bias is very often in evidence in research concerning problems relating to labour law, family law and criminal law. These problems are so closely linked with social ones that it is impossible to study the former without paying attention to the latter.

The limits set to this study make it impossible for us to find room for a multitude of variations of detail. Hence we do not intend to enter into every change of opinion, or even of basic conception, that may have occurred in the various branches of legal study, but will confine ourselves to trends which, even if manifest only in the solution of specific problems or in a single sector of legal science, none the less have an importance that transcends the problem or sector in question and therefore merit inclusion among 'current trends in legal research' in the fullest sense of that expression.

We regard this preliminary warning as necessary in order that the reader may know forthwith not only that this study does not provide a strictly descriptive survey of the current state of legal science, but that it does not even mention certain extensive and undoubtedly respectable branches of that science, leaving them aside where none of the important new trends in research is discernible in them. On the contrary, we deal in detail with several questions which have not yet materially influenced world legal science, but within which certain new trends and perspectives are nevertheless sharply visible.

2. The question remains, what is the origin of these new trends? Prevailing opinion has for a long time accepted – ever since legal science became a *social science* [10] – that trends in legal science are in no way fortuitous or simply determined by the progressive enlargement of the human spirit; they stem from social evolution on which they themselves exert a reflex action.

It is clear, however, from what has already been stated that the prevailing opinion referred to is far from being homogeneous and that the social factors governing the trends in legal science are assessed on a different basis according to the differing conceptions of the development of society itself. We must also be careful not to over-simplify this fundamental notion. It is

---

the series of studies on comparative public law, 'Cases in Comparative Public Administration', published by the International Institute of Administrative Science, Brussels, and the Symposium organized by the International Association of Legal Science at Moscow in September 1970 on the influence of federalism on the development of legal systems (see *Le Fédéralisme et le développement des ordres juridiques/Federalism and the Development of Legal Systems*, 1971). See also *Livre du centenaire de la Société de Législation comparée*, Vol. I, Sec, IV ('Les apports du droit comparé au droit public français', 1969).

9. Lawson, Comments.

10. See below, Section Two, sub-section II,3, pp. 960 sqq., and Section Three, sub-section II,1, pp. 975 sqq.

not therefore enough, even in the Marxist-Leninist view, to assert that the trends in legal science are determined by the social basis [11] since this determines only the main features of the social superstructure. We must also inquire whether within this general framework there are not present other elements which help to determine the various trends of research.

Such elements do, in our view, exist. Being in their turn also a function of the social basis they are somewhat varied, indeed heterogeneous.

Politics, first and foremost, condition the form and content of laws and of legal science itself. Here we may quote the replies to the questionnaire drawn up by the Institute of State and Law of the Academy of Sciences of the USSR, where it is stated that as regards principles one may even detect 'a unity of politics and law in respect of some of their content and aims, the identity of certain of their functions directed to the organization of the social system, a certain coincidence of their spheres of action, etc.' [12]

Law and legal science, and their development, are also influenced by the various philosophical conceptions (philosophic framework of the notion of law, etc.), or in some countries by religion, by ethnological considerations, by the upsurge of the social sciences, even, as already mentioned, by that of the natural sciences, and so forth.

In the second and third sections of this study we shall return to some of these factors in so far as they influence the major theoretical currents and the main trends of contemporary legal science.

3. Before going any further, we should like to say a few more words concerning the reflex action of trends in legal science on social evolution. While research trends in legal science are determined, when all is said and done, by social factors, they in turn are directed towards certain social aims and affect not merely the development of legal thought but also, whether deliberately or otherwise, the development of the State and of law (the existing system from the point of view of democracy, legislation, application of law, and so forth).

It is clear that in a world divided into different social and economic systems, even trends in the development of legal science reflect the social and ideological differences or antagonisms besetting the world today.[13] (We shall be obliged to make frequent reference to this point in the course of the present study.)

Regardless of the fruitfulness or barrenness that these trends may reflect at the strictly scientific level, we can differentiate from the point of view of their social impact between those promoting the progress of society and those which hamper or slow it down. This means that certain trends in the development of legal thought can aid the improvement of society, whereas others

11. See Section Two, sub-section III, pp. 966 sqq.

12. See Replies, p. 8 of original Russian manuscript.

13. This idea is familiar to Marxists but is also found in non-Marxist literature. We read, for example, in CARBONNIER, *Flexible droit* (1969), p. 123, that 'in an economy which is liberal and capitalist, the legal system is bound to be capitalist and liberal ...' ('dans une économie libérale et capitaliste, l'ordre juridique est forcément capitaliste et libéral ...').

tend to preserve what is opposed to social progress. It is clear, however, that even assessment of what is progressive and what is conservative or reactionary hinges on the opposing points of departure to which reference has already been made and that the Marxist-Leninist position [14] is, in this connexion, hostile to any viewpoint linked to a non-Marxist conception of legal science.[15]

4. We have, however, already pointed out that polemics were debarred from this study. For this reason – and without disguising our adherence to the Marxist-Leninist viewpoint – we shall do no more here than draw the reader's attention to the social impact of trends in legal science and to the differing assessments made concerning them in this connexion, owing to the opposition between the different views of society as a whole and of its development.

Before tracing the major currents and main trends of research in legal science today, there are still two prior questions to be examined which have an important effect on the genesis and orientation of trends in legal science within different contexts, and which are, in turn, a reflection of these trends: the notion of legal science itself and what is sometimes called the 'geography of law'.

II. THE NOTION OF LEGAL SCIENCE

1. The notion of legal science is far from clear and we must recognize that we cannot give it 'either a definite content or exact boundaries'.[16]

Even the terminology is far from standardized: writers refer to legal science sometimes in the singular (legal science), sometimes in the plural (legal sciences) – a discrepancy which, even if ascribable in some cases to carelessness of expression or mere linguistic conventions, may frequently betray a genuine divergence of conception.

Where there is such a divergence, we must understand by the plural term *legal sciences*:

(a) either the various sciences which fall wholly within the domain of law, but approach it from different standpoints: we must thus distinguish the dogmatics, the policy and the sociology of law, etc;.

(b) or the disciplines relating to various branches of law (e.g., constitutional, administrative, civil, penal, international, etc.), to the philoso-

14. One of the aims of Marxist-Leninist legal science is an ideological campaign against reactionary trends in legal science – cf. Čhikvadze, Comments (par. 1): '... for the progressive development of [legal] science it is extremely important to identify not merely the positive trends of this development, but also what is hampering it, where opposition lies, such as the old against the new, the reactionary against the democratic, etc.'. We shall come back to this problem at a later stage (Section Two, sub-section III, 6(a) (iv), p. 971) when reviewing the major trends which, from the Marxist-Leninist viewpoint, are regarded as acting as a brake on social progress.

15. Cf., for example, M. Rheinstein's conception of the action of law in furthering social progress; in particular, see Section Three, sub-section IV, 2(a), pp. 1002-1004 below.

16. Bobbio, 'L'objet de la science juridique', a special contribution.

phy (or alternatively to the general theory) of law, to comparative law, and so forth;

(c) or again the ensemble constituted by legal science and the science of the State (*Staatslehre*),

it being agreed that the singular *legal science* is a general term covering collectively all the special legal sciences enumerated, as the case may be, in (a), (b) and (c) above.

In this study we shall use the singular term 'legal science' to designate the general science which incorporates all types of special legal research.

2. The obstacles encountered in attempting a definition of legal science are many. They derive first and foremost from the prevalent uncertainty regarding the subject-matter of this science.

(a) There is a considerable diversity of opinion concerning the proper rôle of the kind of knowledge which legal science proposes to foster; in other words, and simplifying a little, is the purpose of legal science confined to the analysis and interpretation of the law in force (positivist conception), or does it include modification of the law in force, thus implying that legal science is not content to accept the law in force as a given fact but proposes to work for its evolution and transformation, particularly by studies *de lege ferenda*, changes in interpretation and so on (i.e., a sociological conception)? We shall see later that it is the latter conception which has become predominant in contemporary legal science.

(b) There has also been for a considerable time a wide divergence of opinion concerning the *extent* (or limits) to be set for legal science, with particular reference to the question whether legal science is a discipline wholly concerned with the knowledge of law or whether its competence extends to cover knowledge of the State also – in which case it would become the science of State and law.[17]

Attempts to define legal science encounter yet another difficulty, namely the ambiguity of the word 'science' itself. It is well known that the meaning of the French word *science* corresponds only very approximately to that of the German word *Wissenschaft*, to that of the Russian word *naukà* or to that of the English word 'science'. While the terms *Wissenschaft* and *nauka* signify undoubtedly and with no risk of ambiguity knowledge in general, or 'learning', the French word *science* remains loaded with a degree of ambiguity to the extent that it is used to refer both to the sciences generally and to the exact and natural sciences in particular. As to the English word 'science', it appears to be so closely linked to the concept of the natural sciences or natural science that the term 'legal science' falls under suspicion of connoting the view that this sector is a 'science' *stricto sensu*, i.e., a desire to inject into it the self-same methods that have brought success in the natural sciences. (This is why Max Rheinstein, in his replies to the questionnaire, chose not to use the term legal science' and substituted the term 'legal learning'.)

The difference between the 'Continental' conception of legal science and

17. See 1(c) above.

that of the Common Law countries is even more marked, however, and springs to a large extent from the different conceptions of law itself in the two systems.[18] In countries following the 'Continental' tradition, legal science is regarded as a discipline reserved for the 'professional' legal scholars (law professors, research specialists) whereas the practitioners (judges, lawyers, administrators) are presented with the results of 'professional' scientific research and, for the purpose of accepting or rejecting them, refer to the opinions expressed in scientific publications, which thus exert an influence on practice (on legislation and the application of law). In the Common Law countries, on the other hand, it is in the first instance the judges, barristers and civil servants who develop legal thought and, hence, legal science – the function of the academic lawyers is, apart from the teaching of law, to aid the practitioners by criticism and by suggesting new approaches.[19] The rôle of the academic lawyers in relation to the practitioners is thus, in the Common Law countries, very different from what we find in the 'Continental' law countries; and the frontier between legal science and legal practice is much less marked in the Common Law countries than in those where legal science has developed under the influence of continental European traditions.

We may note however that, as Rheinstein reminds us,[20] this difference between the position of legal science in the Common Law countries on the one hand and the Civil Law countries on the other, while still very great, has diminished in relatively recent times and is continuing to diminish. In the Common Law countries, legal science is taking a remarkable interest in the study of positive law (dogmatic jurisprudence), a type of research which traditionally has been less developed than the philosophy, general theory, or history of law.[21] This interest on the part of legal science in the study of positive law is undoubtedly also one of the reasons why legal science has become more important in the Common Law countries, including the developing countries. (We shall return later to the importance attributed in those countries to the teaching of law.)

Conversely, there has been a noticeable lessening of the rôle played by the science of law in the legal life of the countries following the 'Continental' tradition where that rôle has hitherto been preponderant.

We shall not, therefore, in dealing with trends in legal science, limit ourselves to 'academic' legal science, but attempt to identify the main currents of legal thought irrespective of their origin.

3. Regarding the question whether the subject-matter of legal science is exclusively law (positive or 'natural', or even both), or whether its competence extends to cover the study both of law and the State, the following positions have long been taken:

(a) According to the first, legal science tacitly also includes research

18. See II,4 below, pp. 942 sqq.

19. Lawson, Comments. Cf. RHEINSTEIN, 'Die Rechtshonoratioren und ihr Einfluss auf Charakter und Funktion der Rechtsordnungen' (1970).

20. Rheinstein, Observations on the new statement, pp. 1 and 2.

21. Cf. Lawson, 'Doctrinal writing: a foreign element in English law?' (1969).

concerning the State. This conception is particularly prevalent in the French tradition,[22] but is also met with in countries which have been influenced by that tradition, as well as by the German tradition under which *allgemeine Staatslehre* is and has been generally regarded as an integral part of legal science;

(b) A second doctrine holds that knowledge of law on the one hand, and knowledge of the State on the other, are no doubt very closely linked but that neither includes the other, so that what is being built up is a *science of State and law*. This is the typical position in the Socialist countries,[23] although the terminology is not uniform nor invariably crystal-clear. In the USSR, for example, the current term is *pravovaja nauka* (legal science) but there can be no doubt that what is meant is the science of *State* and law.[24] On the other hand in Bulgaria, Czechoslovakia and Hungary among others, the term 'science of State and law' is in constant use. However, disregarding differences in terminology, we can conclude that the Marxist-Leninist position is that the science of State and law is a single science, reflecting the unity of State and law in social life, and considering them from the standpoint of their reciprocal relations; [25]

(c) According to a third conception, research in regard to the State is either set apart from legal science and regarded as the subject-matter of the political sciences (political science, political philosophy, and so forth), or is shared between these sciences and legal science itself.

It is in England, in particular, and in countries where such study has developed under English influence, that research relating to the State is completely separated from legal science; but we should also note that in Italy the general doctrine of State (*dottrina generale dello stato*), though manifestly influenced by the German conception of *allgemeine Staatslehre*, is nevertheless viewed as an autonomous science set apart from legal science. Other examples could be quoted of this separation of the two disciplines: the Scandinavian countries, etc. It is even more common for research in regard to the State to be shared between legal science and the political sciences; indeed, as we shall see immediately, this represents, above all in the 'Continental' law sector, one of the contemporary trends. Of the Common Law countries, it is in the United States that this tendency is particularly prevalent; research relating to the State is split among several branches of the social sciences, problems of the State being studied with particular attention by political scientists, who

22. Cf. René David's replies to the questionnaire: 'Knowledge of the State is, according to French tradition, part of legal science.' ('La connaissance de l'Etat fait, selon la tradition française, partie de la science juridique.'); cf. also 'Zur Einheit der Rechts- und Staatswissenschaften' (1967).

23. Cf. Aržanov, *Gosudarstvo i pravo v ih otnošenii* (1960).

24. Evidence of this fact is provided in the full official title of the institute in the USSR Academy of Sciences that specializes in legal research, viz. 'Institute of State and Law of the Academy of Sciences of the USSR'.

25. Cf. reply of the Institute of State and Law of the Academy of Sciences of the USSR to the questionnaire, pp. 1, 3 of original Russian manuscript.

examine the political life of a given society, by political theorists, who attempt to formulate the ideal State, and by legal specialists.

These three positions subsist in the contemporary world. But the centre of gravity is shifting, as we have stated, towards the third. The modern State has long since ceased to play the rôle of mere 'night-watchman'; its functions and actions within society are gaining momentum, particularly at the level of internal political and economic life, to say nothing of the field of external relations. The social activity of the State is becoming more and more complex, to the extent that legal science, even defined as the science of State and law, has become too narrow to encompass the full complexity of functions and actions which nowadays fall to the State. This fact has given rise to an increasingly strong tendency to carry out multidisciplinary research with respect to the State, its institutions and their functions, etc., to which we shall return in our Epilogue; for the same reasons, research concerning the State quite often (especially in the Western countries) tends to acquire a degree of autonomy, more or less marked, in relation to legal science; at the same time it tends quite often to play an active part even in political life. Knowledge of the State continues in general to be regarded as falling within the orbit of legal science – or, as the case may be, of the science of State and law – in those countries where this has been the rule; but it is also becoming the subject-matter of political science, of which the importance and status are undoubtedly rising. Political science is thus beginning to take shape as an autonomous science even in countries where, in the recent past, the notion of such a science was still not prevalent. (Political science has become a discipline even in some Socialist countries although, since they regard political research as a part of historical materialism, these countries have in general been very reluctant to recognize the autonomous existence of political science.)

III. GEOGRAPHY OF LAW

1. We have already seen that trends in the development of legal thought are strongly influenced by what we may term (and what is sometimes called) 'geography of law'.

Before proceeding, however, we think it advisable to clarify the meaning of this rather equivocal expression. It could relate either to what is properly conveyed by the term 'geography of law', i.e., the *fact* (we shall come back to this straightaway) that law varies from one country to another (or rather from one group of countries to another), or to a chart ('*géographie*') of the various conceptions of law, different legal theories and so on.

What interests us primarily here is the 'geography of law'. Since the two aspects overlap, however, and although the expression *géographie juridique* is more common in French-speaking countries (see the special contribution by René David) we shall, throughout the English text, use the term 'geography of law' without attempting to give it an absolutely precise connotation.

The importance of the geography of law for research trends in legal science derives essentially from the fact that, leaving aside international law (and especially its general principles), law is always found to be related to a specific territory and a specific State authority.[26] We should note, however, that this can give rise to very different interpretations, depending on the position taken in regard to legal science itself. We share Szabó's view that in the positivist conception for example, legal science, which for the positivist is restricted to interpretation of the law in force, constitutes a science which is 'strictly and typically national in the sense that its subject-matter has always been the national law of a given State'.[27]

Contemporary legal science, which has become in the main a social science (see below), accepts – without on that account questioning the territorial character of law[28] – generalization starting from the legal systems of the different countries and classification of these legal systems on the basis of social factors, a classification which is very often influenced by political and economic interests.

These same political, and even economic, pressures have long encouraged a tendency to strive towards the unification of law; however, as we shall see below (in connexion with comparative law) the results achieved remain so far fairly modest. The history of law provides us with ample evidence of the fact that law, being determined by the territory subject to the power of a State, tends to evolve in the direction of reciprocal differentiation of laws previously harmonized rather than towards unification of differing systems. This is instanced by the case of the French and Belgian civil laws: though based on the same Civil Code, the laws as applied in effect in the two countries have drawn sharply apart on a variety of issues through the action of

26. This is of course only true of *positive* law. Natural law, in so far as it is held to exist, is unrelated to territorial limits. Another exception is constituted by *canon* law. It would, however, be going beyond the frontiers of the present survey to attempt to solve the theoretical problem of the character of canon law.

27. '... étroitement et typiquement nationale en ce sens que son objet a toujours été le droit national d'un Etat donné' – Comments (cf. also Professor Rheinstein's view quoted in the next footnote).

28. Even theories concerning the existence of a supra-national – or infra-national – law (ideas which are in no way shared by the author of this study) do not dissociate the reality of law from that of a specific territory. The connexion between law and a given territory or State power also affects legal science, though it has a less pronounced influence on it than on positive law. Rheinstein remarks in this regard, in his 'Observations on the Revised Report' (par. 6b), that 'until recently legal science has almost exclusively been nationally isolated. There have been as many legal sciences as there have been legal systems. Recent years are witnessing the growth of a new supra-national legal science ...'. We only partly share this opinion. The relation established between legal science and a given legal system is only fully valid for legal positivism (see also the view of Szabó in the preceding note), not for legal science in its neo-Kantian, 'natural', or phenomenological variants. On the other hand, it should be remembered that the movement towards a supra-national legal science is not only retarded by the existence of different conceptions of legal science, the roots of which still run deep (see below, Section Two, I, II), but is restricted by the ideological differences which separate countries of different socio-economic structure.

the courts. The same process may be noted in Czechoslovakia in relation to the 1811 Austrian Civil Code (ABGB) which was in force in that country until the close of 1950, and also in relation to Polish law (the unified Family Codes remained in force in Czechoslovakia and Poland for less than fifteen years, from 1950 to 1964; but the law actually applied by the courts in Czechoslovakia on the one hand and in Poland on the other had already diverged widely during that relatively brief period).

On the other hand, as Lawson [29] points out, Common Law and Equity were remarkably uniform throughout the whole British Commonwealth and continue to be so even after attainment of independence by the Commonwealth countries. In contrast, *legislation* between one Commonwealth country and another differs considerably in the sector of law which, in Continental terminology, is called 'public', though uniformity is to a large extent maintained in 'private law'.

2. However, even taking into account everything we have just said, we can base the survey which follows on the idea with which we introduced this sub-section, namely the fact that law is related to a specific territory.

This fact has given rise to many efforts by legal scholars to construct a 'geography of law', in other words to divide the world into several major 'legal families'.

It is not proposed here to recapitulate all those efforts nor to add yet another to the hundred or so existing marshallings of major legal families. We will merely offer a few comments in an attempt to explain our ideas about such classification and especially in order to draw certain conclusions regarding the influence and importance of the 'geography of law' in terms of current trends of research in the field of legal science.

(a) It must first be stated that the very notion of a 'legal family' or a 'major legal family' is imprecise, as are the criteria for the delineation of such entities. It is true that most authors who have sought to identify 'major legal families' have taken as a starting-point the origins of law and confined themselves to private law, disregarding public law altogether. Yet public law would also permit of a grouping of States into 'families' and even into 'major families', although such groups would be very different from those it has been possible to identify on the basis of private law.[30] For example the Latin American countries, in any classification drawn up from the point of view of the sources of law and of private law, undoubtedly belong to the major family of Continental law, the Civil Law, and more particularly to the family known sometimes as Roman law, sometimes Romano-Germanic law, etc., in opposition to the major family constituted by Common Law. On the other hand, according to any classification drawn up from the standpoint of public

29. Comments.

30. Cf. David, R., *Les grands systèmes de droit contemporains*, 4th edn. (1971), p. 23; Rozmaryn, 'Les grandes controverses du droit comparé' (1967); etc. Cf. also the somewhat reserved attitude of Malmström with respect to an exact classification. He writes, in 'The system of legal systems' (1969): '... it is impossible to establish a uniform system of classification which is ideal from every point of view and implies a clear distinction between "families" or "groups".'

law, and especially of constitutional law, many of these countries would belong to the public law family represented by the United States of America, even though they had no connexion at all with that family in terms of private law.

(b) The rules of law have of course existed in a more or less developed form throughout the world from time immemorial. Law forms part of the civilization of all lands and peoples. It is nevertheless a fact that in the realm of private law, the 'exportable' type has been European law, whether Continental or English (Common Law), to which should be added, in respect of public (constitutional) law, that of the United States. Law in continents outside Europe has been influenced by European law, either, depending upon the political situation, by the 'Continental' or the English Common Law type. This is true of the United States also in the private law sector.

The transplantation of European law to the other continents took place in several different ways:

either in such a way as to result in a complete break between the former, traditional law of the land and the new law, in those cases where the colonists introduced the law of their home-country or else created a new law modelled on it, the new law totally replacing the traditional law of the land;

or in such a way that the traditional law remained in existence as the law of the natives, while the law imported from Europe represented, to use a highly simplified formula, the law of the colonists;

or, again, in such a way that certain matters came to be regulated by the 'imported' law, which, in this particular domain, completely superseded the former law of the land, while other matters continued to be regulated by the traditional law which, in that domain, underwent no foreign influence;

or, finally, in yet another way which we shall take up hereafter in regard to the Republic of South Africa, Japan, etc., for example.

In the first of the above-mentioned cases the law became adapted to the political, economic, geographical, etc., conditions of the receiving country, and while retaining the characteristic features of the law in the country of origin, took on relatively different features in matters of detail.

In the second case a process of acculturation took place under which the former traditional law of the land, being less elaborate and less suited to the solution of the increasingly complex problems arising in modern society, gradually lost ground before the law imported from Europe – which for its part also adapted itself gradually in a manner similar to that noted in the first case above. This process continues and has given rise at the present time to one of the major problems of legal policy facing countries that have recently attained independence, thereby bringing into being a highly important trend of research in legal science, to which we shall revert in greater detail below.

The third case referred to above concerns the countries of Moslem law, which we shall deal with in greater detail in the sub-section devoted to the acculturation of law.[31]

3. The criteria governing classification into 'major legal families' differ considerably. The history of attempts to classify these groupings goes back to the beginning of the twentieth century. One of the first classifications was made by Esmein [32] who grouped the legislations (or customs) of different peoples according to their 'historical formation, general structure and distinctive features' ('la formation historique, la structure générale et les traits distinctifs'). The classification proposed by Lévy-Ullmann [33] is based on the different conceptions of the sources of law; that of Arminjon-Nolde-Wolff [34] on the original character of law and relationships based on derivation or resemblance; Schnitzer [35] attempts to classify law according to 'major cultural spheres' (*grosse Kulturkreise);* and Zweigert [36] on the basis of legal 'styles' (determined by the historical origin of law, the specific way of thinking, significant institutions, the nature of the sources of law and their interpretation and, last, by ideological considerations). Malmström, in an effort to make the classification more flexible, begins by classifying law into basic families, and then subdivides them.[37] The classification commonly accepted in the Western world today is that of René David [38] which is based on the ideological criterion and that of legal technique, applying the two criteria in conjunction.[39]

31. Cf. Section Three, VI, 2, below, pp. 1016 sqq.

32. 'Le droit comparé et l'enseignement du droit' (1905).

33. 'Sur les communications relatives au droit privé dans les pays étrangers' (1922).

34. *Traité de droit comparé* (1950), I, pp. 47 sqq.

35. *Vergleichende Rechtslehre* (1945, 1961), pp. 86 sqq.

36. 'Zur Lehre von den Rechtskreisen' (1961). The author develops his idea in a much more detailed way in ZWEIGERT & KÖTZ, *Einführung in die Rechtsvergleichung*, Vol. I (1971), pp. 67 sqq.

37. 'The system of legal systems' (1969) (see also p. 940, n. 30).

38. *Traité élémentaire de droit civil comparé* (1950), pp. 222 sqq.; *Les grands systèmes de droit contemporains*, 4th edn. (1971), pp. 23 sqq.

39. In this connexion it is stated in *Les grands systèmes de droit contemporains*, first edition (1964), p. 16, that 'Two legal systems cannot be considered as belonging to a single family, even if they use the same notions and techniques, when they are founded on opposing philosophical, political, or economic principles and seek to bring about two entirely different types of society'. ('Deux droits ne peuvent être considérés comme appartenant à une même famille, même s'ils utilisent les mêmes concepts et les mêmes techniques, lorsqu'ils s'appuient sur des principes d'ordre philosophique, politique ou économique opposés et qu'ils cherchent à réaliser deux types entièrement différents de société.') In the fourth edition (1971) of *Les grands systèmes*, however, R. DAVID refrains from defining the criteria of classification and adds: 'These discussions have caused a good deal of ink to flow, but they are not very meaningful. There is no real biological foundation for the notion of "legal families"; it only serves a didactic purpose ...' ('Ces discussions ont fait couler beaucoup d'encre; elles n'ont pas pourtant beaucoup de sens. La notion de "famille de droits" ne correspond pas à une réalité biologique; on y recourt seulement à une fin didactique ...') (p. 22). We share this author's view about the amount of ink that has been used up in these discussions, but we should like to point out that the

It is evident from the brief survey of the criteria used to classify legal 'families' that we have just given the reader, that their authors either seek criteria exclusively in the legal domain, or (as David and Zweigert, in particular) take into account even social and ideological considerations, applying these, however, in conjunction with criteria which belong to a very different order of things.

Marxist-Leninist science is opposed to such classification methods. To demonstrate these objections let us take the most widely adopted classification today, that of René David, which distinguishes the following 'families':

(a) Romano-Germanic,
(b) Socialist,
(c) Common Law,
(d) Religious and traditional law.[40]

But how does Socialist law fit into this pattern? It is by no means the product of the vagaries of legal history (such, for example, as the fact that the French colonists in Canada settled in Quebec and not in Saskatchewan). It is a product of the Socialist revolution and is distinguished from the *totality* of non-Socialist law not so much by the form – in certain cases, there are quite similar features – as, first and foremost, by its social origin and purpose. The distinction between 'family' (b) and 'families' (a), (c) and (d) is therefore, in our opinion, qualitatively different from that between 'famlies' (a),(c) and (d).

For this reason Marxist-Leninist legal science takes the differentiation of legal systems according to their *socio-historical type* as the basis for its classification of legal systems.

We should therefore classify law on two planes, first distinguishing between socialist and non-socialist law.[41] Once this classification is made on the basis of the socio-historical criterion, the two groups can then be subdivided according to the nature of the sources of law, the approach to it, the legal apparatus, and so forth.

4. Let us note, however, that this differentiation is much more pronounced within the area of non-socialist law, where the following 'major families' or even 'major systems' of law can, in our opinion, be distinguished:

---

notion of 'legal families', while it naturally has nothing in common with realities of a biological order, does correspond to a social reality which is no less objective than the biological reality. As will become clear from the rest of the article, this is what constitutes the premise of our conception.

40. *Les grands systèmes* . . ., 4th edn. (1971), pp. 23 sqq. Compare the classification made by ZWEIGERT & KÖTZ (cf. p. 941, n. 36), who distinguish between Roman, German, Anglo-American, Nordic and Socialist spheres of law (*Rechtskreise*) and others (the law of the Far East, Islamic law and Hindu law).

41. We realize full well that the 'non-Socialist' group is quite heterogeneous since it includes both capitalist countries and those which have recently attained independence, the latter in turn offering a wide range of political colours. See in particular, ČHIKVADZE *et al., Marksistsko-leninskaja obščaja teorija gosudarstva i prava*, Vol. II (1971), Chaps. IX and X. We shall return to this point in a more detailed fashion in the sections below.

(a) The countries of 'Continental' law, i.e., law which sprang up on the continent of Europe. The system is also termed, as we have seen, Romano-Germanic law ('Civil Law' in England), but the most accurate definition would be: the system of law that has developed on the basis of the reception of Roman law.[42]

(b) The Common Law system, i.e., the system characterized by the existence of judge-made law covering, in addition to England, the countries where law has developed on the basis of the direct reception of English Common Law (and, where appropriate, Equity), or under the influence of English law and legal thought.

(c) The system of Moslem law which, because of its highly elaborate and specific nature, deserves, in our opinion, to be regarded as an autonomous system of law, in the same way as 'Continental law' and 'Common Law'.

(d) Finally, those countries in which traditional laws based on ancestral customs, primitive religions, etc., are still in existence.[43]

The two pillars on which this system rests are *Continental law* and *Common Law*. (We shall see below that traditional or religious law is, as a general rule, influenced either by Continental law or Common Law.)

The essential difference between the Continental law system and Common Law does not lie only in the fact that the former is typically a system of written law and the latter one of judge-made law. According to Rheinstein, the decisive aspect is the difference in the two methods of legal thinking, Continental law being founded primarily on the logic of the syllogism while Common Law is based primarily on the logic of analogy.[44] This is, of course, somewhat of an oversimplification and, in a given country, legal thought is not always so univocal. For example, even in the Common Law countries, application of syllogistic logic is doubtless necessary in questions of written law (as related to Statute Law). Our own experience confirms, however, that even where the application of written law is concerned, the reasoning of the Common lawyer is very different from that of the continental jurist and the distinction put forward by Rheinstein is indeed one of the *general* distinctions which have to be made between the continental conception of law and that of Common Law. (This distinction could, moreover, make it easier to classify, either among the Continental law countries or

42. EHRENZWEIG (*Psychoanalytic Jurisprudence* (1971), pars. 88 sqq.; 'Malmström's "System of legal systems"' (1972), pp. 74, 75) questions the distinctive character of Roman law; see also p. 944, n. 46.

43. The reader will not have failed to notice that in regard to non-socialist law our classification is not dissimilar to René David's.

44. Rheistein, Comments. On the other hand, the idea that two systems of law can be distinguished by their manner of thinking is questioned by some authors (cf., for example, MALMSTRÖM, 'The System of legal systems' (1969), p. 143; EHRENZWEIG, 'Malmström's "system of legal systems"' (1972), pp. 75, 76). Regarding the distinction between Continental law (Civil Law) and Common Law, see also DAINOW, 'Le Droit civil et la *Common Law*' (1965).

the Common Law countries, those States with a mixed legal system which are referred to above.)[45]

Let us note also that one of the greatest differences separating the Continental and the Common Law types of legal system lies in the administration of justice.[46] This difference, of course, does not belong to the domain of *legal science*, yet it springs from a clear duality in the concepts of law which without doubt contributes also to the differentiation between 'Continental' legal thinking on the one hand and that of Common Law on the other.

Among the essential features that distinguish Common Law from the Continental system, mention should also be made of the different positions in society filled by the lawyer in these two groups of countries respectively, particularly in regard to the relationship between 'academic lawyers' and the practitioners.[47]

5. The preceding comments do not, however, dispose of all the problems involved in selecting criteria. The replies received to the questionnaire have convinced us that the difference between Continental (so-called Romano-Germanic) law and Common Law is, on the plane of private law, extremely deep – indeed irreconcilable – if we contrast continental Europe with England and the United States; on the other hand, however, it is far less sharp in countries that have recently attained independence. The legal problems which those various countries face, even in private law, are very similar in each; they affect in much the same manner both the countries whose current law took shape under the influence of Continental law and those belonging to the sphere of influence of Common Law, i.e., regardless of the fact that, from the standpoint of a classification based on history, countries of this type must necessarily be entered under two different 'legal families'.

6. None of the 'major legal families' is in any way homogeneous. Thus within the family known as 'Continental' we have to distinguish (in regard to private law) at the least the following zones of influence: the French zone (area of influence of the French Civil Code of 1804); the German zone (Civil Code – BGB –of 1896); the Austrian zone (Civil Code – ABGB – of 1811), of which the history and even the philosophical sources were very different from those of the French Civil Code and indeed of the German BGB; and the Swiss zone (Civil Code of 1907 and Code of Obligations of 1911 which came into force on 1 January 1912) – each of these having given rise even to a distinct manner of legal thinking. It is well known that the viewpoints and preoccupations of legal science, especially in France, Germany and Austria, were – and remain – quite distinct. Lastly we must not overlook (as is too often the case) the Scandinavian countries which constitute, without any doubt, a specific zone of 'Continental' type law. Be-

45. See 2(a) above, pp. 939-940.

46. This is also by and large the view of EHRENZWEIG (*Psychoanalytic Jurisprudence* (1971), par. 112; 'Malmström's "system of legal systems"' (1972), p. 76) who however, differing in this from the majority opinion, considers that the crucial difference lies in the judge's position and practically rejects all other generally recognized criteria.

47. See above, II, 2, pp. 934-935.

cause of the affinity of legal traditions in the Scandinavian countries, even legal thinking has developed on the same lines in these countries.[48] They have left their mark in the history of legal science through the so-called school of 'Scandinavian realism' or the 'Uppsala School', represented by the familiar names of Hägerström, Alf Ross, Lundstedt, Olivecrona, etc.,[49] and even at present, the trends in legal science (trends that remain broadly positivist, although reflecting very marked sociological influences) are still, if not identical, at least very similar throughout all Scandinavian countries.[50]

In regard to Common Law, we have to realize first of all, as we have noted in another connexion, that in Common Law countries, it is written law or legislation (Statute Law) that gains in relative importance and that legislation varies considerably from one Common Law country to another.

On the other hand, Common Law exhibits far fewer differences; however, although it is relatively stable, we can nonetheless detect within it certain rather notable differences. The greatest difference is that which separates the two geographical zones of Common Law; the zone of *English Common Law*, which has for some time been spread throughout the entire *British Commonwealth*, and that of the *Common Law of the United States*, a zone in which Common Law has, in the course of centuries, become quite remote from its English model,[51] and where there is now a preference for using the notion of 'Anglo-American Law' or 'Anglo-American laws' instead of that of 'Common Law' *largo sensu*, the notion of Common Law being used only *stricto sensu*, i.e., in its original meaning, in contrast to Equity, etc. Although there are also certain differences in the zone of English Common Law, it is more homogeneous than that of the Common Law of the United States, a country where, after a long discussion which was practically settled in 1938 by a Federal Court decision (*Erie Railroad Company* v. *Tompkins*), it was recognized that each of the federated States had its own form of Common Law, and where, as Lawson notes,[52] the difference between the laws of the various states is more pronounced than that which exists between the Common Law of the various countries belonging to the British Commonwealth.

As for the system of *Islamic* law, it must be pointed out that the geographical pattern of the countries belonging to this system is not always identical with that of the countries of Islam, nor even with that of Moslem

48. This subdivision of the 'Continental' system is often overlooked in scholarly literature. A distinction of this kind is especially to be found in the above-mentioned work of ZWEIGERT & KÖTZ (1971) (see p. 941, n. 36 and p. 942, n. 40 above); it is difficult however for us to agree with the authors who class Swiss law and particularly Austrian law in the category of German law (*Rechtskreis*).

49. A brief and pertinent description of 'Scandinavian realism' is found in FRIEDMANN, *Legal Theory* (1960), pp. 258 sqq.

50. Eek, Replies, pp. 8-13; cf. also EEK, 'Evolution et structure du droit scandinave' (1961); von EYBEN, 'Inter-Nordic legislative co-operation' (1962); JØRGENSEN, 'Grundzüge der Entwicklung der skandinavischen Rechtswissenschaft' (1970); MUNCH-PETERSEN, 'Main features of Scandinavian law' (1927); etc.

51. Cf., for example, KAHN-FREUND, *Parallelen und Gegensätze im englischen und amerikanischen Privatrecht* (1960).

52. In his Observations concerning the second version of our survey.

ethnic groups. Some peoples, as they became converted to Islam, adopted its faith but not its law, as, for example, the Berbers of Northern Africa who have maintained their tribal customs (even in respect of the status of the family) or the Yorubas of Western Nigeria who, although Moslem by religion, do not apply Moslem law either. A similar phenomenon occurred in Indonesia, a country which became converted to Islam while maintaining to a large extent its customary law (*Adat*).[53]

The geography of countries following Islamic law is determined by their adherence to one of the schools of Islamic law. It is well known that Moslem law is at present divided into four *Sunnite* (orthodox) schools (rites), namely the Hanafite, Malikite, Shafi'ite and Hanbalite schools. Besides the orthodox schools there is also a Shi'ite law (similarly divided into several schools), which is, however, much less widespread and of lesser importance than Sunnite law.

The various orthodox legal schools have long existed side by side within the same territory; in our time, however, adherence to the respective schools determines the inner geographical pattern of the region of Moslem law. Those countries which were subjected to Turkish rule became, from the standpoint of positive law, *Hanafite* (as did Pakistan, Bangladesh and the Moslem part of India); the Moslem countries of North Africa (Morocco, Algeria, Tunisia and Libya) and of West Africa apply the *Malikite* law; the Islamic law – as distinct from *Adat* law – that is applied in Indonesia is a law of the *Shafi'ite* school; lastly, the *Hanbalite* school prevails in Saudi Arabia.

The last system we have to consider, that which we have termed the 'Family' of *traditional law*, is totally heterogeneous. Moreover this family and both the Continental ('Romano-Germanic') and Common Law families overlap. Almost all the countries with traditional legal systems are at the same time [54] countries having written laws of either the Continental type, or Common Law. But in spite of this distinction, due to historical influences, there are, as we have already pointed out, problems common to countries that have recently attained independence and this community of interest justifies the existence of a special category, namely the family of countries having traditional, or religious, legal systems.

It should, however, be noted that since these countries have attained independence, development of their current law has not been influenced by Continental law or Common Law alone. Many newly independent States have turned (as we shall see further on) to Marxist-Leninist principles of

53. Linant de Bellefonds, 'Les tendances actuelles du droit comparé en pays d'Islam' (Special contribution), p. 7. In Indonesia Moslem law of the Shafi'ite school (see below) is only applied with respect to certain personal status institutions, particularly *waqf* (mortmain property) and, to some extent, marriage, whereas other matters are governed by customary law (*Adat*). The relationship between the application of Moslem law and *Adat* law varies from one region to another. All this places Indonesia in a very special position in the world of Islamic law. Cf. Ter Haar, *Adatrecht in Indonesia* (1962).

54. Cf. 2(b) above, p. 940.

law, i.e., the law of the socialist countries, in constructing their own legal system. This means that in some of these countries the dividing line between socialist and non-socialist law is less marked than in States where the old legal tradition holds good.

7. We could pursue further this examination of 'major legal families' but this is not the object of the present study. Our sole purpose has been to point out to the reader that the 'geography of law', as a historical, socio-political and ideological factor, has a considerable influence on present trends in legal science throughout the world. This is due, in the first place, to the presence of two major socio-historical types of law (socialist and non-socialist), but also to the existence within non-socialist law of several different legal families which, notwithstanding their sometimes rather vaguely defined structure and classification criteria, constitute a real factor which cannot be overlooked when speaking of the main trends of research in contemporary legal science.

SECTION TWO: MAJOR CONTEMPORARY THEORETICAL CURRENTS

I. INTRODUCTION (GENERAL OBSERVATIONS)

1. Clearly, the major theoretical currents in contemporary legal science derive from remote or recent past.[55] The most interesting historical period from this point of view was the second half of the nineteenth century, which – leaving out of account the births of the Marxist conceptions of law that will be dealt with below – was marked by the existence of two rival trends: that of *legal positivism*, limited more or less to interpretation of the law in force (positive law), and that of the *speculative philosophy of law*.[56]

Legal science wavered between these two trends. Philosophy of law, then primarily represented by *neo-Kantianism* which became (at least in Germany) the reigning theory for several decades, took little interest in positive law and was mainly focused on the ontological, epistemological and methodological problems of law. It had, in consequence, correspondingly little influence in the field of research relating to positive law or that of legal practice, these fields being dominated at the time by positivism which was strongly opposed to any form of neo-Kantianism (and, at the same time, to theories of 'natural law'). It can thus be said without any exaggeration that this was a period when neo-Kantianism blossomed in the shadow of positivism.

55. For an overall survey of the history of legal thought, reference may be made, for example, to BRIMO, *Les grands courants de la philosophie du droit et de l'Etat* (1968); FRIEDMANN, *Legal Theory* (1968); KEČEKJAN & FEDKIN, *Istorija političeskih učenij* (1960); and the bibliographies listed in these works.

56. Szabó, Comments.

However, neo-Kantianism, though at the philosophical level it triumphantly fought off the assaults of all its rival theories, succumbed to the passage of time, to the effects of age and to the evidence of its own sterility. The path of neo-Kantianism has become a dead-end allowing neither further progress nor a broadening of the horizons of legal science. Since the Second World War it has in effect vanished from the scene, and with it the paramount rôle played by the philosophy of law in legal science in general.

Legal positivism, although more tenacious than neo-Kantianism – we shall come back to this later – also suffered defeat in regard to both doctrine and methodology. The abandonment of legal positivism was the most significant event in the history of legal science in the period covering the close of the nineteenth century and the first third of the twentieth, an event which determined to a large extent the main trends in the contemporary development of our discipline.[57]

The fatal disease afflicting positivism was its conception of the subject-matter of legal knowledge. This conception, limiting legal science purely to a knowledge of laws as created by the legislator, the judge, custom, etc., is so narrow as to exclude the knowledge of law as a social phenomenon and restrict legal science to interpretation of the law in force, while preventing it from taking part in the process of its creation.

The powerful anti-positivist current that set in – and according to which it was seen as the duty of legal scholars not simply to confine their knowledge to the law in force, but to regard law as a social phenomenon and work towards its development and, in this way, towards social development in general – is that which, broadly speaking, we can term the sociological current. Legal science, which previously wavered, as we have noted, between positivism and speculative philosophy, broke away essentially from these two extremes and became a *social science*.[58] The sociological current spread through Europe and was particularly prevalent in Germany where it was represented by the sociological school of Eugen Ehrlich, by the *Freirechtschule*, by *Interessenjurisprudenz* – a certain revival of which can be seen today in the Federal Republic of Germany – by various attempts to construe law in materialistic terms (Radbruch, Stammler, etc.), or define it by reference to its aims (*Zweck im Recht,* Rudolf von Ihering), by the typology of Max Weber and so forth. In France, sociologically-based concepts are found in the work of François Gény and Léon Duguit, among others.

The transformation of methodology was also the event that had the most decisive effect on legal science in the United States at the beginning of the twentieth century, a period in which this science, naturally much younger than its European counterpart, took its place in the universal history of legal thought.

Positivistic at the outset, it was represented by two remarkable names: J. C. Gray and O. W. Holmes. Holmes, in particular, had already introduced

57. Rheinstein, Comments.
58. Rheinstein, Comments; Szabó, Comments.

certain sociological elements into legal methodology. The following period witnessed the birth and triumph of a very powerful anti-positivist current which was generally termed 'sociological jurisprudence' or, more simply, the 'new jurisprudence'. This trend was represented principally by Roscoe Pound, B. N. Cardozo, K. Llewellyn, J. Frank and others (the two last-named, at least for a time, called their doctrine 'legal realism').[59]

Evidence of the universal nature of the sociological current in legal science is the existence of the Japanese sociological school represented by names such as Suehiro and Wagatsuma. (We may note also the direct influence of American 'legal realism' on Japanese legal science.) [60]

A historical event of far-reaching significance – and which, in the author's view, of all others has been of decisive importance in this 'methodological revolution' – was the birth of *Marxist-Leninist* legal science. Its roots go back to works written by Marx and Engels just before the middle of the nineteenth century (e.g., *Kritik des Hegelschen Staatsrecht,* 1843; *Zur Kritik der Hegelschen Rechtsphilosophie, Einleitung,* 1843, 1844; *Das philosophische Manifest der historischen Rechtsschule,* 1842; but most important of all, *Manifest der Kommunistischen Partei,* 1848) and developed by Lenin in the first quarter of the twentieth century.[61] Set out as a coherent doctrine, on the basis of Marxist-Leninist philosophy, by Soviet jurists following the First World War, it spread, after the Second World War, throughout all Socialist countries and even beyond.

Marxist-Leninist legal science takes issue with all other legal theories. It is a militant doctrine which, in contrast to any other contemporary legal theory, is founded on the notion of a society divided into social classes (we shall refer to this in more detail below) and consciously seeks to serve the interests of the working class throughout the world.

2. How are we to classify the major trends that are discernible in contemporary legal science and that have emerged from the historical development which we have briefly traced? Where are we to find the criteria for such classification? It is clear that these criteria are in no way accidental but fol-

59. CARDOZO, *The Nature of the Judicial Process* (1921); by the same author, *The Paradoxes of the Legal Science* (1928); CASPER, *Juristischer Realismus und politische Theorie im amerikanischen Rechtsdenken* (1967); FRANK, *Law and the Modern Mind* (1930, 1963); HOWE, *Justice Oliver Wendell Holmes* (1957-63); LERNER (ed.), *The Mind and Faith of Mr. Justice Holmes* (1945); LLEWELLYN, 'A realistic jurisprudence – the next step' (1930); by the same author, *Jurisprudence, Realism in Theory and Practice* (1962); POUND, 'The need for a sociological jurisprudence' (1909); by the same author, 'Soziologische Jurisprudenz in Amerika' (1925); REICH, *Sociological Jurisprudence und Legal Realism im Rechtsdenken Amerikas* (1967); Rheinstein, 'Karl Nickerson Llewellyn, 1893-1962' (1962). Cf. also 'Symposium on social science approaches to the judicial process' (1966).

60. Aomi, Contribution of the National Commission of Japan for Unesco, pp. 21 sqq. (cf. AOMI, 'Japan' (1970), pp. 389-390).

61. Cf. also ČHIKVADZE, 'Rol' V. I. Lenina v sozdanii sovetskogo socialističeskogo prava' (1957); by the same author, 'Lenin, Oktjabr' i juridičeskaja nauka' (1967); ROZIN, *Vozniknovenie marksistskogo učenija o gosudarstve i prave* (1967); etc.

low from the philosophical premises adopted at the outset by those engaged in the classification.

It is generally granted today that the criteria for classification are to be found in the history of legal thought. However, it is precisely in the evaluation of this history that opinions divide and that the philosophical basis adopted becomes a decisive factor.

On the one hand, the major rift between past and present legal science is seen in the abandonment of positivism and the transformation of legal science into a social science.

This is, essentially speaking, the position adopted by Max Rheinstein [62] who, like many other Western writers, places even Marxist-Leninist legal science among the sociological theories, as though it were just one among several.[63]

On the other hand, the Marxist-Leninist conception of the development of legal thinking views even sociologically inspired (non-Marxist) legal science as the result of the development of bourgeois legal thinking in Europe and the United States in the nineteenth and twentieth centuries.[64] Although fully aware of the differences separating the positivist, neo-Kantian and other schools of law from the sociological school (and all its variants), as well as of the historical and social origins of these differences, it regards all these schools as the heritage of the same *de cujus*, namely, the bourgeois legal mind, and it is between Marxist-Leninist legal science and all the others that it draws the great, insuperable dividing line.

It is on the basis of this classification – which corresponds, broadly speaking, to the political division of the contemporary world ('broadly speaking' because we are fully conscious of the specific problems in the development of newly independent countries) – that we shall identify, in the following study, the main theoretical currents which can be seen in non-Marxist legal science, namely those which prevail in the countries which are, somewhat inaccurately, called Western, and the major currents which are visible in Marxist-Leninist legal science, representing the predominant theoretical conception of law in Socialist countries.

Before coming to our description of the major theoretical currents, however, we should like to state, in order to avoid any misunderstanding, that the deep and unbridgeable ideological chasm separating Marxist-Leninist legal science from the others in no way implies that communication between them cannot take place at the theoretical level. Experience, to which we shall again

62. See Section Three, sub-section II, 1 below, pp. 975 sqq.

63. 'The development and spread of the Marxist-Leninist theory of law is of course an event of great importance. But is it not that that theory constitutes a particular branch of the sociological theory ...? The rise of the sociological approach to the law has been the most important trend in modern legal science, in its Marxist-Leninist form as well as in its American and in its European-Continental variants.' (Comments, p. 5.) Cf. also Brimo, who places the Marxist doctrine in the same current as sociological and even positivist doctrines, etc. (Cf. *Les grands courants* ... (1968), pp. 217 sqq.)

64. Čhikvadze, Comments, item 3.

have occasion to refer, shows that although Marxist-Leninist legal science and the other schools are irreconcilable on the ideological plane, such opposition does not exclude either mutual understanding of motivations and standpoints (which obviously does not entail their acceptance), or international co-operation in legal science and legal thinking (congresses, symposia, joint undertakings, etc.), particularly in the vast sphere of comparative law, the legal basis of foreign trade and legal problems in regard to peace, as well as the legal implications of the great advances made in science (for example, space law, legal problems arising from the peaceful use of atomic energy, those raised by organ transplants, the possibility of applying cybernetic methods to legal science) and so forth.

II. WESTERN COUNTRIES

In classifying the major theoretical currents which subsist in legal science today we shall take as a starting-point the solution given to the fundamental epistemological problem, that of the relation between law and human society. On this conception depends the reply to the question as to how law is to be viewed: as a social phenomenon (that is to say through knowledge of the social relationships to which law is bound) or independently of the given social relationships.

This is without doubt the fundamental epistemological question. But in order the better to identify the major theoretical currents discernible in contemporary legal science it will be more useful to pose the question differently and to combine it with the 'genetic' (and indeed older) question, which is: where does law come from?

From this angle we can marshal the various conceptions of the notion of law, and on that basis the major theoretical currents, into several categories. Starting with this criterion of origin we can distinguish in the following manner the most significant theoretical currents that motivate contemporary legal science in the Western countries:

(a) The oldest and most persistent is represented by *dualist theories*, that is to say by theories which display a certain recognition of 'natural law' or which, in other words, accept the hypothesis of the dualism of 'given law' (positive law) on the one hand and 'natural law' on the other.

(b) *Theories which limit the subject-matter of legal knowledge to the law in force* and which oppose both dualism involving natural law and any 'meta-juridical' research, that is to say any attempt to relate scientific knowledge of law to knowledge of the social data which determine law and its social impact.

(c) *Sociological theories,* which vary widely but which all have as common denominator that they regard law as a social product, an order arising out of human society and acting within it.

Let us note, however, that this categorization is far from exhausting all the innumerable theories which have ever been held in the past or which are advanced today; its sole object is to impose a classification upon those theories which are still active, which exert a significant influence upon the current evolutionary trends of legal science and which are themselves manifestations of those trends.

1. *Dualist theories (natural law)*

(a) Theories of natural law are concerned with the origins of law – not that law, however, which men have given themselves in fact but another law, a higher law. It follows that they seek to find the origins of law outside society: in the Theory of Ideas (Plato), in Divine creation (the theological version of natural law), in the power of the human reason (the rationalist version of natural law) to give only a few typical examples.

These theories are distinguished by their remarkable tenacity: a hundred times vanquished, they have a hundred times reappeared in the arena of legal science; and there is every justification for speaking of the 'eternal return' of natural law.[65] The different conceptions of natural law change in aspect, of course, but what they retain in common (and this explains their tenacity) is a determination to complete, even to correct, the law in force which, as age-old experience has shown, is indeed often imperfect, incomplete and inequitable, by law that is better, more complete and more just. Nevertheless, within this common framework, innumerable and important differences have separated – and continue to do so – the various schools of natural law; moreover, the notion of natural law itself is rather imprecise. Thus Zippelius [66] holds, in contrast to the rather broad view of natural law which we have just indicated, that in order to be considered a theory of natural law it is not sufficient for a theory to recognize the validity of principles of justice which are independent of positive law; it must also lay down a criterion for determining what is just.

(b) One of the most powerful schools of the eighteenth and early nineteenth centuries was the *rationalist* school of natural law. The conception that it defended, representing at the time an undoubted step forward in legal

65. ROMMEN, *Die ewige Wiederkehr des Naturrechts* (1947). Cf. BRIMO, *Les grands courants ...* (1968), pp. 395 sqq.; HAINES, *The Revival of Natural Law Concepts* (1930); PÉTERÍ, 'O nekotoryh čertah doktriny "vozroždennogo" estestvennogo prava' (1963, 1969); SEIDLER, 'Odrodzenie prawa natury v amerikańskiej jurisprudencji' (1961); SERRANO VILLAFAÑE, 'L'apport du droit naturel au droit positif' (1957); etc.; cf. also 'Works on the survival of natural law' in *Bibliography of German Law* (1964), pp. 58 sqq. After the Second World War, natural law exerted an attraction even for certain writers who had previously been quite remote from this doctrinal position. Thus, for example, G. RADBRUCH, one of the best known representatives of the theory of values (of the so-called neo-Kantian school of Southern Germany), is considered, primarily because of his study 'Gesetzliches Unrecht und übergesetzliches Recht' (1946), as representative of the rebirth of natural law after the Second World War.

66. ZIPPELIUS, *Das Wesen des Rechts* (1969), p. 72.

thought, was however not highly conducive to creative action in legal science. An illustration of this aversion to the creative rôle in legal science will be found in the history of that science in Austria in the early nineteenth century, as represented in the first place by that brilliant jurist Franz von Zeiller.[67] The school of natural law inspired by von Zeiller was the creator of the Austrian Civil Code (ABGB) of 1811, a remarkable monument of the legal thinking of its day. Hence von Zeiller's rationalist school was itself an eminently creative force while denying any such intention. To measure this paradox it is enough to refer to the Austrian Civil Code itself, in particular to Articles 16 and 7 which can be regarded as its 'profession of faith'. According to Article 16 everyone has 'inborn rights' (*angeborene Rechte*) recognizable *by reason alone*; Article 7 is concerned with gaps in the law and lays down that a judge who finds himself unable to settle a case by relying on the text of the law, either by recourse to its natural meaning or even by analogy, shall reach his verdict by reference to *natural* principles of law; at that stage he accordingly administers natural law. It follows that neither legal science nor the judge create law, or ever can create it: they can only recognize it. This conception, after the adoption of the 1811 Code, led to a complete rejection or negation of legal science for a considerable time. Based on knowledge of '*inborn rights*', therefore resting on 'reason alone', the Code was regarded as *ratio scripta* and legal science replaced by a mere exegesis of legal texts.

It is accordingly not surprising that the rationalist conception of natural law, which played such a dominant rôle in Europe towards the end of the eighteenth century and the beginning of the nineteenth, should have vanished without trace after accomplishing its historic task. Nothing in our judgement leads us to think that there is the least prospect of its revival.

(c) Although, however, contemporary opinion looks with disfavour on the rationalist conception of natural law, and not merely on that alone, but on the doctrine of natural law in general, we can nevertheless detect some new signs that seem to indicate the re-emergence of this latter doctrine and reflect a growing interest in natural law. Thus 'natural law' was one of the subjects discussed at the VIIth International Congress of Comparative Law (Uppsala, 1966),[68] and one of the three themes of the Symposium held in 1963 in New York by the New York University Institute of Philosophy.[69]

It seems to us, however, something of an exaggeration to say, as Thieme does in his general report to the Uppsala Congress, that 'the era of natural law is one of the most recent, and still holds sway in our time'.[70] Natural law, although its influence is still appreciable, is far from playing a dominant rôle in current trends in legal thought.

67. *Das natürliche Privatrecht* (1819).

68. *Rapports généraux au VIIe Congrès international de Droit comparé, Uppsala, 1966* (1968) (cf. THIEME, 'L'apport du droit naturel au droit positif', pp. 74 sqq., and the bibliography given in this article).

69. HOOK (ed.), *Law and Philosophy* (1964), pp. 105 sqq.

70. *Rapports généraux* (1968), p. 79.

It is true that the rebirth of natural law we are witnessing is a reaction against legal positivism 'which has failed, which has been incapable of finding a solution for the evils brought about by its own doing'.[71] However, the real vanquisher of positivism was not, in fact, natural law but, as we shall see below, the sociological conception, which is opposed just as much to positivism as to natural law.

It must also be realized that natural law is the subject of criticism by many authors and is rejected not merely by supporters of the sociological conception of law, but by those holding different points of view.[72]

In addition it may be said that certain current conceptions of natural law are very far removed from the notion of natural law in its classical acceptation (namely, law superimposed on 'human' law) and that what we today call 'natural law' sometimes bears such a slight resemblance to classical dualism that we may well ask if certain dualist conceptions we shall refer to below can properly be regarded as natural law doctrines.

(d) The rebirth of natural law (in the fairly wide sense that we have adopted) can be noted in particular in the following trends:

(i) It is almost exclusively as *Catholic* natural law that natural law, in its 'pure' form, is recognized in contemporary legal science, the strongest current in this domain being the neo-scholastic (particularly in the Thomist acceptation).[73] The only phenomenon that can be regarded as denoting a genuine revival of the classic doctrine of natural law is a certain renascence of the Catholic conception of natural law, in which a contrast is made on sharply dualistic lines between law as created by man and the Divine natural law. Even so this is not a 'renascence' in the full sense, since the Catholic church has never surrendered this dualistic conception of law. We can therefore only speak of a revival of that conception in the sense that its influence is currently felt in countries such as the United States,[74] the Federal Republic of Germany,[75]

71. Serrano Villafañe, 'L'apport du droit naturel . . .' (1957), p. 73.

72. Cf., for example, Rheinstein, Replies, 4: '. . . to the bulk of American legal scholars the term [i.e., natural law] is suspect. It is too reminiscent of past abuses.' See also criticism levelled at natural law by Bobbio, *Giusnaturalismo e positivismo giuridico* (1965) and the bibliography listed there, and by Zippelius, *Das Wesen des Rechts* (1969), pp. 79 sqq. It is clear that natural law, whatever its acceptation, is necessarily rejected by Marxist-Leninist legal doctrine. Cf., for example, Seidler who regards the rebirth of natural law in the United States as support for imperialist theories of law (*Doktryny prawne imperializmu*, 1957); see also Naschitz, '"Le problème du droit naturel" à la lumière de la philosophie marxiste du droit' (1966); and Péteri, 'Influence of natural law on positive law' (1966); and so on.

73. Serrano Villafañe, *op. cit.* (1957), p. 73; Rheinstein, Replies, p. 4 (cf. Rheinstein, 'United States of America' (1970), p. 450); Aomi, Contribution of the Japanese National Commission, p. 10 (cf. Aomi, 'Japan' (1970), p. 383); etc.

74. Cf. Rommen, 'In defense of natural law' (1964); cf. also *Natural Law Institute Proceedings* (1947-1951). See further, Rheinstein, Replies, p. 4 and works quoted there (cf. Rheinstein, 'United States of America' (1970), pp. 450-451 and footnotes [61] and [72]).

Spain,[76] Italy,[77], Austria,[78] and so on. Traces of this influence are also found in Japan [79] and, of course, in many other countries.

(ii) In countries imbued with a traditional respect for written law (especially France and Italy) we can observe evidence of some reaction against the omnipotence of the *lex scripta* and a tendency to subject it to critical evaluation. But it is not possible to undertake such an evaluation of positive law starting from within itself. (Positive law is a tautological term: only what corresponds to positive law is just.) [80] Hence the tendency to seek criteria of assessment which lie outside positive law and to show at the same time that law is not only what has been created by the legislator but something incorporating other legal norms such as those set up by the courts, customary norms, currently accepted principles of human behaviour, legal standards, etc. – the whole also being sometimes termed 'natural law'.

This point of view, which, moreover, comes fairly close to the sociologically inclined conceptions of law, is not very widespread; its strength lies not in the number of writers who uphold it but in the regard in which they are held from the standpoint of scientific merit. Among these authors are, for example, in France, André Tunc and René David. The former recalls the current trend towards recognition of the general principles of justice (such as the widespread conviction of the individual dignity of man, equality among men, etc.),[81] whereas the latter relies on natural law in rejecting positivism and acknowledges *expressis verbis* that natural law belongs to the domain of morality.[82]

A similar position is taken in Italy by Norberto Bobbio, who writes: 'If, when we use the term "natural law", we mean to refer to all norms the validity of which does not flow directly from their dependence vis-à-vis an authority delegated to issue norms for a given system, we are fully entitled to claim that the subject-matter of legal science, viewed as research into living law and not only as the interpretation of established law, is also "natural law".' [83]

76. Serrano Villafañe, *op. cit.* (1957).

77. Passerin D'Entrèves, *The Natural Law* (1950); also Giannini, Replies, p. 1 (cf. Giannini, 'Italie' (1970), pp. 462 sqq./'Italy', pp. 422 sqq.).

78. Cf. Knoll, *Katholische Kirche und scholastisches Naturrecht* (1962).

79. In the writings of Kotaro Tanaka quoted by Aomi, Contribution, p. 10 (cf. Aomi, 'Japan' (1970), p. 383).

80. The author of the present study has made a detailed examination of this problem in his work *Filosofické problémy socialistického práva* (1967).

81. Comments. We should also note that the nature of the general principles of law was the subject of discussion at the VIth International Congress of Comparative Law, Hamburg, 1962 (cf. *Rapports généraux*, 1964).

82. Replies.

83. 'Si, en recourant à l'expression "droit naturel", on entend indiquer toutes les normes dont la validité ne dérive pas directement de leur dépendance à l'égard d'une autorité déléguée pour produire les normes du système, on est pleinement autorisé à dire que l'objet de la science juridique, entendue comme une recherche du droit vivant, et non seulement comme l'interprétation du droit déjà établi, est aussi le "droit naturel".' (Special contribution: 'L'objet de la science juridique').

The Scandinavian countries have recently seen the re-emergence of the idea of natural law in the work of F. Castberg.[84] However, although he speaks explicitly of natural law, Castberg's theory too would seem to be more closely allied with the conceptions of R. David, A. Tunc and N. Bobbio than with the 'classic' theories of natural law, his central idea being the recognition and applicability of value judgements in the field of law (and, of course, of legal thinking), particularly in regard to questions on which positive law is silent ('gaps in the law').[85] A similar conception is expressed in the United States by, for example, Fuller.[86]

(iii) Natural law and, where applicable, natural justice play a highly important rôle in the current processes of acculturation of law in many countries, first and foremost in those having recently attained independence but also in others where a process of adaptation is necessary in legal matters.

At the same time, as we shall see later, we are not here concerned with natural law in the classic sense (conforming, that is, to the general concept of natural law as understood in the history of European law) but with natural law (or natural justice) *sui generis*. In cases of this sort what we term natural law is in fact nothing else than the application of the general principles adopted, or even created, by the courts and taken as having the character of sources of law. It is clear that legal science can play a highly active rôle in the elaboration of such general principles.

(iv) We shall therefore conclude, by way of recapitulating the above review, that as far as natural law (natural justice) is concerned the following trends are visible today in legal science:

an appeal to natural law by way of reaction against positivism and against belief in the omnipotence of the *lex scripta*, either by a revival of the influence of the Catholic (especially neo-scholastic) doctrine of natural law or by recourse (expressed in rather vague terms so far) to the general principles of morality as sources of law;

recourse to natural law as an important medium of legal acculturation, particularly during the construction of the legal systems of newly independent countries.

(e) We have so far considered 'natural law' as the subject-matter of current trends in research in our discipline. But the concept of law is not only a theme for an essay on the trends of research; it is also one of the presuppositions or conditions of such trends, and the concept of natural law is no exception to the rule. We must therefore put to ourselves the following question: are current theories of natural law capable of stimulating, and if so to what extent, certain new trends in legal research in general?

84. Cf. Castberg, *Forelesninger over rettsfilosofi* (1965); by the same writer, *Naturrett og menneskerettigheter* (1967) (translated into English as 'Natural law and human rights', 1968); also, in French, *La philosophie du droit* (1970).

85. Opsahl, in Eek, Replies, pp. 6-7.

86. Fuller, *The Morality of Law* (1969).

What can be said on this issue amounts to little. The Catholic doctrine of natural law, which has been the subject of uninterrupted research for centuries, is, in our view, unlikely to contribute anything really new. We may attach a certain degree of importance to the opinions mentioned under point (d) (ii) in so far as they favour an extension of the horizons and interests of legal science and act as a stimulus to research on the different kinds of legal instrument or on those social factors which fall outside the purview of legal science to the extent that the latter is confined to knowledge of written law. (This extension appears to us, however, to depend more upon the growth of a sociological conception of law – to which we shall revert – than on the theoretical views mentioned under (d) (ii).)

It is primarily the concept of natural law referred to under (d) (iii) that can be regarded as one of the important sources of new trends in legal research. This movement has so far taken a rather empirical form; but it has already led to an undoubted increase of interest in the new formulations of legal ethnology and anthropology. It has also injected a number of new factors and interests in the field of comparative law. Much more might, however, be expected of it if it were handled in a more theoretical and complex manner, and less closely subordinated to the solution of concrete cases and problems. One could in the first place expect it to make a contribution of considerable importance to the complex study of 'development law', which represents, as we shall see, one of the foremost trends in contemporary legal science.

2. *Theories which limit the subject-matter of legal knowledge to the law in force*

(a) This category includes theories which, at the risk of appearing unduly severe, we do not hesitate to describe as not in the least concerned with the origins of law but simply with the fact that law exists, or, possibly, that it is valid. It is clear – and no comment is needed – that from the epistemological point of view such theories cannot reckon with social factors – not being concerned with the social origins of law they cannot, for the purpose of studying it, have recourse to social factors. It is surprising (but in fact quite logical) that within this category diametrically opposed and indeed irrevocably antagonistic theories join hands, such for example as legal positivism on the one hand, and on the other the extreme form of legal neo-Kantianism or, in other words, in its various forms, Kelsen's pure doctrine of law (*reine Rechtslehre*), also sometimes called the 'normative theory'.

These two theories were at one time highly novel and important, powerful waves created by new forms of thinking in legal science, namely positivism in the second half of the nineteenth century as a reaction against neo-Kantianism, and neo-Kantianism itself (in its 'normative' form) in the first third of the twentieth century as a reaction against positivism. Both were opposed to the infiltration of sociologically inclined trends in legal science.[87]

87. As regards the 'pure doctrine of law', cf. the chief work of this school,

However, it soon became apparent that the range of the 'pure doctrine of law' was somewhat limited, the mighty wave it had created tended to ebb and *reine Rechtslehre* was soon worn out and lifeless, so that today it no longer figures among the major trends in modern legal science.[88]

Legal positivism, the roots of which can be traced back to J. Austin's work (first third of the nineteenth century), is much more tenacious and maintains, at least in some countries (the Scandinavian and Latin American countries, for example, among many others), a fairly strong position. Legal positivism, whose links with the positivist philosophy of Auguste Comte are very loose, does not greatly favour theoretical reasoning. It tends to interpret the law in force rather than deal with the general problems of law or even legal science. It is also worth noting that positivism – we may venture to say – has not since Bergbohm[89] given rise to any representative theoretical work which can be regarded as providing a theoretical or philosophical platform for contemporary positivism.

It is very interesting, however, to note that the last few years have seen the rebirth of a positivist movement in the country where defeat of legal positivism seemed the most thorough-going and complete, namely the United States. This movement was ushered in by a widely noted article by R. S. Summers, setting out more or less the programme, or even the manifesto, of a new school.[90]

The group known as the New Analytical Jurists, represented, in addition to R. S. Summers, by H. L. A. Hart, R. Dworkin, M. Cohen, etc., is fairly heterogeneous. It can be said, however, that this group, notwithstanding all the differences of conception that separate the writers within it, takes its stand on the classical positivism of J. Austin.

In spite of this evidence of a certain revival of positivism in the United States, we can see that present-day positivism constitutes a way of treating law, or a certain attitude in the scientific study of it, rather than a school of thought in the true sense of the term. Although it still retains, as we have said, some fairly important strongholds and even gains fresh vantage ground

---

Reine Rechtslehre by Hans Kelsen (1934); cf. also, by the same writer, *Grenzen zwischen juristischer und soziologischer Methode* (1911).

88. It is not dead however. Traces of it are to be found, and there are even signs of a certain revival of this concept, in countries where the dominant current is quite different as, for example, in the Federal Republic of Germany where legal science clearly tends towards 'sociologism', in the Latin American countries where legal science is essentially still positivistic, etc. In the Federal Republic of Germany it is especially in the review *Rechtstheorie* published since 1970 in West Berlin that there is a detectable trend towards the revival of the 'pure doctrine of law' (cf. the article by Walter, 'Der gegenwärtige Stand der Reinen Rechtslehre' in the first volume of that periodical); as regards Latin America cf., for example, Carvalho, *A posição filosofico-jurídica de Kelsen* (1953); Feliú, *El mundo jurídico visto por Kelsen* (1956); Gutiérrez, *La teoría pura del derecho de Hans Kelsen* (1958); Peralta García, *En torno al idealismo crítico. Un ensayo sobre Kelsen* (1957); Pino Muñoz, *El estudio sobre la teoría del derecho* (1959); etc.

89. Bergbohm, *Jurisprudenz und Rechtsphilosophie* (1892).

90. Summers, 'The New Analytical Jurists' (1966).

(e.g., the New Analytical Jurists), it is, in our opinion, on the decline, being driven back by the sociological or sociologically inclined conceptions of law; and it is unlikely that we can expect a new contribution to legal science from legal positivism.

(b) We may also place in this category the 'legal method' drawn up in Italy and regarded in contemporary literature (especially in the legal literature of the United States) as a specific 'Italian style'.[91]

The choice of the term 'legal method' was a deliberate assertion of a desire to endow legal science with greater rigour, along lines suited to its nature, by separating it more meticulously both from sociological, economic and political science studies and from that complex of scientific studies which was known in Italy, in the last century, as the 'science of legislation' (following Southern thinkers of the eighteenth century, Genovesi, Filangieri, Pagano and others), whereas today the preference is to describe it as the study of 'legislative policy'.

More explicitly, the adherents of this new methodology concentrated their criticism on French legal science in which they saw a mixture of legal and non-legal factors. They also entered into polemics with the legal science of other countries, but arguing on different lines. With regard for example to German science (which was at that time in a dominating position) they held that criticism could be levelled at its propensity to indulge unduly in abstract concepts to which it ascribed an allegedly universal value; whereas they condemned the contrary vice in the science of Common Law countries, namely that of relying too readily upon empirical concepts linked to the contingent character of historical and factual situations.

This new methodological approach led to a fundamental recasting of all legal disciplines in Italy, for it ended by being accepted by research workers specializing in the study of positive law in its various manifestations. And indeed the denunciations most commonly uttered by lawyers in their mutual controversies and polemics were those of 'sociologism', 'politicism', etc. Such accusations were of course frequently unfounded, but the fact remains that the risk of incurring strictures of this kind acted as a powerful factor of intimidation and dissuasion in the evolution of Italian legal science and that, resulting from this, legal research has been continuously confined within the frontiers of a closely supervised scientific rigour.

As will be readily understood, however, scientifically precise determination of the framework of the 'legal method' presented a far harder task. Whereas on a purely empirical plane the drawing of distinctions between legal science, sociology, political science, economics and the study of legislative policy may appear relatively simple, the situation is very different if we

91. The text which follows has been taken, with slight changes and summarizing certain arguments, from the reply to the Questionnaire made by the Italian Consiglio nazionale delle Ricerche (author: Professor M. S. Giannini) (cf. GIANNINI, 'Italie' (1970), pp. 463-464/'Italy', pp. 423-424). Regarding the 'Italian style' (although conceived of in a slightly different way) cf. also CAPPELLETTI, MERRYMAN & PERILLO, *The Italian Legal System* (1967) (Chaps. 5, 6 and 7).

seek to raise the whole issue, as has been the case here, to the epistemological level. It is in fact the exact delimitation of the frontiers between legal and other sciences, and the precise definition of the means the lawyer must use in studying the reality which is the subject-matter of his science, which have occupied the lengthy discussions devoted to the 'legal method'.

It must immediately be said that there has never been any agreement on these issues, and, in consequence, even within the 'legal method' several tendencies have crystallized. Or rather, to be quite accurate, we should say that although the 'legal method' was accepted by Italian legal science as a whole, unanimity later broke up into separate currents all claiming the 'legal method' as their common denominator.

The issue that proved most controversial turned on the problem of what is commonly termed 'deontology', namely, what a jurist should know and what he should do. In this context there is a first line of thought according to which the jurist is not expected to keep himself informed of developments in sociology, politics, etc., nor is he bound to know the techniques of legislative policy. If he nevertheless takes an interest in these matters also, that may be an advantage to him as an individual, provided always that as a jurist he forgets as completely as possible what he happens to know of them. The result, according to the adherents of this line of thought, is that the phenomenon constituted by law consists in a *quid* which can be sought in a complete scientific vacuum, in other words is totally unrelated to any discipline other than legal science itself.

This current finally resulted in a position close to the 'pure doctrine of law'. The majority of Italian jurists, however, grouped themselves in two other currents both of which assert that links with other disciplines are to be maintained. They can be distinguished as follows: according to the first a jurist should be 'acquainted' with the other disciplines, both theoretical (sociology, political science, etc.) and practical (e.g., legislative policy), but should refrain from resorting to them in his legal work; according to the second, on the contrary, the jurist must necessarily and simultaneously conduct research in the other disciplines also; nevertheless, and in continued recognition of the rules of the 'legal method', he must refrain, in his work as a jurist, from mingling the disciplines.[92]

3. *Sociological theories*

(a) Unlike the positivist and dualist conceptions of law, the sociological doctrine sees the law as an essentially *social phenomenon*, both a product of social life and a force making itself felt in society. According to this conception there is no unbridgeable, epistemological barrier between what 'is' (*Sein*) and what 'ought to be' (*Sollen*): each term reacts on the other, and knowledge of the first is a pre-condition of knowledge of the second.

As we know, the sociological conception of law is marked by a somewhat

92. This concludes the passage borrowed from Professor Giannini's Replies.

lengthy and varied history to which we have referred above. Being a reflection of the great change in the social 'climate' which came about in Europe in the middle of the nineteenth century and the rise of industrial capitalism, the sociologically inclined trend has become the strongest and most widespread in contemporary legal science. But – except of the United States where a continuous link can be traced from the New Jurisprudence of the twenties and thirties down to the present time – there can be no question of a direct succession in the former sociological schools or a simple legacy handed down by their representatives.

There is no sociological *school* of law in contemporary legal science. We have here a genuine *trend* without any settled programme or major work such as Eugen Ehrlich's *Grundlegung der Soziologie des Rechts*.[93] It is rather a spontaneous trend, the nature of which is determined by the needs of a modern society divided into two major socio-economic systems and characterized by a process of decolonization completed in the space of one generation; a society that is witnessing the unexampled blossoming of what it is agreed to call the scientific and technological revolution; in short, a society which cannot be content with law as the passive mirror of social relationships but which deliberately uses law as an instrument for the purpose of achieving a transformation of society.

(b) The notion of the sociology of law is somewhat broad and lacking in precision. There is much discussion as to whether this type of study constitutes an autonomous branch of legal science (i.e., one of the 'legal sciences'), or a branch of general sociology, or whether it should be considered solely a method of legal research, etc.

Without wishing to get involved in these debates we shall concentrate in this study on the *methodological* aspects of the question.

From the methodological point of view the sociological trend stems from the hypothesis that the scientific knowledge of law can never be confined to knowledge or analysis of legal norms, but that:

(i) legal science approaches knowledge of law (having regard to the various shades of sociological tendencies) through knowledge of the social data

93. This does not mean that there are no works dealing with theoretical problems of the sociology of law. Many exist and they have a fairly wide geographical range. For example: ALLEN, *Law in the Making* (1927, 1958); CARBONNIER, *Flexible droit* (1969); by the same author, *Sociologie juridique* (1972); ECKHOFF, 'Sociology of law in Scandinavia' (1960); FAROOQI, 'Sociological factors and the law' (1969); GEIGER, *Vorstudien zu einer Soziologie des Rechts* (1947); GORIĆAR, 'Filosofska in sociološka problematika v naši teorii države in prave' (1953); LÉVY-BRUHL, *Sociologie du droit* (1961). Works on the sociology of law are, of course, found in other countries as well, e.g., in Italy, studies by N. Bobbio, R. Orestano and M. S. Giannini in the review *Jus*, 1957; in Japan, works by Takeyoshi Kawashima and Michitaka Kaino; and so forth. Mention should also be made of the Colloque international de Sociologie du Droit et de la Justice (International Symposium on the Sociology of Law and Justice) held at Brussels in April 1969 (see the '*Actes*' of the Symposium, 1970). Cf. POCAR & LOSANO, *Sociology of Law 1960-1970. A Bibliographical Survey* (1970). See also KULCSÁR, 'Sociologizm v buržuaznoj teorii prava' (1963, 1969).

which determine law, the social surroundings from which law draws life or in which it exists as 'living law', etc.;

(ii) the sociological study of law requires a parallel examination of the impact of law on society.

Thus legal sociology regards law as a living force of which we must not only acknowledge the existence (and, as the case may be, the validity), but also determine the origin and the effect on society. There is accordingly an unceasing and fruitful tension between the 'is' and the 'ought to be', between society and the law which is born in society and acts upon it, a mutual tension the knowledge of which is an essential preliminary to the knowledge of law.

This is, of course, only a very general definition of the contemporary sociological current. Within this current there are several conceptions which can be broadly divided into two main categories: *theoretical sociology* of law, which attempts to interpret the general problems of law in sociological terms (i.e., to elaborate a sociological theory of law), and *applied legal sociology* which, without seeking to frame such a theory, treats positive law by recourse to sociological methods.

This distinction (far from exhaustive, of course) finds its corollary in the United States where *sociological jurisprudence* (see above), as a method of legal thinking and a legal methodology which can be applied just as much in legal science as in legal practice, is distinguished from *sociology of law*, which is a branch of scientific activity observing and studying the law as a social phenomenon.[94] Those favourable to the sociological treatment of law in the United States sometimes go so far as to accept – in the words of Max Rheinstein – 'the postulate of the integration of the law with the social sciences as voiced by such protagonists as Huntington Cairns or Jerome Hall'.[95]

Sociologically inclined trends, often influenced, consciously or unconsciously, by Marxism (even if its philosophical premises are not accepted) [96] are widespread. They are, for example, very pronounced in Japan. They are also manifest in the Federal Republic of Germany, where the interrelation of constitutional law and of social factors is one of the main subjects of re-

94. See p. 949, n. 59.

95. Replies, 7 (cf. RHEINSTEIN, 'United States of America' (1970), p. 449 and n. [49]).

96. Cf. for example, CARBONNIER, *Flexible droit* (1969), p. 151: 'Even though we resist it, even if we do not realize it, there now hangs over our conceptions of property the great legislative shadow from the East'. (See *ibid.*, p. 183: '. . . an analysis with which Marxism has made us familiar'). ('. . . une analyse avec laquelle le marxisme nous a rendus familiers'). See also B. S. Murty, Replies, p. 2, who states: '. . . Marxian socialist thinking, which is likely to become considerably popular in the decades to come . . .'; HAZARD, 'Marxist models for West-African law' (1969); by the same author: 'La familia de sistemas jurídicas de inspiración marxista en Africa' (1967). We should note, however, that interest in Karl Marx goes back, in the case of legal science, to the inter-war period (e.g., in Sweden, Hägerström, Lundstedt) and even to the early twentieth century (Krčmář in Bohemia, for example).

search in the field of *allgemeine Staatslehre* (general theory of the State) [97] and sociologically inclined tendencies are also evident in the context of *genetische Interessenjurisprudenz* and in the doctrine known as *kausales Rechtsdenken* – causal legal thought.[98] The legal science of the German Federal Republic, according to Rheinstein, also displays at the present time another current, which is connected with the problem of judicial law-making; we shall return to this shortly. In France, the sociological study of law has hitherto been generally conducted by philosophers who were not lawyers, and in language which lawyers have difficulty in understanding. Even in France, however, an evolution is now in progress in this field, lawyers having become conscious of the need to study the sociological aspects of law.[99] The trend towards sociology is less marked in Italy, even though the 'Italian style' itself is, as we have seen, not entirely devoid of a sociological element – Norberto Bobbio, in much the same way as the American authors referred to above,[100] regards as one of 'the most significant manifestations of the "vocation" of legal science today ... the elimination of the dividing line between the lawyer's task and that of the sociologist, resulting in the tendency to turn the lawyer into a "social scientist" '.[101]

One of the sociologically inclined trends is linked, as Rheinstein tells us, to increased participation of judges in the creation of law in countries where written law is paramount. 'Judicial law-making necessarily implies the danger of arbitrariness. Efforts are therefore underway to work out directives for the guidance of courts which have so far not been accustomed to large-scale judicial law-making.' [102] According to Rheinstein, this current is most noticeable in the Federal Republic of Germany.[103]

This current however has given rise in that country, as Rheinstein says,[104] to a *Methodenstreit* – a discussion about methods – in which the current which has just been mentioned is set against a current which on the contrary

97. Zippelius, Replies, p. 2.

98. LARENZ, *Methodenlehre der Rechtswissenschaft* (1960, 1969), pp. 47 sqq.

99. David, Replies, p. 3. The sociological current in French legal science at present is represented primarily by the work of H. LÉVY-BRUHL and J. CARBONNIER.

100. See p. 962, n. 95.

101. Bobbio, Special contribution: 'L'objet de la science juridique', p. 12: '... expressions les plus importantes de la "vocation" de la science du droit de nos jours ... la rupture de la ligne de démarcation entre l'activité du juriste et celle du sociologue, avec, pour conséquence, la tendance à transformer le juriste en "*social scientist*" '.

102. Rheinstein, Observations of the Revised Report, I, 7a.

103. Rheinstein cites as examples LARENZ, *Methodenlehre der Rechtswissenschaft* (1960, 1969); ESSER, *Grundsatz und Norm in der richterlichen Fortbildung des Privatrechts* (1956); RAMM, *Einführung in das Privatrecht. Allgemeiner Teil des BGB* (1969); Marxist-Leninist legal science however subjects this trend to critical analysis and considers it a sign of the decadence of the principle of legality in capitalist countries. (See Section III, sub-section (a)(iv), p. 971 and n. 127 below.)

104. In the lecture which he gave before the Society of Legal Studies of the Supreme Court of the Federal Republic of Germany (quoted from manuscript notes supplied by the author).

stresses *Gesetzestreue* – fidelity to the law or, we should say, legality – which therefore confronts a possibly over-emphasized volitional element with the hermeneutic element in judicial decisions, thus tending to create a new form of legal dogmatism.

(c) In addition to what has already been said concerning sociological jurisprudence in the United States, a general observation remains to be made in regard to the sociological method in the Common Law sector, where specific problems arise.

The Common Law system stands in a paradoxical situation: it is at the same time favourable and unfavourable to the sociological trend in legal science, according to the viewpoint adopted.

It is true that judge-made law is more dynamic than written law and that the barrier between 'law in the making' and positive law is much less sharp in Common Law than in written law. This is what enables the judge, under the Common Law dispensation and as the creator of law, to react with greater suppleness and sensitivity to social factors. It is doubtless equally true that 'the question of using law consciously as an instrument of social change is an important part of the discussion of the doctrine of precedent'.[105] Such, in very brief terms, are the elements that under Common Law militate in favour of the penetration of sociologically inclined trends.

What is opposed to these trends is the conservatism of Common Law as mainly exemplified in the rule of *stare decisis*. It is true that this rule has been less rigid in the United States, where the Federal Supreme Court has never hesitated to modify, if necessary, its own precedents, and that even the English courts, which, in the Common Law world, were the most conservative of all, have always found, despite the rule of *stare decisis*, ways of keeping the evolution of law in step with social development. The fact remains that because of the rule of *stare decisis* Common Law, especially in its English form, has over the centuries become a far more set type of law than French law, for under the latter, and despite the well-known Article 5 of the Civil Code which forbids judges from pronouncing on cases submitted to them by 'decisions having a general and regulatory character' ('par voie de disposition générale et réglementaire'), court judgements and even legal science play a very important part in the process of adapting law to social realities. On this point also we must agree with Lawson who, speaking of Common Law, writes as follows: 'Law is certainly determined largely by social data and by their historic evolution, but much that is not easily explained by modern social conditions has survived from earlier ages in which it had a rational basis. In other words quite a fair amount of law is somewhat out of date.'[106] Scientific understanding of social change, and particularly of changed conditions of an ideological and technical character, as determinants of law, has been one of the reasons why American courts, and especially the Supreme Court of the United States, have abandoned rigid enforcement of the rule of *stare decisis:* the American courts are now inclined in

105. Lawson, Replies, p. 2.
106. Lawson, Replies, p. 1.

case of need to depart even from very long-standing precedents.[107]

The rule of *stare decisis* has necessarily been modified in Common Law countries having recently attained independence, owing to the fact that the inevitable process of acculturation of law, making possible the emergence of a new system of law adapted to each country's needs, militates against rigid application of the rule inherited from the former mother-country.

A last but not negligible point is that social necessities have had the effect of weakening the rule of *stare decisis* even in its country of origin, England, where it is no longer binding, at least in respect of judgements of the House of Lords.

The reasons for this change are highly interesting from the standpoint of this analysis, for they reflect on the one hand the respect accorded to acquired rights and the importance attached to stability in jurisprudence, and on the other the concessions made to the evolutionary needs of law.

These reasons were stated by the Lord Chancellor on 26 July 1966 in the following terms:

> Their Lordships regard the use of precedent as an indispensable foundation upon which to decide what is the law and its application to individual cases. It provides at least some degree of certainty upon which individuals can rely in the conduct of their affairs, as well as a basis of orderly development of legal rules.
>
> Their Lordships nevertheless recognize that too rigid adherence to precedent may lead to injustice in a particular case and also unduly restrict the proper development of law. They propose therefore to modify their present practice and, while treating former decisions of this House as normally binding, to depart from a previous decision when it appears right to do so.
>
> In this connexion they will bear in mind the danger of disturbing retrospectively the basis on which contracts, settlements of property and fiscal arrangements have been entered into and also the especial need for certainty as to the criminal law.
>
> This announcement is not intended to affect the use of precedent elsewhere than in this House.[108]

As evidence of the fact that even in England (and Canada and Australia) it is no longer held that traditional methods suffice for the purpose of adapting the law to meet social needs, mention should be made of the very wide terms of reference of the official Law Commissions in these countries which are, as far as we know, empowered to propose codifications.

(d) In countries having recently attained independence, sociologically inclined trends usually manifest themselves in connexion with problems of acculturation in the legal field. On the other hand problems of the 'acculturation of law', while their nature remains sociological, do not arise solely in this group of countries; their 'geography' is broader and includes certain countries which have long enjoyed independence and even, in some respects, countries in which the 'modernization' of law has for many years been successfully accomplished.

The problem of the acculturation of law is of an entirely specific character and is dealt with separately in Section Three.

107. See above, p. 964.

108. [1966] 1 W.L.R. (= Weekly Law Reports) 1234 and [1966] 3 All E.R. (= All England Law Reports) 77.

III. SOCIALIST COUNTRIES (THE MARXIST-LENINIST THEORY OF LAW)

1. Marxist-Leninist legal science is based on Marxist-Leninist philosophy. It looks upon both State and law from the viewpoint of the Marxist-Leninist conception of the world and employs the method of dialectical and, more particularly, historical materialism.

It is however a fact that neither Marx nor Engels, nor Lenin ever set themselves to work out a new legal theory. Nevertheless all three (Marx and Lenin in particular since they were both, as is well-known, trained in the law) were frequently occupied, in various political and economic contexts, with the philosophical problems of law, so that their works provide a solid basis for a new conception of law and a new legal methodology, in short for the building of a Marxist-Leninist legal science.

This new science had its birth in Soviet Russia and became generally recognized first in the USSR and, after the Second World War, in the other socialist countries. A fairly vigorous Marxist school of legal science also exists in Japan,[109] while the influence of the Marxist conception of law is likewise manifest in various forms in the legal science of several European countries, as well as in the development of law in certain African States.[110]

2. It follows from what has just been said that the starting-point of Marxist-Leninist legal science rests upon a philosophical conception of human society. Marxism-Leninism regards this as a complex structure in which the chief rôles are played by social classes and their interrelationship as it has been built up in the course of history, and as it determines the character of society. It must be added however that this concept of human society and its development necessarily includes a recognition of the historical rôle of the proletariat (i.e., the working class).

Both State and law are viewed by Marxist-Leninist legal science as a product of society in the sense that they are both created by the ruling class (or, at a given stage of evolution of socialist society, by the people as a whole).

From this postulate derives the historical approach which characterizes the Marxist-Leninist conception of law. Law is held to evolve in line with history, not as an independent creation of the human intellect but as an element in the evolution of society, which means that the historical development of law as a factor in the social *superstructure* is determined by the evolution of the social *basis,* i.e., in the final analysis, by the evolution of economic relationships within society.[111] Such are, briefly, the philosophical conceptions underlying law from the Marxist-Leninist point of view.

3. One of the characteristic features of Marxist-Leninist legal science is the indissoluble epistemological link between knowledge of the State and knowledge of law (witness the terminology already alluded to, which frequently has recourse to the expression 'science of State and law').

109. Represented by names such as Fujita, Hirano, Numata, etc.
110. See p. 962, n. 96.
111. See 5 below, pp. 967-968.

This is in no way fortuitous but results from the fact that according to Marxist-Leninist thinking, law invariably derives from the State, is always State law. From this point of view Marxist-Leninist legal science is strictly monistic and hostile to any notion of natural law having its origins outside the State (or, *a fortiori*, outside human society). As just stated, Marxism-Leninism regards the State as an organization emanating from the ruling class (or, at a certain stage in the evolution of socialist society, from the people as a whole). In a society divided into antagonistic classes the State is a means in the hands of the ruling class for the suppression of the hostile class (or classes as the case may be). Even after the political and economic liquidation of the *bourgeoisie* as a social class in socialist society, the State remains the mechanism used by the ruling class (or by the whole people, as the case may be) to protect socialist society against such subversive elements as may have survived in the country or may enter it from outside; in addition, as stands to reason, the socialist State is also an instrument for protecting society against anti-social individuals in the criminal sense of the term. An element of force, of compulsion, is therefore always implicit in the Marxist-Leninist concept of the State.

Law is thus a means of State compulsion. As such it is conceived as the expression of the will of the ruling class – a will, it goes without saying, which is not indeterminate but which has been shaped by the material conditions of life in the given society.[112] This is the starting-point of the Marxist-Leninist conception of law. It follows that from a methodological and epistemological point of view law, as an object of scientific knowledge, is not something independent; it can only be recognized as an element in the interaction between social reality and legal reality, through an understanding of the social facts which determine it and of its feedback action on society.

4. Being monistic in the above sense, Marxist-Leninist legal science is nevertheless not confined to positivistic research into the law in force (positive law). It perforce involves the study of the relationship between the 'is' and the 'ought to be', viewing that relationship not only from the standpoint of logic but in broader contexts. This relationship between 'is' and 'ought to be' is also regarded as one between the existing and the possible, between present and future, etc. Furthermore, the notion of 'ought to be' incorporates 'the ideal of justice as a social and ethical yardstick for the evaluation of law and its institutions'. Yet the link between 'is' and 'ought to be' is always, in the final analysis, determined by the links between the basis and the superstructure of society (see below), and ethical principles and ideals (e.g., that of justice) always derive from 'social reality and not from an eternal and immutable "natural law" '.[113]

5. We have just mentioned the relationship between the social basis and the superstructure, which represents the backbone of the Marxist-Leninist

112. Marx-Engels, *Manifest der Kommunistischen Partei*, Chap. II.

113. The quotations are taken from the Replies (p. 5) of the Institute of State and Law of the Academy of Sciences of the USSR.

conception of human society and of its historical development.

Law as an element of the social superstructure is ultimately determined by the social basis and its development, that is to say, if we may have recourse to simple terms, by the material conditions (economic relationships) peculiar to the given society, which involve also the fundamental relationships between classes.

The social factors – and, in the end, the social basis – which determine law are the motive power which promotes the historical development of law and its adaptation to the changes in the social factors by which it is determined. The law, in turn, as a component of the social superstructure influences the society which has moulded it, and its historical evolution.

A moment of great significance in this reciprocal action is that of the social revolution when the revolutionary forces destroy the current structure of society and replace it by a new structure matching their own interests, entailing also the destruction of the preceding legal system (even if it remains formally in force) [114] and the emergence of a new kind of law. An obvious example of this process is the proletarian revolution which, having eliminated the old law, creates a new law of a socialist type.

6. These explanations have appeared necessary in order to set out the unvarying foundations of the Marxist-Leninist legal conception of law [115] and to make intelligible the limits within which new trends appear within the Marxist-Leninist science of State and law.

The significant trends which can be discerned within this science in the USSR – and, for the most part, in other socialist countries as well – are the following: [116]

(a) *Problems within the general theory of State and law.* From the outset, Marxist-Leninist legal science was opposed to the idealist philosophy of law and built up for itself a *general theory of State and law.* This theory, based on Marxist-Leninist philosophy, covers the study of the general problems of State and law, including their philosophical, sociological, ethical and other aspects, as well as the study of questions of a methodological character.[117]

It is possible to detect, within the general theory of State and law, a number of problems which are attracting the attention of the researchers and whose study can legitimately be regarded as providing evidence of current trends in this field:

114. ENGELS, *Ludwig Feuerbach*, Chap. IV.

115. See ALEKSEEV, *Obščaja teorija socialističeskogo prava* (1963); BRATUS *et al., Obščaja teorija sovetskogo prava* (1966); ČHIKVADZE *et al., Marksistsko-leninskaja obščaja teorija gosudarstva i prava* (4 vols., 1970-1971-1972-1973). DENISOV *et al., Teorija gosudarstva i prava* (1967); ROMAŠKIN *et al., Teorija gosudarstva i prava* (1962); SZABÓ, *A jogelmélet alapjai* (1971) (French version, *Les fondements de la théorie du droit*, 1973); etc.

116. In describing these trends we have drawn to a large extent on the Replies of the Institute of State and Law of the Academy of Sciences of the USSR, and Čhikvadze's Comments.

117. Concerning the conception of the general theory of State and law, cf., for example, CHIKVADZE *et al., Marksistsko-leninskaja obščaja teorija gosudarstva i*

(i) One very marked trend that has been evident for a number of years past is towards the intensive study of methodological problems.[118]

It is a well-known fact that the general and fundamental methodology of Marxism-Leninism is dialectical materialism, and more especially historical materialism, the latter being the application of dialectical materialism to an understanding of human society.[119]

Within the context of this general methodology, however, we can see among the special sciences (including legal science) a vigorous movement towards a solution of the methodological problems which relate specifically to each, and the discovery of new methods.

One of the most important trends in this connexion is a mounting interest in the solution of problems relating to the application of the sociological method to research on the science of State and law or, to adopt the expression which Soviet authors in particular prefer, the 'concrete sociological study of reality'.[120]

The main purpose of the concrete sociological study of reality is to investigate the effectiveness of legislation, the methods of State and

---

prava, Vol. I (1970), Chap. III; DENISOV *et al.*, *Teorija gosudarstva i prava* (1967); IOFFE & ŠARGORODSKIJ, *Voprosy teorii prava* (1961); KERIMOV, 'O predmete obščej teorii gosudarstva i prava' (1957); NASCHITZ, 'Nekotorye soobraženija o sootnošenii filosofii prava i obščej teorii prava' (1968); ROMAŠKIN *et al.*, *Teorija gosudarstva i prava* (1962); SZABÓ, 'Gegenstand der marxistisch-leninistischen Rechtstheorie' (1968); WRÓBLEWSKI, 'Zagadnenia przedmiotu i metody teorii państwa i prawa' (1961); etc.

118. Cf. ALEKSEEV, KERIMOV & NEDBAJLO, 'Metodologičeskie problemy pravovedenija' (1954); EHRLICH, S., 'Kilka uwag v sprawie metodologii nauk prawnych' (1954); KAZIMIRČUK, *Pravo i metody ego izučenija* (1965); KERIMOV, 'Grundfragen der Methodologie zur Erforschung staatlicher und rechtlicher Erscheinungen' (1957); OPALEK, *Metodologiczne problemy prawoznawstwa* (1962); STROGOVIČ, 'Metodologičeskie voprosy juridičeskoj nauki' (1965); *Metody badania prawa. Materialy sympozjum Warszawa 28-29 IV 1871* (LOPATKA, ed., 1973); etc.

119. The literature on dialectical and historical materialism is extremely abundant. By way of examples, we may cite the following: SPIRKIN, *Kurs marksistskoj filosofii* (2nd edn., 1966); MAKAROV, POPOV & TERJAEV (eds.), *Marksistsko-leninskaja filosofija. Istoričeskij materializm* (1970); VOSTRIKOV *et al.*, *Marksistsko-leninskaja filosofija. Dialektičeskij materializm* (1970); KELLE & OSIPOV, 'Matérialisme historique, théorie sociologique et recherche sociale en U.R.S.S.' (1968). See also, in the present work, 'Philosophy' by Paul Ricœur, particularly Sec. III, 'Man and social reality', sub-section B, par. 2, pp. 1296-1304.

120. This expression is preferred in order to avoid having to place sociology on the same plane as historical materialism as a fundamental social science and to emphasize that sociology is only one of the methods used within the framework of historical materialism. Concerning the *concrete sociological study of reality*, cf. ČHIKVADZE *et al.*, *Marksistsko-leninskaja obščaja teorija gosudarstva i prava*, Vol. I (1970), III/3; KAZIMIRČUK, *op. cit.* (1965); AFANAS'EV, *Naučnoe upravlenie obščestvom* (1968), pp. 332 sqq.; etc. Legal literature in the USSR however also speaks of sociological research in the field of law: cf. VEJNGOLD, *Pravo kak sociologičeskaja kategorija* (1962); by the same author, *Nekapitalističeskoe razvitie i social'naja volja* (1970). Cf. also KELLE & OSIPOV, 'Matérialisme historique, théorie sociologique et recherche sociale en U.R.S.S.' (1968).

social administration, and so forth. It also covers research relating to specific problems of the reciprocal action of rules of law and society, and the study of such questions as the function of the popular representative bodies [121] and the activities of their members (deputies), the administration of the national economy in general and that of enterprises in particular, agricultural organization and so forth.

In addition to the concrete sociological study of reality, one of the foremost methodological problems with which Marxist-Leninist legal science is concerned is the applicability of methods borrowed from the natural sciences, especially cybernetics. Wide use is also made of the comparative method.[122]

(ii) The evolution of Marxist-Leninist legal science since 1958 has also seen a revival of interest in the problem of the system of law and in particular in the question whether there exists an economic law as an autonomous branch of law, independent of civil law. Controversy regarding the system of law and especially the existence of economic law broke out in the USSR in the thirties and discussion, which was very heated at the time, ended in 1938 with rejection of the hitherto prevalent view in favour of the autonomy of economic law. In 1958 the controversy flared up again in the USSR [123] and spread to most other European socialist countries. This dispute, which illustrated one of the liveliest developments between 1958 and 1964, slowly faded away; but as it did not end in a generally acceptable solution it is far from extinct and still deserves mention as one of the current trends is Marxist-Leninist legal science.

It may be added that the discussion about the problem of economic law has had certain effects on practice, chiefly in regard to legislation, since to recognize an economic law independent of civil law requires a completely new conception of the civil code, and the creation of an economic code distinct from the civil code (as was the case in Czecho-

121. The supreme representative bodies (the Supreme Soviet in the USSR, the Sejm in Poland, the Federal Assembly in Czechoslovakia, for example) are the counterparts of parliaments in Western countries; though they are quite different in character the local representative bodies are the State organs of authority and administration in the territorial units – regions, districts, counties, towns, communes, etc. – all of which are elected by the people and are responsible to their electors.

122. See Section Three, sub-section II, below, pp. 974 sqq.

123. Discussion recommenced in the Autumn of 1956 in the Moscow review, *Sovetskoe gosudarstvo i pravo*. A comprehensive survey of this discussion is provided in the work of SVOBODA, *La notion du droit économique* (1966) (see also the bibliography listed in that work). The notion of 'economic law' is however not limited to the socialist countries alone. It is also found, though in quite a different sense, in the Western countries; see *Begriff und Prinzipien des Wirtschaftsrechts* (1971). See also the above-mentioned work by SVOBODA (1966). Concerning general problems of the system of law, cf. ALEKSEEV, *Obščie teoretičeskie problemy sistemy sovetskogo prava* (1961); see also works cited in *Literature on Soviet Law. Index of Bibliography* (1960), pp. 34 sqq. Cf. also ROT, *Jednošć i podialy systemu prawa socjalistycznego* (1971).

slovakia in 1964). Despite these consequences, it seems to us that somewhat excessive attention has at times been focused on problems of the system of law.

(iii) Recent years have also seen a renewal of interest in the relation between philosophy and legal science [124] and even in the philosophical problems of law themselves, particularly those concerning the notion of law,[125] the existence or even the validity of law, knowledge of law (epistemological problems), the value of law (axiological questions), etc.[126]

(iv) Since Marxist-Leninist legal science is, as we have said, a militant science, it has an inherent tendency to combat non-Marxist legal theories, and also those which seek to deform Marxism-Leninism in legal science.[127]

As regards non-Marxist theories, Marxist-Leninist legal science is not content to criticize their points of departure and theoretical conceptions: it also directs its attention to their practical implications and brings critical analysis to bear on theories which, either directly or indirectly, tend to undermine democracy, legality, etc. – for example, theories which in the written law countries, recalling the *Freirechtschule* of the early twentieth century underestimate the law and plead in favour of judge-made law, those which seek to justify types of extraordinary legislation, those which undermine the rôle of Parliament in order to give greater power to the Executive branch, and so forth. Particular attention is devoted to counter-attacking openly anti-Communist theories expressed in Western legal writings.

On the other hand, Marxist-Leninist legal science regards it as one of its major tasks to expose all deformations of Marxism-Leninism, or indeed deviations from it, in the field of legal science; these deformations or deviations (especially in the form of revisionism) not only occur

124. Cf. STROGOVIČ, 'Filosofija i pravovedenie' (1965); ČHIKVADZE *et al.*, *Marksistsko-leninskaja obščaja teorija gosudarstvo i prava*, Vol. I (1970), Introduction, par. 1.

125. The representative work in this field published during the last few years is by SZABÓ, *A szocialista jog* (1963) (Russian version, *Socialističeskoe pravo*, 1964).

126. KNAPP, *Filosofické problémy socialistického práva* (1967), and the article by the same author, 'La philosophie du droit dans les pays socialistes' (1971); see also the bibliography listed in that article.

127. Cf., by way of illustration: *Kritika buržuazno-reformistskih i revizionistskih teorii v voprosah gosudarstva i prava* (1960); KULCSÁR *et al.*, *Kritikai tanulmányok a modern polgári jogelméletről* (1963, 1969); ČHIKVADZE & ZIVS, *Protiv sovremennogo reformizma i revizionizma v voprose o gosudarstve* (1959); ČHIKVADZE *et al.*, *Marksistsko-leninskaja obščaja teorija gosudarstva i prava*, Vol. II (1971), Chap. XI; IVANOV (ed.), *Gosudarstvenno-monopolističeskij kapitalizm i buržuaznoe pravo* (1969); KERIMOV, *Staatslehre und Revisionismus* (1959); SEIDLER, *Doktryny prawne imperializmu* (1957); TUMANOV, *Kritika sovremennoj buržuaznoj teorii prava* (1957); ZIVS, *Krizis buržuaznoj zakonnosti v sovremennyh imperialističeskih gosudarstvah* (1958); ALEKSANDRENKO, *Buržuaznyj federalizm* (1962); POPOV, *Kritika na savremenija buržuazen praven normativizm* (1964); etc.

in Western literature, but infiltrate from time to time even into legal writings in the socialist countries.

(b) *Study of the socialist State*

(i) One of the most important trends appearing in Marxist-Leninist legal science in the USSR, and in the other socialist countries, is research concerning the development of *socialist democracy*.[128] Moreover, in so far as the basic principles and problems are concerned, this work is the subject of multilateral co-operation among research institutions in the socialist countries of Europe.[129]

We shall come back in more detail to problems of democracy,[130] merely stating here that the study of socialist democracy is conceived in very broad terms. It covers study of specific problems of socialist democracy, such as those concerning the administration of the national economy (e.g., legal processes employed in planning, and the management of enterprises, their legal status in relation to socialist ownership, and so on). It also includes the legal, social and political implications of the scientific and technological revolution [131] in both the internal and external relations of the socialist countries (as, for example, the effect of the scientific and technological revolution on the development of the socialist State and socialist law, on the development of the forms and methods of democratic practice in the socialist countries, on legal regulation of the national economy, on legal problems in foreign trade and economic co-operation, etc.).

(ii) Marxist-Leninist legal science seeks to give to State administration a scientific basis. At this time of great scientific and technological ad-

128. ČHIKVADZE *et al.*, *Marksistsko-leninskaja obščaja teorija gosudarstva i prava*, Vol. II (1971), Chap. IV; ČHIKVADZE, *Gosudarstvo, demokratija, zakonnost'* (1967); ČHIKVADZE (ed.), *The Soviet State and Law* (1969); KOTOK, 'Socialističeskij demokratizm sovetskogo gosudarstva' (1957); LOOSE, *Zur Dialektik der Höherentwicklung der Arbeiterklasse und der sozialistischen Demokratie* (1971); POLAK, *Zur Dialektik in der Staatslehre* (3rd edition, 1963); SZABÓ, 'La démocratie socialiste et le droit' (1966); VINTU, 'La démocratie représentative en Roumanie' (1972); WEICHELT, *Der sozialistische Staat – Hauptinstrument der Arbeiterklasse zur Gestaltung der sozialistischen Gesellschaft* (1972); etc.

129. At a meeting of their Directors held in Prague, 3-5 March 1970, the institutes of legal sciences of the Academies of Sciences of the USSR, Bulgaria, Czechoslovakia, Hungary, Poland and Romania, as well as the 'Walter Ulbricht' Academy of Sciences of State and Law in the German Democratic Republic, selected '*Socialism and Democracy*' as one of the subjects for co-ordinated research.

130. Cf. Section Three, sub-section V, pp. 1007 sqq.

131. Cf. for example, *Pravovye voprosy naučno-tehničeskogo progressa v SSSR* (1967); see also BOBOTOV, 'Vlijanie naučno-tehničeskogo progressa na pravovoje regulirovanie' (1972); RYBICKI, 'L'administration publique dans l'Etat socialiste face à la révolution scientifique et technique' (1972). It should also be mentioned that the meeting of Directors of Institutes referred to in footnote 129 selected as themes for joint study two further subjects which stand in close relation to the preceding theme and, like it, demonstrate a very marked trend in the current development of Marxist-Leninist legal science. These were '*Legal questions relating to economic reform*' and '*Socialist State and law and the scientific and technological revolution*'.

vance, the tasks of the socialist State in ensuring the political, economic, social and cultural development of socialist society are becoming increasingly complex; this also complicates the rôle of State administration and makes it essential that a firm scientific basis be provided for it. Efforts to this end, because of their interdisciplinary nature, also involve legal science and represent an important trend in its current development.

(c) *Socialist legality and problems concerning the legal system.* Marxist-Leninist legal science continues to concern itself with socialist legality [132] in all its manifold general aspects arising under the general theory of State and law, the study of human rights and the study of socialist democracy,[133] as well as in its specific aspects such as, for example, the war against crime, legality in economic relationships, and so forth.

A very marked trend in research to socialist legality is represented by the interest shown by legal science, particularly in the USSR, in studies concerning the legislation of the Soviet Union and the Union republics – the perfecting and development of the system, and its codification.[134] This research embraces the study of ways of further improving legislative procedures, studies concerning the promulgation of legal norms, legal terminology, the legislative process, and so forth. It also deals with work on the development of a scientific system of legal norms in the Soviet Union and the Union republics [135] (resembling the study concerning theoretical problems of the system of socialist law referred to above); study of the relation between federal (All-Union) legislation and legislation at republic level ;[136] examination of questions relating to further improvement of the laws in force in

132. Kerimov, *Obespečenie zakonnosti v SSSR* (1955); by the same author, 'Fragen der Theorie der sozialistischen Gesetzlichkeit' (1957); Nedbajlo, 'Sovetskaja socialističeskaja zakonnost'' (1954); Piontkovskij, 'Ukreplenie socialističeskoj zakonnosti' (1955); Strogovič, *Osnovnye voprosy sovetskoj socialističeskoj zakonnosti* (1966); Strogovič (ed.), *Pravovye garantii zakannosti v SSSR* (1962); Opalek & Zakrzewski, *Z zagadnień praworządnosci socialistycznej* (1962); etc. As regards the difference between socialist legality and bourgeois legality, cf. Weyl, *La justice et l'homme* (1962).

133. See Section Three, sub-section V, below, p. 1008.

134. There is a specialized institute which deals with these questions, namely the Federal Institute of Research into Soviet Legislation (Vsesojuznyj naučno-issledovatel'skij institut sovetskogo zakonodatel'stva), Moscow. Problems concerning the creation and improvement of law have however also been studied by legal science in other socialist countries. Cf., for example, Arlt, 'Zur marxistisch-leninistischen Theorie von der Schaffung und Verwirklichung des sozialistischen Rechts' (1971); Ceterchi, 'Le rôle de la dialectique matérialiste dans la pratique du droit' (1971); Knapp, 'La technique de la réforme du droit dans les pays socialistes' (1970); etc.

135. Bratus *et al.*, *Teoretičeskie voprosy sistematizacii sovetskogo zakonodatel'stva* (1962); Čhikvadze (ed.), *Voprosy kodifikacii* (1957); Kerimov, *Kodifikacija i zakonodatel'naja tehnika* (1962); by the same author, *Fragen der Gesetzgebungstechnik* (1958); Kerimov (ed.), *Zakonodatel'naja tehnika* (1965). See also the collection *Voprosy kodifikacii sovetskogo prava* (1957).

136. Cf. Samoščenko *et al.*, *Sootnošenie obščesojuznogo i respublikanskogo zakonodatel'stva* (1967).

various fields, having regard to the needs and consequences of economic reform (these are mainly problems of the civil law, the so-called economic law, labour law, agrarian law, laws governing the legal status of waterways, forests and mines, etc.); the analysis of experience in regard to the laws of foreign, particularly socialist, countries; and so on.

(d) The above exposition of current trends in Marxist-Leninist legal science is by no means exhaustive. The range of interests dealt with by that science is much larger and includes practically all of the questions that will be referred to below. Reference is made by way of example in this section to only a few of the problems of major importance. In the following developments, the reader's attention is drawn again to the trends which manifest themselves in Marxist-Leninist legal science.

SECTION THREE: MAJOR TRENDS

I. INTRODUCTION (GENERAL OBSERVATIONS)

The major theoretical currents mentioned in the foregoing section provide the general framework of the main trends of scientific research in contemporary legal science and themselves in turn reflect those trends.

In particular we can now classify these main trends, broadly speaking, in accordance with three criteria, namely:

1. trends arising from the needs of developing legal science itself, especially those trends which are apparent in methodology and legal history;
2. trends influenced by political and economic interests;
3. trends caused by the rapid development of the exact sciences and the natural sciences.

There is, of course, no need to add that this classification is far from precise; the influences which give birth to these various trends, or which stimulate them, are superimposed, and the classification has no other purpose than to help to shed light on the main trends of contemporary legal science and facilitate an understanding of them.

II. METHODOLOGY

(a) The methodology of legal science is one of the fields in which current trends of research in our discipline can be seen in a particularly marked way.[137] It is, however, essentially in this domain that the general framework produced by the major theoretical currents plays an important part. It can be said without exaggeration that what has distinguished virtually every theo-

137. Interest in methodological problems is, however, typically 'continental'. Lawson points out (Comments, p. 1) that 'English lawyers at least find methodology extremely repellent.'

retical current in legal science has been the solution it proposed in regard to methodological problems and the use it made of certain methods.

For the sake of the logical arrangement of this survey, the basic trends which are apparent in methodology have already been dealt with in Section Two, where we reviewed the major theoretical currents of contemporary legal science. Consequently, in order to avoid repeating here what has already been said, we shall confine ourselves to certain selected methodological problems, but on the other hand, we shall not limit ourselves, in the present section, to methodological problems *stricto sensu*, that is, to problems which arise at the level of the *theory of methods*. At the same time, we shall be concerned – in some cases primarily – with problems concerning the methods themselves and their application. In other words, under the heading 'Methodology', we shall not restrict ourselves to sketching problems of a purely *methodological* order but shall at the same time endeavour to give the reader a picture of the main trends as regards the *application* of the methods.

(b) Starting from this point, we propose, in the present section, to deal with the application of sociological methods (methods borrowed from empirical sociology), the comparative method and methods borrowed from the exact sciences. These methods are not all on the same plane and do not have the same scope. Obviously they all fall under the major prevailing methodological conceptions and depend upon them, but sociological methods are *general*, whereas the other two are *specific*, hence the sociological methods cover the others. Furthermore, the use of methods borrowed from empirical sociology is of especial significance for the most important theoretical current in contemporary legal science in the Western countries,[138] whereas the other two kinds referred to do not of themselves determine any major current (in the sense in which we have used the notion 'theoretical currents' in Section Two above).

1. *The application of sociological methods*

(a) Even though the application of sociological methods, that is, methods borrowed from empirical sociology, has an essential significance for the sociological current of legal science,[139] it does not coincide exactly with that current. As we have already seen, a clear distinction must be drawn between the *theoretical conception* of human society and the *methods* used in research concerning social data and facts. It is therefore necessary to distinguish between the philosophical, methodological and other starting-points of research and the methods which the research specialist uses. It follows that the sociological *methods* which legal science makes use of can be – and often are – similar, though the theoretical conclusions deduced from the facts so established may be completely different depending upon the theoretical con-

138. That is why, even in the present section, following Rheinstein's recommendation, we touch briefly on sociological methods.

139. See Section Two, II, 3 above, p. 961.

ception of society. An example of this is the difference which separates the sociological theories which we discussed above [140] from the Marxist-Leninist theory.

The use of methods borrowed from empirical sociology thus goes beyond the sociological theories to which we have referred. For that matter, it is possible to imagine sociological methods applied even to theoretical currents other than those. There is no theoretical obstacle, for example, to prevent theories of natural law, phenomenological theories, etc., from having recourse, within the framework of their principal method, to methods taken from empirical sociology. (On the other hand, it is impossible to imagine the use of sociological methods in legal positivism, or in the 'pure doctrine of law', for example, since those theories are clearly and by definition 'anti-sociological'.)

(b) Let us return however to the *sociological theories* of law which we spoke of earlier, and for which the application of sociological methods has an essential meaning.

For these theories, sociological methods are not methods subordinate to another principal method: they are themselves *the* principal method of this theoretical current and their use represents a true *methodological* event. In this connexion we may quote Rheinstein who, in recommending that we remind the reader, even in the present section, of the importance of recourse to sociological methods, points out:

1. that the most significant and most far-reaching trend in present legal learning is the shift toward sociological jurisprudence, and
2. that this shift is universal and quite particularly marked in the United States, Scandinavian countries, the Federal Republic of Germany and Japan.[141]

Consequently, what characterizes the sociological conception of law is work that implies the use of methods borrowed from general sociology, such as the study of statistical reports, interviews, research on the spot, a comparison of social facts, etc. – work which is called in English terminology '*legal fact research*'.

(c) We have seen [142] that even Marxist-Leninist legal science is making increased use of sociological methods (methods used in the 'concrete sociological study of reality'). The methods applied by this type of study are, for example: the analysis of statistical data; the analysis on the spot of certain activities (those of representative bodies, enterprises, etc.) or, in some cases, the building and analysis of 'models' of such activities; the organization of

140. In Section Two, II, 3, pp. 960-965.

141. Rheinstein, Observations on the new statement, p. 13. We have no doubt that the sociological trend – or at least sociologically inclined – is indeed the most important trend in legal science in the Western countries and that in those countries it is very widespread; we regret however that the source materials which we have available do not enable us to share fully the second opinion expressed by Rheinstein above. (See Section Two, II, pp. 962-965, where we have dealt with this.)

142. Section Two, III, pp. 966 sqq.

surveys; the analysis of public opinion; etc. The concrete sociological study of reality also comprises (especially in the USSR) the use of the social test,[143] for example during the preparation of drafts of legal rules; before the introduction of new methods of administration; etc.

(d) For the remainder, we refer the reader to Section Two, II, 3 and III.

2. *Comparative law*

(a) *General problems*

(i) The origins of 'comparative law' go back a century or so: the Paris 'Société de Législation comparée' celebrated its hundredth anniversary in 1969.[144] But modern study of comparative law only dates back to the early twentieth century (the first Congress of Comparative Law was held in Paris in 1900), a period when its foundations were laid by the Frenchmen Raymond Saleilles and Edouard Lambert.[145] It began to win the interest of legal scientists between the two World Wars (Lévy-Ullman in France, Rabel in Germany, Gutteridge in England, Yntema in the United States, among others) and it is only since the Second World War that it has come into its own on a really world-wide scale.[146] One of the causes of this development,

143. It is obvious that the social test raises many theoretical problems: cf. the above-mentioned work by KAZIMIRČUK, *Pravo i metody ego izučenija* (1965); cf. also GOLOVKO, 'Naučnyj èksperiment v gosudarstvennom stroitel'stve' (1966); NIKITINSKIJ, 'Značenie èksperimenta v normotvorčeskoj dejatel'nosti' (1967); PANICKOV, 'Social'nyj èksperiment i naučnoe rukovodstvo razvitiem socialističeskogo obščestva' (1967); etc.

144. The 'Société de Législation comparée', for its 100th anniversary, published a collection in two volumes entitled *Livre du Centenaire de la Société de Législation comparée.* (1969-71.) Evidence of the comparative study of law, however, goes back as far as antiquity (cf. DAVID, R., *Les grands systèmes de droit contemporains*, p. 3 in the 4th edition (1971): 'The comparison of legal systems, from the point of view of their geographical diversity, is as old as jurisprudence itself' ('La comparaison des droits, envisagés dans leur diversité geographique, est chose aussi ancienne que la science du droit elle-même.').

145. Cf. ANCEL, 'La tendance universaliste dans la doctrine comparative française au début du XXe siècle' (1954); by the same writer, *Utilité et méthodes du droit comparé* (1971). Cf. also CONSTANTINESCO, 'Les éléments déterminants en tant que critères fondant la science des droits comparés' (1967), which takes the origin of comparative law back to the first third of the nineteenth century (p. 235); see also TILLE & FAJZIEV, 'Iz istorii sravnitel'nogo pravovedenija' (1970). Cf. likewise 'Buts et méthodes du droit comparé' (in press).

146. The reader will find a comprehensive survey of contemporary comparative law scholars and of the subjects that are currently claiming their attention in several volumes of miscellanies published in honour of the most prominent of these scholars. We may quote, as an illustration, a few of those published during the last decade: *XXth Century Comparative and Conflicts Law* (in honour of Hessel Yntema), Leiden, 1961; *Essays in Jurisprudence in honor of Roscoe Pound*, Indianapolis, 1962; *Liber amicorum Baron Louis Fredericq*, Ghent, 1965; *Libro-Homenaje a la Memoria de Roberto Goldschmidt*, Caracas, 1967; *Etudes juridiques en l'honneur du Professeur Traian R. Ionasco*, Bucharest, 1968; *Ius Privatum Gen-*

as A. Tunc [147] reminds us, has undoubtedly been the rapid expansion of the scientific and technical revolution since the Second World War and especially the development of communications which, by 'shortening' distances, has greatly accelerated the transfer of information, facilitated relations between States and consequently also raised new problems involving the relations between their respective rights. In our view, however, it is essentially the political changes that have occurred since the Second World War – in the first place the co-existence of socialist and non-socialist systems of law and the decolonization process – that have enriched and continue to enrich comparative law with new insights and new horizons.[148] The political issues bound up with the relations between the two great socio-political systems of the present-day world (with their need to understand one another, especially in the field of foreign trade and, at the same time, with their ideological struggle); [149] with the process of decolonization and the birth of dozens of new States, with the formation of regional groupings of States – economic and political – and so on, have set the stage for what is one of the main trends in current legal research, namely the unprecedented upsurge of comparative law.[150]

---

tium (in tribute to Max Rheinstein), Tübingen, 1969; *Miscellania W. J. Ganshof van der Meersch*, Brussels, 1972; *Mélanges de droit comparé en l'honneur du Doyen Å. Malmström*, Stockholm, 1972. Mention should likewise be made of two older works: *Introduction à l'étude du droit comparé. Recueil d'études en l'honneur d'Édouard Lambert*, Paris, 1938, and *Festschrift Ernst Rabel*, Tübingen, 1954. An idea of the subjects of interest in contemporary comparative law is also provided by such meetings as the International Congresses of Comparative Law, the symposia organized by the International Association of Legal Science, etc., to which we shall frequently refer, and by collections published for the anniversaries of institutions concerned with comparative law, as, for example, *Problèmes contemporains de droit comparé* (for the anniversary of the Japanese Institute of Comparative Law), Tokyo, 1962; *Recueil des travaux relatifs au droit étranger et droit comparé* (10th anniversary of the Institute of Comparative Law of Belgrade), Belgrade, 1966; a special issue of the *Boletín Mexicano de Derecho Comparado* (for the 25th anniversary of the Mexican Institute of Comparative Law), Mexico City, 1968; *Livre du Centenaire de la Société de Législation comparée*, 2 vols., Paris, 1969-71.

147. Tunc (in his comments on the second version of the present work) considers these factors to be the main cause of the increased interest in comparative law since the Second World War.

148. Cf. *Buts et méthodes du droit comparé* (1973); Ancel, 'Les buts actuels de la recherche comparative' (1972); Graveson, 'Philosophy and function of comparative law' (1972); Rodière (ed.), *Introduction au droit comparé* (1967); Szabó, 'Sravnitel'noe pravovedenie' (1963, 1969); by the same author, 'Law theory and comparative law' (1972); Szabó, Comments: etc.

149. The rôle of comparative law as a weapon in the struggle against bourgeois ideologies is stressed particularly by Zivs, 'O metode sravnitel'nogo issledovanija v nauke o gosudarstve i prave' (1964), pp. 23 and 31 sq.

150. Interest in comparative law brought about by economic needs in contemporary society is growing more and more in the practical world as well. A. Tunc speaks (in his Comments on the second version of the present work) of an 'incitement to comparative studies which, springing as it does from the existence of the European Economic Community, appears even in professional journals (*La Gazette*

The sustained interest which is now shown in research into comparative law appears concretely in the establishment and activities of numerous institutions, either specialized [151] (such as the Institut de Droit comparé of the Law Faculty of the University of Paris, the Max-Planck-Institut für ausländisches und internationales Privatrecht in Hamburg, the British Institute of International and Comparative Law in London, the Belgrade Institute of Comparative Law, the Institut für ausländisches Recht und Rechtsvergleichung der Deutschen Akademie für Staats- und Rechtswissenschaft 'Walter Ulbricht' at Potsdam-Babelsberg, the Instituto de Derecho comparado in Barcelona, etc.) or engaged in legal research in general but devoting a considerable part of their time to comparative law, e.g., in particular the Institutes of State and Law of the USSR Academy of Sciences and of the Academies of Sciences of Czechoslovakia, Hungary, Romania, etc. These institutions, together with the Chairs and Institutes of Comparative Law which are found in many legal faculties throughout the world, and with a special faculty (the International Faculty for the Teaching of Comparative Law, Strasbourg), constitute the institutional framework of research and teaching in this sphere. Further evidence of this interest can be found in the existence of the International Academy for Comparative Law and of the International Committee of Comparative Law (the latter also serving as executive committee for the International Association of Legal Science), both of which stimulate research in this field by organizing symposia, round tables, lectures, etc. Mention may also be made of the preparation, now in hand under the guidance of the International Committee of Comparative Law, of a basic work in this discipline, the *International Encyclopaedia of Comparative Law,* which is to extend to no fewer than seventeen volumes.

(ii) The interest shown in research in the field of comparative law therefore in itself constitutes one of the most striking current trends in world legal science. Comparative law, however, is not a homogeneous or uniform discipline; the interests of its exponents vary and we can detect a certain diversity of tendencies within the discipline itself.

Before undertaking a review of these tendencies it is however necessary to raise a prior question, that of deciding what constitutes comparative law. The expression is peculiar to the French (*droit comparé*) and English languages, and languages the legal terminology of which was formed under French influence. On the other hand it does not exist in German (one cannot say *vergleichendes Recht* or *verglichenes Recht;* the equivalent of 'comparative law' is either *Rechtsvergleichung* or *vergleichende Rechtslehre)*; such a term would be equally meaningless in Russian, as also in the language of the au-

---

du Palais)'. See also *Livre du Centenaire de la Société de Législation comparée,* vol. I, *Un siècle de droit comparé en France. Les apports du droit comparé au droit positif français* (1969), section VII: 'Les apports du droit comparé à la pratique juridique française'.

151. It may be recalled that the International Association of Legal Science held a symposium in 1954 on the rôle and functions of comparative law institutes and centres. Cf. also David, R., 'La coopération internationale en matière de droit comparé' (1972).

thor of this study, in which the only acceptable formula is 'the comparative science of law'.

What then is comparative law? What does the notion signify? It is, of course, easier to indicate what it does not signify. It is beyond doubt that the term is not used to convey the notion of 'comparative law' as an entity, since there is no such thing. We have French law, English law, Soviet law and so forth, each of which can be compared with the others, but this does not constitute 'comparative law' as a subject for legal research. Comparison is a *placing into relationship;* that is to say it assumes, as in any relationship, a minimum of two elements, a *comparatum* and a *comparandum* mutually linked by a *tertium comparationis*. Even from the standpoint of the subject-matter of knowledge, therefore, we are not faced with one 'comparative law' but with (at least) two laws which we compare one with the other.

To be precise, the term 'comparative law' means nothing more than the *comparative science of law* or, more simply, *comparative research* or *method* in the legal field, that is to say a type of research which uses as its principal means the comparison of legal systems or orders, of the institutions and fundamental principles of law, etc., as between various countries (or, indeed, within the same country: comparative municipal law) or within the same system of law.[152] This, in broad terms, would appear to be what is meant by 'comparative law'.

It must also be said, however, that 'comparative law' raises many methodological problems [153] and that it is by no means an isolated method. The comparative method is applied, as we have stated, within the general framework of the methodological conception of legal science. We can therefore imagine a positivist 'comparative law' [154] (even though positivism is not very favourable to the comparative method),[155] a 'comparative law' within the

152. The problem of 'comparative municipal law' is extremely important in the USSR where it receives constant attention from legal science. Cf. *Sostojanie i zadači sravnitel'nogo izučenija zakonodatel'stva sojuznyh respublik* (1970). Cf. also Knapp, 'Quelques problèmes méthodologiques dans la science du droit comparé' (1968); Linant de Bellefonds, Special contribution, p. 1.

153. Cf., for example, Ancel, 'Méthode comparative et droit comparé' (1965); by the same author, *Utilité et méthodes du droit comparé* (1971); Zivs, 'O metode sravnitel'nogo issledovanija v nauke o gosudarstve i prave' (1964); Čhikvadze & Zivs, 'Sravnitel'noe pravovedenie v praktike meždunarodnogo naučnogo sotrudničestva' (1966); by the same authors, 'L'évolution de la science juridique et du droit comparé en U.R.S.S.' (1971); Constantinesco, 'Les éléments déterminants en tant que critères fondant la science des droits comparés' (1967); Kazimirčuk, *Pravo i metody* ... (1965), pp. 91 sqq.; Knapp, 'Quelques problèmes méthodologiques dans la science du droit comparé' (1968); Mayda, 'Quelques réflexions critiques sur le droit comparé contemporain' (1970); Schmitthoff, 'The science of comparative law' (1949); Szabó, 'La science comparative du droit' (1964); bŷ the same author, 'Law theory and comparative law' (1972); Tille & Fajziev, 'Sravnitel'nyj metod v pravovoj nauke' (1969); Tille & Švekov, *Sravnitel'nyj metod v juridičeskih disciplinah* (1973); etc.

154. Positivism played an important part in the birth of comparative law; one representative of the positivist conception of comparative law, in our opinion, has been the English jurist Gutteridge.

155. See above.

framework of the conception of natural law,[156] a 'comparative law' conceived from the – currently prevalent – sociological viewpoint (i.e., use of the comparative method in sociological methodology) [157] and, of course, a Marxist-Leninist comparative law, which represents 'a particular concrete achievement of the application of the method of dialectical materialism in research relating to matters of State and law'.[158]

(iii) It follows from what has just been said that the interpretation of the notion of 'comparative law' is not invariable. We are moreover not concerned with a mere difference in terminology but with a question which has been debated for a considerable time. This question is whether 'comparative law' is an autonomous branch of legal science distinguished by its use of the comparative method, or whether it is in fact nothing other than the comparative method itself; in other words, whether 'comparative law' constitutes a branch of legal science alongside other branches, such, for example, as the philosophy of law (or, as the case may be, the general theory of State and law) – or those disciplines within legal science which deal with constitutional, administrative, civil, criminal law, etc. – or whether 'comparative law' is merely a method employed by all branches of legal science.

The prevailing opinion is that 'comparative law' is no more than a method. This opinion is shared by the majority of Marxist-Leninist authors.[159] As for non-Marxist legal science, though it has not shown such concern with the theoretical aspects of the question, there can be little doubt that it is also of the same opinion. This was quite clear, for example, at the international symposium held in Budapest [160] from 9 to 11 September 1969 where most of the participants, whether from the East or the West, upheld the opinion that

156. Cf. CALEWAERT, 'Natuurrecht en rechtsvergelijkening' (1965).

157. Cf. DAVID, R., *Les grands systèmes de droit contemporains*, 4th edn. (1971), pp. 14 sqq.; DROBNIG, 'Rechtsvergleichung und Rechtssoziologie' (1953); CARBONNIER, 'L'apport du droit comparé à la sociologie juridique' (1969); KALENSKÝ, 'Les méthodes de la recherche sociologique en droit comparé' (1970); TERRÉ, 'Les méthodes de recherche sociologique en droit comparé' (1970); 'Les méthodes de recherche sociologique en droit comparé' (VIIIth International Congress of Comparative Law (Pescara, 1970); rapporteur général: R. Lukić) – see *Rapports généraux au VIIIe CIDC*, to be published. Cf. also Replies of the Institute of State and Law of the USSR Academy of Sciences, p. 40: 'Comparison of legal institutions can take place only by means of combined legal and sociological analyses.'

158. ZIVS, 'O metode sravnitel'nogo issledovanija v nauke o gosudarstve i prave' (1964), p. 25. It is interesting to note that ENGELS makes mention of comparative law as early as 1891 (in his letter to Kautsky dated 13 June 1891): cf. FAJZIEV, 'F. Engels o sravnitel'nom metode issledovanija estestvennyh i obščestvennyh nauk' (1970).

159. Cf. especially the works of ZIVS, of ČHIKVADZE & ZIVS, and SZABÓ cited in footnote 153 on page 980, etc. The same position was adopted by the Soviet rapporteurs (Professors Zivs and Krutogolov) at the round table organized in Paris on 3 May 1963: see LYON-CAEN, 'Table ronde sur les études et les recherches de droit comparé en U.R.S.S.' (1964), p. 70 ('Rejet du droit comparé comme branche du droit autonome').

160. Cf. 'Tendencies and functions of comparative law in contemporary society' (1971), and also the report on the round table 'Perspectives et limites du droit comparé' (1970).

'comparative law' is no more than a method.

To arrive at a perfectly exact interpretation of the notion of 'comparative law' it would be necessary first to take into account the circumstances and the particular opinions of each author and then to speak sometimes of the comparative science of law, and sometimes of the application of the comparative method in legal science. For the purposes of the present study, however, no such elaborate precision is needed; we shall therefore continue to speak of 'comparative law', henceforward without quotation marks.

(iv) A few words should be added regarding the field of application of comparative law. The question whether and to what extent comparative law also covers the study of foreign law gives rise to frequent discussion.

At first sight an affirmative reply would seem to be self-contradictory: whether we define comparative law as the use of comparison as the principal method or whether we regard it as the comparative method itself, in neither case can comparative law at the same time be made to include the straightforward study of foreign law without recourse to any comparison.

The views in this connexion of the exponents of comparative law reflect, however, widely differing positions. Some make a clear distinction between comparative law in its strict sense and (without underestimating its importance) the study of foreign law, whereas the majority assert that comparative law also incorporates research concerning foreign law. This position, however, gives rise to a number of methodological problems. Unless we are content to define comparative law in terms of the old quip '*lucus a non lucendo*' we must seek an element of comparison even in the study of foreign law.

Before doing so, however, and before reproducing the answer which is habitual in this case, we must explore the notion of foreign law itself. From whose standpoint and for whom is the law being studied to be regarded as foreign? Would, for example, treatises on Czechoslovak, Polish, and Italian law, etc., drawn up by Czechoslovak, Polish and Italian authors, published in French or English and designed for foreign readers or, as the case may be, treatises on United States and English law written by French authors and intended for French readers,[161] and so on, be considered as works dealing with 'foreign law' and accordingly, following the majority opinion just quoted, as belonging to comparative law?

Clearly, the problem is not easily resolved. The answer that is usually given to the question posed above is that foreign law as an object of scientific research is defined by its double relationship with the author and with the reader; 'foreign law' must therefore be foreign both from the point of view of the author and from that of the readers for whom the work is intended. It is maintained that in such a case, the element of comparison resides in the fact that the reader is acquainted (or at least is held to be acquainted) with

161. CAPPELLETTI, MERRYMAN & PERILLO, *The Italian Legal System* (1967); DAVID, R., *Le droit anglais* (1965); HENRICH, *Einführung in das englische Privatrecht* (1971); LAWSON, ANTON & NEVILLE BROWN (eds.), *Amos and Walton's Introduction to French Law* (1967); LUBY (ed.), *Le droit civil tchécoslovaque* (1969); ROZMARYN (ed.), *Introduction à l'étude du droit polonais* (1969); TUNC, *Le droit des Etats-Unis* (1964); etc.

the law of his own country; a work dealing with the law of a foreign country therefore enables him to carry out a comparison between the two laws; in other words it is assumed that he will carry out comparisons in the place of the author, the latter having facilitated the process for him. It is presumed that the author, trained in a given legal milieu, tradition or culture, has treated the foreign law from his own point of view and transposed it into his own native legal language.

Even, however, if it is admitted, as it is by the majority, that comparative law extends to cover study of 'foreign law', it is of course necessary to distinguish instances in which the author deals in depth with foreign law, making an implicit comparison, from those which provide merely a simple description of foreign law or of its institutions; or, indeed, taking the argument to the point of absurdity, from the footnote so well-known in discussions bearing upon the 'comparative-law/foreign-law' issue, in which an author from country A, writing about the law of his own country, informs the reader that a provision of the Civil Code of country B is similar to one in the Civil Code of country A – which of course has nothing to do with comparative law.

The author of the present study does not wish to conceal from the reader that he supports a clear-cut distinction between 'comparative law' in its strict sense and study of 'foreign law' [162] (the value of which he is far from underestimating), though he admits that the two fields overlap. The fact remains, however, that according to majority opinion, undoubtedly representing a current trend, comparative law also includes, for the reasons expounded above, the study of foreign law.

(b) *Current trends*

Current trends in comparative law are quite varied; [163] we shall mention here only the most important ones.

(i) One of the main trends in contemporary comparative law is the comparative study of the laws of socialist and non-socialist countries, with particular attention to the laws which, in these countries, govern foreign trade (research relating to the problems of 'East-West Trade').[164]

Although slightly less marked in recent years, this trend still makes itself

162. Cf. KNAPP, 'Champ d'application du droit comparé' (1972).

163. Cf. 'Tendencies and functions of comparative law in contemporary society' (1971) and also the report of the round table 'Perspectives et limites du droit comparé' (1970).

164. Cf. the documents of the International Association of Legal Science symposia held at Rome (1958), Helsinki (1960), Trier (1961), London (1962) and New York (1964): studies published in *Aspects juridiques du commerce avec les pays d'économie planifiée* (1961); *Some Problems of Non-Performance and Force Majeure in International Contracts of Sale/Problèmes de l'inexécution et [de] la force majeure dans les contrats de vente internationale* (1961); 'La Comparaison d'institutions juridiques en matière de contrat et de propriété entre pays à structure économique et sociale différente' (1961-62); *The Sources of the Law of International Trade with special reference to East-West Trade* (1964); *Unification of the Law Governing International Sales of Goods* (1966).

felt, and this type of study remains one of those that attract the largest number of workers in comparative law. One of the reasons for the persistence of this interest is the fact that the legal problems of 'East-West Trade' and of comparisons between socialist and non-socialist law lead to a further problem of the highest theoretical significance, namely that of the comparability of laws as between countries belonging to different socio-economic systems.

This comparability is generally accepted by Western authors without demur. A quotation from René David will illustrate the point:

> Any serious study in comparative law must envisage law in the social milieu which that law is designed to regulate, with due regard to the diversity of political and economic structures and to traditions. In these circumstances 'comparability' always appears as a question of degree, and it is in no way apparent to us that it should prove impossible, as a matter of principle, for example to compare the law of countries having liberal economies with the law of Marxist-Leninist countries having collectivized economies. The two types of society are evidently different, but similar questions arise in one and the other: the fight against crime, the organization of public authority, the control of the administration, the stability of the family and of employment, in short, the adaptation of man and institutions to a civilization dominated by other techniques. The answers to these questions will differ in the two types of society, but this fact in no way excludes comparability, on the contrary it greatly increases its interest. Comparability of systems can only be rejected if one places oneself on the plane of 'micro-comparison 'and remains closely attached to the concepts applied under the various legal systems.[165]

On this point Marxist authors maintain a more guarded position resulting from their Marxist-Leninist conception of law. We must nevertheless, before explaining the reasons for certain hesitations on the part of Marxist-Leninist science in this connexion, take note of a misunderstanding which very frequently arises in so far as there is confusion between the question of the *comparability* of the legal systems of countries having different socio-eco-

165. 'Toute étude sérieuse de droit comparé doit envisager le droit dans le milieu social qu'il est appelé à régir, en tenant compte de la diversité des structures politiques et économiques ainsi que des traditions. La "comparabilité" nous paraît être toujours dans ces conditions une question de degré – et il ne nous paraît nullement que, par principe, il soit impossible par exemple de comparer le droit des pays d'économie libérale et celui des pays marxistes-léninistes d'économie collectivisée. Les deux types de sociétés sont assurément différents, mais des questions semblables se posent dans les uns comme dans les autres: lutte contre le crime, organisation des pouvoirs publics, contrôle de l'administration, stabilité de la famille ou de l'emploi, bref, adaptations de l'homme et des institutions à une civilisation dominée par d'autres techniques. La réponse donnée à ces questions sera différente dans les deux types de sociétés; mais cette considération n'exclut en aucune façon la comparabilité; elle en accroît bien au contraire l'intérêt. La comparabilité des systèmes ne peut être niée que par qui se place sur un plan de "microcomparaison" en s'attachant étroitement aux concepts mis en œuvre dans les différents droits.' (Replies to the questionnaire.) Cf. also JAKOBS, 'Zur Methodik der Zivilrechtsvergleichung zwischen Rechten aus verschiedenen Wirtschaftsordnungen' (1961); LOEBER, 'Rechtsvergleichung zwischen Ländern mit verschiedener Wirtschaftsordnung' (1963); in addition, RHEINSTEIN, 'Comparative law and legal systems' (1968); TUNC, 'Comparative law, peace and justice' (1961); by the same author 'Le contribution possible des études juridiques comparatives à une meilleure compréhension entre nations' (1964); etc.

nomic structures, and that of the *results* to be expected from such a comparison. Authors sometimes rule out *comparability* when they are actually thinking of a *resemblance* between the laws under consideration (fearing that by admitting *comparability* they may commit themselves to acknowledging a *resemblance* in what is compared).

There is here, of course, nothing but a complete misunderstanding. The problem of *comparability* involves nothing more than the question of the existence of a *tertium comparationis* of a plausible and reasonable nature. A comparison undertaken on the basis of such a *tertium comparationis* may doubtless result in some cases in acknowledging a resemblance between the objects compared, but it can just as well end by recognizing their profound dissimilarity.

From a strict standpoint of comparability (and naturally, without prejudging the outcome of the comparison) we may accept René David's belief noted above that even as between countries with differing socio-economic systems there exist some general notions susceptible of comparison. From the point of view of Marxist-Leninist legal science, however, the question appears much more complex.[166]

As we have already explained,[167] law in the Marxist-Leninist conception is an instrument of power in the hands of social classes and the methodological basis of Marxist-Leninist legal science consists in approaching the study of law in the light of its origins and social impact. The same principle naturally applies in the field of comparative law, which is the reason for certain doubts regarding the comparability of the legal systems of countries belonging to different socio-economic systems.

Law being, in the Marxist view, no more than the formal manifestation of various social relationships, we cannot compare the law of different countries, and especially of those belonging to differing socio-economic systems, unless we take into account the dialectical links between law as form (phenomenon) and its content (substance). It follows that according to the Marxist-Leninist conception of comparative law, we must necessarily take up position on several planes simultaneously if we are to make a comparison and even if we are to establish the *tertium comparationis* which is the essential condition of comparability. We must of course compare the existing legal institutions (positive law), but comparison cannot be confined to the level of positive law: to arrive at a scientific finding we must compare at the same time the social conditions which determine the legal institutions under comparison, and the impact which these institutions themselves have on society.

The outcome of such a comparison may prove to differ considerably from that resulting from a direct comparison of the institutions of positive law, or even of the general principles of law, when examined without reference to

166. Cf. the works by Zivs and by Čhikvadze & Zivs referred to in footnote 153 on page 980. Cf. also Szabó, 'Sravnitel'noe pravovedenie' (1963, 1969); Knapp, 'Zur Frage der Vergleichbarkeit sozialistischen und kapitalistischen Rechts' (1962 [1963]); Eörsi, 'Jogösszehasonlítás és békés együttélés' (1964); etc.

167. Cf. Section Two, sub-section III, pp. 966 sqq.

their dialectical relation to the social data mentioned above. It may, and frequently does, happen that viewed in themselves, the legal institutions of countries belonging to different socio-economic systems appear quite similar; they are however extremely different as soon as they are viewed as the formal or phenomenal manifestations of specific social relationships, in other words as soon as a comparative examination is pursued on the different levels which have to be taken into account under the Marxist-Leninist conception of legal comparison.

Starting from this conception of the comparability of laws between countries belonging to differing socio-economic systems we may, in some cases, arrive at a conclusion contrary to that of René David: there are some legal systems which, from a socio-economic point of view, lend themselves more naturally to 'micro-comparison' than to 'macro-comparison'. For example, the transfer of property, the grounds of its acquisition and similar topics can be more readily studied comparatively than the institution of property itself. We are however in the fullest agreement with René David in believing that from the scientific standpoint the most important form of comparison is indeed that of 'macro-comparison'; remembering at the same time in this context, of course, the ancient and very wise maxim '*bene docet qui bene distinguit*'.

We may therefore conclude that Marxist-Leninist legal science does not deny the comparability of socialist and bourgeois law and is not opposed to it. But it approaches the comparison of laws in accordance with a methodology differing from that employed in the Western countries. It does not underrate the utility of such comparisons, especially in the field of foreign trade; on the other hand it does not overrate the significance of such resemblances as may exist between legal institutions or legal principles viewed as such and in isolation.

(ii) A very marked trend, related to some extent to the foregoing, is found in the persistent interest attached, since the close of the Second World War, to the study of Soviet law and, equally and in growing measure, to the general study of the legal systems of socialist countries, especially European.

This trend is evident both in the socialist countries themselves and in the so-called Western countries, although the objectives are naturally very different as between the two groups.

In the socialist countries, students of Soviet law are not faced with a legal system different from their own from the socio-economic point of view. It is a system of the same quality and one which, moreover, in spite of all the differences, sometimes substantial, which distinguish it from the legal systems of other socialist countries, was, as the first socialist law in history, at the origin of those systems and furnished a model for them in respect of fundamental principles.

The study of Soviet law in other socialist countries was accordingly first conceived as the study of a model-system, the object of the study being to derive benefit from knowledge of law in the USSR with a view to the development of law in each country concerned. Ever since the birth of the socialist countries after the last war, Soviet law has been studied there as foreign law without, however, any attempt at transposing its institutions into the language of the new countries but on the contrary by seeking to adapt the legal language of the latter to the notions at work in Soviet law.

There can be no doubt that in the process of building up a new legal system involving a profound modification of law in the socialist countries born after the Second World War, this method proved efficacious, in spite of the fact that some issues were then viewed in simplified terms. It was, however, only suited to a transitional period. The present trend, which has been evident in the socialist countries for some fifteen years past, consists in approaching the study of law in the USSR and in other socialist States preferably from the standpoint of comparative law in its strict sense. A new discipline has been created which might be labelled *comparative socialist law* [168] and which takes the form of comparative research into the legal systems of the USSR and the other socialist countries, seeking to contribute, by this process of mutual comparison, to a deeper understanding of the law of one's own country, and its development, as well as to a knowledge of the fundamental principles governing the evolution of socialist law in general. It is worth adding that *comparative socialist law* is not solely concerned with 'private law', but also to a broad extent with comparison in the field of 'public law'. (For example the Institute of State and Law of the USSR Academy of Sciences, in its replies to the questionnaire, regards research relating to the 'theoretical problems of the development of democracy, law and the administration of socialist countries', and to the problem of intensifying cooperation between them, as one of the main trends of modern legal science in the USSR.[169] The same is true for the other socialist countries.)

A very different picture emerges from the study of Soviet law (or, as the case may be, of socialist law in general) in the Western countries. Such study is, on the one hand, motivated by the desire to widen the horizons of scientific knowledge, but, in our view, mainly by a political concern to know the law of the other 'camp', in order the better to combat its influence, as well as in order to facilitate the development of international trade with the socialist countries, and so forth. These studies are generally conceived of as belonging to the field of study of *foreign law*.[170]

Nevertheless, close study of Soviet law has taught even Western jurists that the legal systems of the various socialist countries, although resting on the

168. See KNAPP & SZABÓ (eds.), *Introduction aux systèmes socialistes* (1971). Cf. also ZIVS, 'O metode sravnitel'nogo issledovanija v nauke o gosudarstve i prave' (1964), as well as the views of the Soviet rapporteurs at the round table in Paris on 3 May 1963 (Professors Zivs and Krutogolov): See LYON-CAEN, 'Table ronde sur les études et les recherches de droit comparé en U.R.S.S.' (1964), p. 70.

169. Replies, pp. 21, 22.

170. Among the numerous works of this type we may cite those of BERMAN, *Justice in the USSR* (1963); DEKKERS, *Principes nouveaux de droit soviétique* (1962); GEILKE, *Einführung in das Sowjetrecht* (1966); HAZARD, SHAPIRO & MAGGS, *The Soviet Legal System* (1969); etc. See the criticism of bourgeois 'sovietology' in ČHIKVADZE *et al.*, *Marksistsko-leninskaja obščaja teorija gosudarstva i prava*, Vol. III (1972), Chap. XI.

same basic principles, developed by legal science and by Soviet legislation, are in no way uniform, nor are they the result of mere reception of Soviet law. Recognition of this diversity is no doubt the reason why research into socialist law, while continuing to accord pride of place to Soviet law, is broadening its range among Western scholars and, as we have seen, is beginning to concern itself also with law in the other socialist countries. Hence the study of socialist law in Western countries has led to a situation where the study of foreign law and comparative law *stricto sensu* overlap to a certain extent.[171]

(iii) A quite recent trend in research into comparative law is a fast-growing interest in the law of countries having recently attained independence, in the first place the new African States. This trend is even, in the Western countries, beginning to outstrip that relating to the socialist countries. The latter is well-established and based on an institutional structure of considerable size, whereas the study of African law is only at its beginnings, though its development is proceeding very rapidly and there are already a considerable number of works in existence dealing with the legal problems of the newly independent countries.

Study of the law of these countries, however, while classified, as we have just said, under the heading of comparative law[172] is only put in that category as a rule, as being part of the study of 'foreign law' and quite often it falls completely outside the limits of comparative law.[173] In this connexion

171. Cf. HAZARD, *Communists and their Law* (1969); see also *Les formes de l'Etat socialiste* (Proceedings of the Symposium held at Strasbourg, 1-2 Apr. 1966, publ. 1968); DAVID, R., *Les grands systèmes de droit contemporains* (1964, 1971); etc.

172. For example, the following themes were selected by the International Academy of Comparative Law for the VIth, VIIth and VIIIth International Congress of Comparative Law: 'Les tendances constitutionnelles des Etats ayant accédé récemment à l'indépendance' (VIth Congress, Hambourg, 1962, Section IV A 1); 'Modernization of Law in the New States' and 'Le régime juridique des investissements étrangers' (VIIth Congress, Uppsala, 1966, Sections I C 1 and II B 1, respectively); and, lastly, 'Les rapports entre le développement social et économique de la société et le développement du droit' (VIIIth Congress, Pescara 1970, Section 1 B 3). For the documents dealing with these topics, see *Rapports généraux au VIe, au VIIe* and *au VIIIe CIDC* (1964, 1968, and forthcoming).

173. Cf. for example: ALLOTT, *Essays in African Law* (1960); by the same author, *New Essays in African Law* (1970); BARATAŠVILI, *Novye gosudarstva Azii i Afriki* (1966); GONIDEC, *L'Etat africain* (1970); GRAFSKIJ & STRAŠUN, *Federalizm cain* (1970); ČHIKVADZE *et al.*, *Marksistsko-leninskaja obščaja teorija gosudarstva i prava*, Vol. II (1971), Chaps. IX and X; CURRIE (ed.), *Federalism and the New Nations of Africa* (1964); ELIAS, *Africa and the Development of International Law* (1972); ENTAN, *Nacional'naja gosudarstvennost' narodov zapadnoj i central'noj Afriki* (1966); GONIDEC, *L'Etat africain* (1970); GRAFSKIJ & STRAŠUN, *Federalizm v razvivajuščihsja stranah* (1969); HAZARD, 'Classifying Africa's law' (1966); by the same author, 'Marxist models for West-African law' (1969); by the same author, 'La familia de sistemas jurídicos de inspiración marxista en Africa' (1967); LIONS, *Constitutionalismo y democracia en el Africa recién independizada* (1964); RHEINSTEIN, 'Problems of law in the new nations of Africa' (1963); by the same author, 'Law and social changes in Africa' (1962); TUNC (ed.), *Les aspects juridiques du développement économique* (1966); ULJANOVSKIJ (ed.), *Nekapitalističeskij put' raz-*

it may be recalled that contemporary legal science continues to interest itself in these countries from the standpoint of legal ethnology or ethnography[174] and – this represents a fairly important trend in legal science to which we shall also refer later – in relation to study of the 'acculturation' of law in the newly independent countries, both trends falling entirely outside the framework of comparative law (even taken in the broad sense, that is, including research on 'foreign law').

The most significant feature, however, about the current interest in the law of these countries is the tendency to build up a complex study of the problems of State and law which arise there, to create a development law[175] which also goes beyond the limits of comparative law and to which we shall refer more fully in the Epilogue to this study.

(iv) A durable and persistent trend (probably the oldest of all these trends) relates to research concerning the conditions and possibilities of a unification of law at the international level – and here a particular tribute should be paid to the work of the International Institute for the Unification of Private Law (UNIDROIT, Rome) – or at least a rapprochement of the laws of the different countries.[176]

---

vitija Afriki (1967); ULJANOVSKIJ (ed.), *Pravo v nezavisimyh stranah Afriki* (1969); ZIVS, 'Rol' prava v obespečenii nekapitalističeskogo razvitija stran Afriki' (1966); etc. In contrast, there are relatively few works on comparative law written by African authors; we may quote as one example the work edited by M'BAYE, *Le droit de la famille en Afrique noire et à Madagascar* (1968).

174. See below, pp. 1006-1007.

175. See p. 1038, n. 291.

176. A distinction must, of course, be made between the unification of law at the international level and its unification within a country – an internal problem experienced by countries where law has grown up under the effect of various historical influences (cf., for example, BLAGOJEVIĆ, 'L'unification du droit et notamment du droit privé en Yougoslavie' (1958), and, by the same author, 'La signification sociale contemporaine de l'unification du droit sur le plan international' (1972) – similar problems have arisen also in Czechoslovakia, Poland, etc.), and still experienced today, for example, by certain African countries (cf. particularly the works of ALLOTT cited at the end of this footnote), by India (cf. RAMA RAO, *Codification and Unification of Indian Law*, 1963), etc. Problems of this kind can even be detected in certain federations, as, for example, in the United States where, as Rheinstein tells us (Observations on the Revised Report, I, 3), unity of law has been maintained mainly through the influence of 'the universities, their professors and their writings'.

As regards the unification of law at the international level, apart form UNIDROIT, it is principally the International Academy of Comparative Law and the International Association of Legal Science (IALS) that encourage research in this area; cf. the following topics: 'Inventaire critique des matières qui ont fait l'objet de réalisations ou de tentatives en vue de l'unification du droit' (VIth Congress of the International Academy of Comparative Law, Hamburg, 1962, Section I G 1); 'Le problème de la coordination des mouvements d'unification du droit' (VIIth Conference of the Academy, Uppsala, 1966, Section I F 1) – for the relevant documents, see the *Rapports généraux* (1964 and 1968) for these two Congresses (CIDC); *Unification of the Law Governing International Sales of Goods* (symposium organized by the IALS in New York, 1964, documents published under the same title in Paris, 1966); etc. See also, for example, ANCEL, *Utilité et méthodes du droit*

This trend has been powerfully influenced and even modified by the development of international politics since the last war. One of the characteristic features of international life in recent years has been the emergence of numerous regional organizations of States entailing more or less developed forms of economic integration and also in some cases of political integration, and the breaking-up of the great pre-war empires, in particular the disappearance of the British Empire as such and its transformation into the Commonwealth.

This development gives rise to a number of new problems in the sphere of international law [177] and represents one of the significant new trends in its evolution; while the same international realities are also responsible for pointing research in comparative law in the direction of the unification of the legal systems of the different countries.

However, study of the present situation and prospects in regard to a worldwide unification of laws has led, taking into account the rather limited results so far achieved,[178] to a recognition that it is virtually out of the question to expect such unification in the near future, except perhaps on a fractional basis and limited to regional units. The centre of gravity of this line of research has therefore shifted noticeably towards problems concerning the *rapprochement* of the legal systems of the different countries, and particularly towards a study of the conditions of and prospects for *regional* rapproche-

---

comparé (1971); by the same author, 'Les aspects nouveaux de l'unification des droits' (1972); von Caemmerer, 'Rechtsvergleichung und internationales Privatrecht' (1966); David, R., 'The international unification of private law' (1972); by the same author, 'The methods of unification' (1968); Graveson, 'The unity of law' (1968); by the same author, 'L'étude du domaine de l'unification du droit' (1964); Houin, 'Pour une codification européenne du droit des contrats et des obligations' (1964); Limpens, 'Le problème de la coordination des mouvements d'unification du droit' (1968); by the same author, 'Relations entre l'unification au niveau régional et l'unification au niveau universel' (1964); by the same author, 'L'étude du droit comparé envisagée comme moyen de rechercher les matières susceptibles d'unification sur le plan international' (General report to the Vth International Congress of Comparative Law, Brussels, 4-9 August 1958, publ. 1960); by the same author, 'Inventaire critique des matières qui ont fait l'objet de réalisations ou de tentatives en vue de l'unification du droit' (General report to the VIth International Congress of Comparative Law, Hamburg, 30 July-4 August 1962, publ. 1964); Lukić, 'Uporedno pravo i univerzalno pravo' (1966); Nadelmann, 'Uniform legislation versus international conventions revisited' (1968); Zweigert & Kropholler, *Sources du droit uniforme international/Sources of International Uniform Law* (1971); etc. The question of the unification of law at regional level has become one of great topical interest for the African countries. Cf. Alliot, 'Problèmes de l'unification des droits africains' (1967); Allott, 'Towards the unification of the laws in Africa' (1965); by the same author, 'The unification of laws in Africa' (1968), and so on.

177. See Section Three, sub-section VII (b)(ii) below, pp. 1021-1022.

178. Mention can be made of such things as the relatively widespread unification of legislation concerning bills of exchange and cheques; international conventions serving as a basis for the unification of the fundamental principles of copyright and industrial property; the 1964 Hague Conventions on international sales, and a number of regionally-based unifications.

ments of legal systems, or closer harmonization – possibly even partial unification – of various groups of States such as the European Economic Community (EEC), the Council for Mutual Economic Aid (COMECON),[179] the Council for African Economic Co-operation, Benelux, etc. Under the quickening influence of the trend towards the universalization of comparative law, such research is not confined to groups of European States but is also taking an interest in groups of African and Latin-American States.[180] This, then, is the new direction which an already old-established tendency has recently taken.

One group of States, unique of its kind, which has been studied from this point of view is the British Commonwealth, and there is reference in this context to '*comparative Commonwealth law*'.[181]

A very specific field of comparative law is represented by Moslem law. Moslem law, though a single whole, is divided, as we have noted,[182] into several orthodox schools (rites). Given the theoretical immutability of Moslem law as such, both scientific research and the legislative policy in Moslem countries – in theory, there is no *legislative* power in these countries as far as Personal Status is concerned – seek, if the need arises, to 'graft' the rules and principles of one orthodox school (rite) onto the *positive law* based on another orthodox school, that is, to incorporate into a given system of positive law, rooted in a particular school of Moslem law, the rules and principles borrowed from another such school.[183] This process whereby various schools (rites) of Islamic law cross-fertilize one another as regards positive law is facilitated, as Linant de Bellefonds shows, by the study of 'comparative Moslem law'.[184]

179. Very strong emphasis has been placed on rapprochement and even on the consecutive unification of the respective legal norms of the Member States through the Complex programme of continued expansion and improvement of co-operation and development of the economic integration of the Member States of COMECON adopted at the 15th meeting of the Council.

180. We may recall the symposium on *Problèmes actuels du rapprochement régional des droits nationaux* organized by the International Association of Legal Science at Copenhagen in Summer 1969. Cf. also ALLOTT, 'The unification of laws in Africa' (1968); HAY, 'The Common Market Preliminary Draft Convention on the recognition and enforcement of judgments' (1968); HELLNER, 'Unification of law in Scandinavia' (1968); KRÜGER, 'The Council of Europe and unification of private law' (1968); MÁDL, 'Pravovye problemy složenija èkonomičeskoj integracii v stranah SEV' (1970); etc.

181. Simmonds, Replies, p. 9. Opinion on this point is, however, far from being unanimous. Lawson (Comments, p. 3) asserts, on the contrary, that comparison between the laws of the various Commonwealth countries can hardly be regarded as comparative law and that, from the point of view of the Common Lawyers, the notion of comparative law covers only comparison of Common Law with Civil Law (continental law), socialist law, etc.

182. Cf. Section One, sub-section III, 6, pp. 945-946.

183. See Section Three, VI, 2 below, pp. 1016-1018.

184. LINANT DE BELLEFONDS, *Traité de droit musulman comparé* (1965); cf. also his special contribution.

3. *Study of the applicability to the field of legal science of methods borrowed from the exact sciences* [185]

(a) *General problems*

(i) Contemporary society demands that its administration should rest upon scientific bases. But the social sciences,[186] which are called upon to provide these bases, have been suffering for some considerable time from a degree of methodological exhaustion (especially as compared with the thrust and vitality of the exact and natural sciences). They are therefore in danger of finding themselves in a backwater and of falling still further behind the exact and natural sciences.

To avoid this danger most of the social sciences are attempting to inject new vigour into their methods and for this purpose to direct research towards the possibility of applying in their own spheres the exact methods which have proved themselves in the natural sciences.

(ii) Trends aiming to give new vigour to methods of research, and particularly to demonstrate the applicability of exact methods, are visible in a number of countries even in the field of legal science.

As we have already noted, however, the notion of legal science cannot be regarded as firmly established: it varies from country to country, and even within a single country one may find several conceptions of it.[187] We may also note the existence of a tendency favouring the recognition – either as a branch of legal science or as an autonomous science – of administrative science, the subject-matter of which is the administration of the State or, as the case may be, of society in general. (The subject-matter of administrative

185. The following text is taken with some modifications from the special contribution which the author of this study prepared in collaboration with V. Vrecion, entitled 'The possibility of applying cybernetic methods to law and administration'. The reader is invited to refer (in the *International Social Science Journal, op. cit.* (1970), under KNAPP & VRECION) to the extensive bibliography listed there, to which we may add the following works: KUDRJAVCEV (ed.), *Voprosy kibernetiki i pravo* (1967) and *Problems of Cybernetics and the Law* (1968); BAADE (ed.), *Jurimetrics* (1963); HAFT, *Elektronische Datenverarbeitung im Recht* (1972); LOSANO, *Giuscibernetica* (1969); ŠLJAHOV, 'De l'application de la cybernétique au droit en U.R.S.S.' (1970); 'L'informatique juridique et les droits de l'homme' (round table, 1970); etc. Cf. also the bibliographies *Computers, Cybernetics and Law* (1970) and *JUDAC* (1971). The number of references given below will accordingly be kept to the strict minimum. It should also be noted that H. Eek refers in his comments to the prospects of applying methods borrowed from cybernetics to international law, particularly as regards international systems and problems of mediation within them, international organizations, decision-making in international affairs, etc.

186. The author employs the notion of *social sciences* in the sense in which it is used in his country (and in the socialist countries in general), giving it a wider sense than in France, for example. The notion of 'social sciences' used in this study thus covers not only the social sciences *stricto sensu* but also what are in France normally termed the 'human sciences' (*sciences humaines*).

187. See Section One, sub-section II, pp. 933 sqq.

science is therefore far broader than that of administrative law, and appreciably different.) [188] We shall return to this in the Epilogue.

It has appeared necessary to recall this variety of existing conceptions as a warning to the reader, in the first place, that the problem of the applicability of cybernetic methods cannot be viewed from the standpoint of legal science alone; and, secondly, that our attempt to throw light on the problems relating to the applicability of these methods in our domain does not rest on some hard-and-fast conception of legal science and that we are referring here, quite simply, to legal science in general and the study of the administration of society in particular.

(iii) The exact methods which are now beginning to be applied in the field of the social sciences in general, and in that of legal science in particular, are a series of methods borrowed from mathematics (e.g., graph theory, set theory, game theory, etc.), the experimental method (social testing) [189] and especially cybernetics, the methods of which appear the most apt for conferring increased accuracy on the knowledge of society and the study of its administration.

It is, however, well known that the concept of cybernetics is itself not very clear and that definitions vary widely.

We have adopted as a starting point the definition of cybernetics devised by A. A. Ljapunov and S. V. Jablonskij [190] according to which it is an exact science dealing with the general laws governing the structuration of *complex* administrative systems and the processes of administration. Systems of administration are entities of which the structure and elements can be determined and within which it is possible to define the co-ordinates governing the distribution of the elements in the structure, the forces of an internal or external origin which operate within the structure and finally its actual function.

In cybernetics the notions of processes and systems of administration are more generalized than in the usual intuitional figurations.

Despite this generalization, and although one of the works of the founder of cybernetics, N. Wiener, is devoted to the relationship between it and society,[191] cybernetics concentrates on technical problems and relegates to the background research relating to the administration of society.

We must also note that cybernetics, although a science concerned with systems and processes of administration, is not *ex professo* a science dealing with the administration of *society*. Human society is therefore one specific

188. We should note that the idea of an administrative science is nothing new: it goes back to the end of the eighteenth and the nineteenth centuries, when the idea of such a science existed in Germany under the name '*Verwaltungslehre*'. It was however neglected and lost sight of during the rise of legal positivism.

189. The traditional notion of testing (i.e., experimentation) in the legal field (cf. for example Offner, 'Experiment im Recht', 1931) has been revived, on quite different bases, especially in the legal literature of the socialist countries; see p. 977, n. 143.

190. 'Theoretičeskie problemy kibernetiki' (1963).

191. *The Human Use of Human Beings. Cybernetics and Society* (1950, 1954).

subject of cybernetic research, hence the applicability of cybernetic methods to the field of the administration of society gives rise to a number of specific problems of a methodological character.

It follows from the definition given above that cybernetics is only concerned with research bearing upon *complex* administrative systems. If then cybernetics undertakes the study of a given administrative system, including the objects and processes which relate to the law and administration of society, numerous special inquiries must first be carried out to allow a determination of the character of the system under consideration. In determining the degree of complexity of any system of administration whatever, one ascertains whether it is practicable and logically justifiable, in the study of that system, to have recourse to the whole complex mechanism of cybernetic investigations.

Research of this type has shown that the phenomena associated with the administration of society and with law, such as the legal order and all the processes of creation and enforcement of law, generally possess the character of complex social phenomena or objects qualifying them as fit subjects for cybernetic research.

(iv) We have already pointed out that the fundamental problems of the applicability of cybernetic methods to law and the administration of society were of a *methodological* type. In this field above all it was absolutely necessary and indispensable, before deciding upon the *use of new methods*, to examine a whole series of problems related to the *theory of methods* – problems resulting from the fact that, as we have just noted, human society and its administration constitute a highly specific subject of cybernetic research.

It was therefore necessary first to solve certain prior problems. For example the attempt has been made (and, we venture to say, made successfully) to demonstrate that cybernetic research is capable not only of apprehending problems of quantity but also – through quantity – problems of quality (both terms being taken here in their philosophic sense); and that cybernetic research and the practical application of cybernetic methods could be developed even on the basis of the logic of normative precepts – which is very important from the standpoint of the logical structure of law. Finally, and this is of great significance in Marxist science, it has been shown in particular that even dialectical relationships and processes can be analysed by cybernetic means.

(b) *Current trends*

We shall now attempt to describe briefly the main trends in the application of cybernetic methods, or the exact methods linked with cybernetics, in the field of law and the administration of society.

(i) A relatively self-contained as well as rewarding field where the theory and methodology of cybernetics have been applied is that of the philosophy of law or, as the case may be, of the general theory of State and law. Here the use of notions and methods borrowed from cybernetics with a view to the elucidation of legal phenomena or of the impact of State and law upon society has contributed many powerful impulses to legal science. It has, for example, rendered more precise the concept of the fundamental mechanisms of administration in society as well as the concept of law as a social regulator; it has brought to light the part played by feedback in the administration of society; has made it possible to identify the rôle of informational relationships in society and has furnished theoretical bases for its sociological examination.

It follows that the application of cybernetics to law and administration carries with it teachings of a theoretical and methodological character which provide a solid basis for making these branches of scientific activity systematically more exact.

This methodological research, both theoretical and general, is a direct sequel to the work of Wiener.[192] The most important scientific schools in this context are, among others, the Soviet, United States, French and Czechoslovak.[193] Within methodological research we may also include the theoretical and methodological concepts of what is known as *systems engineering*.[194]

(ii) Since complex cybernetic research devoted to the administration of society, and more especially that bearing upon the influence exercised by the State and the law upon society, represents a particularly difficult scientific task usually conceived on a long-term basis, the results achieved so far in the systematic cybernetic analysis of these problems relate chiefly to either a precise analysis of the relatively least complex phenomena or a partial analysis of the complex phenomena. For example intense research has been carried out concerning the application of the theory of information to law and the administration of society.[195]

Many schools have been founded and a great deal of work has been done or planned with a view to establishing cybernetic or mathematical models of the organizational or functional components of State, political and, of course,

192. See footnote 191 on page 993.

193. Cf. KUDRJAVCEV (ed.), *Voprosy kibernetiki i pravo* (1967) and *Problems of Cybernetics and the Law* (1968). See also the works of KERIMOV (1963, 1964 – refer to Bibliographical Annex, II); reports by Deutsch, Rotekau and Rummel in MERRIT & ROKKAN (eds.), *Comparing Nations. The Use of Quantitative Data in Cross-National Research* (1966); see also ALKER, *Mathematics and Politics* (1965); DAVID, A., *La cybernétique et l'humain* (1965); KNAPP, *O možnosti použiti kybernetických metod v právu* (1963) (Russian version, *O vozmožnostijah ispolzovanija kibernetičeskih metodov v prave*, 1965); KNAPP & VRECION, 'Research of the applicability of methods of cybernetics in law conducted in Czechoslovakia' (1970); LÉVY, 'Le droit, l'information et la machine' (1967); etc.

194. See especially GOODE & MACHEL, *System Engineering* (1957); HALL, *A Methodology for Systems Engineering* (1963), MCCORMICK, *Human Engineering* (1957).

195. These applications are based particularly on the following works: BRILLOUIN, *Science and Information Theory* (1956, 1962); CARNAP & BAR-HILLEL, *An Outline of a Theory of Semantic Information* (1952); KNAPP, 'La gestion de la société socialiste et la théorie de l'information' (1963); LANG, *Staat und Kybernetik* (1966); VRECION, 'On some possibilities of the theory of information in the sphere of law and management' (1967).

economic mechanisms. These models are established primarily on the basis of the mathematical theory of multiple command, the mathematical theory of graphs, the theory of games, of dynamic programming, etc. Exact models of the organizational systems have been constructed which operate rationally and carry out certain combined operations; and it has been found possible to draw up exact descriptions of administrative relationships within relatively complex State and political systems. It is even possible to provide precise data for a scientific analysis of complicated power relationships within the State. In most cases research of this type does not consist in merely applying existing mathematical or cybernetic processes but in adapting, often fundamentally, these general processes to the distinctive nature of the organizational and functional components of the socio-political, administrative and other systems which are being analysed.

Cybernetic analysis of the organizational and functional components of systems for the administration of society has yielded positive results in regard to certain theories of limited scope bearing upon the scientific organization and administration of society. The application of these results may even be extended to cover the complex questions involved in State and political mechanisms.

(iii) The most extensive research and the most important practical application of cybernetic methods relate to projects for placing on an automated basis the updating, collection and analysis of information in the fields of legislation, law enforcement and administrative practice. Many methods and systems of this type have already been developed in these fields.[196]

It would in our view be desirable to standardize, on the international level, at any rate the fundamental methods of checking and analysing collective data, in order that it may be possible to organize an international exchange of information without encountering undue technical difficulties and to subject to the necessary comparisons and analyses data relevant to law and administration which are particularly germane to collective phenomena at the international level.

As the methods so far applied in the statistical study of collective phenomena are not generally accompanied by causal or other systematic analyses of these phenomena, we believe that future research will be concentrated on the application of automated methods of logical (or logico-statistical) analysis in this field. The experience so far acquired shows that such methods have fundamental advantages.

(iv) There are numerous projects for automatic processing of the content of the legal order; all these projects aim at automated systems for processing judge-made law, precedents and the interpretation of legal concepts. (We must note in particular the importance attaching to the automatic processing of precedents in Common Law countries.) [197]

196. Attention is called in particular to the joint project directed by Aurel DAVID, described in 'Plan de recherche concernant la documentation juridique automatique' (1968).

197. Cf. for example KEHL, *Automatic Data Processing for the Legal Profession*

A number of projects have also been worked out (we may refer as an example to the Czechoslovak project) which would allow us to hand over to a computer the task of identifying fundamental contradictions in the existing or planned legal order. For instance, the Czechoslovak method rests on automatic processing of the content of legal rules, the computer checking, by means of a direct coding system, even the most basic components of the text. Thanks to the use of rather complicated processes the computer is able to discern semantic divergences even of a relatively complex nature.[198]

(v) Existing projects for automation in the field of the application and enforcement of law are aimed particularly at operations of a relatively special character. Automation is current and of a relatively uncomplicated nature in the fields of taxation and pensions; but the formulation of adequate projects for determining the level of maintenance for minors is a far more complex task from the mathematical point of view.[199]

We must take note of the projects, very important not only from the methodological standpoint but because of the grave issues involved, which relate to automation of the various types of expert evidence encountered in criminology, with particular reference to graphology, finger-print identification, identification of persons from photographs, etc.

(vi) Many special problems arise in the various administrative sectors with regard to the application of cybernetic methods and the automation of planning and decision-making operations. A number of small-scale projects have already been carried out with a view to placing very narrowly circumscribed decision-making operations on a logical, automated basis. They consist essentially in applications of linear programming, dynamic programming and game theory. Ambitious theoretical studies, but also some practical projects, aim at the rationalization of planning in the various administrative sectors, using the methods of the mathematical theory of graphs. For the purposes of planning and keeping close track on the movement of particular indices, simply as a check or to ensure conformity with the planned levels, recourse is made to differential calculus, etc.

The application of cybernetic methods and automation are also relevant to professional training. For instance, projects are being drawn up, theoretically and experimentally, for applying cybernetic techniques to the fundamental teaching process in schools specializing in the training of lawyers and future administrators.

(1963); Tapper, 'Legal information and computers: Great Britain' (1968). An analogous system has been used to check the cases cited in legal literature: cf. Borkowski *et al.*, 'Structure and effectiveness of the citation identifier' (1970).

198. For further details see Vrecion, 'Metoda použití matematických metod a samočinných počítačů při evidenci, analýze a tvorbě právních norem' (1968).

199. In Czechoslovakia for example, at the Institute of State and Law of the Czechoslovak Academy of Sciences, an automated method has been worked out for determining the amount of maintenance, using somewhat complicated processes based on the mathematical theory of graphs. See 'Mathematical methods of determining alimony for minors' (1968) and Vrecion, 'Automated method for determination of alimony to legally entitled persons' (1968).

Cybernetic techniques can also be applied in carrying out aptitude tests such as specialist candidates for administration and law undergo. The method used consists in making a comparison between the candidates' answers to minor problems of administration and decision, chosen for the purposes of these tests, and the 'optimum' answers to the same questions as given by the computer working on a suitable algorithm.

(vii) The problems relating to the application of cybernetic methods to the fields of law and the administration of society, of which we have just described the general features, throw light on a very prominent feature of current research: legal science as well as legal practice and the administration of society are tending to develop more exact information. Despite some outstanding results already achieved, this research is only in its beginnings and it would be a mistake to overestimate the prospects. It nonetheless points, in our view, to a well-attested and typically modern phase in the evolution of the present-day social sciences.

III. HISTORY OF LAW

1. *The subject-matter of legal history*

The first question that arises is that of the subject-matter of the history (or more precisely the historiography) of law. We may view this question from several points of view.

(a) The first aspect of this problem is similar to the problem of the subject-matter of legal science in general, which we have already examined: the question is to determine whether legal history is confined to a historical study of law or whether it is to include also a historical study of the State.

No new trends have emerged on this point. It has long been recognized by most authors that legal history is in reality the history of State and law, the history of the State constituting a specific part of general historiography. For reasons which have already been expounded, the close links between the history of State and the history of law are particularly evident in Marxist-Leninist legal science, which regards the history of both State and law as a specific sector of general history, that is to say of the history of a society divided into social classes. (From a terminological standpoint it is interesting to note, as pointed out above, that Marxist-Leninist science frequently uses the simple formula 'legal science', whereas in the field of historical studies the expression used is invariably 'history of State and law'.)

(b) The question of the subject-matter of legal history also involves that of knowing whether and to what extent legal history deals with the development of State and law viewed apart from the history of the development of political theories and legal thought (legal theories), or whether it also deals with the latter.

Generally speaking, these studies, and especially the study of political theories, have traditionally been regarded as an autonomous branch of legal science (or sometimes of political science) pursued independently of the history of State and law. We may nevertheless venture to affirm that contemporary legal science reveals a tendency towards a merging of the history of State and law and the history of political and legal theories, and even towards the inclusion of the latter in the history of State and law.

(c) Before the Second World War legal history was almost exclusively concerned with European law, namely:

(i) first and foremost with Roman law, its revival in Italy and its reception in France, Germany, etc.;

(ii) with mediaeval law, that is to say Canon law on the one hand and, on the other, German, French, Slavic law, depending on geographical zone;

(iii) with English Common Law which, even from the standpoint of historical research, raises a particular problem which we will consider briefly below.

(d) The evolution of the modern world, and in particular the fact that interest in the future prevails sharply over interest in the past, in the social sciences as well as in the natural sciences, political life and so forth, obviously also affects the history of State and law. Observable tendencies today are broadly speaking the following:

(i) a decline in interest in Roman law, and even more in Canon law;

(ii) a broadening of the field of vision of legal history to include the history of law in the African countries and those of South and South-East Asia;

(iii) an increased interest in the history of State and law in recent times (nineteenth and early twentieth centuries).[200]

2. *Roman law*

We have just indicated that the study of the history of Roman law, formerly the centre of gravity of legal history, is now neglected; this is indeed the most striking feature noticeable today in legal history. It must be added that this loss of interest in Roman law is evident not only at the level of scientific research but also at that of university teaching, in which almost everywhere knowledge of the institutions of Roman law is regarded as something of a luxury, the teaching of Roman law being curtailed or even wholly abandoned in favour of the teaching of foreign law, comparative law and other branches.

This falling-off in historical research concerning Roman law is in our view not only the consequence of a universal tendency in the contemporary world towards a concentration of social interest on the future rather than the past; it is also due to the historiography of classical Roman law being largely worked out. This is the oldest branch of legal history and it has relied on the

200. A picture of the general direction taken by the historiography of law can be gained from the major bibliographies, such as: GILISSEN, *Introduction bibliographique à l'histoire du droit et à l'ethnologie juridique* (1964); KORANYI & KORANYIOWA, *Bibliografia historico-iuridica annorum 1937-1947* (1953-59).

same sources for a very long time. Although the methods for the study of Roman law have considerably developed since the beginning of this century, it is nevertheless understandable that as far as the classical period of Roman law is concerned, there remain few unexplored problems to attract the attention of research workers. Another reason for the decline in interest in Roman law is undoubtedly the well-nigh universal decrease in its relevance to the understanding of law in the contemporary world and the fact that its value as an introduction to the teaching of law has, as a result, considerably diminished.

A trend apparent for some time now towards extending the field of study of the history of Roman law to cover the laws of classical antiquity in general (and particularly the laws of Egypt and of the Middle East) derives from the same reasons, as does also a move towards pursuing the study of Roman law itself into its mediaeval sequels.

3. *Historia magistra vitae*

One of the striking features of legal history in the contemporary world is the decline in the influence, formerly very great, of the maxim '*Historia magistra vitae*'.

Roman law, as we have just seen, has almost entirely lost its pedagogic value; even the history of national legislations is now generally studied without being linked with contemporary law or contemporary political realities.

There are two important exceptions:

(a) *Marxist-Leninist* legal science, which employs the method of historical materialism, approaches knowledge of State and law, by definition, from a historical angle. The aims of the Marxist-Leninist historiography of State and law are always related to the needs of contemporary socialist society. Through a critical assessment of the past, through the analysis of progressive and revolutionary elements in history, etc., the study of the history of State and law serves to enrich the knowledge required for the scientific management of the development of modern society. Marxist-Leninist historiography of State and law is accordingly invariably inspired by the maxim '*Historia magistra vitae*';

(b) The second exception, of a very different nature, relates to the *Common Law* zone, in which the function of legal history (in relation to the law of the country) is very different from that assigned to legal history in Civil Law countries. In the Common Law countries, especially in England but also to a certain extent in the United States, the separation between past and present in law is not the same as in Civil Law countries; Common Law, being a law of precedents, is always resorting to the past; hence under Common Law, it is knowledge of the history of the country's laws that opens the door to knowledge of the law in force at present.

IV. LAW AND MAN

1. *Introduction (general observations)*

Law is always a social, that is to say a human, creation. This thesis is incontrovertible even from the standpoint of theories of natural law, which do not deny the human origin of 'given' law. It is also beyond doubt that law addresses itself to men and that the legal order is a system of rules governing human conduct in a particular society. Law is therefore linked to man twice over, through its origin and through its destination.

We may also note that law is never addressed to man in isolation from society: in his double relationship with law man is always a *zôon politikon.* The relation between man and law is accordingly a species of the general relation between individual man and society.

It follows that scientific knowledge of law might contribute to a knowledge of man's place in society, especially to an understanding of the relationship between man and the law and between man and the State. But legal science has long neglected these prospects offered by knowledge of law. Even when tackling the problems of man, such as human rights and the rights of the citizen, legitimate interests (*jura quaesita*), etc., it was only concerned with analysis of the rules of the law in question (positivist legal science), analysis of the ideal content of law (neo-Kantian theories), etc. It is only with the transformation of legal science into a social science, and especially with the advent of the Marxist-Leninist science of law, that a definite trend has emerged to attain a deeper understanding of man's place in society through knowledge of the law.

As is well known, one of the most marked trends in contemporary philosophy in Western countries has been the upsurge of interest in philosophical problems relating to man. Research of this type is often termed 'philosophy of man', or philosophical anthropology, or again, with a slightly different meaning, 'social anthropology'. Unlike scientific anthropology,[201] it aims at a coherent philosophical explanation of man, achieved in the first place by solving the question of his essential nature and existence.

The increased interest in the whole problem presented by man is, as we have said, also manifest in legal science, which is particularly concerned, in a manner that is both deeper and more complex than before, with problems of the freedom of man as a citizen, the problem of legitimate interests (*jura quaesita*), the rôle of State and law in the moulding of the human personality, etc., and, also naturally primarily with the general problems relating to the rights of man and of the citizen.

201. Study of the problems of scientific anthropology in the field of law (following on Lombroso's well-known theory) has long been a thing of the past and is now only encountered occasionally, e.g., in connexion with the study of juvenile delinquency organized by the European Co-ordination Centre for Research and Documentation in the Social Sciences, Vienna (cf. Péteri, Replies, p. 14).

This last question in particular has given rise to a very marked trend in modern legal science at the levels of both domestic and international law.[202] (There can be no doubt that this trend has been greatly stimulated by the International Human Rights Year proclaimed by the United Nations in 1968.)

It is nevertheless true that as a general rule, problems of human rights or, as the case may be, of the rights of the citizen, are being studied from the standpoint of constitutional and international law rather than from that of philosophy.

2. *Law at the service of man*

From a philosophical point of view it is interesting to note the sustained interest on the part of legal science in the whole matter of the usefulness of law for man, i.e., that of *law at the service of man*. This interest is all the more significant in that it very clearly bears witness to the fact mentioned earlier at the beginning of this sub-section that legal science as a social science can never be divorced from the theoretical or even political conception of human society in general.

Although the question of the rôle that law can play in the service of man is of interest both to Marxist-Leninist and non-Marxist legal science, it is plain that the basic assumptions and axioms differ greatly in the two cases, non-Marxist jurists taking as their starting-point a belief in the perfectibility of liberal (bourgeois) society, while Marxist jurists are fundamentally convinced of the need for revolutionary change in this society and its replacement by socialist society.

(a) Of the so-called Western countries, as we are informed by two of our associate rapporteurs (Professors Rheinstein and Tunc), it is in the *United States* that the greatest interest, from the point of view of both legal science and practice, is displayed in the problems of the rôle of law at the service of man.

202. Cf. BURDEAU, *Libertés publiques* (3rd edn. 1966); COLLIARD, *I ibertés publiques* (1968); FIX ZAMUDIO, 'Los derechos humanos y su protección ante las jurisdicciones nacionales' (1972); GELLHORN, *Ombudsmen and Others. Citizen's Protectors in Nine Countries* (1967); *Problèmes de protection internationale des droits de l'homme* (Vol. I (1969) of the collection *Amicorum discipulorumque liber René Cassin*); etc. In the literature of the socialist countries cf., for example ČHIKVADZE *et al.*, *Marksistsko-leninskaja obščaja teorija gosudarstva i prava*, Vol. I (1970), Chap. XII and Vol. III (1972), Chap. VIII; *Razvitie prav graždan SSSR i usilenie in ohrany na sovremennon ètape kommunističeskogo stroitel'stva* (1962); see also the works cited in *Literature on Soviet Law, Index of Bibliography* (1960), pp. 87 sqq.; etc. Cf. also GLOS *et al.*, *Práva občanů* (1969). It is worth mentioning that as far as Marxist-Leninist legal science is concerned, Hungarian jurists in particular have made a very sustained study of these problems: cf. SZABÓ & KOVÁCS (eds.), *Socialist Concept of Human Rights* (1966); see also the works of SZABÓ cited in Bibliographical Annex, II. Cf. also HANEY, *Sozialistisches Recht und Persönlichkeit* (1967).

This trend has been expressed with remarkable clarity by Rheinstein, who asserts that:

(...) American [legal] scholars as well as judges and lawyers [...] interest themselves in the law as a means of social engineering, of social reform. Such a need existed already in the early days of the Republic when the received English Common Law had to be adapted to the thoroughly different climatic, geographical, political, social and economic conditions of the new country. But the need of changing and adapting the law became even more burning when, in the years of the New Deal, that is in consequence of the Great Depression of the early 1930s, there was started that thoroughgoing and far-reaching movement of social reform which has been aptly labelled the Second American Revolution. It has been recognized that certain groups of the population could not attain the way of life of the American Dream without active support by governmental power. Under the New Deal this power was set into motion in order to bring about the unionization of labor and in consequence thereof the spectacular change in the social and economic position of the skilled part of the labor force. Now the great effort is made to achieve similar effects for unskilled labor, especially in so far as it is recruited from racial minorities (Negroes, Mexicans, Indians). Vigorous efforts at prevention of monopolies, at controlling the issuance of commercial securities, operations of the stock and produce exchange, general control of large-scale operations, and most recently also of foundations and the consequential far-reaching transformations of the social structure in general have made necessary fundamental changes in the law.[203]

Mention should also be made of the particular attention which *Italian* legal scientists have paid to the problem presented by man. Their work, which is based on ancient traditions – although subject to various foreign influences – represents a noteworthy trend within this school of legal science.

The position taken is that law is an instrument in man's service – a view which goes back to eighteenth-century authorities but is still prevalent today. This basic standpoint of the Italian school of legal science is manifested today above all in a whole series of research undertakings devoted to 'non-State law', that is to say to institutions and legal norms emanating from bodies of a collective nature – whether recognized at law or not – claiming the position of law-makers. Examples of such bodies are local organizations, trade unions, domestic or international employers' associations, sport associations and supra-national bodies of a religious character.

To understand the range of this series of studies it must be recalled that most Italian jurists accept the theory known as the 'plurality of legal orders' (*la pluralità degli ordinamenti giuridici*), a theory elaborated by Santi Romano,[204] which has had some very interesting applications in Italian legal science. We can accordingly understand how, through the study of various

203. Rheinstein, Comments, p. 2. In his comments, Tunc also emphasizes the importance accorded by legal science in the United States to the idea that law can be used to mitigate social injustice, for the advancement of underprivileged groups in the population, as a weapon against poverty, etc.

204. Santi ROMANO, *L'ordinamento giuridico* (1947); cf. also BOBBIO, *Teoria dell'ordinamento giuridico* (1960); etc. In this connexion we cannot conceal our feeling that the Italian concept just referred to recalls the view of law as '*die innere Ordnung der geselschaftlichen Verbände*' – the internal order of social associations – propounded by Eugen EHRLICH (*Grundlegung der Soziologie des Rechts*, 1913).

phenomena of an associative character which give rise to legal provisions, it is proposed in the first place to study man in the various functions which allow him to give expression to his personality; and to see how the instrumentality of the legal institution stands in relation to bodies of law different from State law.

Interest in problems that may be placed under the heading 'Man and Law' is not of course restricted to legal science in the United States and Italy. Traces are found over a fairly wide geographic area: in France, England, the Federal Republic of Germany, Mexico, etc.[205]

Let us pause for a moment to consider the *French* viewpoint as presented by Brimo, since this is rather unusual. When dealing, in his work *Les grands courants de la philosophie du droit et de l'Etat,* with the 'humanistic current' (legal humanism), he notes that 'as far as method is concerned, humanism leads to methodological syncretism' ('Sur le plan de la méthode, l'humanisme conduit au syncrétisme méthodologique') and comes down in favour of the humanistic method defined as a method 'made to the stature of man, like law, the very paradigm of a human and social phenomenon' ('à la mesure de l'homme comme le droit, type même de phénomène humain et social').[206] Although Brimo traces the humanistic current back to Montesquieu he believes (and in that we share his viewpoint) that the humanistic trend in legal science represents a relatively recent and anti-positivist current. His conception of the humanistic current however is very broad and extremely heterogeneous, since it not only embraces the schools represented by Hauriou, Gény, Ripert and Dabin but also the above-mentioned German schools of *Interessenjurisprudenz* (a school which bases its view of law on interests) and *Freirecht* (school of free law) and even those of Roscoe Pound and Max Weber. [207]

(b) On the other hand, Marxist-Leninist legal science, geared as it is to militant activity designed to bring about the proletarian revolution and the establishment of the socialist society, is necessarily opposed to any idea of the perfectibility of bourgeois society.[208] It is concerned with the rôle of law as directed to the perfecting of *socialist society* and its transformation into communist society, to which we shall revert later. The impact of law serving the interests of man is thus conceived of by it from the point of view of socialist humanism (it considers law as one of the means for the protection and free development of the human personality in socialist society), and,

205. Cf. especially BRIMO (see next footnote); GRAVESON, 'The unity of law' (1965); JAHRREIS, *Mensch und Staat* (1957); KLUG, 'Tesis para un análisis de la idea del hombre dentro del derecho' (1967); PELLOUX, *Le citoyen devant le droit* (1955); RECASÉNS SICHES, *Vida humana, sociedad y derecho* (1945); WEYL, *La justice et l'homme* (1962); etc. It should be noted also that the problems of man and law have been discussed by the International Congresses of Comparative Law (see, for example, VIIIth Congress, Pescara 1970, Section IV C, which dealt with public freedoms).

206. BRIMO, *Les grands courants de la philosophie du droit et de l'Etat* (1968), p. 308.

207. BRIMO, *ibid.*

208. Hence the author of the present study disagrees entirely with the views mentioned in (a) above.

more especially, from the viewpoint of socialist legality. Founded on the conception of society as divided into classes whose mutual relationships determine the historical character of a given society, it considers that its vocation and its purpose is to act in the socialist society on the creation of law and its application in such a way as to serve the interests of the dominant class (i.e., the working class), and hence the interests of the individual in socialist society.

Thus it never seeks to perfect an abstract society, conceived of in general terms, but always a given society, defined by the particular relationship between the classes; in other words, as we have just noted, it aims at the perfecting of socialist society and the utilization of law in the service of man within that society.

Legal science in the USSR devotes special attention to the question of 'law and man'.

In this connexion we may quote the Reply to the Questionnaire furnished by the Institute of State and Law of the USSR Academy of Sciences:

The problem of man, and the idea of man's primordial social value are immanent in the Marxist-Leninist conception of the world. Soviet social science has always considered the study of questions relating to man, his interests and his needs, as a very important current in social research. Soviet science of State and law is characterized by a perfectly clear standpoint: the rules of law, legal relationships, rights and obligations as well as all other phenomena in regard to State and law do not exist in and for themselves; nor do they constitute a subject for separate study; they only give rise to scientific research by virtue of their rôle and their place in social dealings involving men. In the Marxist-Leninist conception the essence of social processes, including legal processes, is profoundly humanist.

The science of State and law draws up its own model of personality, which is a synthesis of the political and legal attributes of the social individual. Soviet legal science is based on materialist teaching regarding the human personality, the subject-matter of its research being the real man forming part of a system of social relationships and acting effectively within that system. However, legal science is more particularly interested in man from the point of view of his specific qualities, namely the political and legal qualities which characaterize him and which are evident in the field of political and legal relations.

A specific relationship exists between the individual and the Soviet State based on the objective socio-economic principles of socialism; and the character of this relationship determines the political and legal position of the human being in society, it following from the very nature of socialism that man represents the highest value in society. For socialist State and law, a hierarchy of social relationships in which the idea of man represents the goal of social progress is sacrosanct and inviolable. With regard to citizens, the State has a number of important duties and legal obligations, for example to ensure the satisfaction of their material and cultural needs, to create the requisite conditions for the free development of their personalities, to guarantee the inviolability of their persons, and to protect their honour and dignity, etc.[209]

209. Replies, pp. 29-30. See also *Razvitie prav graždan SSSR* (1962); ČHIKVADZE *et al., Marksistsko-leninskaja obščaja teorija gosudarstva i prava*, Vol. I (1970), Chap. 12; KOTOK, TADEVOSJAN *et al., Ličnost', obščestvo i gosudarstvo* (1966); MALCEV, *Socialističeskoe pravo i svoboda ličnosti* (1968); by the same author, 'Gosudarstvo I ličnost pri socializme' (1970); PATJULIN, 'Ličnost', eë prava i svobody' (1966); etc. The literature of the other socialist countries is also devoting increased

3. *Legal ethnology*

Legal anthropology viewed in the ethnological sense of the term is nothing more than legal ethnology or, we might say, legal ethnography.[210]

The subject-matter of legal ethnology (or ethnography) is made up of the traditional or ancestral customs of specific ethnic groups. This discipline does not study traditional rules of human behaviour in the context of the *country's* laws but in the *ethnic* context, that is to say as an element in the civilization of the ethnic group in question. The ethnological standpoint is therefore very different from the legal standpoint, which is probably the reason why the ethnological study of law has traditionally been – and continues to be – regarded as an integral part of general ethnology rather than as a discipline falling within the scope of legal science.

It should however be noted that jurists themselves have long been interested in legal ethnology, particularly in countries that formerly possessed colonies, such as Great Britain for example, where a remarkable work on legal anthropology appeared as early as 1861,[211] and Germany, where the review *Zeitschrift für vergleichende Rechtswissenschaft*, which accorded considerable space to legal ethnology, was founded by Josef Kohler long before the First World War,[212] etc.

The interest in the newly independent African and Asian countries that has developed for political and economic reasons since the Second World War has again turned the attention of jurists to legal ethnology and at the same time modified the goals of this discipline. The study of anthropological (or ethnological or ethnographic) data is usually regarded as falling under the sociology of law, as being an elucidation of the rôle of law in society. A typical example of the sociological conception of legal ethnology (or anthropology) is provided by the works of the French sociologists of law, H. Lévy-Bruhl [213] and recently J. Carbonnier.[214] In the United States a very interesting example of this view is to be found in the study of 'trouble situations' among

attention to the problems of 'law and man': see the works cited by way of illustration in footnote 202 on page 1002.

210. 'The terms *ethnology, ethnography* and *anthropology* are often used synonymously', states J. Poirier ('The current state of legal ethnology and its future tasks' (1970), p. 476). Marxist-Leninist science prefers to use the term *ethnography*. On the contrary French usage, as we know, regards ethnography as a descriptive discipline, and therefore separate from ethnology. English and American terminology prefers the term *anthropology*, or, more exactly, *social and cultural anthropology*; this preference appears to be gaining ground in Western Europe with a consequent lesser use of the term *ethnology*.

211. Maine, *Ancient Law* (1861).

212. However this current came to an abrupt end in Germany after the First World War and only general ethnological research continues which has no influence on jurisprudence (Zippelius, Replies, par. 11).

213. See bibliography in Poirier (ed.), *Ethnologie générale* (1968), pp. 1236 sqq.

214. The latter, however, draws a distinction between legal ethnology and legal anthropology. Cf. *Sociologie juridique* (1972), pp. 29 sqq., 36 sqq.

the American Indians,[215] the object of which was not only to deepen knowledge of Indian ethnic groups inhabiting the country, but also to open up, through the study of these quite specific situations, various new horizons to research into the social rôle of law in general.

However, despite the steady interest in legal anthropology (in the ethnological or even ethnographical sense of the word,[216]) the attention which, doubtless by reason of the political and economic interests involved, contemporary legal science focuses on the law of newly independent countries, particularly African countries, goes far beyond the limits of legal anthropology (ethnology, ethnography) and forms part of the same broad current as the sociologically inclined trend, the trend to develop research in the acculturation of law, the rapid growth of comparative law, and, in general, what is sometimes termed 'development law'.

V. LAW AND DEMOCRATIC PRACTICE

1. It is well known that the notion of 'democracy' has no clearly defined outlines. The meaning of this historic conjunction of the words '*demos*' and '*kratos*' has taken many different forms through the centuries. The democracy of the ancient Greek *polis* bears very little relation to the democracy of current times; even in contemporary society the meaning of democracy varies considerably, largely dependent as it is on the relations between social classes.

The present division of the world, albeit it rests on profoundly differing conceptions of democracy, in no way detracts from, but rather encourages and stimulates, the interest taken in the problems of democracy.

We may note since the Second World War a growing taste for study of the philosophical, politico-social, sociological as well as legal problems of democracy, the universal aim of such studies being to improve the functioning of democratic practice in a given country, or rather the functioning of a given system of democracy, and to demonstrate its advantages.

Although research relating to democracy and its practical operation is typically interdisciplinary in character, legal science (including the science of the State, *Staatslehre*) plays a highly important rôle in it, bearing as it does, for each different type of democracy, on such varied questions as, for example, its foundations, the theoretical problems of its development, the nature of the institutions which ensure its functioning and the guarantees of democratic practice.

We shall not, however, attempt in the present sub-section to deal with legal research relating to democracy in any general perspective, but rather concentrate on one specific question, that of research into the rôle played by *law* in democratic practice.

215. HOEBEL & LLEWELLYN, *The Cheyenne Way* (1941); Rheinstein, Replies, p. 11 (cf. RHEINSTEIN, 'United States of America' (1970), p. 452).

216. See the section on 'Ethnologie juridique' in POIRIER (ed.), *Ethnologie générale* (1968) and the bibliography on pages 1236 sqq.; and cf. in the present work

2. Research relating to the development of socialist democracy represents, as we have already mentioned, one of the main trends of Marxist-Leninist legal science in general and in the USSR in particular.[217] Study of the problems of socialist democracy forms part of the general study of the political organization of socialist society and is intended to perfect it.

A very large part of this research is devoted to studying the specific channels through which the Communist Party exercises its leading rôle in the socialist State and in society at large, as well as the means by which corporate bodies such as trade unions, etc., participate in the administration of a socialist society.

Among the problems of greatest current concern, mention should be made of the study of the transition from the State as characterized by the 'dictatorship of the proletariat' to a State of the whole people and, in the USSR, the study of the essence, form and functions of the latter.

Specific problems arise in socialist countries with a federative structure (questions concerning matters of sovereignty between the Federation and the Federated Republics; the relationship between socialist federalism and the principles of proletarian internationalism and the self-determination of nations, as well as democratic centralism; and the influence of federalism on the development and the structure of legislation).[218]

A question of far-reaching consequence for democratic practice in socialist countries is the participation of the popular masses in State administration, the modes of participation, the relationship between the specialization of the administration of society and the participation of the masses, etc.

Another problem of major importance deserving of mention is that of socialist legality; however, for the sake of logical order of exposition this will be dealt with separately in the following paragraphs.

3. In the Western countries – at least judging by the replies to our questionnaire – as far as the contribution of law to democratic practice is concerned, legal science is above all interested in the problems regarding the

---

pages 42 to 45 of the chapter 'Social and cultural anthropology' by Maurice Freedman, and the works cited therein.

217. ČHIKVADZE *et al.*, *Marksistsko-leninskaja obščaja teorija gosudarstva i prava*, Vol. I (1970), Chap. 8; KOTOK, 'Socialističeskij demokratizm sovetskogo gosudarstva' (1957); by the same author, 'V. I. Lenin i razvitie sovetskoj socialističeskoj demokratii' (1960); LUNEV, *Demokratizm sovetskogo gosudarstvennogo upravlenija* (1967); etc.

218. Cf. for example, KRUTOGOLOV, 'Le fédéralisme et la structure de l'Etat multinational soviétique' (1972); ŠAFIR, *Kompetencija SSSR i sojuznoj respubliki* (1968); JOVIČIĆ, *Savremeni federalizam. Uporednopravna studija* (1973); etc. It is moreover by no means the socialist countries alone that are affected by problems of federalism; they are of concern in other countries too, particularly Canada: cf. TRUDEAU, *Le fédéralisme et la société canadienne-française* (1967); MEEKINSON (ed.), *Canadian Federalism* (1968); etc. Cf. also *Le Fédéralisme et le développement des ordres juridiques/Federalism and the Development of Legal Systems* (Proceedings of the Symposium organized in Moscow in 1970 by the International Association of Legal Science, publ. 1971). Cf. also GRAFSKIJ & STRAŠUN, *Federalizm v razvivajuščihsja stranah* (1969); MACKINNON, *Comparative Federalism. A Study in Judicial Interpretation* (1964); etc.

relationship between the individual and society (between the citizen and the State).[219]

Contemporary society is aware of the importance of law as one of the guarantees of the democratic system against the abuse of power, against violations of the fundamental rights of man, etc. This awareness of the important social function of law has emerged since the Second World War as a general trend towards developing the study of the relations between law and the power of the State, a relation known as 'rule of law' by British and American authors,[220] as '*primauté de la loi*' in France and by the notion of '*Rechtsstaat*' in the legal science of the Federal Republic of Germany.[221] On the other hand, in the socialist countries it is 'socialist legality' which, as we mentioned earlier and shall take up again, is receiving increased attention from Marxist-Leninist legal science.

The common denominator of this trend is apparently the categorical requirement to observe the law as a guarantor of democratic practice, and further to observe the procedures legally prescribed by legislation. (In recent years we can also detect a tendency to foster the study of what might be termed the sociology of legislation.)

Nevertheless, and in spite of this 'common denominator', the notions of 'socialist legality' on the one hand, and of 'rule of law', '*primauté de la loi*' and '*Rechtsstaat*' on the other, differ profoundly. The difference between them stems from the differing conceptions of law itself,[222] namely the fact that socialist legality, while guaranteeing civil rights and liberties, is conceived of from the point of view of the interests of the ruling class, in other words as the internal order of socialist society that simultaneously protects individual interests, the interests of socialist society and the power of the working class, in their rôle of ruling class, or, as the case may be, of the people as a whole.

4. Law plays a most important rôle as a buttress of democratic practice in countries having recently attained independence. As former colonies these countries do not as a rule enjoy democratic traditions of their own and are seeking their way in political terms, some of them inspired by socialism,

219. The bibliographical references available show, however, that even in Western countries there are quite a number of works dealing with democracy in general terms, though these works are usually considered as pertaining to the domain of political science. By way of example we may cite: ABRAHAM, *Elements of Democratic Government* (1958); ARON *et al.*, *La démocratie à l'épreuve du XXe siècle* (1960); BURDEAU, *Traité de science politique* (1949-57, 1966-71); WHEARE, *Modern Constitutions* (1966); etc.

220. The notion of the 'rule of law' is not new; it was invented in England by DICEY in 1885 (*The Law of the Constitution*); see also KEIR & LAWSON, *Cases in Constitutional Law* (1927, republ. 1967).

221. We may recall the symposia organized by the International Association of Legal Science in Chicago in 1957 on the theme *Rule of Law as Understood in Western Countries*, and in Warsaw in 1958 on *Le concept de la légalité dans les pays socialistes* (for publication of these works (1959 and 1961), see their respective titles in the Bibliographical Annex, Section II, pp. 1063 and 1050.

222. See Section Two, sub-section III above, pp. 966 sqq.

others leaning on their former metropolitan countries, others again endeavouring to find a specific formula of political and economic development, and so forth. Whatever their different attitudes, however, most are seeking to establish a democratic order which will protect them against disorder, the excess of power of tribal chiefs or the great land-owners, etc., against coups d'état, abuse of power by officials and so forth. The replacement of the former colonial law and, at the same time, a gradual replacement of particular traditional systems of law by a democratic legal order meeting the needs and the social situation of the people is one of the chief means of serving the democratic evolution of these countries.[223]

VI. ACCULTURATION OF LAW

One of the most important problems in contemporary legal science relates to the study of what may be styled the 'acculturation of law'. Problems of acculturation in the legal field are generally regarded as problems of the sociology of law. For non-sociological theories of law, the problem of acculturation is of a meta-juridical nature and accordingly does not arise within the frontiers of legal science. From both the positivist and normativist points of view law (the legal rule, the legal norm) neither adapts itself nor rebels: it is valid or it is not valid; it expresses an 'ought to be' without the slightest regard for an 'is' of any kind, the two being separated by an unbridgeable epistemological abyss. It goes without saying that the theories we have just mentioned (and others of a similar nature) do not deny the processes of acculturation; they are simply not concerned with them, leaving the solution of such problems to general sociology, to the political sciences, etc.

It follows that the growing interest shown today by legal science in problems of legal acculturation not only reveals in itself a current trend of research in legal science; it also bears witness to the sociologically inclined trend which is increasingly prevalent in contemporary legal science and to which we have already referred.

Acculturation of law is one form of the social action of law, i.e., a conscious use of law as an instrument of social transformation. This is the starting-point. But before undertaking a study of the trends which are evident in the study of problems of legal acculturation and deserve inclusion among the genuinely noteworthy trends of current legal science, we think it necessary to define and elucidate a terminology which is not yet firmly established. The expressions most often used in this context are 'acculturation' and 'adaptation' of law.

We would prefer to reserve the term 'adaptation' of law for the general and unceasing process of adaptation of law to the development of society, which is a social phenomenon that goes beyond the bounds of 'acculturation' of law – a general phenomenon which we examined in greater depth when we dealt with the Marxist-Leninist concept of law.

223. See also the discussion on 'development law' below (Epilogue, point 1).

We must also draw a distinction between acculturation of law and what is meant by rapprochement in legal matters, or even unification of law. It is true that in connexion with certain measures of economic and sometimes political integration, we may witness a process of rapprochement, or even mutual assimilation of the law of different countries belonging to the group thus integrated, brought about either by the converging action of the courts in the various countries thus integrated (as is the case, for example, in the Benelux countries), or through instruments falling within the province of international law (as, for example, the General Conditions of Delivery of COMECON), or by efforts in the direction of a regional rapprochement of legislation, or the unification of law within groups of States. Nevertheless, study of the scientific problems connected with the rapprochement and unification of law is very different from the study of acculturation and belongs to comparative law, which we discussed earlier.

By the term 'acculturation' of law we understand cases in which two or more differing legal traditions meet and influence one another in such a way that the law does not only adapt itself to social realities but that one legal tradition adapts itself to another, though naturally on the basis of social realities.[224]

Cases of acculturation of law have been far from uncommon in legal history. The reception of Roman law on the European continent was nothing less than a magnificent process of acculturation of law in the course of which ancient laws surrendered to Roman law, the latter having in the meantime become 'acculturated' to the needs of the various societies. On the other side of the world we may mention the genesis of South African law, which took the form not of the defeat of one legal system by another, but of a mutual acculturation of two systems, the 'Continental' (the ancient law of Holland, the Roman-Dutch Law) and the Common Law, leading to the emergence of a composite system of mixed origins. Mention may also be made of the acculturation which took place under the influence of proximity to a legal system which happened, in the given situation, to be the stronger – a phenomenon which occurred in Louisiana (and, to a lesser extent, in Canada), where the legal system of French origin, once plunged into the midst of Common Law, either adapted to it or blended with it; the relations between Scots law and English Common Law were characterized by a comparable process.

224. The notion of acculturation is relatively new. According to J. CARBONNIER (*Sociologie juridique*, 1972, p. 166), the word 'acculturation' was invented by the American ethnologist Powell in 1880; the term was first used by sociologists and later adopted by sociologically inclined legal scientists. We may point out, however, that the meaning of legal acculturation varies from one author to another. Thus for example, H. LÉVY-BRUHL makes a distinction between legal *borrowing* and legal *acculturation*, the latter being a wholesale transformation of one legal system as a result of contact with different systems (cf. 'Note sur les contacts entre les systèmes juridiques', 1956). ALLIOT, for his part, distinguishes between three degrees of legal acculturation, namely 'submission, assimilation, reinterpretation' (in 'L'acculturation juridique' (1968), p. 1185). CARBONNIER takes the term 'acculturation' to mean 'any grafting of one culture on another' ('toute greffe d'une culture sur une autre') (*Sociologie juridique*, p. 166); and so forth.

In the modern world acculturation of law takes place in very different conditions, of which the most significant and interesting have been due to the process of decolonization and the accession of numerous African and Asian States to independence. And it is with these countries that we shall begin our brief examination of the concrete problems of the acculturation of law.

1. *Acculturation of law in countries having recently attained independence*

(a) Law undoubtedly raises particular problems in countries having lately attained independence. In almost all these countries law in the modern sense was imported from the former mother-countries (in the form of Common Law or law of a Continental type) and was brought face to face with the customary, traditional or religious law of the country. The latter law, being infinitely less elaborate than law imported from England or the European continent, was nevertheless a far from negligible element in terms of its influence, the strength of its roots and its social impact. There was therefore a mutual interaction between the two legal systems and a reciprocal process of acculturation.[225]

This interaction, which took place in both the former British and French colonies, naturally dates from the period prior to the acquisition of independence by these former colonies and it continues to exert an influence on the character of the legal system of those countries which have meantime become independent. A highly interesting example of such an encounter between the older legal system (or systems) of a country and the legal system imported from Europe is afforded by British colonial legislation which usually included, as regards the legal system of the colonies, clauses governing the relationship between the traditional law of the country and English law. Two examples of such clauses (taken from the national reports received) have to do with the encounter between traditional law (Islamic, Hindu, Chinese) and English law in Malaysia and Singapore.

In Malaysia the complex of traditional laws encountered English law when the country was colonized in the nineteenth century. With regard to the relationship between English law (Common Law and the rules of equity), which predominates, and the traditional laws of the country, the Civil Law Ordinance of 1956 is extremely interesting for it acknowledges the application of English law, but 'so far only as the circumstances of the States and settlements comprised in the Federation and their respective inhabitants

225. Cf. ALLOTT, *Essays in African Law* (1960); by the same author, *New Essays in African Law* (1970); CHABAS, 'Transformation du droit local et évolution économique' (1962); *The Future of Customary Law in Africa/L'avenir du droit coutumier en Afrique.* (Amsterdam Symposium, 1955, publ. in 1956); HOOKER, 'The interaction of legislation and customary law in Malaya' (1968); M'BAYE, 'Droit et développement en Afrique francophone de l'Ouest' (1966, 1967); SY, *Recherches sur l'exercice du pouvoir politique en Afrique noire (Côte d'Ivoire, Guinée, Mali)* (1965); TUNC (ed.), *Les aspects juridiques du développement économique* (1966); VANDERLINDEN, 'La règle de droit dans l'Afrique traditionnelle' (1971); etc.

permit, and subject to such qualifications as local circumstances render necessary'.[226]

In Singapore we find Moslem, Hindu and Chinese law on the one side and English law on the other, the latter being regarded as fundamental. But here too its application is limited: 'The modifications of the law of England which obtain in the colony in the application of that law to the various alien races established there, arise from the necessity of preventing the injustice or oppression which would arise if that law were applied to alien races unmodified'.[227]

These two examples, and particularly that of Singapore, are very instructive in that they reveal the interaction of the country's two legal systems and the acculturation, not only of the ancient law to the modern (as in the case of Nigeria, which we shall examine below), but also of the European law itself (e.g., the 'modifications of the law of England' in Singapore).

We are not therefore certain that we can entirely agree with René David's affirmation that as regards countries having recently attained independence, 'the question which arises is not so much that of the adaptation of the laws inherited from the colonial régime as that of their application in under-administered countries where there is a shortage of lawyers and where the laws frequently constitute a breach with tradition. To inject notions of law, State, equality and freedom among populations living according to a mode of life and in communities where such notions are unknown and sometimes shocking, is the task of a restricted minority, enamoured of Western ways, readily tempted by power and in constant danger of becoming strangers in their own land.'[228]

We must of course acknowledge the reality of many of the *political* problems mentioned by René David; nevertheless from a strictly legal point of view, there can be no doubt that different legal systems meet in the countries in question and that in the modern world at least, the problems which arise are not only those of applying the 'imported' law but also, as we have attempted to show, those of the mutual acculturation of two legal systems coexisting in one and the same country. The problems involved in applying the 'modern' law moreover constitute only one of many types of problem arising in connexion with its acculturation.

We may also note that the situation in this group of countries has now altered to a considerable degree, and is today very different from that which

226. Ibrahim, Replies, p. 5.

227. *Ibid.*, p. 4. (It is hardly necessary to point out that the decision quoted goes back to colonial times.)

228. '... la question qui se pose est moins celle de l'adaptation des lois héritées du régime colonial que celle de leur application, dans des pays sous-administrés, dépourvus de juristes, et où les lois constituent souvent un rupture avec la tradition. Faire pénétrer les notions de droit, d'Etat, d'égalité, de liberté dans des populations vivant selon un mode de vie et dans des communautés où ces notions sont inconnues et souvent choquantes, est la tâche d'une minorité restreinte séduite par l'Occident, facilement éprise de pouvoir et risquant sans cesse de devenir étrangère dans son propre pays' (Replies, p. 10).

prevailed at the time when European law in one or other form was transplanted to them. Whereas at that time European law meant the law of the metropolitan country, a European law for the use of Europeans – the colonial powers, as a rule, being little concerned to introduce their own legal system for the purpose of regulating mutual relationships between natives, questions concerning their civil status, etc. – the law inherited from the former mother-countries has in most countries that have recently attained independence become for the time being *their* law, acting as the starting-point for the building up of a modern law of their own which they are endeavouring little by little to substitute both for the primitive law or laws of the land and for the law bequeathed to them by the former mother-country.[229]

There can be no doubt that these endeavours do encounter the obstacles and dangers to which René David alludes; but it is equally certain that in newly independent countries a process of acculturation of law is in course which represents both a highly important political phenomenon and one of the most significant trends in contemporary legal science. It would therefore seem that we have to agree with Seydou Sy when, in his replies to the questionnaire, he expresses the notion that in countries having recently attained independence 'the active rôle of law' is particularly striking in the solutions put forward to settle the numerous conflicts which arise between traditional law and modern law. 'Legislation', he states 'is in this respect regarded as the instrument *par excellence* of social change.' [230]

(b) It is interesting to observe the important rôle played by the notion of natural law or natural justice in the process of acculturation of law in newly independent countries [231] (and, as we shall seek to explain below, in the process of acculturation of law in general).

The significance of natural law in the process of acculturation has already been briefly mentioned; [232] it is necessary here to dwell on the question further and to underline a paradox which is highly characteristic of contemporary legal science, namely that the most important field of application of the notion of natural law (or natural justice) is linked with the operation of a purely sociological trend. These two mortal enemies, the doctrine of natural law and legal sociology, accordingly meet on the same battleground not as foes but as allies. It must be added that this reconciliation, this paradox in the current tendencies of legal science, provides fertile ground for both scientific knowledge and the practical realization of the process of acculturation of law in general, most particularly in newly independent countries.

229. Cf. Rheinstein, Observations on the new statement, II, 8: '... the majority of the African countries are mostly interested in having a "modern" legal system rather than to preserve indigenous traditions ...'.

230. 'Le rôle actif du droit ... est particulièrement frappant dans les solutions proposées pour trancher les conflits multiples entre droit traditionnel et droit moderne. Le législation est à cet régard considérée comme un instrument par excellence du changement social' (Replies, p. 2).

231. Cf. Marshall, *Natural Justice* (1959); Allott, *New Essays in African Law* (1970), pp. 159 sqq.

232. In Section Two, sub-section II, p. 956.

What we are witnessing in those countries is, however, not so much the renaissance of an earlier theory of natural law as resort to natural law (or, as the case may be, natural justice) in a wholly different sense. For in this group of countries the fundamental legal problem (i.e., the problem not only of legal science, but above all of everyday legal practice) is that of the unavoidable co-existence of customary or religious (traditional) law and modern law either taken over from the former mother-countries or newly created. Obviously, modern law is beginning to gain sway in these countries; but this law, whether founded on the so-called continental concept or on Common Law, undergoes a complex process of acculturation in the course of which it frequently leaves the regulation of many important issues to the old, traditional law.[233] We therefore witness in these countries a collision between two differing legal cultures the incompatibility between which becomes steadily sharper. Traditional law, matching the primitive stages of evolution of these countries, is less and less adapted to the forward surge which has resulted from their attainment of independence, and acquires the character of an archaic, conservative and retarding element. Yet experience shows that it is not possible to abandon traditional law at one blow. The need is felt to amend and supplement it whenever it is either silent or incapable of providing a satisfactory solution in regard to issues which modern law leaves to it to regulate; and for the purpose of thus amending or supplementing traditional law, it is precisely to natural law (or natural justice) that the courts of newly independent countries have recourse in case of need.

Of this, Nigerian judicial practice furnishes two examples. In Oruborama Charlie Amachree v. Chief Inko, alias Goodhead, the appeal judge declared in confirmation of the decision by the 'native courts': 'I . . . certainly cannot uphold the contention that the existing arrangement is repugnant either to natural law or humanity.'[234] The same principle was stated even more categorically by the Nigerian judge in answer to the appeal against the 'native courts' in Ekpenyong Edet v. Nyon Essien when he declared: 'But even assuming that the native law and custom as alleged by the appellant had been definitely established I am inclined to think it should properly have been overruled in this case as being repugnant to natural justice and good conscience'.[235]

233. See above.

234. *Nigerian Law Reports*, Vol. 4, p. 101 (quoted by A. Morgan, Replies to the Questionnaire).

235. *Ibid.* Vol. 11, p. 47 (also quoted by Morgan, *ibid.*). The problems on which the judge ruled in the two cases were as follows: In the former the issue was to know whether the illegitimate child born after the death of its mother's husband, to whom she had been married according to native law and custom, should become a member of the family to which the mother belonged by marriage or of the family to which she belonged by birth. In the second case, the appellant was married to his wife in accordance with native law and custom. She had left him and gone to live with another man by whom she had two children. The appellant claimed that in accordance with native law and custom the children belonged to him since their father had not paid back to the appellant the money settled by him as dowry upon the wife.

In countries having recently attained independence, natural law is accordingly of great importance in relation to the acculturation of law. We are of course not concerned with natural law of a classical type as known in the history of European legal thought, but rather with the application of an ethical norm (as established in the judge's thinking) as a corrective or complement to the legal norm provided by the customs or traditons of a country or province.

2. *Specific problems in the acculturation of Islamic law*

The problem of acculturation arises in a quite specific form in Moslem countries.

It is well known that, since the year 656 of the Hegira (1258 A.D.), Islamic law (the *fiqh*) has in theory been absolutely immutable. The *taqlid*, i.e., the unquestioning acceptance of Islamic law, allows of no creative action whatever, either in case-law or in legislation.

Before proceeding further we must also recall that Islamic law, albeit one, is divided into four orthodox Schools (not counting *Shi'ite* law) representing legal systems which differ considerably from one another as far as the detailed regulations they prescribe are concerned.[236]

Despite the immutability of the law, social and economic life continues to develop in Moslem countries as it does elsewhere, and finds itself in revolt against a legal system that has remained set and unchanging for seven centuries. As in England with regard to the *stare decisis* rule,[237] it was therefore necessary to find some means of developing a legal system which *by definition* precluded all development.

Different countries have chosen different ways to achieve this purpose:

The most radical solution was adopted in Turkey where the Kemal régime simply abandoned the *fiqh* as a whole and replaced it by a system of law derived from European (Continental) sources.[238]

The other Moslem countries have chosen a much more complex solution, based on a distinction drawn between Personal Status and Real Status, the two notions being taken in a broad sense. In its Moslem acceptation, Personal Status embraces not only personal status strictly speaking, but also the whole of family law, legitimacy, maintenance, gifts *inter vivos*, inheritance and wills and, in some countries, even religious or private interest foundations, etc.

236. Cf. above, Section One, III, p. 946.

237. See Section Two, sub-section II, 3 above, pp. 964-965.

238. It is hardly necessary to point out that Islamic law has been completely abandoned and replaced by socialist law in the former Moslem countries which form part of the Soviet Union. Moslem law has also been abrogated in Yugoslavia. While it is true that the law of 1946 on the non-validity of legal norms laid down before 6 April 1941 and during the enemy occupation does not expressly exclude the applicability of the principles of Moslem law, nevertheless decisions handed down by the Yugoslav courts since the enactment of that law do not include any case in which Moslem law has been applied.

The remainder (real property, goods and chattels, contracts and obligations, commercial law, judicial law, the whole of public law, etc.) comes within the purview of Real Status.[239]

It was in the domain of Real Status that Moslem law (the *fiqh)* was least suited to the requirements of modern life, and it was precisely in this domain that Moslem law capitulated to European inspired law giving way either to a codified law of the continental type, or in some places to Common Law, both of which remained entirely separate from Moslem law. It is therefore in the domain of Real Status that an acculturation in the strictest sense of the word [240] occurred: what happened in this case was a wholesale transformation of the legal regime of Real Status when the ancient and immutable *fiqh* came into contact with the then modern systems of law which responded infinitely better than Moslem law to the needs of present-day society.

On the other hand, Personal Status, in the very broad acceptation that is proper to it, continues to be governed by Moslem law, which on account of its unchanging character, precludes any acculturation. The codifications of Personal Status carried out in some Moslem countries amount to no more than an arrangement, in the organized and numbered form of European codes, of the rules of Moslem law drawn from the works of classical authors i.e., the founders of the Schools and their immediate disciples.

The law of the Moslem countries today is therefore strictly dualistic: it is divided into two parts separated from one another by an impassable frontier. In one (the domain of Real Status), the process of acculturation has long since been accomplished; in the other (the domain of Personal Status) there is no trace of European law and (apart from a few rare exceptions which will be mentioned later) no sign of any acculturation.

But even the rules pertaining to Personal Status have become too narrow for contemporary social life, so that Moslem countries are forced to find ways to adapting the *fiqh*, opposed though it be to any such adaptation. The methods applied do not, however, consist in acculturation based on European models. Moslem legal thinking has devised two different methods of changing the law without violating its immutability:

(a) The first method is contingent on the unity of Moslem law. The principle of immutability is only valid in fact for Moslem law as a whole: it does not apply to the different schools individually. It is therefore permissible to

239. On Moslem law, see, for example, ANDERSON, *Islamic Law in the Modern World* (1959); CHARNAY, 'Amplitude sociale et logique de la norme musulmane' (1971); CHEHATA, *Etudes de philosophie musulmane du droit* (1965); by the same author, *Droit musulman, applications au Proche-Orient* (1970); by the same author, *Etudes de droit musulman* (1971) (see especially Chapter I: 'Qu'est-ce que le droit musulman?'); LINANT DE BELLEFONDS, *Traité de droit musulman comparé* (1965); MILLIOT, *Introduction à l'étude du droit musulman* (1953); PEARSON (ed.), *Index Islamicus 1906-1955* and *Suppléments 1956-1960* and *1961-1965* (1958, 1962 and 1967); SCHACHT, *An Introduction to Islamic Law* (1964); etc. Cf. also *VIIe Congrès international de Droit comparé (Uppsala, 1966)*, Section I, D: 'Middle East Law. The link between law and religion' (General Rapporteur: S. HABACHY) (publ. in 1968).

240. See p. 1011, n. 224.

incorporate into the positive law based on the system of a given school rules of law borrowed from another, with the sole proviso that the latter also be one of the orthodox schools.[241]

(b) The second method tends to modify the applicable law indirectly, by recourse to procedure, mainly by virtue of the inadmissibility of certain actions at law. The State (the sovereign) in Islam has no power to modify Moslem law but is free to regulate the judical procedure; thus it – or he – may also determine the conditions under which legal actions are allowable and by this means prevent the holder of a right from claiming his right, without the right being abolished or abrogated.[242]

The further development of these methods moreover represents the main current in present-day Moslem legal thinking.

In conclusion let us add a few remarks on the exceptional cases in which Moslem law has undergone a process of acculturation resulting from its contact with European law:

(a) First we should mention the Tunisian Code of Personal Status of 1956, which to a certain extent underwent an 'acculturation' of this kind, the results of such acculturation, incidentally, coming in for very serious criticism on the parts of jurists in the other Moslem countries.

(b) In Egypt, on the other hand, an attempt was made to reincorporate into the Civil Code of 1949 (which had replaced the old Code of 1883) some twenty or thirty clauses based on Moslem law – concerning, of course, Real Status. This re-acculturation, however, is of little importance, according to Linant de Bellefonds,[243] and in no way alters the 'Western' character of the Civil Code.

(c) In the Moslem countries which were formerly British colonies, justice was at that time dispensed, as a general rule, even in the domain of Personal Status, namely the domain par excellence of Moslem law, by English magistrates who impressed on Islamic law the indelible stamp of their Common-Law way of thinking.[244]

(d) Lastly, Pakistan, the only country to have rejected all such 'legal acrobatics', has attempted to achieve the necessary reforms in Islamic law by means of a direct re-interpretation of the texts of the Koran and the Sunna. Whether such a procedure is legitimate from the standpoint of the Islamic conception of law is, however, very debatable.[245]

241. See Section One, III, item 6 above, p. 946.

242. Cf. the above-mentioned work by LINANT DE BELLEFONDS (1965) and, by the same author: 'Immutabilité du droit musulman et réformes législatives en Egypte' (1955).

243. LINANT DE BELLEFONDS, 'Le droit musulman et le nouveau Code civil égyptien' (1954, publ. in 1956); see also the contributions by Abdel-Wahab, Abou Taleb, Anis, El-Attar and El-Badraoui collected by the National Commission for Unesco of the Arab Republic of Egypt (cf. the summary by ABDEL-HAMID & TAGELDIN, 'Le monde arabo-islamique', 1970).

244. Cf. SCHACHT, *An Introduction to Islamic Law*, 1st edn. (1964), Chap. 14: 'Anglo-Muhammadan law in India'.

245. LINANT DE BELLEFONDS, 'Le Code du Statut personnel irakien' (1960); COULSON, 'Reform of family law in Pakistan' (1957); also Brohi, Replies. Mention

3. *Specific problems of the acculturation of law in Japan*

Another country in which the acculturation of law has caused specific problems is Japan, noted for its highly developed legal culture and a high level of legal science.[246]

The most interesting aspect of the Japanese experience is that a process of acculturation has taken place there twice.

(a) The first time was when the national law (the traditional law) of the country encountered modern European law. Japan's contemporary legal structure took form at a time when continental European law, and particularly French and German law, had come to be regarded as legal perfection itself. English Common Law was then no more than an 'article for export' to countries under the direct political domination of the British Empire; as for American Common Law, it had at that time (except for constitutional law) no influence outside the territory of the United States. Japan, which was beyond the range of British influence, turned for the Europeanization of its law to the famed patterns of continental written law, achieving its own codification on the German model and, in part, the French. This law, however, which drew on European sources came into conflict not only with Japanese legal traditions but particularly with the Japanese social and cultural 'climate', of which the traditional law was simply a reflection. The new legal order in Japan therefore had to take into account the traditional customs of the country and, coming as it did from a quite different social and cultural environment, it had to adapt to the social and cultural environment of Japan. This acculturation process, though it was completed a long time ago as regards problems of principle – Japan has long since become a continental law country – nevertheless continues to pose certain problems even at present.[247]

What is of interest in our study, is the fact that in order to solve these problems, Japanese legal literature and especially court practice frequently have recourse to the application of natural law, as is shown, for example, by the arguments advanced in support of the decision of the Japanese Supreme Court dated 11 October 1950: '... the morality which controls the relation between husband and wife, parents and children, and brothers and sisters belongs to the fundamentals of human morality, universal ethical principle of humanity accepted all through history and the world, i.e., so-called natural law as recognized in legal theory'.[248] Even in this case, however, we shall find that it is a question of the application of ethical norms to the law rather than natural law in the classical meaning of the term.

---

should also be made of the fact that in drafting these paragraphs on Moslem law we have drawn extensively on the 'special contribution' by Y. Linant de Bellefonds.

246. Cf. Noda, *Introduction au droit japonais* (1966); von Mehren (ed.), *The Law in Japan* (1963).

247. Cf. Kitagawa, *Rezeption und Fortbildung des europäischen Zivilrechts in Japan* (1970); Taniguchi, 'La loi et la coutume au Japon' (1964); etc. Cf. also Rheinstein, Observations on the Revised Report, II, 7.

248. Ukai, Replies.

(b) The second period in the acculturation of Japanese law began after the Second World War. Following the military defeat in 1945, Japan found itself included in the political sphere of influence of the United States and this produced – as it is still producing – a certain 'Americanization' of Japanese law. The encounter of two major legal systems made their mutual acculturation imperative. Here we no longer have a case of the acculturation of 'modern' Japanese law to the traditions of the country but one of its acculturation to another 'modern' law or, in other words, a case of the acculturation of a legal system of the continental (civil law) type already adapted to the social and cultural environment onto which it had been grafted, to American Common Law. This process has gone quite far and its consequences have been appreciable enough for Aomi to be able to state that although the country has theoretically remained a civil law country, '. . . it is no longer strictly correct to characterize Japan as a typical Civil Law country, which it definitely was before 1945'.[249]

4. *Native enclaves*

A problem of great interest, although of more restricted scope than the preceding ones, relates to the fact that there survive, in a number of countries where law has long since been codified according to European patterns, small native enclaves, extremely backward and living in conditions of virtual segregation, some of which have retained their ancestral customs. These are therefore communities which have not been integrated with the main community governed by 'modern' law and which, without regard to that law, regulate their social relationships *de facto* by recourse to ancestral customs.

This phenomenon is met with in several Latin American countries (Mexico in particular) and in Asia. It is of occasional interest to ethnological or ethnographic research but does not hold the sustained attention of jurists. It is true that we are witnessing today a revival of interest in legal ethnology,[250] but it would be an exaggeration to claim that the study of the acculturation of traditional customs in native enclaves deserves to be regarded as a current trend in legal science.

In Mexico the problem is viewed rather from the standpoint of the protection of native communities (with particular reference to land-tenure); [251] in India study of the problem has led to the suggestion that various common features of the legal systems of Asian societies should be examined with a view to comparing the forms of their modification and acculturation when exposed to contact with the major Western legal systems.[252]

At the same time we have no evidence that such studies are currently under way.

249. Aomi, Contribution, p. 1 (cf. Aomi, 'Japan', 1970, p. 378).

250. See IV, 3 above, pp. 1006-1007.

251. Cf. *Legislación indigenista de México* (1958); cf. also *Realidades y proyectos* (1964); etc.

252. Sharma, 'Common Law of the Common Wealth' (1966).

VII. CURRENT TRENDS IN THE FIELD OF INTERNATIONAL RELATIONS

1. *International law and current problems of international politics*

International law, one of the oldest disciplines in law and in legal science, and one of the legal disciplines most closely related to political life, has since the Second World War undergone a number of changes of great significance. While pursuing the study of traditional problems, international law in the contemporary world finds itself faced with numerous new problems caused in part by the evolution of international politics and in part by the rapid development of the natural sciences.[253]

(a) The origins of some of the new trends that are manifest in this field go back a long way. Present research is only an extension of what was being done before; however, our justification for speaking of the appearance of new trends is, generally speaking, the emergence of a novel approach to the problems, the growing interest in certain fields of international life and the markedly greater thoroughness of research work.

(i) The most significant trend in this area of research is the unparalleled surge of scientific interest in the *problems of peace*, especially from the legal point of view.

The study of peace goes back to the beginning of the twentieth century, but it has developed most systematically and most vigorously in recent years. There are in the world today some 70 institutes and institutions engaged in research relating to peace.

The study of peace is of necessity multidisciplinary; but international law plays a highly important rôle in it and it would not be too much to state that the growth of this type of research represents one of the outstanding trends in modern international law.

Peace studies also involve the study of the opposite condition, namely war. (Moreover we may note that the scientific bodies engaged in the study of peace include a number of institutes of polemology.) But the study of war in this context and in terms of international law bears no relation to the ancient *jus belli*: it consists essentially in the study of the legal means of preventing armed conflict. Among the main avenues of such research mention must be made in particular of the part played by international law in the formulation of a general theory of conflict, study of the legal problems involved in disarmament and the control of armaments, examination of the rôle of the United Nations in the maintenance of peace, the formulation of peaceful

253. BOBROV, *Sovremennoe meždunarodnoe pravo, ob"ektivnye predposylki i social'noe naznačenie (1962); Entwicklungstendenzen des Völkerrechts* (1965); FRIEDMANN, *The Changing Structure of International Law* (1964); KAPLAN & KATZENBACH, *The Political Foundations of International Law* (1961); TUNKIN, *Ideologičeskaja bor'ba i meždunarodnoe pravo* (1967) (which also contains a reaction to the above-mentioned work by FRIEDMANN); etc.

measures for the solution of international conflicts,[254] etc.[255]

(ii) Another trend relevant to international law is the development of research concerning the *codification* of various branches of *international law*.[256]

The work of codification itself advances relatively slowly; nevertheless, over the past two decades it has accomplished progress unparalleled in preceding periods. Let us for instance recall the Geneva Convention relative to the Protection of Civilian Persons in Time of War signed in 1949, the Vienna Conventions on diplomatic relations and on consular relations signed in 1961 and 1963 respectively, the Vienna Convention on the law of treatises signed in 1969, and a number of treaty laws of a more specific nature. Negotiations are under way regarding various other branches of international law (e.g., responsibility of States, etc.).

In most cases, however, it involves the codification of international law that is already in force, i.e., a simple transformation of hitherto uncodified international law into codified law, hence the formulation of already valid norms rather than a process *de lege ferenda* in the true sense of the term.

Nevertheless even a project of codification of this type raises a number of questions that call for research not only during the codification process but also, frequently, afterwards. For this reason the development of research relating to the codification of international law deserves to be included among the major current trends in this branch of legal science.

(b) In certain cases, although here too the problems were already known and had been studied previously, the development of international politics since the Second World War has so altered the whole situation that one can fairly speak of trends that have genuinely come to light since the end of the war.

The principal result of this development was to broaden considerably the horizons of international law and the field of its application. Whereas in its classical acceptation it was only thought of as the law of States, today international law also encompasses the study of organizations and entities other than States: for example the study of international organizations, or of the status of nations and even individuals in international relations.

(i) Characteristic of our time is the unprecedented growth of *international organizations*. The first international organizations were indeed founded as early as the second half of the nineteenth century, and their political importance increased considerably after the First World War; however it is only in recent years that they have really assumed world-wide prominence as demonstrated both by their proliferation and by the development of their rôle in international relations.

This development creates a number of new problems for international law: it calls for studies in depth of the legal problems facing the international organizations themselves (for example those of the United Nations and its

254. See (b) (i) below.
255. See p. 1041, n. 300.
256. Cf. for example, HOLLOWAY, *Modern Trends in Treaty Law* (1967); Lord McNAIR, *The Law of Treaties* (1961); TUNKIN, Voprosy teorii meždunarodnogo

organs, particularly those recently established; those of the Specialized Agencies; those of regional or continental organizations and other organizations covering a limited geographical area; those of semi-governmental and non-governmental bodies, etc.) as well as research on certain related questions, such as the influence exerted by the international organizations upon the general principles of international law, international jurisdiction, the peaceful settlement of international disputes, etc.[257] The mere existence of international organizations also gives rise, as we have seen, to a number of problems relating to comparative law.

Apart from the increase in the number and importance of international organizations in the normal sense of the term, the political development of the contemporary world is increasingly marked by the ever more frequent cases of *international integration* representing an entirely new form of grouping of States which sometimes comprises elements of a confederative if not federative character. This process gives rise to numerous legal problems unknown to pre-war international law, problems which pertain to public international law and private international law alike; these problems also raise many new questions concerning the relationship between international law and the municipal law of the countries comprised in these groups.[258]

We may recall once again in this connexion that the existence of organizations, or even integrations, of socialist States (particularly the Warsaw Pact, for example, and the Council for Mutual Economic Assistance – COMECON) has been one of the contributing factors in the rise of a *socialist international law*, side by side with universal international law; the study of the problems of socialist international law and its elaboration into a coherent system may

---

prava (1964); BOWETT, *The Law of the Sea* (1967); CAHIER, *Le droit diplomatique contemporain* (2nd edn., 1964), etc.

257. ALEXANDROWICZ, *World Economic Agencies* (1962); BOWETT, *The Law of International Institutions* (1970); COLLIARD, *Institutions internationales* (4th edn., 1967); GANSHOF VAN DER MEERSCH, *Organisations européennes* (1966); GERBERT, *Les organisations internationales* (1966); JENKS, *The Proper Law of International Organizations* (1962); MEYRET (ed.), *Les organisations régionales internationales* (1971); MORAWIECKI, *Organizacje międzynarodowe* (2nd edn., 1969); MOROZOV, *Meždunarodnye organizacii* (1969); 'Nature juridique des actes des organisations et des juridictions internationales et leurs effets en droit interne' (VIIIth International Congress of Comparative Law, Pescara, 1970, Section IV A 1 – see *Rapports généraux au VIIIe CIDC*, to be published); SCHERMERS, *The Law of International Organizations* (1972); SEIDL-HOHENVELDERN, *Das Recht der internationalen Organisationen* (1967); SZALAI, 'The future of international organizations' (1971); VIRALLY, 'De la classification des organisations internationales' (1972); *Legal Aspects of Economic Integration/Les aspects juridiques de l'intégration économique* (RIDEAU (ed.), 1973); etc.

GOODRICH, *The United Nations* (1959); KELSEN, *The Law of the United Nations* (1951); KRYLOV, *Istorija sozdanija Organizacii ob"edinennyh nacij* (1960); EL-MANDJRA, *The United Nations System. An Analysis* (1973); *Les Nations Unies face à un monde en mutation* (Proceedings of the symposium organized in 1970 by the University of Geneva, publ. 1970); etc.

258. Cf. the collection *Les organisations régionales internationales* (1971); PINTO, *Les organisations européennes* (1963); ROBERTSON, *European Institutions* (1959); etc. As regards the socialist integrations, cf. KALENSKY, 'Radá vzájemné hospo-

be considered as constituting one of the current trends in legal science in the socialist countries.

(ii) International law in our time is increasingly concerned with the problems of *nations* as natural human communities and their international profection (apart from the sovereignty of States, Marxist-Leninist legal science in particular has begun to recognize also the sovereignty of nations, even if the term 'sovereignty' is here given a slightly different meaning); hand in hand with this interest in the rights of nations, contemporary international law is concerned with problems affecting the international protection of human rights.[259] This twofold tendency (respect for the rights of nations and for the rights of man) leads to recognition of the right of self-determination of nations as one of the fundamental principles of peaceful co-existence, and results in research being devoted to this question and studies being undertaken with a view to formulating universally applicable principles regarding the political, economic and social rights of both man and citizen.

But there is no need to point out that it is precisely in this domain that the ideological differences that exist in the contemporary world come strongly into play and decisively affect the various basic conceptions and the solutions that stem from them.

(c) We may note yet another current trend on the level of international law. This consists, as we have just seen, in the birth and rise of *socialist international law* [260] which, being based on Socialist internationalism as the fundamental principle of the community of Socialist States, is now developing side by side with universal international law, as an international law that is qualitatively different and governs the mutual relations between Socialist States whose policies are rooted in this principle.

(d) Lastly, the discoveries of science and new technical developments have made it possible for mankind to expand the horizon of his existence and his activities. This leads in particular to an unprecedented increase in the power of the State in areas and spheres that have been mastered thanks to innovations in science and technology and also engenders entirely new and unheard of legal relationships, giving rise to problems which in turn necessitate research – and indeed this research is one of the most pronounced features of international law today.[261]

---

dářské pomoci' (1962); USENKO, *Formy regulirovanija socialističeskogo meždunarodnogo razdelenija truda* (1965); etc.

259. Cf. the collection *Amicorum discipulorumque liber René Cassin* (1969 to 1971); CASSIN, 'La Déclaration universelle et la mise en œuvre des droits de l'homme' (1951); GANJI, *International Protection of Human Rights* (1962); LAUTERPACHT, *International Law and Human Rights* (1950); NEDBAJLO, 'Meždunarodnaja zaščita prav čeloveka' (1968); OSTROVSKIJ, *OON i prava čeloveka* (1969); WEIL, *The European Convention on Human Rights* (1963); etc. See also footnote *202* on p. 1002.

260. BOBROV, *Sovremennoe meždunarodnoe prava, ob"ektivnye predposylki i social'noe naznačenie* (1962); by the same author, *Osnovnye problemy teorii meždunarodnogo prava* (1968); TUNKIN, *Ideologičeskaja bor'ba i meždunarodnoe pravo* (1967); by the same author, *Teorija meždunarodnogo prava* (1970); etc.

261. See also VIII, 1 below, pp. 1027 sqq.

2. *Adaptation of municipal law to the norms of international law*

Scientific solution of the problem of adapting the norms of municipal law to the norms of international law is in a large measure dependent upon our theoretical conception of the relation between international and municipal law.[262]

The current trend is characterized by a firm rejection of the former dualist theory and by an increasingly clear recognition of the direct influence of international law upon municipal law, tending in Western countries towards a gradual recognition of the primacy of international law. This last trend has sometimes led to a conception – roundly contested in the socialist countries – of *international* law as being a genuine *trans-national* law which, in cases of conflict, might even prevail over municipal law. These trends are also reflected to a certain extent in the constitutional laws of certain countries – for example, in Europe, the Federal Republic of Germany,[263] the Netherlands, France, Italy, [264] etc.; the same is true of a number of States in Africa and elsewhere.

Be this as it may, it is recognized in the contemporary world that the norms of municipal law must not be in contradiction with the international obligations undertaken by the State concerned. This is also true of the socialist countries, where one finds a very clear expression of the relationship between international law and State power, as conceived from the Marxist-Leninist viewpoint, especially in the 1968 Constitution of the German Democratic Republic.[265] Should there arise a contradiction between the municipal law and international law, the latter compels the State to adapt its municipal law to its international obligations. Failure to carry out this duty is regarded as an international offence.

Authors on international law further reveal a fairly strong tendency to recognize that States have a duty to adapt their municipal law even to the

262. Cf. Bliščenko, *Meždunarodnoe i vnutrigosudarstvennoe pravo* (1960); Van Bogaert, 'Les antinomies entre le droit international et le droit interne' (1968); cf. also 'Les conflits de la loi nationale avec les traités internationaux' (*VIIth International Congress of Comparative Law, Uppsala, 1966*, Section IV A 2; General rapoprteur: Ch. Fragistas – publ. 1968).

263. Article 25 of the Constitution of the Federal Republic of Germany states: *'Die allgemeinen Regeln des Völkerrechtes sind Bestandteil des Bundesrechtes. Sie gehen den Gesetzen vor und erzeugen Rechte und Pflichten unmittelbar für die Bewohner des Bundesgebietes.'* (The general rules of international law shall form part of Federal law. They shall take precedence over the laws and create rights and duties directly for the inhabitants of the Federal territory); cf. Leibholz & Rinck, *Grundgesetz, Kommentar*, 3rd edn. (1968), pp. 290 sqq.

264. Art. 66 of the Constitution of the Netherlands (1956 amendment); see also the French Constitution of 1958 (Art. 55), the Italian Constitution (Art. 10), etc.

265. Article 8, par. (1) of the Constitution of the G.D.R. states: *'Die allgemein anerkannten, dem Frieden und der friedlichen Zusammenarbeit der Völker dienenden Regeln des Völkerrechts sind für die Staatsmacht und jeden Bürger verbindlich . . .'* (The rules of international law which are universally recognized and serve peace and peaceful co-operation between nations are compelling for the State power and for every citizen . . .).

norms of common international law, in particular those brought together in the Charter of the United Nations, and to the fundamental principles of customary international law that are mandatory in character.[266]

These trends, aimed at the harmonization of municipal and international law, may also be observed in an indirect manner in the sphere of private international law, where there are frequent cases of adaptation of the norms of municipal law to match the draft conventions drawn up by The Hague Conference on Private International Law – cases which occur not infrequently even where the State concerned has not adhered to the convention in question.

VIII. NEW TRENDS DUE TO THE UPSURGE IN THE NATURAL SCIENCES AND THE TECHNIQUES THEREBY MADE POSSIBLE

The discoveries of the exact and natural sciences, and their exploitation at the technological, medical and other levels, have at all times influenced legal science by presenting it with fresh problems. We need only quote as examples the inventions and scientific discoveries which made it possible to build railways, aeroplanes, etc., and which were the prime cause of the appearance and development of legislation to regulate transport by rail, air and road, industrial safety provisions, maritime law and so forth, together with all the problems to which these various regulations give rise in terms of legal science.[267] In our day, the progress in the exact and natural sciences and in the techniques thereby made possible also confront legal science, either directly or indirectly, with a number of new problems. In this connexion, for example, let us recall the legal problems posed, particularly in the sphere of international law, by the exploration of outer space and other explorations,[268] or those resulting from progress in the biological and medical sciences (organ transplants, protection against epidemics, birth control, artificial insemination, etc.), the development of communication and information media, the spread of popular culture, the advent of data-processing techniques and automation, and all the consequences of scientific and technological progress (industrial and agricultural) which threatens man's environment (regulations to combat pollution, etc.).

Since most of these subjects are multidisciplinary in scope they will be dealt with in the Epilogue.

In the following pages we shall deal with two sets of problems which have,

266. Cf. 'Les conflits de la loi nationale avec les traités internationaux' (*VIIth International Congress of Comparative Law, Uppsala, 1966*, Section IV A 2, General Rapporteur: C. FRAGISTAS) (1968).

267. Cf. von STEPANITZ, *Wissenschaft und Recht* (1970); HAZARD & WAGNER, *Legal Thought in the United States of America under Contemporary Pressures* (1970). Concerning international law, see BOURQUIN, 'Pouvoir scientifique et droit international' (1947); EARLE, 'The impact of scientific discovery and technological change on international relations' (1946); ROUSSEAU, 'Progrès scientifique et évolution du droit international' (1954); etc.

268. See item 1 below.

indisputably, given rise to veritable trends in contemporary legal science, namely:

(a) exploration of outer space, and

(b) organ transplants.[269]

As regards cybernetics, the dynamic growth of this new discipline affects in particular the methodology of legal science, hence our treatment of the subject earlier in the sub-section on methodology.

1. *Current trends affecting international law and resulting from developments in the natural sciences: space law*

Rapid progress in science and technology has given rise to certain new trends in research in the field of international law. We may instance the legal problems relating to supersonic flights,[270] the exploration of outer space, of the sea-bed,[271] of Antarctica,[272] the peaceful use of nuclear energy,[273] etc.

Of all these problems, the ones that have given rise to the most pronounced trend in contemporary legal science are the problems resulting from the exploration and utilization of outer space.[274] We propose to outline here at least the most salient ones.

The first launchings of artificial satellites brought about a number of legal

269. The same problems arise with regard to tissue transplants. For the sake of simplicity we shall refer below to organ transplants only, it being understood that tissue transplants involve the same problems and are susceptible of the same solutions.

270. Cf. 'Les responsabilités pour les dommages résultant de vols supersoniques' (VIIIth International Congress on Comparative Law, Pescara 1970, Section III D 1, General rapporteurs: L. Baudouin and M. Eliesco – to be published); HEERE, *International Bibliography of Air Law* (1972).

271. Cf. for example, BROWN, 'Deep sea mining – the legal regime of inner space' (1968); DENORME, 'La question des fonds marins et son examen par les Etats-Unis' (1969); EEK, *Världshavens frihet och fred* (1971); PARDO, 'Who will control the sea-bed?' (1968); VOLKOV, 'Meždunarodnopravovye voprosy èkspluatacii kontinental'nogo šelfa' (1964-65). See also *Belgrade World Conference on World Peace through Law* (1971), Sec. 3: 'The exploration and exploitation of the ocean bed' (in press).

272. Cf. for example, HANESSIAN, 'The Antarctic Treaty 1959' (1960); MENCER, *Mezinárodně právni problémy Antarktidy* (1963); MOVČAN, 'Pravovoj statut Antarktiki' (1959); SIMMONDS, 'The Antarctic Treaty 1959' (1960); etc.

273. Cf. for example, BOURQUIN, 'Pouvoir scientifique et droit international' (1947); KRUSE, *Legal Aspects of the Peaceful Utilization of Atomic Energy* (1962); RODIÈRE, *Aspects du droit de l'énergie atomique* (1965). See also *Problèmes juridiques nés du développement et de l'utilisation de l'énergie atomique* (VIIth Congress of the International Association of Democratic Lawyers) (1960).

274. Cf., for example, GÁL, *Space Law* (1969); HALEY, *Space Law and Government* (1963); JENKS, *Space Law* (1965); KOPAL, 'Treaty on principles governing the activities of states in the exploration and use of outer space' (1968); by the same author, *Dnešní základy a další vývojové tendence kosmického práva* (1971); LACHS, *The Law of Outer Space* (1972); MCDOUGAL, LASSWELL & BRADLEY (eds.), *New Frontiers in Space Law* (1969); MANDL, *Das Weltraumrecht* (1932); SEARA VÁSQUEZ, *Introducción al derecho internacional cósmico* (1961); VEREŠČETIN, 'Osnovy kos-

problems relevant to international law. A start towards their solution was provided in Resolution 1721 (XVI) adopted by the General Assembly of the United Nations in 1961, under the terms of which international law, including the Charter of the United Nations, was declared to apply equally to outer space and to celestial bodies. This same resolution also proclaimed the principle of freedom of exploration and of exploitation of outer space and of celestial bodies by all States in accordance with international law and with the doctrine that neither outer space nor celestial bodies are open to seizure by one nation. In 'terrestrial' legal language, these areas have become *res extra commercium.*

The bases of the contemporary outer space régime were laid down in the first instance in the Declaration adopted by the General Assembly in 1963 (Resolution 1962 (XVII)) and subsequently in the Treaty on principles governing the activities of States in the exploration and use of outer space including the moon and other celestial bodies, dated 27 January 1967 and based on recognition of the common interest of all mankind in the progress made in the exploration and utilization of outer space for peaceful purposes.

Legal research relating to space problems has given birth to a new branch of law which has been given the names of space law or law of extra-atmospheric space or again 'cosmic' law.[275]

Space law is, in general, defined *stricto sensu* as the body of principles and norms of international law which govern the activities of States relating to the exploration and utilization of outer space; in a broader sense space law may be understood to mean the totality of legal rules of either international or municipal origin which relate to cosmic activities.

The major problems arising, which at the same time correspond to the most important trends of research in the field of space law, are the following:

(a) Delimitation of the scope of application of space law, constituting as it does a new branch of legal science, the confines of which remain somewhat imprecise, is the subject of continuing debate. According to one of the current theses space law is based upon a definition of the notion of space. The supporters of this thesis accordingly seek to establish in legal terms the border between atmospheric and extra-atmospheric space. Another opinion holds that space law must be defined in terms of the *activities* involved, and that it regulates activities aimed at the cosmos, leaving aside any delimitation of cosmic space. The difference between these two concepts of space law becomes very acute in regard, for example, to the launching of spacecraft. According to the first view, the craft only falls under the operation of space law

mičeskogo pravoporjadka' (1968); VLASIC, 'The growth of space law 1957-1965' (1967); ŽUKOV, *Kosmičeskoe pravo* (1966) (German version, *Weltraumrecht*, 1968). See also SMITH, *International Telecommunication Control: International Law and the Ordering of Satellite and Other Forms of International Broadcasting* (1969); and the publications of the (French) Centre national de la Recherche scientifique (CNRS): *Les télécommunications par satellites. Aspects juridiques* (1968), and *L'utilisation des satellites de diffusion directe* (1970).

275. Still other expressions are sometimes used, for example 'law of interplanetary space': cf. VALLADAO, 'The law of interplanetary space' (1959).

at the moment when it has left the atmosphere and reached extra-atmospheric space; under the second doctrine, and because the launching of the spacecraft is an activity directed towards the exploration of space, the entire operation is governed by space law. Clearly, delimitation of the scope of application of space law is not merely a theoretical problem; its solution also has considerable practical implications (for example, with regard to damage caused,[276] etc.).

(b) One of the problems of space law is that of establishing an international mechanism for the rescue of astronauts in peril. The Agreement on the rescue of astronauts and the return of astronauts and the return of objects launched into outer space was concluded on 22 April 1968. It provides a sound basis for the development of efficient co-operation.

(c) Another task facing space law is that of finding a solution on the plane of international law to the question of liability for damage caused by objects launched into cosmic space. This problem involves a multiplicity of issues. Damage may be caused to persons or property located on the surface of the earth; if the applicability of space law is defined in terms of the character of the activity, this damage must be dealt with in accordance with space law, even though in such a case space law might prove to be in conflict with the municipal law of the country concerned. Or again damage may be caused by an object in space to another object in space or to persons or property within that object, either within the atmosphere or in outer space; or damage may be caused by objects launched at a spacecraft; we may even imagine that in the near future, damage could be caused as a result of activity carried on on the surface of the moon or of another celestial body; and so forth.

It is clear that compensation for damage of these types raises a great many problems which must be studied by space law. Among such problems are: the designation of the party liable (it is generally admitted that the party liable is the State or the international organization which has launched the object into space); the identity of the victim and the conditions of liability (the general view is that liability should be irrespective of fault, i.e., absolute or objective liability); what damage should be compensated for; the extent of compensation; the tribunal that should deal with the case; the procedure (it is generally conceded that such cases should be judged in accordance with the same procedures as those used for the settlement of disputes between States); and so on.

(d) The next group of legal problems consists of those arising out of the use of telecommunication satellites, meteorological satellites and other systems of functional orbiting devices. With regard to the first, several systems of telecommunication satellites are already in existence. One of the far-reaching issues which arise in this connexion is that of the legal aspects of mutual relationships between telecommunication systems of universal, regional or national usage, of co-operation and co-ordination between them, etc.

(e) Among other important subjects for legal study are the problems in-

276. See (c) below.

volved in the placing of permanent stations either in earth orbit or on the moon, etc., including their use, their mutual relations, the conditions of their occupancy and the questions relating to international co-operation in this field.

(f) The last group of problems we should like to mention here as affecting space law (although there are of course others) relates to the studies to be undertaken with a view to creating a legal régime to govern the exploitation of the resources which might be discovered on the moon, or ultimately on other celestial bodies. Though the international community has, as already noted, debarred the taking possession of outer space or celestial bodies, this naturally does not imply a bar on the exploitation of lunar and other resources.

2. *Legal implications and problems in connexion with organ transplants*

(a) Questions affecting human health and indeed medical questions as such have been the source of a great number of problems on the legal plane, both for legal science and for legislators.[277] These have given rise to the formulation of a variety of legal solutions concerning such matters as liability for injury suffered through medical treatment, consent to surgery, the detention of the mentally disturbed, the prohibition of child labour and of night work for women, and so forth. The development of medicine has posed a number of new questions for legal science, such as the legal problems that arise in relation to organ transplants (similar, in certain respects to those raised by blood transfusions), the legal consequences of artificial insemination, etc. But whereas these latter give rise only to marginal problems, organ transplants carried out in the last few years have aroused the keenest interest on the part of jurists and have produced a very noticeable trend affecting research in this field,[278] some of the legal problems involved in organ transplants are ancient problems and the only difference is the new light in which they are presented today. Thus the legal problems of death are not entirely the same from the standpoint of laws of inheritance and criminal law on the one hand, and from that of organ transplants on the other. It has always been acknowledged that the human body (including of course its various parts) as well as the

277. MALHERBE, *Médecine et droit moderne* (1969); KORNPROBST, *Responsabilités du médecin devant la loi et la jurisprudence françaises* (1957); 'L'incidence de la biologie et de la médecine moderne sur le droit civil' and 'Influence des progrès de la biologie et de la médecine sur le droit pénal' (VIIIth International Congress of Comparative Law, Pescara, 1970, Sections II A 2 (General Rapporteur: P. J. Zepos) and V A 1 (General Rapporteur: M. A. Samuels) respectively; see *Rapports généraux*, to be published); etc.

278. It is true that, even from the point of view of legal science, it is the spectacular (and thus far rather unsuccessful) heart transplants which have attracted the greatest attention. The legal problems of which we are speaking here, however, are by no means limited to heart transplants; they have to do with organ and tissue transplants in general, and have not diminished in practical importance simply because of the failure of the heart transplants carried out thus far.

human corpse (and its parts) are *extra commercium*, that is to say that they can at no time be the object of any legal disposal whatsoever (even without payment), whereas organ transplants necessarily entail the disposal either of a part of one's body (in case of a transplant *ex vivo)* or of one's corpse (in case of a transplant *ex mortuo*).

(b) The examples just given cover only a few of the legal problems resulting from organ transplants. Let us endeavour to review briefly the most important ones in order to illustrate the trend in contemporary legal science analysed in this section.[279]

We can distinguish three different cases, according to where the transplanted organ comes from:

(i) replacement of a discarded organ by an *artificial* organ;

(ii) *auto-transplantation,* that is, transplantation of organs (or in this case tissue) from and to the patient himself;

(iii) *allo-transplantation,* i.e., transplantation in the habitual sense of the term, that is to say replacement of a discarded organ by an organ removed from another person.

The most important and most difficult legal problems arise in the case of allo-transplantations, and it is these that we shall discuss in greater detail.

The issues raised by allo-transplantations fall into three groups:

(i) The first is concerned with the *consent of the recipient.* It is accepted without demur that an organ transplant can only be carried out with the consent of the recipient. Somewhat difficult problems nevertheless arise when the recipient is unable (either physically or legally) to give his consent, that is if he is under age or mentally unfit, if he is in a condition of unconsciousness, if he is a deaf-mute incapable of apprehending the question put to him, etc.

The first problem is to decide whether the consent of the recipient can be replaced by that of a third party authorized to represent him, or whether what is required is a strictly personal act of the recipient; in other words, whether the consent of the recipient is an act under civil law (capable there-

279. Cf. Dworkin, G., 'The law relating to organ transplantation in England' (1970); Gorelik, 'Uslovija pravomernosti izjatija transplantata ot neživogo donora' (1971); Luby, 'Transplantácie tkanív a orgánov a vývoj právnych názorov' (1969); Malherbe, *op. cit.* (1969), pp. 35 sqq.; Nizsalowsky, 'L'incidence de la biologie et de la médicine modernes sur le droit civil' (1970); by the same author, *A szerv- és szövetátültetések joga* (1970); Savatier, '*Et in hora mortis nostrae.* Le problème des greffes d'organes prélevés sur un cadavre' (1968). Cf. aslo Dierkens, *Les droits sur le corps et le cadavre de l'homme* (1966); Kemmer, 'Das Recht am Leichnam' (1932); Geilen, *Einwilligung und ärztliche Aufklärungspflicht* (1963); etc. The problem of organ transplants is beginning to arouse interest also among the legislators of certain countries, even at the level of the codification of civil law. It has been proposed to embody provisions covering organ transplants into the section dealing with the rights of the human person in the new civil code of the Province of Quebec, which is now being prepared. The author of the present article wishes to express his thanks to the 'Office de Révision du Code civil' of Quebec for this most interesting piece of information.

fore of being replaced by the act of a third party under the general rules of civil law or, as the case may be, of family law) or whether this is an act falling within the sphere of *iura merae facultatis* and therefore not replaceable by the act of another.

It may be asked, in a similar train of thought, whether in the case where the recipient is physically unable to give his consent (e.g., because he is in a state of unconsciousness) his consent may be presumed. In such cases some legal authors speak of a presumption of consent where no reasonable motive for the patient's refusal of such consent is apparent.

Finally, a problem close to those which have just been listed is that of deciding whether transplantation can be carried out against the expressed wish of the recipient (for example, where the recipient has expressed in advance his refusal *in eventum* or in reply to an explicit question, and later becomes unconscious at a moment when the transplantation is so clearly necessary that in any other case there would be little hesitation in assuming the consent of the patient).

It is clear that except in the case where the recipient has given his explicit consent, we have here a theoretical problem of the utmost significance, namely the conflict between the principle of the autonomous will and that of *favor vitae*. Although the replies suggested in legal literature and in legislation vary, the principle of autonomy of the will manifestly prevails.

The last question we should like to mention in this context is the question whether the consent of the recipient, considered as a necessary condition of the transplant, is also a sufficient condition. The answer is firmly in the negative. The consent (including the wish) of the recipient is never a sufficient condition of transplantation and cannot absolve the doctor of his responsibility. To justify a transplant it is always necessary that in addition to the subjective conditions (the consent of the recipient and, if appropriate, that of the donor) the objective condition, that is to say a medical justification, should be met.

(ii) The second group of issues relates to the *consent of the donor*. This question falls into two parts, according as we are dealing with transplants *ex vivo* or *ex mortuo*.

1. The legal problems arising out of the donor's consent in the case of transplants *ex vivo* are, *mutatis mutandis*, somewhat similar to those which relate to the consent of the recipient.

Further questions arise in the case of a donor enjoying full capacity in the exercise of his rights: there is first the question whether he may only consent to the removal of the organ without payment or whether alternatively he may exact remuneration; secondly, there is the question whether he can subordinate his consent to a condition, which in practice means whether he can make his consent subject to an assurance that the organ to be removed will be used only for the benefit of a specified person.

Certain specific problems relating to the donor's consent are very closely linked with the legal problems of death. The first case that comes to mind is one where the organ can only be removed for transplant at the cost of causing the donor's death. There can be no doubt that any consent given to the removal of an organ on such terms would be null and void. But that is not the problem.

The difficulties arise when it has to be decided whether the organ can be removed (a) from a dying person; (b) from an individual in the intervening stage between clinical death and biological death ('post-coma'), therefore already dead although still biologically living; it being understood in both cases that it had been established beyond any doubt that there was no possibility of saving the life of the individual in question.

It is clear in such cases that the principle of *favor vitae* becomes self contradictory: two lives are in question, and the law must decide in favour of which of them the issue has to be resolved.

The first case, (a), is equivalent to a form of euthanasia applied to the proposed donor, and the majority view is that such action is not permissible. The second case, (b), is linked with the legal problems relating to artificial maintenance of life, to which we shall revert below.

2. As regards transplants *ex mortuo,* the first problem encountered relates to the philosophy of law, namely the problem of the protection of the personality, and more particularly whether such protection continues to operate even *post mortem.*

Legal science offers several solutions. One of them holds that the protection of the personality never ceases, and that since, after death, there is no living title-holder enjoying the capacity to exercise rights, it becomes absolute; this view, which denies to anyone the right to dispose of a corpse, to all intents and purposes precludes any organ transplant *ex mortuo* and is rejected by most legal experts as inhibiting medical progress.

The diametrically opposed theory holds that protection of the personality only applies to the living being and ceases at death: consequently the corpse enjoys no form of protection of personality and the removal from it of organs required for transplant does not call for the consent of the relatives or for the prior consent (*mortis causa*) of the deceased. But in this case for reasons of a traditional nature (respect for the human corpse, for the feelings of relatives, etc.), this view too is rejected by the majority.

In between these two extreme positions are found theories which seek a reasonable and acceptable solution.

One of these requires as a necessary condition for a transplant *ex mortuo,* the *prior consent* (*mortis causa)* of the future donor. This theory, which endeavours to assimilate the legal régime of transplants *ex mortuo* to that of transplants *ex vivo,* nevertheless gives rise to a number of problems not easily solved, e.g., that of the legal character of such consent *in eventum,* whether it is revocable or irrevocable, etc.

The other theory requires the consent of the closest relatives. This is the most widespread theory but also the least justifiable from the legal point of view. From what source springs this mysterious right of the next-of-kin to consent to the use of their relative's corpse? One thing is certain, namely that no such right can derive from the law of inheritance, for it is well known that a corpse cannot be the subject of legal succession; nor, by the furthest stretch of the imagination, can the right of the *de cujus* to give his consent to the removal or organs from his body for transplant purposes be transmitted by legal succession. Where then does this alleged right of the relatives derive from? And who are to be regarded as the next-of-kin? The argument could continue, but these are questions to which the theory we are dealing with provides answers based more on morals, traditional opinions, etc., than on legal reasoning. At the same time it cannot be denied that this theory is so far the most plausible from a practical standpoint.

(iii) The third group of problems relates to *death* and has already been mentioned. The main legal problems from the point of view of organ transplants derive from the distinction between *clinical* death and *biological* death. The first relevant question is fundamental and its solution will determine the legal provisions to be applied, namely the question whether the removal of an organ during the interval separating clinical death from biological death is a removal *ex vivo* or *ex mortuo;* in other words, whether the notion of death from the standpoint of organ transplants is to be identified with the notion of clinical death or with that of biological death.

This problem is all the more important in that the interval between clinical and biological death can be prolonged artifically through the now current techniques of resuscitation.

Prolonged resuscitation itself is a legal problem of great significance, particularly from the point of view of organ transplants.

Many questions arise in this context. We shall mention only two:

1. Is the doctor compelled to continue the application of resuscitation techniques up to the moment of biological death even though it has become certain that they cannot bring the patient back to life, or is he entitled to bring the resuscitation process to an end as soon as it is established that there is no further hope of restoring life to the patient?
2. The second question is the converse of the first; is the doctor authorized to prolong resuscitation, with a view to making a transplant possible, even when there is no doubt whatever that it can no longer save the life of the patient, in other words is he entitled to prolong artificially the functioning of the organic substance which is still biologically living, but which is already dead as a human being?

It will clearly be seen that the solution of these questions is of the highest importance in terms of organ transplants and of the progress of medicine in general. The answers so far furnished by legal science vary considerably, and the studies which they are prompting represent a very important trend in current legal science.

EPILOGUE

1. *Conclusions and prospects*

(a) Such appear to us to be the most important current trends of research in the field of legal science. It is of course unnecessary to stress once again that this survey of current trends offered to the reader is the outcome of a selection and that it does not exhaust, nor could it exhaust, the entire range of present-day legal science in all the countries of the world, in every school of thought, etc.

However, we think it safe to say that the chief characteristic of the main trend in the development of legal science today is that our discipline has become a *social science* in the true sense of the word and, despite the tenacious hold of legal positivism in some countries, there is no doubt that the sociologically inclined tendency in the study of law is likely to be maintained and become still more pronounced in future.

Nevertheless it should be noted that this increasingly clear conception of legal science as a social science represents no more than a very general common denominator reflecting the very widely held conviction among legal scholars (and practitioners) today that, as A. Tunc says,[280] 'The study of law cannot be isolated from that of its application or from that of the social rules which are its framework' and that, at the same time, 'law is a technique at the service of justice in society' or, as we should say, that law is a means available for achieving a conscious transformation of society and that the task of the scholar in this field is not only to interpret the rule of law, but also, by studying the law in its social context, to exercise an influence on the creation of law and to work towards the effective use of law as an instrument of social change.

In view of the ideological differences [281] which we mentioned at the beginning of this work, it is however self-evident that within the common framework which has been described, the rôle and function of legal science in society – and, consequently, the actual trends of its development – cannot but differ profoundly in line with the theoretical and political conceptions of society itself. The tenor of these various conceptions and the divergences

280. In his Comments on the second version of the present work: 'L'étude du droit ne peut être isolée de celle de son application et de celle des règles sociales qui l'encadrent. [...] Le droit est une technique au service de la justice dans la société.'

281. We have several times introduced the notion of *ideology* without so far defining the meaning we were giving to this notion which is often used in very different ways. May we say here that the term *ideology* in the Marxist-Leninist sense means a system of opinions, ideas, theories, etc., whereby a social group (especially a social class) expresses its economic and social position and interests, becomes aware of what these are and defends them on theoretical grounds: in short, it denotes the theoretical consciousness of social classes. (Cf. for example, *Stručný filosofický slovník* (1966), p. 188). Thus philosophy, like legal science, etc., constitutes only part of a given ideology.

that separate them will have been obvious to the reader in the foregoing sections.

Nonetheless, we hope that the reader will also bear in mind this other consideration to which we likewise drew attention, namely that there is a difference between compromise on the one hand and communication or even mutual understanding on the other and that the unbridgeable gulf, precluding any compromise, which separates Marxist-Leninist legal science from all other schools of thought in no way implies the impossibility of communication, comparison and co-operation in the international sphere.[282] The present work, undertaken under Unesco's auspices, is itself intended to demonstrate that this is so.

(b) The outline presented in this survey seems to show that apart from trends of a methodological nature which cover the whole field of legal science, the most important trends today are to be found in the non-conventional fields of legal science rather than in its more traditional branches; the latter being, as they have always been, more conservative than branches of human knowledge which have either developed more recently or are still in process of formation.

We certainly do not wish to suggest that the traditional branches of legal science are inactive or exhausted: they too are adapting to the needs of society (one might for example mention, as A. Tunc [283] reminds us, the current renewal of interest in divorce and criminology, the research in progress regarding the law of accidents in general and traffic accidents in particular, etc.); nevertheless the new horizons opening up before legal science chiefly affect research on legal problems involved in such things as the exploration of outer space or the ocean floor, organ transplants, etc. In this context we think mention should also be made of a field of technological development which deserves permanent attention from legal science, namely the peaceful use of *atomic energy*. Even though many of the legal problems in this field have already been solved, we nevertheless believe that the legal effects of the peaceful use atomic energy are bound to offer considerable attraction even for the legal science of tomorrow, and that full development of this use will necessarily bring about, in a more or less distant future, many new legal problems falling within the purview of both municipal and international law.[284]

(c) The political and social phenomenon which, as we have seen, has given rise to one of the most significant trends in legal science since the Second World War is the process of decolonization and the emergence of dozens of new States in Africa and Asia.

We have referred to the legal problems of the developing countries from

282. See Section Two, sub-section I, p. 951.

283. Tunc, Comments; cf. also Rheinstein (Observations on the Revised Report) who associates himself with Tunc's views and adds certain examples of the development of legal notions brought about by the innovating action of court decisions in the United States.

284. See VIII, 1 above, p. 1027 and footnote 273.

several different points of view (as regards acculturation of law, comparative law, etc.). Yet it seems that these provide but a very narrow and incomplete framework for the problems which arise in connexion with the decolonization process and the emergence and development of new States. We have here in the field of State and law a grandiose process of inevitable adjustment and change in which elements of a legal order are superimposed on political, economic and other elements, a process which affects both public and private law. This involves the construction of a new legislative and executive apparatus, legal problems in regard to nationality, a radical transformation of the land-tenure system and rural law, the establishment of an effective system for the control of foreign investments, the creation of a banking infra-structure, the adjustment of family law, property law and the law of inheritance, the formulation of a labour code reflecting the needs and interests of the population, etc.[285] Newly independent countries are also faced with a number of problems in the domain of international law (both public and private) such as questions relating to their sovereignty and independence or the succession to rights and obligations under conventions, treaties, etc., concluded by the former metropolitan countries,[286] legal problems of foreign trade and so forth.

The building up of new States and the transformation of their law also entail many other problems that have to do with the implementation of desirable changes and adjustments; [287] particularly as regards the formulation of appropriate techniques with a view to modernizing the law, recasting the administrative and legal framework, etc.

This in turn gives rise to many theoretical problems in connexion with the training of lawyers 'capable *of understanding* and applying a new kind of law intelligently and effectively'.[288] As a result of increased interest in legal structuralization, problems concerning the teaching of law in the African countries are more closely related to those of scientific research. An example of this may be found in the attention given not only to the problems of teaching, but to those of research, by the conference on the teaching of law in Africa organized in 1968 at Addis Ababa by the Haile Selassie I University.[289] This conference observed that 'trends in African legal research and

285. Cf. TUNC (ed.), *Les aspects juridiques du développement économique* (1966). It is true, as Rheinstein points out (Observations on the Revised Report), that some of these trends were already observable in the colonial period. They have, however, since the attainment of independence by those countries, taken on quite a new and very different quality from that which they formerly had. Whereas before the attainment of independence it was a matter of creating 'native' institutions under the colonial regime, measures are now taken by independent States (which naturally may even decide to use for their own purposes institutions which date from the Colonial régime).

286. Cf. MOCHI-ONORY, 'Les aspects récents du problème de la succession aux traités' (1968).

287. Cf. TUNC, *op. cit.* (1966), pp. 18 sqq.

288. *Ibid.*, p. 21.

289. *Proceedings of the Conference on Legal Education in Africa held in Addis Ababa, Ethiopia, Haile Selassie I University, 20-24 October 1968* (1969); we may

writing seem to be organized around five major centres of interest: the knowledge of African law, the "scientification" of African law, the use of African law as a tool for development, the teaching of African law, and the unification of African law'.[290]

It is clear that all the problems just mentioned cannot be grasped from a single standpoint, whether it be the 'acculturation' of law, comparative law, etc., but that they require a complex scientific approach. This explains why some authors are beginning to speak of *development law*.[291]

The decolonization process and the access of new countries to independence has generated yet another new trend in legal science. Hitherto it was usual to consider legal research on developing countries as exclusively within the purview of European or American legal science, hence as research pertaining to the study of 'foreign law'. There is no doubt that European and American legal science retains its interest in the legal problems of newly independent countries or that it exercises a very strong influence on the development of State and law in these countries; [292] what should be stressed, however, is that the formulation of solutions to the problems of 'development law' devolves, and will necessarily devolve increasingly, on the jurists who are emerging in the developing countries themselves; thus the centre of gravity of research on 'development law' will increasingly shift from the countries that 'export' law to the countries whose State and law are the subject of this research.

2. *Interdisciplinary links*

(a) One of the outstanding traits of legal positivism was the total isolation of legal science. Having as its sole aim analysis and interpretation of the law in force, our science thus considered felt no need to communicate with the other

---

note also that the International Association of Legal Sciences organized an international symposium at Pont-à-Mousson in 1966 on the subject of 'La formation des juristes dans les pays en voie de développement'.

290. VANDERLINDEN, 'Trends and priorities in African legal research and writing' (1969), p. 77 in *Proceedings of the Conference on Legal Education in Africa, op. cit.*

291. GRANGER, 'Pour un droit du développement' (1961); by the same author, *Contribution à l'étude du droit du développement* (1970); RARIJAONA, 'Le droit du développement à la recherche de son expression' (1968); M'BAYE, 'Droit et développement en Afrique francophone de l'Ouest' (1966, 1967). Cf. also the Comments of A. Tunc (editor of the collection (1966) containing the article quoted above) who also proposes that legal research regarding developing countries be treated as a complex and independent field of legal research, as 'development law'. A similar opinion is expressed by Čhikvadze in his Comments.

292. Obviously this influence has its political overtones; Čhikvadze recalls (Comments, item 8) that in the field of research relating to newly independent countries '. . . differing tendencies are displayed by Western jurists, especially those of the former metropolitan countries, and by Socialist jurists'. See ČHIKVADZE *et al.*, *Marksistsko-leninskaja obščaja teorija gosudarstva i prava*, Vol. II (1971), Chaps. IX and X.

sciences. The transformation of legal science into a social science has put an end to this isolation.

Legal science in this new meaning, while it preserves its independence within the system of sciences, merely constitutes one of the many sciences geared to the study of human society. It is defined by its subject-matter: State and law as social phenomena, in other words, the organization of society by means of State and law; but the complexity of contemporary social life, within which it is sometimes very difficult to isolate the social action of the State, or of law, from the action of other social forces, makes it necessary that legal science should co-operate closely with neighbouring sciences.

Now that legal science does not only aspire to the knowledge of positive law (that it is no longer content to accept the law as it stands, confining itself to its interpretation), it is closely allied, as regards study of the general problems of State and law, with *philosophy*. Philosophy being the most general science determines the place of the State and of legal phenomena within the framework of a given conception of the world; it also determines the epistemology and methodology of legal science which as we have seen, differ in character depending on the nature of the philosophic conception on which legal science is based.

Legal science necessarily has recourse to philosophy to solve problems of general scope, such as those relating to the origin, essence and development (including revolutionary changes) of the State and law, the relationship between the individual and the State, or even those relating to law, liberty and the equality of men and citizens, or legitimate interests (*iura quaesita*), as well as the problem of justice, problems relating to guilt (including questions that have to do with the contradictory notions of determinism and indeterminism in connexion with the human will), punishment and sanctions, the problem of causality and so on. These are but a few examples of the philosophical problems which are of concern to legal science (be it the philosophy of law, the general theory of State and law or the various branches of legal science constitutional law, civil law, criminal law, etc.) and their solution will depend on the general philosophical background to the science of State and law.

Legal science is also very close to *sociology* and *political science*. But its relationship with these disciplines varies with the general (philosophical) conception of legal science. The history of legal thinking comprises numerous theories, and even schools of thought, which make a very sharp distinction between the subject-matter and methods of legal science and those of sociology. The most typical examples of this are legal positivism, which moreover is equally anti-philosophical, the schools of Natural Law, the 'pure theory of law', legal phenomenology, etc.[293]

293. Let us once again call to mind the well-known work of the founder of the *'Reine Rechtslehre'* ('pure theory of law'), Hans KELSEN, *Grenzen zwischen juristischer und soziologischer Methode* (1911), let us also quote the distinctly anti-sociological position of KNOLL (*Katholische Kirche und scholastisches Naturrecht (1962), p. 13): 'Soziologie überschreitet [...] ihre Aufgabe wollte sie der Versuch*

On the other hand, according to the sociologically-inclined conceptions of legal science, which as we have seen constitute the strongest trend in our discipline today, sociology is looked upon as the source of basic data on the birth, development and rôle of the State and law in society; it gives jurists an understanding of the social action of the State and law, the efficacy of legal rules, their adaptability to the requirements of social development, etc.; lastly, it provides them with a particular methodology, as we have shown in detail.[294] We have also described [295] the place of sociology in the Marxist-Leninist science of State and law; all we need here is recall that it too makes wide use of sociological methods for the concrete study of reality, within the general framework of dialectical materialism and historical materialism.

The above remarks on the relations between legal science and sociology also apply *mutatis mutandis* to the relationship between legal science and *political science*. Although the boundaries of the latter are still somewhat vague, there is no doubt that research relating to various phenomena in political life (both domestic and international), such as the place and function of political parties in society, the electoral system, the political rôle of trade unions and other social organs, the fight against crime in general and against juvenile delinquency in particular, the social function of international organizations and so on, goes beyond the limits of the science of State ad law and is divided between the latter and political science.

It should also be borne in mind that in many countries some of the abovementioned questions do not strictly speaking pertain to legal science at all (even from the sociological standpoint) but are in practice solely the concern of political science.[296]

The solution of a great many fundamental problems of legal science involves close co-operation between the latter and *economics*. This becomes evident as soon as one considers that a vast sector of legislation is concerned with regulating economic life and that the legal notions which apply in this field are but the corollary of the corresponding economic notions – e.g., property law and the law of inheritance, the law of contract (and in particular purchase and sale, leases, liability for damage, etc.), banking law, fiscal law, economic law (commercial law), rural law, transportation law, investment law, the laws governing waterways, forests, mines, foreign trade and so on.

In addition we should remember that economics plays a very special rôle in the Marxist-Leninist science of State and law, which is founded, as we have seen, on the distinction between social base and superstructure, State

---

sein, Geistesgebilde, wie [. . .] Recht usw. zu soziologisieren, das heisst ausschliesslich gesellschaftlich verstehen zu wollen. Das ist ein Missbrauch der Soziologie [. . .]' ('Sociology would be overreaching [. . .] itself if it were to try to sociologize concepts such as [. . .] law, etc., that is, seek to understand them purely from the sociological point of view. This is an abuse of sociology [. . .]').

294. See Section Two, sub-section II, 3, pp. 960 sqq.

295. See Section Two, sub-section III, pp. 966 sqq.

296. See Section One, sub-section II, p. 936.

and law forming part of the superstructure and in turn reacting on the social base.[297]

It is obvious that the interdisciplinary links and affinities of legal science go beyond its co-operation with philosophy, sociology, political science and economics. To confine ourselves to the social, or even human, sciences, legal science must very often draw on general *history* (of which the history of State and of law is but one branch), on *psychology, ethics,* the *logic* of norms,[298] etc.

(b) It is, of course, necessary when embarking on this type of co-operation to guard against the danger of syncretism, i.e., methodological confusion of the different sciences (a danger which could be quite easily engendered by an exaggerated legal 'antipositivism'); nevertheless this risk does not and cannot hinder the development of the interdisciplinary trend which is increasingly evident in contemporary legal science (and the social sciences in general).

At the lowest level, this interdisciplinary co-operation operates on a 'one-way' basis: what we have here is essentially a monodisciplinary (legal) study which borrows certain elements from neighbouring disciplines (for example in the realm of family law it borrows from sociology, in the realm of property law it borrows certain basic data from economics, and so forth).

There is, however, an increasing tendency for a form of interdisciplinary co-operation of a higher type to spread, especially in the form of a genuine *multidisciplinary research,* combining for the purposes of co-ordinated study a number of different social sciences, or even, as we shall see, joining in double harness social sciences and natural sciences. Lastly, it is worth noting that certain types of multidisciplinary research on a complex subject tend to constitute independent branches of science (*hybrid* or *composite* sciences), as has long been the case, as a matter of course, in the development of the exact and natural sciences.

For example we may mention:

(i) Research on *peace* and *polemology* (regarded as the study of means of preventing armed conflicts), which we have already referred to at some length[299] and which demands the co-operation of political science, sociology, legal science (international law), etc.[300]

(ii) Research on the development of *democracy,* which combines political science, sociology, the science of State and law, economics, history, etc., and

297. See Section Two, sub-section III above, p. 968.

298. For recent research in ethics and the logic of action, especially deontic logic, see, in the present work, 'Philosophy', by Paul Ricoeur, especially section Five, 'Man and action', pp. 1380 sqq.

299. In Section Three, sub-section VII, I, pp. 1021 sqq.

300. Cf. ARON, *Paix et guerre entre les nations* (1963); BOUTHOUL, 'L'apport de la polémologie à la solution des conflits' (1967); ČHIKVADZE, *Bor'ba za mir – neodolimoe dviženie sovremennosti* (1969); etc. See also the works cited in the bibliography *Literature on Soviet Law* (1960), under 'Peaceful means of settling international disputes', 'War and international law' and 'The struggle of the USSR for peace' (pp. 229 sqq.); GALTUNG, 'On the future of the international system' (1967); by the same author, 'Friedensforschung' (1968); *idem*, 'Peace research' (1968).

which we referred to above from the point of view of legal science.[301]

(iii) Multidisciplinary study of the *human environment,* which encompasses the problems of protecting (against pollution, etc.) the atmosphere, the earth, waterways and nature as a whole, protection against noise, protection of working surroundings (including safety at work), of cultural monuments, etc. This research, which embraces not only several social sciences (including legal science) but in the first place several natural sciences, tends to constitute an independent branch of science (a composite science) in which, it is to be regretted, the legal element does not yet play the rôle it should.[302]

(iv) *Urbanism,* which in many countries has already become an independent science (coinciding in large measure with research on the environment), where legal science has only just begun to penetrate.[303]

(v) *Administrative science,* which we have already mentioned, and which deals with problems in the administration of the State and, in a wider sense, the administration of society at large, where administrative law plays an important rôle alongside sociology, management theory, information theory, etc.[304]

(vi) Lastly, we would mention the study of the future of society, a complex

301. Section Three, sub-section V, pp. 1007-1010.

302. Cf. CALDER, *The Environment Game* (1967); CAPONERA, 'Towards a new methodological approach in environmental law' (1972); DOXIADIS & DOUGLAS, *The New World of Urban Man* (1966); EEK, 'The hydrological cycle and the law of nations' (1965); FORKOSCH, 'Administrative conduct in environment areas' (1970); HARGROVE (ed.), *Law, Institutions and the Global Environment* (1972); KAZANCEV & KOLOTINSKAJA, *Pravovaja ohrana prirody v SSSR* (1962); MAYDA, *Environment and Resources: From Conservation to Ecomanagement* (1968); SAX, 'Legal redress of environmental disruption in the United States: the rôle of courts' (1970); WALINE, 'La responsabilité des préjudices causés par le bruit des avions aux abords des aérodromes' (1972); cf. also *Control of River Pollution by Industry,* in the series 'Cases in Comparative Public Administration' published by the International Institute of Administrative Sciences (1964); cf. *La défense de l'homme contre les pollutions – air, bruit, eau* (Proceedings of the symposium organized in 1970 by the Association for the Development of World Law, publ. 1971); see also *Belgrade World Conference on World Peace through Law* (1971), Section 2: 'Law and environment' (in press).

303. ROGON, *Les cités de l'avenir* (1968); 'Réglementation au sujet de l'administration des grandes agglomérations urbaines', VIth International Congress of Comparative Law, Hamburg 1962, Section IV A 2 (see *Rapports généraux au VIe CIDC,* 1964); 'La ville et l'aménagement de l'espace' (Symposium organized by the Institut universitaire d'Etudes européennes, Turin, 1970–still unpublished); *Belgrade World Conference on World Peace through Law (1971),* Section 10: 'Urban problems and the lawyer' (publication announced).

304. AFANAS'EV, *Naučnoe upravlenie obščestvom* (1968); AFANAS'EV (ed.), *Naučnoe upravlenie obščestvom* (1967); DAVIS, *Discretionary Justice* (1969); GLADDEN, *The Essentials of Public Administration* (2nd edn., 1968); GOURNAY, *Introduction à la science administrative* (1966); JAMPOLSKAJA, 'K metodologii nauki upravlenija' (1965); LANGROD *et al.*, *Traité de science administrative* (1966); LUNEV *et al.*, *Naučnye osnovy gosudarstvennogo upravlenija v SSSR* (1968); MORSTEIN-MARX, *Verwaltung* (1965); MORSTEIN-MARX (ed.), *Element of Public Administration* (1968); ŠORIN (ed.), *Pravovye i social'no-psihologičeskie aspekty upravlenija* (1972); STAROSCIAK, *Zarys nauki administracji* (1966); by the same author, *Elementy nauki upravlenija* (1965); WAGENER, *Neubau der Verwaltung* (1969); etc.

multidisciplinary study which tends to constitute an independent science often loosely called *futurology*. Although the outlines of this science are very vague and the conceptions of its purpose differ significantly, it is indubitable that the future development of society constitutes a focus of interest of the first importance for the social sciences, including legal science. Viewed as a social science, legal science does not indeed deal only with the past or the present, with the State as it is today and the law as it has come into being; it focuses on the future and endeavours to foresee the development of State and law in a given society (for example the legal implications of the revolution in science and technology which have already been mentioned) and to foster this development. This is of special importance for Marxist-Leninist legal science, one of whose tasks is to study and identify the mechanisms involved in the transition from socialist statehood to public communist self-government in the course of the withering away of the State as foreseen by Marxist-Leninist social philosophy and, simultaneously, those involved in the transition from socialist law to non-legal rules of human behaviour, in other words rules shorn of the element of compulsion by the State.[305]

3. *Epilogue to the Epilogue*

Before closing we would wish to recall that this study is only concerned with current *trends* and does not attempt to review the current *state* of research in legal science. On the other hand the material collected in the course of the work of preparing this chapter (replies to the questionnaire, special contributions, and the contributions of the National Commissions) are more concerned, for the most part, with the current state of research in legal science than with trends. Hence it was not possible to draw full benefit from the wealth of information contained in these documents, except in a very fragmentary manner. As it is unfortunately most unlikely that the replies to the questionnaire will be published *in extenso*, much of the information they contain must remain buried in the archives.

The replies of the national correspondents follow the plan set out in the questionnaire and fall under five headings as follows:

I. General theory of law
II. History of law
III. Law and society (adaptation of legal norms)
IV. Law and man (legal anthropology)
V. Comparative law

The sixth section of the replies is in most cases devoted to a brief bibliography corresponding to each heading.

The reader who is interested not only in current trends but in the present state of research in contemporary legal science is advised to consult these documents in the Unesco archives. The reader in search of more detailed

305. See ČHIKVADZE *et al.*, *Marksistsko-leninskaja obščaja teorija gosudarstva i prava*, Vol. III (1972), Chap. X.

bibliographical information on the various trends and the various countries should do likewise.[306]

BIBLIOGRAPHICAL ANNEX

I. Legal Bibliographies

A. General

1. NATIONAL*

Canada: Bibliographie du droit canadian/A Bibliography of Canadian Law, R. BOULT (ed.), Montreal, 1966.

Czechoslovakia: Bibliography of Czechoslovak Legal Literature 1945-1958, V. KNAPP (ed.), Prague, 1959.

France: French Bibliographical Digest (Law), R. DAVID (ed.), New York, 1952.
Bibliographie du droit français (1945-1960), R. DAVID (ed.), The Hague-Paris, 1964.

Federal Republic of Germany: Bibliography of German Law in English and German, F. BAUR (ed.), Karlsruhe, 1964; Supp. 1964-1969, Karlsruhe, 1969.

Hungary: Bibliography of Hungarian Legal Literature 1945-1965, L. NAGY (ed.), Budapest, 1966.

India: A Bibliography of Indian Law, C. H. ALEXANDROWICZ (ed.), Madras, 1958.

Israel: Israel Legal Bibliography in European Languages, E. LIVNEH (ed.), Jerusalem, 1963; Supp. 1965, Jerusalem, 1965.

Lebanon: Bibliography juridique libanaise, A. NASSIF (ed.), Beirut, 1958.

Luxemburg: Bibliographie du droit luxembourgeois, R. MAUL *et al.* (eds.), Brussels, 1967.

Poland: Bibliographie juridique polonaise 1944-1956, W. CZACHÓRSKI (ed.), Warsaw, 1958.

Romania: Bibliographie juridique roumaine, D. RUSU (ed.), Bucharest, 1969.

Scandinavian countries: Scandinavian Legal Bibliography, S. IUUL *et al.* (eds.), Stockholm-Göteborg-Uppsala, 1961.

Spain: Bibliografía jurídica española, F. DE SOLA CAÑIZARES & D. FERNANDEZ DE VILLAVICENCIO (eds.), Barcelona, 1954.

United Kingdom: A Bibliographical Guide to the Law of the United Kingdom, the Channel Islands and the Isle of Man, London, 1956; 2dn edn., 1973.

United States: The Law in the United States of America. A selective Bibliographical Guide, G. L. ANDREWS *et al.* (eds.), New York, 1965, 1966.
Bibliographie des études publiées sur le droit des Etats-Unis en des langues autres que l'anglaise, M. COING (ed.), Paris, 1967.

USSR: Literature on Soviet Law, Index of Bibliography, P. S. ROMAŠKIN *et al.* (eds.), Moscow, 1960.

Yugoslavia: Bibliographie juridique yougoslave, B. BLAGOJEVIĆ (ed.), Belgrade,

306. Bibliographical information (in the form of attached lists or footnotes) will be found in the national reports submitted by Mr. Brohi (Pakistan), the Yugoslav Inter-Academy Committee, the National Research Council of Italy, and Professors Eek (Scandinavia), Fix Zamudio (Mexico), Ahmed bin Ibrahim (Singapore and Malaysia), Morgan (Nigeria), Rheinstein (United States), Sharma (India), Seydou Sy (Senegal), Ukai (Japan) and Zippelius (Federal Republic of Germany). The reader's attention is also drawn to the legal bibliographies listed in the annex.

* Published under the auspices, or with the assistance, of the International Association of Legal Science.

1959; Supp. 1961; 2nd edn., Belgrade, 1968.

2. REGIONAL

VANDERLINDEN, J. (ed.), *Bibliographie de droit africain, 1947-1966/Bibliography of African Law, 1947-1966*, Centre for African Legal Development (Faculty of Law, Haile Selassie I University, Addis Ababa, and Institut de Sociologie, Université libre de Bruxelles), Brussels, 1972; Annual supplements in the *Annual Survey of African Law* (N. N. RUBIN & A. COTRAN, eds.), London.

B. Other bibliographies

1. HISTORY OF LAW

GILISSEN, J. (ed.), *Introduction bibliographique à l'histoire du droit et à l'ethnologie juridique*, 1964, 6 vols.

KORANYI, K. & KORANYIOWA, J., *Bibliografia historico-iuridica annorum 1937-1947*, Toruń (Poland), 1953, 1959, 2 vols.

2. SOCIOLOGY OF LAW

POCAR, V. & LOSANO, M., *Sociology of Law 1960-1970. A Bibliographical Survey with KWIC Index*, University of Milan, Milan, 1970.

REHBINDER, M. (ed.), *Internationale Bibliographie der rechtssoziologischen Literatur* (= International Bibliography of Literature in the Field of Legal Sociology), (West) Berlin, 1972.

3. COMPARATIVE LAW AND FOREIGN LAW

SZLADITS, Ch., *A Bibliography of Foreign and Comparative Law-Books and Articles in English*, New York, 1955: Supp. 1953-1959, New York, 1962; Supp. 1960-1965, New York, 1968.

4. LAW AND CYBERNETICS

Computers, Cybernetics and Law. New Titles in Foreign and International Law, University of Illinois, Urbana (Ill.), 5 (8), 1971.

SCHUBERT, W. & STEINMÜLLER, W., *JUDAC (Jurisprudence – Data Processing – Cybernetics)*, Munich, 1971.

5. AIR LAW

HEERE, W. P., *International Bibliography of Air Law*, Leiden, 1972.

6. ADMINISTRATIVE SCIENCE

DRAGO, R. (ed.), *Bibliographie de la science administrative pour la période 1967-1970*, Centre de Documentation Sciences humaines du Centre national de la Recherche scientifique (CNRS), Paris, 1971; up-dated quarterly, beginning January 1971, in the CNRS *Bulletin signalétique* (Abstracting Journal), under the heading 'Science administrative'.

II. List of works cited*

ABDEL-HAMID, M. S. & TAGELDIN, Y. M., 'Le monde arabo-islamique', in *Tendances de la science juridique, op. cit.* (1970), pp. 405-415 (parallel publication in English transl., 'The Arab-Islamic World', in *Trends in Legal Learning, op. cit.* (1970), pp. 367-377).

* Abbreviations:

ADAS	–	*Annuaire du droit aérien et spatial*, Montreal.
AJCL	–	*The American Journal of Comparative Law*, Ann Arbor (Mich.), U.S.A.
CIDC	–	Congrès international de Droit comparé [International Congress of Comparative Law].
CNRS	–	Centre national de la Recherche scientifique, Paris.
Ežegodnik	–	*Sovetskij ežegodnik meždunarodnogo prava* (Soviet Yearbook of International Law), Moscow.
IALS	–	International Association of Legal Science.

Abraham, H. J., *Elements of Democratic Government*, New York, 1958.
'*Actes* du Colloque international de Sociologie du Droit et de la Justice (Bruxelles, 9-12 April 1969)' (S.-C. Versele, ed.) = *Revue de l'Institut de Sociologie* (Université libre de Bruxelles), 1970, No. 2.
Afanas'ev, V. G., *Naučnoe upravlenie obščestvom* (= Scientific Administration of Society), Moscow, 1968.
Afanas'ev, V. G. (ed.), *Naučnoe upravlenie obščestvom* (= Scientific Administration of Society), Moscow, 1967.
Aktuelle Probleme der marxistisch-leninistischen Staats- und Rechtstheorie (= Current Problems in the Marxist-Leninist Theory of the State and Law), collection of essays, Budapest, 1968.
Aleksandrenko, G. V., *Buržuaznyj federalizm* (= Bourgeois Federalism), Kiev, 1962.
Alekseev, S. S., *Obščie teoretičeskie problemy sistemy sovetskogo prava* (= General Theoretical Problems of the Soviet System of Law), Moscow, 1961.
—, *Obščaja teorija socialističeskogo prava* (= General Theory of Socalist Law), Sverdlovsk, 1963.
—, Kerimov, D. A., & Nedbajlo, P. E., 'Metodologičeskie problemy pravovedenija' (= Methodological problems of legal science), in *Izvestija vysših učebnyh zavedenii – Pravovedenie* (= Reports by Higher Education Establishments – Legal Science) (Moscow-Leningrad), 4, 1954, pp. 15 sqq.
Alexandrowicz, C. H., *World Economic Agencies*, London, 1962.
Alker, H., *Mathematics and Politics*, New York, 1965.
Allen, C. K., *Law in the Making*, Oxford, 1927, republished 1958.
Alliot, M., 'Problèmes de l'unification des droits africains', *Journal of African Law* (London), 11 (2), Summer 1967, pp. 86-98.
—, 'L'acculturation juridique', in Poirier (ed.), 'Ethnologie juridique', *op. cit.* (1968), pp. 1180-1236.
Allott, A., *Essays in African Law*, London, 1960.
—, 'Towards the unification of the laws in Africa', *International and Comparative L.Q.* (London), 14, 1965, pp. 366 sqq.
—, 'The unification of laws in Africa', in Hay (ed.), 'The international unification of law', *op. cit.* (1968), pp. 51-87.
—, *New Essays in African Law*, London, 1970.
Amicorum discipulorumque liber René Cassin, Paris, 1969-71, 3 vols.
Ancel, M., 'La tendance universaliste dans la doctrine comparative française au début du XXème siècle', in *Festschrift für Ernst Rabel, op. cit.* (1954), pp. 17 sqq.

ISSJ	–	*International Social Science Journal*, Unesco, Paris.
LJ	–	*Law Journal.*
LQ	–	*Law Quarterly.*
LR	–	*Law Review.*
Rabels Z.	–	*Rabels Zeitschrift für ausländisches und internationales Privatrecht*, (West) Berlin-Tübingen.
Recueil Belgrade	–	*Recueil des travaux relatifs au droit étranger et droit comparé*, Belgrade, 1966.
RIDC	–	*Revue internationale de Droit comparé*, Paris.
RISS	–	*Revue internationale des Sciences sociales*, Unesco, Paris (edition of *ISSJ* published simultaneously in French).
RRSS-SJ	–	*Revue roumaine des Sciences sociales, Série des sciences juridiques.*
S.g. i p.	–	*Sovetskoe gosudarstvo i pravo* (= Soviet State and Law), Moscow.
SSL	–	*Scandinavian Studies in Law* (annual), Stockholm.
Z	–	*Zeitung, Zeitschrift* (= review, periodical).

—, 'Méthode comparative et droit comparé', in *Liber amicorum Baron Louis Fredericq, op. cit.* (1965), Vol. I, pp. 67 sqq.
—, *Utilité et méthodes du droit comparé*, Neuchâtel, 1971.
—, 'Les aspects nouveaux de l'unification des droits", in *Miscellanea Ganshof van der Meersch, op. cit.* (1972), Vol. II, pp. 695 sqq.
—, 'Les buts actuels de la recherche comparative', in *Mélanges de droit comparé en l'honneur du Doyen Å. Malmström, op. cit.* (1972), pp. 1 sqq.
ANDERSON, J. N. D., *Islamic Law in the Modern World*, London, 1959.
ANDREEV, N. D. & KERIMOV, D. A., 'O vozmožnostijah kibernetiki pri rešenii pravovyh problem' (= On the applicability of cybernetics to the solution of legal problems), *Voprosy filosofii* (= Problems in Philosophy) (Moscow), 1960, No. 7,, pp. 106 sqq.
AOMI, J., 'Japan', in *Trends in Legal Learning, op. cit.* (1970), pp. 378-392.
ARLT, R., 'Zur marxistisch-leninistischen Theorie von der Schaffung und Verwirklichung des sozialistischen Rechts' (= On the Marxist-Leninist theory of the creation and exercise of socialist law), in *Le Raisonnement juridique, op. cit.* (1971).
ARMINJON, P., NOLDE, B. & WOLFF, M., *Traité de droit comparé*, Paris, 1950, 1951, 3 vols.
ARON, R., *Paix et guerre entre les nations*, Paris, 1963 (English translation by R. Howard & A. Baker Fox, *Peace and War. A Theory of International Relations*, London-New York, 1967).
— *et al.*, *La démocratie à l'épreuve du XXe siècle*, Paris, 1960.
ARŽANOV, M. A., *Gosudarstvo i pravo v ih otnošenii* (= State and Law in their Mutual Relationships), Moscow, 1960.
Aspects juridiques du commerce avec les pays d'économie planifiée (Symposium of the International Association of Legal Science, Rome 1958), Paris, 1961.

BAADE, H. W. (ed.), *Jurimetrics*, New York, 1963.
BARATAŠVILI, A. J., *Novye gosudarstva Azii i Afriki i meždunarodnoe pravo* (= The New States of Asia and Africa and International Law), Moscow, 1968.
Begriff und Prinzipien des Wirtschaftsrechts. Siebzehn Landesberichte zu einem internationalen Symposium (= Notion and Principles of Economic Law. Seventeen National Reports Presented at an International Symposium) (G. RINCK, ed.), *Arbeiten zur Rechtsvergleichung* (= Works on Comparative Law), Frankfurt am Main-(West) Berlin, 1971.
Belgrade World Conference on World Peace through Law (1971), announced.
BERGBOHM, K., *Jurisprudenz und Rechtsphilosophie* (= Legal Science and the Philosophy of Law), Leipzig, 1892.
BERMAN, H., *Justice in the USSR*, Cambridge (Mass.), 1963.
BIPOUN-WOUM, J.-M., *Le droit international africain*, Paris, 1970.
BLAGOJEVIĆ, B. T., 'L'unification du droit et notamment du droit privé en Yougoslavie', in *L'unification du droit, Annuaire de l'UNIDROIT 1957*, Rome, 1958.
—, 'La signification sociale contemporaine de l'unification du droit sur le plan international', in *Miscellanea Ganshof van der Meersch, op. cit.* (1972), Vol. II, pp. 717 sqq.
BLIŠČENKO, I. P., *Meždunarodnoe i vnutrigosudarstvennoe pravo* (= International and Municipal Law), Moscow, 1960.
BOBBIO, N., *Teoria dell' ordinamento giurdico* (= Natural Law and Legal Posi-Turin, 1960.
—, *Giusnaturalismo e positivismo giuridico* (= Natural Law and Legal Positivism), Milan, 1965.
BOBOTOV, S. V., 'Vlijanie naučno-tehničeskogo progressa na pravovoe regulírovanie' (= The impact of scientific and technological progress on legal regulation), *S.g. i p.*, 1972, No. 10, pp. 89 sqq.
BOBROV, R. L., *Sovremennoe meždunarodnoe pravo. Ob"ektivnye predposylki i*

social'noe naznačenie (= Contemporary International Law. Its Objective Prerequisites and Its Social Purpose), Leningrad, 1962.

—, *Osnovnye problemy teorii meždunarodnogo prava* (= Basic Problems in the Theory of International Law), Moscow, 1968.

BOGAERT, E. Van: see VAN BOGAERT, E.

Boletín Mexicano de Derecho Comparado, Instituto de Investigaciones Jurídicas, Universidad Nacional Autónoma de México, N.S., 1st Yr., No. 1, Jan.-Apr. 1968, 466 pp. (Studies in Spanish, French and English.) Special issue commemorating the 25th anniversary (7 May 1965) of the Mexican Institute of Comparative Law.

BORKOWSKI, C. *et al.*, 'Structure and effectiveness of the citation identifier; an operational computer program for automatic identification of case citations in legal literature', *Journal of the American Society for Information Science* (Baltimore), 21 (1), 1970, pp. 8 sqq.

BOURQUIN, M., 'Pouvoir scientifique et droit international', in *Recueil des cours de l'Académie de Droit international*, The Hague, 1947, Vol. I, pp. 335-406.

BOUTHOUL, G., 'L'apport de la polémologie à la solution des conflits', in *Guerre et paix*, Vol. 2, Paris, 1967, pp. 4 sqq.

BOWETT, D. W., *The Law of the Sea*, Dobbs Ferry (N.Y.), 1967.

—, *The Law of International Institutions*, London-New York, 1970.

BRATUS, S. N. *et al.*, *Teoretičeskie voprosy sistematizacii sovetskogo zakonodatel'stva* (= Theoretical Questions on the Systematization of Soviet Legislation), Moscow, 1962.

BRATUS, S. N. & SAMOŠČENKO, I. S. (eds.), *Obščaja teorija sovetskogo prava* (= General Theory of Soviet Law), Moscow, 1966.

BRILLOUIN, L., *Science and Information Theory*, New York, 1956, 2nd edn., 1962.

BRIMO, A., *Les grands courants de la philosophie du droit et de l'Etat*, Paris, 1968.

BROWN, E. D., 'Deep sea mining – the legal régime of inner space', in *22nd Yearbook of World Affairs*, London, 1968, pp. 165-190.

BURDEAU, G., *Traité de science politique*, Paris, 1949-57, 7 vols., Vols. I-VI republished 1966-71.

—, *Libertés publiques*, 3rd edn., Paris, 1966.

Buts et méthodes du droit comparé (Istituto di Diritto comparato A. Sraffa, Milan; M. ROTONDI, ed.), Padua-Dobbs Ferry (N.Y.), 1973 (English version: *Comparative Law: Tasks and Methods*, Dobbs Ferry (N.Y.), 1973).

CAEMMERER, E. von, 'Rechtsvergleichung und internationales Privatrecht' (= Comparative law and private international law), in *Problem des europäischen Rechts* (= The Problem of European Law), Frankfurt am Main, 1966, pp. 63 sqq.

CAHIER, Ph., *Le droit diplomatique contemporain*, 2nd edn., Geneva, 1964.

CALDER, N., *The Environment Game*, London, 1967.

CALEWAERT, W., 'Natuurrecht en rechtsvergelejkening' (= Natural law and comparative law), in *Liber Amicorum Baron Louis Fredericq, op. cit.* (1965), Vol. I, pp. 215 sqq.

CAPONERA, D. A., 'Towards a new methodological approach in environmental law', *Natural Resources Journal* (N. Mex.), 2, 1972.

CAPPELLETTI, M., MERRYMAN, J. H. & PERILLO, J. M., *The Italian Legal System (An Introduction)*, Stanford (Calif.), 1967.

CARBONNIER, J., *Flexible droit*, Paris, 1969.

—, 'L'apport du droit comparé à la sociologie juridique', in *Livre du centenaire de la Société de Legislation comparée, op. cit.*, Vol. I (1969).

—, *Sociologie juridique*, Paris, 1972.

CARDOZO, B. N., *The Nature of the Judicial Process*, New Haven (Conn.), 1921.

—, *The Paradoxes of Legal Science*, Westport (Conn.), 1928.

CARNAP, R. & BAR-HILLEL, Y., *An Outline of a Theory of Semantic Information*,

Cambridge (Mass.), 1952.

CARVALHO, W. V. de, *A posicão filosofico-jurídica de Kelsen*, Minas Gerais (Brazil), 1953.

'Cases in Comparative Public Administration', a series of monographs by various authors, International Institute of Administrative Sciences, Brussels.

CASPER, G., *Juristischer Realismus und politische Theorie im amerikanischen Rechtsdenken* (= Legal Realism and Political Theory in American Legal Thinking), Berlin, 1967.

CASSIN, R., 'La déclaration universelle et la mise en œuvre des droits de l'homme', in *Recueil des cours de l'Académie de Droit international*, The Hague, 1951, Vol. II, pp. 241 sqq.

Cassin, René, Amicorum discipulorumque liber: see *Amicorum discipulorumque liber René Cassin.*

CASTBERG, F., *Forelesninger over rettsfilosofi* (= Lectures on Philosophy of Law), Oslo, 1965.

—, *Naturrett og menneskerettigheter*, Oslo, 1967 (English version: 'Natural law and human rights', in *Nobel Symposium 7, International Protection of Human Rights*, Stockholm 1968, and *Revue des Droits de l'Homme/Human Rights Journal* (Paris), I (1), Mar. 1968, pp. 14-39).

—, *La philosophie du droit*, Paris, 1970.

'Centenaire de la Société de Législation Comparée' (International Comparative Law Week, Paris, 9-15 Oct. 1969), *RIDC*, 21st Yr., No. 4, Oct.-Dec. 1969, pp. 703-739.

ČERNILOVSKIJ, Z. M.: see CHERNILOVSKY, Z. M.

CETERCHI, I., 'Le rôle de la dialectique matérialiste dans la pratique du droit', in *Le raisonnement juridique, op. cit.* (1971).

CHABAS, J., 'Transformation du droit local et évolution économique', *Annales africaines* 1, 1962, pp. 151 sqq.

CHARNAY, J.-P., 'Amplitude sociale et logique de la norme musulmane', in *La Règle de droit, op .cit.* (1971).

CHEHATA, Ch., *Etudes de philosophie musulmane du droit = Studia islamica* (Paris), serial XXIII, 1965.

—, *Droit musulman, applications au Proche-Orient*, Paris, 1970.

—, *Etudes de droit musulman*, Paris, 1971.

CHERNILOVSKY (= ČERNILOVSKIJ), 'Contemporary views in the USSR on the origins and rôle of the State', in *Trends in Legal Learning, op. cit.* (1970), pp. 459-475.

ČHIKVADZE, V. M., 'Rol' V. I. Lenina v sozdanii sovetskogo socialističeskogo prava' (= Lenin's rôle in the building of Soviet socialist law), in *Voprosy sovetskogo gosudarstva i prava* (= Questions concerning the Soviet State and Law), Moscow, 1957.

—, 'Lenin, Oktjabr' i juridičeskaja nauka' (= Lenin, October and legal science), in *Oktjabr' i naučnyj progress* (= October and Scientific Progress), Moscow, 1967, pp. 508 sqq.

—, *Gosudarstvo, demokratija, zakonnost'. Leninskie idei i sovremennost'* (= The State, Democracy, Legality. Lenin's Ideas and the Present Day), Moscow, 1967.

—, *Bor'ba za mir – neodolimoe dviženie sovremennosti* (= The Struggle for Peace – An Invincible Movement of the Times), Moscow, 1969.

— *et al.*, *Marksistsko-leninskaja obščaja teorija gosudarstva i prava* (= General Marxist-Leninist Theory of the State and Law), Moscow, 4 vols.: Vol. I, *Osnovnye instituty i ponjatija* (= General Institutes and Basic Notions), 1970; Vol. II, *Istoričeskie tipy gosudarstva i prava* (= Historical Types of State and Law), 1971; Vol. III, *Socialističeskoe gosudarstvo* (= The Socialist State), 1972; Vol. IV, 1973.

— & ZIVS, S. L., *Protiv sovremennogo reformizma i revizionizma v voprose o gosudarstve* (= Against Contemporary Reformism and Revisionism as regards

Questions of the State), Moscow, 1959.
—, 'Sravnitel'noe pravovedenie v praktike meždunarodnogo naučnogo sotrudničestva' (= The comparative science of law and practical international scientific co-operation), *S.g. i p.*, 1966, No. 2, pp. 12 sqq.
—, 'L'évolution de la science juridique et du droit comparé en URSS', in *Livre du Centenaire . . ., op. cit.*, Vol. II (1971), pp. 581 sqq.
ČHIKVADZE, V. M. (ed.), *Voprosy kodifikacii* (= Questions of Codification), Moscow, 1957.
—, *The Soviet State and Law*, Moscow, 1969.
COLLIARD, C. A., *Institutions internationales*, 4th edn., Paris, 1967.
—, *Libertés publiques*, Paris, 1968.
'Colloques' (of IALS): see 'Symposia' of the IALS.
'La Comparaison d'institutions juridiques en matière de contrat et de propriété entre pays à structure économique et sociale différente' ('Rechtsvergleichung zwischen Ländern mit verschiedener Wirtschaftsordnung': Symposium of the IALS, organized by its German National Committee, the Gesellschaft für Rechtsvergleichung, Trier, 1961), *Rabels Z.*, 26th Yr. (1961), serials 2, pp. 201-254, 4, pp. 604-628; 27th Yr. (1962), serials 1, pp. 73-111, 3, pp. 478-518.
Le Concept de la légalité dans les pays socialistes (Symposium of the IALS, Warsaw-Wrocław, 10-16 Sept. 1958) = *Cahiers de l'Académie polonaise des Sciences* (Warsaw), Vol. XXI, 1961.
CONSTANTINESCO, L., 'Les éléments déterminants en tant que critères fondant la science des droits comparés', in *Akrothinia P. G. Vallindas* (= Tribute to P. G. Vallindas), Faculty of Law, Salonika (Greece), 1967.
'La Contribution possible des études juridiques comparatives à une meilleure compréhension entre nations' (Symposium of the IALS, Rome, 27 Sept. 1963), *RIDC*, 1964, No. 1 (Report by A. TUNC, pp. 47 sqq.; discussion, pp. 63 sqq.).
Control of River Pollution by Industry (IALS & International Institute of Administrative Sciences; general rapporteur: J. LITWIN), Brussels, 1964, series 'Cases in Comparative Public Administration' (International Institute of Administrative Sciences).
COULSON, N. J., 'Reform of family law in Pakistan', *Studia Islamica* (Paris), serial VII, 1957, pp. 135-155.
CURRIE, D. P. (ed.), *Federalism and the New Nations of Africa*, Chicago, 1964.

DAINOW, J., 'Le droit civil et la *Common Law*' in *Liber amicorum Baron Louis Fredericq, op. cit.* (1965), Vol. I, pp. 257 sqq.
DAVID, A., *La cybernétique et l'humain*, Paris, 1965.
— *et al.*, 'Plan de recherche concernant la documentation juridique automatique', *Law and Computer Technology* (World Peace Through Law Center's Section on Law and Computer Technology, Washington, D.C.), 1 (1), Jan. 1968, pp. 11-13, and 1 (6), June 1968, pp. 13 sqq. (English abstracts at the end of articles).
DAVID, R., *Traité élémentaire de droit civil comparé*, Paris, 1950.
—, *Les grands systèmes de droit contemporains*, Paris, 1964; 4th edn. (rev., with add.), Paris, 1971. (For an English version, see DAVID, R. & BRIERLEY, J. E. C., *Major Legal Systems in the World Today: An Introduction to the Comparative Study of Law*, London, 1968.)
—, *Le droit anglais*, Paris, 1965.
—, 'The methods of unification', in HAY (ed.), 'The international unification of law', *op. cit.* (1968), pp. 13-27.
—, 'The international unification of private law', in *International Encyclopedia of Comparative Law*, Vol. II, Tübingen-The Hague-New York, 1972, Chap. 5.
—, 'La coopération internationale en matière de droit comparé', in *Mélanges de droit comparé en l'honneur du Doyen Å. Malström, op. cit.* (1972), pp. 58 sqq.
DAVIS, K. C., *Discretionary Justice. A Preliminary Inquiry*, Baton Rouge (La.),

1969.
La Défense de l'homme contre les pollutions – air, bruit, eau (Proceedings of the symposium organized in May 1970 at Royan, France, by the Association pour le Développement du Droit mondial), Paris, 1971.
DEKKERS, R., *Principes nouveaux de droit soviétique*, Brussels, 1962.
DENISOV, A. I. *et al.*, *Teorija gosudarstva i prava* (= Theory of State and Law), Moscow, 1967.
DENORME, R., 'La question des fonds marins et son examen par les Etats-Unis', *Chronique de politique étrangère* (Brussels), 1969, pp. 407-431.
DICEY, A. V., *The Law of the Constitution*, London, 1885, latest edn., 1961.
DIERKENS, R., *Les droits sur le corps et le cadavre de l'homme*, Paris, 1966.
DOXIADIS, C. A. & DOUGLAS, T. B., *The New World of Urban Man*, Philadelphia, 1966.
DROBNIG, U., 'Rechtsvergleichung und Rechtssoziologie' (= Comparative law and sociology of law), *Rabels Z.*, 18th Yr., 1953, pp. 295 sqq.
DWORKIN, G., 'The law relating to organ transplantation in England', *The Modern L.R.* (London), 33 (4), July 1970, pp. 353-377.

EARLE, E. M., 'The impact of scientific discovery and technological change on international relations', in *Proceedings of the 8th Conference of Teachers of International Law and Related Subjects*, Washington (D.C.), 1946.
ECKHOFF, T., 'Sociology of law in Scandinavia', *SSL 1960*, Vol. IV, pp. 29 sqq.
EEK, H., 'Evolution et structure du droit scandinave', *Revue hellénique du Droit international* (Athens), 14, 1961, pp. 33-51.
—, 'The hydrological cycle and the law of nations', *SSL 1965*, Vol. IX, pp. 51-91; simultaneously published as No. 22 of the *Acta Univ. Stockholmiensis, Studia Juridica Stockholmiensia*, Stockholm, 1965.
—, *Världshavens frihet och fred* (= Freedom and Peace on the High Seas), Stockholm, 1971.
—, 'Some reflections on the development of the human rights law and on the impact on that law of scientific developments', *Revue des Droits de l'Homme/Human Rights Journal* (Paris), IV (2-3) (double issue), July 1971, pp. 443-452.
L'Egalité, studies published by H. BUCH and others, Brussels, 1971.
EHRENZWEIG, A. A., *Psychoanalytic Jurisprudence*, Leiden, 1971.
—, 'Malmström's "system of legal systems": an unsystematic comment', in *Mélanges de droit comparé en l'honneur du Doyen Å. Malmström*, *op. cit.* (1972), pp. 734 sqq.
EHRLICH, E., *Grundlegung der Soziologie des Rechts*, Munich-Leipzig, 1913 (English translation by W. L. Moll with an introduction by Roscoe Pound, *Fundamental Principles of the Sociology of Law*, New York, 1962).
EHRLICH, S., 'Kilka uwag w sprawie metodologii nauk prawnych' (= Some thoughts on the methodology of legal science), *Państwo i prawo* (= The State and Law) (Warsaw), 1964, No. 11, pp. 641 sqq.
ELIAS, T. O., *Africa and the Development of International Law*, Leiden, 1972.
ELMANDJRA, M., *The United Nations System. An Analysis*, London-Hamden (Conn.), 1973.
ENGELS, F., *Ludwig Feuerbach und der Ausgang der klassischen deutschen Philosophie*, republished in (East) Berlin, 1951 (English transl. *Ludwig Feuerbach and the Outcome of Classical German Philosophy*, New York, International Publishers Company, 1941).
—, Letter to Kautsky dated 13 June 1891, in MARX, K. & ENGELS, F., *Werke*, Vol. 38, (East) Berlin, 1968, pp. 114 sqq.
ENTAN, L. M., *Nacional'naja gosudarstvennost' narodov zapadnoj i central'noj Afriki* (= The National States of the Peoples of Western and Central Africa), Moscow, 1966.
Entwicklungstendenzen des Völkerrechts (= Development Trends in International

Law) (Proceedings of the International Symposium organized by the Evangelische Akademie in Loccum, Fed. Rep. of Germany, 25-27 Sept. 1965) = *Loccumer Protokolle*, No. 17, Loccum, 1965.

EÖRSI, Gy., 'Jogösszehasonlítás és békés együttélés' (= Comparative law and peaceful co-existence), *Állam-és Jogtudományi* (Budapest), 1964, No. 3, pp. 390 sqq.

ESMEIN, A., 'Le droit comparé et l'enseignement du droit', in CIDC, *Procès-verbaux*, Vol. I, 1, Paris, 1905.

Essays in Jurisprudence in Honor of Roscoe Pound, Indianapolis (Ind.), 1962.

ESSER, J., *Grundsatz und Norm in der richterlichen Fortbildung des Privatrechts* (= Principle and Norm in Private Law as Developed in the Courts), Tübingen, 1956.

Etudes juridiques en l'honneur du Professeur Traian R. Ionasco = *RRSS-SJ* XII (1-2), 1968.

Etudes juridiques offertes à L. Julliot de la Morandière, Paris, 1964.

EYBEN, W. E. von, 'Inter-Nordic legislative co-operation', *SSL 1962*, Vol. VI, pp. 63-93.

FAJZIEV, M., 'F. Engels o sravnitel'nom metode issledovanija estestvennyh i obščestvennyh nauk' (= F. Engels on the comparative method of research in the fields of the sciences of nature and the social sciences), *Uzbekistonda iztimoij fanlar/Obščestvennye nauki v Uzbekistane* (= The Social Sciences in Uzbekistan), 1970, No. 11, pp. 60 sqq.
[See also TILLE & FAJZIEV.]

FAROOQI, Waheed, 'Sociological factors and the law', in *Proceedings of the 5th Annual Conference of the Pakistan Sociological Association*, Karachi, 1969.

Le Fédéralisme et le développement des ordres juridiques/Federalism and the Development of Legal Systems (Symposium of the IALS, Moscow, Sept. 1970), Brussels, 1971.

FELIÚ, J. M., *El mundo jurídico visto por Kelsen*, Santiago (Chile), 1956.

Festschrift für Ernst Rabel (= Tribute to Ernst Rabel), (West) Berlin-Tübingen, 1954, 2 vols.

FIX ZAMUDIO, H., 'Los derechos humanos y su protección ante las jurisdicciones nacionales', in *Miscellanea Ganshof van der Meersch, op. cit.* (1970), Vol. III, pp. 107 sqq.

— *et al.*, 'Mexico', in *Trends in Legal Learning, op. cit.* (1970), pp. 393-421 (translated from the Spanish).

FORKOSCH, M., 'Administrative conduct in environment areas', *South Texas L.J.*, 1970, No. 1.

Les Formes de l'Etat socialiste (Proceedings of the symposium organized in Strasbourg on 1 and 2 Apr. 1966 by the Centre for Research on the USSR and Eastern Countries set up by the Faculty of Law, Political Science and Economics, Strasbourg University), Paris, 1968.

FRAGISTAS, Ch., 'Les conflits de la loi nationale avec les traités internationaux', in *Rapports généraux au VIIe CIDC, op. cit.* (1968), pp. 361 sqq.

FRANK, J., *Law and the Modern Mind*, Magnolia (Miss.), 1930, new edn., 1963.

FRIEDMANN, W., *Legal Theory*, London, 1960.

—, *The Changing Structure of International Law*, London, 1964.

FULLER, L., *The Morality of Law*, New Haven (Conn.), 1969.

The Future of Customary Law in Africa/L'avenir du droit coutumier en Afrique (Colloque/Symposium organized in Amsterdam in 1955 by the Afrika Instituut-studiecentrum of Leiden in co-operation with the Royal Tropical Institute of Amsterdam), Leiden, 1956.

GÁL, G., *Space Law*, Budapest, 1969.

GALTUNG, J., 'On the future of the international system', *Journal of Peace Research*

(Oslo), 1967, No. 4, pp. 305-333.
—, 'Friedensforschung' (= Peace research), in E. Krippendorff (ed.), *Friedensforschung*, Cologne-(West) Berlin, 1968, pp. 519 sqq.
—, 'Peace research', in *The Social Sciences. Problems and Orientations/Les sciences sociales. Problèmes et orientations, op. cit.* (1968), pp. 192-208.
Ganji, M., *International Protection of Human Rights*, Geneva, 1962.
Ganshof van der Meersch, W. J., *Organisations européennes*, Brussels, 1966.
Ganshof van der Meersch Miscellanea: see *Miscellanea W. J. Ganshof van der Meersch.*
Geiger, T., *Vorstudien zu einer Soziologie des Rechts* (= Prolegomena to a Sociology of Law), Copenhagen, 1947.
Geilen, G., *Einwilligung und ärztliche Aufklärungspflicht* (= Consent and the Doctor's Duty to Inform the Patient), Bielefeld, 1963.
Geilke, G., *Einführung in das Sowjetrecht* (= Introduction to Soviet Law), Darmstadt, 1966.
Gellhorn, W., *Ombudsmen and Others. Citizens' Protectors in Nine Countries*, Cambridge (Mass.), 1967.
Gerbert, P., *Les organisations internationales*, Paris, 1966.
Giannini, M. S., 'Italie', in *Tendances de la science juridique, op. cit.* (1970), pp. 462-473 (parallel publication in English transl., 'Italy', in *Trends in Legal Learning, op. cit.* (1970), pp. 420-433).
Gladden, E. N., *The Essentials of Public Administration*, 2nd edn., London, 1958.
Glos, J. *et al.*, *Práva občanů* (= Citizens' Rights), Prague, 1969.
Golovko, A. A., 'Naučnyj èksperiment v gosudarstvennom stroitel'stve' (= Scientific experiment in the development of the State), *S.g. i p.*, 1966, No. 11, pp. 97 sqq.
Gonidec, P.-F., *L'Etat africain*, Paris, 1970.
Goode, H. H. & Machel, R. A., *System Engineering*, New York, 1957.
Goodrich, L. M., *The United Nations*, New York, 1959.
Gorelik, I. I., 'Uslovija pravomernosti iz"jatija transplantata ot neživogo donora' (= Legal conditions for the taking of graft tissue from a deceased donor), *S.g. i p.*, 1971, No. 11, pp. 105 sqq.
Goričar, J., 'Filosofska in sociološka problematika v naši teorii države in prave' (= Philosophical and sociological problems in our theory of the State and law), *Pravnik* (= The Jurist) (Ljubljana), VIII, 1953, pp. 370 sqq.
Gournay, B., *Introduction à la science administrative*, Paris, 1966.
Grafskij, V. G. & Strašun, B. A., *Federalizm v razvivajuščihsja stranah* (= Federalism in Developing Countries), Moscow, 1969.
Granger, R., 'Pour un droit du développement', in *Dix ans de conférences d'agrégation: études de droit commercial offertes à J. Hamel*, Paris, 1961.
—, *Contribution à l'étude du droit du développement* (3 articles, including the foregoing), mimeogr., Faculty of Law and Economic Science of Madagascar, Tananarive, 1970, to be published later in printed form.
Graveson, R. H., 'Philosophy and function of comparative law', *The International and Comparative L.Q.* (London), Vol. VII, Pt. 4, 1958, pp. 649 sqq.
—, 'L'étendue du domaine de l'unification du droit', *RIDC*, 1964, No. 1, pp. 7 sqq.
—, 'The unity of law', in *Liber amicorum Baron Louis Fredericq, op. cit.* (1965), pp. 507 sqq.
—, 'The international unification of law', in Hay (ed.), 'The international unification of law', *op. cit.* (1968), pp. 4-12.
Gutiérrez, C. J., *La teoría pura del derecho de Hans Kelsen*, Costa Rica, 1958.

Habachy, S., 'Middle East law. The link between law and religion', in *Rapports généraux au VIIe CIDC, op. cit.* (1968), pp. 670 sqq.
Haft, F., *Elektronische Datenverarbeitung im Recht* (= Electronic Data-processing in the Field of Law), (West) Berlin, 1972.

HAINES, G. C., *The Revival of Natural Law Concepts*, Cambridge (Mass.), Harvard University Press, 1930.

HALEY, A. G., *Space Law and Government*, New York, 1963.

HALL, A. D., *A Methodology for Systems Engineering*, Princeton, 1963.

HANESSIAN, J., 'The Antarctic Treaty 1959', *The International and Comparative L.Q.* IX, 1960, pp. 436 sqq.

HANEY, G., *Sozialistisches Recht und Persönlichkeit* (= Socialist Law and the Person), (East) Berlin, 1967.

HARGROVE, J. L. (ed.), *Law, Institutions and the Global Environment*, Leiden-Dobbs Ferry (N.Y.), 1972.

HAY, P., 'The Common Market Preliminary Draft Convention on the recognition and enforcement of judgments – some considerations of policy and interpretation', in HAY (ed.), 'The international unification of law', *op. cit.* (1968), pp. 149-174.

HAY, P. (ed.), 'The international unification of law', *AJCL* XVI (1-2) (double issue), 1968, pp. 1-218.

HAZARD, J., 'Classifying Africa's law', in *Recueil Belgrade 1966*, pp. 217 sqq.

—, 'La familia de sistemas jurídicos de inspiración marxista en Africa' in *Libro-Homenaje a la memoria de Roberto Goldschmidt, op. cit.* (1967), pp. 131 sqq.

—, *Communists and Their Law*, Chicago-London, 1969.

—, 'Marxist models for West-African law', in *Ius privatum gentium. Festschrift für Max Rheinstein, op. cit.* (1969), Vol. I, pp. 285 sqq.

—, SHAPIRO, I. & MAGGS, P. B., *The Soviet Legal System*, New York, 1969.

— & WAGNER, W., *Legal Thought in the United States of America under Contemporary Pressures*, Brussels, 1970.

HELLNER, J., 'Unification of law in Scandinavia', in HAY (ed.), 'The international unification of law', *op. cit.* (1968), pp. 88-106.

HENRICH, D., *Einführung in das englische Privatrecht* (= Introduction to English Private Law), Darmstadt, 1971.

HOEBEL, E. A. & LLEWELLYN, K. N., *The Cheyenne Way*, Norman (Oklahoma), 1941.

HOLLOWAY, K., *Modern Trends in Treaty Law*, London, 1967.

HOOK, S. (ed.), *Law and Philosophy. A Symposium*, New York, 1964.

HOOKER, M. B., 'The interaction of legislation and customary law in Malaya', *AJCL* XVI (3), 1968, pp. 415-428.

HOUIN, R., 'Pour une codification européenne du droit des contrats et des obligations', in *Etudes juridiques offertes à L. Julliot de la Morandière, op. cit.* (1964), pp. 223 sqq.

HOWE, M. D., *Justice Oliver Wendell Holmes*, Cambridge (Mass.), 1957-63, 2 vols.

'L'Information juridique et les Droits de l'homme' (round table, Paris, 1970, organized by the *Revue de Droit contemporain* of Brussels), *Revue de Droit contemporain*, 1970, No. 1, pp. 47 sqq.

International Congresses of Comparative Law (ICCL): see *Rapports généraux* ...

International Institute of Administrative Sciences: see 'Cases in Comparative Public Administration'.

Introduction à l'étude du droit comparé. Recueil d'études en l'honneur d'Edouard Lambert, Paris, 1938, 3 vols.

IOFFE, O. S. & ŠARGORODSKIJ, M. D., *Voprosy teorii prava* (= Questions of the Theory of Law), Moscow, 1961.

Ius privatum gentium. Festschrift für Max Rheinstein (= Tribute to Max Rheinstein), Tübingen, 1969, 2 vols.

IVANOV, S. A. (ed.), *Gosudarstvenno-monopolističeskij kapitalizm i buržuaznoe pravo* (= State Monopolistic Capitalism and Bourgeois Law), Moscow, 1969.

JAHRREIS, H., *Mensch und Staat. Rechtsphilosophische, staatsrechtliche und völkerrechtliche Grundfragen in unserer Zeit* (= Man and the State. Basic Questions in the Philosophy of Law, Public Law and International Law in Our Day), Cologne, 1957.

JAKOBS, O. W., 'Zur Methodik der Zivilrechtsvergleichung zwischen Rechten aus verschiedenen Wirtschaftsordnungen' (= Methods of civil law comparison between the legal orders of different economic systems), *Osteuroparecht* (Stuttgart), 1963, No. 9, pp. 108 sqq.

JAMPOLSKAJA, C. A., 'K metodologii nauki upravlenija' (= On the Methodology of administrative science), *S.g. i p.*, 1965, No. 8, pp. 12 sqq.

JENKS, C. W., *The Proper Law of International Organizations*, London-New York, 1962.

—, *Space Law*, London, 1965.

JØRGENSEN, S., 'Grundzüge der Entwicklung der skandinavischen Rechtswissenschaft' (= Principal features of the development of Scandinavian legal science), *Juristen Z.* (Tübingen), 25th Yr., No. 17, 4 Sept. 1970, pp. 529-535.

JOVIČIĆ, M., *Savremeni federalizam. Uporednopravna studija* (= Present-day Federalism. A Study in Comparative Law), Belgrade, 1973.

KAHN-FREUND, O., *Parallelen und Gegensätze im englischen und amerikanischen Privatrecht* (= Parallels and Contradictions in English and American Private Law), Bonn, 1960.

KALENSKÝ, P., *Rada vzájemmé hospodářské pomoci* (= The Council for Mutual Economic Assistance), Prague, 1962.

—, 'Les méthodes de la recherche sociologique en droit comparé', *Travaux de droit comparé*, Institute of the State and Law of the Czechoslovak Academy of Science (Prague), 1970, No. 2, pp. 1 sqq.

KAPLAN, M. A. & KATZENBACH, N. De B., *The Political Foundations of International Law*, New York-London, 1961.

KAZANCEV, N. D. & KOLOTINSKAJA, E. N., *Pravovaja ohrana prirody v SSSR* (= Legal Protection of Nature in the USSR), Moscow, 1962.

KAZIMIRČUK, V. P., *Pravo i metody ego izučenija* (= The Law and Methods for Its Study), Moscow, 1965.

KEČEKJAN, S. & FEDKIN, G., *Istorija političeskih učenij* (= The History of Political Theories), Moscow, 1960.

KEHL, W., *Automatic Data Processing for the Legal Profession*, Boston, 1963.

KEIR, D. L. & LAWSON, F. H., *Cases in Constitutional Law*, Oxford, 1927, latest edn., 1967.

KELLE, V. & OSSIPOV (= OSIPOV), G., 'Matérialisme historique, théorie sociologique et recherche sociale en URSS', in *The Social Sciences. Problems and Orientations/Les sciences sociales. Problèmes et Orientations, op. cit.* (1968), pp. 209-222.

KELSEN, H., *Grenzen zwischen juristischer und soziologischer Methode* (= The Frontiers between the Legal and Sociological Method), Tübingen, 1911.

—, *Reine Rechtslehre. Einleitung in die rechtswissenschaftliche Problematik* (= Pure Theory of Law. Introduction to the Problems of Legal Science), Leipzig-Vienna, 1934 (English translation based on the 2nd edn., rev. and enl., ed. by M. Knight, *Pure Theory of Law*, Berkeley, 1967).

—, *The Law of the United Nations*, London, 1951.

KEMMER, F., 'Das Recht am Leichnam' (= The rights over the corpse), *Deutsche Z. für die gesamte gerichtliche Medizin* (= German Review covering all Branches of Legal Medicine), Berlin-New York, Springer Verlag, 1932, pp. 443 sqq.

KERIMOV, D. A., *Obespečenie zakonnosti v SSSR* (= The Guarantees of Legality in the USSR), Moscow, 1955.

—, 'O predmete obščej teorii gosudarstva i prava', (= The subject-matter of the general theory of State and law), *S.g. i p.*, 1957, No. 12, pp. 3 sqq.

—, 'Grundfragen der Methodologie zur Erforschung staatlicher und rechtlicher

Erscheinungen' (= Basic questions concerning methodology of research on state and legal phenomena), *Staat u. Recht* (= The State and Law), (East) Berlin, 1957, No. 8-9, pp. 867 sqq.
—, 'Fragen der Theorie der sozialistischen Gesetzlichkeit' (= Questions on the theory of socialist legality), *Neue Justiz* (= The New Justice), 13, 1957, pp. 385 sqq.
—, *Fragen der Gesetzgebungstechnik* (= Questions of Legislative Technique), (East) Berlin, 1958.
—, *Staatslehre und Revisionismus* (= The Science of State and Revisionism), (East) Berlin, 1959.
—, *Kodifikacija i zakonodatel'naja tehnika* (= Codification and Legislative Technique), Moscow, 1962.
—, 'Kibernetika i pravo' (= Cybernetics and law), *S.g. i p.*, 1962, No. 11, pp. 98 sqq.
—, 'Pravo i kibernetika' (= Law and cybernetics), *S.g. i p.*, 1964, No. 9, pp. 86 sqq.
KERIMOV, D. A. (ed.), *Zakonodatel'naja tehnika* (= Legislative Technique), Moscow, 1965.
[See also: ALEKSEEV, KERIMOV & NEDBAJLO; ANDREEV & KERIMOV.]
KITAGAWA, Z., *Rezeption und Fortbildung des europäischen Zivilrechts in Japan* (= Reception and Development of European Civil Law in Japan), Frankfurt on the Main, 1970.
KLUG, U., 'Tesis para un análisis de la idea del hombre dentro del derecho' (= Thesis for an analysis of the idea of man within the law) in *Libro-Homenaje a la Memoria de Roberto Goldschmidt, op. cit.* (1967), pp. 75 sqq.
KNAPP, V., 'La gestion de la société socialiste et la théorie de l'information', *Social Science Information/Information sur les sciences sociales* (Paris), II (3), 1963, pp. 70 sqq.
—, 'Zur Frage der Vergleichbarkeit sozialistischen und kapitalistischen Rechts' (= Concerning the question of the comparability of socialist and capitalist law) = Part I of V. KNAPP, 'Verträge im tschechoslowakischen Recht' (= Contracts under Czechoslovak law), *Rabels Z.*, 27th Yr., 1962, serial 3 (May 1963), pp. 495-501 (cf. 'La Comparaison d'institutions juridiques ...', Symposium of the IALS, *op. cit.*).
—, *O možnosti použití kybernetických metod v právu* (= Concerning the Applicability of Cybernetic Methods in the Field of Law), Prague, 1963 (Russian translation, *O vozmožnostijah ispolzovanija kibernetičeskih metodov v prave*, Moscow, 1965).
—, *Filosofické problémy socialistického práva* (= Philosophical Problems of Socialist Law), Prague, 1967.
—, 'Quelques problèmes méthodologiques dans la science du droit comparé', *RRSS-SJ* XII (1), 1968 = *Etudes juridiques en l'honneur du Professeur Traian R. Ionasco, op. cit.*, pp. 75 sqq.
—, 'La technique de la réforme du droit dans les pays socialistes', *RIDC*, 1970, No. 1, pp. 5 sqq.
—, 'La philosophie du droit dans les pays socialistes', in R. KLIBANSKY (ed.), *Contemporary Philosophy. A Survey/La philosophie contemporaine. Chroniques*, Vol. IV, Florence, 1971, pp. 156-169.
—, 'Champ d'application du droit comparé (Le droit comparé interne et le droit étranger)' in *Mélanges de droit comparé en l'honneur du Doyen Å. Malmström, op. cit.* (1972), pp. 132-140.
—, & VRECION, V., 'L'applicabilité des méthodes de la cybernétique au droit et à l'administration', in *Tendances de la science juridique, op. cit.* (1970), pp. 501-508 (parallel publication in English transl., 'The possibility of applying cybernetic methods to law and administration', in *Trends in Legal Learning, op. cit.* (1970), pp. 495-501).
—, 'Research of the applicability of methods of cybernetics in law conducted in Czechoslovakia', *Law and Computer Technology* (Washington, D.C.), 3 (7-8)

(double issue), July-Aug. 1970, pp. 154-162.

KNAPP, V. & SZABÓ, I. (eds.) *Introduction aux systèmes socialistes*, Budapest, 1971.

KNOLL, A. M., *Katholische Kirche und scholastisches Naturrecht* (= The Catholic Church and Scholastic Natural Law), Vienna, 1962.

KOPAL, V., 'Treaty on principles governing the activities of states in the exploration and use of outer space, including the moon and other celestial bodies', *ADAS*, 1968, pp. 463 sqq.

—, 'Dnešní základy a další vývojové tendence kosmického práva' (= Contemporary foundations and future development trends in space law), *Právník* (Prague), 1971, No. 11, pp. 887 sqq.

KORNPROBST, L., *Responsabilités du médecin devant la loi et la jurisprudence françaises*, Paris, 1957.

KOTOK, V. F., 'Socialističeskij demokratizm sovetskogo gosudarstva' (= Socialist democracy in the Soviet State), in *Voprosy sovetskogo gosudarstva i prava* (= Questions concerning Soviet State and Law), Moscow, 1957.

—, 'V. I. Lenin i razvitie sovetskoj socialističeskoj demokratii' (= V. I. Lenin and the development of Soviet socialist democracy), in *Lenin i nauka* (= Lenin and Science), Moscow, 1960, pp. 212 sqq.

—, TADEVOSJAN, V. S. *et al.*, *Ličnost', obščestvo i gosudarstvo* (= The Person, Society and the State), Moscow, 1966.

Kritika buržuazno-reformistskih i revizionistskih teorij v voprosah gosudarstva i prava (= Critique of Bourgeois Reformist and Revisionist Theories concerning Questions of the State and Law), collection, Moscow, 1960.

KRÜGER, H. C., 'The Council of Europe and unification of private law', in HAY (ed.), 'The international unification of law', *op. cit* (1968), pp. 127-148.

KRUSE, H., *Legal Aspects of the Peaceful Utilization of Atomic Energy*, (West) Berlin, 1962.

KRUTOGOLOV, M., 'Le fédéralisme et la structure de l'Etat multinational soviétique', in *Miscellanea Ganshof van der Meersch, op. cit.* (1972), Vol. III, pp. 177 sqq.

KRYLOV, S. B., *Istorija sozdanija Organizacii ob"edinennyh nacij* (= History of the Creation of the United Nations), Moscow, 1960.

KUDRJAVCEV, V. N. (ed.), *Voprosy kibernetiki i pravo*, Moscow, 1967 (English version, *Problems of Cybernetics and the Law*, Moscow, 1968).

KULCSÁR, K., 'Sociologizm v buržuaznoj teorii prava' (= Sociologism in the bourgeois theory of law), in KULCSÁR *et al.*, *Kritika sovremennoj buržuaznoj teorii prava, op. cit.* (1969, Hungarian original 1963), pp. 31 sqq.

— *et al.*, *Kritikai tanulmányok a modern polgári jogelméletről* (= Critical Essays on the Contemporary Bourgeois Theory of Law), collection, Budapest, 1963 (Russian version, *Kritika sovremennoj buržuaznoj teorii prava*, Moscow, 1969).

LACHS, M., *The Law of Outer Space*, Leiden, 1972.

LANG, E., *Staat und Kybernetik* (= The State and Cybernetics), Salzburg, 1966.

LANGROD, G. *et al.*, *Traité de science administrative*, Paris, 1966.

LARENZ, K., *Methodenlehre der Rechtswissenschaft* (= Methodology of Legal Science), (West) Berlin-Göttingen-Heidelberg, 1960, 2nd edn., 1969.

LAUTERPRACHT, H., *International Law and Human Rights*, London, 1950.

LAWSON, F. H., 'Doctrinal writing: a foreign element in English law?', in *Ius privatum gentium. Festschrift für Max Rheinstein, op. cit.* (1969), Vol. I, pp. 191 sqq.

LAWSON, F. H., ANTON, A. E. & NEVILLE BROWN, L. (eds.), *Amos and Walton's Introduction to French Law*, 3rd edn., London, 1967.

[See also KEIR & LAWSON.]

Legal Aspects of Economic Integration/Les aspects juridiques de l'intégration économique (J. RIDEAU, ed.), The Hague, 1973.

Legislación indigenista de México (= Mexican Legislation for the Indians), Mexico City, 1958.

LEIBHOLZ, G. & RINCK, H. J., *Grundgesetz, Kommentar* (= Fundamental Law, Commentary), 3rd edn., Cologne-Marienburg, 1968.

LERNER, M. (ed.), *The Mind and Faith of Mr. Justice Holmes*, New York, 1945.

LÉVY, S., 'Le droit, l'information et la machine', in *Annales du 5e Congrès International de Cybernétique*, Namur, 1967.

LÉVY-BRUHL, H., 'Note sur les contacts entre les systèmes juridiques', in *Symbolae R. Taubenschlag dedicatae*, I, *EOS* 48, 1956, serial 1, Wrocław-Warsaw.

—, *Sociologie du droit*, Paris, 1961.

—, 'L'ethnologie juridique', in POIRIER (ed.), 'Ethnologie juridique', *op. cit.* (1968), pp. 1111-1179.

LÉVY-ULLMANN, H., 'Sur les communications relatives au droit privé dans les pays étrangers', in *Les transformations du droit dans les principaux pays depuis cinquante ans*, Vol. I, Paris, 1922.

Liber amicorum Baron Louis Fredericq, Ghent, 1965.

Libro-Homenaje a la memoria de Roberto Goldschmidt (= Tribute to the Memory of Roberto Goldschmidt), Caracas, 1967.

LIMPENS, J., 'L'étude du droit comparé envisagée comme moyen de rechercher les matières susceptibles d'unification sur le plan international' (General Report to the Fifth International Congress of Comparative Law, Brussels, 4-9 Aug. 1958), in *Rapports généraux au Ve CIDC, op. cit.* (1960).

—, 'Inventaire critique des matières qui ont fait l'objet de réalisations ou de tentatives en vue de l'unification du droit' (General Report to the Sixth International Congress of Comparative Law, Hamburg, 30 July-4 Aug. 1962), in *Rapports généraux au VIe CIDC, op. cit.* (1964).

—, 'Relations entre l'unification au niveau régional et l'unification au niveau universel', *RIDC*, 1964, No. 1, pp. 13 sqq.

—, 'Le problème de la coordination des mouvements d'unification du droit', in *Rapports généraux au VIIe CIDC, op. cit.* (1968), pp. 95 sqq.

LINANT DE BELLEFONDS, Y., 'Immutabilité du droit musulman et réformes législatives en Egypte', *RIDC*, 7th Yr., No. 1, Jan.-Mar. 1955, pp. 5-34.

—, 'Le droit musulman et le nouveau Code civil égyptien' (Report presented at the CIDC held in Paris, 1-8 August 1954), *Annales juridiques, politiques, économiques et sociales, revue de la Faculté de Droit d'Alger*, published by the *Revue algérienne*), 2nd Yr., 1956, pp. 223-235.

—, 'Le Code du Statut personnel irakien', *Studia islamica*, serial VII, 1960, pp. 135 sqq.

—, *Traité de droit musulman comparé*, Paris, 1965, 2 vols.

LIONS, M., *Constitucionalismo y democracia en el Africa recién independizada* (= Constitutionalism and Democracy in Newly Independent Countries in Africa), Mexico City, 1964.

Livre du centenaire de la Société de Législation comparée, 2 vols.: Vol. I, *Un siècle de droit comparé en France. Les apports du droit comparé au droit positif français*, Paris, 1969; Vol. II, *Evolution internationale et problèmes actuels du droit comparé*, Paris, 1971.

LJAPUNOV, A. A. & JABLONSKIJ, S. V., 'Teoretičeskie problemy kibernetiki' (= Theoretical problems of cybernetics), *Problemy kibernetiky* (= Problems of Cybernetics) (Moscow), 1963, No. 9.

LLEWELLYN, K. N., 'A realistic jurisprudence – the next step', *Columbia LR* (New York), 30, 1930, pp. 431 sqq.

—, *Jurisprudence, Realism in Theory and Practice*, Chicago, 1962.
[See also HOEBEL & LLEWELLYN.]

LOEBER, D. A., 'Rechtsvergleichung zwischen Ländern mit verschiedener Wirtschaftsordnung' (= The comparison of law in countries with different economic systems), *Rabels Z.*, 26th Yr., 1961, serial 2, pp. 201-229 (cf. 'La Comparaison d'institutions juridiques ...', symposium of the IALS, *op. cit.*).

LOOSE, W., 'Zur Dialektik der Höherentwicklung der Arbeiterklasse und der sozialistischen Demokratie' (= On the dialectic of the development of the working class and socialist democracy towards a higher level), *Staat und Recht*, 1971, pp. 724-740.

LOSANO, M., *Giuscibernetica* (= Cybernetics of Law), Turin, 1969.

LUBY, Š., 'Transplantácie tkanív a orgánov a vývoj právnych názorov (= Tissue and organ transplants and the development of legal opinions), *Právnické štúdie* (= Legal Studies) (Bratislava), 1969, No. 2, pp. 233 sqq.

LUBY, Š. (ed.), *Le droit civil tchécoslovaque*, Bratislava, 1969.

LUKIĆ, R., 'Uporedno pravo i univerzalno pravo' (= Comparative law and universal law), in *Recueil Belgrade*, 1966, pp. 335 sqq.

LUNEV, A. E., *Demokratizm sovetskogo gosudarstvennogo upravlenija* (= Democracy in the State Administration of the USSR), Moscow, 1967.

— *et al.*, *Naučnye osnovy gosudarstvennogo upravlenija v SSSR* (= Scientific Bases of the State Administration in the USSR), Moscow, 1968.

LYON-CAEN, G., 'Table ronde sur les études et les recherches de droit comparé en URSS', *RIDC* 1, 1964, pp. 69 sqq.

MCCORMICK, R. J., *Human Engineering*, New York, 1957.

MCDOUGAL, M., LASSWELL, H. D. & VLASIC, I. A., *Law and Public Order in Space*, New Haven (Conn.), 1963.

MACKINNON, V. S., *Comparative Federalism. A Study in Judicial Interpretation*, The Hague, 1964.

MCNAIR, A. D., Lord, *The Law of Treaties*, Oxford, 1961.

MCWHINNEY, E. & BRADLEY, M. A., *New Frontiers in Space Law*, Leiden, 1969.

MÁDL, F., 'Pravovye problemy složenija èkonomičeskoj integracii v stranah SEV' (= Legal problems of economic integration in the COMECON countries), *Acta Iuridica Academiae Scientiarum Hungaricae* (Budapest), 12, 1970, pp. 103 sqq.

MAINE, Sir Henry, *Ancient Law*, London, 1861, new edn., 1930.

MAKAROV, A. D., POPOV, S. I. & TERJAEV, G. V. (eds.), *Marksistsko-leninskaja filosofija. Istoričeskij materializm* (= Marxist-Leninist Philosophy. Historical Materialism), Moscow, 1970.

MALCEV, G. V., *Socialističeskoe pravo i svoboda ličnosti* (= Socialist Law and the Liberty of the Person), Moscow, 1968.

—, 'Gosudarstvo i ličnost' pri socializme. K kritike antikommunističeskih koncepcij' (= The State and the person in socialism. Contribution to the critique of anti-communist conceptions), *S.g. i p.*, 1970, No. 8, pp. 13 sqq.

MALHERBE, J., *Médecine et droit moderne*, Lyons, 1969.

MALMSTRÖM, Å., 'The system of legal systems: notes on a problem of classification in comparative law', *SSL 1969*, Vol. XIII, pp. 129 sqq.
[See also *Rapports généraux au VIIe CIDC*.]

Malmström, Mélanges en l'honneur de: see *Mélanges de droit comparé en l'honneur du Doyen Å Malmström.*

MANDL, V., *Das Weltraumrecht. Ein Problem der Raumfahrt* (= Space Law. A Problem relating to Space Travel), Mannheim-Berlin-Leipzig, 1932.

MARSHALL, H. H., *Natural Justice*, London, 1959.

MARX, K. & ENGELS, F., *Manifest der Kommunistischen Partei* (= Manifesto of the Communst Party) (1848), reprint, (East) Berlin, 1970.

'Mathematical methods of determining alimony for minors', *Law and Cybernetics* (bulletin published by the Institute of State and Law of the Czechoslovak Academy of Sciences, Prague), 2, 1968.

MAYDA, J., *Environment and Resources: From Conservation to Ecomanagement*, Puerto Rico, 1968.

—, 'Quelques réflexions critiques sur le droit comparé contemporain', *RIDC*, 1970, No. 1, pp. 57 sqq.

M'BAYE, K., 'Droit et développement en Afrique francophone de l'Ouest', in

TUNC (ed.), *Les aspects juridiques du développement économique, op. cit.* (1966), pp. 121-165, and *Revue sénégalaise de droit*, 1967, No. 1, pp. 23 sqq.

M'BAYE, K. (ed.), *Le droit de la famille en Afrique noire et à Madagascar*, Paris, 1968.

MEEKINSON, J. P. (ed.), *Canadian Federalism: Myth or Reality*, Toronto, 1968.

MEHREN, A. von (ed.), *The Law in Japan. The Legal Order in a Changing Society*, Cambridge (Mass.), 1963.

Mélanges de droit comparé en l'honneur du Doyen Å. Malmström, Stockholm, 1972.

MENCER, G., *Mezinárodně právní problémy Antarktidy* (= Problems of International Law concerning the Antarctic), Prague, 1963.

MERRITT, R. L. & ROKKAN, S. (eds.), *Comparing Nations: The Use of Quantitative Data in Cross-National Research*, New Haven (Conn.), Yale University Press, 1966.

Metodologičeskie problemy sovetskoj juridičeskoj nauki (= Methodological Problems of Soviet Legal Science), collection, Kiev, 1965.

Metody badania prawa. Materialy sympozjum Warszawa 28-29 IV 1971 (= The Methods of Legal Science Research. Documents of the Warsaw Symposium, 28-29 Apr. 1971) (A. LOPATKA, ed.), Warsaw, 1973.

MEYRET, J. (ed.), *Les organisations régionales internationales*, Paris, 1971.

MILLIOT, L., *Introduction à l'étude du droit musulman*, Paris, 1953.

Miscellanea W. J. Ganshof van der Meersch. Studia ab discipulis amicisque in honorem egregii professoris edita, Brussels, 1972, 3 vols.

MOCHI-ONORY, A. G., 'Les aspects récents du problème de la succession aux traités', *Revue générale de Droit international public* (Paris), 1968, pp. 565-655.

MORAWIECKI, W., *Organizacje międzynarodowe* (= International Organizations), 2nd edn., Warsaw, 1969.

MOROZOV, G. I., *Meždunarodnye organizacii. Nekotorye voprosy teorii* (= International Organizations. Some Questions of Theory), Moscow, 1969.

MORSTEIN-MARX, F., *Verwaltung. Eine einführende Darstellung* (= An Introduction to Administration), (West) Berlin, 1965.

MORSTEIN-MARX, F. (ed.), *Elements of Public Administration*, New York, 1948.

MOVČAN, A. P., 'Pravovoj statut Antarktiki – Meždunarodnaja problema' (= The legal régime of the Antarctic – an international problem), *Ežegodnik*, 1959, pp. 342 sqq.

MUNCH-PETERSEN, H., 'Main features of Scandinavian law', *LQ Review* (London), 43 (171), July 1927, pp. 366-377.

NADELMANN, K. H., 'Uniform legislation *versus* international conventions revisited', in HAY (ed.), 'The international unification of Law', *op. cit.* (1968), pp. 28-50.

NASCHITZ, A., '"Le problème du droit naturel" à la lumière de la philosophie marxiste du droit', *RRSS-SJ*, 1966, No. 1.

—, 'Nekotorye soobraženija o sootnošenii filosofii prava i obščej teorii prava' (= Some comments on the relationship between the philosophy of law and the general theory of law), in *Aktuelle Probleme der marxistisch-leninistischen Staats- und Rechtstheorie* (= Current Problems in the Marxist-Leninist Theory of State and Law), *op. cit.* (1968), pp. 140 sqq.

Les Nations Unies face à un monde en mutation (Proceedings of a symposium organized by the University of Geneva on 30 and 31 Oct. 1970; M. VIRALLY, ed.), Geneva, 1970.

Natural Law Institute Proceedings, College of Law, University of Notre Dame, Notre Dame, Ind., 1947-51.

NEDBAJLO, P. E., 'Sovetskaja socialističeskaja zakonnost', eë osnovnye principy naznačenija (= Soviet socialist legality, its basic principles and goals), *Annales de l'université de Lvov, Série juridique*, 1954, No. 2, pp. 5 sqq.

—, 'Meždunarodnaja zaščita prav čeloveka' (= International protection of human

rights), *Ežegodnik*, 1968, pp. 34 sqq.
[See also ALEKSEEV, KERIMOV & NEDBAJLO.]

NIKITINSKIJ, V. I., 'Značenie èksperimenta v normotvorčeskoj dejatel'nosti' (= The meaning of experiment in legislation), *S.g. i p.*, 1967, No. 6, pp. 26 sqq.

NIZSALOVSKY, E., 'L'incidence de la biologie et de la médecine modernes sur le droit civil', in *Droit hongrois – Droit comparé*, Budapest, 1970, pp. 107 sqq.

—, *A szerv- es szövetátültetések joga* (= The Law with Respect to Organ and Tissue Transplants), Budapest, 1970.

NODA, Y., *Introduction au droit japonais*, Paris, 1966.

OFFNER, J., 'Experiment im Recht' (= Experiment in law) in *Recht und Gesellschaft* (= Law and Society), Vienna-Leipzig, 1931.

OPALEK, K., *Metodologiczne problemy prawoznawstwa* (= Methodological Problems in Legal Research), Warsaw, 1962.

— & ZAKRZEWSKI, W., *Z zagădnień praworzadności socjalistycznej* (= Problems of Socialist Legality), Warsaw, 1962.

Les Organisations régionales internationales, collection, Paris, 1971, 2 vols.

OSTROVSKIJ, R. A., *OON i prava čeloveka* (= The United Nations and Human Rights), Moscow, 1969.

PANICKOV, N. V., 'Social'nyj èksperiment i naučnoe rukovodstvo razvitiem socialističeskogo obščestva' (= Social experiment and scientific administration of the development of socialist society), in *Naučnoe upravlenie obščestvom* (= Scientific Administration of Society), Moscow, 1967.

PARDO, A., 'Who will control the sea-bed?', *47 Foreign Affairs* (New York), 1968, No. 1, pp. 123 sqq.

PASSERIN D'ENTRÈVES, A., *Natural Law*, London, 1950.

PATJULIN, V. A., 'Ličnost', eë prava i svobody v sovetskom socialističeskom gosudarstve' (= The person, his rights and liberties in the Soviet socialist State), *S.g. i p.*, 1966, No. 2, pp. 99 sqq.

PEARSON, J. D. (ed.), *Index islamicus, 1906-1955* (pp. 101-141 relate to Moslem law); *Supplement 1956-1960* (Nos. 876-1294 deal with Moslem law); *Second Supplement 1961-1965*, Cambridge, 1958, 1962, 1967.

PELLOUX, R., *Le citoyen devant le droit*, Paris, 1955.

PERALTA GARCÍA, A., *En torno al idealismo crítico. Un ensayo sobre Kelsen*, Mexico City, 1957.

'Perspectives et limites du droit comparé' (Budapest Round Table, 9-11 Sept. 1969), *RIDC*, 1970, No. 1, pp. 85 sqq.

PÉTERI, Z., 'Influence of natural law on positive law', in *Etudes de droit comparé*, Budapest, 1966.

—, 'O nekotoryh čertah doktriny "vozroždennogo" estestvennogo prava' (= On some aspects of the 'renaissance' of natural law), in KULCSÁR *et al.*, *Kritika sovremennoj buržuaznoj teorii prava, op. cit.* (1969, Hungarian original 1963), pp. 118 sqq.

—, 'Hungary', in *Trends in Legal Learning, op. cit.* (1970), pp. 434-442.

PINO MUÑOZ, J., *El estudio sobre la teoría del derecho*, Concepción (Chile), 1959.

PINTO, R., *Les organisations européennes*, Paris, 1963.

PIONTKOVSKIJ, A., 'Ukreplenie socialističeskoj zakonnosti i nekotorye voprosy teorii sovetskogo prava' (= Reinforcement of socialist legality and some problems in the theory of Soviet law), *Voprosy filosofii* (= Problems in Philosophy) (Moscow), 1955, No. 1, pp. 17 sqq.

POIRIER, J., 'Introduction à l'ethnologie de l'appareil juridique', in POIRIER (ed.), 'Ethnologie juridique', *op. cit.* (1968), pp. 1091-1110.

—, 'Situation actuelle et programme de travail de l'ethnologie juridique', in *Tendances de la science juridique, op. cit.* (1970), pp. 509-527 (parallel publication in English translation, 'The current state of legal ethnology and its future tasks', in *Trends in Legal Learning, op. cit.* (1970), pp. 476-494).

POIRIER, J. (ed.), 'Ethnologie juridique', in J. POIRIER (ed.), *Ethnologie générale*, Encyclopédie de la Pléiade, Paris, 1968, pp. 1091-1246 (articles by POIRIER, LÉVY-BRUHL and ALLIOT, *op. cit.*; and Bibliography, pp. 1236-1246).

POLAK, K., *Zur Dialektik in der Staatslehre* (= On Dialectics in the Science of the State), 3rd edn., (East) Berlin, 1963.

POPOV, P. N., *Kritika na savremenija buržuazen praven normativizm* (= A Critique of Contemporary Bourgeois Legal Normativism), Sofia, 1964.

POUND, R., 'The need for a sociological jurisprudence', *Green Bag* (Boston, Mass.), 19, 1909, pp. 607 sqq.

—, 'Soziologische Jurisprudenz in Amerika', *Jahrbuch für Soziologie* (Karlsruhe), 5, 1925, pp. 88 sqq.

Pravovye voprosy naučno-tehničeskogo progressa v SSSR (= Legal Questions relating to Scientific and Technical Progress in the USSR), collection, Moscow, 1967.

Problèmes actuels du rapprochement régional des droits nationaux (Symposium of the IALS, Copenhagen, 1969), publication announced.

Problèmes contemporains de droit comparé (collection of articles published by the Japanese Institute of Comparative Law, Chuo University), Tokyo, 1962, 2 vols.

Problèmes de protection internationale des droits de l'homme = Vol. I (1969) of *Amicorum discipulorumque liber René Cassin, op. cit.* (1969-71).

Problèmes juridiques nés du développement et de l'utilisation de l'énergie atomique (Proceedings of the Seventh Congress of the International Association of Democratic Lawyers), Brussels, 1960.

Proceedings of the Conference on Legal Education in Africa held in Addis Ababa, Ethiopia, Haile Selassie I University, 20-24 October 1968 (J. VANDERLINDEN, ed.), mimeogr., Centre for African Legal Development, Faculty of Law, Addis Ababa, 1969.

Protiv sovremennoj pravovoj ideologii imperializma (= Against the Contemporary Legal Ideology of Imperialism), collection, Moscow, 1962.

RADBRUCH, G., 'Gesetzliches Unrecht und übergesetzliches Recht' (= Legal injustice and supralegal law), *Süddeutsche Juristen Z.* (= Legal Review of Southern Germany) (Heidelberg), 1946, No. 1, pp. 105 sqq.

Le Raisonnement juridique/Legal Reasoning/Die juristische Argumentation, Actes du Congrès Mondial de Philosophie du droit et de Philosophie sociale/ Proceedings of the World Congress for Legal and Social Philosophy, Bruxelles, 30 août-3 septembre 1971 (A. HUBIEN, ed.), Brussels, 1971.

RAMA RAO, T. S., *Codification and Unification of Indian Law*, Rome, UNIDROIT, 1963.

RAMM, T., *Einführung in das Privatrecht. Allgemeiner Teil des BGB* (= Introduction to Private Law. General Part of the Civil Code), Munich, 1969.

Rapports généraux au Ve CIDC (Bruxelles, 4-9 août 1958), Brussels, 1960.

Rapports généraux au VIe CIDC (Hamburg, 1962), Brussels, 1964.

Rapports généraux au VIIe CIDC (Uppsala, 6-13 août 1966) (Å. MALMSTRÖM & S. STRÖMHOLM, eds.), Stockholm, 1968.

Rapports généraux au VIIe CIDC (Pescara, 1970), Pescara, to be published.

RARIJAONA, R., 'Le droit du développement à la recherche de son expression', *Rec. Penant 1968*, pp. 539 sqq.

Razvitie prav graždan SSSR i usilenie ih ohrany na sovremennom ètape kommunističeskogo stroitel'stva (= The Development of Citizens' Rights in the USSR and the Reinforcement of Their Protection in the Present Period of the Building of Communism), collection, Saratov, 1962.

Realidades y proyectos, 16 años de trabajo (= Reality and Plans, 16 Years of Work), collection, Mexico City, 1964.

RECASÉNS SICHES, S., *Vida humana, sociedad y derecho* (= Human Life, Society

and Law), Mexico City, 1945.
Recueil des travaux relatifs au droit étranger et droit comparé (on the occasion of the tenth anniversary of the Belgrade Institute of Comparative Law), Belgrade, 1966.
La Règle de droit (studies published by Ch. PERELMAN), Brussels, 1971.
REICH, N., *Sociological Jurisprudence und Legal Realism im Rechtsdenken Amerikas* (= Sociological Jurisprudence and Legal Realism in American Legal Thinking), Heidelberg, 1967.
RHEINSTEIN, M., 'Law and social changes in Africa', *Washington Univ. LQ*, 1962, pp. 443 sqq.
—, 'Karl Nickerson Llewellyn, 1893-1962', *Rabels Z.*, 27th Yr., serial 4 (Dec. 1963), pp. 601-605.
—, 'Problems of law in the New Nations of Africa', in C. GEERTZ (ed.), *The Quest for Modernity in Asia and Africa*, New York, 1963.
—, 'Comparative law and legal systems', in D. L. SILLS (ed.), *International Encyclopedia of the Social Sciences*, Vol. 9, New York, 1968, pp. 204 sqq.
—, 'Die Rechtshonoratioren und ihr Einfluss auf Charakter und Funktion der Rechtsordnungen' (= Leader groups in law and their influence on the character and function of legal systems), *Rabels Z.* 34 (1), 1970, pp. 1-13 (adapted English version, 'Leader groups in American law', *University of Chicago LR* 38 (4), Summer 1971, pp. 687-696).
—, 'United States of America', in *Trends in Legal Learning, op. cit.* (1970), pp. 443-458.
Rheinstein, Max, Festschrift für (= volume in tribute to): see *Ius privatum gentium. Festschrift für Max Rheinstein.*
ROBERTSON, A. H., *European Institutions*, London, 1959.
RODIÈRE, R., *Aspects du droit de l'énergie atomique*, Paris, 1965.
RODIÈRE, R. (ed.), *Introduction au droit comparé*, Barcelona, 1967 (published in French and Spanish).
ROGON, M., *Les cités de l'avenir*, Paris, 1968.
ROMANO, Santi, *L'ordinamento giuridico* (= The Legal Order), Florence, 1947.
ROMAŠKIN, P. S. *et al.*, *Teorija gosudarstva i prava* (= Theory of State and Law), Moscow, 1962.
ROMMEN, H., *Die ewige Wiederkehr des Naturrechts* (= The Eternal Reappearance of Natural Law), Munich, 1947.
—, 'In defense of natural law', in HOOK (ed.), *Law and Philosophy, A Symposium, op. cit.* (1964).
ROT, H., *Jedność i podialy systemu prawa socjalistycznego* (= The Unity and Division of the Socialist Legal System), Wrocław, 1971.
ROUSSEAU, Ch., 'Scientific progress and the evolution of international law', *Impact of Science on Society* (Paris, Unesco), V (2), 1954, pp. 71-92.
ROZIN, E., *Vozniknovenie marksistskogo učenija o gosudarstve i prave* (= The Origin of the Marxist Science of the State and Law), Moscow, 1967.
ROZMARYN, S., 'Quelques questions de la théorie des constitutions socialistes', in *Liber amicorum Baron Louis Fredericq, op. cit.* (1965), Vol. II, pp. 888 sqq.
—, 'Les grandes controverses du droit comparé', *Annuaire de l'Institut universitaire d'Etudes européennes*, Turin, 1967.
ROZMARYN, S. (ed.), *Introduction à l'étude du droit polonais*, Warsaw, 1969.
Rule of Law as Understood in Western Countries/Le concept de la légalité dans les pays occidentaux (Symposium of the IALS, Chicago 1957) = Annales de la Faculté de Droit de l'Université d'Istanbul, IX (12), 1959.
RYBICKI, Z., 'L'administration publique dans l'Etat socialiste face à la révolution scientifique et technique', in *Miscellanea Ganshof van der Meersch, op. cit.* (1972), Vol. III, p. 641 sqq.

ŠAFIR, M. A., *Kompetencija SSSR i sojuznoj respubliki* (= The Competence of the

USSR and of the Federated Republic), Moscow, 1968.

SAMOŠČENKO, I. S. *et al., Sootnošenie obščesojuznogo i respublikanskogo zakonodatel'stva* (= The Mutual Relationship between the Legislation of the Union and that of the Republics), Moscow, 1967.

SAVATIER, J., '*Et in hora mortis nostrae.* Les problèmes des greffes d'organes prélevés sur un cadavre', *Rec. Dalloz-Sirey* 16, 1968, pp. 91 sqq.

SAX, J. L., 'Legal redress of environmental disruption in the United States: the rôle of courts', *Social Science Information/Information sur les Sciences sociales* (Paris), IX (5), 1970, pp. 7 sqq.

SCHACHT, J., *An Introduction to Islamic Law*, Oxford, 1964.

SCHERMERS, H. G., *The Law of International Organizations*, Leiden, 1972, 2 vols.

SCHMITTHOFF, C., 'The science of comparative law', *The Cambridge LJ* VII, 1941, pp. 102 sqq.

SCHNITZER, A. F., *Vergleichende Rechtslehre* (=The Comparative Science of Law), Basel, 1945, 2nd edn., 1961.

SEARA VÁSQUEZ, M., *Introducción al derecho internacionel cósmico* (= Introduction to International Space Law), Mexico City, 1961.

SEIDLER, G. L., *Doktryny prawne imperializmu* (= The Legal Doctrines of Imperialism), Cracow, 1957.

—, 'Odrodzenie prawa natury v amerykańskiej jurisprudencji' (= The Renaissance of natural law in American legal science), *Państwo i prawo* (= The State and Law) (Warsaw), 1961, No. 6, pp. 873 sqq.

SEIDL-HOHENVELDERN, I., *Das Recht der internationalen Organisationen einschliesslich der supranationalen Gemeinschaften* (= The Law of International Organizations including Supranational Communities), Cologne, 1967.

SERRANO VILLAFAÑE, E., 'L'apport du droit naturel au droit positif' (summary), in *Travaux du Ier Congrès international de Droit comparé de l'AISJ (Barcelone, Sept. 1956) = Revista del Instituto de Derecho Comparado* (Barcelona), 1957, No. 8-9, pp. 73 sqq.

SHARMA, G. S., 'Common Law of the Common Wealth', *Jaipur LR* 34 (6), 1966.

SIMMONDS, K. R., 'The Antarctic Treaty 1959', *Journal du droit international* (Paris), 1960, pp. 668-701.

ŠLJAHOV, A. R., 'De l'application de la cybernétique au droit en URSS', *Revue de Droit contemporain* (Brussels), 1970, No. 1, pp. 71 sqq.

SMITH, D. D., *International Telecommunication Control: International Law and the Ordering of Satellite and Other Forms of International Broadcasting*, Leiden, 1969.

The Social Sciences. Problems and Orientations/Les sciences sociales. Problèmes et orientations (selected studies written in connexion with the first part of the survey undertaken by Unesco on the main trends of research in the social sciences and humanities), The Hague-Paris, 1968.

Some Problems of Non-performance and Force Majeure in International Contracts of Sale/Problèmes de l'inexécution et [de] *la force majeure dans les contrats de vente internationale* (Symposium of the IALS, Helsinki 1960) = *Studia Iuridica Helsingiensia* (Helsinki), Vol. 2, 1961.

ŠORIN, V. G. (ed.), *Pravovye i social'no-psihologičeskie aspekty upravlenija* (= Legal and Socio-psychological Aspects of Administration), Moscow, 1972.

Sostojanie i zadači sravnitel'nogo izučenija zakonodatel'stva sojuznyh respublik. Tezisy dokladov, rekomenduemaja tematika i bibliografija (= Present State and Problems in the Comparative Study of the Laws of the Union Republics. Summaries of Reports, Recommended Topics and Bibliography), discussion organized in Moscow on 26 and 27 January 1970 by VNIISZ (Federal Institute for Research on Soviet Legislation), Moscow, 1970.

The Sources of the Law of International Trade, with special reference to East-West Trade (Symposium of the IALS, London, September 1962; C. SCHMITTHOFF, ed.), London, 1964.

SPIRKIN, A. G., *Kurs marksistskoj filosofii* (= Course in Marxist Philosophy), 2nd edn., Moscow, 1966.

STAROSCIAK, J., *Elementy nauki upravlenija* (= Fundamentals of Administrative Science), Moscow, 1965.

—, *Zarys nauki administracje* (= Treatise on Administrative Science), Warsaw, 1966.

STEPNITZ, D. von, *Wissenschaft und Recht. Der Einfluss von Naturwissenschaft und Mathematik auf Rechtswissenschaft in zweieinhalb Jahrtausenden. Ein historisches Grundriss* (= Science and Law. The Influence of Natural Science and Mathematics on Legal Thought and Legal Science in the Course of Two and a Half Millennia. A Historical Treatise), (West) Berlin, 1970.

STROGOVIČ, M. M., 'Metodologičeskie voprosy juridičeskoj nauki' (= Methodological problems in legal science), *Voprosy filosofii* (= Problems in Philosophy) (Moscow), 1965, No. 12, pp. 3 sqq.

—, 'Filosofija i pravovedenie' (= Philosophy and legal science), *S.g. i p.*, 1965, No. 6.

—, *Osnovnye voprosy sovetskoj socialističeskoj zakonnosti* (= Basic Problems in Soviet Socialist Legality), Moscow, 1966.

STROGOVIČ, M. M. (ed.), *Pravovye garantii zakonnosti v SSSR* (= The Legal Guarantees of Legality in the USSR), Moscow, 1962.

Stručný filosofický slovník (= Short Philosophical Encyclopaedia), Prague, 1966.

SUMMERS, R. S., 'The New Analytical Jurists', *New York University Review*, Vol. 41, 1966, pp. 861 sqq.

SVOBODA, K., *La notion du droit économique*, Nancy, 1966.

SY, S. M., *Recherches sur l'exercice du pouvoir politique en Afrique noire (Côte d'Ivoire, Guinée, Mali)*, Paris, 1965.

'Symposia' of the IALS: See
— *Aspects juridiques du commerce* ...;
— 'La Comparaison d'institutions juridiques ...';
— *Le Concept de la légalité dans les pays socialistes*;
— 'La Contribution possible des études juridiques comparatives ...';
— *Le Fédéralisme et le développement des ordres juridiques*;
— *Rule of Law as Understood in Western Countries*;
— *Some Problems of Non-performance and Force Majeure* ...;
— *The Sources of the Law of International Trade* ...;
— *Unification of the Law Governing International Sales of Goods* ....

'Symposium on social science approaches to the judicial process', *Harvard LR* 79, 1966, pp. 1551 sqq.

SZABÓ, I., *A szocialista jog* (= Socialist Law), Budapest, 1963 (Russian version, *Socialističeskoe pravo*, Moscow, 1964).

—, 'La science comparative du droit', *Annales Universitatis Budapestinensis – Iuridica* V, 1964, pp. 91 sqq.

—, 'Le citoyen et le droit', *Acta Iuridica Academiae Scientiarum Hungaricae* 8, 1966, pp. 1 sqq.

—, 'La démocratie socialiste et le droit', *Annales Universitatis Budapestinensis – Iuridica* VII, 1966, pp. 101 sqq.

—, *Az emberi jogok* (= Human Rights), Budapest, 1968.

—, 'Gegenstand der maxistisch-leninistischen Rechtstheorie' (= Subject-matter of the Marxist-Leninist theory of law), in *Aktuelle Probleme der marxistisch-leninistischen Staats- und Rechtstheorie, op. cit.* (1968), pp. 18 sqq.

—, 'Les constitutions socialistes et les droits des citoyens', *RRSS-SJ* XII (1), 1968 = *Etudes juridiques en l'honneur du Professeur Traian R. Ionasco, op. cit.*, pp. 119 sqq.

—, 'Sravnitel'noe pravovedenie' (= Comparative jurisprudence), in KULCSÁR *et al.*, *Kritika sovremennoj buržuaznoj teorii prava, op. cit.* (1969, Hungarian original 1963), pp. 165 sqq.

—, *A jogelmélet alapjai*, Budapest, 1971 (French version, *Les fondements de la théorie du droit*, Budapest, 1973).
—, 'Le droit comparé et les droits de l'homme', in *Miscellanea Ganshof van der Meersch, op. cit.* (1972), Vol. II, pp. 925 sqq.
—, 'Law theory and comparative law', in *Mélanges de droit comparé en l'honneur du Doyen Å. Malmström, op. cit.* (1972), pp. 243 sqq.
SZABÓ, I. & KOVÁCS, I. (eds.), *Socialist Concept of Human Rights*, Budapest, 1966. [See also KNAPP & SZABÓ (eds.).]
SZALAI, A., 'The future of international organizations', *Information sur les Sciences sociales/Social Science Information* (Paris), X (3), 1971, pp. 151 sqq.

'Table ronde sur les études et les recherches de droit comparé en URSS' (organized in Paris, on 3 May 1963, by the Section d'études de droit des pays de l'Europe de l'Est of the Société de Legislation comparée and the Institute of Comparative Law of the University of Paris): see LYON-CAEN, G.
TANIGUCHI, T., 'La loi et la coutume au Japon', in *Etudes juridiques offertes à L. Julliot de la Morandière, op. cit.* (1964), pp. 571 sqq.
TAPPER, C. F. H., 'Legal information and computers: Great Britain', *Law and Computer Technology* (Washington), I (1), Jan. 1968, pp. 18-19.
Les Télécommunications par satellites. Aspects juridiques, Paris, CNRS, 1968.
Tendances de la science juridique: see *Trends in Legal Learning*.
'Tendencies and Functions of Comparative Law in Contemporary Society (International Round-Table Conference in Budapest, 9-12 Sept. 1969)', *Acta Juridica Academiae Scientiarum Hungaricae* (Budapest), 13, 1971, pp. 131 sqq.
TER HAAR, B., *Adatrecht in Indonesia* (= *Adat* Law in Indonesia), Djakarta, 1962.
TERRÉ, F., 'Les méthodes de recherche sociologique en droit comparé', *Etudes de droit contemporain (Nouvelle série)*, Paris, 1970.
THIEME, H., 'L'apport du droit naturel au droit positif', in *Rapports généraux au VIIe CIDC, op. cit.* (1968), pp. 74 sqq.
TILLE, A. A. & FAJZIEV, M., 'Sravnitel'nyj metod v pravovoj nauke' (= The comparative method in legal science), *Uzbekistonda iztimoij fanlar/Obščestvennye nauki v Uzbekistane* (= The Social Sciences Uzbekistan), 1969, No. 10, pp. 28 sqq.
—, 'Iz istorii sravnitel'nogo pravovedenija' (= Concerning the history of comparative law), *Uzbekistonda iztimoij fanlar/Obščestvennye nauki v Uzbekistane* (= The Social Sciences in Uzbekistan), 1970, No. 5, pp. 38 sqq.
— & ŠVEKOV, G. V., *Sravnitel'nyj metod v juridičeskih disciplinah* (= The Comparative Method in Legal Disciplines), Moscow, 1973.
Trends in Legal Learning = *ISSJ* XII (3), 1970 (special issue), pp. 363-501 (parrallel publication in French, *Tendances de la science juridique* = *RISS*, same ref., pp. 401-545).
Tribute (Volumes in): see
— *Amicorum discipulorumque liber René Cassin*
— CONSTANTINESCO (*Akrothinia P. G. Vallindas*)
— *Essays in Jurisprudence in Honor of Roscoe Pound*
— *Etudes juridiques en l'honneur du Professeur Traian A. Ionasco*
— *Etudes juridiques offertes à L. Julliot de la Morandière*
— *Festschrift für Ernst Rabel*
— *Introduction à l'étude du droit comparé. Recueil d'études en l'honneur d'Edouard Lambert*
— *Ius privatum gentium. Festschrift für Max Rheinstein*
— LÉVY-BRUHL (*Symbolae R. Taubenschlag dedicatae*)
— *Liber Amicorum Baron Louis Fredericq*
— *Libro-Homenaje a la Memoria de Roberto Goldschmidt*
— *Mélanges de droit comparé en l'honneur du Doyen Å. Malmström*
— *Miscellanea W. J. Ganshof van der Meersch*

— *XXth Century Comparative and Conflicts Law. Legal Essays in Honor of Hessel E. Yntema*
TRUDEAU, P. E., *Le fédéralisme et la société canadienne-française*, Montréal, 1967.
TUMANOV, A. V., *Kritika sovremennoj buržuaznoj teorii prava* (Critique of the Contemporary Bourgeois Theory of Law), Moscow, 1957.
TUNC, A., 'Comparative law, peace and justice', in *XXth Century Comparative and Conflicts Law, op. cit.* (1961), pp. 80 sqq.
—, 'La contribution possible des études juridiques comparatives à une meilleure compréhension entre nations' (Report presented at the CIDC meeting in Rome on 27 Sept. 1963), *RIDC*, 1964, No. 1, pp. 47 sqq.
—, *Le droit des Etats-Unis*, Paris, 1964.
TUNC, A. (ed.), *Les aspects juridiques du développement économique/Legal Aspects of Economic Development* (Study sponsored by the International Association of Legal Science, with Unesco support), Paris, 1966.
TUNKIN, G. I., *Voprosy teorii meždunarodnogo prava* (= Problems in the Theory of International Law), Moscow, 1964.
—, *Ideologičeskaja bor'ba i meždunarodnoe pravo* (= Ideological Struggle and International Law), Moscow, 1967.
—, *Teorija meždunarodnogo prava* (= The Theory of International Law), Moscow, 1970.
XXth Century Comparative and Conflicts Law. Legal Essays in Honor of Hessel E. Yntema, Leiden, 1961.

ULJANOVSKIJ, R. A. (ed.), *Nekapitalističeskij put' razvitija Afriki* (= The Non-capitalist Way of Development of Africa), Moscow, 1967.
—, *Pravo v nezavisimyh stranah Afriki* (= Law in the Independent Countries of Africa), Moscow, 1969.
Unification of the Law Governing International Sales of Goods (Symposium of the IALS, New York, September 1964; J. HONNOLD, ed.), Paris, 1966.
USENKO, E. T., *Formy regulirovanija socialističeskogo meždunarodnogo razdelenija truda* (= Forms of the Regulation of the International Socialist Division of Labour), Moscow, 1965.
L'Utilisation de satellites de diffusion directe, CNRS, Paris, 1970.

VALLADAO, H., 'The Law of Interplanetary Space', in *Proceedings of the IInd Colloquium on the Law of Outer Space*, London, 1959, pp. 156-168.
VAN BOGAERT, E., 'Les antinomies entre le droit international et le droit interne', *Revue générale de Droit international public*, 72nd Yr., No. 2, April-June 1968, pp. 346-360.
VANDERLINDEN, J., 'Trends and priorities in African legal research and writing' = Section V, 'Trends and priorities in research and publication', in *Proceedings of the Conference on Legal Education in Africa, op. cit.* (1969), pp. 77-104.
—, 'La règle de droit dans l'Afrique traditionnelle', in *La Règle de Droit, op. cit.* (1971).
—, 'Aspects de l'égalité juridique dans l'Afrique traditionnelle', in *L'Egalité, op. cit.* (1971), Vol. I.
VEJNGOLD, Ju. Ju., *Pravo kak sociologičeskaja kategorija* (= Law as a Sociological Category), Frunze (USSR), 1962.
—, *Nekapitalističeskoe razvitie i social'naja volja* (= Non-capitalist Development and the Will of Society), Frunze (USSR), 1970.
VEREŠČETIN, V. S., 'Osnovy kosmičeskogo pravoporjadka' (= Bases of the cosmic legal system), *Ežegodnik*, 1968, pp. 116-126.
Vergleichung im öffentlichen Recht (= Comparison in the field of Public Law) = *Zeitschrift für ausländisches öffentliches Recht und Völkerrecht* (= Review of Foreign Public Law and International Law) (Stuttgart), 1964, No. 3.
VINTU, I., 'La démocratie représentative en Roumanie', in *Miscellanea Ganshof*

van der Meersch, op. cit. (1972), Vol. III, pp. 351 sqq.

VIRALLY, M., 'De la classification des organisations internationales', in *Miscellanea Ganshof van der Meersch, op. cit.* (1972), Vol. I, pp. 365 sqq.
[See also *sub Les Nations Unies face à un monde en mutation.*]

VLASIC, I. A., 'The growth of space law 1957-1965: achievements and issues', *ADAS*, 1967, pp. 365 sqq.
[See also MCDOUGAL, LASSWELL & VLASIC.]

VOLKOV, A. A., 'Meždunarodnopravovye voprosy èkspluatacii živyh resursov kontinental'nogo šelfa' (= Problems in international law concerning exploitation of the natural resources of the continental shelf), *Ežegodnik* 65, 1964, pp. 213 sqq.

Voprosy kodifikacii sovetskogo prava (= Problems in the Codification of Soviet Law), collection, Leningrad, 1957, 2 vols.

VOSTRIKOV, A. V. *et al., Marksistsko-leninskaja filosofija. Dialektičeskij materializm* (= Marxist-Leninist Philosophy. Dialectical Materialism), Moscow, 1970.

VRECION, V., 'On some possibilities of the theory of information in the sphere of law and management', *Law and Cybernetics* (bulletin published by the Institute of State and Law of the Czechoslovak Academy of Sciences, Prague), 1, 1967.

—, 'Metoda použití matematických metod a samočinných počitačů při evidenci, analýze a tvorbě právních norem' (= The application of mathematical methods and computers to the verification, analysis and creation of legal norms), *Právník* (= The Jurist) (Prague), 1968, No. 3.

—, 'Automated method for determination of alimony to legally entitled persons', *Law and Computer Technology* (Washington, D.C.), 1 (9), Sept. 1968, pp. 7-9.
[See also KNAPP & VRECION.]

WAGENER, F., *Neubau der Verwaltung* (= Rebuilding the Administration), (West) Berlin, 1969.

WALINE, M., 'La responsabilité des préjudices causés par le bruit des avions aux abords des aérodromes', in *Miscellanea Ganshof van der Meersch, op. cit.* (1972), Vol. III, pp. 729 sqq.

WALTER, R., 'Der gegenwärtige Stand der reinen Rechtslehre' (= The present position regarding the pure theory of law), *Rechtstheorie,* (West) Berlin, No. 1, 1970.

WEICHELT, W., *Der sozialistische Staat – Hauptinstrument der Arbeiterklasse zur Gestaltung des sozialistischen Gesellschaft* (= The Socialist State – The Principal Instrument in the Hands of the Working Class towards the Formation of the Socialist Society), (East) Berlin, 1972.

WEIL, G. L., *The European Convention on Human Rights*, Leiden, 1963.

WEYL, M. & R., *La justice et l'homme*, Paris, 1962.

WHEARE, K. C., *Modern Constitutions*, London, 1966.

WIENER, N., *The Human Use of Human Beings. Cybernetics and Society*, first published in 1950, 2nd edn. rev., New York, 1954.

World Peace Through Law Conference: see *Belgrade World Conference on World Peace Through Law.*

WRIGHT, A., *Study of War*, Chicago, 1965.

WRÓBLEWSKI, J., 'Zagadnienia przedmiotu i metody teorii państwa i prawa' (= (Problems concerning the subject-matter and method of the theory of the State and law), *Państwo i prawo* (= The State and Law) (Warsaw), 1961, No. 11, pp. 751 sqq.

ZEILLER, F. von, *Das natürliche Privatrecht* (= Private Natural Law), Vienna, 1819.

ZIPPELIUS, R., *Das Wesen des Rechts* (= The Concept of Law), 2nd edn., Munich, 1969.

ZIVS, S. L., *Krizis buržuaznoj zakonnosti v sovremennyh imperialističeskih gosudarstvah* (= The Crisis of Bourgeois Legality in Contemporary Imperialist States), Moscow 1958.

—, 'O metode sravnitel'nogo issledovanija v nauke o gosudarstve i prave' (= Concerning the method of comparative research in political science and law), *S.g. i p.*, 1964, No. 3, pp. 23 sqq.

—, 'Rol' prava v obespečenii nekapitalističeskogo razvitija stran Afriki' (= The rôle of law as one of the guarantees of the non-capitalistic development of African countries), *S.g. i p.*, 1966, No. 10, pp. 67 sqq.
[See also ČHIKVADZE & ZIVS.]

ŽUKOV, G. P., *Kosmičeskoe pravo* (= Space Law), Moscow, 1966 (German version, *Weltraumrecht*, (East) Berlin, 1968).

Zur Einheit der Rechts- und Staatswissenschaften (= On the Unity of Legal Sciences and Political Science), *Ringvorlesung der Rechts- und Staatswissenschaftlichen Fakultät in Freiburg i. Br., Wintersemester 1966/67* (= Series of Conferences Held at the Faculty of Law and Political Science in Freiburg in Breisgau; Winter Semester 1966/67), Karlsruhe, 1967.

ZWEIGERT, K., 'Zur Lehre von den Rechtskreisen' (= Concerning the theory of legal spheres), in *XXth Century Comparative and Conflicts Law, op. cit.* (1961).

— & KÖTZ, H., *Einführung in die Rechtsvergleichung* (= Introduction to Comparative Law), Vol. I, Tübingen, 1971.

— & KROPHOLLER, J., *Sources of International Uniform Law/Sources du droit uniforme international*, Leiden, 1971.

4. Philosophy

PAUL RICOEUR

Professor at the Universities of Paris-X and Chicago

FOREWORD

PHILOSOPHY AND PHILOSOPHIES TODAY

How should the main trends of philosophical thought in the world today be set out in a manner conformable with the spirit of the present study?

In the early stages of their work together, the author of the present report and the associate rapporteurs[1] who kindly assisted him in his task agreed on the following rule of method: as far as possible, they have resisted the temptation – which, indeed, would have provided an easy way out of all sorts of problems, albeit at the cost of a loss of significance – to set out the exposition of the main trends of research in philosophy in terms of a geographical framework. Thus, one would have opposed Western thought and non-Western thought (i.e., that of East Asia, South Asia, the Middle East, Africa); within the limits of the Occident, one would have distinguished between the Anglo-Saxon world, Eastern Europe, Western continental Europe, and the Ibero-American world. In order to break up this geographical framework, they have tried to relate the interests of those who are carrying on research in philosophy to a small number of philosophical 'fields' or '*loci*' respectively corresponding to those questions which are at present most alive at the levels of research, publication and discussion.

The object of the present introduction is to lay out the topography of these philosophical fields or *loci* and to establish the method which will be used to examine them. To do this, we shall first of all define and distinguish, in the most general terms, the main conceptions that philosophers of various leanings have themselves formed of philosophy, its essential aspirations and its course, its function and its task within the culture of our time.

1. According to the first conception, the business of philosophy is still to build up a systematic representation of *reality*. According to this school of thought, philosophy is a *Weltanschauung*. Epistemologically speaking, it is firmly committed to some form of realism: there exists 'a reality', at both the natural and the social levels, which is in no way constituted by man's own theoretical or practical activities, and whose laws are binding on his consciousness rather than being produced or projected by it. These laws belong to a logic of reality which cannot be reduced to any form of merely formal logic, symbolic or mathematical. As compared with science, philosophy is deemed a more fundamental, more widely comprehensive form

1. Daya KRISHNA, Professor and Head of the Department of Philosophy and Psychology, University of Rajasthan, Jaipur, India; A. C. R. G. (Alan) MONTEFIORE, Balliol College, Oxford University; Teodor Ilič OJZERMAN (Theodor Ilich OIZERMAN), Professor at Moscow State University, corresponding member of the Academy of Sciences of the USSR.

of knowledge, not only on the grounds that it seeks to link together the results of the various sciences and to contrast and relate with each other their diverse methods, but also in that it seeks to create a unified view of the world out of the facts ascertained by science and the values that emerge from the social experience of mankind. By some of their major characteristics, a number of currently 'living' philosophies correspond to an ideal of this type, in spite of the profound differences which separate them. Marxism, as inherited through Engels and Lenin, pre-eminently presents that all-embracing structure: its inherent realism combines itself with the epistemological aspects of its notion of reality, and with practice, in a vast unified synthesis; but other schools too share in varying degrees the same ideal of philosophic achievement: for one example, those types of thinking which can be traced back to Hegelianism, especially the Hegelianism of the *Encyclopaedia;* for another, the various 'scientists' philosophies' which call for a synthesis of cosmology and anthropology within some kind of all-embracing apprehension; finally, in yet another way, those streams of thought which derive from the Aristotelian-Thomist synthesis.

2. At the opposite extreme from this ambitious ideal, we find deliberately modest modes of philosophizing, according to which science alone provides an image of reality, an image to which philosophy is quite unable to add. The task assigned to philosophy is then to cure philosophy itself of its ill-founded claim to know more and better than science. This reductionist and therapeutic conception is generally linked with an investigation of language and of its grammar, the warping and misuse of which are, in their various forms, held responsible for the illusions of philosophy. In this overall conception, the theory of propositions and of speech-acts holds the place that is held by the theory of reality in the preceding conception. Moreover, philosophers holding this conception tend frequently to make use of analytical techniques, that is, both of the decomposition of propositions into simple (or more simple) ones, and of the reformulation of complex propositions according to strict rules of derivation. This analytical style stands in contrast with the synthetic style of the preceding schools of thought. This second stream is represented most notably by English and American analytic philosophy and by its diverse variations in Europe and throughout the world. These general characteristics allow, nevertheless, for considerable differences, according to whether one expects symbolic logic to provide a model for the reformulation of ordinary language, or whether one takes ordinary language to be characteristically irreducible to any such reformulations; the only justification for setting up an overall contrast between this and the first tendency lies in the general task which it assigns to philosophy – which is to increase not the realm of facts but the understanding which we have of them – and the related emphasis on language.

However, these two attitudes, standing in polar opposition, do not exhaust the entire field of possibles; they merely provide a framework for identifying and locating the other overall conceptions of philosophy, in relation to the alternative thus set up; attention will be given in the present report to those original conceptions that represent something more, and something more significant than just a compromise between the two above-described trends, and constitute original attempts to get beyond the alternative itself. For the sake of schematic clarity, only two or three of those conceptions will be mentioned here.

3. Repudiating both the ambition of the first trend as an unwarrantable pretension and the modesty of the second as an unacceptable abandonment of responsibility, another conception of philosophy stresses the diversity of human experience: perceptual and scientific experience, of course, but also aesthetic, ethical, political and religious experience. Recognizable influences here include those of the young Hegel, of Kierkegaard, of the young Marx, as they extend to certain developments of phenomenology. According to this trend of thought, philosophy cannot hope to embrace all the sciences in one overall vision of reality; its responsibility is, rather, to preserve, to make explicit and to co-ordinate other forms of experience than objective knowledge. Accordingly, its theme is subjectivity, in all its diverse and various modes. Giving up any ambition of offering some all-embracing vision, it aims rather at depth, if necessary at the cost of adopting a method of fragmentary description. The charge of subjectivism and idealism that is commonly laid against it by adherents of the first-mentioned trend of thought is equally commonly met by counter-accusations either of 'positivism' and 'scientism' in so far as reality as conceived by philosophers of this first trend is dominated by models borrowed from the sciences, or of indulging in 'metaphysics' in cases where reality as conceived by the philosopher is deemed to be irreducible to what is allowed by science to count as real.

4. There are, however, other attempts at getting beyond the alternative of dialectical realism and analytical philosophy. Admitting that philosophy cannot embrace the whole of science in a more fundamental and more comprehensive conception of reality, admitting that it is equally unable to restrict itself to a mere critical examination of language, but admitting also that it should not shut up itself within the boundaries of human subjectivity, there remains the task of exploring certain possibilities opened up by meditation on the very breakdown of philosophy and on the end of metaphysics. Presented often as 'post-philosophy' or 'metaphilosophy', this trend is mainly related to the delayed influence of Nietzsche and to the eminently contemporary influence of Heidegger; the vision here offered of philosophy is in complete contrast to the humanism and subjectivism of the preceding trend: seen in its light, subjectivity appears to embody the final triumph as well as the final illusion of metaphysics. Its criticism of subjectivity and its strong sense of the cosmic reality in which man is included seem to bring it close to the first trend of thought; on other occasions, however, the stress that it lays on language tends to make it appear closer to the second. But the reality which it exposes in this way is neither the reality of dialectical

materialism, nor that of the realistic ontologies; nor is the language which is here extolled that which the analysts dissect: both reality and language turn out as more archaïc, more poetical and also more fragmentary than they do as seen by either of the first two trends of thought.

In the light of this general review it is clear that present-day philosophy offers conflicting views of its own mission. It is, therefore, not possible to put forward a synthesis in which the basic differences of opinion which prevail inside philosophy would somehow be concealed. At the same time, it must be admitted from the start that there can be no neutral form of exposition: the very order in which issues are dealt with is bound to reflect a particular style of philosophic thinking. All that one can hope to do is to identify the 'fields' in which the several forms of debate cut across each other. It must also be admitted, however, that certain schools are more at home in certain fields rather than in others: the first-mentioned trend of thought is more at home in the field that we have called 'Natural Reality', while the field labelled 'Language' is more densely occupied by the analytical school. This is the reason why it will be impossible to present the different conceptions of philosophy in the same order with respect to each respective philosophical field or *locus*.

The rapporteur is responsible for the division, here proposed, into six broad fields. Section I deals with contemporary research and reflection concerning the very instrument of thought and the processes of thought, that is to say the recent developments of logic, the questions that it raises for philosophy and the relation of philosophy to its own logic. The following two sections come under the category of reality: natural reality and social reality. Thence one is led, by a process of reflection, to consider two of man's essential operations: speaking and acting. The concluding section is devoted to a general discussion of man and the foundation of humanism.

The rapporteur is likewise responsible for the internal arrangement and the subdivisons of each of these parts. For the first five, he has placed at the beginning of each part a brief reminder of the present state of those sciences and corresponding epistemologies which force the various schools of thought to a constant renewal of their interpretations in the domain in question. He then turns to those broad trends which share each given field of philosophical thought between them; the order in which they are expounded reflects in each case the relative importance of that particular field for the conflicting doctrines within it. The rapporteur also thought that Section VI, to be devoted to man and the foundation of humanism, set out in terms of the major areas of civilization and, within each, in terms of the contrast between religious and secular modes of thinking (or, more exactly, between recourse to the transcendental dimension and the will to keep to the plane of human immanence), could provide an opportunity for a systematic general comparison of the main trends of research in philosophy within the focus of a more precise exposition of the schematic view presented in this Foreword.

It is clear that whatever principles of division had been adopted, many works would call for mention under more than one heading. It is of the essence of philosophy to call into question each new boundary line as soon as it is proposed, to resist all attempts to shut off any of its recognized fields from all intimate connection with any other. It is thus natural that in the pages that follow many authors and many works should appear under more than one heading. From time to time explicit mention is made of such overlaps. To have pointed to each of them, however, would have been an over-cumbersome procedure. The reader will readily appreciate that many possible cross-references have thus been left implicit for him to complete for himself according to his interests and his needs.

It appeared permissible to arrange the body of this work in two main chapters, each comprising three of the 'fields' previously described and entitled respectively 'Modes of thinking and the different classes of reality' and 'Language, action and humanism'.

The overall structure of the work is as follows:

First chapter (Chapter VII): *Modes of thinking and the different classes of reality*

Section I : *Man and his forms of knowledge: thinking*
- A. Philosophy of logic
- B. Logic of philosophy

Section II : *Man and natural reality*
- A. Epistemology of the natural sciences
- B. Theory of natural reality

Section III: *Man and social reality*
- A. The logic of explanation
- B. Social and political philosophy

Second chapter (Chapter VIII): *Language, action and humanism*

Section IV: *Man and language*
- A. Epistemology of linguistics
- B. Philosophies of language

Section V : *Man and action*
- A. Theory of practical discourse
- B. Philosophies of action

Section VI: *Man and the foundation of humanism*
- A. Man according to Eastern modes of thinking
- B. Man according to Western modes of thinking
- C. Philosophy of religion.

The List of Works Cited which is given in the annex refers to both chapters. It provides the reader with additional bibliographical information about each of the publications – volumes or articles – mentioned in the notes or in the body of the work.

The present work has benefited from numerous contributions brought together as a result of international collaboration.

A certain number of Unesco National Commissions supplied the rapporteur with statements obtained at their request from institutions and scholars in their respective countries:

AUSTRALIA	Notes prepared by Professor A. BOYCE GIBSON on the development of philosophy in Australia (with bibliography).
BELGIUM	(Commission of the Royal Academies for the Unesco Study) *Tendances actuelles des disciplines philosophiques en Belgique.*
CHILE	Report by Professor Arturo PIGA, former head of the Department of Psychology of the University of Chile.
Arab Republic of EGYPT	*Phénoménologie et existentialisme*, by Dr. Yéhia HOUEIDI (University of Cairo); *Al-Gouwania, or The Philosophy of Inwardness*, by Professor Osman AMIN (University of Cairo).
FINLAND	*Philosophical Research in Finland*, by Mrs Raili KAUPPI, of the University of Tampere.
INDIA	Reports of the meetings held by the Special Committee (incorporated in the contribution of Professor Daya Krishna).
ITALY	Work of a committee of Professors of the University of Bologna (Preliminary report by Professor Felice BATTAGLIA, summary of the discussion, communications by Professors ANCESCHI-BARILLI, PASQUINELLI-SANDRI, MANFERDINI, BABOLIN) preceded by a note of Professor Marino GENTILE, Director of the Institute of Philosophy of the University of Padua and President of the Società Filosofica Italiana; Statement by Professor Gustavo BONTADINI (Catholic University of Milan); Statement by Professor Pietro PIOVANI (University of Naples).
JAPAN	*The Postwar Trends of Japanese Philosophic Thought*, by Professor MASAAKI Kosaka, Emeritus Professor of Kyoto University.
MEXICO	*Las tendencias actuales de la investigación filosófica en México*, by Dr. Fernando SALMERÓN, Director of the Centre for Philosophical Studies of the independent national University of Mexico.
UKRAINIAN S.S.R.	(National Commission) *Razvitie filosofskoj nauki v Ukrainskoj S.S.R. za 10-15 let* (= The development of philosophical science in the Ukrainian S.S.R. over the last 10 to 15 years).

YUGOSLAVIA (Special inter-academy committee) *Le développement de la philosophie en Yougoslavie depuis 1945.*

The International Council for Philosophy and Humanistic Studies (ICPHS) supplied the rapporteur with a written essay by Professor I. S. KON of Leningrad, as a contribution to the last phase of the work for the Study, on the theme '*Istorija i obščestvennye nauki (Novejšie tendencii v sovetskoj istoriografii)*' (= History and the social sciences – Recent trends in Soviet historiography).

Of the international learned societies which form part of this Council, the International Academic Union and the International Federation of Societies of Philosophy contributed to the work in response to an invitation by the General Secretariat of the Council:

– the International Academic Union in the form of several comprehensive surveys drawn up by Canon G. VERBEKE, of the University of Louvain, member of the Royal Academy of Sciences of Belgium, including an ample account of research trends in the philosophy of religion, with the collaboration of Messrs. A. DONDEYNE and A. VERGOTE and other professors of the University of Louvain;

– the International Federation of Societies of Philosophy by way of remarks formulated on the basis of the preliminary outline by its officers and by Professor J. J. C. SMART of the University of Adelaide (Australia).

Thanks to the assistance and agency of the associate rapporteurs, the author was able to consult a body of documentation consisting of reports prepared from first-hand knowledge of those sectors of philosophical research with which he was least directly familiar.

For instance, the recent achievements and current development of Marxist philosophical thought in the Socialist countries were the subject of an extensive description prepared under the editorship of T. I. OJZERMAN and to a large extent written by him, the sections relating to logic and dialectics, the philosophy of natural reality and the natural sciences, and the philosophy of language being the work of V. A. LEKTORSKIJ, N. F. OVČINNIKOV (Ovchinnikov) and V. S. ŠVYREV (Shvyrev) respectively, of the Institute of Philosophy of the Academy of Sciences of the U.S.S.R.

In the field of analytical philosophy, the advice and mediation of Alan Montefiore made it possible to assemble the following contributions of a specialized nature:

L. Jonathan COHEN (Queen's College, Oxford University):
- *Philosophy as Criticism and Analysis*
- *The Epistemology of Linguistics*
- *The Analytic Conception of the Philosophy of Language.*

Kit Fine (Edinburgh University):

- *Logic and Mathematics. The Question of the Foundation of Mathematics since Frege: the Criticism of Transcendentalism and Intuitionism*
- *Deontic Logic.*

Patrick Gardiner (Magdalen College, Oxford):

- *The Historical Sciences and Their Methodology*
- *Philosophy of History*
- *History of Philosophy.*

Rom Harré (Oxford University):

- *The Epistemology of the Natural Sciences as a Subject for Philosophic Reflection today*
- *Relationships between the Natural and the Social Sciences.*

J. P. Hierro (University of Madrid):

- *The Analytic Conception of the Philosophy of Action. Propositions about Actions. Ethical Philosophy from G. E. Moore to Hare and after.*

A. J. P. Kenny (Balliol College, Oxford):

- *The Analytic Conception of Man and Humanism, and the Concept of Mind.*

John R. Lucas (Merton College, Oxford):

- *Decisions Theory, Game Theory, and Cybernetics, as Subjects for Philosophic Reflection. Philosophy of Technique and Technology.*

Steven Lukes (Balliol College, Oxford):

- *Political Science and the Sociology of Institutions and of Social Dynamics, as Subjects for Philosophic Reflection, especially along 'Analytic' Lines.*

Alasdair MacIntyre (University of Essex, Colchester):

- *The Sociological Concepts and Conceptions of Behaviour and Action, and the Philosophy of the Social Sciences.*

Charles Taylor (McGill University, Montreal):

- *The Psychological Concepts and Conceptions of Behaviour and Action.*

Recent research in the field of general logic and the logic of philosophy was synthesized in a note by Jean Ladrière, of the University of Louvain.

The trends of recent research on the philosophical meaning of the history of philosophy were analysed by Pierre Garniron, of the Centre National de la Recherche Scientifique (C.N.R.S.), Paris.

The treatment of questions concerning the philosophy of religion by 'analytic' philosophical thinkers was the subject of a note prepared by Jacques Poulain, of the University of Montreal.

Recent research and the present evolution of Asian philosophies – particularly in India and Japan – were the subject of a general report by Daya Krishna, for the preparation of which the author was able to draw not only on his own vast experience but also on the valuable information contained in the written contribution prepared at the request of the Japanese National Commission by Professor MASAAKI Kosaka, and on the results of direct consultations with the most representative philosophers, among whom, in Japan, Professors FUJIYOSHI Jikai, TAKEYUSHI Yoshinori, ABE Masao, UMEHARA Takeshi and NAGAO M. Gadjin.

Lastly, the rapporteur had the benefit of two essays by Pierre AUGER (Paris) on the themes 'L'homme et la machine' and 'La morale de l'homme de science', and of an essay by Paulin Jidenu HOUNTONDJI (Dahomey) on the theme: 'Remarques sur la philosophie africaine contemporaine'.[2]

The author of the present analysis expresses his gratitude to the institutions and persons without whose assistance he could not have brought this undertaking to completion, and first and foremost to the associate rapporteurs who, from the early stages, gave him their advice and suggestions, then collected, in their respective fields, the indispensable basic documentation and, finally, by comments and recommendations which they formulated on the basis of a draft, contributed to the final version of the text.

He wishes to express his personal gratitude to the scholars who lent their co-operation to Professor Ojzerman in the completion of his heavy task of working out a comprehensive survey of the present trends of Soviet philosophy, and in the critical examination of the draft.

He also associates himself with Professor Daya Krishna's expression of gratefulness to the philosophers who generously contributed to the elaboration of the latter's report on the present trends of philosophic research in Asia, in particular to the late Professor Masaaki Kosaka, and to the Japanese philosophers who took part in the discussions indispensable for the preparation of that report.

He also thanks all the collaborators who, often at very short notice, accomplished a task of special documentation and synthesis which was of very great value to him; he is particularly grateful to those among them who, in fields where the highest degree of technical precision was required, agreed to cast their contributions in a literary form such as to facilitate their inclusion in the final text with every guarantee of accuracy and authenticity.

Finally he takes pleasure in acknowledging the support and co-operation which he enjoyed from the general rapporteur of the Study at all stages of the work, and the perfect community of views which marked this close and happy association. It is to the general rapporteur that he wants to address his ultimate thanks.

Paul RICOEUR

2. Published in *Diogenes*, No. 71, Fall 1970, pp. 109-130: 'Remarks on contemporary African philosophy'.

CHAPTER VII

Modes of thinking and the different classes of reality

CONTENTS

SECTION ONE: MAN AND HIS FORMS OF KNOWLEDGE: THINKING

SECTION TWO: MAN AND NATURAL REALITY

SECTION THREE: MAN AND SOCIAL REALITY

SECTION ONE

MAN AND HIS FORMS OF KNOWLEDGE: THINKING

SECTION ONE: MAN AND HIS FORMS OF KNOWLEDGE: THINKING

INTRODUCTION

The problems considered in this first section, unlike those in Sections II to V, do not represent philosophy at work. They concern what might be called, after Aristotle, the *organon* of philosophy. Our reason for beginning with this enquiry is that we wanted to take stock of the rational equipment available to philosophy. In other words, the question we are here aiming at is the relation of the logic of philosophy to philosophy itself. Is that logic really a separate instance? Can it be formal in the sense of formal logic? or, at least, formalized?

But that level of questioning cannot be reached till we have considered logic itself and disengaged the properly philosophical questions of justification and foundation which it poses. That is what we propose to do in sub-section A, the relation of which to sub-section B is similar to that which we have found to obtain elsewhere in this study between preliminary epistemological considerations and discussion of the strictly philosophical theses. We have therefore brought together in sub-section A the questions logic puts to philosophy: What relation is there between logic and mathematics? What is the present state of foundational research in the philosophy of mathematics? More broadly, what form does the question of foundations take over the whole logico-mathematical area? To discussion of these three topics we have added a slightly incongruous study of the logic of science, anticipating problems and discussions that will be taken up at greater length in Section II, sub-section A, with regard to the natural sciences. Here we confine ourselves to aspects of the theory of confirmation that concern the status of logic itself.

Sub-section B contains the answers to the question that launched our enquiry: What relation has philosophy to its own logic? We have tried to set forth the issues as they stand at present in a simple framework which, without doing violence to the subtlety of the various positions that are held, will enable non-specialists to find their way. We have taken as guideline the distinction between the two main types of philosophy: analytical and synthetic. Within this broad dichotomy finer distinctions are introduced: for the analytical type, between an elucidatory and a transcendental approach; for the synthetic type, between various kinds of system: categorial, hermeneutic, dialectical. Among the dialectical systems we have given dialectical materialism a place apart because of the number of works on the subject. Thus we have tried to compensate by complexity of execution for the simplicity – not to say the simplifications – inherent in the overall conception.

A. PHILOSOPHY OF LOGIC

§ 1. *Logic and mathematics**

A paradigm case of reason in man is his ability to draw conclusions. Given that all philosophers are clever and that some women are philosophers, he accepts that some women are clever. It is the traditional task of logic to lay bare the principles of reasoning, to present in a systematic way all the correct arguments that come within the scope of reasoning. It was noticed long ago by Aristotle that the correctness of such arguments depends, not on their subject-matter, as indicated by the words 'philosopher', 'women', and 'clever', but only on their form, as indicated by the particles 'all' 'some', and 'are'. So logic could content itself with presenting the forms of all correct arguments.

This presentation has been of two kinds. One can either give rules for deriving all correct argument forms or one can give rules for determining which argument forms are correct and which incorrect. In the one case we have an axiomatization, the classic instance being afforded by Euclid's *Elements,* while in the other case we have an algorithm, a well-known example being the truth-table test for validity in the propositional calculus. The difference between axiomatization and algorithm is the difference between being able to draw certain numbers out of a hat and knowing in advance whether or not a particular number is going to be drawn. Essentially, one is a derivation procedure and the other is a checking procedure.

(a) One can say that to some extent the traditional task of logic is still being pursued. Logicians still try to axiomatize or to find algorithms for correct argument forms. Until recently, logicians had concentrated on form as indicated by the particles that had emerged from the analysis of the 19th century logicians: the quantifiers 'for all individuals x' and 'for some individual x'; and the truth-functional connectives, such as 'or', 'not', and 'if-then'. But recently some logicians have taken a more liberal view of form; it could be indicated by such words as 'necessarily', 'obligatory', 'know' or 'believe', 'will' or 'has', and be applied to questions, commands and other types of non-indicative sentences (examples of such formalization will be found in Section V of this study). One might instance at this point, Lewis' *modal logic,*[1] von Wright's *deontic logic,*[2] Hintikka's *epistemic logic*[3] and Prior's *tense logic.*[4]

* §§ 1 and 2 in sub-section A are taken from a contribution specially written by Kit Fine for inclusion in the present work.

1. Cf. HUGHES & CRESSWELL, *An Introduction to Modal Logic* (1968) and of course the standard work of LEWIS & LANGFORD, *Symbolic Logic* (1932, re-edn. 1951).

2. Cf. von WRIGHT, 'Deontic logic' (1951); 'A note on deontic logic and derived obligation' (1956); 'Deontic logics' (1967); *An Essay in Deontic Logic and the General Theory of Action* (1968).

3. Cf. HINTIKKA, *Knowledge and Belief* (1962).

4. Cf. PRIOR, *Past, Present and Future* (1967).

This extension of the scope of logic is, of course, at one with the trend in analytical philosophy to assimilate philosophy to logic and both to the study of language. So it is not surprising that the study of these new logics has received impetus from, and had great influence on, the corresponding areas in philosophy. For example: imperative logic is associated with Hare's moral philosophy and the view that statements of obligation are a special kind of imperative; modal logic has proved invaluable in recent discussions on the nature of necessity and essence; and tense logic has been connected with the revival of McTaggart's view that the language for events being past, present or future is more fundamental than the language for instants of time being earlier or later than one another.

Despite the importance of these studies, it has to be said that most of the recent work in logic has moved away from its traditional task. There has been a change of perspective. Instead of constructing this or that theory or presentation of logic, logicians now study the theories themselves. Nor do they confine themselves to the theories of logic; they also consider the theories of mathematics and science. Thus instead of trying to axiomatize the predicate calculus, the logic of quantifiers and truth-functional connectives, they will prove that no algorithm for the predicate calculus exists or they may show that a particular theory based on the predicate calculus is consistent, i.e., contains no self-contradictory theorem. From being a particular theory, logic has become a general theory of theories.

This change was partly the result of developments in the foundations of mathematics.[5] First, in order to substantiate their claim that mathematics was continuous with logic, the logicists had to give a formal axiomatization of logic, i.e., to specify with absolute precision the language, axioms and rules of inference that they wished to use. Second, to justify his formalist philosophy of mathematics, Hilbert had to show that a formalization of the kind provided by the logicists was consistent, did not contain a self-contradictory theorem. Thus one might say that logicism provided the material and formalism the motive for a meta-theoretical approach in logic.

As a consequence, logic is now as close to mathematics and the foundations of mathematics as it was to philosophy. The motive for a great deal of modern research is to diagnose, not the nature of reason, but the nature of mathematics. Also, with the increased complexity of its material, logic has become more explicitly mathematical in its methods. This has meant that many topics are considered from a purely mathematical viewpoint. Perhaps a good example of this is the recent interaction between logic and algebra, with the logical treatment of algebra and the algebraic treatment of logic, as in Robinson's *Introduction to Model Theory and to the Metamathematics of Algebra* (1963) and Halmos' *Algebraic Logic* (1962). Of course, in this short and general survey, we shall have to slide over the mathematical core of modern logic.

The first questions in meta-logic were mainly about axiomatizations as

5. Cf. § 2 below, pp. 1094 sqq.

systems of signs, without regard to the meanings of the signs or the truth of formulas under a given interpretation. In other words, logic was predominantly syntax or proof theory. The principal results in this field are Gödel's two incompleteness theorems: according to the first, there is a reasonable formalization of arithmetic which contains an undecidable sentence, i.e., a sentence such that neither it nor its negation is provable; according to the second, in this formalization of arithmetic one cannot prove that the formalization itself is consistent. Recent work in proof theory has become of wider scope and shown greater sensitivity to exactly what requirements a proof in proof theory should meet.

This typical predominance of syntax in logic was partly the result of Hilbert's formalism, but it also arose from a distrust in regard to the notion of truth itself. After all, proof simply involves a configuration of signs, whereas truth seems to involve an obscure relation between the signs and the world. However, in his classic paper of 1933, *'Der Wahrheitsbegriff in den formalisierten Sprachen'*, Tarski showed how the notion of truth in an interpretation could be made mathematically precise, at least for the languages of interest to the logician; and since that date, logicians and philosophers have become increasingly willing to use this notion. Indeed, it is no exaggeration to say that model theory or the theory of truth in an interpretation now enjoys that pre-eminence in logic that was once enjoyed by proof theory.

It is impossible to give an account here of all the topics covered in model theory.[6] However, one very general tendency should be noted. Work on models for languages based on quantifiers over individuals and the truth-functional connectives had made logicians realize that these languages are inadequate in some respects. One of these inadequacies, known in the 1920's, is that it is impossible to formulate in such a language a theory which will commit one to the existence of a greater number of individuals than there are natural numbers. As a result, logicians have studied languages of greater and greater expressive power. For example, they have considered languages in which one can say that a certain, possibly infinite, number of individuals satisfies a certain condition, or that some property or relation, rather than individual, satisfies a certain condition, or languages in which one can express the conjunction of an infinite number of sentences or quantification for an infinite number of individuals.

Experience in model theory has led some philosophers and logicians to the, usually implicit, view that a notion is not coherent unless it can be given a semantics, i.e., unless a formal and adequate account of truth in an interpretation can be given for a language capable of expressing that notion. Thus, whereas before a logician would attempt to axiomatize a new theory, the tendency now is to find semantics for the language of the theory.

This approach can work both for and against a contested notion. Thus Robinson's non-standard analysis, Kripke's possible worlds semantics and

6. Cf. ADDISON, HENKIN & TARSKI (eds.), *The Theory of Models* (1965).

various classical interprctations of intuitionism have helped to rehabilitate the notions of the infinitesimal, of necessity and of intuitionism respectively.[7]

On the other hand, it could be argued that the notion of entailment, as embodied in the system of Anderson and Belnap, and the notion of set, as embodied in Quine's 'New Foundations', are not secure until appropriate semantics for these systems have been given.[8]

There can be no doubt that model theory has often provided great illumination where none was available before. (So it is interesting to note that in this respect formal linguistics and the theory of computer languages have lagged behind logic, although there are signs that they are now catching up.) However, despite the great value of the previously mentioned work, it is in set theory that the model-theoretic approach has led to the most outstanding results. Before one can describe these results, it is necessary to say something about set theory itself. Suppose someone asks a question of the sort: 'How many?' (How many words are on this page? How many angels can dance on a pinhead?) Then it is appropriate to answer by specifying a natural number such as 75 or 0. But now suppose someone asks: how many natural numbers are there? Then clearly it is not appropriate to specify a natural number. So can one still talk of there being a definite number of natural numbers?

Cantor showed in the late nineteenth century that it was possible to extend a large part of the theory of natural numbers to non-natural or the infinite numbers. There were, however, two main difficulties in his theory. Cantor had been able to show that the number of natural numbers was the smallest infinite number (just as 0 is the smallest natural number). He was also able to determine that the number of *sets* of natural numbers was greater than the number of natural numbers. But he was not able to determine how much greater. So he put forward the Continuum Hypothesis, which states that the number of sets of natural numbers is the second smallest infinite number (just as 1 is the second smallest natural number). The first difficulty was to settle whether the Continuum Hypothesis is true or false.

The second difficulty arose as follows. Even to prove that any two sets could be compared for size, Cantor had to assume the Axiom of Choice. This states, in effect, that one can pick out a single person from each compartment of a train (assuming that no compartment is empty). This is clear if there are a finite number of people in each compartment, but the axiom of choice is meant to hold no matter what the number of people in each compartment or the number of compartments. Now the axiom of Choice, unlike the Continuum Hypothesis, enjoys a degree of self-evidence. All the

7. Cf. ROBINSON, *Non-standard Analysis* (1969); HUGHES & CRESSWELL, *An Introduction to Modal Logic* (1968); CROSSLEY & DUMMETT (eds.), *Formal Systems and Recursive Functions* (1964).

8. Cf. *Proceedings of a Colloquium on Modal and Many-valued Logics* (1963); QUINE, 'New foundations for mathematical logic' (1937); the same author's *From a Logical Point of View* (1953).

same, it would be pleasant to be able to derive it from the other assumptions of set theory, for, unlike all these assumptions, it states that a set with a certain property exists without giving a condition of membership for the set.

Now recent work on set theory has produced a startling turn to these two difficulties. First, Gödel showed in 1939 that neither the Continuum Hypothesis nor the Axiom of Choice could be proved to be false on the sole basis of the standard axioms of set theory. Then in 1963, Cohen showed that the same was true for the negations of these propositions. Thus neither the Continuum Hypothesis nor the Axiom of Choice can be settled on the basis of the standard axioms of set theory.

The effect of these two results is to remove the Axiom of Choice and the Continuum Hypothesis from the realm of conventional mathematics, which is essentially equivalent to the standard axiomatizations of set theory. One can no longer regard the truth or falsity of these propositions as a straightforward mathematical problem. What attitude should one take to them then? On the one hand, one can take these results as confirming that set theory does not have a unique subject-matter, contrary to the views of Platonists and others. The situation in regard to the two propositions is then analogous to what it was in the nineteenth century in regard to the proof of independence for the parallel postulate in Euclidean geometry: just as there are non-Euclidean geometries, so equally there should be non-Cantorian set theories. Indeed, work is already being done along these lines. On the other hand, one can take the view that set theory does have a unique subject-matter, but that these results show that further self-evident axioms are needed to settle the Continuum Hypothesis. These new axioms might be formulated in set-theoretic terms, although at the moment this does not look like a very promising approach, or they may be formulated in terms of a new notion or new notions. The discovery of such notions would have tremendous consequences for mathematics. It would undermine an orthodoxy that underlies the modern approach to mathematics; viz., that a mathematical notion is not legitimate unless it can be given a set-theoretic definition. It would also destroy the autonomy of set theory, for there would be propositions which could be formulated in set-theoretic terms but which could only be settled by going outside of set theory or of mathematics as we now understand it.

(b) So far we have considered the study of axiomatizations. We must now turn to the study of algorithms or what is now called recursive number theory. It will be remembered that an algorithm is essentially a checking procedure, i.e., a mechanical procedure which determines in a finite number of steps the answer to a particular instance of a general problem. The problem might be, for example, to determine whether or not a number was prime or a statement in arithmetic was true.

Which general problems are thus solvable? A great step forward was made in the 1930's when a precise characterization was given of the notion of a solvable problem. This characterization was given in various forms, but all the forms were shown to be equivalent. Thus the general theory of solvable problems was able to develop on a firm basis.

The present work in this field is very technical and cannot be described in any detail. Hartley Rogers' *Theory of Recursive Functions and Effective Computability* (1967) covers most of the topics. Suffice it to say that much of this has been concerned to generalize the notion of a solvable problem. For example, we may allow an algorithm to perform an infinite number of steps or we may consider problems about problems, problems about these higher-order problems, and so on. Again, we may consider problems whose solution is already contained, in some sense, in the solution of another problem. This approach leads to the so-called degrees and hierarchies.

This technical work should not blind us to the fundamental importance that the notion of a solvable problem has taken on the philosophy of mathematics. This importance derives from its connection with other important notions in the field. Thus it is through, and perhaps only through, possession of an algorithm that we are capable of *knowing in principle* an infinity of mathematical truths. Again, a solvable problem is essentially one that an *ideal computing machine* can be programmed to solve. So the notion is relevant to the question of which problems can be handed over to the computer programmers. Or again, we require of an *axiomatizaton* that a proof in accordance with its rules be recognizable as such or, in other words, that it be a solvable problem whether or not a given string of symbols constitute such a proof. In this way the notion of a solvable problem has pervaded a large part of the philosophy of mathematics.

This makes recursive function theory relevant to that field as long as the logician's precise characterization of the intuitive notion of a solvable problem is correct. This assumption is known as Church's thesis and is widely accepted. Let us give one example of its application. In recursive function theory we can prove, using Church's thesis, a generalized version of Gödel's first incompleteness theorem. This states that if an axiomatization of mathematics is consistent, contains a reasonable part of arithmetic, and is such that its proofs are recognizable as such, then that axiomatization contains an undecidable sentence, one neither provable nor refutable. It follows that either the body of mathematical truths we can in principle know is axomatizable as above, in which case there are mathematical truths which we cannot in principle know, or else this body of truths is not so axiomatizable, in which case our ability to know mathematical truths cannot be explained on the assumption that we are computing machines.

Recursive function theory is also relevant to automata theory and formal linguistics. By restricting algorithms in various ways, one obtains different kinds of *automata,* some of which correspond to the types of grammar studied by the formal linguist.

Finally, we should mention the constructive trend in mathematical logic. A sketch of the underlying philosophy can be found in § 2 below. Most of the technical work of this school has attempted to give accounts of constructive objects or constructive proofs within classical or conventional mathematics. For example, recursive function theory has been used to define and study constructive real numbers (points, or sets of natural numbers) and set theory has been used to formulate semantics for various parts of intuitionistic mathematics. However, there is still little intuitionistically acceptable material on intuitionistic thought.

§ 2. *Mathematical philosophy**

Mathematical philosophy is that branch of philosophy which deals with the nature of mathematics. It asks such questions as: do numbers exist? are conventional proofs legitimate? what is the connection between arithmetic and counting? A systematic answer to questions of this sort is a philosophy of mathematics. Such a philosophy is, or should be, of great relevance to a more general philosophy of reason, since mathematics is the most perfect example of man's rational endeavour.

(a) The last hundred years have seen the rise of three great philosophies of mathematics: logicism, intuitionism and formalism. The first of these was logicism. This claims that mathematics is continuous with logic, that the difference between the two is one of complexity and not of kind. Mathematics is merely a vast elaboration on logical material.

(i) The logicists attempted to demonstrate this continuity by deriving the whole of mathematics from logic. This does not mean that any mathematical result that had ever been proved had to be derived from logic, for then the task of logicism could only end with mathematics itself. Rather it was supposed that the various parts of mathematics could be derived from a limited number of axioms and that these axioms could, in their turn, be derived from a limited number of logical axioms. To substantiate their claim, the logicists had to formalize logic, to specify its language, axioms and rules with absolute precision, and then to set out the whole derivation in detail. This programme was first attempted by Frege in his *Grundgesetze der Arithmetik* of 1893. Unfortunately, his system was not consistent; it allowed the derivation of Russell's paradox. It was only with the publication of Russell and Whitehead's *Principia Mathematica* from 1910 to 1913 that the paradoxes were avoided and the programme fulfilled in a – so it was hoped – consistent system.

Logicism has an importance quite apart from the correctness of its claims. First, by providing such an important example of a formal theory, it helped to initiate the study of formal theories themselves, a study which has dominated mathematical logic for the last forty years.[9] Second, logicism produced a language which has had a great influence on analytical philosophy for the last sixty years. The language of *Principia Mathematica* or of the

* Taken from contribution by Kit Fine (cf. note * at beginning of § 1 above).
9. Cf. § 1 above, pp. 1089 sqq.

Grundgesetze essentially comes down to this: there are predicates which attach to variables to form open sentences; and there are quantifiers and truth-functional connectives for the construction of complex sentences. Logicism claimed that such a language could say anything mathematical. This was the first step towards saying that such a language, suitably enriched with descriptive predicates, could say anything whatever. Logicism also claimed (if only implicitly) that all logical relations are made perspicuous in such a language. That was a step towards saying that all philosophical matters are made perspicuous in the enriched language, and that, in particular, the relation between language and reality and the significant core of metaphysics are thus rendered perspicuous.

Thus the logicist's language led to the idea of an ideal or universal perspicuous language. Such a language was at the heart of most of the analytical philosophy of the 1920's, principally Wittgenstein's *Tractatus logico-philosophicus*, and it still plays an important, though less pretentious, part in the theory of meaning.

What is the present view on logicism? Its thesis that mathematics is derivable from logic would, I think, be accepted by most philosophers of mathematics, though with certain reservations. Firstly, Gödel's first incompleteness theorem shows that all mathematical *truths* cannot be derived from all logical *theorems*. All the same, it may be possible to derive all mathematical truths from logical truths and all mathematical theorems from logical theorems. Secondly, the logicist's logic is perhaps best regarded as set theory. This is a subtle issue which we need not go into. Thirdly, and somewhat controversially, there may exist notions which do not admit a set-theoretic definition.[10] Thus the thesis that *mathematics* is derivable from *logic* is reasonable if *mathematics* is extant mathematics and *logic* is set theory.

However, despite the plausibility of the logicist's derivability thesis, its philosophical significance is open to some doubt. This significance can only rest on the assumption that the nature of logic is clearer than the nature of mathematics. For otherwise no philosophical progress would be made in going from mathematics to logic; this reduction would shed no more philosophic light than the reduction of one branch of mathematics, say the theory of rational numbers, to another branch of mathematics, the theory of natural numbers. It is clear that the logicists themselves hold this assumption. Indeed, they often 'read off' the nature of mathematics from the nature of logic. For example, Frege assumes that logic is analytic and so concludes that mathematics is; again, Russell, in *The Problems of Philosophy* (1912), assumed that logic is about universals and so concludes that mathematics is too.

Now the assumption of logic's greater perspicuity seems to be quite reasonable. After all, logic constitutes the simplest organized body of necessary truth; also it is not beset by the problems of existence and of infinity which beset mathematics. However, once one looks at the logicists' logic,

10. Cf. § 1 above, p. 1092.

the assumption no longer appears to be reasonable. There seems to be no clear and significant way of demarcating the logicists' logic as analytic. Moreover, their logic has to make assumptions, such as that an infinite collection exists, which raise the same problems of existence and of infinity, as are raised by mathematics itself. This suggests that logicism leaves the fundamental problems of mathematical philosophy as they are. It cannot solve them; it can, at best, only put them into sharper focus.

(ii) The second great philosophy of mathematics is intuitionism. It will be convenient to discuss this philosophy under the general heading of constructivism. Suppose I say: either 10 7's occur in the expansion of π or they do not. According to the constructivist, this is illegitimate, since I do not possess a method for determining whether or not there are 10 7's in the expansion of π. Again, suppose I say: consider the set of all sets of natural numbers. Again, according to the constructivist, this is illegitimate, since I do not possess a method for generating all sets of natural numbers. Thus a constructivist wishes to restrict himself, in this way, to constructive principles of reasoning or to constructive objects, or to both. Accordingly, constructivism is the only philosophy of mathematics which is not conservative. It aims to change mathematics, not merely to interpret it. As such, it is something of an anomaly in present-day analytical philosophy, which enjoins philosophers to leave other subjects alone. Indeed, until the advent of intuitionism, most philosophers would have regarded mathematics as the non-philosophical subject least vulnerable to philosophical criticism.

There are various forms of constructivism, which differ in how radical a change they wish to make in classical or conventional mathematics. The least radical form of constructivism is predicativism. In response to the paradoxes of set theory, both Russell and Poincaré suggested that a set should not be defined in terms of totalities which involve that set itself. This point of view was called predicativism. It was systematically developed by Weyl in 1918; the view was exploited in Gödel's Independence Results in set theory; and it has lately been analysed by Feferman.[11] Note that predicativism limits itself to constructive objects, but uses classical logic.

At the centre of constructivist thought is intuitionism. This modifies both the principles and the objects of classical mathematics. It would be wrong, however, to suppose that intuitionistic mathematics is a straightforward restriction of classical mathematics. It treats of its objects in a fundamentally different way, so sometimes its theories, for example its theory of real numbers (points, sets of natural numbers), are in conflict with the corresponding classical theories.

The intuitionists have developed a mathematics of great beauty and subtlety. Unfortunately, the quality of their philosophical writings is, on the whole, rather poor; and, as a result, many criticisms of intuitionistic doctrine have been based on simple misunderstandings.

Recently, Dummett has freed intuitionism from its worst idealistic ex-

11. Cf. his 'Systems of predicative analysis (1969).

cesses and placed its doctrines in a general setting. He suggests that intuitionism should not be regarded as a particular view on mathematical activity, but as a general theory of meaning in rivalry with the classical theory. According to the latter theory, one understands a language if one understands what it is for the sentences of the language to be true. The truth or falsity of complex sentences is then explained in terms of the truth or falsity of simples sentences, and the truth or falsity of the simplest sentences is explained in some direct way. For example, a disjunction is true if one of its disjuncts is true, and a negative sentence is true if the sentence itself is false. On the other hand, what distinguishes intuitionism is explaining meaning, not in terms of truth-conditions, but in terms of assertion conditions, the conditions under which one is justified in *asserting* the sentence. Thus a disjunction can be *asserted* if one has a method which will lead to an assertion of one of the disjuncts, and a negative sentence can be *asserted* if one has a method which will transform an assertion of the sentence itself into the assertion of any sentence whatever.

This interpretation of intuitionism has great appeal and does not suffer from some of the weaknesses of previous interpretations.

The most radical form of constructivism is Wittgenstein's strict finitism, as expounded in his *Remarks on the Foundations of Mathematics*, published in English in 1956. The difference between intuitionism and strict finitism may be brought out as follows. The intuitionist accepts the proposition that every natural number has a square, since he has a method which will *in principle* convert any number into its square. The strict finitist, however, does not accept the proposition, since he has no reason to suppose that such a method can be carried out *in practice*. Thus the strict finitist substitutes practical possibility for theoretical possibility in the intuitionist's account of mathematics or of meaning.

Strict finitism is but one aspect of Wittgenstein's philosophy of mathematics. He also held novel views on what it is to accept a proof and on the connection between arithmetical truths, say, and certain very general empirical facts about counting. Since these views, like his strict finitism, are very counter-intuitive, no one has taken Wittgenstein's philosophy of mathematics very seriously, although the Russian Vopenka has attempted to use strict finitist mathematics for a proof that set theory is consistent.

(iii) The third great philosophy of mathematics is Hilbert's formalism. Suppose a person has no money and then on three separate occasions he earns £3, spends £6 and earns £9. How much does he have left? We may suppose that the transactions occur in the order stated. So he has successively £3, minus £3 ($3 - 6 = -3$) and then £6 ($-3 + 9 = 6$). Now in my deduction I have used the statement 'the person has minus £3', which is strictly meaningless. (We are supposing there is no credit.) However, it is useful in deducing one meaningful statement from another.

Now in exact analogy to the above example, Hilbert thought that certain statements in mathematics were strictly meaningful or finitary, that the others were strictly meaningless or non-finitary, and that the non-finitary statements were useful in proving finitary statements. The significance of the non-finitary statements is indirect and derives entirely from the rôle they play in proving finitary statements. Roughly speaking, the non-finitary statements involve an illegitimate use of the infinite. So a large part of artithmetic is finitary and a large part of set theory is not.

To show that the use of non-finitary statements in mathematics was justified, Hilbert had to prove that they did not enable one to prove false finitary statements. In particular, he had to prove that non-finitary mathematics was consistent; and he had to prove this within finitary mathematics, since the consistency of non-finitary mathematics was in question. Now Gödel's second incompleteness theorem showed that this could not be done, that even finitary mathematics could not be proved consistent within finitary mathematics. Consequently, Hilbert's strict form of formalism is no longer tenable, although a modified version of his programme still lies behind a good part of the recent research in logic.[12]

Recently, another form of formalism has become popular. According to this view, mathematics is essentially an applied logics. The mathematician makes statements of the form 'this follows from that', where 'this' is an uninterpreted theorem and 'that' is a body of uninterpreted axioms.
'Follows' may even be understood in a way that guarantees that if the axioms are true under a given interpretation, the theorems are too. However, it is no part of the business of the mathematician to assert that there is an interpretation of the axioms which makes them true.

This view does have difficulties, although it is often favoured by the working mathematician.

(b) This concludes our discussion of the three great philosophies of mathematics. We have put aside two topics which do not easily fall into any particular school of thought. They concern mathematical existence and the relation between mathematics and science.

(i) Recently, Quine has given an interesting gloss on philosophical disputes about existence.[13] The nominalist asserts that numbers do not exist; yet the mathematician asserts that prime numbers exist. Since the nominalist is not trying to repudiate conventional mathematics, he must be using 'exists' in a special sense. Quine suggests that the nominalist is asserting that one does not need to quantify over numbers, e.g., use the phrase 'for some number', in order to say what can be said in mathematics; and, quite generally, he thinks that the traditional ontological disputes have been about what one needs to quantify over to say what can already be said. This is perhaps a classic example of the linguistic treatment of traditional metaphysical problems that is so characteristic of present-day analytical philosophy.

12. Cf. LADRIÈRE, *Les limitations internes des formalismes* (1957) and 'Les limitations des formalismes et leur signification philosophique' (1960).

13. Cf. his essay 'On what there is' (1953); see also LADRIÈRE, 'Objectivité et réalité en mathématiques' (1966).

On this construal of the issue, is Platonism correct? Quine reluctantly concedes that it is, since there seems to be no way of expressing a great deal of mathematics without quantifying over numbers or, for that matter, without quantifying over sets.

It is worth noting that Quine's gloss does not do justice to all the traditional ontological disputes, even within the philosophy of mathematics. For example, both the formalist and the intuitionist do not avoid quantifying over real numbers (points, or sets of natural numbers), yet neither accepts the existence of real numbers. For the formalist, this is because statements about real numbers are strictly meaningless; and for the intuitionist, this is because only natural numbers exist – real numbers are to be explained, but *not* defined, in terms of the growth of sequences of natural numbers. Thus both formalist and intuitionist explain real numbers away, but their explanations do not necessarily yield the translation required by Quine. More generally, we may say that both formalist and intuitionist adopt a non-classical semantics and that Quine's gloss is appropriate only if classical semantics is not in question.

Discussion on Platonism has also arisen from the 'Independence Results' in set theory. These state that the Axiom of Choice and the Continuum Hypothesis cannot be settled on the basis of the current axioms of set theory. Roughly speaking, the Axiom of Choice says that one can pick a single representative from each set in a given collection of non-empty sets, and the Continuum Hypothesis says that either one can count the members of a collection with the natural numbers or one can use the members of the collection to count all *sets* of natural numbers. Set theory is the modern equivalent of the logicist's logic; it suffices for the derivation of conventional mathematics. Therefore the Independence Results state that the Axiom of Choice and the Continuum Hypothesis cannot be settled within conventional mathematics.[14]

For the Platonist these results are slightly embarrassing. For he must either say that the Continuum Hypothesis is essentially undecidable or that it can be settled outside conventional mathematics. For example, it has been suggested that it might be settled with the help of new axioms, possibly formulated in terms of new notions. It has also been suggested that although we may not be able to prove the Continuum Hypothesis or its negation, it may be reasonable to accept it, or its negation, as a fruitful hypothesis.

On the other hand, the non-Platonist can draw some comfort from the Independence Results, just as seventy years ago he drew comfort from the paradoxes. He can take the view that set theory is like the theory of groups, say, and does not describe a unique structure of sets but describes a large variety of structures which have something in common. Or he can take the view that set theory does not describe structure or structures at all, that it obtains its significance in some altogether different way. This might, for example, be the view of the formalist.

14. For a more detailed discussion see § 1 above, p. 1092.

(ii) So far we have only considered the nature of mathematics in itself and not in its relation to science. Recently, many philosophers and logicians have suggested that mathematics is like science in certain essential, though not immediately obvious, respects. For example, it has been suggested that proof is not the only source of knowledge in mathematics, that the hypothetico-deductive method, as it is used in science, may also be a source of knowledge. It has also been suggested that the necessary/contingent and *a priori/a posteriori* distinctions are ones of degree and not of kind. Thus mathematical propositions are not 'absolutely' *a priori*; in certain extreme cases, they may be revised in the light of experience. Finally, it has been suggested that mathematics is best regarded as a scientific theory, only to be distinguished from other scientific theories by its greater generality. Despite this recent tendency towards 'scientism', there are still those who hold that in its method, its subject-matter and the status of its propositions, mathemathics is to be distinguished from all other forms of rational enquiry.

§3. *Philosophy of logic: foundational problems**

Logic has played and continues to play a highly important part in research into the foundations of mathematics. But certain developments in logic concern the foundations of logic itself, in that they have as their object elucidation of the concepts and procedures used in the logical theories we might call 'classical' (i.e., the ones which are in one way or another related to the nucleus made up of propositional logic and predicate logic as these were elaborated in Russell and Whitehead's *Principia Mathematica* and later developed on the basis of that work). There are two noteworthy trends in such research: 'natural deduction logic' and 'combinatory logic'.

(a) 'Natural deduction logic' goes back to the work of Gentzen.[15] Its bearing on the problem of foundations has been brought out in recent years, especially by Curry.[16] Gentzen started with the idea of constructing a calculus representing as faithfully as possible the 'natural' way in which thought goes about mathematical demonstrations. But this idea has a quite general bearing. Deductive logic is essentially concerned with the *form* of deductive reasoning, i.e., with the demonstrative links that are evident in deductions, considered independently of the particular content of the statements that actually go to make up the demonstrations. We could express this formal point of view by saying that logic is interested not in the concrete inferences that are made but in inferential situations in general. An inferential situation can be represented by the following schema: From such

* §§ 3 and 4 in sub-section A are taken from a contribution specially written by Jean Ladrière for inclusion in the present work.

15. GENTZEN, 'Untersuchungen über das logische Schliessen' (1934) (French translation, *Recherches sur la déduction logique*, 1955).

16. CURRY, *Foundations of Mathematical Logic* (1963).

and such hypotheses one can derive such and such statements. More precisely: there exists a demonstration that leads from the set of hypotheses $A_1, A_2..., A_n$ to one of the statements $B_1, B_2..., B_m$. If we let the sign $\vdash$ represent the relation of deducibility, we can express this inferential situation by means of the following statement: $A_1, A_2,..., A_n \vdash B_1, B_2,..., B_m$. This statement is to be interpreted as follows: From (A_1 and A_2... and A_n) one can demonstrate (B_1 or B_2... or B_m). The true object of deductive logic is to study the relations between inferential situations, to examine how one can pass from one inferential situation to another legitimately, that is, without destroying the demonstrative link. Such a transition is represented by a schema: If an inferential situation exists leading from a sequence of statements X to a sequence of statements Y, then there exists an inferential situation leading from the sequence of statements X' to the sequence of statements Y', sequences X' and Y' being obtained from sequences X and Y by given transformations. Such a schema expresses in short an invariance of the inferential situation as regards certain transformations. Natural deduction logic studies these schemata and their properties.

Curry has shown that this logic makes it possible to elucidate the significance, from the point of view of inferential possibilities, of logical operators. Thus it is a true 'meta-logic' in that its statements relate to situations that occur in an underlying logical language. Let us show this by an example, that of the implication operator of 'classical' propositional logic. This operator makes it possible to form from two statements A and B the complex statement $A \supset B$. We can interpret this as follows: If in a demonstration there is a deductive path leading from A to B, then one can accept that one has a demonstration of $A \supset B$ and, following statement B, write the complex statement $A \supset B$. In the language of natural deduction logic this situation can be represented by the following schema: If there exists an inferential situation leading from hypotheses X and A to statement B (or, more generally, to the sequence formed of B and of sequence Y), then there exists an inferential situation leading from hypothesis X (only) to statement $A \supset B$ (or more generally to the sequence formed of $A \supset B$ and of sequence Y). Such a schema indicates in short under what conditions the implication operator can be introduced into an inferential situation. The inferential situation at point of departure is expressed by statement I, thus: $X, A \vdash B, Y$. In this statement, sequence X, A is called *antecedent* and sequence B, Y is called *consequent*. The inferential situation at point of arrival is expressed by statement II, thus: $X \vdash A \supset B, Y$. Statements I and II are, in relation to statements A and B, second-level statements. Let us call them inferential statements. As can be seen, the schema above indicates how the implication operator can be introduced into the consequent of an inferential statement. Another schema will show how this operator can be introduced into the antecedent of an inferential statement. These two schemata therefore characterize the implication operator by describing the rôle it can play in an inference.

In the same way, the other operators that figure in 'classical' logical theories can be characterized by appropriate schemata. Natural deduction logic, then, constitutes a genuine formalized theory of operators. Moreover this theory is at the same time a formalized theory of deduction. It is at a higher level of formalization than the 'classical' systems of logic and hence is more successful in isolating the formal aspect, which is logic's proper domain. It founds or grounds, in the sense that it provides a representation of the operations of logic in terms of a pure theory of inferences.

In the same line of research is the work on 'operative logic' developed by Lorenzen.[17] The fundamental concern of operative logic has been to give a constructive interpretation of logical operators. It has elaborated such an interpretation by analysing logical operators in terms of rules of operation; the meaning of a logical operator is fixed by the manner in which it can be introduced into a rule of operation. The constructivist preoccupations of operative logic tie in with those of intuitionist logic and of Hilbert's meta-mathematics. They are to be found in all present-day developments of intuitionist logic and also in the work of A. A. Markov and his pupil M. A. Šanin.[18]

(b) Combinatory logic is due to the work of Schönfinkel, Church, Rosser, Kleene and, above all, Curry.[19] It is a theory that studies the properties of certain objects called combinators. A combinator is an operator characterizable by the way in which it makes combinations among the variables it acts on. Let us take for example the inversion operation. This consists in changing the order in which the arguments of a function are given. It can be represented by combinator C, which is characterized by its mode of action on variables f, x, and y: $Cfxy = fyx$. This formula signifies: Combinator C transforms some functions, f, applied to variables x and y taken in that order into that same function f applied to the same variables taken in inverse order, y and x. Combinatory logic shows how more and more complex combinators can be constructed from a very small number of elementary combinators (such as the permutator just described) which are themselves defined axiomatically. It also shows how a certain number of fundamental concepts of logic and mathematics can be formulated in terms of combinators. (Thus it is that the theory of recursive functions can be formulated in the setting of a theory of combinators.)

From the foundational point of view combinatory logic appears to be the appropriate setting for study of the mechanisms that enter implicitly into the handling of ordinary logical systems. Thus combinatory logic has made it possible to elucidate the nature of the substitution operation that

17. Lorenzen, *Einführung in die operative Logik und Mathematik* (1955).

18. Markov, *Teorija algorifmov* (English version, *Theory of Algorithms* (1961); Šanin, *O nekotoryh logičeskih problemah aritmetiki* (= On some Logical Problems of Arithmetic) (1955).

19. Curry & Feys, *Combinatory Logic*, Vol. I (1958) and Curry & Hindley, *Combinatory Logic*, Vol. II (1973).

A historical account of combinatory logic is given in Curry, 'Combinatory logic' (1968).

comes into all systems where there are variables (these systems include a 'substitution rule' allowing substitution of one variable for another or replacement of a variable by a constant). Above all it has made it possible to elucidate the mechanism underlying certain classical paradoxes. Curry has shown that there exists a combinator Y with the strange property that if Y is applied to any term, X, a term Z is obtained which is itself identical with the expression formed by applying X to Z. If for X we take a term playing the role of negation, we thus get: There exists an expression Z such that this expression is identical to its own negation. If a system contains both this paradoxical combinator Y and a combinator corresponding to the implication operator, it will necessarily be contradictory. The only way to avoid the contradiction is to restrict the implication operator's field of application suitably (so as to keep the operator from being applicable to paradoxical-type expressions). For that, one has to elaborate a theory that will enable one to divide the expressions of the system in question into categories. Among these categories will be one of propositions; and it will be specified that the implication operator may only be applied to expressions in this category (paradoxical expressions belonging to another category). Curry did in fact succeed in constructing in the framework of combinatory logic a theory which does this, and which he calls the *theory of functionality*. (Combinators can be regarded as functions of the variables on which they operate. The fundamental combinator of the theory of functionality makes it possible to characterize the functional type of a combinatory expression, that is interpret it precisely *qua* function. It indicates the category of the expression obtained by applying a given combinator to certain arguments, and does so according to the categories of those arguments.)

Combinatory logic, then, makes it possible to thematize what goes on within known logical systems – not, of course, in their psychological but in their purely operative aspect. This aspect is entirely susceptible of objectivation and lends itself to formal representation. We can say that combinatory logic develops the theory of operations underlying ordinary systems of logic. It is not a metatheory but rather a theory of infrastructures. It studies not the properties of formal systems (non-contradiction, saturation, etc.) but the conditions under which formal systems function. It may readily be seen that combinatory logic can itself give rise to paradoxes and that their mechanism has to be made clear, just like that of the classical paradoxes. This leads to the idea of an underlying theory that would be responsible for thematizing the operating conditions of combinatory logic itself. And it appears plausible that this thematization of the foundations leads to an infinite regress, in other words that there is no ultimate theory capable of providing, by itself, a complete elucidation of its own mode of operation. Combinatory logic, then, gives us a glimpse of a situation which is in a way symmetrical with the one revealed by the well-known incompleteness theorems, particularly Gödel's. What those theorems have shown is that there exists no formal system capable of encompassing its own metatheory (considered both in its syntactical and its semantic aspects) and in particular that there exists no formal system capable of establishing, through its own resources, its non-contradiction. If we accept Hilbert's view that to found a formal theory is to demonstrate its non-contradiction, we can say that it is impossible for a system to be founded on itself and that accordingly there is no way of determining its ultimate ground or foundation. But we can take the expression 'to found' in another, more radical sense. To found a system is to elaborate the theory of the processes through which the system's expressions are formed and transformed, and through which derivations in the system are effected, and to determine, on the basis of such a theory, what conditions will be both necessary and sufficient to arm the system against any intrusion of paradox. A demonstration of non-contradiction only establishes the absence of paradoxes (or more exactly the impossibility of deriving a paradox). Foundational analysis in the sense just spoken of goes much further. It shows how and why paradoxes might appear and what rules must be laid down in regard to the system to ensure that no paradoxes can be formed. The prospects opened up by combinatory logic show that foundational analysis in this sense can be pursued but can most probably never be considered as ended. In this direction too it appears impossible to found a system on itself and therefore impossible to determine its ultimate ground or foundation.

What has been said raises a very far-ranging philosophical problem. It shows that, in the area of formal thought at least, the idea of radical founding or grounding is illusory and has to be replaced by the idea of foundational elucidation (in the sense of combinatory logic). But does this mean that the ultimate foundations of logic (understood as theory of formal systems) are necessarily non-logical? Does not the recent evolution of logic rather invite philosophy to re-examine the very notion of founding or grounding and, with it, certain classical notions like explanation, justification, truth, etc.? In a somewhat curious, yet basically very significant manner, this change in the way in which the problem of foundations in logic presents itself parallels the 'dismantling (*déconstruction*) of metaphysics' which forms a central theme in Heidegger's philosophy[20] and which has been taken up by certain authors who draw part of their inspiration from Heidegger[21]. According to this line of thought the very idea of founding, as used in consideration of the 'foundations' of mathematics and logic, must be interpreted as belonging to a certain form of 'metaphysics', i.e., to a certain conception of the being of being, in Heideggerian terminology. This is the conception that dominates modern times and that first found unequivocal expression in the Cartesian metaphysic, which for the first time determined being as 'objectivity of representation' and truth as 'certitude of representation'.[22] To undertake to get at the 'foundations' of the different

20. *Sein und Zeit. Erste Hälfte* (1927) (English translation, *Being and Time*, 1967), § 6.

21. See for example DERRIDA, *De la grammatologie* (1967).

22. 'Erstmals wird das Seiende als Gegenständlichkeit des Vorstellens und die Wahrheit als Gewissheit des Vorstellens in der Metaphysik des Descartes bestimmt.'

formal theories does mean attempting to establish their validity with certainty. The 'dismantling of metaphysics' shows up the assumptions that govern the whole modern conception of truth, through a more fundamental questioning that reactivates the original processes of ontology (as adumbrated in early Greek thought). It thus points to the existence of other ways in which truth can be constituted. From this point of view, the different light in which problems now present themselves within the framework of formal thought could be interpreted as heralding the advent of a new concept of truth.

§ 4. *The logic of science**

Certain developments in contemporary logic directly concern the problems raised by the scientific approach.

A first set of problems concerns the process of putting hypotheses to the test. The central question here is confirmation: Can a satisfactory, quantitative theory be elaborated that will enable us to give a precise meaning to the idea of 'confirmation of a hypothesis by an item of information of experimental origin'? This question has been approached from different sides by Carnap and by Popper. Whereas Carnap speaks of 'degree of confirmation' (in the sense of the credibility afforded a hypothesis, positively, by empirical data that conform to it), Popper speaks of 'degree of corroboration' (in the sense of the capacity a hypothesis displays for resisting attempts to invalidate it).[23] For Carnap, the problem of confirmation is closely tied in with that of induction on the one side, of the founding or grounding of probabilities on the other. Indeed, he distinguishes two concepts of probability: One is the frequential concept used in physics, the other a purely logical concept which concerns the relation between two propositions and is identical to the 'degree of confirmation' concept. (This latter relates two propositions, one expressing a hypothesis, the other an item of empirical information.) Despite the criticism Carnap's attempts to establish the foundations of an inductive logic have come in for, research into that logic has continued, partly along the lines he explored,[24] and partly in other directions.[25] The problem of induction has been brought into relation with semantic problems (for instance by way of construction of models for inductive systems of logic or by way of analysis of the notion

– HEIDEGGER, 'Die Zeit des Weltbildes' (1938), p. 80 in *Holzwege* (1950) (English translation, 'The Age of the World View', 1951).

* Taken from contribution by Jean Ladrière (cf. note * at beginning of § 3 above).

23. See CARNAP, *Logical Foundations of Probability* (1950) and *The Continuum of Inductive Methods* (1952); and POPPER, *The Logic of Scientific Discovery* (1959).

24. See HINTIKKA & SUPPES (eds.), *Aspects of Inductive Logic* (1966) and LAKATOS (ed.), *The Problem of Inductive Logic. Proceedings of the International Colloquium in the Philosophy of Science, London, 1965*, Vol. 2 (1968).

25. See in particular HACKING, *Logic of Statistical Inference* (1965).

of 'semantic information'). Moreover, this whole set of questions about induction and confirmation has also been developed through use of methods provided by information theory.[26] There has also been study of the relations between these questions and certain epistemic attitudes (e.g., the 'degree of acceptance' of a hypothesis).

A second set of problems concerns the mechanism of explanation. Here there is a theory which is regarded in some sort as a classic theory of explanation, Hempel's 'nomological-deductive' model.[27] But this theory appears inadequate; the simple logical notion of deduction appears inadequate to account for the explanatory link between propositions. It certainly seems that the notion of explanation brings in, in a manner that requires further elucidation, the notion of cause. There are attempts at formalizing this notion. The possibility of using more adequate concepts of implication than the concept of material implication of standard propositional logic is being studied. It would mean accounting for the link which the implication of ordinary discourse establishes, in other words giving the 'intensional' aspect of the concept of implication its due. The work of Ackermann,[28] Anderson and Belnap[29] should be mentioned in this regard. It also seems necessary to call on the logic of questions and answers.

Bound up with this question of the nature of explanation is the question of the characterization of nomological propositions. For every scientific explanation involves propositions that have the character of laws. But just what is it that constitutes this character? Can formal criteria be given? It certainly seems that the notion of 'law', just like that of 'explanation', has 'intensional' aspects. A way of expressing this is to point out that a nomological proposition is linked with certain epistemic attitudes; a law is a proposition which occupies a relatively secure position, it is something one is not inclined to abandon without good reason. Can we suitably explicate the notion of 'law' in the framework of a logic of epistemic modalities? And for the modalities that will necessarily be involved, can we construct a purely extensional interpretation based entirely on the notion of reference by using Hintikka's methods, which have already been mentioned with regard to semantics?

A third set of questions concerns the invention of hypotheses and the transition from one theory to another. There are, of course, psychological and sociological aspects to these processes. But apparently there are also logical aspects which ought to be elucidated in an appropriate theory. This problem is connected with Popper's work on 'the logic of scientific dis-

26. See HINTIKKA & SUPPES (eds.), *Information and Inference* (1970) and WATANABE, *Knowing and Guessing. A Quantitative Study of Inference and Information* (1969).

27. See HEMPEL, *Aspects of Scientific Explanation, and Other Essays in the Philosophy of Science* (1965), especially the chapters 'Studies in the logic of explanation' (taken from the article of the same title by HEMPEL & OPPENHEIM, 1948), and 'Aspects of scientific explanation.'

28. ACKERMANN, 'Begründung einer strengen Implikation' (1956).

29. ANDERSON & BELNAP, 'The pure calculus of entailment' (1962).

covery'. In this context the work of Hanson[30] and Kuhn[31] should be mentioned. Both in elaborating theories and in choosing among rival theories, certain formal criteria like the criterion of simplicity[32] or very general principles such as principles of invariance or of symmetry[33] seem to enter in. Such principles are not properly speaking laws. They are rather prescriptions indicating what criteria laws must obey. As such they have a metatheoretical character and can therefore serve as guidelines in the elaboration of new theories. Here we come back to the question of form. If one can speak of a 'logic of invention', even a 'logic of evolution', it is to the extent that the governing principles involved are formal in nature. But in what sense *are* they formal, exactly? For instance, is a principle of invariance purely logical in nature? Must one not suppose that the principles involved are even more general in nature and concern the way of using metatheoretical principles (such as principles of invariance)? The principles in question would relate to the general form of theories, and by virtue of that fact they could be considered logical principles.

But one may ask whether a 'logic' of content does not have to be invoked as well. When one theory is replaced by another, not only do certain metatheoretical principles enter into play; there is also internal rearrangement of the existing conceptual framework. Now this rearrangement does not take place arbitrarily: the new framework is descended from the old. Besides, the old theory often appears as a particular case within the new theory, which clearly shows that there is a logical relation between them. Thus we are led on to hypothesize an inner sublationary necessity which preordains the principles according to which the conceptual restructuring will have to operate. The theories would not be mutually discrete but connected among themselves by necessities of essence. And the appearance of new ideas would have to be explained not in terms of sudden discoveries but by the immanent development of the possibilities inherent in the concepts. Thus we would be brought to speak of a logic of contents to which we would look for the theory of conceptual restructurings, and for the principles of this immanent necessity underlying the theoretical rearrangements. The idea of such a logic quite naturally suggests the idea of dialectical logic. The chief reference here is to what is said on the subject in the posthumous work by Cavaillès,[34] the last sentence of which is at once a thesis and a programme: 'The motive necessity is not that of an activity but of a dialectic.' (*translation*)

30. Hanson, *Patterns of Discovery. An Inquiry into the Conceptual Foundations of Science* (1958).

31. Kuhn, *The Structure of Scientific Revolutions* (1962, 1970).

32. On this subject see, for instance, Barker, *Induction and Hypothesis. A Study of the Logic of Confirmation* (1957), Chap. 5, 'Induction and simplicity', pp. 91-105.

33. On this subject see Weyl, *Philosophie der Mathematik und Naturwissenschaft*. Pts. I, II (1926) (enl. English edn., *Philosophy of Mathematics and Natural Science*, 1949).

34. Cavaillès, *Sur la logique et la théorie de la science* (posthumously published, 1946): 'La nécessité génératrice n'est pas celle d'une activité mais d'une dialectique'.

§ 5. *Logic and ontology in contemporary Indian and Japanese thought**

A number of Indian and Japanese thinkers have reflected on the ultimacy and applicability of logic that bases itself on the law of contradiction and the law of excluded middle. In this connection, perhaps the most interesting and thoroughgoing work is the presentation by the Indian philosopher Kalidas Bhattacharyya of what he has called the *Logic of Alternation.*[35] This seeks to provide the logical underpinning to the philosophy of alternative absolutes propounded by his father, Krishna Chandra Bhattacharyya.[36] But it is interesting in its own right also. Kalidas' reflection on the logic of alternation may be briefly presented as follows:

The truth-functional logic finds the non-exclusive 'or' to be merely one of the logical connectives emerging in the 16 possible dyadic logical operators which are possible in a two-valued truth-functional logic. The non-exclusive 'or' is defined by the truth table which gives $\begin{matrix}T\\T\\T\\F\end{matrix}$. However, Kalidas has contended that it provides a fundamental clue to philosophical reality, if reflected upon in a certain manner. The relation is of such a nature that if one of the alternatives is asserted as true, then the other alternative may be either true or false; that is, both the possibilities are equally granted. In the *exclusive* 'or' which is defined by the truth table which gives $\begin{matrix}F\\T\\T\\F\end{matrix}$, the other alternative is definitely negated. The judgement in either case does not assert either of the alternatives but *only* that if either is asserted then the other has to be definitely negated (exclusive 'or') or it may be either true or false, that is, one is indifferent towards it (non-exclusive 'or').

Normally, it is contended that the use of the disjunctive judgement is a sign of our *ignorance* about the situation, that is, about what actually is the case. Once this is *known*, the uncertainty ceases and a definite categorical statement emerges. But there *is* no such necessity. It is quite possible that the situation in which man finds himself with regard to knowledge is irremediably of a *disjunctive* kind. And Kalidas' contention is that such is indeed the case. For example, if we reflect that, however long we allow for the knowledge-enterprise of man to fructify, its basic situation will remain the same, we might discover some point in what Kalidas is trying to assert. We would never have known *all* that is to be known, nor would all that is claimed to be known be true in the sense that later knowledge would never

* Written by Daya Krishna for inclusion in the present work.

35. Kalidas Bhattacharyya, *Alternative Standpoints in Philosophy* (1953).

36. K. C. Bhattacharyya, *Studies in Philosophy*, edited by G. Bhattacharyya (1956-58).

reveal its falsity. However, this situation itself may be understood in alternative terms. Either the reality itself is determinate in character and there is something intrinsic in the knowing process which makes it incapable of grasping reality completely, or the nature of reality itself is indeterminate so as to permit or even require alternative formulation of itself. But as there is no way of deciding between these alternatives, the ultimate situation can only be articulated in a disjunctive judgement which, thus, ceases to be provisional in character.

The Japanese philosopher Seizo Ohe, on the other hand, has questioned the applicability of the law of Excluded Middle to realms where we do not have clear-cut dichotomies.[37] It is not merely at the sub-atomic level that we fail to find such clear-cut dichotomies, but at the perceptual level also. But if this is so, then the attempt to use logical thought which assumes such dichotomies in order to understand experience which is not dichotomous would be misleading in character.

Imamichi, on the other hand, has suggested that if one accepts the law of contradiction then one cannot think about the whole or the totality, as the attempt to comprehend totality always leads to paradoxes. Mathematics has been aware of the problem, but, not being interested in ontology proper, has tried to deal with it in an *ad hoc* manner. Yet if totality has to be thought about – and it has to be thought about if one is to do any real metaphysics – then one is forced to give up the law of contradiction and treat the paradoxes as not paradoxical.[38]

The application of the law of contradiction to empirical reality has been questioned from another side.[39] It has been urged that this assumes the point-instant analysis of those objects to which one seeks to apply the law, since in all cases the limiting clause concerning the 'same point of space, and same instant of time' has to be used in order to validate the operation. Yet a strict interpretation of the clause would make the law applicable only in a vacuous sense, as every instant may be treated as a new instant, thus making the law irrelevant to whatever the objects to which one seeks to apply it.

The widespread interest in the problems which logic in general and modern logic in particular raise for philosophy may be seen in the volume containing papers discussed at a seminar devoted to that topic.[40]

37. OHE, 'The multiple structure of our external knowledge' (1964).
38. IMAMICHI, *Betrachtungen über das Eine* (1968).
39. KRISHNA, 'Law of contradiction and empirical reality' (1959).
40. KRISHNA, MATHUR & RAO (eds.), *Modern Logic. Its Relevance to Philosophy* (1969).

B. LOGIC OF PHILOSOPHY

Introduction

What has been said in § 2 and, more particularly, § 3 of sub-section A makes it plain that any foundational analysis leads into far-ranging philosophical problems: Is the idea of radical founding or grounding illusory, at least in the area of formal thought? Is the idea of an ultimate foundation for logic still a matter of logic? Is it not the actual idea of ground or foundation that has to be re-examined?

These same questions come to the fore again when we approach the problems the other way round and no longer consider the philosophical implications of logic, but the logical characteristics of philosophy itself.

In what sense can one speak of a logic of philosophy? Like all discourse, philosophical discourse is obviously dependent on logic in that it seeks to avoid contradictions and attempts to construct itself systematically. But the question is whether there is a specific logic of philosophical discourse. Now this question itself presupposes another: What exactly is the specific characteristic of philosophical discourse? Are these two questions strictly philosophical questions or do they fall outside the competence of philosophy to answer? Do they in fact already form part of that logic to the existence of which the first of them relates? If so, that would mean that the logic of philosophy must itself be able to furnish its own justification. And if not, to what then must we appeal in order to establish the possibility of a logic of philosophy?

Since philosophical discourse is characterized by the form of totality, and since its endeavour can be described as the attempt to make itself adequate to that form, we are prompted to think that philosophy must be capable of providing its own justification itself, and of specifying the law of its own discourse; in other words, that the questions just posed have to be examined inside philosophy and that philosophical logic, if it exists, has to be explicable by its own resources. If philosophical logic exists it is for it, and no other discipline, to answer questions concerning it. This is already enough to show us that a logic of philosophy will necessarily be circular in character.

§1. *Analytical philosophy: the task of elucidation**

As was said in the Introduction, we propose to divide contemporary work on the logic of philosophy into two main classes: philosophies of analytical and of synthetic type.

* §§ 1, 2 and 3 in sub-section B are taken from the contribution specially written by Jean Ladrière for inclusion in the present work (cf. note * at the beginning of § 3 in sub-section A above).

But an analytical philosophy may be *elucidatory* or *transcendental* in its approach.

In the first case, it sets out to clarify experience or, more exactly, the natural language which, for us, accompanies experience and provides us with an immediate interpretation of it. It attempts to analyse the mode of operation of the language and thereby of experience itself, either through use of the language's own resources or through recourse to those of an artificial language (e.g., to the methods furnished by formalized languages).

Contemporary work under the elucidatory approach falls into two groups. The first aims at defining a *'scientific philosophy'* by extending logic to certain forms of expression of the natural languages, the second at defining a *'philosophical logic'* closely related to the philosophy of language.

(a) *'Scientific philosophy'*

Under this head can be put work which is properly logical in character but extends the competence of logic beyond formal systems and sets out to analyse a number of forms of expression used by the natural languages. The basic idea is that natural language is fundamentally well-suited for its task, that it comprises an adequate interpretation of experience but that it involves ambiguities and may give rise to false problems; and that, in order to make it fully effective, it is necessary to eliminate the equivocations and difficulties resulting from misuse of language and accordingly to lay down the rules for its proper functioning.

This attempt at elucidating and purifying natural language called in turn for an appropriate extension of logic itself. While modern logic may generally be characterized as a discipline that studies formal systems, it nonetheless continues, in some of its branches, to treat of the correct forms of reasoning, in accordance with its traditional rôle. But analysis of valid inferences must necessarily be preceded by analysis of the language within which the inferences are formulated. When it began, and till around the middle of this century, modern logic was above all concerned with analysing the language used by mathematics. This led it to evolve systems which even today are considered as a sort of base: propositional logic, logic of classes, first order predicate logic, logic of relations.

But over the last twenty years a whole current has developed which has concentrated on analysing a number of forms of expression used by natural languages. (Actually in certain cases, e.g., the study of modalities, this contemporary research links up with work that dates from before the Second World War, or goes back even earlier to a very ancient tradition indeed.) In this context we should mention modal logic (more exactly the logic of alethic modalities, i.e., modalities based on the notions of possibility and necessity), the logic of action (including deontic logic, logic of norms, logic of imperatives, logic of decision, logic of preference) – to which we shall come back in Section V, *Man and Action* –, chronological logic (logic of time, of change, of processes), ontological logic (logic of existence, of the part and the whole, 'ontology' in Leśniewski's sense)[41] and epistemological logic (logic of questions and answers, of suppositions, of epistemic operators or of propositional attitudes such as assertion, belief, memory, hope, wishing, etc.).[42]

These different logics, usually starting from certain interpretations of the notions under study, try to construct axiomatic systems which will make it possible to characterize these notions. For example, one version of deontic logic interprets obligation thus: To say that *A* is obligatory is to say that if *A* does not take place then necessarily a certain state of things *S* (interpreted as 'sanction') does take place. Using this interpretation as base, Anderson was able to show that the logic of deontic operators (obligation, permission, prohibition) can be reduced to the logic of alethic modalities.[43] In the framework of an axiomatic system of deontic logic, the formal relations between deontic operators are systematically explicated. Therefore such a system constitutes an instrument for analysing these notions. In like manner an axiomatic system of chronological logic constitutes an instrument for analysing the structure of time. Of course, since different axiomatic systems are possible, we find that the ordinary notions branch out into several notions.

But axiomatization is only one aspect of the research we are considering here. It is generally agreed that three areas can be distinguished in the study of language: syntax, semantics and pragmatics.[44] Syntax studies a language's internal structure; it educes the rules for constructing permissible expressions (called 'well-formed expressions') and the relationships that exist among these well-formed expressions. There might for instance be relations of 'definability' among terms (Is such and such a term definable by means of such and such other terms within the language?) or relations of deducibility among propositions (Is such and such a proposition deducible from such and such other ones within the language?). It is in studying relations among propositions (and, through them, among terms) that the axiomatic method proves useful. Semantics studies the relations that exist between expressions in a language and the objects (in the most general sense of the term) to which they can refer. Finally the pragmatics of a language studies the conditions of its use, in other words studies relations between its expressions and the contexts in which these are used by speakers.

41. LEŚNIEWSKI, 'Grundzüge eines neues System der Grundlagen der Mathematik' (1929, 1938); 'Über die Grundlagen der Ontologie' (1930); 'O podstawach matematyki' (= On the foundations of mathematics) (1927-31).

42. For all this see RESCHER, *Topics in Philosophical Logic* (1968), Chap. I, 'Recent developments in philosophical logic', pp. 1-13. This study includes a bibliography of the subject which RESCHER used as basis for pp. 37-40 of 'Recent developments in philosophical logic', one of his three contributions to KLIBANSKY (ed.), *Contemporary Philosophy, A Survey*: I, *Logic and the Foundations of Mathematics* (1968).

43. ANDERSON, 'The formal analysis of normative systems' (1956, 1957).

44. This distinction was proposed by MORRIS, C. W. in *Foundations of the Theory of Signs* (1938). See also his *Signs, Language and Behaviour* (1946).

The work of Tarski[45] and Carnap[46] showed that it was possible to formalize the semantics of formal languages. Formalized semantics developed in the shape of model theory, which turned out to be extremely fruitful both for the study of formal systems and for that of mathematical structures. But it emerges that the methods of formalized semantics can also be used for studying the semantics of natural languages. At any rate they have already been used successfully in the logical theories discussed above that concern what might be called the formal aspects of natural languages. One current trend in research is to extend the applicability of formal semantics methods further and further. Indeed use of semantics seems not only fruitful but indispensable. At least this has been demonstrated as far as modal logic is concerned.[47]

Pragmatics on its side has not as yet had a development comparable to that of syntax and semantics, but at least it, too, is being formalized, above all thanks to the work of Martin[48] and Montague.[49] Indeed pragmatics as conceived by Montague appears as a general formal theory of interpretation covering both semantics and the study of contexts of use.

The methods of (formalized) semantics and pragmatics do not merely contribute (along with the regular syntactical methods) to elucidate the significance of concepts such as: possible, necessary, obligatory, permitted, past, future, etc. They also make possible precise analysis, in relation to determined contexts, of general semantic concepts such as: designation, denotation, connotation, reference, meaning, satisfaction, truth, analytical proposition, synthetic proposition, synonymity, logical equivalence, etc.

In particular semantic research has shed new light on the notion of significance. Since Frege, it has become usual to distinguish two aspects of significance: sense (*Sinn, sens*) and reference (*Bedeutung, référence*). To convey the 'meaning' of a term or a phrase it seems we have to make use of notions of 'intensional' (or connotational) type, as opposed to notions of 'extensional' type; such 'intensional' notions are concept, proposition, intention (or connotation), etc. Now Hintikka and others have shown that it is possible to absorb the theory of 'meaning' in a theory of 'reference', if not absolutely at least for most known problems.[50] The key concept here is that of the 'possible world', which Carnap had already introduced

45. TARSKI's early work on semantics was reprinted in *Logic, Semantics, Metamathematics. Papers from 1923 to 1938* (1969). This collection includes in particular (in English translation) his well-known paper on formalizing the concept of truth, which has been crucial in the development of formalized semantics.

46. CARNAP, *Introduction to Semantics* (1942) and *Meaning and Necessity. A Study in Semantics and Modal Logic* (1947, 1956).

47. MONTAGUE, 'Syntactical treatments of modality, with corollaries on reflexion principles and finite axiomatizability' (1963).

48. MARTIN, R. M., *Toward a Systematic Pragmatics* (1959).

49. MONTAGUE, 'Logical necessity, physical necessity, ethics, and quantifiers' (1960) and 'Pragmatics' (1968).

50. See *inter alia* HINTIKKA's 'Semantics for propositional attitudes' (1969) and 'Logic and philosophy' (1968).

in his work on semantics and the foundations of probability theory. A statement is interpreted not as relating merely to *one* clearly determined universe but as relating simultaneously to *several* possible universes which might be infinite in number. And it is stipulated that in general an individual term (such as a name or the description of an individual) can refer to *different* individuals in different possible universes. A semantics constructed on this basis enables one to interpret the different kinds of modality (alethic, deontic and epistemic) satisfactorily without having to use intensional notions.

Montague's pragmatics uses similar ideas for interpreting linguistic expressions whose meaning depends on context. These are expressions that act as indicators: personal and demonstrative pronouns, adverbs of place and time, tense forms (present, past, future). In fact it seems possible to integrate the interpretation of modalities and of indicators in one single theory. Montague's idea is to construct interpretation of a language by introducing two interpretative sets simultaneously: one is a set of 'points of reference', the other the set of possible individuals (an individual in this second set is not supposed necessarily to exist for *all* points of reference). The set of points of reference will be the set of instants, or of speakers, or of possible worlds, as the case may be. In the context of this theory, theory of meaning can also be absorbed in a theory of reference. More exactly, the notions 'intension of a term' ('meaning of a term') and 'intension of a sentence' ('meaning of a sentence') can be defined on the basis of the notion 'extension of a term'.

As can be seen, study of modal contexts, propositional attitudes, indicators, etc., represents a progressive extension of the applicability of modern logical methods (syntactical as well as semantic) into the field of the natural languages. This research is, however, to be distinguished from scientific linguistics research using methods inspired by those of formal logic or methods that are frankly mathematical in nature (the work of Chomsky and his school; work in mathematical linguistics).[51] The latter research is concerned not with the content of language, considered in all its concrete diversity, but with certain general forms that are applicable to the most varied contents. If linguistics uses formal methods it does so to the extent that it is trying to reveal the *underlying* mechanisms whereby expressions of a language are formed (syntactic viewpoint) and whereby these expressions can be coupled with sounds (phonological viewpoint) and meanings (semantic viewpoint). When one studies modalities or the status of pronouns one is concerned with certain definite expressions which appear *explicitly* in the language but whose precise manner of operation one wishes to determine.

51. See especially CHOMSKY, 'Formal properties of grammars', and CHOMSKY & MILLER, 'Introduction to the formal analysis of natural languages' (both, with a bibliography, in Volume II of the *Handbook of Mathematical Psychology*, edited by LUCE, BUSH & GALANTER, 1963); KATZ & POSTAL, *An Integrated Theory of Linguistic Descriptions* (1964). For mathematical linguistics, see MARCUS, *Lingvistică matematică* (1963) (French version, *Introduction mathématique à la linguistique structurale*, 1967), and *Algebraic Linguistics. Analytical Models* (1967).

Now it so happens that it is just these forms of expression that have been prompting so many questions of a philosophical nature: the nature of modalities, the structure of time, the difference between norms and facts, the nature and scope of propositional attitudes, etc. It also happens that semantic methods, which are particularly important in studying these forms of expression, enable us to subject to rigorous examination the traditional philosophical problems of a semantic nature: the distinction between analytic and synthetic, the nature of the relation of significance, the nature of the relation of truth, etc. Hence all this research can be grouped under the general heading 'philosophical logic', to indicate that it involves the re-examination of problems of a philosophical nature in a theoretical context employing the methods of formal logic. We should make it clear that this is by no means a matter of linking up a particular theory with the 'classical' theories by way of deduction but only of extending the applicability of certain methods. Obviously this extension is linked with the intervention of certain hypotheses as to interpretation of the concepts studied: every attempt at formalization involves a particular interpretation. (Thus, as we have seen, the system of deontic logic constructed by Anderson rests on a modal interpretation of the concept of obligation.) That the methods of logic prove effective in such contexts must be accounted for by the 'formal' character of the concepts concerned. But this raises a knotty problem: in just what sense are those concepts 'formal'? What does a modal operator, like 'it is possible that...,' have in common with an operator in ordinary propositional logic, like 'if... then...'? And what does a modal operator have in common with an indicator, like the personal pronoun 'I'? One might say that all these concepts are representable in the framework of a generalized formal semantics. But that is only an indication. The question is *why* it is so. Perhaps here we are up against a question that can no longer be formalized. And supposing it can be, from one point of view, in the framework of a very general formalism, then the very possibility of such formal treatment would in its turn pose a question of the same order. When one poses the problem of the nature of the formal, it seems that at a given moment one has to have recourse to a language which is no longer, or not yet, a formalized language. In other words, it does not seem possible to conceive an entirely formalized explanation of the actual possibility of formalization.

Be that as it may, what has been said comes down essentially to the fact that certain philosophical problems can be treated by scientific methods, at least in some of their aspects. Nothing prevents us from thinking that in future such methods will become more and more widely applicable, in the general field of 'philosophical problems'. It would therefore be appropriate to speak in this connection of 'scientific philosophy'; the label is preferable to 'philosophical logic', which, as we shall see, can be taken in an altogether different sense. Yet it must be admitted that 'philosophical logic' in the sense of 'treatment of philosophical problems by the methods of formal logic' has already come into widespread use.[52]

52. See Rescher, *Topics in Philosophical Logic* (1968). See also, for example,

(b) *'Philosophical logic'*

We have in fact, at this point, to refer to an entirely different conception of philosophical logic; the one proposed by P. F. Strawson.[53] In the *Tractatus logico-philosophicus*, Wittgenstein suggests that at least in principle all formal logic, or at least its classical core (propositional and predicate logic), could be elaborated from the single idea of proposition in general.[54] Strawson sees in such a suggestion a typical example of an approach that depends not on formal logic but on 'philosophical logic'. Formal logic uses certain notions like proposition, propositional operator, deduction, etc. The significance of these notions may well be queried: and the task of elucidating them depends, no longer on formal logic, but rather on a logic that is philosophical in nature. In this connection it would perhaps be more exact to speak of 'philosophy of logic'.

The question brought up earlier as to expressions of 'formal' character in natural language obviously comes under such a discipline. With a view to educing the content of 'philosophical logic', Strawson draws attention to the following problems: What is the general form appertaining to language expressions which we recognize as having 'propositional' status? What is the difference and what are the relations between reference (the modus of expressions that serve to identify particular aspects of reality) and predication (the modus of expressions that serve to characterize, from a certain general point of view, the aspects of the real of which we are speaking)? What is the exact nature of propositional connectors and in particular of implication, which comes into propositions of conditional form? How should the notion of significance be analysed? The notion of truth? Of analytical truth? How to account for the difference between falsity (which consists in attributing to an object a predicate which does not belong to it in fact but which could belong to it) and absurdity (which consists in attributing to an object a predicate which cannot apply to it, e.g., attributing to a living creature an arithmetical property like 'being a prime number')? The last problem just give rise to a theory of categories: the expressions in a language have to be broken down into classes and rules of predication formulated on that basis which will make it possible to avoid 'errors of category'. It is to be noted that this problem is different from that posited by the theory of functionality. We could express the difference between them by saying that that theory concerns syntactical categories whereas the one we are now considering concerns semantic categories. It concerns categories that involve the 'meaning' of the expressions studied. Work on this problem of categories should be viewed in conjunction with the research that has been developed

DAVIS, HOCKNEY & WILSON (eds.), *Philosophical Logic* (1969).

53. See his 'Introduction' to STRAWSON (ed.), *Philosophical Logic* (1967). See alsc QUINE, *Philosophy of Logic* (1970).

54. WITTGENSTEIN, 'Logisch-philosophische Abhandlung' (1921) (*Tractatus logico-philosophicus*, 1922, 1961): see No. 5.47.

in the field of linguistics under the name 'structural semantics'.[55] On either side there is the search for a general method for identifying the semantic structures underlying the expressions used in a given language. Such a method must make it possible to characterize the relations of synonymy, antonymy, exclusion, compatibility, incompatibility, etc., that exist among the terms of a language considered in their significative content.

Evidently, the problems of 'philosophical logic' are very close to those of the philosophy of language. Whereas linguistics studies concrete languages, the philosophy of language studies the way in which language operates. In particular it asks questions about the nature of the semantic relations (designation, significance, truth, etc.) that are essential to the operation of a language. Even more fundamentally, it asks how, in general, a system of expressions can acquire significative value, or again how semantic relations are possible. Now such questions bear on the formal aspects of language; that is, on the properties a system of material signs has to have in order to function as language, independently of the particular contents proper to concrete languages. But this brings us back to our earlier question: Just what is the criterion of the 'formal'? This question is closely bound up with the following one, which concerns the relations between logic and the philosophy of language. If the problems of the philosophy of language are formal in nature, does that mean that, as such, they belong to logic? But what could be the logic they would belong to? If one thinks these problems can be treated in the setting of a formal logic (though possibly one far more general than 'classical' formal logic), will one not inevitably return to the question we have just recalled: What, then, is the aspect of language that lends itself to analysis within a formal theory? And will one not be led to ask at the same time: what, ultimately, is the specific nature of formalized languages over against the natural languages? On the other hand, if one thinks these problems have to be treated in the framework of a philosophical logic (in the sense suggested by Strawson), will one not be driven to ask: What, after all, differentiates such a logic from formal logic? To elucidate such a question one will inevitably have to ask what is specific to formal theories. So, no matter how one comes at them, the problems mentioned above all refer back to a more fundamental problem: What about form? Can the problem of form be considered merely one of 'philosophical logic'? Does it not rather depend on an overall philosophical approach that must be able to interpret the very status of logic and situate logic in relation to language in general, on the basis of a doctrine of structures valid for structures existing in the world as well as for those found in linguistic systems (formal or not)?

These are the questions with which, to a great extent at least, Wittgenstein's philosophy is concerned.[56] Both the *Tractatus* and the *Philosophical*

55. LYONS, *Structural Semantics. An Analysis of Part of the Vocabulary of Plato* (1969); GREIMAS, *Sémantique structurale* (1966).

56. WITTGENSTEIN, 'Logisch-philosophische Abhandlung' (1921) (*Tractatus logico-philosophicus, 1922, 1961*) and *Philosophische Untersuchungen/Philosophical Investigations* (1953).

Investigations treat of the possibility of language. In the *Tractatus* the problem of form is central, since it is at the very heart of 'picture theory', the theory of meaning peculiar to the *Tractatus*. The *Philosophical Investigations* bring in the notion of 'language game' and correlatively of 'constitutive rules'. But here too the question of form reappears, since a 'language game' is characterized as a 'form of representation' (*Darstellungsform*).[57]

'Philosophical logic' poses another problem. As we have seen, many questions of philosophical nature (like the distinction between 'analytic' and 'synthetic') can be studied by scientific methods. It is highly probable that the field of questions that can be scientifically treated will continue to expand. No limits can be assigned in advance to the applicability of formal analytical methods. However, when a problem is subject to scientific treatment in this way, it is so usually only in certain of its aspects. Thus, as we have seen, formal semantics can throw light on the distinction between meaning and reference. But it still leaves us with the general problem of the nature of significance; for that matter, this problem governs the very possibility of formal semantics. In every single case one would have to be able to specify what aspects of the problem do in fact come within the domain of science and what aspects still appertain to 'philosophical logic'. Moreover every advance towards formalization brings up again, in a new shape, the basic problems of philosophical logic, such as the problems of significance or of form. Thus while the domain of what pertains to science keeps on endlessly expanding, as fast as its frontier is pushed back the properly philosophical problems of logic (and especially the problem of the very possibility of this expansion) seem to crop up again; they are ceaselessly revived by the very progress of the scientific approach.

Among the philosophical problems posed by the development of logic we should, in concluding, call particular attention to the problem of the relation between logic and mathematics. If logic becomes the general theory of formal systems, the boundary between logic and mathematics blurs. On the other hand the logicist theory that mathematics is reducible to 'purely logical' principles encounters difficulties that look to be decisive.[58] Truth to tell, what is involved here is, once again, the nature of logic. If logic plays a leading rôle in mathematics, it is to the degree that mathematics has a formal aspect. But it what sense exactly must one say that logic is the science of form? Besides, it does seem that in mathematics there are specific contents irreducible to pure forms. At all events that is what is suggested by results like the impossibility of adequately characterizing arithmetic by means of an axiomatic system, or the inability of the axiomatic system of set theory, in this present state at least, to characterize appropriately the notion

57. See Wittgenstein, *Philosophische Untersuchungen*, 1st pt., § 122, p. 49: 'Der Begriff der übersichtlichen Darstellung ist für uns von grundlegender Bedeutung. Er bezeichnet unsere Darstellungsform, die Art, wie wir die Dingen sehen'. – 'The concept of a perspicuous representation is of fundamental significance for us. It earmarks the form of account we give, the way we look at things.' (P. 49e of the authorized English translation by G. E. M. Anscombe, *Philosophical Investigations*).

58. On this topic see Church, 'Mathematics and logic' (1962, 1968).

of set. (At any rate that is what is apparently indicated by Gödel's and Cohen's results in regard to the continuum problem: in its present state, the axiomatic theory of sets admits both models in which the continuum hypothesis is true and models in which it is false. An adequate theory ought to make it possible to determine whether it is true or false.)[59] But, if this is the case, one ought to be able to understand why. In other words, one ought to be able to specify what differentiates properly mathematical concepts from formal concepts. This question, however, is strictly correlative with the question of form.

§2. *The analytical type (contd.): the transcendental approach**

Within the analytical type we have distinguished an elucidative and a transcendental approach. By a transcendental approach we mean that whereby an analytical philosophy investigates under what conditions language is possible. (In short it is the kind of undertaking to be found in Wittgenstein's philosophy. But in a sense this is also the way in which Kant saw the problem, at least in the *Critique of Pure Reason.*) An inquiry into what makes language possible, particularly into the nature of the significance relation, is necessarily also an inquiry into the limits of language. Now to lay down limits for language is to refuse to use it in an attempt to 'totalize' experience: it is on the contrary to accept that there is a 'beyond the limit' which on the one hand must belong to the very structure of language (language being possible only on the basis of what limits it) but which on the other hand cannot be thematized within language.

What of logic, then? In the case of a philosophy of elucidation, logic merged with the analytical method and took the form either of a scientific philosophy or of a philosophical logic in the sense considered above. In the case of a transcendental philosophy, the logic will be what Kant called a 'transcendental logic'. As logic, such a discipline is concerned with the formal aspects of language (or experience), that is to say, with determinations that are independent of content (content proper to a particular language or experience). As transcendental, it is concerned with those forms that constitute the very conditions for the concrete (be it an actual language or a determinate content of experience) to arise. These conditions have a character of necessity: without them there can be neither language nor experience. The necessity involved here has a peculiar nature: it concerns not the deductive link, which formal logic treats of, but the status of the *a priori* and its structure.

It is worth noting that there has in recent times been a substantial revival of interest among analytical philosophers in Kantian studies and in the

59. On this subject see, for instance, MOSTOWSKI, 'Recent results in set theory' (1967): 'Such results show that axiomatic set-theory is hopelessly incomplete' (p. 93).

* Taken from contribution by Jean Ladrière (see note * at the beginning of § 1 above), with supplementary material provided by Alan Montefiore.

nature of transcendental arguments. The most obvious name to mention in this connection is probably once again that of Strawson with his *Individuals* (1959) and above all his study of the *Critique of Pure Reason, The Bounds of Sense* (1966). Other influential studies of Kant include Jonathan Bennett's *Kant's Analytic* (1966) and S. Körner's *Kant* (1955). Among books worth mentioning for their interest in or use of transcendental arguments are Stuart Hampshire's *Thought and Action* (1959), especially Chapter I, and Sydney Shoemaker's *Self-Knowledge and Self-Identity* (1963); among articles 'Transcendental arguments' by Barry Stroud (1968) and 'Are transcendental arguments a version of verificationism?' by Peter Hacker (1972). Strawson's contribution to Bryan Magee's collection of radio talks, *Modern British Philosophy* (1971), is also of very considerable interest for the account which it gives of what seems to him to be his own very Kantian philosophical development as he looks back on it.[60]

One task of a transcendental logic is to explain itself, to define its own task and establish that the task is legitimate. If it is truly transcendental it cannot refer to another authority, upon which it would be founded, but must itself elucidate its own foundations and therefore develop a theory of the *a priori* as such. Before showing what the structure of the *a priori* is, it has to show what are its nature and function; it is indeed only after determining the foundations in this way that it will be able to educe the structure of its content. Thus it will be led to investigate in what sense the *a priori* is form; and, as regards the problem of form, it will have to take up a position in relation to formal logic. In this context we should mention Husserl's *Formal and Transcendental Logic*[61] and also the book by Cavaillès which has already been alluded to.[62] In certain respects Wittgenstein's *Philosophical Investigations* also partake of this approach, by virtue of the concept of grammar. The central question that arises here is whether the *a priori* represents only the locus of the most general forms or whether it does not already contribute to determining the content; to put it differently, whether the *a priori* has not a certain 'material' character. If it is truly constitutive, then it seems that it must surely provide not merely very general rules, but an articulated system that must function inside concrete reality as what actually holds it together, i.e., as the sum of the determinations that confer on it its actuality. It is doubtless within the framework of this approach to

60. As one might expect, works such as those of Strawson, Hampshire and Shoemaker find appropriate mention under other headings: respectively, Section IV, *Man and Language*, sub-section B, § 1, 'The "analytic" approach', pp. 1338 sqq., and Section VI, *Man and the Foundation of Humanism*, sub-section B, § 1, 'The analytic conception of man and humanism and the concept of mind', pp. 1434 sqq.; Section V, *Man and Action*, sub-section B, § 1, '"Analytic" philosophy and the problems of human action', pp. 1392 sqq.; and Section VI, sub-section B, § 1, again.

61. HUSSERL, *Formale und transzendentale Logik. Versuch einer Kritik der logischen Vernunft* (1929) (English translation, *Formal and Transcendental Logic*, 1970). See the important study by the French translator of this work, Suzanne Bachelard, *La Logique de Husserl* (1957) (English translation, *A Study of Husserl's Formal and Transcendental Logic*, 1968).

62. *Sur la logique et la théorie de la science* (cf. sub-section A, § 4, p. 1107).

the *a priori* that it will be possible to treat the question we called attention to earlier as to how the possibility of formal logic can be grounded. This question is in fact tied up with that of form, and it is to a theory of the *a priori* that falls the task of pronouncing on the nature of form, defining its modalities and thereby determining, within the limits of the theory itself, the locus of formal logic and where the dividing line between it and transcendental logic is situated.

§ 3. *Synthetic philosophy: non-Marxist systems**

A philosophy of synthetic type purports to be a system: that is, a conceptual whole that is both formally coherent and materially adequate. The criterion of adequacy concerns the system's ability actually to accommodate all experience, or more exactly to provide an interpretation satisfying the condition of completeness. This interpretation must constitute an extension of natural language (and of the interpretation which that extension provides) such that any new extension would necessarily already be contained in it. In other words a system must have a totalizing character, must by appropriate extrapolation yield a fully self-contained experience. It must accordingly be in a position to furnish its own justification and must therefore be circular in character. But philosophies of synthetic type do not all present the same type of systematicity. By way of hypothesis, three main varieties of system could be proposed: categorial, hermeneutic and dialectical. By 'categorial system' we here mean a coherent, closed and connected set of categories providing an interpretative framework for all aspects of experience. (By saying it must be 'connected' we mean that all the categories must be linked together, that it must be possible to pass from any one of them to each of the others.) This is the conception of system found, for example, in Whitehead's speculative philosophy.[63] A hermeneutic system purports to be a progressive deciphering of experience disclosing, in successive layers, the structures which are embedded in what is given and in terms of which there develops an increasingly adequate comprehension of experience through itself. Philosophies of phenomenological style could be cited as belonging to this type: the very idea of phenomenology is that of a progressive unveiling of the phenomenon, i.e., of an exposure, as exhaustive as possible, of all that is implied in it. A phenomenology is a theory of manifestation. Finally, a dialectical system is the complete exposition of the constitutive categories of reality insofar as these are linked together according to a process of necessary generation which is the motive force of the system itself.

What is the position of logic in these different varieties of system? In all three, a logic can be spoken of as the theory of the system's systematicity. For each, its corresponding logic represents within it that which permits it to

* Taken from contribution by Jean Ladrière (cf. note * at beginning of § 1 above).

63. WHITEHEAD, *Process and Reality. An Essay in Cosmology* (1929).

be constituted as a system. In other words, the logic is the theory of the conceptual chains by means of which the system is constructed. This implies that it is not a logic of form only: it must bring out what, in the concepts themselves, necessarily relates them to one another and at the same time points beyond them. In a categorial system, logic has the task of determining how the system's categories form a coherent, closed, connected whole. It is therefore the theory as to how the categories operate; specifically, it must determine what gives them their particular status and enables them first to form an interconnected whole and then to constitute, in and through their interconnection, a medium for the interpretation of experience. It thus assumes the character of a transcendental logic. For categories are not only representations; they are constitutive, forming the framework of reality. In a hermeneutic system, logic is a theory of interpretation. Its task is to show how the 'passing over' process that brings to light the structure of the phenomenon can and must take place, how interpretative concepts develop, and what makes each interpretative stage call forth the next stage. In a way, logic has to bring out the very necessity of the steps that are being taken and, thereby, found the legitimacy of the interpretation that is exposed. In a dialectical system, logic is a theory of the system's self-generation or, what amounts to the same thing, the concept's internal coming into being. In other words it is itself dialectic. As such it is both a theory of discourse and a theory of the real. It is a theory of discourse in that it makes explicit the immanent laws (internal necessity) according to which concepts are linked, and it is a theory of the real in that, the system being materially adequate, the way in which the concepts are linked is also the way in which the corresponding aspects of the real are linked. In this sense logic is identical with the system itself, which is nothing more nor less than a theory of reality in its entirety.

Three references are important for dialectical logic: Hegelian logic, Eric Weil's 'logic of philosophy',[64] and dialectical materialism. In the Hegelian system logic is defined as 'the science of the idea in itself and for itself.'[65] As such it makes up only a part of the system, but a part of cardinal significance. It so to speak delineates in advance the necessary links according to which the real can and must be produced, and according to which therefore the continuation of the system – which has ultimately to link up with the real, even in its most concrete aspects – must also necessarily be constructed. In a sense logic is abstract in relation to the rest of the system; after logic come philosophy of nature and philosophy of spirit, which have to describe the successive stages by which absolute reality goes outside itself, in order then to come back towards itself and find itself again in the presence-to-self of absolute reflection. And together, these successive stages are

64. WEIL, *Logique de la philosophie* (1950).

65. HEGEL, *Enzyklopädie der philosophischen Wissenschaften im Grundrisse* (1830): 'Die Logik, die Wissenschaft der Idee an und für sich' (p. 51 in Nicolin & Pöggeler's edition).

just what constitutes concrete reality in its two great facets, nature and spirit. But in another sense logic already is the entire system and as such is adequate to concrete reality: indeed it provides in advance the categorial design according to which the further movement of the system will be effected – as also, for that matter, the concrete movement of reality. If dialectical logic presents this double aspect, it is because it is not purely formal. It is the law of discourse but at the same time the law of reality, and shows of itself that it has this character by showing how discourse is the very structure of reality. More exactly, it shows how discourse and reality are produced, are constituted, in the movement of '*Aufhebung*' (sublation). But it shows this by accomplishing this movement within itself: thus it is both the theory of the dialectical (sublationary) process and the process itself at work. Hence it has a circular and self-founding character. Inasmuch as it is self-founding it is foundational for the whole system and unites within itself the form of the connection (the dialectical pattern itself) and production of the content. The peculiarity of dialectical logic is to be a form which in developing engenders the content of which it is the form.

Eric Weil conceives of the logic of philosophy as *philosophia prima*, that is as the foundation for all particular philosophies: ontology, ethics, philosophy of nature, political philosophy, etc. It is none other than the progressive constitution of totally coherent discourse, i.e., of coherent discourse which comprehends itself, and rises to explicit recognition of what makes it possible and at the same time fulfils it. This absolutely coherent discourse is therefore necessarily complete: it is made up of the succession of all possible coherent discourses, each of which is possible only through and in this total discourse and has meaning only insofar as it leads in its direction. More precisely, the logic of philosophy is the expostition of a complete system of categories, i.e., a system of categories constituting the necessary formal framework of every concrete action and every particular discourse – in a word, of all experience. The exposition in question is not made from without. It consists in showing how the system generates itself, i.e., how one passes from one category to the next until, finally, the system closes on itself. Now since in a sense the development of a category is what leads to the next category one could say that the exposition of a system is in reality its self-generation. What provides the motive force is the relation between attitude and category. Every category is born of the reflection of an attitude onto itself, but as soon as a category has appeared it makes a new attitude possible, whose development will make the following category appear. In other words, it is in being exercised, actualized, that a category reveals in what way it is sublated and, so to speak, calls on the next category. But this passage through actuality, represented by attitude, is as it were the concretization of the category. In a sense the system of categories is nothing but the form of all possible experience. But it is only by being actualized, in the life of concrete man, that it emerges as a system. It is, as Eric Weil puts it, 'the logos of the eternal discourse in its historicity'.

(*translation*)[66] It is therefore form which only determines itself as form by posing a content, in accordance with the very essence of all dialectical discourse.

§ 4. *The synthetic type (contd.): logic and dialectic in dialectical materialism**

In dialectical materialism, dialectical logic appears as that part of philosophy which deals with the fundamental laws of reality. Since it is taken as axiomatic that reality can be definded as matter in movement, these fundamental laws are the laws of the movement of the whole. And they are valid for thought as well as for nature. Thus dialectical logic, as in any dialectical theory, has really ontological significance. That is, it is not purely formal but constitutes a logic of content, a theory of the self-generation of content. In this sense 'dialectical logic' as a theory of thought is to be regarded as identical with objective, materialistic dialectic and distinct from formal logic. Whereas the latter considers the forms of thought only from the point of view of their structure, dialectical logic considers them from the point of view of their origin and development. But the reason why this is so is precisely because it is a general theory of the laws of development of all reality. In one sense it determines only the general forms of development, of which it educes the laws. But in another sense it is identical with the overall theory of reality, with materialism itself, since in educing the laws of development it defines reality (matter in movement) in its quintessence.

By its very nature dialectical materialism resists every philosophical venture that would separate logic and the formal disciplines. As a scientific-philosophic world-vision it strives towards a complete comprehension of the world and of man's place in that world. The philosophy of Marxism starts with recognizing the unity of principle between the characteristics of objective reality and the knowledge of that reality. Its foundation is the theoretical analysis of social activity, i.e., of that specific and universal human activity which, in the final reckoning, transforms the reality surrounding man, and man himself.

Not only do the formalisms proceed from an abstraction performed on this overall comprehension; theoretical activity itself could not be separated from practice without its own character's being altered. Hence the compulsory starting-point for a study of the logic and gnoseology of dialectical materialism is inevitably their relation to practice.

According to Marxism, philosophical analysis of practice makes it possible to eliminate a series of problems that are traditional and insoluble for classical 'bourgeois' philosophy, to surmount them by treating them as insoluble

66. Weil, *Logique de la philosophie*, p. 77: 'logos du discours éternel dans son historicité'.

* Based (with occasional rearrangement and summarizing) on a paper written by V. I. Lektorskij as part of the general contribution to the present work prepared under the direction of T. I. Ojzerman.

because falsely posed: for example, the problem of the thing in itself in Kant. For Marxism, the relation of consciousness to reality, of the ideal to the real, can be correctly understood only if the ideal is taken not as an independent essence but as a specific expression of the reality of social man. The same holds true for the gnoseological problem of 'primary' and 'secondary' qualities, for the rôle of cognition as original 'mediator' in the relation of subject to object, etc. There are various studies of practice as criterion of truth.[67] In substituting practice for sense data as the criterion of truth Marxism contrasts with philosophical empiricism. It holds the neo-positivist principle of verification to be too narrow and limited to particular lines of investigation. Into the content of this practice there enter not only sense data but instrumental factors; practice functions as criterion of truth inasmuch as it actually constitutes, itself, the historical process of social development. Practice also subsumes the sense experience of social man, as conditioned by social and historical factors. Thus the forms in which we perceive and observe the world, and distinguish in these forms objective characteristics, conditioned as they are by the forms of social reality and language, turn out to be subordinate forms of practice. Only in practice is man in immediate contact with the object; and practice alone gives rise to a multilateral assimilation of reality. It is by virtue of this that it forms the criterion of truth.

It also follows from the historical significance of practice that, in the philosophy of dialectical materialism, the truth of knowledge is not understood as a static characteristic attaching to particular elements; it is understood as dynamic, as the process of an increasingly multilateral and adequate reflection of the real. There is therefore no Marxist understanding of the dialectical nature of truth without the interpretation of practice in its quality as criterion.

However, in recent years Marxist philosophers have begun to be more critically aware that the criterion of truth is inadequately elaborated in Marxist philosophical literature. In particular, it is evident that for any given degree of development of practice there may be a number of corresponding theoretical hypotheses in science. The choice between these rival hypotheses is made not only on the basis of their correspondence to the existing practice but also with the help of certain other criteria which, while deriving from – though not replacing – the fundamental principle of practice, are yet relatively independent (the criterion of simplicity, the criterion of a theory's consistency with previously acquired knowledge, etc.). Moreover, the existence of different types and levels of practice implies that practical success can accompany theoretical constructs that reproduce, not authentic reality, but only its 'transformation' into external forms. Over these last years Marxist philosophers have been concerning themselves more and more with elaboration of this approach to the problem.

But if dialectical materialism counters empiricism and logical empiricism

67. E.g., the collection *Praktika – Kriterij istiny v nauke* (= Practice – Criterion of Truth in Science) (1960).

with the union of theory and practice, on the other hand it also seeks to differentiate its gnoseology from that of earlier materialism, which Marxism characterizes as a non-dialectical understanding of the reflection of the outside world in men's consciousness and a purely psychological interpretation of the process of knowledge. From the point of view of dialectical materialism, knowledge as reflection of the real differs qualitatively from simple apprehension of the outside world (which is also a reflection) in being active, selective, directed to an end. The idea of the activity of the knowing subject, organically linked with the gnoseological stress on practice, is central in the gnoseology of dialectical materialism. The subject of knowledge is not an incorporeal mind that knows itself; it is not the theoretical and practical reason of German idealism but man as total being, whose manifestations in the manifold forms of culture are all subsumed in the single process of real vital activity, for the production and reproduction of life. Seen thus, the process of knowledge is a social one, historically changing, in development; and gnoseology can be seen to be connected with the sociology of knowledge, also. Thus the philosophy of Marxism goes beyond both the non-historical and the naturalistic interpretations of the process of knowledge.[68]

The gnoseological problem of reflection concerns the relations of Marxism not merely with earlier materialism but also with contemporary science. The whole problem of the status of the 'object' in science and philosophy is involved. If, in line with a well-known idea of Lenin's, gnoseology is the theoretical balance-sheet of the history of knowledge, the contemporary stage of that history necessitates a further development of gnoseology.[69]

Lately a whole series of investigators have set themselves the goal of defining the concepts 'reflection' and 'image', particularly from the point of view of using the data of cybernetics and semiotics in philosophical investigation. Several views have been expressed as to the nature of the image. According to one, the characteristics of reflection can be described and explained in terms of the relation of the model to the original, i.e., in terms of isomorphism and homeomorphism.[70] This viewpoint has aroused objections among various writers. Be that as it may, this controversy, in the course of which the basic features of the development of contemporary scientific knowledge have been subjected to gnoseological analysis, has contributed to the further development of dialectical materialism.[71]

68. Cf. RUBINŠTEJN, *Bytie i soznanie* (= Being and Consciousness) (1957); KOPNIN, *Vvedenie v marksistskuju gnoseologiju* (= Introduction to Marxist Gnoseology) (1966); IL'ENKOV, article 'Ideal'noe' (= Ideal) (1962).

69. A special critical analysis of contemporary attacks on the theory of reflection will be found in the Soviet-Bulgarian collection *Leninskaja teorija otraženija i sovremennost'* (= The Leninist Theory of Reflection and the Present Day), under the general editorship of Academician T. PAVLOV (1969).

70. Cf. for example TJUHTIN, *O prirode obraza* (= On the Nature of the Image) (1963).

71. The problems raised by the theory of reflection, and the nature of the image, have received close attention in contemporary Bulgarian philosophical literature, as is evidenced by PAVLOV's basic work *Teorija otraženija* (= Theory of Reflection) (1949).

Marxist gnoseology has also been enriched by new research into the classical methodology of Marxism. In this connection work done on the method used by Marx in *Das Kapital,* quite apart from its historico-philosophical significance, has had an immediate repercussion on the contemporary gnoseological problem.[72]

All Marxists who work on the philosophical structure of *Das Kapital* take as their starting-point recognition of the important methodological implications of Marx's idea that in science concrete knowledge comes as the result of assembling different determinations. Each of these determinations is unilateral, that is, abstractly characterizes the object studied. That is why the movement from particular abstractions to their synthesis, or what Marx called the ascent from abstract to concrete, is the necessary condition for constructing a theoretical knowledge. But as to how to understand the process of rising from abstract to concrete, how to interpret its logical mechanism, there is some disagreement. Some investigators emphasize that the method Marx applied in *Das Kapital* presupposes two different logical movements: one from the concrete datum obtained by observation to certain abstractions, and then one away from those abstractions to the concrete understood as synthesis of different abstract determinations. Others argue that in Marx the two logical movements form one single movement. Be that as it may, this ascent from abstract to concrete has a decisive rôle in the case of theoretical knowledge: it is in the course of this movement that transformation of perceptions and representations into concepts takes place. Certain Marxists interpret the relation of abstract and concrete knowledge as one of the possible relations between different degrees in the development of scientific theory; in this case, the deduction of abstract determinations is connected with the logical procedure of isolation, and the obtaining of concrete knowledge with the logical procedure of concretization. In this sense the characterization of an item of knowledge as abstract concrete is relative: it is abstract or concrete only in relation to another item of knowledge. Other students consider the ascent from abstract to concrete as the specifically dialectical way of constructing theoretical knowledge of global systems. From their viewpoint, the concrete is interpreted as theoretical reproduction of a completed cycle of development. Yet other students, finally, see the method of *Das Kapital* as a generalized expression of the axiomatic method, subject to rigorous formal algorithms and the rules of deduction. For all who work on the method of *Das Kapital,* the important thing is to discover the universal significance of this method.

Work done on Lenin's *Philosophical Notebooks*[73] has brought into the open an analogous methodological trait, viz. the consequence of dialectic,

72. Cf. ROZENTAL', *Voprosy dialektiki v 'Kapitale' Marksa* (= Questions of Dialectic in Marx's *Das Kapital* (1955, republ. 1967); IL'ENKOV, *Dialektika abstraktnogo i konkretnogo v 'Kapitale' Marksa* (= The Dialectic of Abstract and Concrete in Marx's *Das Kapital*) (1960); GRUŠIN, *Očerki logiki istoričeskogo issledovanija* (= Essays on the Logic of Historical Research) (1961).

73. KEDROV, *Edinstvo dialektiki, logiki i teorii poznanija* (= The Unity of Dialectic, Logic and Theory of Knowledge) (1963); KOPNIN, *Filosofskie idei V. I. Lenina*

logic and gnoseology. According to the prevailing interpretation, this consequence means that the materialist dialectic is not only a doctrine of the development of natural and social things but also a theory of the development of knowledge; in this way, the dialectic would comprise gnoseology. The logic involved here is not formal but dialectical logic: it is interested in forms of thinking and in categories only as regards the development of their content. In other words dialectical logic expounds the development of the logical form of knowledge. Thus the principle of the convergence of dialectic, logic and gnoseology reveals the universal character and fundamental aspects of the materialistic dialectic.

This general appreciation of the principle of convergence does not preclude disagreement as to how it should be interpreted. Some Marxists hold that dialectic, logic and theory of knowledge are absolutely identical in their content; while others tend to think that the relation between them is one of only partial coincidence in respect of logical extension or denotation.

Thus the intensive study of the characteristics and structure of Marxist dialectical logic and its inter-relations with theory of knowledge and formal logic which began back in the fifties has been stimulated both by interpretation of Marx's method and by the new orientation of research into Lenin's *Philosophical Notebooks*. Besides the problems of the ascent from abstract to concrete and of the convergence between logical dialectic and gnoseology, interest has extended to the relationship between the logical and the historical, between analysis and synthesis, between induction and deduction, and to the construction of a theoretical system in general.

On this basis various systems of dialectical logic have been built.[74] For some, the methodology of science is identical with that of dialectic, and the elaboration of methodological problems amounts solely to analysing the rôle played by the laws and categories of dialectic in knowledge. According to others, dialectical logic is merely the general methodology of philosophy, and a methodology of science presupposes the elaboration of special disciplines like semiotics, cybernetics and mathematical logic.[75] Work on dialectical logic also goes into the reciprocal relationship of generalization in dialectical logic and in formal logic, the similarities and differences between contradiction in formal and in dialectical logic, and so on.

Allied to this work on dialectical logic is work on the systematic-structural methods used in *Das Kapital* for analysing the system of capitalist productive relations considered as an organic whole. These methods are re-

---

i logika (= Logic and Lenin's Philosophical Ideas) (1969); ROZENTAL', *Lenin i dialektika* (= Lenin and Dialectic) (1963); etc.

74. Cf. for instance KOPNIN, *Dialektika kak logika* (= Dialectic as Logic) (1961); ROZENTAL', *Principy dialektičeskoj logiki* (= Principles of Dialectical Logic) (1960); FOGARASI, *Logik* (1955); de GORTARI, *Introducción a la logica dialéctica* (1956); JOJA, *Studii de Logică* (1960, 1966).

75. Cf. GORSKIJ, *Problemy obščej metodologii nauk i dialektičeskoj logiki* (= Problems of General Scientific Methodology and Dialectical Logic) (1966); the collection *Dialectica metodelor în cercetarea ştiinţifică* (= The Dialectic of Methods in Scientific Research) (1966).

garded as specific investigatory procedures which, at their most highly developed, are modifications of the principles of dialectical logic. Marx's views as to procedures for analysing complex systems thus provide a key to understanding the methods of systematic analysis elaborated in contemporary biology, in cybernetics, in special social disciplines (ethnography, anthropology, etc.). Thus contemporary research into the logical foundations of a general theory of organization is linked with dialectical logic, as are procedures for analysing the structure and ordering of multi-level complex systems, specific methods for analysing systematic aggregates of various types, the inter-relation between structural and genetic methods, etc. Conversely, research in this area provides a specific content for one's understanding of the methodological function of such dialectical categories as development, progress, the qualitative leap and the form of organization of matter. In this connection contemporary structuralism as seen abroad, especially in France, is coming in for special critical analysis.[76]

Along with this work on dialectical logic we can mention other work relating to the dialectical materialist method – e.g., the dialectical movement between the historical and the logical in the history of science, the relationship between explanation and description and between the theoretical and the empirical levels in scientific research, the logical structure of the hypothesis, theory as a procedure for organizing knowledge, the dialectic of form and content in theoretical thought, the problem of the beginning in construction of a theoretical system, the rôle of monistic principle in knowledge and the logic of forming and developing scientific concepts.[77]

All these considerations lead to the problem posed by the assertion that philosophy must be an open system. There is a whole series of books and articles on the particular categories of dialectic.[78] From the fifties on Marxist philosophical literature has been tackling the problem of how to construct a system of categories of materialist dialectic, with thought both to the systematizing principles and to the concrete character of such a system. The starting-point here is the fact that the categories of dialectic cannot but be linked together in some form of single whole, since theoretic knowledge in every field, including the philosophic, can only exist in the form of a

76. FROLOV, *O pričinnosti i celesoobraznosti v živoj prirode* (= On Causality and Hormicity in Living Nature) (1961) and *Genetika i dialektika* (= Genetics and Dialectic) (1968); KREMJANSKIJ, *Strukturnye urovni živoj materii* (= Structural Levels of Living Matter) (1969); BLAUBERG, SADOVSKIJ & JUDIN (eds.), *Sistemnye issledovanija* (= Systems Research) (1969); SEBAG, *Marxisme et structuralisme* (1964); *Structuralisme et marxisme* (special number of *La Pensée*, 1967).

77. Cf. for instance KEDROV, *Razvitie ponjatija himičeskogo èlementa* (= The Development of the Concept of Chemical Element) (1956); TAVANEC (ed.), *Problemy logiki naučnogo poznanija* (= Problems of the Logic of Scientific Knowledge) (1964); POPOVIČ, *O filosofskoi analize jazyka nauki* (= On the Philosophical Analysis of the Language of Science) (1966).

78. Cf. SVEČNIKOV, *Kategorija pričinnosti v fisike* (= The Category of Causality in Physics) (1961); UËMOV, *Vešči, svojstva i otnošenija* (= Things, Properties and Relations) (1963); the collection *Materialismul dialectic şi ştiinţele moderne* (= Dialectical Materialism and Modern Sciences) (1967).

system of concepts. At the same time the system of categories in Marxist dialectic, insofar as it can be built up, is understood not as something closed, completed once and for all, but as an open system – open with regard to the new phenomena that may appear in historic reality and in scientific knowledge. That is, it lends itself to self-criticism, modification and concretization both of particular categories and of the relations between them and, consequently, alteration of the categorial structures. Different variants have been proposed for solving the problem of the system of categories.[79] Though a good deal of work has been done on the problem, no generally acceptable solution has yet been found.

But with all this research into dialectical logic the point of departure – the gnoseological import of practice as criterion of truth – must not be lost sight of. That is why, over the last decades, Marxists have given ever increasing attention to problems of the dialectic of social processes. In this respect, philosophers who accept the viewpoint of dialectical materialism start from the fact that the social sphere is not simply one where the universal dialectical principles operate. A scientifico-philosophical understanding of the underlying general philosophical characteristics of knowledge, of man as the subject of knowledge and activity and of the inter-relation between the cognitive and the axiological planes presupposes an inventory of the social dimensions of human activity. In work on the dialectical mechanism of social development (above all in socialist society), analysis chiefly concerns the relationship of antagonistic and non-antagonistic contradictions in socialism, the correlation between quantitative and qualitative change in the development of socialist society, the dialectic of social progress, etc.[80] More precisely, Marxist philosophers have paid particular attention to the specific problems involved in the dialectic of practical activity. The dialectic of the end, the means and the result of activity, the relationship between alienation, objectivation and reification, between freedom and necessity, the philosophical aspect of the problem of value, the analysis of what are, in the terminology of *Das Kapital*, the 'transformed forms' of human activity – all these and other problems have recently come in for study in Marxist philosophical literature. The authors of these studies stress that for Marxism, by contrast with existentialism, there is a difference in principle between alienation, which characterizes man only in certain social conditions, and objectivation as a necessary structure pertaining to any action. In this respect, the dialectic of practical activity is to be distinguished, according to Marxism, both from naturalistic objectivism, which leads to fatalism and predestination, and from subjectivist voluntarism.[81]

79. Cf. for example Bibler, *O sisteme kategorij dialektičeskoj logiki* (= On the System of Categories of Dialectical Logic) (1958).

80. Fedoseev, *Dialektika sovremennoj èpohi* (= Dialectic of the Present Era) (1966); Cernea, *Despre dialectica construirii socialismului* (= On the Dialectic of Constructing Socialism) (1967).

81. Cf. articles on 'Ob"ektivacija' (= Objectivation), 'Opredmečivanie' (ditto), 'Oveščestvlenie' (= Hypostatization), etc. in *Filosofskaja Ènciklopedija*, Vol. 4 (1967); Laković, *Marksistička teorija otudenja* (= The Marxist Theory of Aliena-

Within the context of this dialectic of practical activity, there are studies of the complex dialectical relation between end and result, end and means of action, and of such general philosophical problems posed by the methodology of the social sciences as dialectic and structuralism, genetic and systematic methods of analysis, generalization and idealization, knowledge and value, knowledge and the formulation of aims in social research.

§ 5. *Logic of philosophy and formalization**

Examination of logic of philosophy of the synthetic type (whether dialectical or non-dialectical) leads to the same conclusions as examination of the analytic type.

Whatever type of philosophy we consider, it comprises a logic, but always as one 'moment' or constituent of the whole. The logic of philosophy is not outside philosophy, does not amount to a really separate authority or tribunal. That is why it is not formal in the same sense as formal logic. Still one may ask if it cannot be made formal, if the methods of formal logic cannot be imported into philosophy and adapted to the logic of philosophy with its own highly individual status. Some attempts have already been made to study the possibility of formalizing dialectical logic.[82] At first sight, any attempt at formalization is doomed to failure. Formalization is in fact a representation; it therefore constitutes, in relation to what it formalizes, a separate authority or tribunal. But a true philosophical logic must have a circular character; it cannot itself become the object of an external tribunal. It has nothing to do with representation. It does not allow of being represented and it is not representation itself. It constitutes its own theory: it shows forth its essence and manner of operation by constructing itself and by operating effectively. And its manner of operation is constitution, not representation; precisely because it is not pure form it, so to say, penetrates into reality from the inside, as a framework, not from the outside as a mere theory of formal links. All this could be expressed by saying that any logic of philosophy has a reflexive character. But it does not, *a priori*, seem inconceivable that a formalization could be elaborated which would present just this character of reflexivity. This would mean a representation that would in a way be self-representation, that would not involve referring back to another authority or tribunal; it would represent itself by operating. Such a formalization would have to take the form of the theory of a struc-

---

tion) (1968); CORNFORTH, *Marxism and the Linguistic Philosophy* (1955); GULIAN, *Problematica omului. Eseu de antropologie filozofica* (= The Problem of Man. An Essay in Philosophical Anthropology) (1966) (French translation, GOULIANE, *Le marxisme devant l'homme. Essai d'anthropologie philosophique*, 1968).

* Based on general views expressed by Jean Ladrière.

82. DUBARLE, 'Sur une formalisation de la logique hégélienne' (1969); DUBARLE & DOZ, *Logique et dialectique* (1972), Pt. I: 'Dialectique hégélienne et formalisation', pp. 1-200 (this first part of the work is by D. DUBARLE).

ture capable of engendering its own image within itself. The work of D. Dubarle indicates that modern algebra might well open up possibilities in this direction.

However, supposing that the attempt at formalizing philosophical logic leads to positive results, we have to ask whether the formalization obtained can ever be complete in character. Doubtless the situation we shall be confronted with here will be similar to that which already obtains in regard to logic, as described above. It will always be necessary to ask what makes the formalization possible, therefore always necessary to raise the question of form. It will also be necessary to specify the conditions governing the operation and use of the proposed system, and to provide it with an interpretation. So it looks as if at a given moment one will always be thrown back on natural language. No doubt the boundaries of what can be formalized can be pushed further and further back, but it is doubtful whether all that is in fact formalized can ever form an enclosed domain. Naturally this impels us to enquire into the reasons for such a situation, into what it is that apparently forces all discourse to take root, in the final analysis, in natural language. The question that emerges here is that of the genesis of meaning. No doubt the genesis is effected in and through discourse; and this it is that justifies transcendental logic and dialectical logic alike. But this advent of meaning in discourse is doubtless possible only in relation to some authority capable of taking under its sway language and discourse, uniting in one single act the movement of discourse and that of existence. Rooted in natural language, discourse is thereby anchored in existence. This connection itself poses an ontological problem. If logic has ontological implications, logic itself needs to be grounded in an ontology. That is exactly why no logic of philosophy can be more than a constituent moment in a self-grounding discourse; nor can it be a theory of the interconnections which constitute the ground except insofar as it is itself an integral part of ontology.

§ 6. *The philosophical significance of the history of philosophy*

No inquiry into the logic of philosophy could conclude without casting a glance at the relationship between the *logic* and the *history* of philosophy. Indeed the history of philosophy, insofar as it is the work not of historians but of philosophers, poses the question of its significance in relation to the idea of philosophical truth professed by the historian-philosopher. Even if, as is very often the case, the latter does not himself create an original philosophical work, and even takes care not to do so, his purpose being to remain a *pure* historian of philosophy, his very activity as a historian *of* philosophy inevitably brings him back to the question of the philosophical meaning of his enterprise. In fact, this question comprises two questions: what is the *interest* of works of the past in relation to present-day *research*? and what is the degree of *truth* (or truth-content) inherent in a philosophical work as such, with regard to hat the philosopher writing the history of philosophy considers to be the *paradigm* of truth?

(a) Analytic philosophy – represented mainly by English-speaking writers – does not often pose the second question, which is on the contrary typical of those philosophies that derive from Hegelianism, or those that oppose it, but oppose it on its own ground. On the other hand, analytic philosophy provides at least an indirect answer to the first question, through the *selection* of works of the past to which the historian-philosopher devoted full case-studies. In this connection the very noticeable widening of the field of interest since 1945 to date reflects a parallel trend in regard to the criteria of what analytic philosophers consider to be meaningful or meaningless.*

During the years immediately succeeding the Second World War, many Anglo-Saxon philosophers considered that their main task consisted in the elucidation of conceptual tensions and tangles presumed to lie at the root of traditional philosophical problems; as a result the approach adopted to previous thought was of a markedly critical and deflationary nature. The term 'metaphysics' still carried the pejorative overtones it has acquired from the Logical Positivists; hence there was an inclination to treat the system born of classical speculation at best as ingenious extravaganzas which, albeit in a mystifying and misleading manner, dramatized certain pervasive features of experience and our thought about it, at worst as products of a pretentious word-spinning devoid of any real content and illustrative of a fatal disregard for the limits of significant discourse. In consequence attention was largely focused upon those philosophers of the past who were seen as having anticipated to some degree the wary attitude of contemporary analytic philosophers in regard to the possibilities of conceptual confusion in our language and thought, and whose major contributions could be said to have been epistemological and methodological, rather than ontological or transcendental in character. Outstanding examples of such thinkers were Hobbes, Locke, Berkeley, Hume, Mills and Peirce.[83]

During the last decade or so, analytic philosophers may be considered to have arrived at a more open, less sceptical conception insofar as the task of rendering an account of past philosophy is concerned. This wider and more sympathetic understanding of the works of the past is due to a combination of factors which have affected the very conception of the philosopher's task as a whole: a more liberal view of meaning and of the boundaries of intelligibility; a less rigid adherence to the canons of 'ordinary

* What follows is taken from a synopsis contributed by Patrick Gardiner.

83. For some of the most important recent studies see PETERS, *Hobbes* (1956); O'CONNOR, *John Locke* (1952); YOLTON, *John Locke and the Way of Ideas* (1956); WARNOCK, G. J., *Berkeley* (1953); ARMSTRONG, *Berkeley's Theory of Vision* (1960); PASSMORE, *Hume's Intentions* (1953); FLEW, *Hume's Theory of Belief* (1961); BRITTON, *J. S. Mill* (1953); GALLIE, *Peirce and Pragmatism* (1952); AYER, *The Origins of Pragmatism. Studies in the Philosophy of Charles Sanders Peirce and William James* (1968).

language'; an increased attention to the fundamental structure of our conceptual scheme as a whole, as contrasted with the 'surface phenomena' of actual usage displayed in particular regions of discourse; a clearer recognition of the historical evolution of basic notions and of the forms of social life in which they are embedded. One way in which this change has manifested itself has been the proliferation of Kantian studies, with emphasis upon the logical and conceptual implications of Kant's doctrines as opposed to the more speculative extensions of Kantian idealism.[84]

From another point of view, a number of recent writers have come to see avowedly metaphysical works as inspired by a sensitivity to the limitations of existing conceptual structures[85]: fresh ways of describing reality are advanced which to some extent loosen the bonds of accepted categories and distinctions, thereby opening up unsuspected avenues of thought which have proved to be fruitful subsequently in the field of empirical investigation.[86] Other works give prominence to the 'revisionary' character of much metaphysical literature, or stress the motives that have led philosophers to question established modes of describing the world and to propose comprehensive alternatives.[87] It is noteworthy in this connection that a number of nineteenth century thinkers, whose stock was for a time very low in the Anglo-Saxon world, have recently attracted greater interest; examples are Hegel, Marx, Schopenhauer, Nietzsche and Bradley.[88] It should be noted that, concurrently with this revival of concern, existentialism and its sources have given rise to a growing number of analyses and studies which reflect, in large measure, the feeling of dissatisfaction with the limitations of moral philosophy as has been pursued in England during the last few decades. Even when they are highly critical, such studies exhibit a much livelier sympathy and understanding with the central preoccupations of existentialism than it received at the hands of Anglo-Saxon philosophers just after the war.[89]

Analytic philosophy also has to its credit numerous works on the history

84. Cf. for example BIRD, *Kant's Theory of Knowledge* (1962); STRAWSON, *The Bounds of Sense* (1966); BENNETT, *Kant's Analytic* (1966). See also above, § 2, p. 1119.

85. Cf. among other authors, WAISMANN, 'How I see philosophy' (1956).

86. Stuart HAMPSHIRE's study on *Spinoza* (1952) showed the extent to which aspects of the thought of an uncompromisingly abstract thinker could be illuminated in this light. A similar treatment was extended more generally in some of the contributions to *The Nature of Metaphysics*, edited by D. F. PEARS (1957).

87. WALSH, *Metaphysics* (1963); LAZEROWITZ, *The Structure of Metaphysics* (1955).

88. Among the most recent studies are FINDLAY, *Hegel. A Re-examination* (1958); WALSH, *Hegelian Ethics* (1969); PLAMENATZ, *Man and Society*, Vol. II (1963) (extended discussions of Hegel and Marx); KAMENKA, *Marxism and Ethics* (1969); DANTO, *Nietzsche as Philosopher* (1965); GARDINER, *Schopenhauer* (1963); WOLLHEIM, *F. H. Bradley* (1959).

89. Cf. WARNOCK, M., *Ethics since 1900* (1960). Sartre, in particular has aroused considerable interest: see for example MURDOCH, *Sartre: Romantic Irrationalist* (1953); CRANSTON, *Sartre* (1962); WARNOCK, M., *The Philosophy of Sartre* (1965); MANSER, *Sartre. A Philosophic Study* (1966).

and sources of its own tradition,[90] or on the history of ethics, including Marxism and existentialism.[91]

(b) The influence of Hegelianism on the general tenor of the history of philosophy in the nineteenth century explains why continental philosophers remained more attentive to the second of the questions we formulated above, in roughly the following terms: what is the truth-value of systems of thought in the philosophical tradition when compared to the idea of truth that a present-day philosopher is likely to entertain?*In tackling this question the contemporary period seems to differ from that running approximately from the end of the nineteenth century to the middle of the twentieth century, which tended to evaluate the work of philosophy as such negatively, viewing it as defective with regard to its own content of meaning. Such is the conception of the historical anthropology of Dilthey,[92] in which the claim of philosophical systems to universality is overthrown by consideration of their dependence on empirical historical consciousness, the sole unity in all these *Weltanschauungen* subjected to all the consequences of history being that pertaining to a typology of a psychological character; similarly, with Bergson,[93] philosophical intuition, which is in itself irreducible to history, is the source of a discourse which is always inadequate; and with an intellectualistic rationalism such as that of Léon Brunschvicg,[94] the positive element of

90. WARNOCK, G. J., *English Philosophy Since 1900* (1968); URMSON, *Philosophical Analysis. Its Development Between the Wars* (1956); PASSMORE, *A Hundred Years of Philosophy* (1957); PEARS, *Bertrand Russell and the British Tradition in Philosophy* (1967); AYER, *Russell and Moore: The Analytical Heritage* (1971).

91. MACINTYRE, *A Short History of Ethics* (1966). Historical studies by authors of the analytic school are to be found in major collections such as *The Encyclopedia of Philosophy*, edited by P. EDWARDS (1967) and *A Critical History of Western Philosophy*, edited by D. J. O'CONNOR (1964).

* The whole of what follows is taken from a synopsis contributed by Pierre Garniron.

92. Cf. DILTHEY, *Weltanschauungslehre. Abhandlungen zur Philosophie der Philosophie* (*G.W.*, VIII).

On Dilthey's view of the history of philosophy, the reader is invited to consult Chapter 1, Section 5, of Raymond ARON, *La philosophie critique de l'histoire. Essai sur une théorie allemande de l'histoire* (1938, reissued 1969), pages 89-100 in the 1969 edition.

The idea of a typology of philosophy, which had its proponents towards the end of the eighteenth and the beginning of the nineteenth century (particularly Grohmann and Fr. Schlegel) and which is to be found again in the nineteenth century with, for example, Cousin and Renouvier, still finds expression today in authors such as Alois DEMPF, who takes as his basis a comparative study of successive philosophical periods in the West and the East, particularly China and India. In this conection the reader should also consult the contribution by DEMPF, *Philosophie de l'histoire de la philosophie*, to CASTELLI *et al.*, *La philosophie de l'histoire de la philosophie* (1956). See also GELDSETZER, *Die Philosophie der Geschichte der Philosophie im 19. Jahrhundert* (1968), pp. 127-128.

93. Cf. BERGSON, 'L'intuition philosophique' (1911, 1934) (English translation, 'Philosophical intuition', 1946); see also *L'évolution créatrice* (1907) (English translation, *Creative Evolution*, 1911), Chap. IV.

94. Cf. BRUNSCHVICG, *Le progrès de la conscience dans la philosophie occiden-*

a philosophy is the mind's activity at work in mathematics, such activity being more or less faithfully reflected by the system, so that philosophical progress always follows, more or less closely, in the train of scientific progress. A similar comment may be made on the subject of the *philosophia perennis* of Etienne Gilson,[95] who sees historically constituted doctrines as imperfect realizations of typical essences of systems which are themselves imperfect in relation to the philosophy of being as such. We may also view from this standpoint the reflections of historians of philosophy such as Emile Bréhier[96] or Henri Gouhier[97] who reproduces Bergson's views (in greater detail and with more shades of meaning), or those of, for instance Ortega y Gasset,[98] or, yet again, the various strands of Italian historicism: Benedetto Croce,[99] Giovanni Gentile[100] and their descendants.

In all these examples – and subsequent writers to date [101] – it is the concrete historical element which is the source of the disparity between the philosophical intention and the actual philosophical work: either it is the positive element which philosophy remains inadequate to express (this is the position of historicism), or else it is the negative element which prevents actual philosophy from adequately expressing a non-temporal reality, or else, as with Brunschvicg, it is positive in one of its aspects – the development of the sciences – and negative in others. Such attitudes might be described as efforts to achieve 'immunization';[102] their purpose is always to go beyond given systems towards a truth which transcends them and explains their essence, the truth which the philosopher-historian wishes – even if only implicitly – to preserve, either as the spirit that gives such systems life or as the reference which explains their obsolescence.

---

tale (2 vols., 1927); and see GUÉROULT *et al.*, *Brunschvicg et l'histoire de la philosophie* (presentation followed by discussion, *Société française de Philosophie*, 1954).

95. Cf. GILSON, *The Unity of Philosophical Experience* (1937); see the review (1938) of this work by Henri GOUHIER, Appendix 2, pp. 127-134 in *La philosophie et son histoire* (1944). Reference may also be made to Joseph MOREAU's view of *philosophia perennis* (cf. *La construction de l'idéalisme platonicien* (1939), §§ 365-366) and the conception, inspired by Hegel, of Eric WEIL (cf. *Logique de la philosophie* (1950), pp. 68 sqq. and 427 sqq.).

96. Cf. BRÉHIER, *Histoire de la philosophie*, Tome I, Vol. 1, Introduction (1926); by the same author, *La philosophie et son passé* (1940), Introduction and Sections I and II (pp. 1-75). For the overall conception of the history of philosophy during this period, particularly in France, the reader should also consult ROBIN *et al.*, *Sur la notion d'histoire de la philosophie* (presentation followed by discussion, Société française de Philosophie, 1936).

97. Cf. GOUHIER, *La philosophie et son histoire* (1944), 'De l'histoire de la philosophie à la philosophie' (1949) and *L'histoire et sa philosophie* (1952).

98. Cf. ORTEGA Y GASSET, *Ideas para una historia de la filosofía* (1942) and 'Dos Prologos' (1944).

99. Cf. CROCE, *Logica* (1909) (English translation, *Logic*, 1917); *Teoria e storia della storiografia* (English translation, *History. Its Theory and Practice*, 1921).

100. Cf. GENTILE, *La riforma della dialettica hegeliana* (1913) and *Se e come è possibile la storia della filosofia* (2nd edn., 1966).

101. Such is still the overall situation reflected, for example, in the essay by Harold R. SMART, *Philosophy and Its History* (1963).

102. SARTRE, *Critique de la raison dialectique* (1960), pp. 55 sqq.

Following on the preceding examples, consideration should be given to recent or contemporary authors such as Karl Jaspers and Ferdinand Alquié: for the former, the conceptual realizations of the philosophical absolute are debasements due to the necessities of social communication,[103] while for the latter 'systems are [. . . .] what remains of philosophical activity as such when it has come to an end'. (translation)[104] But these authors at the same time strongly insist – and this is not easy to reconcile with the distinction each of them makes between that which is philosophical, i.e., timeless, and that which is not – on the individuality and absoluteness of each philosophy, philosophy being thus clearly distinguished from science.

But this eternal aspect of each philosophy can be sought – as is done by Etienne Souriau[105] – in the works themselves, as completed movements; philosophical activity then appears as the task of elevating the totality of the real to the perfection of a single reality, namely the actual philosophical work, thus identified with a work of art. We are thus led to stress the importance of the architectonic processes involved in the elaboration of philosophical systems, and these systems no longer appear, in their plurality, as *theories*, in the manner of the sciences, but as works, whose value consists in their own plenitude. Any idea of progress being thus eliminated, the history of philosophy is in essence different from the history of science; but while it is no longer concerned with anything beyond the works themselves, it still does not come to grips with the works in their specific character as *philosophical* works, but only in their *aesthetic* aspect.

It is this specifically philosophical character of great philosophical works which Martial Guéroult,[106] in the very forefront of contemporary writers on the history of philosophy, tries to identify, seeking to elucidate its significance by the projected elaboration of a science of *dianoematics*[107],

103. Cf. Jaspers, 'Die grossen Philosophen' (2 vols., 1957) (English translation, *The Great Philosophers*, 1962) and 'Essay on my philosophy' (1956); see also Schilpp (ed.), *The Philosophy of Karl Jaspers* (1957).

104. 'Les systèmes sont [. . .] ce qui nous reste de l'activité proprement philosophique lorsque celle-ci a pris fin.' – Alquié, *Signification de la philosophie* (1971), pp. 70-71; see also Alquié *et al.*, *Structures logiques et structures mentales en histoire de la philosophie* (presentation followed by discussion, Société française de Philosophie, 1953).

105. Cf. Souriau, É., *L'instauration philosophique* (1939).

106. Among works by Martial Guéroult, see particularly: *Leçon inaugurale*, Collège de France (1951); *Brunschvicg et l'histoire de la philosophie* (1954), already cited; 'La voie de l'objectivité esthétique' (in *Mélanges d'esthétique et de science de l'art offerts à Etienne Souriau*, 1952); '*Le problème de la légitimité de l'histoire de la philosophie*' (in Castelli *et al.*, *La philosophie de l'histoire de la philosophie*, 1954/1956); 'Logique, architectonique et structures constitutives des systèmes philosophiques' (in *Encyclopédie française*, Vol. XIX, 1957); interventions (pp. 453-463 and 473-475) in the general discussion at the Royaumont symposium on *Descartes* (publ. in 1957); 'The history of philosophy as a philosophical problem' (in the special issue of *The Monist* entitled *Philosophy of the History of Philosophy*, 1969). Among the many important works on the history of philosophy by Martial Guéroult, we shall refer here to only one of the best known, *Descartes selon l'ordre des raisons* (2 vols., 1953).

107. Cf. *Etudes sur l'histoire de la philosophie en hommage à Martial Guéroult*

(from the Greek *dianoema*, doctrine), i.e., 'investigation of the conditions that must be fulfilled for philosophical works to possess indestructible philosophical value' ('science des conditions de possibilité des oeuvres philosophiques en tant qu'elles possèdent une valeur philosophique indestructible'). But he refuses to assimilate them purely and simply to works of art, commenting that philosophy, like science, aims at elaborating a theory in accordance with reality, a truth of judgement (*veritas in intellectu*). It will be the task of dianoematics to investigate 'the manner in which the truths of judgement in individual philosophies are scientifically grounded, in such a way as to ground an intrinsic truth (*veritas in re*) independent of any truth of judgement' ('de quelle façon dans chaque philosophie l'instauration scientifique d'une vérité de jugement rend possible l'instauration d'une vérité intrinsèque [. . . .] indépendante de toute vérité de jugement'). Thus philosophy aims at *veritas in intellectu* but achieves fulfilment in producing *veritas in re*, i.e., a conception of the world which replaces the everyday world and gains acceptance through the illumination it provides.' ('La philosophie vise ainsi une *veritas in intellectu*, mais elle s'accomplit en produisant une *veritas in re*, c'est-à-dire une conception du monde qui se substitue au monde ordinaire et se fait accepter par son éclat propre.') Thus dianoematics is a science which 'treats of the different ways of viewing reality' ('une problématique de la réalité'): each philosophy starts with a *common reality*, of indeterminate nature, and goes on to construct *true reality* of which it represents not a copy but, like a Platonic Idea, the constitutive principle, a free act of grounding, in response to the problems posed by the common reality.[108]

This dianoematic approach is based on a methodology of the study of philosophical doctrines, the *technicology of philosophical systems* or determination of the *demonstrational techniques* which distinguish philosophical works as such and constitute the original structure peculiar to the individual nature of each of them. This structure is manifest in the *sequence of reasoning* actually employed by the philosopher, expressed in an architectonic structure which technicological analysis studies in its relationship to the technique used for proof.

We thus come to the principle governing the survival of philosophical systems: this is in fact nothing other than their 'internal structure' (économie structurale'), a true 'principle of integration, of never-ending assimilation' ('principe d'intégration, d'assimilation indéfinie'), of 'the most diverse contributions' ('apports les plus divers').[109] It is thus understandable that a

(1964), particularly the contribution by Henry DUMÉRY, 'Doctrine et structure'. See also the entire special issue of *The Monist* (1969) already quoted, with the Introduction and Bibliography by L. W. BECK, articles by EHRLICH, 'Principles of a philosophy of the history of philosophy', FAUROT, 'What is history of philosophy?', KUNTZ, 'The dialectic of historicism and anti-historicism', etc.

108. Cf. in *Etudes sur l'histoire de la philosophie*, already referred to, the article by Chaïm PERELMAN, 'Le réel commun et le réel philosophique'.

109. Cf. DUMÉRY's contribution, already cited, to the same collection, p. 171.

metaphysical system can survive the physical, cosmological, psychological and other representations which were originally coupled to it; it is a permanent basic structure peculiar to the presentation of a particular thesis, but which can be employed again and amplified in the presentation of other theses, without being thereby merged with them. This would be one way of restoring in a manner of speaking the objective meaning, so strongly emphasized by Hegel, of the term 'history of philosophy', as a series of the *res gestae* of the 'heroes of the thinking spirit'.[110] The keynote of such a history, however, would be discontinuity.

This explains how M. Guéroult's method as a historian of philosophy should have come to be regarded as akin to contemporary structuralist thought;[111] the sequence of reasoning is not to be sought as underlying the text, which it is not a question of interpreting, but of understanding in its explicit and manifest *expression*, though this is what most easily passes unnoticed by whoever concentrates solely on the substance of *what is expressed*.[112] This understanding – which is not intuitive – of the demonstrational structures used by philosophers in their objectivity makes it possible to elaborate a formal analysis of them, as is done for example by Jules Vuillemin and Victor Goldschmidt.[113] It is also from this point of view that one can discuss the attempt made by Wittgenstein, in his *Tractatus logico-philosophicus*,[114] to determine critically the nature of the reasoning employed in philosophical works, revealing it as a characteristic of the very nature of philosophy and meaningless, but suggestively meaningless. This introduction of increasingly rigorous logical procedures into the analysis of philosophical discourse has led to investigations – for example those of André Robinet[115] in France – into the possibility of using computers to throw light on the subject. However, the philosophy of the history of philo-

110. Hegel, *Vorlesungen über die Geschichte der Philosophie* (lectures first given during the Winter term 1805-06, Jena), *Einleitung.*

111. Cf. Deleuze, 'Spinoza et la méthode générale de M. Guéroult' (1969). See also in the following chapter of this study, Section IV, sub-section B, § 4.

112. Significant in this respect is the attempt by Jean Bollack to rediscover literally the meaning of the texts of the pre-Socratics, disentangled from the many layers of interpretations and projections through which they have reached us. One's mode of access to a text seems in this way to rise to an entirely different plane by comparison with the erudite approach of nineteenth-century scholars. See particularly Bollack, *Empédocle* (4 vols., 1965-69); Bollack, J. & M., & Wismann, *La lettre d'Epicure* [Letter to Herodotus] (1971); Bollack & Wismann, *Héraclite, ou la séparation* (1972).

113. Cf. the contributions by Jules Vuillemin, 'Sur les propriétés formelles et matérielles de l'ordre cartésien des raisons', and Victor Goldschmidt, 'Exégèse et axiomatique chez saint Augustin', to the collection already referred to, *Etudes sur l'histoire de la philosophie* (1964), pp. 43 and 14 respectively.

114. Wittgenstein, *Logisch-philosophische Abhandlung/Tractatus logico-philosophicus* (1921/1922). Cf. the contribution by G.-G. Granger, 'Systèmes philosophiques et métastructures. L'argumentation du *Tractatus*', to *Etudes sur l'histoire de la philosophie, op. cit.* (1964) and the article by Helmut Fahrenbach, 'Die logisch-hermeneutische Problemstellung in Wittgensteins *Tractatus*', in the collection *Hermeneutik und Dialektik* (1970), Vol. II.

115. Robinet, 'La communication philosophique à l'ère des ordinateurs' (1969).

sophy which, in the manner of M. Guéroult's dianoematics, attempts to base the indestructibility of philosophies on a structural analysis of philosophical works gives rise in turn to a fundamental question:[116] is the antimony between the history of philosophy – which notes that the various philosophies forever contradict each other – and philosophy – which as philosophy aims at absolute universality – in fact transcended by a system of philosophy which makes each particular philosophy the free grounding of one Idea of reality? Is not this also just one particular philosophy like the others? Does it not enter into conflict with realist philosophies and, more generally, does it not make them all relative, contrary to the intent of each of them? Further, does not this Platonic plurality of intelligible realities call for a unification of a neo-Platonic type? Does it not throw us back on the transcendental unity of an ineffable infinite plenitude?

It has been noted that the metaphysics of Plotinus was the first 'metaphysics of metaphysics', the metaphysics of the Greek metaphysical *logos*, which came after it, after it had developed its possibilities, the finite limits of which it recognizes.[117] Today it is the line of thought initiated by Heidegger[118] that, avowedly, puts itself forward as the philosophy of Western metaphysics, the philosophy of the history of metaphysics which, coming at the end of metaphysics, identifies its characteristics and goes back to the unconsidered possibilities that are concealed by the contingent nature of its origins. It is thus understandable that this line of thought which has been calling since 1927 for a 'destruction' of our philosophical tradition, still provides inspiration – though with the greatest liberty as regards the detailed manner in which it interprets the philosophical past – for investigations into the history of philosophy, and also for original philosophical developments which discover new threads of meaning in the history and study of thought, such as the work of Jacques Derrida,[119] with whom the 'destruction' of metaphysics takes rather the form of 'dismantling' (*déconstruction*).

This idea of completion, of a *ne plus ultra* reached by a certain cultural tradition – characterized by its philosophical tradition –, seems to find reflection in the importance attached to spatial images in structuralist thought: whereas, for example, at the beginning of the century the notions of genesis, history, evolution and driving force were predominant, today talk is mainly of space, configurations, loci, fields, strata and levels. What we need to do therefore, in the view of authors such as Heidegger and Derrida, is to dig down below this preoccupation with spatial externals to the underlying stratum of temporality, whose presence it reveals and of which it is the

116. Cf. the contribution by Fernand BRUNNER to *Etudes sur l'histoire de la philosophie* (1964), 'Histoire de la philosophie et philosophie'.

117. Cf. AUBENQUE, 'L'auto-interprétation de la philosophie' (1969).

118. Among HEIDEGGER's works see in particular *Sein und Zeit* (1927) (English translation, *Being and Time*, 1967); *Holzwege* (1950) (French translation, *Chemins qui ne mènent nulle part*, 1962); *Hegel und die Griechen* (lecture, 1958; published 1960, 1967) (French translation, 'Hegel et les Grecs', 1968).

119. DERRIDA, for example *L'écriture et la différence* (1967).

culmination: in this way, though without Hegel's optimism, we come back again to the theme of the concluding pages of the *Phenomenology of Mind.*

Again, it is in large measure to the thought of Heidegger – and also, directly, to Husserl's phenomenology[120] – that is linked contemporary hermeneutic thought, in for example authors such as H. G. Gadamer[121] in Germany and Paul Ricoeur[122] in France, in regard to investigations into the significance of the history of philosophy insofar as it explicates in particular the conditions governing the apprehension of philosophical works and statements. Their object is not to destroy but fully to accept, following on Heidegger, the hermeneutic circle of understanding and existence: the understanding of another's thought is understanding oneself, in the sense of precomprehensive existence, and understanding oneself is understanding other thoughts. This leads Gadamer to a theory of the conscious rehabilitation of belonging to history, or to be more precise, to a tradition within which alone the significant works of the past can convey something to us. We show our enlightened adherence to a philosophical tradition by referring for example to a well-known author, whose thought we regard as being of normative value in a particular field; this thought thus appears not as a detour, but as a shortcut towards acceding to oneself and things. It is easy to see how this approach – which affirms the intimate nature of the link uniting philosophy and the appropriation of its past, which there is no question of massing together – converges with P. Ricoeur's conception of the philosophical community, inescapably at war with and divided against itself, the unity of truth being 'a kind of ontological hope' ('une sorte d'espérance ontologique') that inspires the history of philosophy, which can only be comprehended as a dialogue between one individual and another. Lastly, it is clear that for a hermeneutic philosophy of the history of philosophy the significance of philosophical works is inexhaustible; their truth is implicit, forever being explicated; 'it can only be *possessed* in the form of a demand to *go on seeking for it*' (translation)[123]

120. With regard to the problems here discussed, reference can be made, among the works of HUSSERL, to 'Philosophie als strenge Wissenschaft' (1910-11) (English translation, 'Philosophy as rigorous science', 1965); *Erste Philosophie*: Pt. I, *Kritische Ideengeschichte* and Pt. II, *Theorie der phänomenologischen Reduktion* (ms. 1923-24; published 1956-59) (French translation, *Philosophie première*, 1970-72). For the bearing of a philosophy calling for a radically new beginning on the idea of the history of philosophy, see also BELAVAL, *Leibniz critique de Descartes* (1960), pp. 126 sqq.

121. GADAMER, *Wahrheit und Methode. Grundzüge einer philosophischen Hermeneutik* (1960, 1965); by the same author, 'Begriffsgeschichte als Philosophie' (1970).

122. RICOEUR, 'L'histoire de la philosophie et l'unité du vrai' (1953/1954; reproduced in *Histoire et vérité*, 1955, 1964).

Cf. the overall studies by P. FRUCHON, 'Ressources et limites d'une herméneutique philosophique (H.-G. Gadamer, *Wahrheit und Methode*)' (1967) and 'Connaissance et interprétation du passé en histoire des idées' (1966).

123. Cf. the article by L. PAREYSON, 'Originarietà dell'interpretazione', in the collection already referred to, *Hermeneutik und Dialektik* (1970), Vol. II, pp. 353 sqq. It is worthwhile consulting this collection as a whole and in particular, over and

But in emphasizing, with reference to the circle of understanding and existence, the necessary fact of the interpreter's belonging to history, to *a* history, hermeneutic thought rediscovers the Hegelian affirmation that the philosopher is 'a child of his time'. And since the link with a specific past is internal to his thought, it follows that it is the privilege of a specifc moment in history to open on to a particular aspect of the philosophical past, and also on to the philosophical past in general, as was perceived by Hegel when he argued that it was as a child of his time that he was able to integrate into his own thinking earlier philosophies as they actually were, for example the Platonic dialectic or 'the treasury of Aristotle, virtually unknown for centuries past'.[124]

This Hegelian heritage in the philosophy of the history of philosophy (which is inseparable from his philosophy of history) is to be found after a fashion in Marxism, which holds that one's ability to understand a particular aspect of the philosophical past depends on the attitude one adopts to the class struggle.[125]

Some have seen in this conception a link with a form of historicism, but this is now vigorously disputed, particularly, in France, by Louis Alt-

---

above the contributions cited elsewhere, KUHN, in Vol. I, 'Ideologie als hermeneutischer Begriff', LLEDÓ, in Vol. II, 'Lenguaje e historia de la filosofía', etc.

124. HEGEL, *Vorlesungen über die Geschichte der Philosophie*, p. 368 in *Jubiläumsausgabe*, Vol. XVIII. See the essay by BUBNER, 'Philosophie ist ihre Zeit, in Gedanken erfasst', in the collection *Hermeneutik und Dialektik* (1970).

125. For the Marxist conception of the history of philosophy, mainly in the U.S.S.R. and the other socialist countries, the reader should consult, for example: *Symposium der Philosophie Historiker* (*Moscow, 1967*) (1968) with, in particular, the synopsis, 'Übersicht', by KUMPF, and the contributions by KOPNIN, 'Zur Methode der philosophiegeschichtlichen Forschung', BOGDANOV, 'Methodologische Probleme der Philosophiegeschichte', etc.; *Geschichte der Philosophie*, published by the Academy of Sciences of the U.S.S.R., Vol. I (2nd edn. in German, 1962); ARZAKANIN, 'K voprosy o stanovlenii istorii filosofii kak nauki' (1962) (English translation, 'On the problem of the rise of a scientific conception of the history of philosophy', 1962/63); IOVČUK, 'O nekotoryh metodologičeskih problemah istorii filosofii' (= About some methodological problems of the history of philosphy) (1959); MAMARDAŠVILI, 'Nektorye voprosy issledovanija istorii filosofii kak istorii poznanija' (= Some problems of the investigation of the history of philosophy as the history of cognition) (1959) and 'K probleme metoda istorii filosofii' (= Concerning the problem of method in the history of philosophy) (1965); NOVIKOV, 'Predmet i zadači istoriografii filosofii' (1964) (English translation, 'Historiography of philosophy: subject matter and aims', 1964); BUHR & IRRLITZ, *Der Anspruch der Vernunft*, Vol. I (1968); BUHR (ed.), *Zur Kritik der bürgerlichen Ideologie* (a series of fascicles, 1971, etc.); LUKÁCS, *Die Zerstörung der Vernunft* (1954); (French translation, *La destruction de la raison*, 1958-59, 2 vols.); BANU, *Introducere in storia filozofiei* (1957); POSESCU, 'De l'explication dans l'historiographie de la philosophie' (1966). For a view of the conceptions of 'western' Marxists, see for example, in addition to the works of ALTHUSSER referred to below, GOLDMANN, 'Matérialisme dialectique et histoire de la philosophie' (1948, 1959); GOLDMANN and LEFEBVRE, contributions (pp. 463-471 and 476-477) to the General Discussion at the *Descartes* Symposium already cited (1957). For critical accounts by English-language philosophers see, among others, KAMENKA, 'Marxism and the history of philosophy' (1965).

husser who, starting with a fresh analysis of the classical works of both Marxism and Leninism, elaborates a much more structural, coherent conception of social *praxis* and its history, with philosophy appearing as a form of theoretical practice closely dependent on *scientific practice* (which is not a 'superstructure', and has a relatively independent history), the scientific nature of which it reflects against a background of *ideologies* that ignore or obscure it, either combating or upholding them, ideologies being always an expression of the interests of classes struggling for political power.[126] L. Althusser thus outlines a critical theory of traditional philosophical practice which leads to its complete transformation, the 'object' of the history of philosophy then being not the history of systems but the history of concepts organized into complexes of problems corresponding to a given state of science and ideology; here the influence becomes apparent of the works of Gaston Bachelard on the history of science.[127] This conception is linked to a critique of historicism in that it sees each philosophy as the immediate expression of a historic present considered as an all-embracing substantial unity of a 'spiritual' type. Ascribing the origin of historicism to Hegel, Althusser continues his critique of the historicist interpretation of Marxism, particularly in Italy in the works of Antonio Gramsci[128] (influenced by Benedetto Groce), Galvano Della Volpe[129] and Colletti,[130] and finally in Sartre.[131] For the same sort of reasons he also rejects any kind of humanist interpretation, whether linked or not to the foregoing.[132] Leaving aside the discussions to which Althusser's interpretation of Marxism (resembling but distinct from structuralism) gives rise, and despite the fact that it refuses to recognize the autonomous status of the history of philosophy, it nevertheless represents a characteristic attempt to apprehend the specific nature of philosophical activity within history, i.e., the diversified aggregate of rational activities.

One might consider the movement of the philosophy of the history of philosophy which we have tried to outline by analogy with the transition from scepticism to neo-Platonism in Hegel's 'Lectures on the History of Philosophy'[133]: in our case the transition has been from scepticism (the var-

126. ALTHUSSER, 'Le marxisme n'est pas un historicisme', pp. 73-108, in ALTHUSSER *et al.*, *Lire le Capital*, Vol. II (1965); see also by the same author *Pour Marx* (1965) and *Lénine et la philosophie* (lecture, 1968, publ. 1969, 1972).

127. See below, Section II, sub-section A, § 4, p. 1164.

128. Cf. GRAMSCI, *Il materialismo storico e la filosofia di Benedetto Croce* (Vol. II of *Opere*, publ. posth., 1948).

129. Cf. DELLA VOLPE, *Chiave della dialettica storica* (1968).

130. Cf. COLLETTI, Introduction to the Italian edition of Lenin's *Philosophical Notebooks*.

131. Cf. SARTRE, *Critique de la raison dialectique* (1960), particularly the first part, 'Questions de méthode' (1957, 1960) (English translation, *Search for a Method*, 1968).

132. Cf. below, in the following chapter, Section VI, sub-section B, § 2, 'Marxism and the discussion of humanism', pp. 1439 sqq.

133. Cf. HEGEL, *Vorlesungen über die Geschichte der Philosophie*, introduction to neo-Platonism, in *Jubiläumsausgabe*, Vol. XIX.

ious forms of historicism and its immediate negations) with regard to philosophical works as such, tending to absorb them in the flow of history, to the asseveration of their intrinsic intelligible consistency, which allows them to escape from the realm of becoming and its conflicts, and this has been achieved largely as a result of the processes of objectification and 'appropriation' associated respectively with structuralism and hermeneutics. But at the same time it has become apparent that the philosophy of the history of philosophy could not fail to be itself philosophy, a particular brand of philosophy, present among, and against, others; this gives value – or fresh value – to efforts to understand the *history* of *philosophy* as the *Aufhebung* which effectively reconciles these two terms, i.e., which both reunites philosophy with its past and links it to all the living complexity of the aggregate of historical realities acknowledged as intelligible and entitled to consideration as rational phenomena.[134]

134. In our attempt to characterize briefly present trends in the 'philosophy of the history of philosophy' we have tried to analyse a number of works which appear to us especially significant. In addition to the works cited in the foregoing footnotes, the reader may also select for reference, from a copious output particularly in German and the Romance languages, works such as the following (drawn up as an indication, this list is limited to general considerations on the nature, methods and significance of the history of philosophy; the terms of reference and dimensions of the present work did not allow of giving any kind of adequate picture of the output of works on the history of philosophy, a discipline which is still very much in vogue and extremely fertile in the sphere of so-called 'continental' philosophy):

Collections: 'La philosophie et son histoire' (Section IV of *Actes du VIème Congrès des Sociétés de Philosophie de Langue française*, 1952); CASTELLI *et al.*, *La filosofia della storia della filosofia* (1954) or *La philosophie de l'histoire de la philosophie* (1956); LLERA (ed.), *Idea de la historia de la filosofía, de su eficacia didáctica y su importancia en el presente* (Sociedad cubana de Filosofía, 1954); etc.;

Individual works: ABBAGNANO, 'I compiti di una storiografia della filosofia' (1967); ANTONELLI, *Filosofia e storia della filosofia. Idea di storia della filosofia* (1968); BALADONI, 'Filosofia, storia, e storia della filosofia nel marxismo' (1964); BARTOLONE, *Struttura e significato nella storia della filosofia* (1964); BEIERWALTES, 'Geschichtlichkeit als Element der Philosophie' (1968); BELAVAL *et al.*, *L'histoire de la philosophie et son enseignement* (presentation followed by discussion, Société française de Philosophie, 1961, publ. 1962); BERGMAN, 'Philosophy and the history of philosophy' (1968); BLUMENBERG, 'Paradigmen zu einer Metaphorologie' (1960); BRELAGE, 'Die Geschichtlichkeit der Philosophie und die Philosophiegeschichte' (1962); CARPIO, 'El pasado filosófico' (1963); CARRERAS Y ARTAU, *Balance, estado actual y perspectivas sobre la historia de la filosofía española* (1951); CENCILLO, 'Consequencias de un criterio ecuménico en la historia de la filosofía' (1968); CHIODI, 'Filosofia, storia e realta umana' (1965); CHRISTENSEN, 'Philosophy and its history' (1964-65); DAL PRA, 'Storia e verità della filosofia' (1971); DELHOMME, 'Histoire, histoire de la philosophie, philosophie' (1965); DESCHEPPER, 'Le Colloque de l'Institut des Hautes Etudes de Belgique: Les méthodes en histoire de la philosophie' (1972); DUNN, 'The identity of the history of ideas' (1968); EHRHARDT, *Philosophiegeschichte und geschichtlicher Skeptizismus. Untersuchungen zur Frage: Wie ist Philosophiegeschichte möglich?* (1967); EHRLICH, *Philosophie der Geschichte der Philosophie* (1965); FAIN, *Between Philosophy and History. The Resurrection of Speculative Philosophy of History within the Analytic Tradition* (1970); FOUCAULT, *L'archéologie du savoir* (1969); FRANCHINI, *Esperienza dello storicismo* (2nd rev. edn., 1960); GARIN, *La filosofia come sapere storico* (1959, with lengthy bibliography); GAROTTI, 'Filosofia e storicità' (1967); GATES, *Adventures in the History of Philosophy. An Introduction from a Christian Viewpoint* (1961); GELDSETZER, *Was heisst Philosophiegeschichte?* (1968); GOLDSCHMIDT, 'Temps historique et temps logique dans l'interprétation des systèmes philosophiques' (1970) and *Platonisme et pensée contemporaine* (1970); de GRAAF & BAKKER, *De bezinning over goed en kwaad in de geschiedenis van het menselijk denken. Geschiedenis der wijsgerige ethiek. De mondige mens tussen goed en kwaad* (2nd edn., 1967); GROETHUYSEN, 'Philosophie et histoire' (1961); GRUNDER, 'Bericht [. . . .] über das "Archiv für Begriffsgeschichte"' (1967) and 'Perspektiven für eine Theorie der Geschichtswissenschaft' (1971); HEIMSOETH, 'Stand der philosophiegeschichtlichen Forschung' (in WINDELBAND, *Lehrbuch der Geschichte der Philosophie*, 14th edn., 1950); HINNERS, 'The ideological turn and its problem for the history of philosophy' (1969); IRIARTE, 'Teoría de la historia de la filosofía' (1961) and 'Valor filosófico de la historia de la filosofía' (1964); JANNACCONE, *Il pensiero filosofico nella storia* (3 vols., 1967); JOEL, *Wandlungen der Weltanschauung. Eine Philosophiegeschichte als Geschichtsphilosophie* (2 vols., 1928-34; reissued 1965); KRAFT, 'Dreierlei Philosophiegeschichte' (1971); KRISTELLER, 'History of philosophy and history of ideas' (1964); LANDOLT, 'Cos'è storia della filosofia?' (1969); LEE, 'Hypothetic inference in systematic philosophy' (1969); LIBERTINI, *Influsso della leggenda sulla storia della filosofia* (1961); LINDNER, *Der Entwicklungsgang des philosophischen Denkens* (1966); LOMBARDI, *Concetto e problemi della storia della filosofia* (new edn., 1970); LOURENÇO ARANJO, 'História da história da filosofia' (1961-62); LÖWITH, *Von Hegel zu Nietzsche* (4th edn., 1958) and *Permanence and Change. Lectures on the Philosophy of History* (1969); LUBBE, 'Sprachspiele und "Geschichten"' (1960-61) and *Säkularisierung. Geschichte eines ideenpolitischen Begriffs* (1965); MARTY, 'Le rapport de la philosophie à son histoire. L'accès à la vérité pour un être dans l'histoire' (1964); MASSOLO, *La storia della filosofia come problema* (new edn., 1967); MASSON-OURSEL, *La philosophie comparée* (1923); METZ, 'La historicidad de la filosofía' (1966); MICHELI, 'Storia della scienza e storia della filosofia: problemi di metodo' (1967); MONDOLFO, *Problemi e metodi di ricerca nella storia della filosofia* (1969); MORRA, 'Genesi e dissoluzione della categoria di progresso nella storiografia della filosofia' (1968); NEGRI, *Storia della filosofia e attivita storiografica* (1972, with lengthy bibliography); NOTA, 'Geschiedschrijving en wijsbegeerte' (1966); OEHLER, 'Der Entwicklungsgedanke als heuristisches Prinzip der Philosophietheorie' (1963); Van PEURSEN, 'Fasen in de ontwikkeling van het menselijk denken' (1963-64); PIOVANI, *Filosofia e storia delle idee* (1965); POSER, 'Philosophiegeschichte und rational Rekonstruktion. Wert und Grenze einer Methode' (1971); PRINI, *Introduzione critica alla storia della filosofia* (1967, 1969); PRO, 'Problemas de la historiografía de las ideas' (1971); RÁBADE ROMERO, 'Hacia una revisión de concepto histórico de filosofía moderna' (1960); RANDALL, *How Philosophy Uses Its Past* (lecture, 1961, publ. 1964); ROBINET, 'De la vérité en histoire de la philosophie' (1964), 'Dialectique et histoire de la philosophie' (1969) and 'Hypothèse et confirmation en histoire de la philosophie' (1971); ROMERO, *La estructura de la historia de la filosofía, y otros ensayos* (1968); ROSSI, 'Sulla storicità della filosofia e della scienza' (1964) and *Storia e filosofia* (1970); SALERNO, 'Opposizione relativa, progresso e storia del pensiero' (1971); SANDOR, 'Die Entwicklungsgesetze der Geschichte der Philosophie' (1968-69); SASS, 'Philosophische Positionen in der Philosophiegeschichtsschreibung. Ein Forschungsbericht' (1972); SASSO, *Passato e presente nella storia della filosofia* (1967); SCHNEIDER, Helmut, *Das Verhältnis von System und Geschichte der Philosophie als Methodenproblem* (1968); SCHWARZ, *Über das innere Prinzip der Periodisierung der Philosophiegeschichte. Antrittsvorlesung* (1966); SEBBA, 'What is "history of philosophy"?' (1970); TEIXEIRA, 'Filosofia e história da filosofia' (1962); TONELLI, 'Qu'est-ce que l'histoire de la philosophie?' (1962); TOPITSCH, *Vom Ursprung und Ende der Metaphysik* (1958) and *Mythos, Philosophie, Politik. Zur Naturgeschichte der Illusion* (1969); VERDE, 'Un debattito intorno a "filosofia e storia della filosofia"' (1960); VIANO, 'Storia della filosofia e sociologia' (1966); VIDAL MUÑOZ, 'O método comparativo na investigação da história das idéias' (translated from Spanish, 1964); VOLLENHOVEN, 'De consequent probleemhistorische methode' (1961); XIRAU, *Introducción a la historia de la filosofía* (1964); etc.

Attention should also be drawn to the most recent attempt, undertaken by a large team of specialists, to compile a general history of philosophy: CHÂTELET (ed.), *Histoire de la philosophie. Idées, doctrines* (8 vols., 1972-73).

SECTION TWO

MAN AND NATURAL REALITY

SECTION TWO: MAN AND NATURAL REALITY

Few sectors of contemporary philosophy are as difficult to view synoptically as that covering the relationship between science and philosophy. There are several reasons for this: firstly, the subject is vast and rapidly changing; in addition, it would require some kind of generalist, acquainted with both living science and the philosophical tradition, to deal with it. Few people today combine such extensive scientific culture with such a wealth of philosophical knowledge. The list of works meeting this requirement is thus a short one. It includes those of Ernest Nagel, *The Structure of Science* (1961), Wilfrid S. Sellars, *Science, Perception and Reality* (1963) and collections such as those edited by Herbert Feigl and Grover Maxwell, *Current Issues in the Philosophy of Science* (1961), Sidney Morgenbesser, *Philosophy of Science Today* (1967), and Mario Bunge, *Scientific Research* (1967). A more important reason, however, is that there is a time-lag today between philosophy and science, particularly philosophy and physics, which is in singular contrast with the way in which they kept pace with each other from the time of Galileo up to that of Kant. This time-lag operates both ways; philosophers are frequently a scientific revolution behind the times, just as scientists often cling to dead philosophies and are unaware of the changes which have taken place, for example, in logic, semantics, syntactics and throughout the various sectors of the formal theory of discourse. This study will therefore be divided into two parts, the first of which will contain a brief review of epistemological problems as they appear for the purpose of a direct investigation into the scientist's work. Next, we shall consider the changes brought about by these epistemological problems as such, either in the theory of knowledge, or in regard to the theory of reality, or, lastly, in the evaluation of specifically scientific values. This second group of problems are those which may properly be entitled 'the philosophy of science'.

A. EPISTEMOLOGY OF THE NATURAL SCIENCES *

It is commonly agreed that science is defined by its method rather than its object of research, and that in addition, scientific research differs from its applications and from any kind of technology in that it is pursued solely for the sake of knowledge, without any thought of controlling things or minds. Thus the first question that arises concerns the nature of the end sought by such research, and the relationship between scientific method and that end.

The question is all the more justified in that in fact it appears that the

* The working out of this sub-section has benefited from notes and references provided by Rom Harré and by A. Polikaroff, of the University of Sofia, then a member of the Unesco Secretariat.

greatest changes in epistemology concern the actual definition of the end sought; the contemporary scientist is less inclined to insist on the discovery of facts and the descriptive nature of science in relation to a supposedly given reality, than on elevating knowledge to the level of theorizing. This emphasis on theory means that mathematics, or at least mathematical logic, forms an integral part of the actual description; the description is therefore less a description of facts than a description of the models of the object which the scientist constructs.

It is thus less easy to define science as an attempt at discovery, if by this we mean fact-collecting or the systematization of data in the form of purely taxonomic systems. The purpose of a theory is not only to stand up to tests of refutability, but to be coherent and to provide, with other theories, an overall view of nature; going still further, one cannot legitimately speak of coherence unless the mathematical language is not merely a translation medium, but the actual language in which the body of the theory was originally built up. This is the most far-reaching, and also the most radical consequence of the mathematization of nature symbolized by the name of Galileo.

This all-pervading change in the goal of science, making it ever more theoretical, is reflected in the more limited changes we shall now consider in four sectors in which epistemology appears to have made the most significant advances.

§ 1. *Law and theory*

The hallmark of theory is visible in the first place in stable interrelated patterns normally known as laws; it is the search for these laws which constitutes the end for which the method represents the means. The actual conjectures tested by scientific method are conjectured laws, and the entire methodological cycle is subordinated to this search. In *Scientific Research* (1967), Mario Bunge summarizes the cycle as follows: posing well-formulated problems; proposing verifiable hypotheses; deriving and critically examining the logical consequences of those hypotheses; designing techniques to test the conjectures through their consequences; testing the reliability of the techniques themselves, executing the tests and interpreting their results in terms of satisfactorily colloborated theories; evaluating the truth claims of the conjectures and the fidelity of the testing techniques; and finally discussing the solution obtained, determining its domain of validity and the extent to which it supports (or undermines) our previous knowledge, as well as the new problems it raises. The entire cycle hinges on the notion of law, which is thus central to any study of epistemology and reflects in its own development the general trend of science towards theory.

In a science which has not reached the level of theory, law is taken to mean something approximating to an empirical generalization, a summary of observations. The philosophers' view of the concept of law as a regularity which can be observed, or a statement expressing such a regularity, does not willingly rise above this level; the concept is not false if we take into account only the relationship between states, and not the individual states, in particular the initial and final conditions which as such are contingent. The contemporary idea of law stresses the systematic interrelatedness of several laws, which requires that laws previously derived from an empirical generalization be reformulated in terms of theoretical concepts, i.e., concepts which are themselves elucidated or introduced by a theory. Like empirical generalizations, laws at the stage of theory claim to be both general (with the qualifying clause 'to a certain extent', in order to make room for stochastic theories in the field of mechanics or genetics, which combine regularity and change) and objective, in the sense that they claim to be a conceptual reconstruction of a real pattern in nature. They further resemble empirical generalizations in requiring sufficient confirmation (with all the difficulties inherent in the use of the adjective 'sufficient', which we shall consider in §§ 2, 3 and 4 below). Unlike empirical generalizations, however, the notion of law at the stage of theory posits the added requirement of consistency with other theories, thus bringing certain non-empirical requirements into the epistemological definition of the notion of law. A law goes beyond experience not only because it claims to be universal but also because it involves non-observational concepts. Thus one can no longer define a law without defining a scientific theory, i.e., a hypothetico-deductive system. In such a system some of the initial propositions concern the mathematical nature of the symbols used, others their factual meaning, others the 'thing' itself, i.e., the objective pattern of the theoretical entity described by the theory (the status of these theoretical entities will be discussed separately in sub-section B). It is only subject to this threefold condition that one can speak of a law, in the sense of a law-statement, as being a confirmed hypothesis. Thus what is stated to have been confirmed – *pro tempore* – is the entire system of mathematical, semantical and strictly referential hypotheses which together constitute a reconstruction, by approximation, of the basic traits of the thing considered.

An important corollary of the transition to the stage of theory is that the notion of law appears today to be less closely linked with the notion of causality, traditionally associated in former times with the conception of laws in the sense of empirical generalizations. In addition to causal laws, there are stochastic laws (see above) and dynamic laws.

We shall discuss in sub-section B the philosophical implications of this definition of the idea of law in terms of the idea of theory. Let it be said here that any definition of determinism must accommodate the new conception of the lawfulness of events and take account of the variety of types of law. The proposition that there is no such thing as a lawless event itself becomes a proposition at the same level of theory as the notion of law itself. In particular, if some laws are stochastic, the proposition that no event is lawless does not exclude the notion of random or chance events. Nor does it exclude the possibility that a sequence of individual events, each governed by a certain number of laws, when taken together do not 'obey' any law. Any contemporary discussion of a truly philosophical nature on the subject of determinism must take account of the changes in the strictly epistemological concept of lawfulness in the sense of conformity to laws. If the statement of determinism is taken as a proposition relating not to science but to reality itself, i.e., the way in which events really happen, then the actual formulation of this proposition, which purports to relate to reals, is dependent on the way in which science connects the notion of event with that of law, whether it be causal, dynamic or stochastic. When we speak of the production of events, we mean something which is already situated within the theoretical system connecting events with laws.

We shall go further into this discussion of the ontological implications of determinism as a separate question in sub-section B. One reason for postponing it – and here again we are dealing with a strictly epistemological consideration – is that its outcome depends not only on the notion of law, but also on other notions which have not yet been introduced, in particular, that of the 'model', which seems undoubtedly the most suitable, among those we are considering here, for conveying the claim to 'realism' made by the idea of law, taken not only in the sense of a law-statement but also as a pattern of reality or objective regularity. The function of the theoretical model is to provide what we described above as a reconstruction, by approximation, of the basic traits of the thing considered. But just as the model gives descriptive scope to the law, so the law gives the model its theoretical status. If a model is intended to provide a simple explanation of a set of facts, the theoretical status conferred on the model by the notion of law implies that any fact so represented should be non-observable. This theoretical status of the idea of law should not be overlooked in contemporary discussion on the subject of confirmability. This concept is central to the 'logic of explanation', to revive a term from a twenty-year-old work by Hempel and Oppenheim.[1] A large number of recent studies deal with the question whether explanation and prediction are symmetrical concepts, as argued by N. Rescher, R. N. Manson and M. Scriven.[2] A full discussion of the subject is to be found in the collection of articles edited by Feigl and Maxwell.[3]

Nelson Goodman had already in 1947 advanced the theory that the notion of a confirmable hypothesis is linked to the interpretation of what he calls 'counter-factual conditionals'.[4] This theory has had a considerable influence on epistemological thinking, and there are a number of studies on the significance of 'counter-factuals' and their relation to the notion of scientific law, through that of confirmability or 'projectibility'. The interest of this

1. Hempel & Oppenheim, 'Studies in the logic of explanation' (1948).
2. Rescher, 'On prediction and explanation' (1950); Manson, 'On the symmetry between explanation and prediction' (1950); Scriven, 'Explanation and prediction in evolutionary theory' (1959).
3. Feigl & Maxwell (eds.), *Current Issues in the Philosophy of Science* (1959).
4. Article reprinted in Goodman, *Fact, Fiction and Forecast* (1955, republ. 1965).

discussion lies in the introduction of semantic considerations, and the connection established between the epistemology of science and the logic of discourse. The strictly linguistic status of the subjunctive conditional reveals the connection immediately. This is seen in the transition from a statement dealing with a property, capability or disposition ('x is capable of A in B circumstances') to a statement which explains it by introducing, first, the subjunctive conditional ('if x were put into B circumstances, A would happen') and, second, as a particular case of the subjunctive conditional, the counter-factual ('if x – *contrary to facts* – had been put in B circumstances, A would have happened'). According to N. Goodman, only law-statements, as opposed to accidental generalizations, imply counter-factuals, i.e., propositions in the form 'if – contrary to facts – such a thing had such a property, such-and-such would happen'. This obviously presupposes that the class of law-like statements is identified with the class of hypotheses capable of receiving confirmation 'from the evidence describing their instances'.

Once this relation is established between law and confirmation, and confirmation suitably defined, an explanation of the significance of counter-factuals becomes essential if we are to distinguish between a law-statement and an accidental generalization. Part of the discussion has borne on the status of these counter-factual conditionals.

The problem of counter-factuals itself consists, in N. Goodman's view, in 'discover[ing] the necessary and sufficient conditions under which counter-factual coupling of antecedent and consequent is warranted'. Part of the current debate on counter-factuals deals with their relationship with the notion of 'truth-conditions'. There is not only the question what connection exists between the warrant of the counter-factual (in the subjunctive mood) and the truth-conditions of a proposition no longer involving the subjunctive conditional, but also the question what are the truth-conditions of the actual statement of a counter-factual. Some authors, such as Karl R. Popper, in an article in *Mind*[5] and in *The Logic of Scientific Discovery*,[6] considering that the latter question has not been satisfactorily solved, have rejected the theory that only law-statements are capable of supporting or implying counter-factuals.

Others, such as Roderick Chisholm,[7] have interpreted these contrary-to-fact conditionals as elliptic statements which express the deducibility of the consequent (in the indicative mood) from the antecedent (also in the indicative mood), and to which should be added a set of arbitrary but consistent assumptions. In this way the contrary-to-fact inference could be integrated into the theory of hypothetical reasoning.

Another part of the debate has dealt with the question whether counter-

5. POPPER, 'A note on natural laws and so-called "contrary-to-fact" conditionals' (1949).

6. Revised and augmented English version, by the author himself, of his *Logik der Forschung*, published in Vienna in 1934.

7. CHISHOLM, 'The contrary-to-fact conditional' (1946, 1949) and 'Law-statements and counterfactual inference' (1955, 1960).

factuals provide an adequate distinction between laws and accidental generalizations. This introduces the problem of induction, by linking confirmation with projectibility. For N. Goodman, however, the problem of justifying induction is not something over and above the problem of defining valid induction, that is of defining and describing the valid confirmation of a hypothesis. As with many problems in modern logic, analysis has proceeded by the paradox method. This consists in devising intuitively unacceptable, non-projectible hypotheses which are nevertheless equally well confirmed by the same material evidence, and then trying to discover how in scientific practice such hypotheses can be eliminated and valid projections selected. The main interest of the discussion on confirmation paradoxes lies in its having focused attention on the need to go beyond the syntactical and semantic properties of language and have recourse to its pragmatic properties. Thus according to N. Goodman, a projection is to be ruled out if it conflicts with the projection of a 'much better entrenched' predicate.

This discussion on confirmation paradoxes and the solutions proposed, either for solving them, or for avoiding them,[8] brings us to the heart of the problems raised by inductive inference.

§ 2. *Probability*

After reviewing the ideas of law and theory and before proceeding to discuss the disconcerting problem of induction, it seems useful to insert a few remarks on probability. In the first place, the question of probability is one of the keys to the rationale of non-demonstrative arguments. In the second place, a sudden breakthrough in the philosophy of induction occurred when J. M. Keynes related it to questions of probability.[9] Furthermore, unlike the paradoxes which apparently beset all attempts at justifying inductive inference, the theory of probability, from the outset, forms part of a mathematical theory of great technical sophistication, handled in a way which inspires the greatest confidence. Lastly, notwithstanding this technical sophistication, the question of probability has the advantage of keeping us firmly in touch with everyday usage. As will be suggested at the end of this sub-section, it is perhaps by a return to the everyday use of the concept of probability that it is possible to solve, or even, by a kind of linguistic argument, dissolve, the paradox of induction.

Before any question of a 'probability calculus' arises, there is a practical concrete way of taking rational decisions on the basis of non-demonstrative arguments which entitle us to consider as probable the conclusions we draw from the knowledge of certain facts and apply to other unknown facts. S. E. Toulmin and J. P. Day[10] provide a philosophical analysis of the ordinary reputedly correct use of the word 'probable'. As emerges from

8. Cf. Carnap, 'The aim of inductive logic' (1962).
9. Keynes, *A Treatise on Probability* (1921, republ. 1962).
10. Toulmin, *The Uses of Argument* (1958); Day, *Inductive Probability* (1961).

these analyses and that by Max Black,[11] it is primarily in the form of an adverb that probability properly enters into ordinary usage: 'Such and such will *probably* happen.' This adverb modified a kernel sentence which refers to facts. For this reason, probability immediately falls under the heading not of logical but of empirical possibilities; for the same reason, probability concerns a state of affairs held to be true; again for the same reason, notwithstanding the speaker's uncertainty as to the probable outcome, his judgement is in the nature of an assertion which commits him in regard to the truth of the 'kernel' proposition.

The grammatical use of the adverb *'probably'* takes us somewhat further yet. The assertion qualified by the index of probability is related to the knowledge of initial conditions on which is based the material evidence for the assertion: 'Given this, such-and-such will probably happen.' It is here that the inferential nature of the assertion appears, together with the idea of degree of validity of the inferential relationship, hence of relative distance from certainty: something may be improbable, fairly probable, more probable than . . . etc.

To the notion of degree of probability is linked the idea of attributing quantitative values to the relative degree of probability and the distance in relation to the evidence. This is expressed in ordinary language when we say that an event is likely to happen, or is more likely to happen than some other event. Thus ordinary language brings us to the threshold of the problem of rational conjecture, from which point onwards probability is handled in mathematical terms. This threshold is crossed when it is possible to find a number corresponding to the degree of probability. Jacques Bernoulli, in his *Ars Conjectandi* (1713), followed by Laplace in his *Essai philosophique sur les probabilités* (1814), laid the foundations of this purely formal theory. For Bernoulli, probability concerns the degree of certainty of an ideal belief, the number denoting it – the ratio of the number of favourable cases to the total number of cases – measures the strength of the belief of a perfectly reasonable man who derives his conjecture solely from the evidence at his disposal. Such a being is supposed from the outset to be totally indifferent in the event that the reasons admit equally of one or other decision, and it is only in the case of such a being that the notion of equiprobability, and accordingly also the notion of probability calculus, has a meaning.

This transition from usage in ordinary life to logico-mathematical usage concerns both the theory of action and the theory of science, since both the rules for rational action and those for inductive behaviour in regard to reality receive their formal structure from the mathematical theory of probability. We shall not here discuss the implications of the former, postponing such a study to Section V in the following chapter, sub-section A, § 1, which contains an account of how decision and game theory provides a formal structure to the everyday notion of rational preference between

11. Black, article on 'Probability' in Edwards (ed.), *The Encyclopedia of Philosophy* (1967).

several courses of action, and also to the related notion of strategy. In this section we shall deal solely with those aspects of the notion of probability which may elucidate the debate on induction. We shall consider to what extent the theory throws light on the operation of non-deductive arguments which are held to be rational and which relate to the existence of things which have not been observed.

The parallelism between these two studies is in itself highly instructive. In the same way as decision and game theory raises a difficult problem of 'application' in order to pass from the formalism of the purely mathematical systems to specific action situations in real life, it is also a difficult problem – of a non-formal nature – to provide 'interpretations' whereby the formal systems can be linked to the material evidence underlying empirical arguments. It is this connection between a formal system and its interpretations which will also make possible the transition from a formal theory of probability to the problem of induction. It is at this stage that we shall encounter all the enigmas involved in 'justifying' induction, for example, the meaning to be assigned to the reasons we advance for accepting the axioms of a given system as being an acceptable reconstruction either of a rational course of action, or of a well-grounded empirical expectation. A formal theory of probability is free from enigmas of this type, precisely because it steers clear of such questions of application and interpretation. But while the mathematician is entitled to regard problems of the interpretation of formal systems as lying outside his discipline and having no bearing on theory, the philosopher cannot merely dismiss them as subsidiary. A formal theory of probability can only allow of formulating specific judgements concerning non-observed events and non-verified hypotheses if it is possible to give the undefined terms of the theory semantic definitions, i.e., an interpretation; only then will the axioms of mathematical theory lead to well-grounded conjectures about real outcomes.

It is the task of a philosophical theory of probability to provide this bridge between formal theory and its interpretations, and to pose the problem – which precedes that of induction – of the extent to which the theory advocated justifies our confidence in applying probability calculus to non-observed events, i.e., events for the time being unknown.

Philosophers are divided roughly among three trends.

The first trend, deriving from Keynes, which may be called 'logical' to distinguish it from the 'empiricist' and 'subjectivist' trends, keeps the theory of probability within the limits of *a priori* demonstration, and consequently excludes any reference to material evidence drawn from the relative frequency of occurrences, as in empiricist theories, or any reference to the degree of confidence of an ideal reasoner, as in subjectivist theories. Statements of probability in the form '*The probability of P in relation to S is p*' belong to pure mathematical logic. Rudolf Carnap, R. C. Jeffrey and E. Kyburg[12] are the main representatives of this school of thought. If we refuse

12. Carnap, *Logical Foundations of Probability* (1950, republ. 1962); Jeffrey,

to admit that probability varies with the evidence of associated frequencies, or to relate the strength of the evidence to the belief of an ideal judge, the only way out is to consider probability as a logical relation holding good between two propositions and resulting solely from their significance. It was in order to keep the theory of probability within the bounds of this logical relationship that Carnap introduced the notion of relative 'range' (*Spielraum*).

The other trends draw their refutations of logicism in all its forms from the logical gap which separates *a priori* truths from the anticipation of the unknown on the basis of non-demonstrative reasons.

With this practical problem in mind, empiricist theories are mainly frequency theories in the tradition of Locke and Venn (*The Logic of Chance*, 1866). As with common sense, the judgement of probability is based in this case on the knowledge of relative frequencies. Here the logical gap has moved elsewhere; from being between an *a priori* truth and an anticipation, it now lies between a given frequency and an anticipation, and it is necessary to introduce something on the lines of a limiting value to this frequency as it emerges in the long run. Richard von Mises and Hans Reichenbach[13] are the chief representatives of this school; a separate place should be accorded to C. S. Peirce, to whom we shall return in sub-section 3 below, dealing with induction.

Peirce is in fact one of the philosophers who never ceased studying probability and its impact on philosophy in general, for the reason, that, if probability implies something more than the notion of actual relative frequency, namely something on the lines of a *dispositional* relative frequency, then the criterion of meaning, in its most general sense, should be less nominalistic and more realistic, in order to make room for real possibilities. Here we come back to subjunctive conditionals, which, half a century later, were to lead Goodman to the problem of contrary-to-fact conditionals mentioned in § 1 above. It was pragmatism which led Peirce to consider propositions in the form 'If experiment E were performed in circumstances C, the observable result would be R.'. Hence room had to be made for would-be-frequencies as well as actualities. It was against this background that Peirce worked out, before Reichenbach, the concept of weight and applied it to single cases ('the probability of this coin falling heads'). In addition, like Reichenbach but before him, Peirce spoke of probability in terms of the frequency of the truth of arguments rather than the frequency of the occurrence of events. From this point of view probability theory appeared as a method for justifying predictions, and probability thus became the key to the problem of induction. One difference remains,

---

The Logic of Decision (1965); Kyburg, *Probability and the Logic of Rational Belief* (1961).

13. Cf. von Mises, R., *Probability, Statistics and Truth* (rev. English edn., 1957, republ. 1961) and *Mathematical Theory of Probability and Statistics* (1958, rev. 2nd edn., 1964); Reichenbach, *Theory of Probability. An Enquiry into the Logical and Mathematical Foundations of the Calculus of Probability* (English version, 1949).

however, between the two theories. Peirce refused to extend the notion of probability beyond cases in which an argument leads from true premises to true conclusions a proportionate number of times. A scientific hypothesis is not true a given number of times, and for this reason Peirce refused to speak of the probability of a scientific hypothesis.[14]

Generally speaking, empiricist, frequentist or pragmatist theories have considerable cogency in all cases in which one can point to high frequency rates within a large population of events. They are less suitable for interpreting probability in single cases. Further, as has been seen with Peirce, the notion of relative frequency cannot be plausibly applied to a general law-statement. Lastly, recourse to the 'in the long run' argument runs counter to the aims of empiricism and pragmatism by leading us away from the actual conditions of verification or falsification.

As opposed to the 'realistic' trend of Peirce's dispositional concepts, subjectivist theories purport to remain on the firm ground of a comparison between degrees of expectation, belief and confidence. This is the feature common to the work of B. de Finetti, Good, Jeffrey, Kyburg and Smokler, and Savage.[15] For these authors, by and large, probability denotes the degree of confidence a given person may have regarding the truth of a specific proposition, whose truth is unknown to him at the time, given the knowledge he has of the truth of certain other propositions. In posing the 'coherence conditions' with which a personal judgement must comply if it is to be considered reasonable, subjectivist theory thus takes its place unashamedly within the framework of a logic of uncertainty. This is the logic implicit in any art of betting successfully, i.e., of avoiding penalties and not losing one's stake. The notion of coherence proves to be irreducible to that of logical consistency; the use of scales of utility values necessarily introduces psychological preferences, and it is then impossible to define probability itself without reference to utility, the simplest form of decision rule being the maximization of expected utility. When this stage is reached probability tends to be defined as a degree of rectified confidence. We shall see in the following chapter, in Section V, the implications of these ideas for decision and game theory.

§ 3. *Induction*

It is difficult not to be struck by the contrast between the extraordinary sophistication of the logic of probability (and the kind of confidence placed

14. Cf. Moore, F. C. & Robin (eds.), *Studies in the Philosophy of Charles Sanders Peirce*, 2nd ser. (1964).

15. De Finetti, *La probabilità e la statistica nei rapporti con l'induzione secondo i diversi punti di vista* (1959); Good, *The Estimation of Probabilities. An Essay on Modern Bayesian Methods* (1965); Jeffrey, *The Logic of Decision* (1965); Kyburg & Smokler, *Studies in Subjective Probability* (1964); Savage, *La probabilità soggettiva nei problemi della statistica* (1959). See also Levi, *Gambling with Truth* (1967).

in the use of its arguments) on the one hand, and on the other the perplexity, not to say confusion, that besets philosophical discussion when it broaches the well-known problem of the justification of induction.

The actual terms in which the problem is stated indicate the circle in which the discussion seems condemned to turn. 'Induction' is used to cover all cases of non-demonstrative argument in which the truth of the premises, while not entailing the truth of the conclusions, purports to furnish a good reason for belief in it. In the following paragraphs we shall discuss the various questions that arise in the same order as does Max Black in his article on 'Induction' in the *Encyclopedia of Philosophy* (1967).[16]

On the one hand, the specific nature of induction as a mode of argument is stated positively by the notion of *reason for belief* applied to the existence of individuals not yet observed; on the other hand, its logical validity is implicitly called in question by the same definition, which presents the argument as non-demonstrative; thus the problem appears from the outset as a paradox. This has been the case ever since Hume, in his famous analysis of causality,[17] noted the logical gulf between the idea of necessary connection and that of a simple conjunction of events constantly observed in the past. Even if the problem of lawfulness is today no longer limited to causality, and if little support is forthcoming for the psychological solution proposed by Hume – namely the conversion of habit into expectation by means of a belief which is 'an operation of the soul' – Hume's problem remains posed in its most general terms: how can one justify the additional meaning which seems to attach to the idea of *lawfulness*, over and above that of a series of observations which can never be complete? Several authors, including Goodman,[18] have tried to state this famous enigma in the form of a paradox of the kind familiar to theorists, and this paradox has been used by a number of authors as a basis for posing and solving certain philosophical problems (e.g., Russell in his 'theory of types'[19]). But no matter how sophisticated the form in which the paradox of induction is propounded, it is still deductive reasoning which supplies the basis for comparison in proving the logical weakness of induction.

The solutions proposed to this enigma may be classified according to their references to deduction. At one end of the spectrum, we have the solutions which derive induction, in one way or another, from deduction; at the other end, those which try to attribute a special status to the 'reasoning' involved in induction. The most spirited way of relating induction to deduction is to deny that induction denotes a specific mode of reasoning; the most determined anti-inductionists speak, with Karl Popper, of the 'myth of induction', and reduce it to the hypothetico-deductive method. Formu-

16. One may also refer to KATZ, *The Problem of Induction and Its Solution* (1962).

17. Cf. in particular *A Treatise of Human Nature* (1739), Bk. I, 'Of the Understanding', Pt. III, 'Of Knowledge and Probability'.

18. GOODMAN, *Fact, Fiction and Forecast* (1955).

19. RUSSELL, 'Mathematical logic as based on the theory of types' (1908, reprinted in RUSSELL, *Logic and Knowledge. Essays 1901-1950*, 1956).

lating hypotheses, they say, is one thing, testing them factually is another; and this latter part of the method, which consists in verifying or falsifying observable consequences, does not involve any mode of reasoning unknown to deductive and mathematical logic. This position is the most consistent with the scepticism which seems to result from the requirement of necessary connection posited by Hume. Most authors maintain, however, that the original problem remains intact,[20] since only the problem of 'how' has been solved, and not that of 'why'. The 'how' is a problem of analysis and comparison: how to codify the criteria which render some inductive arguments rationally acceptable, and what it is that makes one inductive conclusion preferable to another as better supported or more deserving of trust. The 'why' cannot be reduced to a differential appraisal of arguments. It concerns the justification of all non-demonstrative arguments, a problem which reappears as soon as we do not reject all inductive arguments but make a selection among them.

At this stage some philosophers are tempted to reconstruct induction deductively. Encouraged by John Stuart Mill, who tried to posit a supreme inductive principle (such as 'Every event has a sufficient cause'), they attempt to introduce an extra premise between insufficient premises and the amplifying conclusion. One way or another, such theories are the equivalent of postulating the regularity of the universe, and thus run the risk of circularity in that they demonstrate lawfulness by order.

The most elaborate forms of these theories associate the problem of induction with that of probability. Keynes[21] opened the way to this theoretical investigation by reformulating the conclusion of inductive arguments ('If P, then probably K') in such a way as to bring about the convergence of the two problems. What gives weight to his attempt is the comment that the confirmation of a hypothesis takes the form of an asymptotic convergence in an unbroken series of favourable cases, each independent of the other.

The most celebrated of the theories stemming from this attempt at reconstruction is still the inductive logic of Rudolf Carnap,[22] in which probability is taken to express a logical relation between propositions. Its basic concept is that of the logical 'width' of a given proposition stated in a certain type of language. This concept can be used to define the degree of confirmation conferred on a proposition y by a proposition x as the ratio of the width $x.y$ to the width x. The definition of logical width depends on the class of possible universes which can be expressed in the language in question.

This system of inductive logic is presented and discussed by Popper, Nagel and Putnam in the volume by P. A. Schilpp dealing with Carnap's

20. See the recent discussion of Popper's position by LAKATOS in 'Changes in the problem of inductive logic' (1968) and 'Criticism and the methodology of scientific research programmes' (1969).

21. See above, § 2, p. 1154.

22. CARNAP, 'The aim of inductive logic' (1962).

philosophy.[23] Among those who seek to eliminate the technical deficiencies of Carnap's system mention should be made of the important work of Hintikka.[24] The work of Kyburg, Jeffrey and Leblanc[25] also pertains to this formalist approach based on the axioms of probability calculus.[26]

The opponents of formalism aim to show that any attempt to give analytical form to confirmation statements leads to emphasizing the gap between the analytical nature of the proof and the nature of belief which seems to be inseparable from any law-statement. An 'inductive leap' remains, and the question of the rational nature of the belief involved in the inductive leap itself is still problematical. It is this question of belief which has led a number of authors to pose the problem not in terms of validation or justification – where the reference to deduction still plays a manifestly important rôle – but in terms of 'vindication'. These authors take as their starting-point well-known situations and strategies in everyday life, in which 'reasonable' decisions are taken in default of any certain knowledge of the consequences: the relative frequency already verified in previous similar cases is one of the 'reasonable' arguments used to anticipate an ultimate limiting value. The notion of 'the long run' is the cornerstone of this mode of argument, and all the difficulties concerning the short run result from it.

C. S. Peirce and H. Reichenbach – whose positions should not be confused, even if they can be justifiably compared – are the most important supporters of this 'pragmatic' type of defence. C. S. Peirce does not belong to the period under study, but the publication of his *Collected Papers* between 1931 and 1935 has given rise in recent decades to a full discussion, part of which is contained in E. C. Moore and R. S. Robin (eds.) *Studies in the Philosophy of Charles Sanders Peirce*, 2nd series (1964). Induction is conceived by Peirce as a process of testing statistical hypotheses by examining random samples. Its operation is to be understood in relation to two other processes, statistical deduction (according to which it is highly probable that a sample randomly drawn from a particular population will have the attribute which has a known frequency of occurrence), and abduction (which, as opposed to statistical deduction, creates new ideas and provides generalizations having 'some chance of being true'). The three processes used in combination produce the notion of a self-correcting method which, if followed indefinitely, leads the scientific community as a whole closer and closer to the truth. In such asymptotic convergence to truth lies the validity of induction.

The most controversial point is the 'self-corrective tendency' attributed by Peirce to inductive procedure, which he believed dispensed with the need for any other postulate of principle. A host of other questions arises on

23. SCHILLP (ed.), *The Philosophy of Rudolf Carnap* (1963).

24. In HINTIKKA & SUPPES (eds.), *Aspects of Inductive Logic* (1966).

25. KYBURG, *Probability and the Logic of Rational Belief* (1961); JEFFREY, *The Logic of Decision* (1965); LEBLANC, *Statistical and Inductive Probabilities* (1962).

26. For L. J. COHEN, however (*The Implications of Induction*, 1970), the validity of the inductive inference is based, not on the mathematical calculus of probability, but on a generalization of the concept of logical truth.

all sides. What, for instance, is the significance of Peirce's condition that the samples on which induction is based should be a random selection? Of what assistance is the idea of convergence in the long run for the relative appraisal of given hypotheses in the short run? There has also been discussion as to the conditional or non-conditional nature of Peirce's self-corrective argument. Must we say that the inductive method will disclose the limit 'if there is any ascertainable limit or truth to be discovered'? What is the connection between this theory and the theory of truth? In particular, how does the apparent chimaera of recourse to the long run fit in with the general outlook of pragmatism? Can we speak of a metaphysical basis when we are dealing with the actual constitution of series and the real existence of a limit? Peirce's work accordingly appears today as a vast mine of ideas which has not been completely exploited.

Reichenbach's theory of induction has provided almost as much matter for extensive discussion as that of Peirce. Nowhere else is such emphasis placed on the practical decision-making nature of the belief sustaining the inductive leap. If a reasonable man uses relative frequency as the basis of a provisional estimate of the limiting value, his judgement is not backed by any absolute certainty that the convergence will continue to the limit. This 'strategy' is closer to that adopted in everyday life, which bids one follow that course of action whereby one has 'nothing to lose'. Max Black[27] objects that, as with Peirce, a mode of reasoning based on consideration of the long run and asymptotic convergence towards a limiting frequency yields no criteria for decisions in short-run cases; further, the determination of limiting values for relative frequencies is only one particular case. Thus we are driven back to examining and classifying those inductive arguments which are actually held to be rationally acceptable, and building up on this basis a differential methodology whereby one inductive conclusion is recognized by the scientific community as being preferable to another.

This reduction of the problem of justification to an examination of actual inductive behaviour brings us at one and the same time to the other end of the spectrum of induction theories. Some authors suggest that the problem itself is badly posed in that it is expressed in terms which render it insoluble, since if one expects a non-demonstrative argument to satisfy conditions which are met only by deduction, there exists no mode of reasoning which can render induction valid. Inductive conclusions cannot follow (in the deductive sense of 'follow') from premises. Max Black is among those who consider that the question of justification is a pseudo-problem. He brings all his weight to bear on proving, by linguistic argument, that the 'value' of induction is presupposed by the very use of expressions such as 'have good reasons for . . .', 'it is reasonable to . . .' or 'the knowledge of certain facts justifies'. Anyone using such expressions is 'committed' by the rules of the communication game to interpreting them in a sense which implies the real application to unknown events of knowledge acquired from

27. Black, *Philosophical Analysis* (1963).

certain sequences of events. There remains only the comparative problem (What procedures are more valid than others?) and the analytical problem (What is it that renders such an argument acceptable?) As for the celebrated problem of justification, it should be dissolved rather than solved.

The objection may be raised that ordinary language is full of metaphysical prejudices, and that sooner or later in this philosophical discussion we are bound to break through the linguistic barrier. Max Black's reply is as follows. To say that the use of words such as 'justification', 'reason' and 'reasonable' commits us to an approach involving the inductive leap is not to revert to Hume's view of induction as a habit or customary belief, nor to profess any kind of conventionalism. It is neither by habit nor convention that all reasonable adults consider that any increase in the number of independent confirmatory instances of a law strengthens the probability of the law's truth. It would be more appropriate to call this rule of the game an 'institution'. All adults, according to Max Black, participate in 'a complex system of ways of learning from experience. [. . . .] Like other institutions [. . . .], the inductive institution requires that its participants have mastered a system of distinctive concepts (among them the concepts of good reason, sound argument, and relative likelihood) having both descriptive and normative aspects ... Understanding what people mean by [these concepts] requires acceptance of certain types of situations as paradigmatic of empirical evidence'.[28]

This way of dissolving, instead of solving the problem is typical of the linguistic approach to epistemological problems. It comes up against the same difficulties as any linguistic solution to an epistemological problem, namely that if the institution is a formal crystallization into linguistic rules of general modes of response to the universe inherited from our ancestors, it is difficult to explain why this institution differs from all others in that it is impossible to get away from it. Max Black recognizes that the idea of ceasing to be an inductive reasoner makes no sense. This is true, but in that case a distinction must be drawn between two kinds of institutions: those which can be changed, because the actual change can be expressed by means of language, and those which cannot be changed without destroying language. The idea of a 'commitment' on the part of the speaker as prior to any attempt to call in question the inductive institution merely shifts the enigma to a higher level, that of the institution of language itself.

It then becomes even more of an enigma than before that language should claim, by the successive approximations of the inductive procedure, to attain to reality itself. Neither conventionalism nor pragmatism have accounted for this claim to truth, and it is open to question whether it is any better accounted for by a purely linguistic theory. In any case, it is this claim which makes the problem of induction a continuing enigma for the philosopher.

28. Black, 'Induction', p. 178 (b) in Edwards (ed.), *The Encyclopedia of Philosophy* (1967), Vol. IV.

§ 4. *The logic of discovery*

The question whether there is a 'logic of discovery' is closely related to the questions raised on the one hand by the formalism of the advanced sciences, and on the other by the manner in which induction operates. For example, how can knowledge be extended to fields of experience which become increasingly remote from the familiar sphere of reality and the human scale of experience, tending now towards the infinitely small, now towards the infinitely large, and how, as knowledge so extends, does it rise to the level of increasingly abstract theories, that is to say, theories which are increasingly remote from the description of experience on the human scale? And by what ampliative processes do we, whenever we make even the simplest law-statement, infer the unknown from the known?

These problems were central to the thought of Gaston Bachelard. His epistemological work, situated at the meeting point of logic and the psychology of the scientific mind, belongs chiefly to the period preceding that covered here. Since the war, however, several important works were published,[29] in the period when his wisdom was at its height – a wisdom that drew sustenance from the most varied sources and was constantly attentive to the latest trends and achievements of advancing science – while a recently published posthumous collection[30] has given us, just when it was most needed, a selection of the philosopher's previously published essays and articles.[31]

The twin paths of induction and formalization lead to the question whether there is a separate logic which corresponds to this increased degree of abstraction, this extension and amplification of our knowledge. The logician's natural tendency is to give a negative reply to this question: the very notion of a hypothetico-deductive method implies that as far as logic is concerned the brunt of the matter is the second half of the term – deduction – and that the first half – hypothesis – falls within the province of psychology and sociology, particularly a psycho-sociology of scientific imagination and discovery. Such a reply appears less and less satisfactory with the realization of the magnitude of the change in scale involved in the transition to theory. Hence the idea of a 'logic of discovery', which would take over the question of the conditions for devising and formulating hypotheses. That this

29. *Le rationalisme appliqué* (1949); *L'activité rationaliste de la physique contemporaine* (1951); *Le matérialisme rationnel* (1953).

30. *L'engagement rationaliste* (1972).

31. BACHELARD's most general epistemological works, which have now become classics, are: *Le nouvel esprit scientifique* (1934) and *La philosophie du Non* (1940) (English translation, *Philosophy of No*, 1968). His thought has exerted a profound influence on the philosophy and history of science in France and Western Europe – an influence which remains very much alive – while poetics and literary criticism continue to draw on the theory of material imagination which is the other side of his reflection on experience and the elaboration of experience. On this last point we refer the reader to the two chapters in this work by Mikel Dufrenne on *Aesthetics and the Sciences of Art*. The durability of Bachelard's influence is attested by the recent publication of a special issue of the review *L'Arc* (1970).

is a task for logic proper is confirmed by an important comment which keeps recurring during the discussion on the empirical applications of mathematical formalisms; as such formalisms become more complex and also more arbitrary (in the sense that they depend on freely conceived systems of axioms), it becomes more urgent to introduce separate semantic and syntactical rules for their interpretation, and consequently for their empirical application.

Two classical problems have been given a new lease of life as a result of introducing the semantic interpretation of formalisms: the problem of inductive amplification, known as the logic of hypothesis, and the problem of the rôle played by models in increasing the degree of abstraction of theories.

The first problem was revived to some extent by the publication and discussion of Peirce's *Collected Papers* (1931 to 1935). For Peirce, the way in which hypotheses are formed is not a question pertaining solely to psychology or sociology. The conditions governing a good hypothesis are themselves of a logical nature; the number of consequences that stem from it, the existence of independent verifications, the rôle of counter-examples, the simplicity of the hypothesis itself, and its compatibility with the network of theory already accepted. Some authors, following up the idea of a special logic of hypothesis, have gone back to Peirce's suggestion that abduction is a separate component part of the combination of procedures constituting induction (cf. above); these authors, who include Hanson,[32] insist on the rôle of discordant facts, anomalies and perturbations in the 'backward' proceeding of inference from facts to hypothesis.

The second problem is that of the rôle played by models in expanding the conceptual framework as a result of introducing more highly abstract theories. This problem concerns particularly physics, where models are specially effective in interpreting the unfamiliar and artificial situations created in laboratories (modern atomic theory). When we transfer the pattern of a known real order of events (in the case of atomic theory, the pattern of the solar and stellar system) to another order, it becomes possible, with certain additions, also to transpose to this new order the formalisms which have proved effective in the previously-known order of events. The notion of a *model* is thus comparable to that of interpreting formalisms by means of semantic and syntactical rules, the model providing, in this sense, a partial interpretation. In the case of atomic theory, it is easy to demonstrate the operation of the model: the existence of new particles is only established by 'evidence' which is interpreted with reference to the entire system constituted by both the instrumental and theoretical fields and the hypothesis in question. Since the properties of the new entity cannot be visualized, we relate them to the properties of mechanical bodies which can be visualized. But the rôle of visualization, however vital, should not blind us to the even more important fact that what is at stake is the construction of a theory

32. HANSON, 'Is there a logic of discovery?' (1959). Also TOULMIN, *Philosophy of Science* (1953).

which is more 'abstract', in the sense indicated above. The problem then arises of the correspondence between one theory and the other, the limited nature of the similarity, and the character of the additional hypotheses which must be introduced in order that such correspondence may exist.

The logic of models has been discussed by numerous authors. One question that has been studied is the relation between the epistemological rôle of the model and the use of metaphor in ordinary language[33]: it may well be that there is in both cases analogical transfer of vocabulary. Other authors[34] concentrate their investigations on the different types of analogical relation which a model introduces between situations (movement of electrons and movement of planets) and between descriptions (atomic theory and celestial mechanics). It seems certain that the stumbling-block with all these analogies lies in the apparent contradiction between the model's function of visualization and its function of increasing formalization or, to state the difficulty in other terms, between the partial nature of the interpretation provided by the model, due to the previous situation which it transposes, and the breakthrough to a wider, more comprehensive theory. It is in fact here that the originality of this dual transposition is to be found: transposition from a familiar to a remote experience, and from a theory of a certain degree of abstraction to one that is more abstract yet.

A final problem inherent in the theory of models is the theory's 'realistic' scope. The model's visualization function, which is uncontested, seems to support the view that the purpose of science is to represent nature, whereas the rôle played by the model in making theories more abstract seems to corroborate the system-building character of science.[35] This question, taken in conjunction with those raised in the preceding paragraphs, goes beyond the bounds of methodological considerations and leads us to the confines of a philosophy of science which is at the same time a philosophy of reality.

B. THEORY OF NATURAL REALITY

It is not easy to say where the dividing line comes between epistemology strictly speaking, as we have been considering it in sub-section A, and a philosophical inquiry dealing with natural reality. The answer depends on how philosophers view the relationship of philosophy to science, and on this matter there is a vast range of opinion. At one extreme we have the school which developed from the *Vienna Circle*, known more generally as logical positivism or logical empiricism, which holds that philosophy is indistinguishable from an epistemology that is itself closely dependent on scientific methodology. Then we have philosophies of science which share logical

33. BLACK, *Models and Metaphors. Studies in Language and Philosophy* (1962).
34. HESSE, *Models and Analogies in Science* (1963); THEOBALD, 'Models and methods' (1964). Also ACHINSTEIN, *Concepts of Science* (1968).
35. Cf. HARRÉ, *The Principles of Scientific Thinking* (1970).

positivism or logical empiricism, which holds that philosophy is indistin- to dissociate themselves from neo-positivism. At the opposite end of the spectrum are philosophies of natural reality which, by contrast with those, do not profess to be just philosophies of science, that is, conceptions of reality based on the empirical sciences alone. Marxism, phenomenological ontology and Aristotelian or Thomist philosophies of nature are all to be found at this end of the spectrum.

This superficial classification based on schools leads to the consideration of a more interesting classification which no longer deals just with schools but, if one dare say so, with 'things themselves'. Indeed, each of the schools we have just mentioned highlights a particular dimension of the problem of reality. And these it is which interest us here, over and above the actual schools.

Going back to the first school mentioned above, the logical positivist, we may ask what is meant here by reality. It is no more nor less than the *referent* of scientific discourse. To ask what is real is to ask what science is speaking about. The theory of reality then becomes more or less equivalent to that of the status of theoretical entities compared with the 'observables' of science. In regard to this problem the classic touchstone is the status of the entities of quantum mechanics. This is the first dimension of the problem of reality.

With the philosophical viewpoints which we group together under the name 'non-positivist philosophy of science' two traditional types of problem concerning reality arise. The first I shall call the problem of the categories of reality. The question of entities which we have just mentioned could already be thought of as a kind of first chapter in a modern treatise on *categories* which, like Aristotle's, would open with the question of substance. The subsequent chapters of this treatise would deal with traditional questions like those of time (of events and processes), space (and the world) and causality. The second group of problems where the non-positivist philosophies point to a link with traditional philosophy I shall call the problem of the *degrees* of reality. These concern the relationship between the levels, orders or domains (matter, life, consciousness or mind) among which natural things are distributed or arranged.

We thus have three associated problems concerning three different meanings of the word 'reality': the *referent* of scientific discourse; the *categories* of reality; and its *degrees*.

This third problem, in turn, touches on a new one which may be called that of the *whole of reality* or *things as a whole*. And this will bring us to the theories of reality which no longer put themselves forward as philosophies of science and explicitly link their concept of reality to a criticism of the claims of positive science to provide an exclusive criterion by which to judge what we call reality.

This approximate correlation of schools and problems suggests the following outline:

1. The referent of scientific discourse;
2. Science and the categories of reality;
3. Science and the levels of reality;
4. Reality according to dialectical materialism;
5. Reality according to the philosophies of nature;
6. Reality according to linguistic philosophies;
7. Reality according to ontological phenomenology.

§ 1. *The referent of scientific discourse*

What does a scientific theory speak about? As has just been mentioned, this problem has acquired very great significance with the constructs of modern science.

For logical positivism, the question of the status of theoretical entities is a pseudo-problem. The only question which can be asked is whether it is advantageous for physical theory to adopt a certain form of language. Hence the mixture of phenomenalism and operationalism, inherited from E. Mach, which characterizes logical empiricism. Positivism is a form of phenomenalism in the sense that it takes observables to be in the last resort the yardstick of what is real. Positivism, for this very reason, is also a form of operationalism since if, in fact, only observables are real, constructs can only be auxiliary expressions, i.e., either entities based on observables or ways of calculating observables. In the last analysis, only experimental statements have any meaning.

A statement of these views will be found in the work of Bridgman.[36] The opponents of logical positivism say that positivism is not so much the philosophy proposed by science as a philosophy imposed on science. For it presupposes a theory of meaning which says that only those statements have significance which imply (or are accompanied by) another statement concerning the conditions in which the first could be verified. This is why a non-positivist philosophy of science is based both on a criticism of the verificationist theory of meaning and on greater attentiveness to what science actually does. Radical empiricism has thus been transcended on two fronts – that of semantics and that of epistemology. Quite frequently, however, physicists know nothing about the progress which has been made in the first direction, that of linguistic philosophy, and take it for granted that logical empiricism is the philosophy implicit in their own methodology.

The attempts to link theoretical concepts with observables by means of 'correspondence rules' may be regarded as still falling under the head of positivism, properly speaking. In this way, statements on observables remain, as it were, the court of final appeal, the correspondence rules making it possible to replace all non-observables by observables. Since Carnap's famous

36. See BRIDGMAN, 'P. W. Bridgman's *The Logic of Modern Physics* atter thirty years' (1959).

article in 1956,[37] a large part of the philosophy of science, faithful to the heritage of the *Vienna Circle*, has been based on the trilogy: empirical basis, theoretical apparatus, correspondence rules.[38]

Carnap's view seemed to other theoreticians like a first breakthrough in the direction of the recognition of the non-operational and genuinely explanatory status of theoretical concepts. Juhos[39] sees the need to introduce theoretical entities as 'fictions'.

Breaking right out of the framework of operationalism and phenomenalism of which it is the complement, other writers question the initial hypothesis of logical empiricism, which is that in the last resort science is based on observables. They say that in order to acquire scientific status any phenomenon must surely already be a construct.[40] If this is justified, one may ask whether the conviction that science hinges on observables is not the factor which prevents us from recognizing what is really happening in the cognitive process at work in science. The question of the referent of scientific discourse consequently takes on quite a different meaning. If constructs have some referent in reality, it is not to the extent that they reduce or are linked to observables – which alone would partake of reality: it is as constructs *per se* that they claim to explain reality.[41]

This tendency to give a real significance to constructs themselves, without their requiring the sanction of observables, very often stems from inside criticism of the semantic presuppositions of logical positivism. The work of Sellars, Grünbaum, Beth, Morgenbesser and Bunge[42] is characteristic of the anti-positivist views of many philosophers of science.

This philosophical position, however, has its own difficulties which may be expressed as a paradox: how can what is farthest removed from ordinary experience be said to be the most real? This 'paradox of the theoretical', to use Hempel's words, points the way to a particularly complex type of solution, according to which the explanatory function of a theory resides in its ability to represent one segment of reality, without, however, any of the concepts involved having phenomenal value. The idea of a construct which is, at the same time, a description of reality is characteristic of the 'critical realism' which is tending to take the place of positivism.

The preceding discussion about the notion of model[43] may throw some light on this epistemological discussion. If it is true that the function of

37. Carnap, 'The methodological character of theoretical concepts', published in Vol. I of the *Minnesota Studies in the Philosophy of Science*, edited by Feigl & Maxwell (1956).

38. Cf. Leinfellner, *Struktur und Aufbau wissenschaftlicher Theorien* (1965) and Braithwaite, *Scientific Explanation* (1953).

39. Juhos, 'Die Methode der fiktiven Prädikate' (1960).

40. Bunge (ed.), *Scientific Research* (1967).

41. Cf. Kneale, *Probability and Induction* (1949).

42. Sellars, *Science, Perception and Reality* (1963); Grünbaum, *Philosophical Problems of Space and Time* (1963/64); Beth, 'Semantics of physical theories' (1960); Morgenbesser (ed.), *Philosophy of Science Today* (1967); Bunge, *Foundations of Physics* (1967).

43. See above, sub-section A, § 4, pp. 1165-1166.

models is to interpret logico-mathematical formalisms with the help of semantic and syntactical rules, it can be seen that by interpreting these formalisms in an empirical sense, the models 'apply' them at the same time to a sector of reality. It would then be the function of a highly formalized science to express through its theoretical concepts the deep-rooted aspects of reality. Whereas for the positivists non-observable entities have only a methodological character, to use the title of Carnap's article, for critical realism the reasons for adopting a theory are also the reasons for accepting the existence of the entities which the theory postulates (Bunge).

The collection of articles edited in 1961 by Feigl and Maxwell[44] clearly shows the state of the discussion concerning the 'ontological status of theoretical entities'. (The opposition will be noted between this title borrowed from G. Maxwell and the title used by Carnap in the article quoted above.)

Criticism of logical positivism is more marked in Germany than in English-speaking countries, and stems less from a close study of scientific methodology itself than from a historico-critical elucidation of the sources of logical positivism. Bohm[45] throws light on its metaphysical sources, while K.-F. von Weizsäcker,[46] taking the history of cosmogonies as his theme, sees scientism as secularized Christianity.

This *general* discussion about theoretical entities provides the setting for a similar but clearly circumscribed discussion on the status of the entities of quantum theory. This particular discussion may, indeed, be considered as affording the test case for discussion of the issues we have just been outlining.

It is the originators of quantum mechanics themselves who initiated the great philosophical debate which, far from dying down, continues to give rise to fresh viewpoints.

The kind of philosophy elaborated by Bohr, Heisenberg and Pauli as an extension of their physics does not follow in the path of the critical realism we have just been describing. In fact, the first idea they advance as a philosophical implication of quantum physics is that of a contradiction between two 'images' of what is real – the particle and the wave. This contradiction imposes a new series of constraints on scientific discourse. These constraints operate first of all on the logical plane: here the task is to formulate the contradiction in terms of a principle of complementarity which tears holes in the logic of identity inherited from Aristotle.[47] As to what interests us here, namely 'ontological status', the implications are no less considerable. Once we are obliged to introduce two rival representations of reality and, what is more, accept transformations from discrete element to wave and vice versa, the very identity of the 'thing' starts to vacillate.

44. Feigl & Maxwell (eds.), *Current Issues in the Philosophy of Science* (1961).

45. See in particular Bohm, *Causality and Chance in Modern Physics* (1957), 'Hidden variables in the quantum theory' (1962) and *Problems in the Basic Concepts of Physics* (lecture given at Birkbeck College, 13 February 1963).

46. Von Weizsäcker, K.-F., *Zum Weltbild der Physik* (1958).

47. Bohr, *Atomic Physics and Human Knowledge* (1958).

This is why quantum physics has not only suggested introducing into reality the existence, equally, of particle and wave but has also suggested interpreting this alternation in subjective terms. This subjectivity can itself be understood in a variety of ways. Thus one speaks of the subjective nature of a *preference* as between two equally valid modes of expression, but one also speaks of the subjective nature of the *representations* which alternate in these two modes. In the case in point, the wave function does not represent a physical property like mass and position but acquires from its probabilistic character an aspect of uncertainty which gives it a somewhat subjective tinge.

What more than anything else marks off the philosophical interpretation of the formulae of quantum physics from any form of objectivism or realism is the statement that the observer and the object observed are inseparable at the level of reality at which the theory operates. The subjective element, once introduced into physics, is sometimes taken a long way. The Copenhagen School even saw in the *indeterminism* drawn from Heisenberg's principle of indeterminacy an argument in favour of human freedom. Whatever may be said about this, the micro-object seems indeed to be without autonomous existence. If nothing happens without the intervention of the observer, the micro-object is a micro-event situated at the intersection of the observer and the thing observed.

Certainly, the philosophical interpretation developed by the Copenhagen School tends in a different direction to that taken by critical realism, which claims to be the successor to logical positivism.

This is why it is not surprising that the supporters of critical realism try to separate the theory relating to physics as such from its philosophical implications. These criticisms attempt to show that the above-mentioned scepticism in regard to autonomous reality rests on the same theory of verifiability that logical positivism applies to meaning in general. The Copenhagen School argues, in fact, that the impossibility of carrying out measurements entails the unverifiability of statements about reality. To this it has been objected that the impossibility of predicting an individual event within a statistical whole is not the same as the impossibility of measuring an event which is taking or which has taken place. It is also objected that even if the argument of non-measurability holds good, it is still wrong to conclude that what cannot be measured does not exist. This arrangment of the Copenhagen School on the score of positivism leads certain philosophers of science to try and reduce the gap between critical realism and the type of thinking based on quantum physics.

We shall say very little here about the work being done in the field of *physics* as such with a view to bringing quantum physics back into the fold of classical physics. The aim of this work is to resolve the contradictions in purely physical terms, without having recourse to an irreducible dualism either on the plane of reality, or between physical object and mental reality, or between object and subject. Landé, Destouches and Bunge[48] are attempt-

48. LANDÉ, *New Foundations of Quantum Mechanics* (1965); DESTOUCHES, 'Phy-

ing in various ways to restore a unitary representation of physical reality more favourable to an objectivist and realist interpretation. The evolution of the thinking of Louis de Broglie[49] is particularly striking in this respect.

This criticism on the part of physicists is reinforced by that of certain philosophers who are anxious to eliminate subjectivism from the Copenhagen School's interpretation, while still maintaining its basic formalism.[50] Subjectivism would be completely eliminated if the quantum theory could be stripped of all symbols which stem from an analogy with classical physics – e.g., specifically particles and waves. By thus being made axiomatic the formulation would, at the same time, become perfectly objectivist.[51]

We shall not develop, either, the arguments which the logicians have put forward against the view that physics must be reformulated on some system of logic other than the ordinary calculus of predicates. Indeed, according to Birkhoff and von Neumann, P. Février and Reichenbach,[52] the statements of quantum physics present us with non-commensurable propositions; and if this is the case, the underlying logic cannot be ordinary logic. To this it has been objected that the argument relates to the experimental propositions and only applies to the theoretical statements if the positivist doctrine of verifiability is accepted. Thus we see the would-be defenders of classical mechanics and classical logic joining forces.

We shall spend more time on the arguments which aim to demonstrate that the paradoxes of the Copenhagen School lose some of their force if they are interpreted in the light of a general conception which is less dependent on positivism in regard to the nature and rôle of theory in physics. Is the action of the observer, in so far as it is itself physical – i.e., linked to instruments and not mental – fundamentally different from the combination, in any theory, of experimental situation, instrumental arrangement and the available theoretical apparatus? Do statements which it is difficult or even impossible to verify in terms of prediction as applied to individuals for this reason lose their realist meaning? Does not this equate existence with observability? Does not the argument that the observer disturbs the physical object in fact presuppose its autonomous existence? In the work of Putnam, Suppes and Mehlberg, and also in the collection of essays made in 1967 by M. Bunge,[53] various manifestations will be found of this attempt by critical realism to strip of its philosophical accretions the scientific achievement of the Copenhagen School and to integrate it in a more objectivist and more realist conception of non-observable theoretical entities.

---

sique moderne et philosophie' (1958); BUNGE (ed.), *Quantum Theory and Reality. Studies in the Foundations, Methodology and Philosophy of Science* (1967), Vol. 2.

49. De BROGLIE, *Certitudes et incertitudes de la science* (1966).

50. Cf. FEYERABEND, 'Problems of microphysics' (1967).

51. BUNGE, *Foundations of Physics* (1967).

52. BIRKHOFF & von NEUMANN, 'The logic of quantum mechanics' (1936); DESTOUCHES-FÉVRIER, *La structure des théories physiques* (1951); REICHENBACH, *Philosophic Foundations of Quantum Mechanics* (1944, republ. 1965).

53. BUNGE (ed.), *Quantum Theory and Reality, op. cit.* (1967).

§ 2. *The categories of reality*

As we said above, in the introduction to the present sub-section, a presentation of contemporary problems concerning natural reality can easily adopt the framework of an ancient treatise on categories. The foregoing discussion of theoretical entities can, without doing violence either to ancient tradition or to the problems which arise today, be placed under the heading of substance. Let us now go through the remaining chapters of such a treatise, considering the concepts of time and event, space and cause.

The last two decades have seen the publication of works written by philosophers of science, with titles which once upon a time would have been called metaphysical.[54]

(a) *The notions of event and time*

The notions of event and time are in the forefront of the philosophical discussions which quantum mechanics have set going. The problem of time, however, is in reality co-extensive with physics itself and with reality as a whole. Problems as classical as *time reversal* and the *irreversibility, directionality* or *asymmetry* of time are today treated both on the semantic and physical planes.

In semantic terms, what do we mean when we employ expressions like those we have just used? Are the three latter expressions identical? A 'tense-logic' like that of A. N. Prior[55] seems today to be an indispensable preliminary to any treatment of the philosophical perplexities surrounding the notion of time. These perplexities themselves are no longer the province of linguistic analysis. This, in fact, moves at the level of expressions such as past, present, future – previous and subsequent – to remain, to last, to change – which mix time and the experience of time and rest on an implicit, uncriticized metaphysics. Above all, however, ordinary language with its verbal tenses cannot express the unified concept of space-time implied by the theory of relativity as stated by Hermann Minkovski, following Einstein, in his celebrated article 'Space and time' (1923). This is why W. v. O. Quine asks that space-time science should be expressed in a language without tenses but simply with a timeless 'is'.[56] Costa de Beauregard[57] confirms the need for this from the point of view of modern physics. Smart[58] shows that tenses can be translated into timeless language by mere reference to

54. Cf. Reichenbach, *The Direction of Time* (posthumous work edited by Maria Reichenbach, 1956); Grünbaum, *Philosophical Problems of Space and Time, op. cit.* (1963/64); Gonseth, *Le problème du temps* (1964).

55. Prior, *Time and Modality* (1957).

56. Quine, *Word and Object* (1960).

57. Costa de Beauregard, *La notion du temps. Equivalence avec l'espace* (1963); see also, by the same author, *Le second principe de la science du temps: entropie, information, irréversibilité* (1963).

58. Smart, J. J. C., *Philosophy and Scientific Realism* (1963).

the statement itself: 'now' is simultaneous with '*this* statement'. Sellars,[59] on the other hand, says that a tenseless language can express only the scientific image of the world, not what he calls its manifest image.

If one passes from semantics to the ontology implied by space-time, the problem is now posed in terms of physics. The directionality or asymmetry of time (to use the conventional figure of speech, time's arrow) is a characteristic of the universe attested both by certain processes of a causal nature and by the phenomena of radiation from a point source. Contemporary discussion has centred particularly on the relationship between directionality and increase or decrease in entropy. The notion of *trace* is a good illustration of what this discussion is about. The fact that there is no trace of the future but only of the past (prints, photographs, fossils, various writings and recordings, memoirs) poses the whole problem of the asymmetry of time. Reichenbach, in the posthumous work quoted above, links the formation of a trace with that of sub-systems having a temporarily decreasing entropy (the branch-systems of Reichenbach and Grünbaum). A footprint in the sand is a state of the sand offering temporarily a high level of order. O. Costa de Beauregard, quite independently, gives an interpretation which is very close to that of Reichenbach and Grünbaum. The problem of time, and more especially of time's arrow, is thus posed today principally in terms of statistical mechanics. The close link between information and entropy has given other writers the idea of viewing the problem from the standpoint of modern information theory. The 'branch-systems' phenomenon has also been related to the receding galaxies and the expansion of the universe.[60] The philosophical significance of this research is that it adds nothing to the analysis of expressions like 'previous' and 'subsequent' in ordinary language, but that it addresses itself to the general features of the universe which make it possible for us to use the language of time although we know nothing of entropy and 'branch-systems'. It thus goes beyond the realm of language and falls genuinely within the scope of a theory of reality.

(b) *The notion of space*

That the philosophy of space, considered from the point of view of a theory of reality, is intimately linked with that of time follows from the relativistic notion of space-time, the now classical concept which is the background to all contemporary work, whether, as in most cases, this work takes it for granted, or whether it leads into a criticism of the conceptual framework of relativity theory.

One can obviously not tackle this series of problems without deliberately renouncing the conceptual implications of the term space in ordinary lan-

59. Sellars, 'Time and the world order' (in *Scientific Explanation, Space and Time*, Vol. III of the *Minnesota Studies in the Philosophy of Science*, edited by Feigl & Maxwell, 1962).

60. Gold, 'Cosmic processes and the nature of time' (1966).

guage. Here, space appears as a container which things, considered overall, never cease to occupy and in which things, taken individually, change their place. In ordinary language, too, as well as in daily experience, space is also something which persists in time. In solid geometry, however, a three-dimensional space is already thought of as intemporal and the verbs which express it have no tenses. Intemporality, in the phenomenal and grammatical sense of the term, is the condition under which there is also something like a space-time. This space-time is not a container and is not liable to change, neither, indeed, does it persist.

It is within this conceptual framework and not at the level of daily experience that some authors attempt to derive topological properties – principally dimensionality – from the notion of temporal order thought of as a unidimensional order of succession, and to reduce spatial measurements to temporal ones. This trend can be traced back to an old article by Carnap on the dependence of the properties of space on those of time.[61] Reichenbach[62] likewise pursues the same line. One may nevertheless wonder whether relativity has completely eliminated (as Čapek thinks[63]) the idea of instantaneous co-existence which started with Leibniz.

The discussion about dimensionality, which is a topological problem par excellence, is of particular interest for a philosophical theory of reality. Several writers, among whom may be included Whitrow,[64] have tried to establish a relationship of conformity or congruence between the dimensonal status of space-time and the kind of world in which man exists: stability of the earth's orbit and of Bohr's model of the atom, conditions for the transmission of electro-magnetic signals and for transformations of energy, etc. In the order of metric geometry, a similar discussion has been initiated by the work of Grünbaum[65] as to how far a particular metric can be held to have an empirical warrant. This discussion is taking place within the framework of a general evaluation of conventionalism and empiricism. The question is, in fact, whether physical space is intrinsically amorphous and whether the attribution of a particular metrical structure is simply the result of the application of measurements performed with solid bodies which provide in this way the normal model for congruence (cf. Putnam's examination of Grünbaum's ideas[66]).

The question, par excellence a philosophical one, as to the absolute or relative character of space, a question which goes back to the dispute between the followers of Newton and those of Leibniz as well as Kantian philosophy, inevitably underwent a profound change under the influence

61. CARNAP, 'Über die Abhängigkeit der Eigenschaften des Raums von denen der Zeit' (1925).

62. REICHENBACH, *The Philosophy of Space and Time* (English version by Maria Reichenbach, 1957).

63. ČAPEK, *The Philosophical Impacts of Contemporary Physics* (1961) and 'Relativity and the state of space' (1955).

64. WHITROW, *The Natural Philosophy of Time* (1961).

65. GRÜNBAUM, *Philosophical Problems of Space and Time, op. cit.* (1963/64).

66. PUTNAM, 'An examination of Grünbaum's philosophy of geometry' (1963).

of the theory of relativity. The choice of the word *relativity* itself is significant since a certain conception of absolute space, making space something completely separate from time, is now excluded. But this has not brought the debate to an end.

In fact, the relative character of space is affirmed in much more radical terms in the line of thinking initiated by E. Mach, who considered the inertial properties of space to have their ultimate origin in the matter contained in the universe. A relational space-time theory thus goes much further than affirming the inseparability of space and time. It implies that the curvature of any portion of space-time is produced by the matter contained in it, and if this is so, Newton's absolute space is definitively ruled out. Indeed, relativist cosmology frequently offers a representation of matter consisting simply of regions of special space-time curvature. Nevertheless, other ways of thinking of an absolute space are making their appearance. Some are linked with attempts to think of curvatures and hence space-time structures in the absence of all matter. Others start with the present difficulty of harmonizing physics and the particles of quantum mechanics with the filed theory of general relativity. On this point, J. J. C. Smart concludes: 'One day, we may know whether a particle theory will have absorbed a geometrical field-theory or vice-versa. Until this issue is decided one cannot decide the question whether space (or space-time) is absolute or relational, in other words, whether particles are to be thought of as singularities [. . . .] or whether space-time is to be understood as a system of relations between particles.'[67]

(c) *The idea of cause*

Of the 'categories' of reality, the notion of causality is perhaps the one most discussed by philosophers. It is also the one which lends itself least well to radical revisions because of the very ambiguity of the problem and because of the illustrious authority of the views handed down by tradition from Aristotle, Hume and Kant. It is thus not uncommon for philosophers, especially when they start speaking about human activity and freedom, to argue for or against Hume's notion of cause as if this were still applicable in the physical sciences. They then take for granted that cause and effect are two distinct observables with no logical link of implication, succeeding each other (or even occurring simultaneously) in time and joined together by a mere empirical regularity. As we shall see later, the philosophy of action still conducts its polemics on this terrain.

The persistence of Hume's model of causality can doubtless be explained by an even more fundamental factor inhibiting change. In ordinary language, it is the archaic meaning codified by Aristotle which persists: we ordinarily call cause whatever brings about the event, change or process with which

67. Smart, J. J. C., article on 'Space' in Edwards (ed.), *The Encyclopedia of Philosophy* (1967), Vol. 7; cf. also Smart, J. J. C. (ed.), *Problems of Space and Time* (1964).

we are concerned. Generally speaking, anything in the universe forms part of the conditions leading to the appearance of an event, but among all the candidates for the rôle of agent responsible for a change, the cause is said to be that one which corresponds to a particular interest, i.e., on which one can act either to forestall an undesired result or to obtain a state of affairs postulated as an end of which the cause represents the means. This pragmatic restriction certainly seems to tie up with the meaning of causality on the level of phenomena.

The epistemological changes affecting the notion of law in theoretical science,[68] as well as those affecting the status of theoretical entities,[69] have brought about a radical redefinition of causality which opposes both Hume's and Aristotle's ideas of cause. The relation of causality is no longer valid between observables but between states of a system, such states themselves containing only terms defined by the theory. Causality is then seen as a relation between the states of a system at different times.

Once this epistemological change is accepted, philosophical discussion deals essentially with the universality of the relation of causality.[70] Indeed, not all the changes in the world are either known or explained in *fact*. In particular, causal laws can only be stated in isolated systems. It is therefore legitimate to ask whether one can speak of a *universal* causal link. As is well known, the basic model of an isolated system governed by causal laws was furnished by classical mechanics. In this system, velocity and position are the causal variables of the change called motion. In this ideal condition of isolation, Newton's law of motion provided the model of causal laws for many generations of scientists and philosophers, until the day when quantum mechanics made its appearance. This was the model which Laplace generalized in the celebrated formula which summed up the deterministic conception of the universe. According to this conception, a mathematician of unlimited ability would be able to calculate with absolute precision all future values of all causal variables. The scientific notion of cause is traditionally identified with this formula.

Such is the background to the contemporary discussions on causality; these have, in the main, been stimulated by the appearance of quantum mechanics. Laplace's formula in fact applied to a world consisting of bodies to which position, velocity, size, shape, etc., could be assigned. The perplexities which the notions of position and velocity have given rise to in quantum physics (and which we have mentioned in connection with the ontological status of theoretical entities) were bound to reappear on the causal level. Faced with this difficulty, authors tend to adopt a great variety of views.

At one end of the scale are those who consider the notions of causality and determinism to be indissolubly linked with a notion of the states of systems implying the univocal determinability of the position and velocity

68. Cf. sub-section A, § 1 above, pp. 1150 sqq.
69. Cf., in this sub-section, § 1 above, pp. 1168 sqq.
70. Cf. BUNGE, *Causality* (1959).

of individual particles. This is the theory of Bohr[71] and the Copenhagen School. For these philosophers, quantum mechanics offers sufficient justification for philosophical indeterminism. Von Neumann similarly maintains[72] that the interference of an instrument of observation constitutes an unpredictable and uncontrollable element, even statistically. In this sense, one has to speak of changes of a non-causal kind.

The majority of writers take up a central position and hold a statistical view of causality. They consider that Laplace's formula must be extended to quantum mechanics with a redefined notion of state which would satisfy the probabilist nature of calculation at this level. According to this view, the relation of causality is no longer between singularities but between abstract states having a definite range of probability. As the probability of a particle being observed at a given place in a given state is fixed, the relation between the states of the world can be said to be equally fixed and it is therefore inappropriate to speak of indeterminism.[73]

At the opposite extreme are the defenders of traditional causality. They see no need to treat statistical causality separately. Thus Bohm[74] introduces what he calls 'hidden variables' (meaning inaccessible to observation) which re-establish strict determinism.[75]

Some writers try to make traditional causality subordinate to statistical causality. Thus Born[76] attacks Laplace's model itself. In his opinion, mechanistic determinism has meaning only in an abstract mathematical space, where a point can be represented with absolute accuracy by real numbers. The physical world cannot be determined in this absolute way, since there is an element of indeterminacy in any physical state. The idea of infinite precision excluding the slightest error is thus an illusion. Consequently, even in classical physics, there are only very high probabilities. According to these writers,[77] the division comes between mathematics and physics, and any physical formula implies a 'scientific uncertainty'. Invoking information theory, Brillouin holds that the notion of 'mistake' or 'error' in observation does not pertain to something merely accidental but constitutes an integral part of our knowledge of the world. Consequently, the establishment of a causal link does not permit the absolute precision and essential univocity which the notion of absolute determination seems to require.

71. BOHR, 'Quantenphysik und Philosophie' (1961); cf. BOHR, *Essays 1958 to 1962 on Atomic Physics and Human Knowledge* (1963).

72. Von NEUMANN, *Mathematical Foundations of Quantum Mechanics* (English version, 1955).

73. MARGENAU, *The Nature of Physical Reality* (1951).

74. BOHM, *Causality and Change in Modern Physics* (1957).

75. Concerning this discussion, see FÉVRIER, *Déterminisme et indéterminisme* (1955).

76. BORN, 'Vorhersagbarkeit in der klassischen Mechanik' (1958); by the same author, *Natural Philosophy of Cause and Change* (Waynflete Lectures, 1948) (1949).

77. SCRIVEN, 'The key property of physical laws – Inaccuracy' (in FEIGL & MAXWELL (eds.), *Issues in the Philosophy of Science, op. cit.*, 1961); WAISMANN, 'The decline and fall of causality' (1961); BRILLOUIN, *Scientific Uncertainty and Information* (1964).

Determinism is not the only issue at stake in the philosophical debate on causality; a further question is whether physical causality is the only conceivable form of causality. We alluded earlier to the stubborn persistence, not only of Hume's model of causality but, beyond that, of the Aristotelian notion of efficient cause, which in turn seems to have a particular affinity with a part of human experience. Hence the question arises: has modern thought broken completely with the old notion of efficient cause understood as that which produces something? When one passes from explaining natural things to explaining human activity, this question repeatedly comes back into the forefront of philosophical discussions. As far back as 1940, Collingwood[78] distinguished three meanings of cause: the causation of a voluntary act by an agent; the thing that man acts upon, like a lever, in order to make something happen or to prevent something; lastly, the whole set of conditions which are invariably accompanied by a change. Collingwood held the second one to be the primary meaning. H. L. A. Hart and A. M. Honoré,[79] making a wide-ranging study of jurisprudence, demonstrate that we call cause only the new, unexpected and unusual condition which is added to a whole range of more lasting conditions (the spark which starts the fire) and which, in addition, falls within the sphere of human action (what man can bring about or prevent). The legal reasoning here testifies to a use of the concept of cause which can in no sense be called archaic and which lends itself to very precise codification.

More recently, several theorists,[80] influenced by Wittgenstein's conception of language activity, have maintained that the theory of human action implies a clean-cut distinction between the notions of motive and cause. It is true that these writers take for granted Hume's notion of cause to which they oppose explanation through motives. They say that cause is an antecedent having no logical link with its consequent and joined to the consequent merely by constant conjunction. On the other hand, the grammar of the notion of motive shows between motive and action a relationship of mutual implication which is quite alien to the contingency of cause as understood by Hume. The argument is particularly strong if one takes the notion of motive as synonymous with 'reason for...'. The motive is then something which makes action intelligible and gives it a communicable sense, and which sometimes validates, justifies or excuses an action subject to moral judgement or reprobation.

Other authors attempt to redefine the notion of motive in terms closer to Hume's causality. For this purpose, they relate the notion of motive to desire rather than to reason and also seek to interpret desire and belief as mental events which cause action.[81] Others again, like Charles Taylor,[82]

78. Collingwood, *An Essay on Metaphysics* (1940).

79. Hart & Honoré, *Causation in the Law* (1958).

80. Melden, *Free Action* (1961); Anscombe, *Intention* (1957); Taylor, R., *Action and Purpose* (1966).

81. Goldman, *A Theory of Human Action* (1970).

82. Taylor, C., *The Explanation of Behaviour* (1964). See also below, Section III, sub-section A, § 2, pp. 1238 sqq.

also accept the view that there is no valid opposition between motive and cause but further agree that Hume's causality does not admit of the form of explanation required by goal-oriented or consciously purposive behaviour. In this case only a teleological explanation, in terms of Aristotle's final cause, would be valid. The notion of final cause thus has to be redefined in terms which will not reintroduce some obscure force or call for a reversal of the temporal order of causality. In a teleological explanation, it is the total form of a closed system which constitutes the explanatory principle. According to the teleological explanation, the condition for the appearance of an event is that a state of affairs should materialize such that it will bring about the end in question or such that this event is 'required in order to' bring this end about. In this way, no additional entity is introduced. One is simply describing a form of law, namely the law by which the appearance of an event is acknowledged to depend on the fact that this event is 'required for' some end. Only the teleological explanation would make it possible to link the phenomenological experience of purposive action and the causal explanation used in the behavioural sciences.

Nevertheless, even if certain motives are causes rather than reasons, or even final causes, the question arises whether the notion of an agent does not inevitably reintroduce a still more primitive idea, namely that of *being able* to do. It certainly seems that the notion of cause is invested here with one of its most primitive and most irreducible meanings. This is the meaning implied in the following expressions whose relatedness is obvious: 'I can do this'; or else: 'I could do this if I were asked'; or: 'I could do it if I wanted to'; or: 'I could have acted differently'. Whether in the form of a simple assertion or as a future conditional or as a past conditional, the expression of *being able*, conceived as *my* being able, certainly appears to designate something like a primitive idea (Richard Taylor).

Thus in various ways the philosophy of action has become, with quantum physics, the modern battlefield on which conflicting ideas of causality confront one another, ordinary language being ranged on one side of the field and the language of science, chiefly the behavioural sciences, on the other. The unanswered question is then the following, and it is a dual question: can the modern concept of causality drive out the primitive notion of cause when this goes back to its original source, the experience of the 'I can'? Must philosophy tend towards a single notion of causality, according to the 'unified science' programme of the Vienna Circle?

Or has philosophy to resign itself to the concept of causality necessarily having a variety of meanings, as the semantics of this expression – and also experience itself, divided between practical action and the contemplation of nature – seems to suggest?

§ 3. *Levels of reality*

It is in regard to *life*, and then in regard to *consciousness* considered as a natural reality, that the problem of the levels of reality arises in an acute form.

(a) One of the questions in respect of which the past twenty years or less have brought the most dramatic changes, even in the way they are presented, is the question whether **biological organization** is specific, or whether it is reducible to biochemical phenomena.

By laboratory simulation of the conditions of the appearance of life, biochemistry, first of all, has reached a closer approximation of what the historical phenomenon of the appearance of life may have been. For many theorists, like A. I. Oparin[83] the key to the biological phenomenon is to be found in the chemical evolution which preceded and made possible the organic evolution by producing the chemical substances on which life is built. The emergence of life then seems to be less a question of sudden appearance than a matter of slow preparation. This does not resolve the enigma of the origin of life but transfers it to the phenomenon of convergence by virtue of which the end result of the chemical phenomena was none other than the synthesis of living protein. In terms of probability, the synthesis which created a chemical system like protein appears as a highly unlikely combination. This improbability seems all the more striking as biochemistry reveals the structural complexity of living matter. Is this tendency of chemical systems towards highly structured phenomena a fortuitous trend, as materialist philosophy requires, or must one suppose that there is a preferential direction imprinted on it? The former view, which is similar to the determinism described in the previous paragraph, has to accept the difficulties inherent in the concept of chance, ultimately involving the paradox that 'Life does not exist'.[84] The latter view, for its part, must accept the difficulties inherent in the concept of goal, ultimately involving the paradox that consciousness, the model of goal-directed activity, precedes itself as an organic consciousness immanent in life. One does indeed seem to be faced with a choice between the universal extension of chance and the universal extension of goal-directed activity.[85]

It is genetics, however, which has provided the new ideas that have had the most profound effect on the problem of life. The discovery and the deciphering of the genetic code begun in 1954 (Watson and Crick) marks, in fact, the introduction into biology of semiological concepts such as code, message, information, coding, decoding, etc., whose special sphere of application till then had been human language, i.e., the conscious and intentional production and exchange of messages intended for social communication. The idea of a genetic code represents a dual revolution. On the one

83. Oparin, *Life. Its Nature, Origin and Development* (translated from the Russian, 1961); Oparin *et al.* (eds.), *The Origin of Life on Earth. Proceedings of the International Symposium on the Origins of Life on the Earth, Moscow, 1957* (1959).

84. Kahane, *La vie n'existe pas* (1962).

85. Cf. Ruyer, *La genèse des formes vivantes* (1956), *L'animal, l'homme, la fonction symbolique* (1964), and *Paradoxes de la conscience et limites de l'automatisme* (1966).

hand, the notion of code has to undergo an unprecedented extrapolation beyond the phenomenal situation of human communication in order to be applied to a chemical molecular model and to account for the properties of DNA (deoxyribonucleic acid). On the other hand, the very notion of structure in biochemistry takes on an unexpected significance when molecules become the carriers of genetic information transmitted from one cellular division to another.

This combination of a biochemically-inspired concept of structure with a semiologically-inspired concept of code represents a considerable conceptual revolution, with a direct bearing on the problem of morphogenesis and, hence, the problem of the origin of life. Before the introduction of semiological concepts, the question was whether or not to introduce a specific organizational concept such as force or form. The problem for those who gave an affirmative answer to this question was whether this organizational principle was to be considered as analogous with a soul, like Driesch's entelechy, or as a 'holistic' principle, irreducible indeed, but capable of the same theoretical elaboration as other scientific concepts. Georges Canguilhem writes of the conceptual upheaval as follows: 'At this point appeared a possible definition of life as order and language. The secret of life was no longer to be sought in a hinterland or background since, in its elementary and universal structure, life knows neither depth nor surface. It is contained in a single dimension, in a linear order in which a few amino acids are linked together. The secret of life is the way in which elementary basic units, grouped in threes, succeed each other. The phenomenon of self-reproduction, the peculiar property of what is alive, from being vainly imagined as a mode of construction with or without an architect, now comes to be seen merely as the copy or echo of a message.' (*translation*)[86]

One after another, all the guiding principles of biology are having to be revised. In the same article, Georges Canguilhem notes that the same explanatory scheme makes it possible to account for two hitherto divergent problems, that of organic permanency or constancy, in the tradition of Claude Bernard, and that of variation or difference, following the Darwinian or neo-Darwinian tradition. In fact, if permanency appears to result from the indefinite transmission of genetic information, erratic decoding, which means in the last analysis some modification in the sequence of nucleotides, certainly seems able to account for the smallest conceivable mutation. In

86. 'Or voici qu'apparaissait une définition possible de la vie comme ordre et comme langage. Le secret de la vie n'était plus à chercher dans un arrière-fond ou un arrière-plan, puisque la vie, dans sa structure élémentaire et universelle, est étrangère aussi bien au relief qu'à la surface, puisqu'elle tient sur une dimension unique, en un ordre linéaire d'enchaînement de quelques acides aminés. Le secret de la vie, c'est le sens de succession de bases élémentaires groupées trois par trois. Le phénomène propre au vivant, la reproduction de soi, jusqu'alors vainement imaginé comme une construction avec ou sans architecte, vient à être conçu comme la copie ou l'écho d'un message.' – CANGUILHEM, 'Biologie et philosophie', pp. 388-389 in KLIBANSKY (ed.), *Contemporary Philosophy. A Survey*, Vol. II, *Philosophy of Science* (1968).

this way differentiation is introduced into life and the resulting variations form the basis for the selection process, the cornerstone of the neo-Darwinian explanation of evolution. Thus the phenomena of evolution, at least of micro-evolution, are included under the same all-embracing theory as 'faithful' transmissions of the hereditary message. Admittedly it remains to be seen whether the phenomenon of macro-evolution, which is really the evolution of species, can be explained by the same mechanism.

The physiologically-inspired concept of regulation undergoes similar reinterpretation. J. Monod and F. Jacob have introduced the new concept of teleonomy to designate the mechanism by which a cell employs part of the hereditary message which is carried in its entirety by the chromosomes of each cell. According to these authors, teleonomy replaces teleology. The equilibrium of living systems is to be explained in purely mechanical terms, and the only process at work in sorting out the variations of the genetic regulation mechanism is that of natural selection. Finally, the phenomena of dysfunction appear in a new light. G. Canguilhem shows that 'a new conception of illness as a biochemical 'error' tends to assimilate dysfunction to congenital abnormality and macro-molecular monstrosity. This profundly modifies the meaning and scope of the concepts of 'normal' and 'pathological'. It depends on the relationship of the living thing to its surroundings whether the error of life becomes an evil or not.' (translation)[87]

The age-old debate between mechanism and teleology is thus continuing today under new forms, and Jacques Monod's recent book *Le hasard et la nécessité*[88] has placed this debate in its broad philosophical setting. A Stoic vision of man, which alone gives meaning in a universe of chance, appears as the counterpart of a vision of the world from which it is desired to eliminate all traces of 'animism'. On the other hand, the discoveries which have turned biology upside down have extended our knowledge of structures, and in particular introduced a new concept of order. If the old teleological concepts still have a future, it is to the extent that they take account of the new semiological concept of order. As was said earlier when speaking of the peculiarly biochemical aspects of the problem, the choice seems to lie between universal chance, which is the counterpart of determinism, and universal goal-directed activity, the counterpart of a philosophy of emergence.

(b) The problem of the specific character of **consciousness** is, as was in-

87. '[. . . .] une nouvelle conception de la maladie, "erreur" biochimique, tend à assimiler la dysfonction à l'anomalie constitutionnelle, à la monstruosité macro-moléculaire. Le sens et la portée des concepts de normal et de pathologique s'en trouvent profondément modifiés. C'est dans le rapport du vivant au milieu que l'erreur de la vie se mue ou non en mal.' – CANGUILHEM, 'Biologie et philosophie, *op. cit.*, p. 391; see, by the same author, *Le normal et le pathologique* (1966) and 'Le concept et la vie' (1966) – article reprinted in CANGUILHEM, *Études d'histoire et de philosophie des sciences* (1968).

88. MONOD, *Le hasard et la nécessité. Essai sur la philosophie naturelle de la biologie moderne* (1970) (English translation, *Chance and Necessity*, 1971).

dicated at the beginning of this sub-section, the second source of perplexity for a philosophical reflection about the levels of reality.

The problem of levels of reality created by the emergence of life and consciousness in matter and their relation to the material substrate gets even more complex by the introduction of what Sri Aurobindo has called the levels of consciousness beyond the one usually enjoyed by most human beings and which, to a certain extent, have been known to most spiritual traditions of the world. The sense of reality itself is a correlate of the type of consciousness one enjoys, just as what one regards as the paradigm case of objective reality is a function of what one pre-eminently identifies oneself with as the subject. The reason why the perceptual world continues to be regarded by almost everybody as the paradigmatic example of objective reality lies, as K. C. Bhattacharyya remarks, in the fact that almost everybody identifies his self with the body. All non-perceptual objects of knowledge thus appear as half-shadowy adjectival characters deriving their reality from, and applying ultimately to the perceptually knowable object which alone is substantive in character. Yet 'they are [. . . .] necessarily symbolized as substantive by objective metaphors, being not merely thought, but sought to be believed as though they were substantive objects'. But, 'if they could be known as such, the actual perceived object would be, at best, a particular manifestation of a more essential object'. Yet it is almost impossible to know them as such, 'because as a matter of fact, though not necessarily, we are wedded to the body and to the perceptual object that is organic to it'.[89]

The problem may be seen at another level by asking oneself as to what is the reality of a created art object. It opens up questions of a different sort than the ones concerned with the relation between matter, life and consciousness. The primacy of material reality is justified generally in causal terms in the sense that it is that independent variable whose control gives the control over those which basically are either its result or at least associated with it in an intimate sense. But the art work does not seem related to its material base in this causal sense at all. Nor, for that matter, is reproducibility what we want from our understanding or knowledge about a work of art.*

For Occidental thought, the problem of the specific character of consciousness is that of the status of conscious events in their relationship to cerebral events or processes.

We shall examine here three kinds of solutions recently suggested for the age-old problem of the relationship between consciousness and the brain. The first is a solution of principle in the sense that it resolves the difficulty at the level of the entities themselves –brain and consciousness. The second brings in a technological criterion based on the possibility of constructing machines having all the properties normally attributed to consciousness. The

89. BHATTACHARYYA, K. C., *Studies in Philosophy* (1958), Vol. II, p. 41.

* We owe Daya Krishna the preceding references and reflections.

third is based on the construction of an informational model applicable both to minds and machines.

(i) The first solution, known as the 'theory of identity', is distinguished by its radicalism from the traditional solutions which explain psychical events as constructions, epiphenomena, accompaniments or associated phenomena. As its name indicates, it postulates the identity of the two sorts of phenomena. It can be expressed in these terms: states of consciousness are sub-groups of physical processes in the brain. The theory has variants, it is true, which range from the claim that statements about consciousness can be translated with no loss of meaning into statements about the brain, to the claim that the entities themselves are identical (in the sense in which light is identical with electro-magnetic energy).[90] This theory of principle implies that the distinctive features of mentalistic language conform to no essential necessity, that this language is merely a product of our cultural usage and that by teaching it a new usage, it will be possible to get it to signify cerebral events. More basically still, the duality of conceptual systems which underlies the duality of linguistic processes, a duality by virtue of which what we signify by a sensation is different from what we signify by a cerebral movement, must be regarded as a cultural habit and not as expressing an ontological duality.

We must admit, however, that we have no physiological explanation allowing us to understand psychical phenomena any better than we already do at their own level of conceptuality. Still less do we have an explanation which accounts for the difference between a conscious and a non-conscious process. In this respect, the theory of identity – and this will also be true of the theories which invoke a technological criterion – argues from a future state of science, to anticipate which itself implies the truth of the theory in question.

Moreover, in those cases where we do have sufficient information on the correlation between conscious phenomena and nervous activity, we do not know why the response is conscious, why, for example, we see the world as coloured. It is not certain that better neuro-physiological information would be able to solve this problem. One may even ask whether a complete neuro-physiological explanation of consciousness would prove the theory of identity and whether it would not be compatible with other theories of consciousness, including dualism. It may well be that there is no scientific progress whereby the theory of identity could be either substantiated or refuted.

What, in fact, is the aim of this theory? It does not propose any explanation which increases our knowledge; it merely pleads for an ontological interpretation of the scientific explanations of conscious behaviour. Its

90. Cf. FEIGL, 'The mental and the physical', in FEIGL *et al.* (eds.), *Concepts, Theories, and the Mind-Body Problem* (1958) = Vol. II of the *Minnesota Studies* – discussed by SELLARS in 'The identity approach to the mind-body problem' (1965); RORTY, 'Mind-body identity, privacy, and categories' (1965); PLACE, 'Is consciousness a brain process?' (1956); SMART, J. J. C., *Philosophy and Scientific Realism* (1963).

basic arguments are (a) that reductionism is possible, (b) that the duality of entities is not necessary. The first argument is essentially in the nature of a refutation and draws its main force from all the difficulties of dualism much more than from the progress of our neuro-physiological knowledge. As for the argument that it is not necessary to posit mental entities distinct from their supposed cerebral support, this is an argument of conceptual economy akin to the famous argument called Ockham's Razor. An argument of this kind is always to be recommended if an explanation involving a small number of categories is in fact simpler and no less effective than an explanation which increases the number of entities; and we are told that recourse to mental entities adds nothing to scientific explanation. The question is, however, whether recourse to mental entities has the same explanatory intention as the postulation of neuro-physiological entities and whether postulation of the consciousness will add anything at all to scientific explanation of a naturalist kind. In brief, is consciousness a form of nature?

This is the question which underlies the debate. On the reply to this question depends the interpretation which can be given of 'mental' language and the conceptuality connected with it. Is it a language of observation? Have we learnt its meanings as we have learnt the language in which we speak of cerebral processes? Can cerebral processes become the referent of the language of consciousness merely by changing the mode of linguistic training? Is the meaning of the language of consciousnness just the product of linguistic and cultural custom?

(ii) The appearance of man-made machines capable of simulating operations referable to the purposive activity of ordinary human beings has given a new dimension to the classical problem of the relationship between brain and consciousness.[91]

In asking whether machines can be conscious, the new theoreticians of artificial intelligence are posing a problem which is both theoretical and practical. In the first place, they ask whether there is a logical contradiction between the idea of a machine and the idea of a conscious being. In the second place, they ask whether it would ever be possible, in fact, to build a machine which could with certainty be described as conscious.

The writers who try to decide the matter by a purely technological argument put on one side the question of principle. They say that it is impossible to answer the question of logical coherence without becoming involved in a vicious circle. To say, for example, that a machine cannot be conscious because it cannot be said to act in the sense of a human agent, or to say it cannot be conscious because it cannot be living or dead or because it cannot suffer, all supposes that the problem is solved. In fact, say these authors, ordinary language does not make it possible to decide one way or

91. The following collections of articles may be consulted in this connexion: HOOK (ed.) *Dimensions of Mind. A Symposium* (1960); SAYRE & CROSSON (eds.), *The Modeling of Mind: Computers and Intelligence* (1963) and *Philosophy and Cybernetics* (1967); ANDERSON (ed.), *Minds and Machines* (1963); MELTZER & MITCHIE (eds.), *Machine Intelligence* (1969).

the other whether it is possible for conscious robots to exist.

One thus has to come back to the practical question. Here the difficulty is no less, but is of another order; it is that of assessing the credibility of technological forecasts. What will become technically possible in the development of artificial brains?

One may ask, however, whether the question can be decided in a somewhat isolated fashion, purely on the basis of technological information and excluding all arguments put forward by non-specialists. Indeed, is it not erroneous to suppose that the technologist's special knowledge of machines equips him to recognize a conscious mechanism if he should by chance come across one? Does not this recognition presuppose a clear idea of the difference between consciousness and its absence? It thus seems legitimate to ask whether the question what is a conscious entity can be answered in a purely technological fashion.

Are we to say that it will be up to the machine to tell us whether it is conscious? But even supposing it were possible to indoctrinate a machine so that it could employ our mentalistic terminology, could distinguish truth and falsehood and, lastly, could not lie, such a machine would be faced with the same difficulty as ourselves in knowing whether it was conscious or not. For it still would not have the indispensable criteria for using outside the sphere of human behaviour, the descriptive adjective 'conscious', the whole meaning of which has been built up, taught and learnt within the framework of communication between human beings.

As for the purely factual question whether it will be possible one day to build machines capable of carrying out all the operations which we call conscious, this depends more than at first sight appears on the question of theory. What we have to do is to forecast a future state of technological development, just as earlier, we had to forecast a future state of neurophysiological science. The credibility of the forecast does not reside simply in a few extrapolations from the present state of the art but rests on the very significance of the operations in question. If we say that a robot can simulate a certain operation termed conscious, we still have to know what has been simulated and what simulate means. Let us consider the case of purposive action, as broadly co-extensive with the notion of consciousness. If the purposive nature of an action is defined by its being directed towards a specific aim, it is quite obvious that many mechanical systems have this distinctive feature. Such are rockets which seek out their targets, radar-controlled navigation devices, various automatic temperature control systems, etc. These various systems have to receive control signals from outside, act in relation to their surroundings under the control of these signals and compare the results of their activity with this specific aim. One then speaks of negative feed-back to underline that the functioning of the system is controlled by the margin of error and that the effect of this control is to bring the system closer to this desired aim. Nevertheless, although this simulates in a remarkable way what we call purposive action, there is a difference between the behaviour of these systems and human behaviour.

Action in people may be occasioned by an aim which is not present (a lost, distant or even non-existent object, an undesirable aspect of things seen as having to be got rid of or changed). People may also act without a precise aim simply for pleasure (as in dancing and games). Lastly, they may choose between several courses of action. Where machines are concerned, the aim is necessarily a real part of the surroundings themselves. In addition, the signals which regulate the system's behaviour are themselves factors belonging to the surroundings. Finally, this condition excludes initiatives and a change of intention when faced with alternatives. Here, the technologist is tempted to reply that it is possible to imagine machines programmed in such a way that they are able to choose and such that the choices they make are not purely and simply determined by the surroundings or by the programme or by a combination of the two. All the same, by arguing from a stage of technology which it is admitted has not yet been reached but which is said to be not far distant, is it not implied that the problem is solved? Can faith in a subsequent stage of technology take the place of an argument? Furthermore, if this faith is in the form of a conjecture, do we not fall into that same vicious circle with which we reproach those philosophers who say that it is impossible, *a priori*, to construct conscious machines?

In point of fact, it is difficult to find conclusive arguments in this field because we scarcely know what consciousness is itself. If we say that machines can do all that man can, do we know what we mean when we say that a man can do this or that? When we say that the machine simulates this or that activity, do we know what it is simulating?

Let us just recall here the analyses of Gilbert Ryle in *The Concept of Mind*. Since this book was published in 1949, there has been a tendency among English-speaking philosophers to see consciousness as a disposition or as a form of behaviour which is described most adequately in 'dispositional' terms. Is it possible to credit machines with dispositions, in other words the ability to do certain things *when the need arises*, and tendency to do them in certain circumstances?

With desire, tendency, disposition and ability, we come back to ideas which certainly seem to belong to the same constellation as negation and doubt, something like a potentialization of action which indeed seems to give a specific definition of consciousness.

Another simple way of demonstrating our ignorance about the meaning of the word 'conscious' is to recall the extreme confusion that prevails in regard to the problem of 'other minds'. How could we with certainty attribute consciousness to machines if we do not know what it means 'to attribute consciousness' – for example a pain – to a person other than ourselves? If we admit that two people cannot experience the identically same suffering and that the word 'pain' derives its meaning only from its reference to private feelings or sensations, how am I able to attribute the same pain to other people? The paradox is markedly greater than that of knowing how I know that other people suffer (by inference? by observation? through sympathy?). It concerns the identity in meaning of the words *to suffer* and *suffering* when they are applied to myself and to others. It will be even more difficult to apply these same terms in any significant way outside the context of human behaviour.

(iii) A third theory attempts to elaborate a concept of consciousness in such terms that it can be attributed not only to people with brains but also to machines. The supporters of this theory borrow their model of consciousness from information theory.[92]

This theory is close to the previous one in so far as it is based on the technology of artificial intelligence, but its argument is not technological (what it may be possible to construct one day) but theoretical. What is important to it is the intelligibility of the informational model. On the other side, this theory is akin to the theory of the identity of mind and brain in the sense that it admits the possibility of simulating consciousness itself. It is fundamentally distinct from this theory, however, in that consciousness is identified with function and not with material components. This is in fact why the brain is not necessarily the only support for this function.

The heart of this theory is the concept of information in the mathematical sense of the word. Consciousness is in fact defined as a way 'to process information in the form of patterned responses'.[93] The concept of information is taken in its strict sense as defined by information theory: 'Information is a mathematical characteristic of an event, selected out of a set of possible events, each of which may be assigned a specific probability of selection'.[94] To this mathematical concept of information is joined the concept of an information channel: 'An information channel consists of two sets of possible events, each member of one of which is linked to every member of the other by a specific conditional probability of occurrence'.[95] This is an important notion for determining the ontological status of the media of communication. One of the sets can be thought of as an input alphabet of code symbols and the other as an output alphabet. The two alphabets, input and output, taken independently, are characterized by their own *a priori* entropies. The relations of probability between the input and output alphabets make it possible to define relations of 'indication', of 'equivocation', of 'mutual information', of 'faithful representation' and of 'interference'. It is by the use of these information concepts that the pattern formation required by the informational definition of consciousness is subsequently explained. The heart of the theory is the explanation of visual perception as information processing in the form of structured responses. The principle of the explanation is based on the notion of *identical* information at the two extremities of an interference-free channel. What is the 'same' in terms of information is the lessening of uncertainty concerning the appear-

92. SAYRE, *Consciousness. A Philosophic Study of Minds and Machines* (1969); see also an earlier work by the same author: *Recognition. A Study in the Philosophy of Artificial Intelligence* (1965).

93. SAYRE, *Consciousness*, *op. cit.* (1969), p. 152.

94. *Ibid., loc. cit.*

95. *Ibid.*, p. 153.

ance of an event at the input end. This uncertainty has no connection with the observer's subjectivity and is a function of the probabilist characteristics of the input alphabet.

Having described the notion of 'structured visual response' in these terms, the theory proceeds step by step to reconstruct all the activities involving the discernment of informational convergence, recognition of configurations and storage of data, etc., which in 'mental' terms are called perception, imagination and memory.

If this theory is to be compared with the theory of mind-brain identity, it is important to describe its intention correctly. Whereas the theory of identity had no explanatory pretensions, this one has considerable heuristic possibilities, manifest, in particular, in the way in which it accounts for the distinction between conscious and unconscious phenomena. Nevertheless, the chief difference between the theories resides elsewhere. This theory does not try to describe either machines which might one day be made (in this sense, the argument is not technological) or a real brain – the physical carrier makes no difference. What the theory sets out is a 'model', in the sense of the word as given earlier.[96] The aim of this model is to indicate the broad outlines of the *processes* involved in consciousness. These processes are, of course, constantly defined in quantitative terms, but the model does not require these processes to be brought about through a nervous system. They can also be brought about by an entirely mechanical system (historically speaking, the application to mechanical systems came before the applications to biology and physiology). To the extent that the responses explained in informational terms are the basis of consciousness in men and the higher animals, one may say that machines are capable of consciousness in a similar kind of way. This is because the description is neutral in regard to the kind of system in which the process is produced. All the model requires is that certain informational measures be applied at certain levels of information processing, the physical make-up of the system to which these measures are applied being itself of no importance.

There is thus a great gap separating this theory from the theory of identity. A new kind of duality makes its appearance, not now, it is true, between consciousness and brain but between the non-material, although quantitative, properties of information, as statistical properties of a system with alternative states, and the physical nature of the carrier or support. Within the notion of cerebral process itself, it then becomes necessary to distinguish the material aspect – the series of electro-chemical discharges – from the function which may be simulated by other systems, this function being nothing other than the relationship between the system's input and output. A system simulating the brain would be one which in an equivalent input situation gave an output equivalent to the original system. Whatever the mechanisms or procedures involved, the function can then be simulated by a system which could even consist of nothing more than symbolic nota-

96. See sub-section A, § 4 above, pp. 1165-1166.

tions. The error of identity theory would then consist in confusing information processing with the activity of the nervous system thanks to which the processing takes place. Or again, if we assimilate the events in the cortex to the elements of the code by which information is transmitted, identity theory would be guilty of confusing the code with the information.

The philosophical interest of this discussion goes beyond the problem of the relation between consciousness, brain and machine in information theory. The interest, in my opinion, is twofold.

What is at stake first of all is the modern destiny of materialism. One may say that the theory of identity is a materialistic theory (at least it is interpreted by its supporters in this way), in the sense that the ulimate entities which serve as referent for all existential statements are of a single sort and are entities which can be fully described and explained by the physical sciences. Is the identification of consciousness with the processing of information a materialistic theory? If materialism is defined negatively by the rejection of Cartesian dualism, to assimilate consciousness to the same information as that processed by brains and machines may be considered a variety of monism and, in this sense, of materialism. However, the fact that the informational model re-introduces a kind of dualism, between function and carrier, re-opens the discussion. In addition, if one considers that the mutual information existing between the various configurations of nervous activity is a mathematical concept, information is not then a concept which can be accounted for in terms of physical entity; nor is the processing of information a process which results from the physical properties of the information channels.

This argument is in fact decisive only for a definition of matter which is today out of date. Thermodynamics and quantum physics impose a view of matter in which the mathematical relationships directly constitute the physical reality itself. It is then possible that a materialism modified in a mathematical direction may demand an informational theory of consciousness. But in that event the whole dispute is perhaps merely a question of label.[97] However, an ontological dualism could just as well invoke on its own behalf the distinction between the processing of information and the physical code, between mathematical process and physical entity. In this respect, the situation is the same as with regard to the progress made in neuro-physiology, which can be interpreted with equal plausibility by a monistic or a dualistic theory.

What is still in the balance in this distinction is the very meaning of the word consciousness. The difficulties met with here are perhaps the cause of those which we have just been explaining. What do we mean, in fact, when we say that consciousness *is* a processing of information? Are we invoking the same kind of identity as in the so-called identity theory? The same questions reappear in a more refined form: Does an interpretation of consciousness in terms of information exhaust the meaning of what consciousness is

97. Cf. HANSON, 'The dematerialization of matter' (1963).

or does it only provide a simulacrum, a formal equivalent of these operations? Nothing is gained by substituting a formula which appears to be more ambiguous, as when one says that the same operations can be carried out by a brain or by a machine. Does 'same' mean identical or similar? The difficulties involved in the concept of simulation, as used in robot technology, also reappear: Should a machine carrying out by different means the same information processing operations as those on which consciousness is based be called conscious or quasi-conscious? If we say that it is conscious, why is it not called a person? And if we say that it is only quasi-conscious, in what does the difference reside?

What is disturbing is that one can no more decide this alternative by having recourse to arguments drawn from the information model itself than one can settle it by using arguments taken from the progress (largely still to come) of neuro-physiology or from the (anticipated) success of robot technology. The hesitations of Sayre in this respect are significant: 'Although I have argued that the type of information processing that results in consciousness when performed by the human brain can be performed within an entirely mechanical system as well, there is nothing in this claim which in any way entails that men are machines. There is nothing even to entail that the operations of the brain involved in consciousness are the same as operations by which machines might perform the same functions... In this respect, the [Information Processing] Theory is on a par with the Identity theory. Both leave the issue of mechanism right where it is otherwise, moving it neither closer to nor further from the realm of plausibility. For my own part, I consider mechanism a most implausible thesis. Despite the fact that I think machines are capable of consciousness, I consider that no behaving system with which we are acquainted is further from being a machine than man himself.'[98] This statement is very interesting since it signifies that the reasons why we *believe* or do not *believe* that men are machines are of another order than those which we can draw from an increasingly perfect knowledge, whether of mathematical models, real brains or man-made artifacts.[99]

§ 4. *The concept of natural reality in dialectical materialism**

The chief characteristics of dialectical materialism from the gnoseological

98. SAYRE, *Consciousness, op. cit.* (1969), p. 215.

99. All these problems will be found discussed from a phenomenological standpoint in DREYFUS, 'Why computers must have bodies in order to be intelligent' (1967); see also, by the same author, *What Computers Can't Do. A Critique of Artificial Reason* (1971).

See also, for the examination of these problems from various angles by contemporary philosophy, below, Section III, sub-section A, § 2, 'The logic of explanation in psychology', item (b), 'Physicalism old and new', pp. 1244 sqq.; and Section V, sub-section B, § 1, under item (d), 'Analytic philosophy and the theory of games', pp. 1402-1403.

* Based on a paper written by N. F. Ovčinnikov as part of the general contribution to the present work prepared under the direction of T. I. Ojzerman.

and logical points of view have been described in a previous section.[100] This is the general background against which consideration of the relationship between Marxist philosophy and the theoretical interpretation of the natural sciences should be seen. In the nineteenth century, dialectical materialism already took account of the great scientific discoveries such as thermodynamics, cellular theory and the theory of evolution, and it continues to develop on the basis of the recent discoveries of natural science. This philosophical work sometimes takes the form of correcting errors made by the previous generation in their assessment of scientific discoveries; cybernetics and molecular genetics afford two notable examples.[101]

One dominant theme of Marxist philosophy is the form of scientific development itself. It is true that an overall assessment of the contemporary natural sciences is made more difficult on the one hand by the process of their internal differentiation, to the point of utter severance, and on the other by the appearance of new disciplines such as biophysics, geophysics and astrobiology, which represent a new stage of integration in scientific knowledge but which add an external differentiation to the internal one. The Marxist interpretation of science considered as a system of knowledge deals precisely with the workings of these processes of differentiation and integration. Soviet philosophers have paid particular attention to one essential characteristic of this development, namely the application of natural science to cosmic space. This trend is, of course, an extension of one aspect of classical mechanics, but the discovery of new non-terrestrial objects, the development of astrophysics, astrogeology, space biology, etc., is today giving a new dialectical slant to the process of applying knowledge to non-terrestrial space. This process is, in fact, going on in two opposite directions, firstly towards discovering and studying the identity of cosmic and terrestrial laws, and secondly towards discovering and investigating the specific character of cosmic phenomena (twentieth-century astrophysics has established, for example, the existence of hyper-dense states of matter and hyper-elevated temperatures which cannot be achieved on earth). The unity of identity and difference in regard to terrestrial and cosmic laws thus emerges as a characteristic feature of the contemporary process of extending science to outer space.[102] A further feature leads to a more specifically Marxist interpretation. All the philosophies of science are asking questions about the revolutionary changes affecting scientific concepts and theories. The distinctive approach of Marxism here is to consider science both as a system of knowledge and as a social institution. From the first point of view, science is the object of a theory of scientific knowledge, and from the second point of view, it is the object of a science dealing with science. Three new pe-

100. See above, Section I, sub-section B, § 4, pp. 1124 sqq.

101. Cf. LEWIS, *Science, Faith and Scepticism* (1959); NØRLUND, *Kybernetik og marxisme. Menneskets nye muligheder* (= Cybernetics and Marxism. New Possibilities for Mankind) (1964).

102. ŠKLOVSKIJ, *Vselennaja, žizn', razum* (= The Universe, Life, and Reason) (1962).

culiarities of contemporary science as a social institution characterize the scientific and technical revolution. Firstly, a particular kind of professional activity has made its appearance, which consists in elaborating scientific discoveries and exploring their practical applications. Next, the development of scientific activity in the form of an immense and complex system in which groups of men engaged in the most varied tasks are at work, in its turn demands special machinery for the guidance of such a system. Lastly, as a result of the two previous features, science as a social institution becomes a direct productive force in society.[103] These characteristics are peculiar to contemporary science: nineteenth-century science did not yet have this character of a system of applied activity, had not yet become a great undertaking capable of influencing social processes to a significant degree. And if one leaves aside this aspect of science as a social institution, the revolutionary changes that have occurred at the level of concepts and ideas no longer have any specific character. The uneven development of scientific thought is a constant feature of the history of science.

The second dominant theme of the Marxist philosophy of natural science is the status of scientific entities as objective realities. This problem is directly linked to the previous one. As was shown in the first paragraph of this sub-section, the sudden breaking down of the fundamental concepts of science has led many scientific theorists to emphasize the problematical character of the objective existence of physical objects whose existence is known to us only on the basis of these concepts. Soviet philosophers, reflecting in their turn on the radical changes which the concepts of space and time, mass and energy have undergone as a result of relativity and quantum physics[104] distinguish two levels of examination: (i) that of immediate examination or mere identification of the changes which have occurred in the formal structure; and (ii) that of their meta-theoretical examination. If one remains at the first level, one inclines towards an interpretation whose general tenor is to refuse an objective value to science. It is only at the meta-theoretical level that we see the transformation of fundamental concepts is subordinated to laws of a special kind. Investigation of these laws is among the most important problems of Marxist gnoseology. One of these laws concerns the inner relationship which governs the transition from one conceptual system to another. The complete break with the past is evident only at the superficial level; deeper down, the new development of knowledge leads to the discovery of more profound concepts which themselves do not vary. A table of concepts in the process of change thus gives no more than an external picture of the deep and complex process of change in the structure of knowledge. The concepts of the old theory are

103. For a study of the distinctive features of the scientific and technical revolution, see the work of B. M. KEDROV and the collective works of his collaborators.

104. Cf. OVČINNIKOV, *Ponjatie massi i ènergii v ih istoričeskom razvitii i filosofskom značenii* (= The Concepts of Mass and Energy in Their Historical Development and Philosophical Significance) (1957).

changing, in fact, and are ceasing to be of use in the new. The new theory, however, at the cost of changing the classical concepts, is bringing to light more fundamental magnitudes and elaborating concepts which are largely independent of cognitive processes and more general than traditional concepts. It is enough here to quote as an example the discovery of the invariancy of the space-time interval, which is a constant more profound than the absolute space and absolute time of Newtonian mechanics. From this point of view, the classical idea of the absolute nature of space and time was not an error but was simply an approximation of the absolute and invariant character of the space-time interval.

If we consider the actual content of the process by which the concepts of the natural sciences have broken down we see that the process is not arbitrary but is in accordance with fixed inner laws independent of the theoretician's free choice. The radical change in the concepts takes place in such a way that they reveal deeper levels of reality than that associated with the traditional concepts. In relation to the new theory, the old concepts appear only as aspects of the reality in question, as approaches to it or projections of it which change with the cognitive process used to study it.

A simple description of the cognitive process thus gives only an external picture of the development of knowledge behind which, as this picture is constantly changing, the object of knowledge existing independently of the changes in this cognitive process becomes lost. The choice, therefore, is as follows: either, owing to the continually changing picture of scientific knowledge, there exists no object independent of knowledge; or else it is necessary to find, within scientific knowledge itself, certain features which could represent elements of knowledge that are independent of these changes – and in this case, we shall not remain at the level of meta-empirical description but shall pass on to a more detailed analysis of the mechanism of the cognitive process. Such elements could be made up of certain non-transient magnitudes (constants), properties or relationships which, by their invariant character in comparison with the continual flux of knowledge, would indicate something opposed to this flux and something quite independent of it. And we do, in fact, find such properties as necessary elements in any cognitive process. They it is which make it possible to state that behind this process, different levels of reality are revealed whose existence becomes known to us as a result of bringing into the open these invariant elements of knowledge.

Must we conclude from this that, for Marxism, an immutable reality exists independent of the cognitive process?

Marxism replies to the accusation of naive realism by distinguishing, on the one hand, the empirical and theoretical levels in special scientific research and, on the other hand, the meta-empirical and meta-theoretical levels of methodological research.

At the empirical level of the special field of science, the existence of the object of research is almost postulated by the very content of such research. And although, in general, the problem of the existence of the object does not arise, its existence is a condition and not a result. Even in those cases where it is subsequently discovered that the presupposed object does not exist (as happened in the case of the ether presupposed in the Michelson-Morley experiment), the elimination of the object takes place only at the theoretical level. In fact, it was only the theory of relativity that eliminated the ether, whereas in the Michelson-Morley experiment, it was presupposed by the experimental procedure itself as an object existing in reality.

As for the theoretical level of research, in a special science, it is linked to the introduction of the concept of the abstract object. To provide the basis for developing a theoretical system, it is sufficient just to find such an object. The problem is then that of the relationship of the abstract to the real object. This problem is insoluble within the framework of any given special theory; in fact, the problem cannot be posed. The construction of an abstract object is indeed a slow and complex process in which, in one way or another, all human culture is involved. By its very nature, the process of forming an abstract object falls outside the competence of any given special theory. It is a process which, with its specific laws, lies outside the domain of the laws of natural science. To study it, one has to go outside the framework of the special theory in question and enter the domain of meta-research first of all at the meta-empirical and then at the meta-theoretical level.[105]

A new set of problems then arises which already pertains to the sphere of philosophical knowledge, even if it is once again the scientist who becomes aware of them in the course of interpreting the process of his own theoretical activity. These new problems arise because the meta-empirical examination of the problem of the correlation of knowledge and the object does not go beyond merely identifying the processes that occur in cognition. This examination merely makes it possible to describe a situation, at both the empirical and the theoretical levels. At the meta-empirical level, however, we are unable to inquire into the process of transition from object of knowledge to knowledge itself, and from knowledge to object. This impossibility is formal in character. If a philosopher does not realize this, he is at the mercy of a host of illusory solutions, one of which is absolute relativism with its characteristic interpretation of the relativity and subjectivity of knowledge. To find a way out of this impasse, one has to understand that consideration of the problems has remained confined to the meta-empirical level of examination. It is through not having understood the inadequacy of this theoretical level that one is led to subjectivist and agnostic theories. It is consequently necessary to pass on to examination of the problems at a meta-theoretical level, which presupposes a systematic elaboration of the theory of scientific knowledge.

The first attempts to construct this kind of theory on the basis of a philosophical analysis of the development of contemporary natural science

105. Cf. the collective work *Filosofskie voprosy estestvoznanija* (= Philosophical Questions of the Natural Sciences), Vols. 1-3 (1958-60).

showed the need for an analysis of the process of conceptual change and of the corresponding change in theoretical systems. This analysis shows that the process of conceptual change is subject to principles of a particular kind. One of these principles is well known and may be called the principle of the receivability of knowledge or, if it concerns the development of physical concepts, the principle of correspondence.[106] Previous theories based on traditional concepts are much more than temporary scaffolding to shore up contemporary theory while it is being constructed, and to be thrown away once the theory is complete. The development of a scientific theory resembles rather the development of a living organism in which the basic cells do not die while the complete organism is being formed but become the necessary elements of a larger system. Just as a distinction is made in the development of living organisms between phylogenetic and ontogenetic rules, so the diachronic change of theoretical systems is subject to a specific logic of development. In the real process of this development, the historical engenders the logical and the logical the historical. The difference between phylogenesis and ontogenesis is the difference between the development of scientific knowledge as a whole and the development of a given special theory. Phylogenesis is based on the dialectical interaction of two opposite factors: the new concept negates the old one and at the same time conserves it as a particular form of its own expression. Ontogenesis, for its part, resulting as it does from analysis of the development of a given theoretical system, makes it possible to distinguish the genetic and structural aspects of this development. It is the structure of a theory in process of formation which reveals the dynamic constants sublated in it. And it is these invariant magnitudes which constitute the criteria of reality, in regard not only to the conservation of things, but also to the conservation of properties and relationships.

A particular case which can serve as an example is provided by the category of causality. This philosophical category illustrates particularly well the theoretical situation which has just been described, and a number of Marxist studies are devoted to this subject.[107] Quantum mechanics has made it necessary to delimit the action of causal laws in connection with the study of micro-processes. As was seen earlier, a certain interpretation of the principle of indeterminacy has led to the abandonment of the principle of causality characteristic of traditional physics. For Marxist authors, the discoveries of the theory of relativity and quantum mechanics in no way justify indeterminist ideas: even if in certain fields the form of causality linked with the determinism of Laplace, characterized by a univocal determination of cause and effect, is not applicable, this by no means justi-

106. Philosophical research on this principle was undertaken for the first time in KUZNECOV, *Princip sootvetstvija v sovremennoj fizike i ego filosofskoe značenie* (= The Princple of Correspondence in Contemporary Physics and Its Philosophical Implications) (1948).

107. Mention may be made in particular of the collective work *Problema pričinnosti v sovremennoj fizike* (= The Problem of Causality in Contemporary Physics) (1960).

fies renunciation of the principle of causality but calls for the establishment of new causal forms which are in accordance with the new data of natural science. The various forms of causal relationship are in fact linked to the knowledge of various kinds of reality; in particular, the different levels of research in physics employ essentially different forms of determination. At the empirical level, a statistical representation of causality is necessary. It is only at the theoretical level of research that determination can become dynamic in character; the necessary and homogeneous link posited by Schrödinger's equation between the changing states of the quantum system expresses the causal dependence of the micro-processes. This dependence, however, does not exclude the specifically statistical nature of the quantum process. The problem of the ground for the statistical nature of the laws is linked to the problem of interpretation of the scientific theory and is the object of more detailed examination. All we can say in this connection is that the solution to this problem presupposes, in particular, the analysis of categories such as those of system and structure.

These two ideas of system and structure are at the centre of the most recent work of Marxist philosophers.[108] A system is defined as a set of elements linked together by a pattern of relations displaying fixed properties. The idea of system belongs as such to the language of the natural sciences but it can be compared to the idea of structure at the level of a description of the cognitive process. To bring out the structure of the system is a deeper aspect of research on one and the same object. If systems analysis entails describing the way in which the object can be divided into its elements and the relations and properties characterizing it determined, the structural approach, for its part, consists in bringing to light to invariancy of the system's essential parameters. Explication of the elements of the system and the relations between them is complemented by explication of their integrality. The problem of the relationship between the whole and the parts of a system assumes a particular significance in biological research.

Unlike the traditional natural sciences, whose ideal of scientific explanation was to reveal the causes of phenomena, contemporary natural sciences try to present objects as systems. The aim of scientific explanation is to be found in the search for the structure of the systems under consideration. The causal explanation is not refuted by the new structural approach but is deepened by it and becomes an aspect of it.

Study of the various types of system and their classification have now become an essential preliminary to the search for general principles of systems analysis.

The problem of mathematization has a direct bearing on this formal analysis of systems. Several works deal with the methodological generaliza-

108. Cf. the collective works *Struktura i formy materii* (= The Structure and Forms of Matter) (1967); *Problemy formal'nogo analyza sistem* (= Problems of the Formal Analysis of Systems) (1968); *Sistemnye issledovanija* (= Systems Research) (1969).

tion of the process of mathematization of the natural sciences.[109] These works are fully in accordance with the theory of models as set out by us in § 4 of sub-section A above. They too stress the break between the experimental reality of the world of natural phenomena and the formal constructions of the theoretician; in its ultimate stage, the model is no longer just the quantitative elaboration of empirical data but marks the construction of abstract objects into developed theoretical systems. Marxist theoreticians, however, insist on the fact that the mathematical apparatus, by virtue of its heuristic character, affords the means for discovering natural objects which were not present in any previous experience. The heuristic force of mathematical constructs tends in the same direction as the realism defended above. In particular, it is through the process of interpretation that a mathematical theory receives its real significance. A mathematical theory can be interpreted in another theory. Interpretation is thus a procedure which makes it possible to compare the abstract theoretical constructions of contemporary science with the real world of nature.

§ 5. *The philosophy of nature*

In this sub-section, something will be said of contributions to the theory of reality which relate more to ontology than to the philosophy of science.

The philosophy of nature is firmly rooted in the tradition of Oriental thought, chiefly among the contemporary thinkers of India.* With them, the philosophy of nature may be said to take at least two significant turns. One is found in the work of Sri Aurobindo who reflects on the fact of evolution and draws consequences which usually are not drawn from it.[110] The first and perhaps the most important conclusion that he draws is that if life and mind have evolved out of matter, then they must have already been implicitly involved in it. In other words, if mind and life are evolved forms of matter, then matter is a sort of involved mind or life in principle. The second question that he poses is whether all evolution has stopped with evolution of mind and, if not, what direction may possibly be taken by evolution in the future. In other words, what is involved in the mind now which we cannot quite clearly see but of which we are vaguely aware? The clue, he suggests, may be found in those fleeting moments of consciousness which, when they occur, are felt to be most real, and yet on return to the normal mode of consciousness are felt as illusory or unreal. The first stirrings of life in inanimate matter and the first stirrings of intelligence in life would have seemed the same to any observer, if he had happened to be present then.

109. Cf. AKČURIN, VEDENOV & SAČKOV, *Poznavatel'naja rol' matematičeskogo modelirovanija* (= The Cognitive Rôle of Mathematical Modelling) (1968).

* We owe Daya Krishna the developments which follow.

110. See in particular AUROBINDO, *The Future Evolution of Man. The Divine Life upon Earth* (new edn., 1963).

On the other hand, nature itself is seen as the correlate of the cognitive enterprise of man. The enterprise itself is bound up with man and is essentially unending in nature. It is impossible to think of a time when man would have known all there is to know – and this, however long the time that may be granted to him on this planet. Science is as much the pursuit of an ideal, as any other pursuit of man. The crucial question about nature, therefore, always revolves around the fact whether man is to be wholly included in natural reality without a residuum and, if so, do ideals which man seeks and this seeking itself form part of natural reality? If the answer is 'yes', the nature of natural reality itself is radically transformed from what most of the scientists generally take it to be. On the other hand, if the answer is 'no', a bifurcation is introduced into the heart of reality and nature is understood in a partial and truncated manner.

The paradox of science as a meaningful activity in the pursuit of an ideal value embedded in a reality revealed by it to be meaningless, is brought to a sharp focus in an article entitled 'Science and Zen', published in 1965 by the well-known Japanese philosopher, Keiji Nishitani. The paradox consists in the fact that the activity of the scientist *qua* scientist cannot be accounted for in the world-view presented by science which consists in banishing all teleology from the world. But teleology is the basis not only of all traditoional religions, but also of all activity of man including the scientific one. If what science reveals about the world is to be considered true, then science itself would have to be considered meaningless or perhaps, in a deeper sense than even Nishitani implies, impossible. Nishitani's solution to the paradox is through Zen Buddhism, which reconciles and holds simultaneously in one vision the contradiction which constitute reality.

Another problem that has been of concern to most thinkers is the status of reality as revealed in perception and the exact nature of what is so revealed. In this context, perception is not merely seen as providing the empirical touchstone on which various scientific theories are tested and validated, and in terms of which theoretical concepts get their ultimate meaning, but also as providing the *knowledge* of a world in terms of which we live, feel, act, and share a common experience of an inter-subjective character. The world revealed in perceptual experience, however, is not 'a world of clear-cut dichotomy' to which alone the law of excluded middle is applicable. Thus, as the Japanese philosopher Seizo Ohe points out,[111] it is not only at the subatomic level that the law is not applicable due to lack of clear-cut dichotomies, but also at the perceptual level. To insist on a strict application of the law in such a situation would, as he points out, be misleading in character.

Further, the world revealed in sense perception is not as atomistic and chaotic as is thought by many Western thinkers, requiring principles of structure and organization from some faculty other than itself. Rather, it is always intrinsically structured – and structured not merely in terms of what

111. Cf. OHE, 'The multiple structure of our external knowledge' (1964) and 'Flexibility in human cognitive behaviour towards nature' (1968).

is perceived, but also in terms of significance. Eastern philosophers thus generally feel more drawn to *Gestalt* psychology, though they generally tend to ignore its identification of perceived configurational patterns with the underlying neurological structure. On the other hand, they have paid far more attention to what may be called the *ground* against the background of which every figure emerges, than either the psychologists or the philosophers who have paid attention to the school have ever done. The differentiations are, so to say, within an undifferentiated continuum which itself is as, if not more, real as the differentiations, and which is experienced as directly and immediately as the differentiations themselves. The one important task of philosophy, thus, consists in drawing attention to the ultimate ground which is present in experience and against whose background everything appears and disappears. Also, the moment one concentrates attention on experience, the limitation to sense-experience, on which most of modern Western epistemology is based, begins to appear extremely arbitrary. Philosophy, as a reflection on experience, then, knows no limitations on the type or types of experience it would reflect upon or which it would consider relevant or significant for cognitive purposes.

In Germany, *Naturphilosophie* sometimes has the meaning of philosophy of science, as can be seen in the works of W. Büchel[112] and F. Dessauer,[113] as well as in the relevant dictionary and encyclopaedia entries.[114] This is because the philosophy of science itself tends to become ontology as soon as it no longer confines itself to a logic and methodology of natural science but undertakes to say something about reality itself. Have we not ourselves presented the discussion on the objective status of the theoretical entities of science as being the modern equivalent of a discussion on substance, and have we not considered the treatment of the categories of time, space and cause in modern science as the development of the Aristotelian treatise on categories, which in fact began with the category of substance? Have we not also linked the discussion of the degrees of reality to ontological theories concerning materialism, monism and dualism, and the emergence and hierarchy of being? As for the exposition of dialectical materialism, is it not in its way a statement of a philosophy of nature, especially if one emphasizes the contribution made by Engels to a dialectic of nature?

All this is true. Nevertheless, what distinguishes the philosophy of nature from the philosophy of science and dialectical materialism is its connection with the ontological tradition. We have a 'philosophy of nature' wherever consideration is given to the question how the characteristics of being dealt with in the previous sub-section are related to the intelligibility of being as such, and this constitutes by common consensus the distinctive theme of

112. Büchel, *Philosophische Probleme der modernen Physik. Vorlesungen über Naturphilosophie* (1957).

113. Dessauer, *Naturwissenschaftliches Erkennen. Beiträge zur Naturphilosophie* (1958).

114. Feyerabend, 'Naturphilosophie', in *Fischer-Lexikon. Enzyklopädie des Wissens* (1958); Margenau, 'Naturphilosophie', in *Die Philosophie im XX. Jahrhundert* (1959).

ontology. This ontological tradition, which will be seen to embrace many currents, has a common source in Aristotle. 'There is,' says the Stagirite, 'a science which studies being as such, and the attributes which essentially belong to it.'[115] It is the 'as such' which distinguishes this science. As for its content, Aristotle's greatest single legacy to the West is the idea of the polysemia of being: 'Being expresses itself in a multitude of ways'. Indeed the different meanings of being provide the texture for an exposition which has necessarily the nature of a discourse – the discourse of ontology itself. This multitude of meanings can of course be distinguished from one another, and related to each other, by a variety of ontological criteria. According to one such criterion or trend, the discourse of being follows the succession of categories (in the primary sense of the word, meaning schemes of attribution). Thus the *ousia* – the *substantia* of the Scholastics – is the first term of that series which Aristotle said was 'relative to a first term'; for in order to have quality, dimension, relation, place, time, etc., some thing must be, and be determined. To describe this original unity which groups categories together and which is irreducible to that of a genus, the Scholastics coined the term 'analogical unity'. This analogous character of being will provide modern philosophies of nature with a key: the question will then be how do the categories and entities of reality as described by modern science link up with this discourse of being and its multiple meanings? Aristotelian substance also expresses itself in a multitude of ways, the first term of reference being the 'form'. These 'substantial forms' were, of course, a serious obstacle to Galilean and Cartesian mathematical physics. All the same, even at the beginning of the eighteenth century, Leibniz did not disdain to link his dynamic physics to the tradition of substantial forms. The reason is that Aristotle, while including Platonic Forms in the real and thus making them both a principle of intelligibility and a real substratum of perceptible substances, introduced another distinction as important for the history of ideas as the distinction of being according to categories and as the definition of substance by form, namely the distinction between being as actuality and being as potentiality. On the one hand, the notion of potentiality made it possible to understand movement, and, in general, change and becoming which in this way were no longer a simulacrum or shadow. On the other hand, the notion of actuality, i.e., of accomplished being or of being in entelechy, introduced a principle whereby the different types of being could be placed at different levels, ranging from those with potentiality – the perceptible substances – to those stripped of potentiality, the heavenly intelligences and thought as a pure act. The idea that not all being was actual but that there was also potential being was not exhausted by the medieval 'virtues'. From Leibnitzian dynamism onwards, it reappears in every century. More than one theme of the *Naturphilosophie* of German speculative idealism stems from it, among others Schelling's concept of *Potenz*. The Nietzschean theory of the Will to Power is its most famous modern development.

115. *Metaphysics*, Bk. IV.

But the counterpart to the notion of potentiality, i.e., that of different levels of being corresponding to different degrees of entelechy, is of no less ontological significance. This is perhaps the key idea of a philosophy of nature – not only to situate physical entities in relation to perceptible substance, but to hierarchize the degrees of reality according to the degrees of actualization of substance, i.e., according to degrees of entelechy.

In this way, the philosophy of nature pursues a dual aim: over and against the fragmentation of scientific knowledge, to grasp the generic unity of nature; and over and against the empiricist and materialist reduction to a physical order of being, to reflect the irreducible distinction between the different levels of being – physical, animate, mental and spiritual. What one might call fundamental Aristotelianism provided Western thought both with a unitary theory of perceptible substance and a hierarchical theory of degrees of emergence, according to the principle of the progressive subordination of potentiality to actuality.

This is the dual theme to be found all through the multiple tradition which stemmed from fundamental Aristotelianism. In the past, Scholasticism was its main branch and neo-Thomism – the foremost exponent of which is still Etienne Gilson, with his large and constantly expanding body of work[116] in three spheres, the history of philosophy, ontology and the philosophy of art – is today its most important representative.[117] However, the repeated allusions made to Leibniz, to the *Naturphilosophie* of speculative idealism and even to Nietzsche, lead one to think that the philosophy of nature has many faces besides neo-Thomism.

Even within neo-Thomism itself, there are great divergences concerning the relationship of nature and science. Some hold, like J. de Tonquédec,[118] that the philosophy of nature owes very little to science, the former constituting a part of *philosophia perennis* while the latter results from progress in knowledge; most hold, however, that the philosophy of nature and science form a pair in which the two terms, while distinct, jointly define each other.[119] Conversely, some think that science owes very little to the philosophy of nature, other than the explication of its implicit assertions about nature;[120] for others, a science of nature completely cut off from the philosophy of nature cannot arrive at a real understanding of science.[121]

116. Cf., in the field under consideration, GILSON, *L'être et l'essence* (1951). For the contribution of the neo-Thomists, particularly Jacques Maritain and Etienne Gilson, to the philosophy of art, we refer the reader to the account of the subject given by Mikel Dufrenne in this work: see, in Chapter IV, *Art and the Science of Art Today*, Section II, 'The philosophical scene', p. 551.

117. Cf. the collection of articles *Idée du monde et philosophie de la nature* (1966), containing, *inter alia*, the articles by S. BRETON, 'Monde et nature', and D. DUBARLE, 'Epistémologie et cosmologie'.

118. TONQUÉDEC, *La philosophie de la nature* (3 vols., 1956-59).

119. Cf. in particular SIKORA, *The Scientific Knowledge of Physical Nature. An Essay on the Distinction between the Philosophy of Nature and Physical Science* (1967).

120. Van MELSEN, *The Philosophy of Nature* (1961).

121. Cf. ASHLEY, 'Does natural science attain nature or only the phenomena?',

Here, the philosophy of nature in a strange way links up with dialectical materialism which holds that no science is possible without an overall grasp of the structures of reality.

What underlies this debate is the question whether understanding nature is linked to a single mode of knowledge, namely science or to many. Are the terms being, substance, potentiality and change to be understood *differently* in philosophy and in science? This question of the many ways of knowing nature is at the centre of the work of M. Ambacher,[122] J. J. Kockelmans,[123] R. Ruyer,[124] and V. von Weizsäcker.[125]

This question, although it is not completely identical, partially coincides with that of the status of pre-scientific knowledge. What, though, is to be understood by pre-scientific knowledge? If it is taken as the form of knowledge which historically precedes science, it has to be admitted that this knowledge is an amalgam of prejudice and that science's only possible attitude to it is critical and reductionist. In this respect, the critique of 'secondary' qualities and then of 'primary' qualities is organically a part of the birth of modern science. Is pre-scientific knowledge just a stage in the history of science, however, as it were its pre-history? Is there not also a non-historic, non-genetic sense to be given to the expression 'pre-scientific', one that would designate the characteristics of a primordial experience which is never completely eliminated and from which scientific knowledge itself takes its sense of reality? Man, in fact, has always had some understanding of nature through his daily familiarity with things, through myth and poetry and through the affinity of his own body with bodies in general. And this understanding generates a conceptuality which supports the same kind of intelligibility as is generated, for example, by Aristotelian concepts. Rather than pre-scientific, then, it would be better to say 'meta-scientific'. This would then designate something other than the perceptive level of the knowledge of nature, or the allusions imprinted in experience by myth and outworn cosmologies. This experience of nature would embrace a grasp of being directly akin to perception but furnishing an implicit representation of the intelligible order that Aristotelian discourse had placed under the heading of the science of being-as-such.

One proof that science alone does not give access to reality is provided by the history of science. There is no science which does not make use of tacit presuppositions[126] which are neither dependent on perception nor anticipations of objective knowledge and which bring into play a philoso-

in SMITH (ed.), *The Philosophy of Physics* (1961); and KANE, 'The extent of natural philosophy' (1957).

122. AMBACHER, *Méthode de la philosophie de la nature* (1961).

123. KOCKELMANS, *Phenomenology and Physical Science. An Introduction to the Philosophy of Physical Science* (1966).

124. RUYER, 'La philosophie de la nature et le mythe' (1956).

125. Von WEIZSÄCKER, V., *Am Anfang schuf Gott Himmel und Erde. Grundfragen der Naturphilosophie* (6th edn., 1963).

126. Cf. V. von WEIZSÄCKER, *Am Anfang . . ., op. cit.*

phical conception of nature, of human experience and man's position in the world. These presuppositions are not just 'epistemological obstacles',[127] but 'undemonstrables' which play the part of regulatory *a priori* principles.[128] In the past, the same conception – that of Aristotle for example – may, from a certain point of view, have constituted an obstacle to scientific development and, from another point of, view, preserved the sense of reality on which science depends, in so far as it remains 'an experience' and not just a language game. And even if science is a language game, the question is what are the general features of the universe which enable us to play it. To map out these features is not a task for objective knowledge but for the philosophy of nature. To quote A. van Melsen: 'What kind of nature must we have if science is to be what it is?'[129]

This theory of the philosophy of nature is confirmed by several doctrines which do not really belong to the philosophy of nature, with its philosophical *credo* derived from Aristotelian sources. All that we said at the beginning of this chapter on the rôle of theory in modern science and on the distance separating theory and empirical generalization confirms the view that science proceeds by continual rectification, starting from projections and presuppositions which, to quote Max Scheler, express 'Man's place in nature' for each age.[130] In this sense, the philosophy of science and the philosophy of nature confirm each other. From another point of view, which gets much closer to a theory of culture, M. Foucault[131] demonstrates that only problems and methodologies in accord with a dominant *epistēmē* can come to maturity at any given time. This *epistēmē*, far from being reducible to the achievements of science as a whole, constitutes the major presupposition of the group of sciences which are most advanced at any given time. And this major presupposition dictates *a priori* how the field of knowledge is to be conceptually related. From one *epistēmē* to another, the leap is irrational.

All in all, everything which moves the theory of science away from empiricism helps to substantiate the philosophy of nature without, however, justifying it entirely. For the 'theory' we have spoken about, although it may be mixed with meta-scientific *apriorisms* and metaphysical presuppositions, constitutes a body of propositions completely at one with the instrumental arrangement, the system of laws and the verifications in pro-

127. Gaston BACHELARD.

128. MARGENAU, 'Does physical "knowledge" require *a priori* or undemonstrable presuppositions?', in BRIDGMAN, *The Nature of Physical Knowledge* (ed. L. W. Friedrich, 1960).

129. Van MELSEN, 'The philosophy of nature', p. 154 in KLIBANSKY (ed.), *Contemporary Philosophy*, Vol. II, *Philosophy of Science* (1968); van Melsen refers in a footnote to KATTSOFF, *Physical Science and Physical Reality* (1957), and to his own books: *The Philosophy of Nature* (1961) and *Evolution and Philosophy* (1965).

130. SCHELER, M., *Die Stellung des Menschen im Kosmos* (1928) (English translation, *Man's Place in Nature*), 1961).

131. FOUCAULT, *Les mots et les choses. Une archéologie des sciences humaines* (1966) (English translation, *The Order of Things. An Archaelology of the Human Sciences*, 1970) and *L'archéologie du savoir* (1969) (English translation, *The Archaeology of Knowledge*, 1972).

gress which together make up scientific knowledge. This is why these associated points of view represent for the philosophy of nature a challenge as well as a confirmation. The question still remains whether these regulating meta-scientific principles, impossible though it may be to reduce them to the synthesis of the results of science, can be linked to the themes of *philosophy perennis* following the tradition of the analogy of being. If, on the other hand, this meta-science has a history of its own, more or less the same as the history of science, is this history absolutely contingent and composed of irrational breaks, or does it lend itself to a specific method of approach of a structural-genetic kind?[132] Is this history autonomous or must it be attached to something else? For example, is it an integral part of a history of culture as we are told by the philosophers of the Frankfurt School, with their leanings to Marxism?[133] Or does it mark the outcropping of the 'epochal' character of being itself as Heidegger claims?

This last question brings us to the confines of another philosophical tradition, closer to modern phenomenology than to the fundamental Aristotelianism which has been the guiding thread running through this sub-section.

§ 6. *Language and ontology*

Science is not the only human activity which develops a complex of ontological problems. It is open to question whether one should not seek an even more primitive justification than science itself for the resistance which many theoreticians of science oppose to conventionalism, and for the stubborn conviction of scientists themselves that scientific discourse attains reality, even (perhaps especially) when it elaborates entities whose meaning is exactly defined by a theory, in the full sense of the work. According to critical realism, which seems indeed to be the natural philosophy of science, the reasons for adopting a theory are also, as we have noted above, reasons for accepting the existence of the entities postulated by the theory. The idea that, as scientists, when we build up theoretical entities, we are at the same time describing reality, certainly appears to be a feature common to both scientific and other forms of discourse. It is thus reasonable to explore the hypothesis that the postulate of being is one belonging to discourse as such. It is this hypothesis which leads us to consider the ontological implications of language itself.

The aspect of language involved here has been designated by the general

132. KUHN, *The Structure of Scientific Revolutions* (1962); see also LAKATOS & MUSGRAVE (eds.), *Criticism and the Growth of Knowledge* (1970), in particular KUHN's two contributions to this colloquium.

133. Cf. in particular ADORNO, *Negative Dialektik* (1966); HORKHEIMER, *Eclipse of Reason* (1947) (German version, ed. Alfred Schmidt, *Zur Kritik der instrumentellen Vernunft*, 1967); by the same author, *Kritische Theorie. Eine Dokumentation* (ed. A. Schmidt, 1968); HABERMAS, *Theorie und Praxis. Sozialphilosophische Studien* (1963); by the same author, *Technik und Wissenschaft als 'Ideologie'* (1968), particularly the essay 'Erkenntnis und Interesse'.

term of *reference*. We have already made allusion to this in our description of the theoretical entities of science as *the referent* of scientific discourse. We have now to consider the general theory of reference, without distinguishing between scientific and ordinary discourse. Let us define what we mean by discourse. Emile Benveniste makes a distinction between discourse and language.[134] Language is the combination of phonological, lexical and syntactical codes which characterize a historical community; discourse is the succession of messages sent out by individual members of that community who are in a situation to communicate. As W. von Humboldt said, discourse is an infinite use of finite means. The infinite use is the succession of messages, and the finite means are the codes. Language as such does not exist; it is merely the enabling system, indicating what is permitted and what forbidden, on which the acts of discourse, which alone are real, are built up. Discourse is language put into action. Language and discourse are not based on the same units: the unit of language is the sign, which has the same potential, enabling character as language, while the unit of discourse is the sentence, or temporal, factual 'instance', which brings discourse about, i.e., makes it appear and disappear. Thus the latter kind of unit is not derived from the former. The sign is in fact merely a differential value in the language system; it exists only in contrast with other units in the same system, whereas the sentence is a circumstantial combination of signs intended to express subjects, to communicate and represent reality. These brief remarks are sufficient to explain why there is no problem of reference for language, but only for discourse. In language, all relations are internal, being differences between signs; only discourse can be the subject of something. We can say that language is made up solely of differences, and only discourse involves reference. Thus from now onwards we shall discuss not language but discourse.

To understand fully the complex of ontological problems involved in any kind of discourse, a useful starting-point is the distinction introduced by Frege, in his well-know article in 1892,[135] between *Sinn* (sense) and *Bedeutung* (meaning, reference or denotation). Sense is what is stated in a linguistic expression; the reference is that about which it is stated. Thus, 'the victor in the battle of Iena' and 'the vanquished in the battle of Waterloo' or 'Plato's pupil' and 'Alexander's tutor' both have different senses each time, but designate the same entities, Napoleon or Aristotle. *Sense* can also exist without reference, as for example when one speaks of a speed greater than that of light; thus signifying (having sense) and designating (something) are not coincidental. Sense is ideal (i.e., it cannot be reduced either to the physical reality of things or to the mental reality of the representation), whereas reference purports in addition to capture reality. As Frege states emphatically, we are not satisfied with sense, we also want reference; this 'intention' which presses sense into its service, this impulsion

134. BENVENISTE, *Problèmes de linguistique générale* (1966).

135. 'Über Sinn und Bedeutung' (English translation, 'On sense and reference', in FREGE, *Philosophical Writings*, 1960).

(*Drang*) to 'advance' from sense towards reference, is nothing other than our hankering for truth. In the case of proper names, all we require is that there should be correspondence between the name and the particular thing denominated. In the case of a complete sentence, it is the truth value itself which constitutes the reference; to say the same thing another way, there is 'knowledge' in the full sense of the word, when thought is considered in conjunction with its truth value, i.e., its full reference. Thus a linguistic sign theory is only complete if sign is related to sense, which is not something in the mind, and if in addition one proceeds from sense to reference. It is this forward step of postulating a real which constitutes the ontological implication of discourse.

The same requirement is formulated by Husserl in the first of his *Logical Investigations*,[136] 'Expression and meaning'. It is not possible to have a signifying linguistic expression without an 'act which confers sense'; what gives signification the 'nature of an act' is its power to 'aim at' something or 'direct itself towards' an object. This act of intent may either remain 'empty' or be 'filled' by the presence of something, which may be a categorial relation or something physical in flesh and blood. With Husserl's distinction between empty and full intent we thus come back again, using a different vocabulary, to Frege's distinction between sense and reference. The act of intent as such is merely aiming at the same, identical object; this 'ideal', like Frege's 'sense', is still not that about which one speaks; only the actual 'filling' provides the reference of language. When intuition and sense 'overlap', language transcends itself by going beyond the sign; we speak of the 'object' (*Gegenstand*) where the sign is a name, and of a 'state of affairs' (*Sachverhalt*) where the sign is a sentence.

Thus at the beginning of this century Frege and Husserl posed the philosophy of language with the problem of the relation obtaining between sense and reference, or meaning and 'filling'; this problem contains the seeds of a new complex of ontological problems which owes nothing to metaphysics before Kant.

In this respect the philosophy of Russell is an extraordinary mine of material for an ontology dominated entirely by logic, the theory of knowledge and the philosophy of language. At the time of writing his *Principles of Mathematics* (1903), Russell's ontology teems with life. Every word refers to something; if the word is a proper name, its object is a 'thing' which one can describe as existing; if not, it is a concept, which one can describe only as subsisting. 'Words all have meaning, in the simple sense that they are symbols which stand for something other than themselves'. Russell freely admits into the kingdom of 'things' both instants and points; outside this category of existence, which he attributes only to 'things', are to be found all other entities. 'Numbers, the Homeric gods, relations, chimeras and four-dimensional spaces all have being, for if they were not entities of a

136. HUSSERL, *Logische Untersuchungen*, II: *Untersuchungen zur Phänomenologie und Theorie der Erkenntnis* (1901, 1913), *Untersuchung* 1, 'Ausdruck und Bedeutung'.

kind we could make no propositions about them. Thus being is a general attribute of everything, and to mention anything is to show that it is.' This type of non-discriminating ontology is comparable to that of Meinong, who considers 'being' as embracing even impossible objects. The whole subsequent philosophy of Russell, considered from the standpoint of the development of his views on ontology, is a persevering, long-drawn-out attempt to reduce this teeming population of entities. His well-known theory of 'definite descriptions'[137] leads us to restrict existence solely to objects answering to logical proper names (by 'definite descriptions' are meant expressions of the type 'the so-and-so', i.e., expressions comprising the definite article followed by a common noun or equivalent term which identifies one particular thing and it alone). Russell's strategy consists in elaborating the paradoxes which result from the indiscriminate inclusion of all seeming names in the circle of expressions which denote something; all paradoxes (for example, the paradox in the expression 'the king of France is bald') are eliminated by reformulation of the proposition so as to consign to the predicate all false proper names and recognize only logically proper names as performing the function of logical subject. The apparent subject is then a definite description. ('The king of France is bald' has a meaning only if it is an abbreviated form of a sentence which does not have 'the king of France' as its subject: 'There is an entity which is [now] king of France and is bald'.) The argument presupposes that we were justified in believing in the existence of the things denoted by true proper names. In this sense, the reduction of paradoxes serves to ward off any doubt or scepticism regarding what may be called the ontological commitment entailed by the use of proper names. The reason that Russell is so anxious to eliminate the paradoxes of reference is that he does not have Frege's resource of distinguishing sense and reference. ' "Napoleon" means a certain person' and ' "man" means a whole class of such collections of particulars as have proper names'.[138] This being so, the lack of reference does not differ from the lack of meaning; this is why the king of France cannot be excluded from the kingdom of being without constructing the complicated apparatus of the theory of descriptions.

Definite descriptions are themselves only one case of 'incomplete symbols'; another case is that of classes, the same classes which Russell, in this first ontological fervour, included among things, not concepts. Here again it is the paradox method which sets the argument in motion. For example, the paradox of the class of all the classes which are not members of themselves makes it impossible to identify the class with the sum of all the things included in it; and the solution by means of the theory of types (1908)[139] leads to constructing a hierarchy of levels of discourse and

137. Cf. RUSSELL, 'On denoting' (1905; reproduced in RUSSELL, *Logic and Knowledge, Essays 1901-1950*, 1956).

138. RUSSELL, *The Analysis of Mind* (1921).

139. RUSSELL, 'Mathematical logic as based on the theory of types' (1908, reproduced in RUSSELL, *Logic and Knowledge. Essays 1901-1950*, 1956).

thus excluding classes from the realm of ontology, grouping them instead with incomplete symbols. From here onwards the theory of types – and, with it, classes and all universals – is imbued with an increasingly nominalistic theory of 'propositional functions'. In that case, to what can one attribute being? Here the perplexities begin again. In 'The philosophy of logical atomism'[140] Russell speaks of '[. . .] ultimate simples, out of which the world is built [. . .] and [which] [. . .] have a kind of reality not belonging to anything else. Simples [. . .] are of an infinite number of sorts. There are particulars and qualities and relations of various orders, a whole hierarchy [. . .].'[141] But if it is phrases, not words, which convey the complete meaning, should we not include among things that which corresponds to the phrase and is asserted by it? If, like Russell, we call 'fact' that which answers to a true proposition, then we must say that 'facts belong to the objective world'.[142] The distinction between facts and things thus parallels, in the case of existents, the distinction between phrases and names in the cases of language. Here again the absence of distinction between meaning and reference leads us to seek an entity in the same relation to the phrase as things are to names. This ontology of 'facts' is not easy to combine with the ontology of things, since 'atomic', i.e., simple, facts are complex objects; and complex objects are classes of sense-data which, according to *Our Knowledge of the External World* (1914), are the ultimate atoms.

We can see the distance covered from an ontology which included among its entities 'numbers, the Homeric gods, relations, chimeras and four-dimensional spaces' to an ontology of the external world for which reality is reduced to the sum of sense-data – passing by way of, first, an ontology of objects and, second, an ontology of facts.

It is Russell, whose philosophical activity covers and dominates three quarters of a century, who provides the best possible example not only of the complex of ontological problems developed by language but also of the considerable inherent perplexities. The distinction between objects (answering to names) and facts (answering to phrases), elaborated by Russell even before he met Wittgenstein, is also to be found in Wittgenstein's *Tractatus logico-philosophicus*,[143] without either having borrowed from the other. Wittgenstein pushes the theory of reference to the extreme of postulating a looking-glass relation between true propositions and facts (termed 'states-of-affairs'). This is the well-known theory of the 'picture', which, while not reviving the old theory of *mimesis* which Plato had so much difficulty in discarding, introduces a similarity of structure between discourse and reality. This is probably the most radically simple form of ontology that can be imagined; but its simplicity, which gives it a crystal-clear beauty, heightened still further by the aphoristic style of the *Tractatus*,

140. First published in 1918, reproduced in *Logic and Knowledge* (1956).

141. *Ibid.*, p. 270 in *Logic and Knowledge*.

142. *Ibid.*, p. 183 in *Logic and Knowledge*.

143. WITTGENSTEIN, *Logisch-philosophische Abhandlung* (1921) (English versions, *Tractatus logico-philosophicus*, 1922 and 1961).

brings out most forcibly the paradox on which the *Tractatus* comes to grief. With a kind of daring virtuosity, Wittgenstein begins the treatise with propositions which are plainly ontological; about the world – which is 'everything that happens' – about facts and objects. But since the limits of our language are those of the world, it is not possible to attribute any meaning to this theory of the world without employing the very structure and propositions of language. The result is that the facts of the world are a doublet of the structures of propositions, which are in their turn a picture of facts. Merely to state this reflection of a reflection is considered by Wittgenstein himself as nonsense.

Strawson[144] grappled with Russell's paradox of expressions which, because they refer to a thing and one thing only ('the so-and-so'; 'the king of France') resemble genuine proper names and thus seem to postulate a world of special entities. He rejects Russell's reformulation which consigns false logical subjects to that part of the statement given over to descriptions, i.e., to the predicate, and retains as the subject only real, logical proper names, which alone designate something which is. According to Strawson, this solution has the major drawback of making ontology dependent on the denotational capacity of undiscoverable grammatical entities. For that is what these 'logically proper names', which exclude definite descriptions, are: they are undiscoverable because they should ultimately designate all the 'thises' and 'thats' of individual experience and this implies a vocabulary which is infinite and incommunicable, a presupposition doubly fatal to language. Strawson's solution is both simple and elegant: it is to distinguish between the sense of a sentence and its use. For example, exactly the same sentence – 'The king of France is wise' – uttered at different times, under different reigns, is sometimes true and sometimes false. Depending on the circumstances, we are speaking about different individuals, and it is the circumstances surrounding the use of the sentence which create the conditions for the truth or falsity of the assertion. Taken outside the context of a specific use, a sentence is neither true nor false; this is the case when one quotes a sentence in which one mentions a name without using it, i.e., without applying it to anyone. Referring is thus not something done by an expression: it is something done by someone when using an expression. In that case, what is meaning? A general directive or rule for use in the formation of true or false assertions. Thus we come back, by a roundabout way via the theory of use, to Frege's distinction between sense and reference; and it was, according to Strawson, because Russell overlooked this distinction that he had to have recourse to a complex reformulation of ordinary language and his 'mythical' conception of logically proper names.

But a theory of use does not settle the problem of the ontological commitment entailed in attributing existence to the things about which we speak; it merely pinpoints it by stating that it is only in a specific use that

144. STRAWSON, 'On referring' (1950).

our language refers to something which we consider to exist. But what is 'referring'?

This necessarily takes us back to the functions constituting the predicative relation, i.e., discourse itself. The predicative relation is based on the dissymmetry of two functions, each of which is performed by different grammatical means but has a unity of intent which can be perceived by analysis. On the one hand we carry out identifications, and on the order we classify or attribute characteristics or qualities. The two functions are asymmetrical in the sense that only the first aims at particulars (to identify is to designate a thing, and one thing alone), and only the second aims at universals. For the purposes of our ontological investigation, this asymmetry is the decisive feature, since only the act of aiming at particulars, as part of the process of singularization and identification, entails an ontological claim and commitment. To ask whether, and how, universals exist, is to require of the predicate that which can only belong to the subject; in a word, it is introducing a symmetry into the manner of asking questions in precisely that sphere in which dissymmetry of functions prevails. By recognizing this dissymmetry we do away with a false problem which has encumbered philosophy ever since Plato, that of the mode of existence of universals; we also circumscribe a real problem, that of the ontological commitment which accompanies the identifying function.

This ontological commitment raises many problems and at the same time gives rise to new difficulties:

(a) In the first place, it should be noted that this function is a logical, not a grammatical function; it is performed equally well by the use of proper names, pronouns, demonstratives and Russell's definite descriptions ('the so-and-so'). Here the hunt for logically proper names becomes less relevant, and it is no longer necessary to consign definite descriptions to the predicate if, when a sentence is used in specific circumstances, a genuine identification takes place by means of a definite description. But if the difficulty inherent in specific descriptions is eliminated, it is replaced by another, which relates to the logical function of identification as a whole: does it not sometimes happen that we identify fictitious beings, such as characters in novels or plays? It is not enough to say that even in this case the existence of such beings is posited, but only in fiction: we must then ask how the existential thesis can be neutralized. This is a problem which goes beyond linguistic philosophy; Husserl was able to pose it only on the basis of a theory of reduction.

(b) Next, it should be noted that the ontological commitment can only be understood, at least by the speaker, if we give a complete description of the 'speech-act' in accordance with the requirement stated by Austin in the last few lectures of *How to Do Things with Words*,[145] i.e.,

145. Published in 1962 (edited by J. O. Urmson).

if we take into account not only the logical structure of the proposition constituting the locutionary act, but also the illocutionary force investing it (assertion, wish, command, etc.). It then becomes clear that it is not only performatives (a promise, an order, a wish) which do something 'in' speaking (hence the word 'illocution') but that constatives also do something. The ontological commitment is in fact the illocutionary force of constative verbs: the assertion that something is, entails an 'I believe that' (one cannot say, without self-contradiction as regards the illocutionary act, 'The cat is on the mat and I do not think so'). But recourse to the theory of the speech-act, taken in conjunction with the distinction between identifying and predicating, gives rise to a serious ambiguity about where, in language, the ontological commitment is manifest: is it in the name or in the sentence? It will be seen that this is an old question which goes back to Frege. The theory of the speech-act implies that the illocutionary force invests the entire act of discourse. This means that it is the sentence as a whole which expresses being. If, on the contrary, we adopt the theory of the dual function of discourse, we shall tend to argue that it is the identifying function which supports the existential intent. In other words, to express being is to do two things: to claim that an individual exists which has specific characteristics or qualities, or is a member of such and such a class; and also to claim that the state of affairs constituted by the possession of such characteristic by the individual in question is itself real. This duality brings us back to the difficulty encountered by Frege, Husserl and Russell of deciding whether it is the name or the sentence which carries the primary ontological implication; the theory of identification stresses the name, while the theory of illocutionary force stresses the sentence. Thus it appears clear that language, as discourse, deploys successively or simultaneously an ontology of things or individuals and an ontology of facts or states-of-affairs.

(c) If we now consider the ontological commitment as regards the thing, not the speaker, it is necessary to distinguish between theory of language and theory of the world. To say that discourse has two functions, identifying and predicating, is to state something about language itself. But we state something about the reality outside language when we say, like Strawson in *Individuals* (1963/64), that the two kinds of 'basic particulars' about which we can speak, in the context of human experience, are *bodies* and *persons*. A theory about particular identification remains a linguistic theory; a theory about basic particulars is a theory which goes outside language. It is for this reason that Strawson subtitles his work 'An Essay in Descriptive Metaphysics'. The question then arises how the transition is made from the individuals signified to the basic particulars in reality. It is this transition which forms the hub of the ontological commitment of discourse; in it we

> in fact go beyond language. In his *Speech-Acts* (1969), Searle places Strawson's theory of the dual function of discourse in the context of Austin's theory of the speech-act, and has no hesitation in designating as a 'postulate' the claim that particular identification relates to existing individuals. But if it is a postulate, can it be suspended or changed? And if it is a belief, is it not possible to question it? Russell already encountered this problem in his *Principles of Mathematics* when he wrote: 'Numbers, the Homeric gods, relations, chimeras and four-dimensional spaces all have being, for if they were not entities of a kind, we could make no propositions about them'. We have reduced these entities to basic particulars, but we have not changed the argument whereby we postulate these entities, since what we say also about basic particulars, is that if they were not entities of a kind, we could make no identifications in regard to them, nor could we commit ourselves to the belief which supplies the illocutionary force of even the most elementary constative utterance.

What is this postulate of existence which is entailed by particular identification? What is this belief which characterizes the illocutionary act specific to assertions? It seems very much as if language bases itself on an ontological faith of which it is the vehicle, but not the origin.

§ 7. *Phenomenology and ontology*

A fresh approach to ontology is opened up by phenomenology, the descriptive study of the essential features of experience taken as a whole. We may approach the subject by starting from the preceding discussion: the possibility that language refers to reality presupposes that we are already oriented towards reality and linked to it by various ties which discourse merely subjoins to language. To inquire what is thus subjoined to language is to turn from linguistics to phenomenology.

This transition is not without its difficulties, since it is still within language that we intend to discuss that which precedes language. Is such a language available? The phenomenologists say yes. They argue that it is not an object-language, i.e., a language bearing on things, but a language bearing on our relation to things; nor is it a meta-language, like that spoken of by linguists, i.e., a language about language considered as a set of codes on which our messages are constructed; it is a language which expresses that which precedes language. This language has always existed. It is that which was hammered out in the ontology of the pre-Socratics and the poetry of fundamental thinkers; man has never been without a language to express his inclusion in the world of being and the effacement of language itself when confronted with that which is.

Can we say that this transition from object-language to a language expressing what precedes language constitutes an unjustifiable leap from reference, which is still a fact of language, to the thing, which is a reality outside language, or alternatively from 'being said' to 'being'? And indeed this transition must appear as a leap for anyone who has fenced himself inside a closed world of signs; but once we realize that in language we are still oriented towards that which comes before language, then it becomes plain that it is the fact of language being closed in on what is signified within language which should be seen as entailing the loss of the ontological dimension and making us forgetful of being. It may well be that the hypostatization of language as a closed-in world of signs and the philosophers' enthusiasm for linguistics – though not the linguistics of linguists – are among the symptoms of this forgetfulness. But being is never so entirely forgotten that we cannot still recognize traces of the ontological assertion in the very impulse which carries language on from sign towards sense and from sense towards reference. This impulse can only be accounted for, at the level of a philosophy of language, as a formal requirement or postulate: something must 'be' in order that we may be able to talk about it. We have seen this postulate or requirement stated by Russell and Searle, but we might equally well trace it back to Kant's well-known Introduction to the second edition of the *Critique of Pure Reason*, in which he argued that if nothing existed, nothing would appear in our experience either. Replace 'appear in our experience' by 'be expressed' and you have the formula for the postulate by virtue of which phenomenology marks out for itself a terrain different from that of the philosophy of language.

But how do we know that we are still oriented by language towards that which precedes language, if such is indeed the case?

From the beginning phenomenology has always been an investigation into the structures of experience which precede connected expression in language; but it has not always been an ontology. We must therefore examine a first stage in which the subordination of language to the structures of experience is still interpreted in the framework of an idealistic philosophy of consciousness, and a second stage in which the primacy of experience over consciousness is interpreted in ontological terms. In this sense it is only in the process of phenomenological investigation that problems posed in ontological terms actually emerge, though even in the first stage of phenomenology all the conditions for their doing so are already fulfilled.

The first stage is represented by the phenomenology of Husserl and its continuation in French existential phenomenology, as long as this asserts the primacy of consciousness (we shall return to this later).

The development of the argument in Husserl's *Logical Investigations*[146] exemplifies this first stage of phenomenology. After criticizing the psychologistic interpretation of logical laws, Husserl in his *Prolegomena* to these essays[147] maintains that logical truths are necessary truths. Subsequently,

146. HUSSERL, *Logische Untersuchungen* (1900-1901, 1913-1921) (English translation, *Logical Investigations*, 2 vols., 1970).

147. HUSSERL, *Logische Untersuchungen*, I: *Prolegomena zur reinen Logik* (1900).

in the first four essays,[148] he links the things signified by language with the act of conferring sense, of which the things signified become the objective correlate. Lastly, beginning with the fifth essay,[148] he bases this relationship on a general characteristic of all kinds of consciousness, in the sense of actual experience, namely the property of being related to something. This intentionality was for long understood by Husserl in a sense which took due account of its logical function. Intentionality is not limited to the vague formula 'All consciousness is consciousness of something'. To intend something is to intend something identical, the 'same', which can be repeated and recognized as the same. What we call an 'object' is this kind of unity of meaning; and intentionality, in the full sense of the term, is aiming at identical meaning. But the general formulation of intentionality was to result in phenomenology continually going beyond the frame of reference of its original preoccupation with logic. It begins by asserting the noetic aspect – the capacity of being related to something – of all non-objectifying acts, such as affects, volitions, etc. It then discovers the variety of forms of referential 'filling' (*Erfüllung*) whereby the intent, the taking aim at something, becomes a specific act. It also emphasizes the constructional, thetic aspect, whereby consciousness asserts the being of that at which it aims; this *doxa,* underlying every assertion that such-and-such is, recalls the ontological commitment stressed by analytic philosophy in its description of the speech-acts, though phenomenology tries to root it in pre-linguistic functions. Finally, there is evidence of an increasing tendency to find in perception the repository of all these features of intentionality in action, so much so as to make phenomenology as a whole veer towards a phenomenology of perception. In all this, language has lost its pre-eminence; it is solely the layer of expression, and as such produces nothing, unless it be the transfer into a connected system of signs of that which was already pre-connected in the structure of the noesis (subjective intent) and the noema (objective correlate).

It is nevertheless difficult to say that Husserlian phenomenology is an ontology. In his *Ideas*[149] and again in his *Cartesian Meditations*,[150] Husserl gives an idealistic interpretation of these structures. The theory of constitution is that it is consciousness which constitutes not only the sense but also the characteristics of being which correspond to the doxic consciousness

148. *Id.*, II: *Untersuchungen zur Phänomenologie und Theorie der Erkenntnis* (containing *Untersuchungen* 1 to 5 (1901; rev. edn., 1913) – the 6th *Untersuchung* was added in 1921).

149. HUSSERL, *Ideen zu einer reinen Phänomenologie und phänomenologischen Philosophie*, Erstes Buch: *Allgemeine Einführung in die reine Phänomenologie* (1913) (English translation, *Ideas*, 1931).

150. HUSSERL, *Méditations cartésiennes. Introduction à la phénoménologie* (French version, expanded and edited, of the lectures given at the Sorbonne on February 23 and 25, 1929, published in 1931) (original German text, *Cartesianische Meditationen und pariser Vorträge*, publ. posth., 1950) (English translations: *Cartesian Meditations. An Introduction to Phenomenology*, 1960, and *The Paris Lectures*, 1965/1966).

(for *doxa*, see above). The non-existence of the world, taken as the totality of the correlates of consciousness, is possible; it is only consciousness of which the non-existence is impossible; it is therefore consciousness which has absolute and necessary being; the world, by its contingency, has merely relative being as a phenomenon.

But just as phenomenology developed in such a way as visibly to go beyond its own logical starting-point, it also inwardly transcended its own idealistic finishing point. And from this stems a new ontology.

By discovering what precedes any language, judgement or predicative process, phenomenology discovered the limits of constitution itself, considered as an active process. The pre-predicative embraces passive organizational processes which are no longer within the purview of consciousness and which reveal the existence of receptivity preceding the imposition of any kind of form. A new kind of reflection, involving a backward questioning, discovers the references to the primitive – Merleau-Ponty uses the term *'l'être sauvage'* – implied in the actual work of constitution. Thus phenomenology, starting out from logical considerations, succeeds in expressing in thematic terms a living world – a *Lebenswelt* – which precedes the world of language and logic. It is true that, unlike any form of phenomenalism or empiricism, phenomenology does not go as far as to assert the possibility of recreating in a present experience something on the lines of sense-data. The anterior event always remains the anterior event implied in the backward questioning; it is never a 'given', like a paradise lost and regained. This flight backwards could therefore still be interpreted within the limits of an idealistic theory of constitution.

A truly ontological interpretation could only be produced by an 'overthrow', brought about within and by phenomenology. What is overthrown is the primacy of the subject-object relation, which still dominates Husserl's phenomenology and his investigation into the correlation between *noesis* and *noema*. This primacy can only be questioned if one simultaneously renounces the central position of consciousness, which governs the whole complex of problems that arise in regard to constitution. The idealistic interpretation of phenomenology is not in fact arbitrary, but faithfully transposes, in terms of a long-standing controversy with realism, the primacy of the *cogito* with which the phenomenologists' analyses of the pre-predicative and the *Lebenswelt* are still imbued.

This 'overthrow' of the primacy of consciousness found its exemplary expression in Heidegger's *Being and Time*.[151] Even if this work eventually produces its own overthrow, it is already an upheaval in action. It is enough to consider the development of the argument in *Being and Time* to appreciate its scope. The starting-point is not the *cogito* but the question of being; we proceed from the being which gives rise to the question to the being which puts the question. This sequence, which governs the well-known introduction to *Being and Time*, is in itself significant. It implies that con-

151. HEIDEGGER, *Sein und Zeit, Erste Hälfte* (1927) (English translation, *Being and Time*, 1962).

sciousness is not the measure of all things; man, the questioner, is not designated by a term expressing this consciousness, but by an ontological term, expressing that very being which makes him a questioner of being – *Dasein* or *'Da-sein'* (being-there), being the 'there' or locus where the question of being is put. The analytical description of the *Dasein*, the central theme of *Being and Time* is still phenomenological, but in a new sense; it expresses the appearance of a 'being' whose whole rôle lies in being open to the question of being.

By its inner logic this analysis in turn reveals the primacy of being over knowledge. We start with the global structure of 'being-in-the-world', of which the subject-object relation, even stated in terms of intentionality, is only a derivative form; in analysing this structure we then consider its 'world' aspect, which, even before we emphasize the existential pole, contains signifying relations (for example in man's array of tools and the way each tool is referred to a totality which is a world). From the 'world' structure, we make our way back towards the 'being-in' relation which, before expressing a relation of inclusion which might still be dominated by consciousness, designates a relation of belonging and familiarity which the act of indwelling, or the act of caring for, express better than the intellection of a geometrician, since only they have knowledge who begin by having this close proximity of concern with things. It is this analysis of 'being-in' which contains the seeds of all phenomenology of 'one's own body'.

Only then can one develop a non-idealistic analysis of the *'da'* of *Dasein*. The analysis is non-idealistic in the sense that the theory of statements comes last, though placed first by both logical empiricism and the phenomenology of Husserl's *Logical Investigations*. The analysis begins by pairing 'finding oneself in a situation' and 'purposely directing oneself thither'. It is first necessary to have roots, and to project one's innermost potentiality against the background of this given 'being', before any of the problems involved in understanding and interpreting can arise. Interpretation therefore is not something absolute, resulting from the existence of texts, but stems from the possibility of explicating in a number of ways our understanding of the relationship between our situation and our possibilities. It is this initial bipolarity which creates the hermeneutic situation, in the sense that it is always possible to understand more fully and to interpret in a different way the ontological condition of the existent which we represent. Discourse, as a medium of connection, and the statements which logic systematizes in terms of propositional functions, then constitute merely the derivative forms of these primordial ontological structures. The forgetfulness of being begins with the hypostatization of these derivative forms, as is seen in logicism, logical empiricism, structuralism and all philosophies which immure themselves in the structures of language.

In the same way as Heidegger comes back to ontology from phenomenology, so Gabriel Marcel comes back to ontology from descriptions of a much more existential nature. With Gabriel Marcel a pact is concluded between being and existence, rather than being and phenomenon. By *existence* is meant, as in Kierkegaard, the concrete appearance of the human individual, considered at one and the same time in his physical and social embodiment, in his dramatic relation to a 'thou' and in his power to reject or accept his mortal condition. It was this form of existence which Gabriel Marcel described in an essay dated 1925, 'Existence and objectivity',[152] as preceding thought concerned with objects or representations. If it is true of the *cogito* that it 'stands guard at the threshold of the valid', ('garde le seuil du valable') or even of that which can be stated as a problem, existence, like Heidegger's *Dasein,* opens the door to mystery. But even if the intention is similar, as when the French thinker speaks of '*être-avec*' and '*être-chez*', his analyses, unlike those of Heidegger, attest a keen interest in personal relationships, to which the theatre is more likely to do justice than is philosophy. From this point of view nothing is less abstract than ontology. Basing himself on Max Scheler's interpretation of Nietzsche[153] Gabriel Marcel sees in the spirit of abstraction, directed against life, the expression of the devastating power of 'resentment', which begets fanaticism. In opposition to resentment and rebutting the temptation to despair, the ontological experience *par excellence* is probably the hope expressed by the 'in spite of all'; this it is which gives us the assurance of belonging to 'an invisible world outside which there would be nothing but mad absurdity, and of which the astonishing and priceless first-fruits have been most notably revealed to us by great spiritual figures and great musicians such as Bach, Mozart and Beethoven' (*translation*) ('un monde invisible hors duquel tout ne serait que délirante absurdité et dont les grands spirituels, d'une part, les grands musiciens de l'autre, un Bach, un Mozart, un Beethoven, avant tout, nous ont livré les inestimables et fulgurants prémices').

The position of Jean Wahl may be compared to those of Heidegger and Gabriel Marcel; for Wahl, hearkening to the 'eternal words of philosophy' and reflection on the 'present revolution in concepts' are inseparable from the deepening of a properly metaphysical experience, which is part of existence itself and gives access to being, in the variety of its aspects and the plurality of its levels, only because it is essentially of the order of being, before it is reflected – always imperfectly – in that of thought. Although, for Wahl, the controlled practice of conceptual thinking is a legitimate necessity in an intermediate region where (as opposed to Heidegger) the traditional notion of truth as representation still holds good, it cannot give access to the higher plane, that of the absolute (or absolutes) or the lower plane, that of the immediate presence of the real. At these two

152. Marcel, 'Existence et objectivité' (first published as article, 1925; reproduced as an Appendix to his *Journal métaphysique*, 1927 – English translation, *Metaphysical Journal*, 1952).

153. Cf. Scheler, M., 'Über *Ressentiment* und moralische Werturteile' (first published as article, 1912), republished in a revised and expanded version under the title 'Vom *Ressentiment* im Aufbau der Moralen' (1915, 1919) (English translation, *Ressentiment*, 1961).

levels, the fundamental experience is, in a manner logically inexpressible, both undivided intimacy and irremediable distance. In the quest for a truth which is no longer of the order of representation and is averse to any kind of predicative formulation, philosophy cannot do justice to the dual transcendence – trans-ascendence and trans-descendence – which haunts the existence of man, otherwise than by drawing its light from what – whether by immediate apprehension or by mystical intuition – we know as *poetry*, and by affirming its own relativity and its own dependence.[154]

The work of Merleau-Ponty, with which we close this review of the philosophies which go back to ontology via phenomenology, is also probably that which can most appropriately be described in this way. It can be said briefly that it bears witness to the progressive veering of Husserlian phenomenology in the direction of the ontology of Heidegger, with the adjunction of theories from Marcel such as that of 'one's own body' (*'le corps propre'*), these being fully assimilated to form a new original theory. The publication of *The Visible and the Invisible*,[155] a book begun in 1959 and interrupted by the author's death in 1961, enables us to measure the distance he had covered since his *Phenomenology of Perception*.[156] In this key work Merleau-Ponty waged battle on two fronts, against behaviourism and against intellectualist philosophies of judgement. The object was to reinstate perception, at the level – below that of language – where the subject's 'own body' (*corps propre*) submerges him in the world of living experience. Considered retrospectively, however, this existential phenomenology, which linked the notions of meaning and experience, carried with it the seeds of an ontology close to that of Heidegger, and equally remote from any psychological or biological considerations; moreover the writings of Merleau-Ponty's intermediate period [157] show that he was also investigating art – in particular painting – in search of an ontological sense of wonder. It was however necessary to break with the philosophy of consciousness which had inspired the psychological inquiry in the *Phenomenology of Perception*; it was even necessary to break with the subtle form of the philosophy of consciousness which he himself had worked out in connection with what he called the 'tacit *cogito*' ('*cogito* tacite'), using significations that did not involve language. The break with the psychology of

154. Cf. in particular WAHL, *Existence humaine et transcendance* (1944); *Poésie, pensée, perception* (1948); *Traité de métaphysique* (1953); *L'expérience métaphysique* (1965).

155. MERLEAU-PONTY, *Le visible et l'invisible, suivi de 'Notes de travail'* (edited by Claude Lefort, posthumous publication, 1963) (English translation, *The Visible and the Invisible*, 1969).

156. MERLEAU-PONTY, *Phénoménologie de la perception* (1945) (English translation, *Phenomenology of Perception*, 1962).

157. MERLEAU-PONTY, *Sens et non-sens* (collected essays, 1948) (English translation, *Sense and Nonsense*, 1964): see in particular 'Le doute de Cézanne (originally published in 1945); *Signes* (collected essays, 1960) (English translation, *Signs*, 1964); *L'œil et l'esprit* (dated July-August 1960, publ. 1961, 1964).

experience is not complete unless we no longer take as our starting-point the distinction between consciousness and object, and unless we follow Heidegger in starting from the involvement of the subject in being. Some obscure jottings appended to his posthumous work, on *'l'être des lointains'*, *'l'être de latence'*, *'l'être sauvage'* and the language of being, foreshadow an ontology which abounds in difficulties, and is grappling not only with traditional language but also with its own language. In particular the concept of flesh – 'my flesh is the flesh of the world' ('ma chair est la chair du monde') –, applied henceforth to the visible, to the world, to history, aims at recording palpably the relation with being which becomes, for the philosopher, that to which no name can be given.

This complete attenuation – not to say destruction – of ontology at the hands of those who have been the most active in re-establishing it beyond where metaphysics ends, is probably the most striking paradox of the return to ontology. It is on this theme that we wish to end this account of ontology. It is a theme which also has its starting-point in Heidegger, since the same ontological requirement which earlier led him to overthrow the primacy of consciousness and the subject-object relation, also produces an upheaval inside his own philosophy. *Being and Time* still accorded *Dasein* a central position; and despite the author's clearly-stated intentions in the well-known Introduction of 1927, this reference to the 'being that we are', gave encouragement to an anthropological interpretation of *Being and Time* which seemed to be further borne out by the themes of dread, 'being-to-death' and authentic and purposeful being. In his final philosophy, Heidegger gives up this reference to human being, and attempts to express the forms of emergence of being which do not involve reference to the being that we are. The return to the pre-Socratics and their pre-metaphysical ontology, listening to poets, above all those who write poetry about poetry, meditation on the simple life and its daily wonders, all provide different but converging paths towards a direct ontology which attests the fact that man, consciousness and liberty are a requirement of being itself, regardless of any claim they might make to be supreme or constitute meaning.[158]

In a last attempt to free himself from the most obdurate dictatorship, that of the subject that speaks, Heidegger tries in *On the Way to Language*[159] to subordinate the power of speech, which is our achievement and glory, to that of expressing, which is being itself speaking to us before ever we direct our discourse from ourselves towards things. But is this 'expressing' still within the realms of philosophy, i.e., philosophy as a mode of thought invented by the Greeks, with its own discipline and technique? It seems clear that the short-circuit between being and expression does away with

158. Cf. in particular HEIDEGGER, *Vorträge und Aufsätze* (1954) (French translation, *Essais et conférences*, 1958); *Holzwege* (1950, in particular 'Der Ursprung des Kunstwerkes') (French translation, *Chemins qui ne mènent nulle part*, 1962, in particular 'L'origine de l'œuvre d'art'); *Erläuterungen zu Hölderlins Dichtung* (2nd edn., expanded, 1961) (French translation, *Approche de Hölderlin*, 1962).

159. HEIDEGGER, *Unterwegs zur Sprache* (1959) (English translation, *On the Way to Language*, 1971).

all possibility of discourse, which ever since Plato and Aristotle has been concerned with 'forms' and 'determinations'. Going still further, is not the very idea of *being* challenged by an experience which is so fundamental, and at the same time so primitive and simple, that nothing in the philosophical discourse of Greece and the West is now relevant to it? It is not surprising that in one of his last works[160] Heidegger writes the word 'being' with a cross through it. Must we now speak, not only of that which lies beyond metaphysics, but also of that which lies beyond ontology? 'What is the meaning of this crossed-out *Sein*?' asks Jean Wahl. 'Either no name fits being, or else being is only a name. And both views are tenable.' (*translation*) ('Et que signifie ce *Sein* barré? Soit qu'à l'être ne convient aucun nom, soit que l'être n'est qu'un nom. Et les deux peuvent être soutenus.') Must we say 'the thing' (*das Ding*)? Or the 'occurrence' (*Ereignis*)? And must we say *one* single thing? In the same way as we spoke at the beginning in regard to the Platonic Ideas of a polytheism of being, so one could conceivably speak of a post-ontological polytheism, including Heidegger's tetrad of heaven, earth, mortals and immortals.

Some post-Heideggerians are in fact searching for another name for what lies beyond being. Emmanuel Levinas suspects that the category of being is linked with that of totality, in which all things are reduced to 'the Same'; he contrasts it with that of the Infinite, which has more affinities with the idea of the 'Other' exemplified in each persona (*visage*) one comes across.[161] Michel Henry, on the other hand, sees the idea of being as linked with that of transcendence, that towards which one strives and in so doing goes beyond oneself. This he contrasts with the experience of passivity, accessible to feeling alone, and more primitive than the intentionality of being.[162] Jacques Derrida, pursuing Levinas' idea of the traces left by being – an idea which is admittedly bound up with Levinas' meditation on '*le visage*' – looks for the key to the problem in writing, the origin of which seems to him to have been obscured by the prominence given to speech; but in that case it is the pact between being, logos and voice which must be called into question, and it is Difference which must be placed beyond Being and the Same.[163]

The paradox is that thought, in attempting to think beyond ontology, finds itself brought back to the dialogues in which Plato appeared as a critic both of himself and of ontology.

160. Heidegger, *Zur Seinsfrage* (1956) (English translation, *The Question of Being*, 1958).

161. Levinas, *Totalité et infini. Essai sur l'extériorité* (1961); 'La trace de l'autre' (1963); 'La signification et le sens' (1964).

162. Henry, *L'essence de la manifestation* (2 vols., 1963); *Philosophie et phénoménologie du corps. Essai sur l'ontologie biranienne* (1965).

163. Derrida, *La voix et le phénomène. Introduction au problème du signe dans la phénoménologie de Husserl* (1962); *L'écriture et la différence* (1967); *De la grammatologie* (1967); *La dissémination* (1972). And see Derrida's article on Levinas: 'Violence et métaphysique. Essai sur la pensée d'Emmanuel Levinas' (1964).

SECTION THREE

MAN AND SOCIAL REALITY

SECTION THREE: MAN AND SOCIAL REALITY

A. THE LOGIC OF EXPLANATION

This sub-section deals with the philosophical aspects of the epistemology of the human sciences; what social scientists themselves have to say about their sciences will be found in the relevant chapters of this *Study*;[1] and what is brought out here is in the nature of 'second-order' thinking, from the philosopher's point of view.

In § 1 we shall consider the logic of eplanation in the social sciences, dwelling on those logical problems which are liable to arise in regard to all the human sciences. We shall then deal with the specific problems of psychology (§ 2), psychoanalysis (§ 3) and history (§ 4), in the light of the general mode of approach adopted in the human sciences.

§ 1. *The logic of explanation in the social sciences**

(a) *The subject-matter of the social sciences*

Before tackling the essential problem of the logical structure of explanation, it is worth taking a quick preliminary look at the controversy concerning the 'reducibility' or 'irreducibility' of the subject-matter of the social sciences. After what Durkheim and Lévy-Bruhl had to say about the matter it might hardly have seemed possible to question further that social facts are purely social and are not reducible to the sum of individual interactions; nevertheless, the contrary view – that of what is termed 'methodological individualism' – is put forward by major theoreticians like Hayek[2] and Popper[3] who argue that, in the last analysis, the individual is the referent of all the human sciences.

1. See in particular, in the previously published volume, *Main Trends of Research in the Social and Human Sciences, Part I: Social Sciences*, the chapters on *Sociology* (Paul Lazarsfeld), *Political Science* (W. J. M. Mackenzie), *Psychology* (Jean Piaget), *Economics*, and *Demography* (Jean Bourgeois-Pichat); and, in the present work, the chapters on *Social and Cultural Anthropology* (Maurice Freedman), *Archaeology and Prehistory* (Sigfried J. De Laet), *History* (Geoffrey Barraclough) and *Legal Science* (Viktor Knapp), and the relevant sections of the two chapters on *Aesthetics and the Sciences of Art* (Mikel Dufrenne, with the collaboration of various specialists).

* The discussion of this subject is largely based on contributions, notes and references supplied by Steven Lukes, Alasdair MacIntyre and Rom Harré.

2. Cf. von HAYEK, *The Counter-Revolution of Science. Studies on the Abuse of Reason* (1952), Chaps. IV, VI and VIII; and the same author's *Individualism and Economic Order* (1948-1949).

3. Cf. POPPER, *The Poverty of Historicism* (1944-45, 1957), especially § 7, 23, 24, 29, 31 and 32; and the same author's *The Open Society and Its Enemies* (1945), especially Chap. 14.

This discussion remains purely academic so long as it is posed in terms of ultimate explanations; science is perfectly at liberty to ignore it and confine itself to the 'penultimate' phenomena: groups, and their beliefs and institutions, in so far as the latter can be assigned to groups and not to individuals. But the debate ceases to be a matter of indifference if the question is put in logical terms and one asks of whom or of what the facts or characteristics brought to light by the explanation (whatever it may be) are to be predicated. Are they to be predicated of the individual or of the group? In a word, who or what are the logical subjects of social predicates? When couched in these terms, the problem is incapable of logical solution. How indeed can the controversy be settled, without recourse to a specifically philosophical analysis of the human phenomenon – an analysis which, in the eyes of a scientist, will always have a metaphysical, *a priori* taint? From the purely epistemological standpoint, it is the very nature of the predicates which alone can serve as a criterion: if an explanation appeals to factors which can only be described in 'social' terms, the individualist thesis ceases to be of methodological significance, in the sense of providing directives for research; it becomes merely a question of linguistic reformulation, all statements about society having to be recast as statements about individuals. As Durkheim saw at the beginning of the century, the very possibility of developing sociology as an independent science here plays an *a priori* rôle: in this connection methodological individualism does indeed seem to have disastrous consequences for explanation itself.

On the other hand, the problem of the individual itself becomes a problem for the social sciences, *via* the concept of the individual's social 'rôle'. The crucial question here is how the individual is related to the social rôle which he plays and this is a conceptual as well as an empirical question. If, like some writers, one maintains that the individual's behaviour is dictated by the rôle-structure he inhabits, this can be understood either as a proposition verifiable within social science or as the assumption implicit in the adoption of a sociological standpoint; in the latter case, to say that the individual is socially determined is merely to say that he must be envisaged as such from the standpoint of sociology;[4] in the first case, however, it is to claim that sociology shows that the individual's manifest behaviour is all a matter either of adopting socially prescribed rôles or of following rules which enable such rôles to be played.[5]

It is at this point that some writers refer to linguistics to elucidate the concepts of 'structure' and 'rôle'; structuralism, in its extreme form, contends that what exists are structures; the individual is not the bearer of meaning, but is himself determined by the meaning of the structure; finally, the individual as subject disappears. It is obviously open to question whether this is really what linguistics teaches; for it also teaches that the structures of language are merely the potential conditions of speech which is always

4. Dahrendorf, *Essays in the Theory of Society* (1968).

5. Goffman, *Encounters* (1961); also the same author's *The Presentation of Self in Everyday Life* (1959).

what someone says in a particular situation, and that it is always the speaker's job to find his way among the numerous codes and sub-codes, selecting whichever is most appropriate to the circumstances in which, the people to whom and the subject about which he is talking.

It is difficult, therefore, to settle the question by an argument drawn from linguistics. The disappearance of the individual is a doctrine of philosophical structuralism and not of linguistic science, which always requires a 'speaker', without having to decide on the status of the speaker outside the speech-situation.

In this connection, Karl Popper's methodological individualism is interesting although Popper's original attacks were directed against historicist rather than sociologizing doctrines.[6] The point at issue is: What is the ultimate subject-matter of the social sciences?[7] Reference to socially established norms does not eliminate the problem, but merely shifts it. The nature of the norms and, above all, the way they work inevitably lead us back to a problem similar to that of the relation between the speaker and the structure of the linguistic code or codes; the way in which the individual assumes the rôle prescribed by the social structure, or follows the rules which enable such rôle to be played, is comparable to the way in which, according to Emile Benveniste, each speaker, in using language, takes possession of the 'empty' forms which it offers him. Must we say of 'rôle-structures' what the linguist says of language, namely that 'it is organized in such a way as to allow every speaker to appropriate for his own use the entire code by calling himself 'I' (*translation*)?[8]

But sociology is no more competent than linguistics to solve the problem of what 'I' means and how the concepts of structure and of rôle are related to the meaning of 'I'. It is quite possible that the two views are compatible, but at different levels: in that case, one might say, on the one hand, that the social sciences can and should sidestep the whole question and build entirely on the basic assumption of the sociological standpoint as such – that of an autonomous social subject, endowed with socially established norms; and, on the other hand, that there is another type of inquiry for which this assumption is no longer ultimate, but which can and should call

6. Cf. *The Poverty of Historicism, op. cit.*

7. Recent defences of methodological individualism include: WATKINS, 'Ideal types and historical explanation' (1952) and 'Historical explanation in the social sciences' (1957) – the latter article being a reply to the criticisms of GELLNER (see below); AGASSI, 'Methodological individualism' (1960); and the works of HOMANS, especially *Social Behavior: Its Elementary Forms* (1961) and *The Nature of Social Science* (1967); see also JARVIE, *Concepts and Society* (1972), Chap. I. Among attacks on it we may mention: GELLNER, 'Explanations in history' (1956) (containing a discussion of WATKINS' arguments in 'Ideal types', *art. cit.*, 1952); MANDELBAUM, 'Societal facts' (1955) and 'Societal laws' (1957); DANTO, *Analytical Philosophy of History* (1965); and LUKES, 'Methodological individualism reconsidered' (1968), an article which gives a complete bibliography of the controversy.

8. '[. . . .] il est ainsi organisé qu'il permet à chaque locuteur de s'approprier la langue entière en se désignant comme je' – BENVENISTE, *Problèmes de linguistique générale* (1966), p. 262.

into question postulates which are adequate for an empirical inquiry.

The debate would then again become pertinent, from the standpoint, not of epistemological, but of genuinely transcendental thinking. One can indeed ask whether, to understand 'we', it is not first of all necessary to know the meaning of 'I'. But such a derivation of 'we' from 'I', as sketched by Husserl in his fifth *Cartesian Meditation*,[9] is not a scientific hypothesis: it concerns the origin of meaning though it is in no way empirical; it consists in saying that one cannot understand the meaning of an institution, a group belief or a collective symbolism unless these phenomena are related to a 'we'. But one can speak of a 'we' only if every member of it can say 'I'; thus meaning proceeds from 'I' to 'we' and thence to the group, the beliefs, the institutions. In this sense, Hayek's and Popper's thesis might be justified, but it would no longer be the thesis of a *methodological* individualism.

Human phenomena can accordingly be regarded as irreducibly social for the purposes of description and explanation; their reduction into terms of an 'I' is relevant to an analysis of a different sort, which has nothing to do with the construction of a theory of social phenomena.

This distinction between the construction of an empirical theory and reflection on the ultimate conditions of meaning may prove to be useful when we go on to discuss the crucial question of the logical structure of explanation.

(b) *Explanation*

Here we shall consider only those discussions which concern the status of explanation in the social sciences as a whole, and for everything relating to demarcation of the various disciplines within the social sciences we shall refer the reader to the accounts mentioned above, especially in regard to sociology, psychology and history; the concern of philosophers with methods is not with their usefulness in a specific branch but with the logical structure of explanation in general. The overriding question here is the relationship between the principles of explanation in the natural and in the social sciences respectively.[10]

9. 'Détermination du domaine transcendantal comme "intersubjectivité monadologique"'/'Enthüllung der transzendentalen Seinssphäre als monadologische Intersubjektivität' (1931/1950).

10. There are a number of collections of essays in English which cover the whole field of explanation in the social sciences. Useful references are: BRODBECK (ed.), *Readings in the Philosophy of the Social Sciences* (1968); and, amongst earlier collections: BRAYBROOKE (ed.), *Philosophical Problems of the Social Sciences* (1965); GROSS (ed.), *Symposium on Sociological Theory* (1959); LAZARSFELD (ed.), *Mathematical Thinking in the Social Sciences* (1954); NATANSON (ed.), *Philosophy of the Social Sciences. A Reader* (1963). See also, in German: ALBERT (ed.), *Theorie und Realität. Ausgewählte Aufsätze zur Wissenschaftslehre der Sozialwissenschaften* (1964); TOPITSCH (ed.), *Logik der Sozialwissenschaften* (1965).

And see, amongst general works: ARON, general introduction ('Leçon I'), 'De la sociologie', in *Dix-huit leçons sur la société industrielle* (1962) (English translation, *Eighteen Lectures on Industrial Society*, 1967); BOURDIEU, CHAMBOREDON & PASSE-

Numerous writers, especially in Britain and America adhere to a viewpoint that may be traced back to John Stuart Mill who argued for the basic unity of the two types of science; both consist in the observation of phenomena and the endeavour to establish laws, primarily causal laws. Explanation is closely linked with the possibility of prediction: observation, explanation and prediction constitute an 'objective' strategy which is common to the natural and the social sciences. Carl Hempel's concept of a 'covering-law model'[11] is the clearest contemporary expression of this 'unitary' conception; on similar lines, Popper[12] and Ernest Nagel[13] assert that the differences between the natural and the social sciences relate to the techniques of research and not to its logic or its justification.[14]

But the opposite view, which was characteristic of German philosophy of the social sciences at the end of the nineteenth and the beginning of the twentieth centuries, has also found supporters among British and American writers,[15] particularly under the influence of the later Wittgenstein, whose

---

RON, *Le métier de sociologue, I* (1968); BROWN, *Explanation in Social Science* (1963); GEHLEN, *Studien zur Anthropologie und Soziologie* (1963); GOLDMANN, *Sciences humaines et philosophie* (1952, 1966); GRANGER, *Pensée formelle et sciences de l'homme* (1960, 1967), *Essai d'une philosophie du style* (1968), and 'L'explication dans les sciences sociales' (1971); GUSDORF, *Introduction aux sciences humaines* (1960, 1974) and *Les sciences de l'homme sont des sciences humaines* (1967) (see also the same author's monumental historical work, *Les sciences humaines et la conscience occidentale*: 6 vols. published to date, 1966-73); HABERMAS, *Zur Logik der Sozialwissenschaften* (1967); HANDY, *Methodology of the Behavioral Sciences. Problems and Controversies* (1964); KAPLAN, *The Conduct of Inquiry* (1964); LAZARSFELD, *Qualitative Analysis. Historical and Critical Essays* (1971); by the same author, *Language of Social Research* (1971); by the same author, *Philosophie des sciences sociales* (choice of basic texts in French translation, 1970); MERTON, *Social Theory and Social Structure* (1957); PALMADE, *L'unité des sciences humaines* (1961); SARTRE, *Critique de la raison dialectique, I: Théorie des ensembles pratiques* (1960), especially the Introduction: 'Questions de méthode' (1957); STRASSER, *Phänomenologie und Erfahrungswissenschaft vom Menschen. Grundgedanken zu einem neuen Ideal der Wissenschaftlichkeit* (1964) (French translation, *Phénoménologie et sciences de l'homme. Vers un nouvel esprit scientifique*, 1967); VIET, *Les méthodes structuralistes dans les sciences sociales* (1965); von WRIGHT, *Explanation and Understanding* (1971), especially Pts. III and IV. Cf. also the controversy (1962) between POPPER, 'Die Logik der Sozialwissenschaften' and ADORNO, 'Zur Logik der Sozialwissenschaften'. For the Marxist points of view, see KON, *Der Positivismus in der Soziologie* (1968), especially Chaps. VII-XI; the collection *Očerki metodologii poznanija social'nyh javlenij* (= Essays on a methodology for the understanding of social phenomena) (1970); etc.

11. HEMPEL, *Aspects of Scientific Explanation, and Other Essays in the Philosophy of Science* (1965).

12. Cf. *The Poverty of Historicism*, especially § 29.

13. Cf. NAGEL, E., *The Structure of Science* (1961), especially pp. 401-428.

14. Some of the strongest advocates of this position in recent years have been the following: RUDNER, *Philosophy of Social Science* (1966); GIBSON, *The Logic of Social Enquiry* (1960); AYER, 'Man as a subject for social science' (1967); HOMANS, *The Nature of Social Science* (1967). See also BRODBECK (ed.), *Readings in the Philosophy of the Social Sciences* (1968).

15. This view is found for instance in COLLINGWOOD: cf. *The Idea of History*

talk of a multiplicity of 'language-games' each applicable to mutually irreducible speech-situations[16] has encouraged the idea of a comparable pluralism in matters of scientific methodology. The central argument of those who advocate a duality of methods is not unlike that of Max Weber: social behaviour has an aspect of 'meaningfulness' which is not contributed or projected by the observer, but which is simply the behaviour itself when regarded as a social phenomenon; it is this meaningfulness which makes it possible for other people to understand it. Meaningfulness is bound up with being governed by rules; but to perceive the rule which governs a phenomenon is not the same as to ascribe to it a cause. We may be told that causes are themselves regular relations between phenomena. But rules are not regularities: they do not merely conjoin logically independent phenomena, but weld phenomena together in unique configurations, each comprised within a unity of 'sense' or meaning.

It will be seen that this discussion connects up with the one we shall turn to in another section on the difference between human action and natural events[17]: the contrast between reasons or motives on the one hand, and causes on the other[18] is common to both discussions and can be dealt with in either context. If we accept that the concepts of 'meaningfulness' and 'rule' are irreducible, two courses are open to us: either, like Max Weber, we can try to combine certain features of *Verstehen* with certain features of causal explanation not merely selecting those features which suit our purposes best but achieving a real dialectical synthesis of the principles involved; or else we can take the bull by the horns and frankly accept the irreducibility and uniqueness of social science methods.

Turning to the latter position, it may be useful to consider first its most extreme exponent: P. Winch[19], who goes so far as to argue that the social sciences are not scientific, but philosophical, so that the central problem of

---

(1946). Among recent defenders of the notion of the specific nature of social science methodology, we may mention: OAKESHOTT, *Rationalism in Politics and Other Essays* (1952); for the most extreme expression of this view, cf. WINCH, *The Idea of a Social Science and Its Relation to Philosophy* (1958); and, in decreasing order of radicalness: COWLING, *The Nature and Limits of Political Science* (1963); LOUCH, *Explanation and Human Action* (1966-67); RUNCIMAN, *Social Science and Political Theory* (1963). Von HAYEK himself defends the idea of a certain methodological independence of the social sciences in relation to the natural sciences: cf. *The Counter-Revolution of Science* (1952).

16. Cf. in particular the *Philosophical Investigations* (1952).

17. Cf. HAMPSHIRE, *Thought and Action* (1959); von WRIGHT, *Explanation and Understanding* (1971), Pt. III. See below Section V, *Man and Action*.

18. As regards the identification of motive and cause, cf. DAVIDSON, 'Actions, reasons and causes' (1963, 1968). The view – which seems to be gaining ground at present, even in the English-speaking world – that it is necessary to distinguish methodologically the motive or reason from the cause of an action is set out for example in the seminal article by SCOTT & LYMAN, 'Accounts' (1968, 1970).

19. Cf. in particular WINCH, *The Idea of a Social Science* (1958). See also LOUCH, 'The very idea of a social science' (1963) and *Explanation and Human Action* (1966-67).

sociology is not that of method, but of subject-matter, because the subject matter determines the method. But what are the social sciences about? One might say: they are about the relations between human beings. But what are these relations but 'ideas' embodied in action? To understand 'the ideas which men's actions embody' and to explain social relations are one and the same. It is this reference to 'ideas' which brings philosophy into sociology. In arguing thus, Winch has no intention of imposing a philosopher's viewpoint on the human sciences; on the contrary, it is the positivist sociologists who impose their viewpoint by describing social phenomena in their own terms and not in those used by the social agents themselves; for how the agents describe their own actions and culture in constitutive of what those actions and culture are.[20]

In support of combining *Verstehen* with causal explanation, it can be argued that they are not on the same level. *Verstehen* is involved in grasping and identifying social phenomena, whereas causality belongs to the logical structure of explanation. Thus Alfred Schutz,[21] whose ideas were very close to Husserl's phenomenology, regarded *Verstehen* as the starting-point for seeking and giving a causal explanation; on that point, he agreed with Hempel and Nagel, but the social phenomenon must first be recognized as such, and in such identification the scientist's *Verstehen* is essentially the same as our everyday understanding of each other – our intentions, motives, feelings and beliefs. Without such identification, there would be nothing to study, let alone to explain. In other words, *Verstehen* is a condition of there being any social phenomenon at all, whereas causal explanation is a condition for the objective, scientific knowledge of such phenomena.

A somewhat more moderate version of this position seems acceptable to many writers who put forward similar methodological views concerning, in particular, the interpretation of and criteria for causal judgements in sociology.[22] But here too the discussion should be approached from the standpoint of methodological efficiency, though this question is more directly raised by the discussion of the concept of *rules*. Has the word 'rule' more

20. This view is violently disputed by some: cf. GELLNER, 'Concepts and society' (1963, 1970) and 'The new idealism' (1968). Louch himself, whose position is close to that of Winch on certain points, diverges from him as regards the most extreme consequences of his position (cf. (d) below, 'Rationality').

21. Cf. SCHUTZ, *Collected Papers* (publ. posth.), especially the following articles: in Vol. I, *The Problem of Social Reality* (1962), 'Common-sense and scientific explanation of human action', pp. 3-47; 'Concept and theory formation in the social sciences', pp. 48-66; 'Phenomenology and the social sciences', pp. 118-139; etc.; and, in Vol. II, *Studies in Social Reality* (1964), 'The social world and the theory of human action', pp. 3-19; 'The dimensions of the social world', pp. 20-63; and 'The problem of rationality in the social world', pp. 64-88. For an earlier stage in SCHUTZ's thinking, see his *Der sinnhafte Aufbau der sozialen Welt: eine Einleitung in die verstehende Soziologie* (1932); the republication of this work in 1960 and its recent translation into English (*The Phenomenology of the Social World*, 1967) are evidence of the increasing influence of this theorist.

22. Cf., for example, BLALOCK, *Causal Inference in Nonexperimental Research* (1964); GALTUNG, *Theory and Methods of Social Research* (1967); etc.

meanings than one?[23] To say it has only one meaning amounts to saying that when we speak of the 'regularity' of natural and of human events we are using the word in the same sense; to say there are two means that the 'rules' governing conduct are of a different kind from the regularities of natural phenomena.

It cannot be denied that the scientist's own behaviour – what he expects from his research – is involved in this debate: the merging of the human in the natural, or the accentuation of man's specificity.

At this point we may introduce the allied notion of *norm*; if human action is explained in terms of 'norms', it is perfectly possible to regard the explanation as a form of causal explanation, and not as an alternative to it. This is what Talcott Parsons does; in his work, the system of norms, interpreted as a self-balancing system, supplies the terms in which individual phenomena are explained. Where abstract and schematic thinking sets up an opposition between static, and as such irreconcilable, terms – 'rules' and 'causes' – the *sociology of action* establishes a conceptual complementarity between explanation in terms of norms and causal explanation; and the notions of conflict and equilibrium are used in passing from one to the other.[24]

The debate then focuses on the relationship between norms, on the one hand, and conflict and equilibrium on the other. In these discussions in the general area of 'functionalism', one finds a rather surprising parallelism between sociologists of the English or American schools[25] and Marxist writers like Althusser.[26] Both use 'structures' and their 'dynamism' to reconcile the ideas of 'cause' and 'norm'.

The point at issue lies at the heart of the discussions concerning 'functional' explanations: these are particularly relevant to general sociology, but also to social anthropology and political science.[27] To give a clear meaning to

23. Cf. BLACK, 'Notes on the meaning of "rule"' (1958) and, on a particular point of application, RAWLS, 'Two concepts of rules' (1955). And regarding the importance of the concepts of rule, purpose and meaning in psychology, cf. HARRÉ & SECORD, *The Explanation of Social Behaviour* (1972).

24. Cf., from among Talcott PARSONS' numerous publications, *Essays in Sociology* (1949); *The Social System* (1951); *Social Structure and Personality* (1954); *Structure and Process in Modern Societies* (1960); *Sociological Theory and Modern Society* (1967); *Politics and Social Structure* (1969); etc. See also: PARSONS & SHILS, *Toward a General Theory of Action* (1951) and PARSONS, SHILS, NAGELS & PITTS (eds.), *Theories of Society. Foundations of Modern Sociological Theory* (1961). Cf. also BOURRICAUD, 'En marge de l'œuvre de Talcott Parsons: la sociologie et la théorie de l'action' (an introduction to selected writings of Talcott PARSONS translated into French under the title *Eléments pour une sociologie de l'action*, 1955) and TOURAINE, *Sociologie de l'action* (1965), especially pp. 75-82 (see also the same author's recent work, *Production de la société*, 1973).

25. See, for example, DAHRENDORF, *Class and Class Conflict in Industrial Society* (1959).

26. ALTHUSSER, *Pour Marx* (1965); ALTHUSSER *et al., Lire le Capital* (1965).

27. For the most general problems, see, from among the vast literature that has appeared in English, the following: NAGEL, E., 'A formalization of functionalism' (1957) and *The Structure of Science* (1961), pp. 247-283; HEMPEL, 'The logic of

this idea – without merely saying, as some writers do, that everything in a given system has a function – one has to consider the various types of causal and/or teleological claim which may be implied in the notion of 'function'.[28]

In this connection, it is particularly worth mentioning the work of Max Black[29] as an example of philosophical thinking on methodology, which cuts through a somewhat confused mass of sociological theorizing with the tools of analytic philosophy.

(c) *Neutrality and value*

The foregoing debate on the relation between 'explanation' and 'understanding' as well as the debate on the rôle of evaluation can be described as concerned with the 'objectivity' of the social sciences: the former the 'objectivity' of their subject-matter, and the latter the possibility of 'objectivity' or 'neutrality' on the part of the social scientist himself.

It was again Max Weber who was the first to raise the question of *evaluation*, which is distinguished from the preceding question purely for reasons of expository convenience; for if social phenomena are meaningful, their meaning is essentially practical, i.e., a value; and if understanding is understanding of meaning, it is *ipso facto* also perception of value. Thus the question inevitably arises: how far are explanations based on value-judgements? Do these affect merely the selection of problems, or do they also determine our choice of causes and the relative weights we attach to them? Or do they even determine the conceptual framework used in the explanations?

It is hardly surprising that the thesis of axiological neutrality should be maintained by those who, for the reasons given above, underline the word

---

functional analysis' (1959); EMMET, *Function, Purpose and Powers* (1958). For a more complete bibliography the following may be consulted: ISAJIW, *Causation and Functionalism in Sociology* (1968) and MARTINDALE (ed.), *Functionalism in the Social Sciences* (1965). A critical assessment of functionalism, from a Marxist point of view, will be found in Chapter X, 'Der Funktionalismus in der amerikanischen Soziologie', of the previously mentioned work by KON, *Der Positivismus in der Soziologie* (1968).

For the problems of general sociology, see, for example: HUACO, 'The functionalist theory of stratification. Two decades of controversy' (1966); ANDERSON & MOORE, 'The formal analysis of normative concepts' (1957), 'Some puzzling aspects of social interaction' (1962) and 'The structure of personality' (1962); MACINTYRE & EMMET, 'Sociological theory and philosophical analysis' (1970).

28. Cf., on this last point, DAVIS, 'The myth of functional analysis as a special method in sociology and anthropology' (1959); GRUNER, 'Teleological and functional explanations' (1966); RUDNER, *Philosophy of Social Sciences* (1966); and, for an overall view of the question, DEMERATH & PETERSON (eds.), *System, Change and Conflict. A Reader on Contemporary Sociological Theory and the Debate over Functionalism* (1967).

29. BLACK (ed.), *The Social Theories of Talcott Parsons* (1961), especially his personal contributions to this collection, entitled: 'Some questions about Parsons' theories'.

'science' in 'human sciences';[30] those who stress the rôle of beliefs in social phenomena and regard values as irreducible elements in explanation itself tend to join the other camp.[31]

The argument is lively not only in *social anthropology* – in endeavouring to understand and explain societies culturally remote from the observer;[32] it is just as heated – but for opposite reasons – in *political science*,[33] where it is cultural nearness which constitutes the obstacle. Is empirical research into theories of democracy as neutral as it pretends to be?[34] The discussion is conducted at two levels, the conceptual and the methodological; at the first level, we find discussions in political philosophy about the concepts of 'power' and 'authority', 'élite', 'liberty', 'equality', 'justice', 'interest'.[35] At the second level are discussions on the nature of political argument:[36] if the method of political argument is held to be *sui generis*, can it be used in a theory of behaviour? Are apolitical politics possible at all?[37] The discussion at both these levels calls into question the very status of political science as a social science, as sociology. Can political sociology be distinguished from political philosophy? And can the latter criticize itself by formulating a theory of the ideological factors which contaminate it? Can a sociological study of ideology itself be ideology-free?[38]

30. For example, NAGEL, *The Structure of Science* (1961); RUDNER, *Philosophy of Social Science* (1966); BRODBECK (ed.), *Readings in the Philosophy of the Social Sciences* (1968); see also BERGMANN, 'Ideology' (1951).

31. Cf., for example, MYRDAL, *Values in Social Theory* (1958); TAYLOR, C., 'Neutrality in political science' (1967); and among practising social scientists, Wright MILLS, *Sociological Imagination* (1959).

32. Cf. NAGEL, *The Foundations of Social Anthropology* (1951); JARVIE, *The Revolution in Anthropology* (1967). On the question of the limits of comparison of one culture to another, especially in respect of institutions, cf. the views of ALMOND & VERBA, *The Civic Culture. Political Attitudes and Democracy in Five Nations* (1963) and the opposing views of GOLDSCHMIDT, *Comparative Functionalism. An Essay in Anthropological Theory* (1966); cf. also, in *Part I: Social Sciences* of the present *Study of Trends in Research* (1970), the chapter by ROKKAN: 'Cross-cultural, cross-societal and cross-national research'.

33. Cf. STRETTON, *The Political Sciences* (1969); RUNCIMAN, *Social Science and Political Theory* (1963); LUKES, 'Alienation and anomie' (1967). See also, on a related problem, RUNCIMAN, 'Sociological evidence and political theory' (1962).

34. Cf. DAHL, *A Preface to Democratic Theory* (1956); DUNCAN & LUKES, 'The new democracy' (1963).

35. Cf., as examples, BACHRACH & BARATZ, 'Two faces of power' (1962); DAY, 'Authority' (1963); PARRY, *Political Elites* (1969).

36. Cf. BARRY, *Political Argument* (1965) (a work which also deals in part with the traditional concepts of political theory); see also PERELMAN, *Eléments d'une théorie de l'argumentation* (1968).

37. Cf. MCCOY & PLAYFORD (eds.), *Apolitical Politics. A Critique of Behavioralism* (1967).

38. Cf., *inter alia*, the following works: ADORNO, *Negative Dialektik* (1966); ALTHUSSER, 'Idéologie et appareils idéologiques d'Etat: notes pour une recherche' (1967); ARON, 'Fin de l'âge idéologique' (conclusion of *L'opium des intellectuels*, 1955, 1968 – in English translation, *The Opium of the Intellectuals*, 1962); *Trois essais sur l'âge industriel* (1966) (English translation, *The Industrial Society. Three Essays on Ideology and Development*, 1968); *De la condition historique du socio-*

(d) *Rationality*

Lastly, it is the status of rationality in the human sciences which has been called into question. Not, of course, in these abstract terms, but as far as the social sciences are concerned more specifically in the form of the question whether the criteria of the rationality of an explanation are 'context-dependent'. Do models of rationality differ from one culture to another?

This problem is particularly serious in social anthropology. We find it already discussed by Frazer and Tylor in England, Durkheim and Lévy-Bruhl in France; more recently, it has come to the fore again with Evans-Pritchard,[39] who attacks the intellectualist approach and interpretation prevalent among British and American writers on religion and magic.

But, in varying degrees, the discussion concerns all the social sciences: one finds it in the sociology of knowledge, in particular in the work of Karl Mannheim;[40] in linguistics, in the work of B. L. Whorf;[41] and even in the history of science.[42]

As far as it relates to rationality in the social sciences, the question of rationality links up with the question of the logic of explanation, already considered above; hence one finds much the same line-up of antagonists. We have already seen that Peter Winch suggests that a culture should be described in terms of its own beliefs; but if social institutions are a func-

---

logue (1971); *Etudes politiques* (1972); GADAMER, 'Rhetorik, Hermeneutik und Ideologiekritik' (1967) (French translation, 'Rhétorique, herméneutique et critique de l'idéologie', 1971); GOLDMANN, *Sciences humaines et philosophie* (1952, 1966), especially Chap. II; HABERMAS, *Theorie und Praxis: Sozialphilosophische Studien* (1963); *Technik und Wissenschaft als 'Ideologie'* (1968); HALPERN, '"Myth" and "ideology" in modern usage' (1961); HORKHEIMER, *Eclipse of Reason* (1947) (German version, *Zur Kritik der instrumentellen Vernunft*, 1967); KELSEN, *Aufsätze zur Ideologiekritik* (1964); LICHTHEIM, 'The concept of ideology' (1965); LUKÁCS, *Schriften zur Ideologie und Politik* (1961); MACINTYRE, *Against the Self-Images of the Age. Essays on Ideology and Philosophy* (1971); MARCUSE, *One-Dimensional Man: Studies in the Ideology of Advanced Industrial Society* (1964); MONTEFIORE, 'Fact, value and ideology' (1966); MUKERJI, *Implications of the Ideology Concept* (1966); PARTRIDGE, 'Politics, philosophy, ideology' (1961) and 'Political philosophy and political sociology' (1965); RUNCIMAN, *Social Science and Political Theory* (1963); SCHAAR & WOLIN, 'Review essay: Essays on the scientific study of politics – A critique' (1963); TOPITSCH, *Sozialphilosophie zwischen Ideologie und Wissenschaft* (1961, 1966). For earlier works up to 1960, see the previously mentioned essay by BERGMANN, 'Ideology' (1951) and for an overall view: BIRNBAUM, *The Sociological Study of Ideology (1940-1960): A Trend Report and Bibliography* (1962), which includes a treatment of the subject and a classified and annotated bibliography.

39. Cf., in particular, *Nuer Religion* (1956) and *Theories of Primitive Religion* (1965).

40. Cf., among the works by MANNHEIM, *Ideologie und Utopie* (1929) (English translation, *Ideology and Utopia*, 1936); *Mensch und Gesellschaft im Zeitalter des Umbaus* (1935) (English translation, *Man and Society in an Age of Reconstruction*, 1940); *Essays on the Sociology of Knowledge* (1952); *Essays on the Sociology of Culture* (1956).

41. WHORF, *Language, Thought and Reality* (1956).

42. KUHN, *The Structure of Scientific Revolutions* (1962).

tion of a set of socially conditioned ideas, the models of rationality – including those of 'truth' and 'logic' will themselves depend on the context, so that rationality will be reduced to conformity with standards internal to the society, truth becomes relative, and logical relations depend on social relations.[43]

The upholders of universal objectivity and rationality reply to this that in order to understand a religion one does not have to be a believer, and that an essential prerequisite of a scientific attitude is for the observer to stand back from his subject-matter and to 'put into brackets' his own convictions and prejudices.[44] Thus the debate on rationality becomes a debate on the *criteria* for according the social sciences scientific status. A large number of books, collections of articles, symposia, etc., have been devoted, during the last ten years, to rationality and its limits, the idea of a social science, the notion of Western science, etc. The crucial point of the debate remains: how far are the criteria of rationality context-dependent? Is it possible to find any that are not? And, on the basis of these, to determine the precise nature of the remainder and to what extent they for their part are dependent?

Our own view is that this problem can be solved to the satisfaction of scientists if the question of the ultimate identification of the subject-matter of the human sciences is kept separate from that of its scientific treatment. The latter question rests on procedures of observation and explanation where the criteria and models of rationality are not essentially different from those of the natural sciences. *Verstehen,* on the other hand, concerns the identification of the subject-matter or 'object' of knowledge, without which there can be no explanation and no problem of objectivity of rationality.

The two questions are distinct. The first concerns the ultimate nature of social phenomena, and it is to this that Max Weber's *Verstehen* applies: it is the act of grasping social phenomena in their inner essence, and from this standpoint social phenomena are: (a) reducible in the last resort to

43. The theses exposed by WINCH in *The Idea of Social Science* (1958) have been further elaborated, in respect of this particular issue, in his article 'Understanding a primitive society' (1964, 1970).

44. Cf. the strong criticism which is directed against Winch's relativism in the following, among others: MACINTYRE, 'Is understanding religion compatible with believing?' (1964, 1970); the series of three articles (1967): LUKES, 'Some problems about rationality'; HOLLIS, 'The limits of irrationality'; and TORRANCE, J., 'Rationality and the structural analysis of myth'; Chap. II of JARVIE, *Concepts and Society* (1972). See also the debate between MACINTYRE and BELL about Winch's ideas, entitled 'The idea of a social science' (1967). Two recent collections are also worth consulting: WILSON (ed.), *Rationality* (1970) (in which several of the above-mentioned essays are reprinted) and HORTON & FINNEGAN (eds.), *Modes of Thought* (1973). On the related but separate problem of the value of the notion of the rationality of behaviour in explaining behaviour, cf. BENNETT, *Rationality* (1964) and GARFINKEL, *Studies in Ethnomethodology* (1967). The influence of G. H. Mead and Schutz is manifest in these discussions which are also of interest to the sociology of knowledge. Schutz's point of view is applied to this field in a philosophically interesting way in BERGER & LUCKMANN, *The Social Construction of Reality* (1966-67).

individual agents; (b) essentially significant or meaningful; (c) themselves imbued with axiological characteristics, reflecting norms, rules, attitude, value, belief; (d) intelligible in terms of their own norms. Couldn't we say that *social* science (with the accent on the word 'social') derives its subject-matter from an act of understanding, but that it becomes a *science* as a result of the scientist's efforts to apply procedures of observation and explanation, where the criteria and models of rationality are not fundamentally different from those used in the natural sciences? It is by a process of 'methodological transfer' that social phenomena can be treated in the same way as natural phenomena. Once this transfer has been made, we can say of social phenomena – in contradistinction to what can be said of them from the standpoint of *Verstehen* – that they are: (a) not reducible to some other order of phenomena; (b) essentially observable, in the logical empiricist's sense of the term; (c) amenable to scientific investigation as free from axiological bias as self-criticism of ideological preconceptions allows; and (d) imbued with the same rationality as is universally involved in all the sciences.

(e) *The distinctive problem of knowledge in the social sciences**

One of the key issues relating to the problem of knowledge in the social sciences arises from a reflection on the phenomenon which is usually known amongst social scientists as 'the self-fulfilling prophecy'. The essence of the phenomenon may be said to lie in the fact that if some situation is *defined* in a certain way, and the definition is *believed* to be real, then it is real in its consequences. Also, the fact whether a belief is true or false is irrelevant to the question whether it has actual consequences or not. The phenomenon is generally defined in terms of an initially 'false definition of the situation evoking a new behaviour which makes the originally false conception come true'.[45] But it is obvious that the phenomenon has nothing to do with the falsity of the belief as such. Rather, it is rooted in the fact that men can become aware of what is thought or postulated about human reality, and this awareness can affect or influence it in a certain direction. The phenomenon is not found with respect to the world of non-human reality, for what we think about nature does not affect it in any sense, except the purely technological one. But men are capable of self-consciousness, and thus can become aware of what is thought about them. Accordingly, the absence or presence of the capacity for self-consciousness on the part of that which is the object of study lies at the foundation of the fact that the phenomenon of 'self-fulfilling prophecy' is absent from our encounter with natural reality, and not quite so absent from our encounter with social reality.

* Reproduced from a contribution specially written for the purpose by Daya Krishna.

45. MERTON, *Social Theory and Social Structure*, rev. edn. (1957), p. 423. See also POPPER, *The Poverty of Historicism* (1944-45, 1957), §§ 5, 6.

But if what we think about social reality tends to affect it in a certain way, then how can we study it in the same way as we study objects in the natural sciences? It is because what we think about nature does not affect it in any sense, that we have the freedom to form whatever hypothesis we want to make about it. We can do so, for we are certain that the phenomenon about which we are forming the hypothesis will remain as it is and will in no way be affected by what we think or postulate about it. But if this certainty is absent, even to the least possible extent, in the social sciences (and the phenomenon of self-fulfilling prophecy means just this), then how can the usual postulational-deductive-verificational method be applied in this field?

On the other hand, once the effective rôle of belief in the *creation* of social reality is accepted, the question arises as to exactly what it would mean for a statement to be true with respect to such a reality. For if a false definition of the situation evokes a behaviour which makes the originally false conception come true, then we have to consider the cognitive implications of the fact that what is described by a *true* proposition in the present may be the consequence of the fact that a 'false' proposition was entertained as 'true' in the past. The truth of the proposition describing the present actuality would thus become, in an essential sense, dependent upon the contingent fact of some 'false' proposition having been entertained as 'true' in the past. Also, if the beliefs are considered as a constituent part of social reality in the sense that a change in them would change the nature of social reality itself, then how can we talk of 'truth' about this reality independently of the beliefs that are held about it and the way it is conceived to be?[46]

§ 2. *The logic of explanation in psychology*

So far we have been considering the most general problems of the logic of explanation in the human sciences, giving particular attention to the social sciences. Here we shall consider the special forms which the debate as to whether or not the human sciences can be reduced to the natural sciences takes in psychology. This singling out of psychology is partly justified by the long association between it and philosophy in university departments; but it was dictated more especially by recent developments in psychology – mainly in the schools of genetic structuralism and psychoanalysis – which have raised new points concerning the logic of explanation that did not emerge from consideration of the human sciences in general.

The question of 'reductionism' can be tackled on two levels: one which might be described as purely methodological, and the other, more boldly ontological. At the first level, the defence of behaviourism in psychology

46. Daya KRISHNA, '"The self-fulfilling prophecy" and the nature of society' (1971). See also, by the same author, *Considerations towards a Theory of Social Change* (1965) and *Social Philosophy – Past and Future* (1969).

largely coincides with the doctrines of logical empiricism in philosophy; but behaviourism, by stopping at the concept of 'behaviour', from which it takes its name, in no way implies any explicit doctrine about the reduction of that concept to the order of physical movements or even to that of neuro-physiological systems. 'Physicalism', in this respect, might be called a reductionist doctrine in an ontological sense, in so far as it asserts the possibility, in principle, of translating all propositions containing mental terms into propositions containing only physical terms. This was Rudolf Carnap's thesis in *Psychology in Physical Language* (1931).[47]

These two discussions will be presented separately. We shall then set forth the views of genetic structuralism and of psychoanalysis, in so far as these two schools of thought have given a new turn to the discussion on the logic of explanation.

(a) *Behaviourism**

Behaviourism is a reductionism with respect to ordinary language and the philosophies which adopt its concepts. The stimulus-response (S-R) model excludes any reference to purposes, intentions or motives as explanatory principles; on this point behaviourism and logical empiricism agree.

Broadly speaking, one can agree with Hull that 'it is the primary task of a molar science of behaviour to isolate the basic laws or rules according to which various combinations of stimulation, arising from the state of need on the one hand and the state of environment on the other, bring about the kind of behaviour characteristic of different organisms'.[48] What is ordinarily called 'consciousness' ceases to be the key to the solution of any problem; it is itself a problem which demands a solution in virtue of the principle that 'data-language' should only contain terms belonging to the language of physical things. One must therefore refrain from speaking of a 'situation' as perceived by a living being, or 'what it "means" to an organism'. Skinner lays down the rule that 'the independent variables must also be described in physical terms. [. . .] The events affecting an organism must be capable of description in the language of physical science'.[49]

In his view, an attempt to distinguish between a psychological world of experience and the physical world violates the fundamental principle of method and also reflects a confusion between dependent and independent variables. The reason why is clear: in a teleological explanation, the situation (the supposedly independent variable) is described in terms of its relation to the action (the supposedly dependent variable), inasmuch as it calls forth the action. It is precisely this which is debarred by the theory;

47. This extreme reductionist position has given way to subtler views: cf. SKLAR, 'Types of inter-theoretical reduction' (1967-68).

* Based on contributions, notes and references supplied by Charles Taylor and Rom Harré.

48. HULL, *Principles of Behavior. An Introduction to Behavior Theory* (1943), Chap. II, p. 19.

49. SKINNER, *Science and Human Behavior* (1953), Chap. III, p. 36.

in consequence, it also excludes from the internal conditions all concepts 'expressing a disposition to' behave in a certain way. It is further forbidden to name certain sequences of action according to the goals aimed at: that sort of talk is only justified in ordinary practical usage; in science, it must be cut out. To quote Hull again: 'An ideally adequate theory even of so-called purposive behavior ought, therefore, to begin with colorless movement and mere receptor impulses as such, and from these build up step by step both adaptive behavior and maladaptive behavior.'[50]

This relationship between 'receptor impulses' and 'colourless movements' is the essential principle of the S-R theory. It rejects, therefore, the claims of ordinary language theories that a model of explanation is given in ordinary language, and it excludes the model in question. At best, it agrees to retain the popular ideas of intention and motive as secondary principles, on condition that they can be deduced from the primary, elementary and objective principles.

If adaptation to a purpose is not an explanatory principle, but something to be explained, the whole weight of the explanation must rest on the concept of 'learning'; it is easy to see, therefore, that this problem has not been merely one amongst others, or a particularly popular field for experimentation, but the touchstone of the whole behaviourist system.

The controversy about behaviourism has taken (at least among philosophers) two clearly distinct forms.

On the one hand, the British and American philosophers of the ordinary language school have opposed the implicit logic of our language of action to the explicit logic of the science of behaviour. We shall deal with these theories of action-language in another place;[51] here we shall examine only the impact of this logic on psychological explanation, which, conversely, will not be dealt with at all in Section V.

These writers (E. Anscombe, A. I. Melden, A. Kenny),[52] faithful to Wittgenstein's contention that there were many kinds of 'language-games',[53] tend to regard the distinction between the two logics as establishing a simple dichotomy between ordinary language and the language of science. To talk, on the one hand, of 'projects', 'intentions', 'motives', 'desires' and even of 'actions' and, on the other, to construct causal chains of movements are to inhabit two different 'universes of discourse': an action is a slice of behaviour characterized, at least partly, by what it aims at or is designed to produce; intentions and desires are characterized, partly, if not entirely, by the actions to which they tend to lead. This explanation of their actions by the agents themselves, as a form of practical knowledge which is mingled with the action, is opposed – at least at this level of discourse – to the mechanistic or physicalistic description of behaviour.

50. HULL, *Principles of Behavior*, Chap. II, p. 25.

51. See below, Section V, *Man and Action*, pp. 1380 sqq.

52. ANSCOMBE, E. *Intention* (1957); MELDEN, *Free Action* (1961); KENNY, *Action, Emotion and Will* (1963).

53. Cf. *Philosophical Investigations* (publ. 1952).

But most of the philosophers of this school do not draw the conclusion that the two explanations are mutually exclusive; they refer each of them to different levels of discourse which in no way clash with each other. G. Ryle in *Dilemmas* (1954) held a similar position.

This means that the discussion on the logic of action-language has no implications for psychological explanation. Unfortunately, however, the discussion too often proceeds without any reference to developments going on within behaviourism or any contact with the theories current in psychology departments. Very few of the philosophers of this school enter into any real dialogue with the psychologists. In this respect, the most interesting writers are those who, like R. S. Peters,[54] attack behaviourism on its own territory and accuse it of conceptual confusion, taking as their cue Wittgenstein's remark: 'In psychology, there are experimental methods and *conceptual confusion*.[55]

To demonstrate this, however, one must go into behaviourism's own language-game and defeat it on its home ground. But the ordinary-language philosophers have rarely paid attention to developments in behaviourism itself or to the slow retreat from positivism in experimental psychology; during this period, in fact, the attempt to get beyond the level of facts and laws and to arrive at a genuine deductive theory, within which the laws are co-ordinated, tended to make psychologists think in terms of 'models', very different from Hume's idea of causality as consisting merely of regular sequences of atomic facts. These 'models', as Piaget explains in Part I of this *Study*, raise science to the level of genuine explanation in so far as they are able 'to represent the real processes and to express them in the form of deductive operations, the aim being achieved when the deductive operations are matched by the actual changes that have taken place.'[56]

This comment leads us to the second form of the debate about behaviourism, which consists in really putting to the test its capacity to explain and seeking its inherent limitations.

One theme which offers an excellent terrain for discussion of the inherent limitiations of behaviourism is the question of linguistic behaviour. (We shall consider the discussions about language in another connection;[57] here we shall deal, not with language as such, but with linguistic behaviour, i.e., that part of behaviour which is determined by the use of signs and discourse. Thus the limited problem which we shall consider here is: can language, as a specific type of behaviour, be explained within the canons of behaviourism? That is the only question which interests us for the moment.)

In his *Verbal Behavior* (1957), Skinner specifically suggests a functional

54. PETERS, *The Concept of Motivation* (1958).

55. *Philosophical Investigations*, Pt. II, subsec. XIV, p. 232 e.

56. PIAGET, 'Psychology', especially pp. 234 and 263-269 in *Main Trends of Research in the Social and Human Sciences, Part I: Social Sciences* (1970): cf. p. 234 for our quotation.

57. Cf. below, Section IV, *Man and Language*, pp. 1323 sqq.

analysis of language in which all the variables that govern this behaviour can be described in terms of concepts like 'stimulus', 'reinforcement', 'privation' – all derived from experiments with animals. As early as 1959, Noam Chomsky, in his 'Review of B. F. Skinner's *Verbal Behavior*', claimed that 'the magnitude of the failure of this attempt to account for verbal behavior serves as a kind of measure of the importance of the factors omitted from consideration, and an indication of how little is really known about this remarkably complex phenomenon'.[58] Sinner's characterization of verbal behaviour 'as "behavior reinforced through the mediation of other persons" [. . .] is clearly much too broad';[59] it seems difficult to distinguish between requests, orders, advice, etc. on the basis of the hearer's behaviour disposition. But the biggest difficulties arise from the question of sentence structure: 'It is evident that more is involved in sentence structure than insertion of lexical items in grammatical frames.'[60] The composition and enunciation of an utterance cannot be reduced to laying end-to-end a series of responses under the control of an external stimulus and intra-mental association: syntactical organization is not something that is directly and simply represented in the physical structure of the utterance itself; behind the manifest sequences, there are a vast number of integrating processes which can only be inferred from their terminal effects. Integrating processes, imposed patterns, selective mechanisms – these are all among the problems raised by the acquisition of linguistic competence and which behaviourism seems incapable of fitting into the exclusive rules of its functional analysis. A behaviourist theory can doubtless account for some important aspects of language – i.e., all those which involve the same procedures as learning by reinforcement; but the mastery of grammar seems to raise a problem of competence of a different order. 'It is reasonable to regard the grammar of a language *L* ideally as a mechanism that provides an enumeration of the sentences of *L* in something like the way in which a deductive theory gives an enumeration of a set of theorems. (*Grammar,* in this sense of the word, includes phonology.) Furthermore, the theory of language can be regarded as a study of the formal properties of such grammars, and, with a precise enough formulation, this general theory can provide a uniform method for determining, from the process of generation of a given sentence, a structural description which can give a good deal of insight into how this sentence is used and understood. In short, it should be possible to derive from a properly formulated grammar a statement of the integrative processes and generalized patterns imposed on the specific acts that constitute an utterance.'[61] Thus to learn a language is to acquire the overall competence corresponding to the formal properties of such a grammar, including their integrative processes and generalized patterns.

58. Chomsky, 'A review of B. F. Skinner's *Verbal Behavior*' (1959), p. 549 in Fodor & Katz (eds.), *The Structure of Language* (1964).

59. Chomsky, *ibid.*, p. 565.

60. Chomsky, *ibid.*, p. 574.

61. Chomsky, *ibid.*, p. 576.

A second topic which leads into the same discussion of the inherent limitations of behaviouristic explanations is that of directed, oriented, intentional behaviour. Charles Taylor in *The Explanation of Behaviour* says not merely that the ordinary language of action is teleological, but that scientific language itself ought to be if it is to do justice to action. But it is necessary to understand clearly what teleological explanation involves. The logic of explanation only too often confines itself to consigning teleological explanation to metaphysics, purposes being dismissed as non-observable entities, like the soporific property of opium; conversely, the logic of explanation has been content to work with a very meagre concept of causality, inherited from Hume, according to which cause and effect are two events without any internal logical connection, but purely contingent and associated only by regular succession. In Taylor's view, the 'purposive' character of animal and human behaviour requires a form of explanation which can be described as 'teleological'; but a teleological explanation is not the contrary of a causal explanation, but a particular form of it, that in which the resulting order is itself, in some way, also that by which it is actualized. 'Now when we say that an event occurs for the sake of an end, we are saying that it occurs because it is the type of event which brings about this end. This means that the condition of the event's occurring is that a state of affairs obtains such that it will bring about the end in question, or such that this event is required to bring about that end. To offer a teleological explanation of some event or class of events, e.g., the behaviour of some being, is, then, to account for it by laws in terms of which an event's occurring is held to be dependent on that event's being required for some end.'[62]

Thus to explain an event in terms of intention is not to invoke a non-observable entity, but to make a statement about the form or type of laws which govern this system; for example, to say that an animal is 'stalking' its prey means that in its repertoire of available action-patterns, this is the action 'required' to achieve its goal (satisfy its hunger). No hidden entity controlling the behaviour from inside is postulated here: one simply describes the fact that, in the case of beings whose acts are intentional, the sufficient condition for the production of an event (in this case, the 'stalking' behaviour) is that it should be 'required' to bring about the goal. It is not a separable characteristic, but a property of the system as a whole.

Charles Taylor thinks that, in this way, he has freed teleological explanation from the usual charges brought against it by logical positivism (recourse to nonobservable entities, *petitio principii*, etc.). He also accuses logical positivism of trying to impose its doctrine of a single model of

62. Taylor, C., *The Explanation of Behaviour* (1964), p. 9. Taylor's thesis is discussed, for example, by Noble, 'Charles Taylor on teleological explanations' (1967). For related problems, see Davidson, 'Actions, reasons and causes' (1963, 1968) and Scott & Lyman, 'Accounts' (1968, 1970), both already cited under § 1 above.

explanation in the sciences, for reasons which have nothing to do with the investigation of intentional behaviour.

One might add to Taylor's arguments that when behaviourist psychology excludes explanation in terms of the form of the system, this exclusion does not occur as the outcome of a process of reasoning, in connection with the derivation of laws, but at the very time the experiments are organized; the atomism is implicit in the way problems are divided up and the choice of an experimental procedure; it is at that time that the experimenter, by deciding what the stimulus in a given experiment is to be, favours a corresponding class of responses.

This defence of teleological explanation – as a theory of the form of systems or of 'order' – is very different from the argument from ordinary language and the opposition between a 'cause' and a 'motive' derived there from; that distinction leads in fact to the separation of two 'universes of discourse': that in which we treat an action as a physical event in the world, and that in which we normally speak of action, either to inform other people or in replying to questions about our intentions or our motives. For this school of thought, there tended to be a 'split' between ordinary language and science. Charles Taylor is arguing for an asymmetry, within science and its causal explanations, between teleological and atomistic explanations; it is causality itself which splits in two according to the nature of the 'ordering' relationship between antecedent and consequence. In this way, conscious intentions, rational motives (reasons), personal responsibility are reintroduced as part of the teleological explanation, which involves only the form of the system. 'Intentionality' in the phenomenological sense, denotes a type of system where, in addition to its general explanatory form, the origin of the action can be assigned to an 'internal' centre of responsibility. Ordinary action-language would appear to be the appropriate medium of expression for this type of system; but ordinary language itself would not be understood unless the system of intentional action can be subsumed under the system of teleological behaviour characterized by the purely formal idea of self-materializing order.[63]

(b) *Physicalism old and new*

Reductionism in the ontological sense was for a long time synonymous with physicalism. As we said above, it is not necessarily implied by behaviourism for which the concept of 'behaviour' is the fundamental concept of psychology; for methodological behaviourism, translation into physical language is an ontological thesis.

Anti-reductionism has long sought to distinguish different degrees of organizational complexity, thus reviving the doctrine of emergent evolution and 'levels of reality'. Piaget helps to clarify the discussion by distinguishing between two problems: that of the 'relations between con-

63. For a general critique of Skinnerian behaviourism, see also Malcolm, 'Behaviourism as a philosophy of psychology' (1964).

sciousness as such and its nervous concomitant' and that of 'the interaction between conduct (which includes consciousness but goes beyond it) and physiological processes'.[64]

In regard to the first problem, the general trend today is to recognize a degree of isomorphism but not interaction between states of consciousness and those of the 'concomitant'; but to deny interaction between consciousness as such and its nervous concomitants does not in any way entail denying the existence of interactions between conduct and physiological processes. In this way, to the reductionist tendency which aims at purely and simply identifying the mental process (conceived as a simple phenomenological expression) and its organic concomitant (conceived as constituting the former's underlying reality, or at least a direct explanation of it) one can oppose a trend that can be called relational or dialectic and which 'involves distinguishing between many scales of phenomena, both in the organism or nervous system and in behaviour or conduct, and discerning the interaction or feedback between processes of different scales, so that there is no longer any reduction from the higher to the lower but closer and closer solidarity'.[65] Psychology today offers many examples of this influence of higher-level 'scale' on lower-level phenomena, including the recent history of the conditioned reflex and conditioning processes. This dialectic of levels, in which the higher and the lower levels form an interacting whole, is then extended to the entire field of relationships between behaviour and organic processes, in such a way as to suggest the construction, storey by storey, of an edifice representing all the functions, but with each stage commencing with a *re*-construction of what has already been built or achieved at a lower level.

But there was implicit in the physicalism of a Fechner or a Köhler another sort of reductionism, which has come into its own today on the basis of electronic models. This was a physicalism which established a direct relationship between the physical world and the most objective and depersonalized aspects of perception. This physicalism immediately split away from mental atomism and, under the influence of *Gestalt* psychology, moved towards a theory of structures, seeking at this level isomorphism, if not identity, between the physical and the mental *Gestalt*. This configurationist physicalism then took over the idea of a control system conforming to cybernetic models: in this way, 'the theory of information, which springs from essentially human considerations, has found itself partly converging – but in a way remarkable for its formal and mathematical apparatus – with the fundamental equations of thermodynamics relating to entropy (it being possible to define information as a negative entropy); and the theory of decision or of games, the specific field of which is economics, has found physical applications (such as Maxwell's demon interacting with entropy)'.[66]

The result is a new reductionism: the identity of the physical and the

64. PIAGET, 'Psychology', *op. cit.* (1970), p. 235.
65. PIAGET, *ibid.*, p. 235.
66. PIAGET, *ibid.*, p. 242.

mental is now modelled on digital computers. This reductionism goes hand in hand with the appearance of non-behaviouristic forms of mechanism which, in many respects, provide real alternatives to classical behaviourism. These take the form of theories of competence: this instrument enables psychologists to admit once more into their explanatory universe to commonsense distinction between capacity and performance without breaking with mechanistic principles; the development of programming gives them a new language in which to elaborate their models as well as the assurance of their mechanistic respectability.

More recently, this development has affected philosophical discussion. Many philosophers, sympathetic in principle to mechanism, but convinced of the inadequacy of behaviourism, revived the out-and-out reductionist thesis on the basis of this new form of physicalism. The group of Australian materialists pushed their reductionism to an ontological theory of identity between mental and neurophysiological states, so that one could do away with the mental altogether.[67]

Other philosophers, anxious to defend materialism, and hence reductionism, do not necessarily want to embrace this identity theory. The concern of some of them is rather to show how, in spite of the weakness of behaviourism and the consequent necessity of a 'mentalist' psychology, the properties of our psychological model can in turn be seen as embodied in neurophysiology, so that a complete science of man might eventually be built on this indirect reduction of mentalist concepts to neurophysiology. Hilary Putnam and Jerry Fodor[68] belong to this group. Their work leans heavily on the computer analogy: a computer programme can provide an adequate account of what a machine does on one level, while capable of instantiation in many different ways in hardware on a more fundamental level. The moral, in general, is that psychology should proceed quite freely in elaborating theories in mentalistic terms, and not concern itself with the reduction of these to neurophysiology till afterwards. The impetus for this non-behaviourist yet mechanistic view comes partly from the advances of computer technology, but also from the influence of Chomsky and his anti-behaviourist account of language-learning.

Thus the main discussions in philosophical psychology are coming to be centred round the following questions: First, to what extent are digital computers really a reliable model of intelligent conduct? One of the most interesting challenges to the view that they do comes from Hubert Dreyfus.[69] Second, what is to be said about reduction itself? In opposition to the reductionist thesis, other writers argue that we cannot understand

67. Cf. SMART, *Philosophy and Scientific Realism* (1963); ARMSTRONG, *A Materialist Theory of the Mind* (1968).

68. PUTNAM, 'Minds and machines' (1960); FODOR, *Psychological Explanation* (1968).

69. Cf. the most recent work by DREYFUS on this question: *What Computers Can't Do. A Critique of Artificial Reason* (1971). And cf. above, Section II, sub-section B, § 3, 'Levels of reality', under (b), pp. 1183 sqq.; also below, Section V, sub-section B, § 1, (d), 'Analytic philosophy and the theory of games', pp. 1402-1403.

human conduct without introducing the notion of levels of behaviour such that the higher levels cannot be reduced to the lower levels. Michael Polanyi, for example, has defended this view.[70] Norman Malcolm has also argued against reductionism on grounds of logical conceivability.[71]

(c) *Genetic structuralism*

The idea that psychology deals only with comprehensive and quantifiable 'observables' is an idea no longer accepted unreservedly by psychologists today. In the first place this was a controversial idea associated with the divorce from philosophy which was supposedly concerned with essences or confined to introspective activity by the subject. Furthermore, it was a guideline, contemporary with early empiricism, which invited the research worker to study forms of conduct which respond to stimuli known to the experimenter and which are expressed in behaviour chains themselves susceptible of experiment and calculation. But such guidelines become prohibitions and inhibitions when they prevent consciousness from being considered as an aspect of action itself, or when the description of fully constituted mechanisms conceals the processes and stages of development, i.e., their genesis; or when exclusive concentration on passive experience blocks access to intellectual operations; or, lastly, when a purely analytical concept of observation dooms to failure 'the search for structures of behaviour or structures of thought resulting from the gradual interiorization of actions'.[72]

All these prohibitions are dispelled by the *genetic structuralism* of the Geneva School. Its initial hypothesis is that of 'overall forms comprising their self-regulation or their operators, in contrast to interpretations of an atomistic type'.[73]

This involves querying the most general hypothesis of those theories of learning which would make learning the primary phenomenon and mental development its derivative. If, on the contrary, we start from the hypothesis that mental development obeys its own laws, then learning in a particular and determined situation no longer constitutes anything but a more or less artificially defined sector of such development. We must then try to find out what active forms of co-ordination govern the progressive equilibration of development. Thus it was that the Geneva school brought to light the logic or algebra underlying these constructions. The growing child shows a greater capacity than 'the degenerate white rat'[74] for grasping underlying logical structures even while learning contingent and arbitrary facts. This grasping of logical structures does not come from external reinforcements (i.e., success or failure recognized according to results), but by the generaliz-

70. POLANYI, *Personal Knowledge* (1964) and *The Tacit Dimension* (1966).
71. MALCOLM, 'The conceivability of mechanism' (1968).
72. PIAGET, 'Psychology', *op. cit.*, p. 228.
73. PIAGET, *ibid.*, p. 252.
74. PIAGET, *ibid.*, p. 254.

ation and differentiation of logical or pre-logical structures. This operative logic which governs the organization of actions and their strategy is situated at the point where biological facts and reflective logic meet. On the one hand, since it is responsible for the general co-ordination of actions, it forms one of the self-regulating and self-correcting factors well known to biologists; on the other hand, it lends itself to those mental and reflexive interiorizations from which explicit logic derives.

Thus development appears as a real construction and a construction of structures, not as an additive accumulation of isolated acquisitions. From the child's point of view, the universe is not organized unless he re-invents its organization step by step under the guidance of an operative logic. The same is true for social norms: here too he only learns them by reconstructing their models. In all these activities, experience is not what empiricism claims; it is from the outset an action upon objects and, contiguous with action, such an unfolding of intellectual – particularly logico-mathematical – operations (putting together, counting, changing the order of objects, counting them again and thus deriving order and enumeration from the action itself). Furthermore, these operations are to be found bound up with one another within overall systems which, in turn, are based on definite mathematical structures.

The philosopher cannot fail to be interested in this conception of natural structures.

In the first place, genetic structuralism enables him to make a completely new approach to the problem of the specific nature of psychic phenomena. Psychology is often caught between two types of reductionism – aiming at reduction to the organic and the social respectively – and has always had some difficulty in accounting for the existence of structures common to all individuals except as carbon copies of organic structures or as the individual expression of social structures. As is shown by the development of the child and the adolescent, socialization is not limited to reflecting external constraints but implies a partial re-invention of the rules by the individual in a highly complex interplay of inter-individual dependencies. And as the comparative method shows, this self-structuring of the individual, which runs parallel to his moulding by the group, brings into play dynamic structures of socialization different from the cultural traditions and codes characteristic of special groups. The most remarkable of these dynamic structures are those which underlie the general co-ordination of action and the thought processes. The problem of reductionism thus tends to be replaced by a relational analysis in which social structures and mental structures are linked by structural homologies, themselves governed by broad structures of general algebra, 'so that the sociological explanation thus coincides with a qualitative mathematization similar to that which occurs in the setting up of logical structures'.[75]

There is another philosophical implication: the theory of natural struc-

75. Piaget, *ibid.*, p. 248.

tures unexpectedly finds common ground between the phenomena of self-regulation, in the biological and cybernetic senses, and the various kinds of logic – all of them consistent though without any direct relation to one another – which have sprung up from modern mathematics. The psychology of intelligence thus places itself midway between the two by showing how these forms of logic are constituted spontaneously, along with the operations themselves, building upon more elementary structures, while at the same time extending the range of the strictly biological phenomena of equilibration.

In turn, these natural structures are formed at a non-conscious level in the individual who makes use of them. The individual as subject is only aware of the outward results of these operative structures at the level of performance, and therefore of the particular operations which he performs. This does not mean that the subject is negated, but that the processes of consciousness, reflection and interiorization are themselves brought back within the ken of operative logic.

But psycho-genetic structuralism, as distinct from that which prevails in phonology and semiology, is from the outset a constructivism. Thus it escapes the dilemmas – structure versus genesis, synchrony versus diachrony – to which the taxinomic forms of structuralism lead. Here structure and construction go hand in hand, and we stand closer to the transformational models of the Chomsky school of linguistics than to the structuralism of the French semiologists and semanticists.

In conclusion, whereas for positivism abstract models did not attain to the real – which it reduced to the observable – but only constituted a convenient language, that is to say, one that was simple and useful in making predictions, for psycho-genetic structuralism models have an explanatory value that is real 'in so far as the operative transformations of the model match the real transformations of the phenomenon that has to be explained'.[76] This involves an important decision with respect to the very notion of *reality*, for if it is true that the model 'offers glimpses of real but still hidden processes accounting for observables and if it thus facilitates explanation',[77] then it must be admitted that the very notion of real ceases to match that of the observable and adapts itself to that of causal interpretation.

It is with all these implications that philosophy will henceforth have to enter into debate.

§ 3. *Psychoanalysis and the human sciences*

The advent of psychoanalysis to the field of the human sciences raises a specific problem with regard to the logic of explanation, which is the sole point of view being considered here.

76. Piaget, *ibid.*, p. 264.
77. Piaget, *ibid.*, p. 265.

As a general rule, epistemologists of the logical positivist school have questioned the scientific character of psychoanalysis. If, they say, psychoanalysis is a 'theory' in the same sense as the molecular theory of gases or the theory of genes in biology, i.e., a set of propositions which systematize, explain and predict certain observable phenomena, it should satisfy the same logical criteria as theory in the natural or social sciences. In the first place, it should be capable of empirical verification. To that effect, it must be possible to deduce from its propositions specific consequences, otherwise the theory itself has no definite content. Furthermore, there must be specific procedures (whether they are called matching rules, co-ordinating definitions or operative definitions), which make it possible to relate this or that theoretical notion to definite, unambiguous facts. It seems, however, that nothing precise can be deduced from the notions of Freudian energetics because they are so vague and metaphorical. They are perhaps suggestive notions, but they are not susceptible of empirical validation. Moreover, attempts at co-ordination with the facts of behaviour come up against their invincible ambiguity. Thus it is impossible even to say on what ground the theory might be refuted. Then too, empirical validation must satisfy the logic of proof if it is to be considered evident. It is said that interpretation is its principal method (in addition to its confirmation through study of the child's development and through ethnology). But on what condition is an interpretation valid? Is it because it is coherent, or because the patient accepts it, or because it helps someone who is sick? But such an interpretation must first of all have an objective character, and this means that it must be possible for a series of independent specialists to have access to the same material gathered together under carefully codified circumstances. There must then be objective procedures for deciding between rival interpretations. Furthermore, the interpretation must lead to verifiable predictions. But, say the logical positivists, psychoanalysis is not in a position to meet these demands: its material stems from the individual relationship between the analyst and the person analysed, and for want of a comparative procedure and statistical research it is impossible to eliminate the suspicion that the interpretation may have been imposed on the facts by the interpreter. Lastly, the psychoanalysts' claims to therapeutic efficacy do not meet the most elementary rules of verification: since it is impossible to establish or even define strictly the rates of improvement, the therapeutic efficacy of psychoanalysis cannot be compared with that of another type of investigation or other treatment, or even with rates of improvement in spontaneous cures. Hence the criterion of therapeutic success is useless.

These epistemologists are supported by those of the behaviourist school who, like Skinner, demand that every human science conform to the rules of an operational language. According to this strict operationalism, the psychoanalytic theory and all the concepts which revolve round the idea of mental apparatus necessarily appear to be dangerous metaphors of the 'phlogiston' type. From the epistemological point of view, the psychoanalytic theory does not represent any decisive advance on animism and its various substitutes. 'Freud's explanatory scheme', Skinner writes, 'followed a traditional pattern of looking for a cause of human behavior inside the organism.'[78] This 'traditional fiction of a mental life' – what Ryle called 'the ghost in the machine'[79] – leads to positing something which is not observable and on which it is impossible to act. Operationalism, however, only takes into consideration changes in an organism related to variables in the environment. Skinner goes so far as to accuse Freud of only being interested in those aspects of behaviour which must be considered expressions of mental processes, and of having thus considerably narrowed the field of observation. He reaches the conclusion that the idea of the mental apparatus which Freud imposed on psychoanalysis retarded the inclusion of that discipline in the body of science properly so called. Does not an operational definition of all the terms of psychoanalysis in fact beg the question whether psychoanalysis is reducible to a psychology of behaviour, and the thinking involved in analytical interpretation to construction of an empirically verifiable theory?

Other epistemologists reach less rigorously exclusive conclusions. What, after all, does operationalism demand? For a statement to be meaningful in operational terms it must be linked at some point with observables. Hence there is one requirement which is irreducible: it must be possible, in some way or other, to confirm a statement or a hypothesis, that is, they must be linked in a significant manner or correlated with some sort of verifiables. Such confirmation however can be made indirectly through a whole network of intercalary constructions containing abstractions of different degrees; it is in this connexion that we may introduce what have been called 'intercalary variables' or 'dispositional concepts'. If therefore psychoanalysis is to satisfy the demands of logical positivism, it must reformulate its principles in a language entirely derived from two basic observables: perception and response. In order to satisfy these criticisms and meet these requirements, some psychoanalysists have actually tried to recast the theory in terms that would be acceptable to psychologists of the Skinnerian type. Thus the facts of psychoanalysis could, they thought, be classed with the observables of scientific psychology; psychoanalysis would only add a latent character to the notion of behaviour. The functioning of the Freudian topographical systems (id, ego, superego) according to specific laws of their own would represent a cluster of 'separate models' governing 'partial' modes of functioning similar to the reflex model; the economic point of view would fit quite well into the tension/relaxation entropy model, while the theory of stages would presumably represent a genetic point of view; lastly, it would be possible to see Freud's theory in terms of a Jacksonian model, where the various systems formed a hierarchy of integration, with the higher inhibiting and controlling the lower.

The price that must be paid for bringing psychoanalytic theory into line

78. SKINNER, 'Critique of psychoanalytic concepts and theories', p. 79 in Vol. 1 of *Minnesota Studies in the Philosophy of Science* (1956).

79. RYLE, *The Concept of Mind* (1949), Chap. I, § 2 and *passim*.

with general psychology is twofold. On the one hand, the dual functioning of the psychic apparatus according to the pleasure principle and the reality principle as posited by the fundamental theory, is reduced to a mere loop or detour in the process of adaptation, itself subject to the basic S-R (stimulus-response) pattern. On the other hand, psychoanalysis moves in the direction of an *egopsychology*. These two changes go hand in hand: the adaptive point of view is only valid if the reality principle is itself interpreted in terms of pre-adaptation for the ego. This principle then becomes a variant of social determination.

How then do matters stand with psychoanalysis if these changes are made?

Is Skinner's rejection not more enlightening than all the attempts at reformulation? This rejection leads us to ask whether psychoanalysis should not remain a parallel discipline in relation to behaviouristic psychology which alone fully meets the requirements of operationalism. Instead of trying to redefine psychoanalysis in terms which both distort it and satisfy poorly the methodological requirements of a science of observation, would it not be better simply to give up the attempt to class it with the sciences of observation of behaviour and put it with the semiological and historical sciences? Efforts to make psychoanalysis into a science of behaviour inevitably end up by regarding it as an inferior form of observatonal theory and its hypotheses as metaphors from the pre-scientific age. A distinction is either to be drawn initially or not at all. It is our personal conviction that whereas psychology is a science of observation dealing with facts of behaviour, psychoanalysis is an exegetic science dealing with relationships of meaning as between the substitute objects and the original lost objects of instinct. It is from the outset, in regard to the primordial notions of 'fact' and 'inference from facts', that the two disciplines diverge. To what, for example, in analytic theory do the psychologists' 'environmental variables' correspond? For the analyst these are not facts as known to an outside observer; the analyst is interested in the dimensions of the environment as the individual 'believes' them to be. What is relevant for him is not the fact but the meaning that the fact has acquired in the individual's life. It is therefore incorrect to say that 'early punishment of sexual behaviour is an observable fact that undoubtedly leaves behind a changed organism'.[80] The subject of the analyst's study is the meaning the individual confers on those events which the psychologist considers as an observer and terms environmental variables.

Hence behaviour does not represent for the analyst a 'dependent variable', observable from without; rather it is the expression of changes of meaning in the individual's life, as they come to the surface in the analytic stuation. We may likewise speak of 'changes in the probability of action'. Thus a psychoanalytic patient may also be treated in terms of the psychology of behaviour, but it is not in those terms that the facts of behaviour are re-

80. Skinner, *op. cit.*, p. 81.

levant to analysis. Their value is not as observables but as significant elements in the development of desires. It is precisely this significance that Skinner casts into outer darkness, amid the general welter of theories on mental life and pre-scientific metaphors. This idea of steps in development, however, does not imply a less advanced stage along the one and only road to behaviourism: speaking absolutely, there are no 'facts' in psychoanalysis because it does not observe but interprets.

This, in our view, is the analyst's only reply to the behaviourist. If he accepts a methodology already constituted on the basis of the axioms of behaviourism, and if he begins to conceive of his research in terms of the 'probability of certain responses', he is doomed either to be rejected as non-scientific or to have to beg for partial rehabilitation by means of what Skinner calls the simple expedient of an operational definition of terms. The line of defence is along the outposts and the issue is decided on the basis of the prior question: what is relevant in psychoanalysis? If the answer is human reality as it can be described in operational terms of observable behaviour, then condemnation inevitably follows. If the specific nature of questions of meaning and double meaning is not recognized, and if these questions are not related to that of the method of interpretation whereby the questions come to light, the psychic reality of which psychoanalysis speaks can never be more than a supererogatory, redundant 'cause' in relation to what the behaviourist quite rightly describes as behaviour. In the last analysis, it will be no more than a ritual form or a kind of mental alchemy, to use Skinner's damning expression.[81]

It would seem however that it is an irreducible characteristic of psychoanalytic explanation to associate concepts of an energetic nature, such as repression, displacement, condensation, cathexis, discharge, etc., with concepts of an exegetic character such as censorship, double meaning, overdetermination, object-choice, interchangeable objects, symbolism, etc. Presumably this is not because of any equivocation or lack of conceptual clarification, but on the contrary out of faithful recognition of the very status of the unconscious. According to the Freudian doctrine, the unconscious mind is a reservoir of instincts. These instincts cannot be known in their biological existence, but only through their 'representatives' or delegates, which are of the nature of ideas or affects. Furthermore, these primary representatives themselves become accessible only after a laborious deciphering of the effects of meaning which rise to the surface of consciousness. These offshoots of the unconscious are like the traces or documents on which the historian or semiologist works. Thus it would seem that the specific subject-matter of psychoanalysis is itself situated at the point where force and meaning intersect and that the theoretical concepts used to depict it must necessarily combine the two universes of discourse in which the notions of force and meaning are deployed. It is the task of a 'semantics of desire' to develop systematically all the implications of this

81. In his contribution to the symposium *Psychoanalysis, Scientific Method and Philosophy*, edited by Sidney Hook (1959).

dual origin. Thus it has been possible to match the concepts of condensation and displacement, characteristic of 'dream-work' and expressed in the language of force, with rhetorical devices such as metaphor and metonymy which are situated at the level of the effects of meaning. But this matching operation does not mean that it is possible in turn to reduce the Freudian economy to language structures. There are two reasons for this: first, because the economy cannot be replaced by rhetoric without destroying the very subject-matter of psychoanalysis, namely instinct in its representative and affective delegates; but mainly because transcription into linguistic terms overlooks the fundamental fact that language is itself organized on the level of pre-conscious and conscious systems and that what appears as 'work' in the language of neurosis and dreams is rather a distortion of language itself, due to its being disoriented in relation to its signifying function. It is for this reason that the use of Saussure's terms of 'signifying' and 'signified', as applied to the unconscious, is always apt to jeopardize the great Freudian discovery, namely that in the primary process there is neither doubt nor negation nor passage of time. These functions however are essential to the birth of linguistic signs, which are based specifically on the distinction between the signifying and the signified, on the separateness of sign and thing. Consequently psychoanalysis cannot be reduced to semiological terms any more than it can to biological. It is irremediably a hybrid, with all the ambiguity that entails.

Far from eliminating the problems of the boundaries between it and psychology, this hybrid character of psychoanalysis only makes them more complicated. This is so for a number of reasons.

To begin with, psychoanalysis, like any explanation, only gets to grips with its subject-matter by means of a *theoretical* structure.[82] The question then arises how this structure can vindicate itself both in the face of other theories and also with respect to its own content, i.e., what it asserts about the motives underlying all discourse. As Jean Ladrière points out in *L'articulation du sens*: 'Either it must be able to furnish a complete clarification of its own presuppositions, or else it must show itself independent of the unconscious. In the former case it is necessary to establish that it is possible to keep watch over all the workings of the unconscious, in other words for the unconscious to be fully reflected in the conscious mind – which constitutes a highly ambitious programme, to say the least. In the latter case it is necessary actually to prove that psychoanalytical discourse is itself independent of the unconscious.'[83]

82. Cf. *inter alia* Hook (ed.), *Psychoanalysis, Scientific Method and Philosophy* (1959); Farrell, 'The status of psychoanalytic theory' (1964); MacIntyre, *The Unconscious* (1958).

83. 'Ou bien elle doit être capable de fournir une élucidation complète de ses propres présuppositions, ou bien elle doit se montrer indépendante de l'inconscient. Dans le premier cas, il s'agit d'établir qu'il est possible d'obtenir un contrôle complet de l'inconscient, autrement dit de réaliser une réflexion totale de l'inconscient dans le conscient, ce qui constitue un programme à tout le moins fort ambitieux. Dans le second cas, il s'agit de prouver effectivement que le discours psychana-

This brings us to the question of the form of the theory and its model. Thus a link-up may be established with the new forms of mechanism and physicalism by attempting to formulate the theory in cybernetic terms; it would then be at the level of non-behaviourist models that psychology and psychoanalysis might meet. It is also possible to conceive of psychoanalytic discourse being rigorously formalized by means of the algebraic and topological models of modern mathematics, as is beginning to be done in certain structuralistic interpretations of psychoanalysis; this would then lead to a type of discourse which would not necessarily be physicalist but which would in any case have the same logical form as physical theories. Here again the overlapping with psychology of behaviourist inspiration takes place at the level of the theoretical structure.

A second area of contact and possible conflict is with 'psychogenetic structuralism'. That school looks upon psychoanalysis, despite its genetic programme (theory of the stages of the libido, theory of the formation of the super-ego by identification, etc.), as a rival, and in large measure opposing, theory. Compared with Piaget's constructivism, psychoanalysis seems one type of reductionism. 'In its original Freudian form', Piaget writes, 'psychoanalysis presented the remarkable example of a doctrine that explained an individual's present by his past, therefore the adult by the child, and which in this sense had a definite genetic intention behind it, but which conceived genesis not as a continuous process of building up but merely as the development of certain initial tendencies, so that the present was reduced to the past and the different phases of development reduced merely to the shifting of the points of application of the initial pulsive energy.'[84] This is indeed a form of reductionism, though this time it is 'not from the mental to the organic or to the social, but in fact from higher psychic forms to elementary forms that live on throughout life underneath the former and in the "unconscious" '.[85]

This difference deserves further consideration instead of being passed over summarily. The strength of Freudianism lies in having brought to light what *The Interpretation of Dreams* (1900) called 'indestructible desires' and, generally, the regressive tendencies of the human psyche, including its libidinal clinging to the archaic objects of desire, whereby the individual's history takes on the appearance of repetition. The keynote of destiny represents perhaps Freud's most profound insight into man's libidinal make-up. On the other hand, it is rather in logical operations and their progressive development that Piaget's school finds the principle of psychic construction. Thus psychoanalysis and genetic psychology see two different facets in man and it is no easy matter to fuse them. What appears to Piaget as an integral reductionism stems from a reading of the palimpsest of human desires, while what he proclaims as a constructivism

lytique est lui-même indépendant de l'inconscient.' – LADRIÈRE, *L'articulation du sens. Discours scientifique et parole de la foi* (1970), p. 44.

84. PIAGET, 'Psychology', *op. cit.* (1970), p. 249.

85. PIAGET, *loc. cit.*

is situated in the area that Freud calls the 'reality principle', of which the ego is the correlative psychic agent. Here the constructivism of the genetic psychology school joins hands with the less markedly reductionist forms of psychoanalysis represented by the trend found in H. Hartmann[86] and especially D. Rapaport.[87] For the latter, thought is a system of mechanisms which makes it possible to withdraw from areas of conflict so as to concentrate on cognitive advances. Hence mental effort is no longer entirely sublimation or defence mechanism, and there is room for a real genesis of the ego.

The two major trends in contemporary psychology remain separated however to the extent that genetic constructivism is concerned with an area other than that explored by Freudianism and does not come to grips with affectivity itself. It is specifically in regard to the theory of stages and observation of the young child that confrontation is currently developing. Interdisciplinary research, which causes psychologists and clinical specialists to co-operate and compete, is alone capable of finding a way out of the present impasse.

There is one more area of contact and possible conflict to be considered. Psychoanalysis necessarily links up with social psychology in its theory of the superego. It was stated above that the subject-matter of psychoanalysis is desire as represented in effects of meaning, but a semantics of desire is complete only if desire is considered initially in its clash with social – originally parental – forces which set up barriers and prohibitions against it and push it back into the unconscious. This confrontation between desire and social force puts psychoanalysis in a new relation, not only with the psychology of behaviour, but with all the social sciences. It is no accident that psychoanalysis has ventured into the fields of ethnology, morality, aesthetics and religion; its interest in cultural phenomena represents no sideline or chance digression beyond its field of competence, but follows from its most fundamental aim. Psychoanalysis does not aspire to be merely a therapy; from the outset it sought to be something more, namely an interpretation of human reality in its entirety. Freud's *Letters* to Fliess, published quite recently,[88] show that at a very early period an interpretation of the Greek tragedy of Oedipus and the Elizabethan tragedy of Hamlet was bound up with the interpretation of dreams and symptoms, and this initial link must be based on something more than chance; the reason for it is in fact that the very subject-matter of psychoanalysis is not instinct, but the relation between desire and culture – this is the point

86. Cf. HARTMANN, *Psychoanalysis and Moral Values* (1960); *Ego Psychology and the Problem of Management* (1964); *Essays on Ego Psychology. Selected Problems in Psychoanalytic Theory* (1965); HARTMANN, KRIS & LOEWENSTEIN, *Papers on Psychoanalytic Psychology* (1965); LOEWENSTEIN, 'Rapport sur la psychologie psychanalytique de H. Hartmann, E. Kris et R. Loewenstein' (1966).

87. Cf. RAPAPORT, *The Structure of Psychoanalytic Theory. A Systematizing Attempt* (1967).

88. Under the title *Aus den Anfängen der Psychoanalysis* (1950) (English translation, *The Origins of Psychoanalysis*, 1954).

where all analysis comes to bear. It is for this reason that psychoanalysis cannot be confined to the terrain of instincts, dreams and neurosis; everything relating to the connective links between desire and culture falls within its purview.

This can easily be seen from the premisses of psycho-analysis themselves. Analysis is never concerned with the energy of desire in its biological roots. From its very first emergence, instinct is placed in a cultural situation – usually an antagonistic one. Censorship, in the theory of dreams, is a cultural factor which plays an inhibiting rôle in regard to the most deeply rooted desires; and therewith the interpretation of dreams evokes the same picture as that revealed by anthropology, in terms of the prohibition of incest. In *Three Essays on the Theory of Sexuality* (1905), the same antagonistic factor appears under the guise of the barriers which channel libido towards adult genitality. In the papers on *Metapsychology,* which were written about 1915, the three 'localities' (unconscious, pre-conscious and conscious) are considered to be in a dialectical relationship in which the untamed unconscious is always faced with the pre-conscious – the area of language – and the conscious – the area of access to the outer world (the world of things and the human world). The three 'areas' are the topological representation of that dialectic. What is true of the primary system (unconscious, pre-conscious, conscious) is even more true of the secondary system (ego, superego and id), namely that these represent rôles which cause the anonymous, the personal and the supra-personal to alternate in culturally determined situations. Thus in many ways psychoanalysis is constantly being confronted not with desires only but also with their alternates.

This being the case, how shall we define the psychoanalytical point of view on culture, in relation to the psychological and sociological points of view? It is not possible to parcel out the proper fields of competence of the various sciences, as if one were competent to deal with affectivity, another with sociality, etc. The social sciences are not distinguished according to their fields but according to their theories, i.e., their guiding hypotheses, methodology and practice. If this is true, the psychoanalytic explanation must be said to be limited not by its subject-matter, which is man in his entirety, but by its point of view, which is the construction of a theory to account for the effects of meaning that desires exert in a cultural situation of conflict. That is the way psychoanalysis has chosen, and that is why, in all its explanations bearing on art, morality or religion, it is no use looking for anything other than the direct or indirect ramifications of a semantics of desire in a cultural situation. All cultural phenomena fall within its purview in so far as they can be projected on to this clearly defined plane of the semantics of desire; at the same time they must be subjected to prior testing against the theoretical hypothesis underlying the psychoanalytical approach.[89]

89. Cf., in the present work, the essay on 'The psychoanalytic approach' (to art and literature) by Jean-François Lyotard, pp. 636-652, in Chapter IV, *Art and the Science of Art Today*, under the editorship of Mikel Dufrenne.

The limited character of this interpretation – limited but valid within the limits of the established rules – becomes even clearer if we consider that psychoanalysis approaches the vast field of the effects of meaning starting from an initial model and seeking counterparts to it in all branches of culture. This model, as we know, is constituted by the dream-neurosis correlatives. Freudian psychoanalysis furnishes not only a type of interpretation which I have tried to characterize in terms of the relation between meaning and force,[90] but also a very specific model of distortion, the *Verstellung* referred to in Chapter III of *The Interpretation of Dreams*. This transposition and distortion, which are characteristic of disguised wish fulfilment, provide a guiding line through the labyrinth of the effects of meaning which we call works of art, legends, folklore, myths, etc.

The level at which psychoanalytic interpretation can be confronted with other types of explanation is becoming clear. The psychoanalytic interpretation of cultural works does not seek to bring to light the instincts or even the buried conflicts of infancy, but to discover the very structures of distortion and the laws of transformation which govern the disguises of desire. In relation to these laws of transformation, the fact of instinct appears as a quasi-mythological reference: our 'earliest desires', our desires which are in some way 'immortal', act as the backmost limit in relation to interpretation. What is important is what is said not about desire but about the procedures that may appear as counterparts to displacement, condensation, secondary revision and acting out. It is at this level, the level where meaning is produced, that other explanations might be co-ordinated with psychoanalysis, but this would be on the basis of other working hypotheses susceptible of playing, in relation to cultural objects and their effects of meaning, the same organizing rôle as the semantics of desire. Thus in the field of literary criticism, as in that of the explanation of myths, the specific subject-matter of psychoanalysis is the study of the forms of distortion capable of being treated as counterparts of that which is operative in dreams and neuroses. To put it in another way, the essence of psychoanalytic interpretation consists in the relation between the semantics of desire and the syntax of distortion. It is because of this that Freud was able from the outset to apply to *Oedipus Rex* and *Hamlet* an explanatory schema which was not exclusively concerned with the world of neurosis. The structures of distortion have indeed their original model in dreams and neuroses, but the relatively formal character of these structures makes possible their analogical transposition within no limits, other than those set by the disguised expressions of desire.

Thanks to this dual limitation on psychoanalysis – namely that it is from beginning to end a semantics of desire and that it extends by analogy the initial model of the effects of meaning of dreams and neuroses – it is in a position to collaborate with other modes of explanation, based on dif-

90. Cf. Ricoeur, *De l'interprétation. Essai sur Freud* (1965) (English translation, *Freud and Philosophy. An Essay on Interpretation,* 1969), and the bibliography cited in the notes.

ferent theories and different initial models. It is not only the mutual limitation of explanations which is important, however, but their co-ordination; and it would seem that it is in regard to the syntax of distortion that psychoanalytic theory intersects and overlaps other types of explanation. What Freud called dream-work, the work of mourning and the work of neurosis bring into play transformation procedures which lend themselves to a comparison with other explanatory models much more readily than the quasi-mythological theory of instincts. At this level, psychoanalysis no longer encounters these other models in a fortuitous manner and, as it were, outside itself, but of itself both presupposes and entails them.

§ 4. *Epistemology of the historical sciences*

Epistemologists all agree that history has a place apart in a survey of the problems raised by the human sciences. But, as can be seen from the chapter on historical research which Professor Barraclough contributes to this *Study*, none of the human sciences has gone through such a radical rethinking of its own methodology in the last two decades. This methodological revolution – carried out by the historians themselves – is of interest to philosophers in their epistemological reflection in so far as it is directed against premisses which historians in the first half of the century borrowed from philosophy instead of taking from the practice of their own discipline. These historians, in point of fact, were influenced in their actual work by a debate which had not originated with them but in which their whole work was at stake, namely the debate between neo-Kantian idealism and positivism: this debate concerned history in that it gave rise to a schism within the historical discipline itself, as between its theory which was idealistic and its practice which was positivistic. From the idealism of writers like Dilthey and Simmel, followed by Troeltsch and Meinecke, in Germany, or Croce in Italy, the historian derived his conviction that history was not a science, if by science is meant a nomothetic discipline, that is, one which subordinates facts to laws, but an art, or at least an idiographic discipline dealing with unique events, outstanding individuals or remarkable minds. But it was from positivism that history took its sense of the true fact, its critique of prejudices and unilateral judgements. Practice alone provided a fragile compromise between the conflicting requirements, assigning to them two distinct phases, in one case the sifting of documents and in the other the interpretation of facts. An unstable equilibrium between comprehension and explanation seemed the best that historical knowledge could hope to achieve.

It was from that very practice that doubt was to arise, doubt which affected first the aim – the idea of historical truth progressively worked out through mutual rectification of errors – and then the means – the establishment of the facts. For if idealism is right, the impossibility of reaching an objective interpretation, independent of the observers, necessarily implies also the impossibility of attaining to the true facts. It was with the incipient crumbling of the concept of a historical fact that the word 'historicism' began to take on the derogatory connotation which it now commonly has, at least in the West. Was it reasonable to try to revive the past? Did it even make sense to endeavour to relive the past for its own sake and to reconstruct what had really happened? Was the goal not placed beyond the reach of historical research by the very presupposition of idealism, namely that historical knowledge is irremediably dependent upon the outlook of the present and the point of view of the historian? But the doubt went further: if it had been possible for the historian to recognize in idealism the philosophy of his own practice,[91] was it not because he himself had defined his discipline as a science of events, i.e., of what only happens once? And was this predilection for events not related to a predilection for individual action, principally by great men? And was not that in turn related to the historian's traditional predilection for political events, and consequently for the State, considered as the principal agent of historical change?

These doubts on the part of the historians had repercussions on philosophy itself and on the most basic premiss of historicism, namely the opposition between the natural world and the spiritual world, conceived by Hegel as the counterpart to their dialectical continuity, by virtue of a dialectical logic issuing in a totality of meaning.

Historicism was Hegel's problem without Hegel's solution. The problem inherited from Hegel was the question of the relation between two unlike spheres: nature and spirit. But once the Hegelian key of absolute knowledge and the logical nature of historical change itself had been lost, the two spheres were obliged to evolve without any definite relation between them. Hence the bitter opposition between explaining and understanding, between abstraction and event, between general and individual, between anonymous and personal, and all the dichotomies which keep history from being a science. It is understandable that since the historian as such had not really pondered on Dilthey's problem in depth as the German philosopher had done, he ultimately only perceived the adverse practical consequences of historicism: subjectivism and relativism, idolatry of the past as such, fascination with the problem of great individuals and virtually exclusive emphasis on political events, the whole concept of a science of history being eventually undermined by that very epistemological hypothesis on which it was supposed to be founded.

It is worthy of note that Soviet historiography in the same period was confronted with similar problems starting from initial positions which were quite different and even opposite. It was the 'schematism' and 'dogmatism' imposed on historical research by the over-simplified interpretations of

91. Cf. ARON, *Introduction à la philosophie de l'histoire. Essai sur les limites de l'objectivité historique* (1938) (English translation, *Introduction to the Philosophy of History. An Essay on the Limits of Historical Objectivity*, 1961) and *Essai sur la théorie de l'histoire dans l'Allemagne contemporaine. La philosophie critique de l'histoire* (1938, 1969); MARROU, *De la connaissance historique* (1954, 1962).

Marxism-Leninism in the Stalin era which here fostered the doubt and internal critique of the new generation of historians. Thus the initial positions of the East and West were diametrically opposed, but the ultimate doubt was similar. On the one hand one had subjectivism, on the other the dogmatism of official explanations; on the one hand the quest for the event, the individual act, the subjective experience, on the other the general search for the laws of social development, with emphasis on the rôle of the masses and reference to the economic and social infrastructures, and against that background, the attempt to view world history in terms of a breakdown into clearly defined periods. But the difficulties inherent in each initial position caused similar embarrassment to historians on both sides. The result was a remarkably parallel situation. For it was also in practice that the Marxist historians discovered the difficulties of principle which – for reasons that were the opposite of those that arose in the case of historicism – impeded strictly scientific work. The problem of 'periodization' has often been the touchstone for a methodology which was governed by directives rather than criteria.

Thus it is that the years from the mid-1950's on may be characterized broadly speaking by the critical re-examination of the theses of historicism in the West, and dogmatism and schematism in the East.

Since Professor Barraclough has endeavoured to show the contribution of historians themselves to 'The search for new concepts and a new methodology',[92] we shall dwell at greater length here on the epistemological thinking of philosophers. It will be taken for granted therefore that the reader is fully informed about 'the impact of the social sciences' on history,[93] including the taking over from sociology of quantitative methods and methodological concepts such as structures and models, as well as of the concepts relating to 'group analysis', the transference from sociology to history of problems formerly concealed by political history (mass behaviour, acculturation, urbanization, the rôle of élite groups, etc.) and, above all, the extension to history of a renewed belief in the possibility of applying scientific method to the human sciences.

On the other hand, the subjects retained for philosophical discussion will include the very idea of shifting the accent from the particular to the general, from events to uniformities, from narration to analysis. Were it possible to carry this idea to its logical conclusion, the uniqueness of historical facts would cease to be an epistemological obstacle; but would there still be any *historical* facts left? The individual would cease to be ineffable; but can structures be defined apart from their function of acting as a supporting framework for events and individuals? The question of determinism would cease to be a labyrinth, since history would appear less as a series of events linked by causality than as a system of correlations between structures; but can interest in structures replace the narrative aspect? The

92. Cf., in the present work, the section so entitled (Section II), pp. 256-273, in Chapter III on *History*.

93. Ibid., Section III, 'The impact of the social sciences', pp. 273-322.

rôle of wars, treaties and divisions would vanish before the long-term effect of social forces; but is it really possible to reduce the drama played out on the political stage to a compound equation made up of reproducible components? Quantification may be substituted for impressionism and for the platitudes of ordinary psychological and moral judgement; but does quantitative analysis still make any sense if it is divorced from qualitative concepts such as Max Weber's ideal types – which are, in the last analysis, configurations of social change, that is, rules for constructing historical sequences (e.g., bureaucracy, towns, etc.) in type fashion? In a word, can social reality and basic structures be made to constitute the substance of history without taking into account the problem of time, and thus the action of individuals on the durable conditions of their own history? Or, to put the question in other terms, is it possible to take to a logical conclusion and actually carry out the project of substituting systematic analysis for narrative description?[94] One small word presents a difficulty in the definition of history which Professor Barraclough borrows from Frédéric Mauro: 'history is the projection of the social sciences into the past'.[95] What is the meaning of *past*?

These questions, arising from historical methodology itself, have their counterpart in the epistemology of philosophers reflecting on history.

Philosophers generally reply by applying to the peculiar situation of history a general conception of intelligibility which they attempt to make specific. Epistemological reflection on history thus finds itself at the point where two movements meet, one of which starts from history itself, as we have just shown, and the other from the theory of science. It is this second line that we shall now follow.

(a) The **English-speaking philosophers** have approached the problem of history with the same mistrust of speculation as they display towards the philosophy of nature.* But it has been harder for history to disentangle itself from the philosophy of history than for physics to free itself from the philosophy of nature, so deeply is the desire for totalization rooted in the approach of philosophers to history.

In so far as analytic philosophers have discussed speculative theories of history, it has been chiefly in a critical fashion, with a view to exhibiting the logical and conceptual difficulties believed to be inherent in the doctrines propounded by Hegel, Marx, Toynbee and Spengler, though some writers have also stressed the positive rôle of such theories as a source of procedural suggestions which have influenced the development of scientific history. Among the critics of speculative theories are such names as K. R.

94. *Ibid.*, Section III, proposition No. 18, p. 289, under the title 'The contribution of sociology and anthropology'.

95. *Ibid.*, proposition No. 21, p. 291.

* Based on a report by Patrick Gardiner, whose text we have followed very closely.

Popper, I. Berlin, W. H. Dray, A. Danto, W. H. Walsh and P. L. Gardiner.[96] For a comparison between British or American and 'Continental' approaches to historical theorizing of the speculative type, see the article by Charles Taylor, 'Marxism and empiricism' (1966).

Rather than exhaust itself in rectifying and correcting speculative theories, the analytic philosophy of history has set out to explore the logical structures and epistemological presuppositions of the thought and practice of working historians. Its aims have been descriptive rather than revisionary; in some cases, however, the analyses offered have been implicitly critical of accepted concepts and criteria.

A key work, which has formed the starting-pont of much recent discussion, is R. G. Collingwood's *The Idea of History* (1946). Collingwood was not an analytic philosopher in the contemporary sense; his general philosophical outlook derived largely from British idealism, and in his approach to history he owed much to Continental writers like Dilthey and Croce. Yet by stressing the autonomy of history as a branch of study, with its own procedures and methods of inquiry, he did much to make British philosophers aware of problems relating to the human sciences which they had previously overlooked or ignored.

A second important source of current interest in the philosophical analysis of history was C. G. Hempel's seminal article, 'The function of general laws in history', first published in 1942. Approaching the subject of history from a philosophical position diametrically opposed to Collingwood's – namely, logical empiricism – Hempel maintained that no differences in principle separated historical modes of explanation from those typical of the natural sciences and severely criticised *Verstehen* or 'empathy' models of historical understanding as involving unwarrantable mystification. Following Hume, Hempel claimed that the historian's use of a causal vocabulary – like its use in the context of other empirical inquiries – carried an implicit reference to covering laws, such laws being in princple subject to experiential test. This claim ran directly counter to Collingwood's assertion that the word 'cause' was used in history in a special sense; to refer historical events to their causes was, for Collingwood, essentially a matter of presenting them as the expression of rational processes of thought and calculation, and not of subsuming them under empirical regularities.

The question of historical explanation has thus become the focal topic of the analytical philosophy of history. The Hempelian thesis that the explanations provided by historians presuppose general statements which can be elucidated by logical analysis has been defended (though often with drastic qualifications and reservations) by, amongst others, Morton White, Robert

96. Popper, *The Poverty of Historicism* (1944-45, 1957); Berlin, *Historical Inevitability* (1954) and 'History and theory. The concept of scientific history' (1960); Dray, *Philosophy of History* (1964), Chaps. 5-8; Danto, *Analytical Philosophy of History* (1965), Chap. 1; Walsh, *Introduction to Philosophy of History* (1951); Gardiner (ed.), *Theories of History* (1959).

Stover, M. Scriven, M. Mandelbaum and P. L. Gardiner.[97] The most severe attack upon the 'covering law' theory is to be found in W. H. Dray's *Laws and Explanation in History* (1957); this book not only offers a series of arguments designed to demonstrate the unacceptability of a Hempelian-type analysis, but also puts forward an alternative model of 'rational explanation' which, the author suggests, fits some (though not all) of the standard ways in which historians seek to render the human past intelligible. Acknowledging a debt to Collingwood, Dray argues that the rôle of much historical explanation is one of showing the actions of historical figures to have been appropriate in the light of their particular beliefs, aims and principles; such a consideration was, he claims, uppermost in the minds of those theorists who have emphasized the part played by imaginative or empathetic understanding in historical thought, but it has been largely overlooked or misinterpreted by their positivistic opponents. Dray's position on this score has found both supporters and critics; representative arguments on both sides are contained in the collection edited by Sydney Hook, *Philosophy and History* (1963). Other problems raised by historical explanation are dealt with in articles by I. Berlin, 'The concept of scientific history' (1960); P. H. Nowell-Smith, 'Are historical events unique?' (1956–1957); and A. Donagan, 'Historical explanation. The Popper-Hempel model reconsidered' (1964). We shall not revert here to P. Winch's critique of the positivist conception of explanation in the social sciences,[98] which was discussed in § 1 above.

The debate with regard to explanation in history has brought back into the foreground a question to which we have already alluded above, namely the notion of importance in history. Analytic philosophers approach this from the point of view of the criteria used in selecting causes, or, to avoid prejudging the answer, the criteria according to which historians select from amongst the relevant conditions the one considered responsible for the occurrence of an event. Collingwood himself, in attempting to answer this question in a general treatment of the problem of causality, argued that we tend to identify as cause that condition which we are able 'to produce or prevent'.[99] More recently, in the context of a discussion on the concept of legal responsibility, it has been suggested that the principal criteria for selecting causal conditions involve an appeal to: (a) 'abnormality' (what is in some way 'unexpected' or deviates from the 'ordinary course of events'); (b) the notion of voluntary human activity.[100] Both these interpretations have proved influential in discussions dealing with historical causation. The part played by practical interest in the selection of causes, along with other contextual factors, is considered by Gardiner in *The Nature of Histori-*

97. White, M., *Foundations of Historical Knowledge* (1965); Stover, *The Nature of Historical Thinking* (1967); Scriven, 'Truisms as the grounds for historical explanation' (1959); Mandelbaum, 'Historical explanation. The problem of covering laws' (1961); Gardiner, *The Nature of Historical Explanation* (1952).

98. Winch, *The Idea of a Social Science* (1958).

99. Cf. Collingwood, *Essay on Metaphysics* (1940).

100. Cf. Hart & Honoré, *Causation in the Law* (1959)

cal Explanation;[101] while, in his *Foundations of Historical Knowledge*, Morton White lays stress upon the criterion of abnormality, with the important reservation that what is regarded by a given historian as abnormal within a certain context is subject to considerations deriving from the point of view or interest of the historian in question and that his judgement may not therefore be accepted by other historians whose points of view are different. The suggestion that an element of 'subjectivity' may thus be inherent in many causal statements, rendering some disputes concerning the 'real causes of events' undecidable on purely empirical grounds, has been accepted by a number of contemporary writers: Dray, for example, has persuasively argued that moral considerations play an ineliminable rôle in various well-known historical controversies (e.g., about the origins of the American Civil War), the historian's identification and weighting of causal factors frequently depending essentially upon a moral standpoint which may not be shared by some of his professional colleagues.[102] Other philosophers, however, have tried to show that, in certain cases at least, historical judgements concerning the relative importance of causes may be interpretable in a fashion that renders them straightforwardly open to objective empirical test. A careful analysis along these lines is undertaken by Ernest Nagel in *The Structure of Science* (1961).[103] Despite the work done in this field, many substantial problems still await adequate treatment; in particular, concepts like those of chance and accident have received only cursory attention. A thoroughgoing investigation of such topics would be useful, both as throwing light upon puzzles experienced by practising historians themselves when they reflect upon the nature of their discipline, and also as being relevant to the assessment of avowedly 'monistic' theories of historical development like the Marxian.

These issues bring us back to the problem of objectivity and the rôle of evaluation discussed above in the general context of the social sciences. All the difficulties referred to there take on an acute character where history is involved. A clear statement of the problem is to be found in Walsh, *Introduction to Philosophy of History* (1951), Chapter 5, entitled 'Can history be objective?' It has been argued[104] that general questions concerning the objectivity of history as such cannot sensibly be raised; answerable questions concerning objectivity only arise in contexts where some specific piece of historical work is in dispute, and here there are criteria recognized within the field of historical criticism which can be specified and applied. This short way with the problem has, however, not met with general acceptance. It has been pointed out[105] that perplexity concerning historical objectivity becomes most acute for many people (historians among them) when history is compared with other forms of inquiry, e.g., physics:

101. See pp. 99 sqq.
102. Cf. DRAY, *Philosophy of History* (1964), pp. 47 sqq.
103. See pp. 582 sqq.
104. By BLAKE, for example, in his article 'Theories of history' (1955).
105. By PASSMORE, for example, in his article 'The objectivity of history' (1958).

when set beside such inquiries the historian's procedure, the way in which he presents his material and the forms of argument he adopts, may seem to be guided by personal or culturally determined predilections. One central topic which has engaged the attention of analytic philosophers in this connection has been the part played by evaluation, and specifically moral judgement, in historical description and narration. To some it has appeared evident, from a consideration of the historian's subject-matter – human beings acting with particular motives and purposes – and also of the language he customarily uses, that history cannot avoid being to some degree 'value-charged': a thesis of this type has been propounded by I. Berlin, in his book *Historical Inevitability* (1954). What was said above about judgements of importance and the part they play in the selection of material points in the same direction; for the attribution of importance may itself depend upon presupposed values, which further enhances the rôle of the evaluative component in historical work.[106]

Other discussions of historical objectivity and the issue of relativism are to be found in the works of A. Danto, M. White, J. W. Meiland and E. Nagel.[107]

The question of causality necessarily brought the analytic philosophers face to face with the more general problem of determinism and responsibility, which every philosophy of history encounters sooner or later. In general, analytic philosophers have been highly critical of 'inevitabilistic' theories which see the course of history as governed by inexorable forces or identifiable 'laws of historical development'. To deny validity to such theories is none the less not the same as denying determinism as such; their rejection is perfectly compatible with the view that, for any given historical event, a complete causal explanation in terms of sufficient conditions could in principle be given.[108] What is less clear is whether historical determinism in this latter sense is consistent with the actual practice and terminology of history as these are ordinarily understood. According to I. Berlin,[109] it is not: there is an intimate bond between the concepts intrinsic to the historian's interpretation of human thought and action and the 'libertarian' assumptions that underlie everyday conceptions of moral responsibility; consequently, subscription to a thorough-going determinism would involve sweeping and perhaps unimaginable changes in the accepted categories of historical description. Berlin's view rests upon an explicit rejection of all attempts to analyse the concepts of freedom and moral responsibility in such a way as to make them compatible with a deterministic interpretation of human behaviour. This, together with other features of his argument, has been challenged by a number of writers, notably by E. Nagel, M. White,

106. See, for example, WALSH, 'The limits of scientific history' (1961, 1966).

107. DANTO, *Analytical Philosophy of History* (1965); WHITE, M., *Foundations of Historical Knowledge* (1965); MEILAND, *Scepticism and Historical Knowledge* (1965); NAGEL, E., *The Structure of Science* (1961), Chap. 13, and 'The logic of historical analysis' (1952, 1959).

108. See, for example, NAGEL, E., *The Structure of Science* (1961), pp. 592 sqq.

109. In *Historical Inevitability* (1954).

R. Stover and E. H. Carr.[110] A rejoinder to some of his critics is to be found in Berlin, *Four Essays on Liberty* (1969).[111]

These discussions – on causality, objectivity and determinism – all come back to the same point, namely the juxtaposition of history with the other human sciences and with the sciences of nature. It is this very rapprochement which is called in question in works dealing with the notion of narrative in history. W. B. Gallie claims that 'narrative is the form which expresses what is basic to and characteristic of historical understanding';[112] he considers that if the logical and pragmatic aspects of story-telling are taken seriously, many of the models in terms of which analytic philosophers have been prone to interpret historical explanation, particularly those inspired by scientific procedures, appear inappropriate and misconceived. Danto, in his *Analytical Philosophy of History* (1965), holds that a proper appreciation of the rôle played by narrative throws a new light on the standard difficulties concerning objectivity and the place of prediction in history.[113]

(b) **Soviet philosophy** shows a growing interest in the philosophical problems arising from the theory of historical knowledge and in the logical structure of historical methodology.* A large number of studies have been devoted to these problems.[114] These various works demonstrate how historians are collaborating with philosophers and logicians.

Discussion about the theory of historical knowledge is characterized in opposition to two extremes. On the one hand, Marxist philosophy has never accepted the neo-Kantian distinction between idiographic and nomothetic

110. NAGEL, *The Structure of Science* (1961), Chap. 13; WHITE, *Foundations of Historical Knowledge* (1965), Chap. 7; STOVER, *The Nature of Historical Thinking* (1967), Pt. II; CARR, *What is History?* (1962).

111. See particularly the Introduction to this work.

112. GALLIE, *Philosophy and the Historical Understanding* (1964).

113. In addition to the titles already mentioned, a general discussion will be found in the article by MINK, 'Philosophical analysis and historical understanding' (1968).

* Based on contributions by I. S. Kon, especially his essay, 'Istorija i obščestvennye nauki (Novejšie tendencii v sovetskoj istoriografii)' (= History and the social sciences (Recent trends in Soviet historiography)) (unpubl.).

114. Cf. in particular: KON, *Filosofskij idealizm i krizis buržuaznoj istoričeskoj mysli* (= Philosophical Idealism and the Crisis in Bourgeois Historical Thinking) (1959) (rev. and enl. version in German, *Die Geschichtsphilosophie des 20. Jahrhunderts. Kritischer Abriss*, 1964, 1966); GRUŠIN, *Očerki logiki istoričeskogo issledovanija: process razvitija i problemy ego naučnogo vosproizvedenija* (= Essays on the Logic of Historical Research: the Process of Development and Problems in Its Scientific Reconstruction) (1961); and the collections *Istorija i sociologija* (= History and Sociology) (1964); *Metodologičeskie i istoriografičeskie voprosy istoričeskoj nauki* (= Methodological and Historiographical Questions in Historical Science) (1963-66); *Voprosy metodologii istoričeskoj nauki* (= Questions of Methodology in Historical Science) (1967); *Filosofskie problemy istoričeskoj nauki* (= Philosophical Problems in Historical Science) (1969); *Istočnikovedenie. Teoretičeskie i metodologičeskie problemy* (= The Study of Sources. Theoretical and Methodological Problems) (1969).

disciplines which justifies keeping history at the descriptive level where it is concerned with events; to regard history as opposed to the social sciences, and the human sciences in general as opposed to the sciences of nature, is incompatible with the materialist conception of history.[115]

On the other hand, Marxism has always taken position against positivist naturalism and abstract sociologism which do not give sufficient recognition to the epistemological and methodological specificity of history as a science. This second aspect of the question is often misunderstood in the West where Marxism is interpreted in a grossly naturalistic sense, particularly by the neo-Kantian critics and others at the beginning of the twentieth century. Below, in sub-section B, 'Social and Politicial Philosophy', we shall examine the general conception of social phenomena in Marxist theory, particularly the place accorded to superstructures and subjectivity, and the general discussion of the concept of reflection;[116] we may simply recall here the general theme according to which history appears, on the one hand, as a historico-natural process subject to laws and, on the other, as a universal historical drama in which human beings are both actors and authors. This conception of history as a human activity justifies the importance attributed to the subjective factor, i.e., the conscious aims and motivations of historical figures, etc. By the same token, the simplified models of physicalist type employed in the natural sciences are obviously inadequate here. In this paragraph we shall consider the strictly methodological aspects of the discussion. Several works have been written in the U.S.S.R. on the methodological discussion which has developed in the West – particularly in the review *History and Theory* – concerning the Popper-Hempel thesis referred to above. The need to find specific criteria for the objectivity of historical knowledge has led to a critical examination of the notion of historical fact and other concepts relating to historical cognition, such as the concept of historical time.[117]

For an observer comparing from without the dominant problems in the tradition of British and American analytic philosophy with those most readily dealt with by Soviet scholars, it seems clear that among the former stress is laid on problems arising from juxtaposition with the sciences of nature and from the philosophical tradition of logical empiricism, whereas among Soviet scholars the epistemology of historical science is dominated by problems involving total comprehension of the historico-social phenomenon. This feature is especially prominent in the earliest work devoted to the epistemological-methodological specificity of historical science, namely the work by B. A. Grušin, *Essays on the Logic of Historical Investigation*.[118] Basing himself on a systematic analysis of the works of Marx and on experience in existing disciplines – only to a limited extent in historiography

115. On this point, see the work of Kon cited in the previous footnote.
116. See below, sub-section B, § 2, 'Historical materialism', item (a) pp. 1296-1297.
117. See Gurevič, 'Vremja kak problema istorii kul'tury' (= Time as a problem in the history of culture) (1969).
118. *Očerki logiki ...*, *op. cit.* (1961).

as such – B. A. Grušin lays stress, in the scientific reconstitution of any historical phenomenon, on 'discovery of the *laws* governing the *process* of its *development* as a *system*'. This thesis has several implications. In the first place, the research worker, faced with a phenomenon characterized by a complex internal structure, has the task of reconstituting it not in its isolated components but in the integrality of their structural interdependence, that is, as an organic, structured and functional whole. But this is only one side of the matter, for what must be reconstituted is a process. The investigator is faced with factors which have succeeded one another in time and he must therefore reconstitute all the historical links between those components. Thirdly, the process to be reconstituted is that of a phenomenon's development; what must be discovered therefore is not the sundry changes observable in time, but the qualitative changes which affect the structure of the system as a whole. Lastly, it still remains to attain to the notion of a law governing the process. The historian must discover and reconstitute, not only a series of qualitatively differing historical situations in the life of a phenomenon, but the laws governing the transition from one given historical situation to another.[119]

As is clear, this approach is definitely directed towards the elaboration of 'theoretical history'. Later works by A. I. Rakitov, V. I. Stoljarov and E. P. Nikitin exhibit the same philosophico-epistemological character. It must be admitted however that the professional historians, who have a poor understanding of epistemological terminology as applied to the concrete problems of their science, do not take much account of these discussions on the epistemological nature of history.

Historians themselves are divided into two major groups when they theorize about historical knowledge. Some, like Kedrov,[120] put particular emphasis on the elements common to history and the other social sciences, and either question whether there is any such thing as an epistemological specificity of historical science or else consider such specificity as relatively unimportant. Others, like A. V. Gulyga,[121] A. Ja. Gurevič and I. S. Kon, deliberately avoid intuitivist and ideographic conceptions of history, though they recognize that 'historical generalization constitutes a unique synthesis of theoretical and aesthetic modes of knowledge about the world'.[122] Thus Gulyga studies in detail the relations that exist in historical research between statistical laws and various typological procedures and models of an aesthetic nature. History as a process and history as drama admit of different modes of representation: 'History has two facets, the sociological and the human. One is turned towards mass processes, and the other towards personality. These are two connected and interdependent worlds, but they do not coincide, and the task of the historian is to produce a synthesis by

119. Grušin, *Očerki logiki . . ., op. cit.*, p. 18.

120. B. M. Kedrov *et al.*, in the collection *Istorija i sociologija* (1964).

121. Gulyga, 'O haraktere istoričeskogo znanija' (= On the nature of historical knowledge) (1962).

122. Gulyga, 'O haraktere . . .', p. 37 in *Voprosy filosofii*, 1962, No. 9.

penetrating into and recreating these two worlds, the sociological and the human, in their organic unity.'[123] The same argument leads the scholar to look for another kind of typology, for the study of historical events and the history of culture, than that employed in the history of economic processes, where statistical methods are fully adequate.[124]

The problem of historical knowledge may be called meta-historical and be of interest only to scholars working on the most general problems of historiography. Professional Soviet historians are taking considerable interest in the practical problems relating to the interaction between history and the social sciences. Whereas for a long while research was concentrated on the problems of delimiting scientific areas, more recent work has laid greater stress on co-operation and mutual enrichment between different branches of study. In its most elementary form, co-operation of this type implies the mastery by historians of research techniques long familiar to sociologists and specialists in the other social sciences. This has been the case with structuralist methods: the possibilities and limits of structural analysis in history are today the subject of heated discussions.[125] Various authors have a tendency to exaggerate the importance of structuralist methods without always taking into account the difficulties inherent in the combination of the structural and functional approaches to history. Others, on the contrary, question from the outset the applicability and effectiveness of the structuralist approach to history. As Ju. A. Levada observes,[126] the opposition between history and structure and the inclusion of structures in processes involve the whole problem of a structuralist conception of historical processes as such. To solve the methodological difficulties bound up with the historicism and evolutionism of the nineteenth century, it will be enough simply to choose such quantitative indications and isolated statistics as fit one's purpose and add them to the traditional historical descriptions; historical research must have at its disposal scientifically based

123. Gulyga, 'Istorija kak nauka' (= History as science) p. 35 in the collection *Filosofskie problemy istoričeskoj nauki, op. cit.* (1968).

124. Kon, 'K sporam o logike istoričeskogo ob"jasnenija: shema Poppera-Gempelja i eë kritiki' (= Contribution to the discussion on the logic of historical explanation: the Popper-Hempel thesis and its critics), pp. 293-295 in the collection *Filosofskie problemy istoričeskoj nauki, op. cit.* (1969); by the same author, 'Istorija', pp. 369-376 in *Filosofskaja ènciklopedija*, Vol. 2 (1962) and 'Istorija i sociologija' (1970) (French translation, 'Histoire et sociologie', 1971).

125. Cf. Barg, 'O nekotoryh predposylkah formalizacii istoričeskogo issledovanija' (= Some prerequisites for a formalization of historical research) (1967); Štaerman, 'K probleme strukturnogo analiza v istorii' (= On the problem of structural analysis in history) (1968); Danilov, 'K voprosu o metodologii istoričeskoj nauki' (= On the question of methodology in historical science) (1969); Barg, 'Učenie ob obščestvenno-èkonomičeskih formacijah i konkretnyj analiz istoričeskogo processa' (= Study of socio-economic formations and concrete analysis of the historical process), in the collection *Očerki metodologii poznanija social'nyh javlenij* (= Essays on a Methodology for the Understanding of Social Phenomena) (1970).

126. 'Istoričeskoe soznanie i naučnyj metod' (= Historical consciousness and the scientific method), pp. 220-221 in the collection *Filosofskie problemy istoričeskoj nauki, op. cit.* (1969).

concepts regarding the structure of the phenomena being studied. This is precisely what is supplied by the Marxist theory of the socio-economic formation as an integral social entity.

One branch of study to which history has drawn notably closer in recent years is psychology. For many years Soviet scholars tended to neglect this aspect, partly because social psychology was not sufficiently developed and the historian found nothing that he could borrow from general psychology. The situation has changed considerably in recent years. B. F. Poršnev's *Social Psychology and History*,[127] on the general principles of the relation between these two disciplines, was published in 1966, and a collection entitled *History and Psychology*[128] was published in 1970. This was the first time that historians and psychologists co-operated to study problems of common interest (for example, the theory of the national character). The founders of Marxism adopted a rather critical attitude towards psychological explanations of history; not because they questioned the importance of motivations, but because the explanations in question were usually arbitrary and subjective, founded on conceptions that were generally accepted at the time; furthermore, it is absolutely impossible to explain mass social processes by the motives of isolated historical figures, as is so often done in traditional historiography. Soviet historians also adopt a sharply critical attitude towards the conceptual schemas of psycho-analysis which are essentially extra-historical.[129] In order to enable the historian to make real use of the concepts and methods of psychology, it is first necessary to constitute a social psychology, and secondly to 'historicize' the actual thinking of psychologists. This is the task undertaken by the authors of the already mentioned *History and Psychology*, published jointly by the Institute of World History and the Institute of Philosophy of the U.S.S.R. Academy of Sciences. This new conception of historical psychology, heralded on the side of psychology by the works of several Soviet writers – e.g., L. S. Vygotskij, A. R. Lurija and A. N. Leont'ev – has also been influenced by French scholarship represented by such writers as H. Wallon, I. Meyerson and J.-P. Vernant; and Soviet specialists have high regard for the work done in historical psychology by the '*Annales*' school, particularly Robert Mandrou and Jacques Le Goff. The ideas set forth by Poršnev in *Social Psychology and History*, concerning the formation of the concepts 'we' and 'others', the specific nature of group consciousness, the relations between processes of unification and particularization and the whole range of concepts and terms of a socio-psychological character, are playing a considerable part in current historical research. This research has gone beyond the stage of discussions about principles and has produced remarkable results at the level of concrete research. Of particular interest in this

127. Poršnev, *Social'naja psihologija i istorija* (1966).

128. Poršnev & Anciferova (eds.), *Istorija i psyhologija* (1970).

129. Cf. Tokarev, 'Načalo frejdistskogo napravlenija v ètnografii i istorii religii' (= The beginnings of the Freudian orientation in ethnography and the history of religions), in the collection *Istorija i psihologija*, cited in the previous footnote.

regard are the works of the mediaeval scholar A. Ja. Gurevič[130]: using the most recent semiotic and other methods to study the widest variety of source materials, Gurevič recreates with talent the socio-psychological structures of man in the Middle Ages, his ways of thinking and feeling, his conception of time, etc. His work raises a series of methodological questions regarding the deciphering of the symbolic language of a given historical period, stereotypes of social behaviour, their genesis and their social function as part of a larger socio-historical whole.

The same desire to grasp the historical phenomenon as a structured whole pervades the work done on the – for Soviet science – classic problem of the periodization of history, with its corollary notion of historical epochs. For Marxist science, the basic structural category, for the macro-analysis of social systems, is the socio-economic formation, whereas the problem of periodization is that raised by the multiple forms the historical process takes. That process involves the development of internal structures and geographical regions as part of the formation. In this regard, great importance is attached to the characteristics of the historical development of the Eastern countries and the so-called 'Asiatic mode of production'.[131] The same central problem of the repetition and originality of historical phenomena is the subject of spirited discussions among historians of culture, especially with regard to the concept of Renaissance. M. A. Konrad in a work which had considerable success, entitled *The West and the Orient*,[132] interprets Renaissance as a concept of universal historical application and claims that it has equivalents and analogies in the cultures of the East. Other writers disagree with this.[133] The discussions which have followed from this have also dealt with the general principles of the methodology of the history of culture, the relations between system analysis and genetic analysis in this branch of history, the typology of cultures and means of recognizing their individuality.[134]

130. See the following articles by GUREVIČ: 'Nekotorye aspekty izučenija social'noj istorii: obščestvenno-istoričeskaja psihologija' (= Some aspects of the study of social history: socio-historical psychology) (1964); 'Social'naja psihologija i istorija. Istočnikovedčeskij aspekt' (= Social psychology and history: the 'study of source materials' aspect), in the above-mentioned collection, *Istočnikovedenie ...* (1969); 'Vremja kak problema istorii kul'tury' (= Time as a problem in the history of culture), *op. cit.* (1969); and 'Predstavlenie o vremeni v srednevekovoj Evrope (= The conception of time in mediaeval Europe), this last in the collection already referred to, *Istorija i psihologija* (1970).

131. Cf. the collections *Obščee i osobennoe v istoričeskom razvitii stran Vostoka. Materialy diskussii ob obščestvennyh formacijah na Vostoke (Aziatskij sposob proizvodstva)* (= General and particular aspects of the historical development of the Eastern countries. Material for discussion on social formations in the East (the 'Asiatic mode of production')) (1966); DANILOVA (ed.), *Problemy istorii dokapitalističeskih obščestv* (= Problems in the history of precapitalist societies), Vol. 1 (1968).

132. KONRAD, *Zapad i Vostok* (1966).

133. ÈJDLIN, 'Idei i fakty. Neskol'ko voprosy po povodu idei kitajskogo Vozroždenija' (= Ideas and facts. Some questions with regard to the idea of a Chinese Renaissance) (1970).

134. BATKIN, L. M., 'Tip kul'tury kak istoričeskaja celostnost'. Metodologičeskie

Such then are the general problems of methodology and epistemology arising from the influence of Marxism-Leninism on historical reflection.

(c) **French-language writers** on the epistemology of the historical sciences take their cue specifically from the methodological thinking of those historians who, in the wake of Marc Bloch and Lucien Febvre, centre round the French review *Annales*, founded in 1929.[135] As is commonly known, in place of event-oriented history this school substituted the integral history of societies and men of the past, embracing geography, economics, sociology, psychology and semiology. Instead of trying to rediscover the past as it was, it preferred to reconstitute scientifically the historical 'object', in the sense in which the 'object' of a science is always determined, as regards its status, by the theory under consideration. In place of the over-simplified notion of the unique event, it substituted a more complex view of several concurrent time-streams or processes: the 'long process' (*longue durée*) of history, in regard to major geographical regions, the 'intermediate process' (*moyenne durée*) of the history of structures, with its recurrent periods and cycles, and the 'short process' (*courte durée*) of 'conjunctures', akin to individual time but dealt with by quantification.[136] It is therefore not by accident that Marc Bloch, in his *Apologie pour l'histoire, ou Métier d'historien*,[137] uses the language of the other sciences, and especially the sciences of nature, and calls his first chapter *Observation*, by which he means that the reconstruction of facts on the basis of documents is an observational effort quite comparable to that deployed in the sciences of nature. It is noteworthy, moreover, that his second chapter is entitled *Analyse* (Analysis), and not synthesis; the establishment of retrospective sequences is a task of analysis in the double sense of breaking down and getting back to principles. The critical understanding of history becomes a matter of the systematic analysis of historical knowledge.

All the questions which we raised above are here thrown into acute relief. In the first place it is certain that the very notion of a science of men in society singularly weakens the antithesis between nomothetic and idiographic disciplines, which was itself, in German idealism, the corollary of the opposition between the world of nature and the world of the

zametki v svjazi s ital'janskim Vozroždeniem' (= The cultural type as a historical entity. Methodological notes on the Italian Renaissance) (1969).

135. Cf. GLÉNISSON, 'L'historiographie française contemporaine: tendances et réalisations', in the collective work *La recherche historique en France de 1940 à 1965* (1965).

136. The differentiation between the various kinds of historical time, and the relationships between them, were, as is well known, to be fully developed by Fernand BRAUDEL (who attributed greater intelligibility to *longue durée*). Cf. his theoretical studies since 1949 gathered together in the collection *Ecrits sur l'histoire* (1969), particularly pp. 9-122.

137. A posthumous work published in 1949 (English translation, *Historian's Craft*, 1954).

mind. But does it do away with the antithesis entirely? As soon as one clarifies the new definition of history as a science of men by adding that it is a science of men in time,[138] it becomes evident that the problem of the autonomy of history is shifted but not eliminated. Naturally the notion of time is not identified merely with that of event: the 'virtually motionless' time of *longue durée* makes the historical subject, as observed and analysed, akin to the physical object, which is also not outside time. Yet history is not sociology: the relation between the geographical time of long-range history, the social time of structures and the individual time of 'conjunctures' indeed seems to be specific to history. No matter how close a relationship is established between history and the social sciences, therefore, the two cannot be confused.[139] We see clearly how the social sciences can broaden the historian's horizon, provide him with tools for analysis and at the same time deliver him from common-sense psychology and rhetoric, but it is not so clear how they can eliminate the narrative character of history.

That the objectivity claimed for historical knowledge is not simply a particular case of the objectivity claimed for the social sciences in general is amply clear from the debates on the notion of importance in history; this notion was the general topic of a series of study meetings organized in Brussels by the 'Centre National de Recherches de Logique' from 1960 to 1963.[140] As Jacques Havet points out, 'If every historical work stems from a choice and remains linked to an outlook, on what bases does the historian select the elements of becoming which he considers 'important' and on which he proposes to build his intelligible reconstruction of the past? And which of these elements is he entitled to regard as determining, i.e., to set up as legitimate factors in explanation? And to what extent does he have a strict method at his disposal for these purposes, and how many historians do, in point of fact, explicitly define their criteria for selecting and relating historical facts?'[141]

These questions lead us back to the crux of the debate: is it possible totally to discard the traditional idea that the subject-matter of history is necessarily marked by irreducible singularity and inexhaustible concreteness, and that historical knowledge furthermore is irremediably separated from its subject-matter by differences in time, outlook and interests? This traditional conception of the historical 'object' no doubt once served as a basis and justification for a method of understanding the past the purpose of which was seen as sympathetic reconstitution, without the benefit of any rules, in which the concerns and understanding of the present time were necessarily projected. In this former approach, the effort to achieve his-

138. Bloch, M., *Apologie pour l'histoire, op. cit.* (1949), p. 4.

139. This is also the conclusion reached, after a thorough examination of most of the epistemological problems of contemporary historical science, by the Marxist philosopher Adam Schaff, in his work *Histoire et vérité* (1971).

140. Cf. Perelman (ed.), *Raisonnement et démarches de l'historien* (1963).

141. Havet, critical review of the last-mentioned work, p. 348 in *Revue de Métaphysique et de Morale*, No. 3, 1969.

torical objectivity could only be successfully applied to the criticism of documents and the establishment of historical facts, before they were arranged in intelligible sequences; but the assimilation of history to the human sciences should not make us forget that what was a weakness in the eyes of scientific method borrowed from the natural sciences became a strength in so far as history remained the privileged terrain of human awareness and its expressions throughout the ages. By the same token, history, promoted to the rank of source and focus of all the human sciences, became *par excellence* the seat of knowledge about man. And maybe this represents an ultimate, abiding characteristic of history, even after it has been enriched by the social sciences, with their quantification techniques and scientific concepts.[142] It would seem that the very attempt to bring historical knowledge into line with the other forms of scientific knowledge unfailingly brings us back to the essential characteristics of historical objectivity, the very ones, indeed, of which the problem as to judgements of importance keeps reminding us; it is at the very heart of historical objectivity that the 'hermeneutic circle' arises between the interpreter and the interpreted, though this quite specific bond between the historian and his subject-matter or 'object' does not justify the intrusion of a non-scientific subjectivity. Historical objectivity, by very reason of the explicatory effort called for, always consists in rising above hatred and anger. The struggle against emotive self-projection, against resentment, against hatred and the spirit of indictment, debars the historian forever from a subjectivity uninformed and uncorrected by methodology. Thus the relativist or perspectivist aspects of historical consciousness, which certainly seem to be part and parcel of historical knowledge, must be raised to the level of a systematic methodology. But while it is true that quantification and mathematization have proved more or less undeniably successful in historical demography, economic history, the study of comparative changes in prices and wages and, generally speaking, all spheres which lend themselves to statistical treatment – provided data are not flagrantly lacking – is it reasonable to expect that these methods can be applied with the same success to the exploration of that which is qualitative, that which belongs to the realm of the mind, and so make it possible for the subjective factor to be brought within the orbit of historical intelligibility? This difficulty is increased by the circumstance of the historian himself belonging to a concrete historical whole which has its own structure different from that to which the historical 'object' under consideration belongs. It must be admitted that the historian does not at present have at his disposal an objective and rigorous theory for passing from one structure to another, basically heterogeneous structure;[143] unlike ethnology, history so far lacks a mathematized and

142. Cf., *inter alia*, the collection of articles by Raymond ARON, *Dimensions de la conscience historique* (1961).

143. Cf. LEBRUN, 'Structure et quantification. Réflexions sur la science historique', p. 44 in PERELMAN (ed.), *Raisonnement et démarches de l'historien, op. cit.* (1963).

formalized theory of transformations, and in this regard is still confronted with a multiplicity of individual cases which it has either to interpret or compare.[144]

It is for this reason that many historians still doubt whether quantified history can ever come to grips with 'apparent' history and actually serve as a basis for its study. Is it possible to unite in a single form of history several specialized and partial histories, resulting from an operation of abstraction? Is not the concept of *longue durée* once again a matter of convention, or in any case abstraction? What do we know, in the last analysis, about the '*mechanisms* which characterize collective behaviour patterns'? All these interrogations throw an all too revealing light on the still largely craft-like character of the historian's work.[145]

The question which the philosopher may raise is more radical: can a history which seeks both to resemble the other human sciences and to create its own schemas of intelligibility, its own 'models', still come to terms with the historical 'object'? It claims to eliminate the subjectivity of 'sympathetic understanding', but is it not imposing its own intellectual interests, its own mental structures, under the pretext of coherence and formal precision? More basically, such a confrontation between the logic of explanation and the historical 'object' presupposes that value is attributed to reflection on the historical being of man and the meaning of history – a type of reflection which is not regulated *a priori* by epistemology, but which leads back from epistemology to general philosophical thinking.[146]

144. See however an interesting but isolated attempt at logical construction: MORAZÉ, *La logique de l'histoire* (1967) and 'Logique et histoire dans l'événement et l'expression' (1968).

145. Cf. DHONDT, 'Histoire et reconstitution du passé', especially pp. 95-105 in PERELMAN (ed.), *Raisonnement et démarches ..., op. cit.* (1963); and, by the same author, 'L'histoire récurrente' (parallel English version, 'Recurrent history') (1971).

VEYNE, in *Comment on écrit l'histoire. Essai d'épistémologie* (1971) and 'Contestation de la sociologie' (parallel English version, 'A contestation of sociology') (1971), gives some powerfully reasoned (and rather caustic) arguments against the claims of 'scientific' historians who try to follow social science models; see also in this connection the major critical review of the first of these two works, by Raymond ARON: 'Comment l'historien écrit l'épistémologie. A propos du livre de Paul Veyne' (1971).

146. With regard to philosophical – or para-philosophical – thinking on history and historicity, see, as examples, the following: BULTMANN, *History and Eschatology* (1955, publ. 1957); CROCE, *La storia come pensiero e come azione* (1938) (English translation, *History as the Story of Liberty*, 1941); FOUCAULT, *L'archéologie du savoir* (1969) (English translation, *The Archaeology of Knowledge*, 1972); GADAMER, *Wahrheit und Methode* (1960); GOLDMANN, *Sciences humaines et philosophie* (1952, 1966), especially Chap. I, and *Recherches dialectiques* (1959), Pt. I; GOUHIER, *L'histoire et sa philosophie* (1952); HUSSERL, *Die Krisis der europäischen Wissenschaften und die transzendentale Phänomenologie* (1936, 1954), and 'Die Frage nach dem Ursprung der Geometrie als intentional-historisches Problem' (publ. posth., 1939) (French translation, *L'origine de la géométrie*, 1962); JASPERS, *Vom Ursprung und Ziel der Geschichte* (1949) (English translation, *Origin and Goal of History*, 1953), especially Pt. III; LACROIX, *Histoire et mystère* (1962); LÖWITH, *Meaning in History. The Theological Implications of the Philosophy of History* (1949/1950); MANNHEIM,

B. SOCIAL AND POLITICAL PHILOSOPHY

Introduction

In this portion of our survey we distinguish *social and political philosophy* from the *logic of explanation* in the human sciences. In sub-section A above, which was devoted to the latter subject, an attempt was made to isolate the philosophical aspects of the epistemology of the human sciences. In so far as this bore on the thinking of philosophers rather than that of the sociologists, psychologists, historians and so on themselves, problems which were strictly philosophical were already anticipated, as, for example, the question whether human phenomena must be considered irreducibly social in nature, or whether the subject-matter of the social sciences is ultimately reducible to the personal subject which says 'I'. This reflection on the *ultimate conditions of the meaning* of the human phenomenon is now faced squarely here in sub-section B.

1. We begin with a review of trends in the *philosophy of law* and *political philosophy*. And tradition itself left us no choice in the matter. Ever since Plato and Aristotle, philosophers have based their assessment of 'the political animal' on theories about law, right and the State. For Leibniz, legal science was a discipline as essential to philosophy as mathematics, physics or theology. Kant called the first part of his *Metaphysic of Morals* the 'Doctrine of Right'. Hegel extracted from his *Encyclopaedia of the Philosophical Sciences* the great chapter on Objective Mind or Spirit and built it up into the *Principles of the Philosophy of Right,* where an ethical system derived from Kant was set in a framework constituted by a theory of abstract right, contracts and punishments, and by a theory of 'objective morality' which culminated in a theory of the State. These great precedents are recalled here not in order to take shelter behind tradition but to stress the

---

Mensch und Gesellschaft im Zeitalter des Umbaus (1935) (English translation, *Man and Society in an Age of Reconstruction*, 1940); MARITAIN, *On the Philosophy of History* (1957); MARROU, *Théologie de l'histoire* (1968); NIEBUHR, Reinhold, *The Nature and Destiny of Man* (2 vols., 1941) and *Faith and History* (1949); POPPER, *The Poverty of Historicism* (1944-45, 1957); RICOEUR, *Histoire et vérité* (1955, 1964) (English translation, *History and Truth*, 1965); SARTRE, *Critique de la raison dialectique* (1960); WEIL, *Essais et conferences*, Vol. I: *Philosophie* (1970), Chaps. IX and X, pp. 199-231.

Considerations in regard to historicity run through the entire development of HEIDEGGER's ontological and aesthetic thought. Reference would here have to be made to a large portion of his work from *Sein und Zeit* (1927 – English translation, *Being and Time*, 1967) on; see however, in particular, 'Nietzsches Wort "Gott ist tot"' (1950) (French translation, 'Le mot de Nietzsche "Dieu est mort"', 1962) and *Nietzsche* (2 vols., 1961) (French translation, 1971).

For the structuralist points of view on diachrony and historicity, and the hermeneutic approach to the problems of historical understanding, see below, in Section IV, *Man and Language*, sub-section B, § § 4, 'Philosophical structuralism' and 5, 'The "hermeneutic" current: the interpretation of language' respectively, pp. 1354-1364, and also the works cited therein.

difference in nature between the philosophical reflection which follows and the more properly epistemological study which went before. With the philosophy of law and political philosophy, we are no longer considering 'second-order' discourse concerning the 'first-order' discourse afforded by the social sciences; the philosophy of law and political philosophy themselves comprise a realm of 'first-order' discourse, one in which an essential part of what is human in man determines itself directly.

2. We have put in second place a separate study on *historical materialism*. In a sense, it might have figured among the sociological theories of law and of the State. But besides the fact that it includes theory of law and theory of the State in a single discipline, the Marxist theory of superstructures brings into play a comprehensive conception of the history of mankind and its place in the total process of reality. In this regard, the present exposition of historical materialism will also serve as an introduction to the Marxist conception of language,[147] of action,[148] and of humanism.[149]

3. In the third and last part, devoted to considerations of a more ontological nature, we have endeavoured, in a more personal style, to relate the foregoing analyses to the structures of a *philosophical anthropology*.

§ 1. *Philosophy of law and political philosophy*

(a) *Philosophy of law*

Reference should first be made, in the present *Study*, to the account given by Professor Knapp of the main current trends of research in the field of Legal Science. From the very first pages Professor Knapp brings out the ambiguity of the word *science* as applied to the sphere of law (not to mention the fact that it is in singular – 'legal science'). The ambiguity, we are told, is manifold: Who engages in this science? Who develops legal thought? Is it scholars and professional jurists, as in the Western (the so-called Romano-Germanic) tradition of law, or is it the practitioners – judges, barristers, administrators, – as in the Common Law countries ('judge-made law')? And is the function of legal science merely to describe the law as it exists within the limits of a given national territory or a particular State? Or is its function to evaluate existing law from the point of view of a normative law, one less limited in space and less bound up with a particular State power? Is the subject-matter of legal science exclusively the law (whether it be positive, natural or both), as is the case in England and those countries where such research has developed under English influence? Or does its competence extend not only to the study of law but also to that of the State, as in the Soviet Union and the socialist countries?

147. Cf. below, Section IV, sub-section B, § 3, pp. 1348-1354.
148. Cf. below, Section V, sub-section B, § 3, pp. 1416-1426.
149. Cf. below, Section VI, sub-section B, § 2, pp. 1439-1449.

In other words, what are the relations between legal science and political science?

These questions obviously raise philosophical issues. To ask who makes the legal rules, to ask whether legal science is confined to describing the law in force or whether it measures it against a more basic law, and to ask whether law and the State constitute separate spheres, is to raise in three different ways a more fundamental issue, namely, by virtue of what do we call a norm legal.

The various replies given by theorists to this basic question are what determine, even today, the major currents of research in legal science. In a broad sense, the name *philosophy of law* can be applied to the body of teaching in which an answer is given to this question regarding the specific properties or essence of legal rules.

(i) In the opinion of non-philosophers, the philosophy of law is identified with **the natural law**. Such an identification however is not justified. The so-called 'natural law' school is itself clearly delimited in history: stemming from Grotius, Pufendorf and Thomasius, it developed with Leibniz and Wolff. The philosophy of right of the neo-Kantian school, which reached its peak in the first third of the 20th century, in no way arises from it. The most that can be said is that the 'philosophy of natural law' and 'neo-Kantianism' share a desire to measure the law in force, incomplete, imperfect and unjust as it is, against a more basic law, one that is better, more complete and more just. But what characterizes the natural law tradition, as compared with that of Kantianism and neo-Kantianism, is that it is more ambitious in deducing the law, down to legislative details, and especially the fact that it bases the 'ought to be' of legal propositions on man's being, on his essential inner nature. This is the crux of the matter: neither legal science nor the judge creates or ever can create law; they merely recognize it. This recognition of what is makes it impossible to set up an opposition, in the hard-and-fast manner of Kantianism, between the *prescriptive* domain and the *descriptive* domain. The constant resurrection and re-emergence of natural law corresponds precisely to a rejection of the barrier between 'ought' and 'is' and an attempt to find the level of rationality where the validity of the principles of justice is manifest independently of positive law. At this radical level, such principles are in no way 'posited' but belong to the very structure of man's being: because mankind itself has certain characteristics, there are certain rights which belong to it; and the task of the philosophy of law or right is to attain, by a method which is both abstractive and intuitive, to this inner core of what is and what ought to be.

This demand for radicality is so insistent as to have survived the decline of the rationalist school of natural law, which was bound up with a certain conception of human nature. It must be admitted that whereas critical examination of the idea of 'human nature' from a historical or cultural point of view inevitably takes us away from the idea of natural law, each time not only philosophers or jurists, but ordinary citizens invoke the idea of inborn rights, imprescriptible rights, inalienable rights – or simply human rights – we are inevitably brought back to it. How would it be possible, for example, to invoke human rights against the abuses of the State if the only kind of law were a positive law? It is this paradoxical situation which undoubtedly explains what has been called 'the eternal return' of natural law.[150] Legal positivism, and also the sociological conception of which we shall speak later, to the extent that they fail to take account of this age-old appeal from the law to a higher idea of what is just, thereby add fuel to these periodic revivals. As some critics have said, positive law is tautological: only what corresponds to positive law is just. But must not the criteria of assessment be sought elsewhere than in the *lex scripta*? Does the lawmaker himself not feel the constraint of other legal norms which he has not created? Norms established by judges, customary norms, general principles of law and generally recognized principles of human behaviour do not necessarily constitute a supra-human tribunal nor do they necessarily give rise to legal dualism, as positivism, sociologism and Marxism claims, but map out a constituent element of the human condition underlying the whole sphere of right; and this sphere is not purely the creation of the lawmaker, still less does it stem from decrees by the State power. Thus the concept of natural law may well take different forms than under 18th century rationalism.[151]

(ii) It was **the neo-Kantians** who, along lines prepared by Kant's *Rechtslehre,* developed the question of the status of legal propositions in all its abstract nakedness. In their view, the divorce between 'ought' and 'is', formalized in Kant's second *Critique,* governs the entire legal structure. The notions of moral duty, a moral obligation to perform the duty, the attribution of responsibility for human acts and the moral freedom of the individual serve to distinguish legal propositions, inasmuch as they are normative, not only from all the descriptive propositions of science, but from the so-called propositions of natural law;[152] the school of natural law, completely rationalistic though it was, still sought a foundation for its constructions in what might be called a residual experience, that of 'human nature' abstracted from any social context. This experiential residuum of the systems of the natural law school is inconsistent with a pure theory of 'the imperative'. But, at the same time, the neo-Kantian school was less ambitious than the school of natural law in what is sought to deduce from principles.

150. Cf. ROMMEN, *Die ewige Wiederkehr des Naturrechts* (1947); HAINES, *The Revival of Natural Law Concepts* (1930); THIEME, 'L'apport du droit naturel au droit positif' (Rapport général au VIIe Congrès international de Droit comparé, 1966) (1968); HOOK (ed.), *Law and Philosophy. A Symposium* (1964).

151. Cf., in the present work, Chapter VI, *Legal Science*, by Viktor Knapp, Section Two, II, 1, pp. 952-957.

152. 'A merely empirical system that is void of rational principles is like the wooden head in the fable of Phaedrus, fine enough in appearance but unfortunately wants brain.' – Kant, Introduction to the *Science (Doctrine) of Right,* § B.

It was no longer considered possible to deduce everything, as mathematics does, or to convert *all* law into a rational system. The practical reason dictates only the forms of all legal thought, the very definition of law and of its principal notions, and the structure of its language – in short, what many treatises call the 'general theory of law'; the content of the law, the forms it may take in time and space, belong to the realm of positive research. This was the door through which legal positivism was to rush. What is left to jurists is interpretation of existing law, while to philosophers belongs the task of reformulating the whole general part of law in the language, categories and conceptual framework of an authentically philosophical *Rechtslehre*.

Positivism was to find another form of support – or at least another cue – in neo-Kantianism, by virtue of the rôle which the latter assigned, following Kant, to coercion by the State. The neo-Kantian school never came near confusing law and morality. The *Rechtslehre*, though cut off from the natural sciences and even from politics and economics, has remained, since Kant's time, distinct from the *Tugendlehre* (Doctrine of Virtue). Kant introduced his *Doctrine of Right*, it may be recalled, in the following terms: 'The Science (Doctrine) of Right has for its object the principles of all laws which it is possible to promulgate by external legislation'. By 'external legislation' he meant first of all the rule of justice applied to the practical relation of one person to another, in so far as they can influence each other. Legislation therefore, as a regulatory system for ensuring the coexistence of freedoms, is something external: it merely proscribes the harmful effects due to my external action. The universal law of right is expressed as follows: 'Act externally in such a manner that the free exercise of thy will may be able to coexist with the freedom of all others, according to a universal law.' Since the law forbids that which is an outward hindrance to freedom, it follows that resistance to what acts as a hindrance, i.e., constraint or compulsion, also enters into the definition of what is just. 'According to the logical principle of contradiction, all right is accompanied with an implied title or warrant to bring compulsion to bear on anyone who may violate it in fact.' This inclusion of constraint in the definition of right implies that the jurist must look for his answers solely to legal texts, since only the rules of positive law are backed up by compulsion. In this way neo-Kantianism made its adversary – 'legal positivism' – its indispensable complement.[153]

(iii) With **legal positivism,** the notion of philosophy of law changes radically. It does not consist in a body of legal theory distinct from the rules of law embodied in the various codes; legal positivism is even the negation of the 'philosophy of right' conceived of as a 'general theory of right',

153. WIEACKER, *Privatrechtsgeschichte der Neuzeit* (2nd edn., 1967); CERRONI, *Kant e la fondazione della categoria giuridica* (1962); VILLEY, *La formation de la pensée juridique moderne* (1968); by the same author, *Seize essais de philosophie du droit* (1969).

logically anterior to law itself. Philosophy here does no more than limit legal science to a knowledge of the only rules of law there are, those established by the lawmaker.[154]

This does not mean that, so far as legal positivism is concerned, legal propositions are descriptive propositions. It is legal science which *describes* and *interprets* the propositions of the law in force, but they constitute a body of rules separate from all other social phenomena. Consequently, while legal positivism, in the name of *legal facts*, will have no truck with the kind of questions relating to the ground or foundations of law as they arise from legal Kantianism and theories of natural law, it just as stoutly resists the reduction of legal fact to social factors other than the law, and in a general way opposes the sociologically-inclined trends which pervaded legal science in the first third of the twentieth century.[155] Hence the complexity of its relations with the neo-Kantian tradition: it combats the deductive pretensions of that tradition on the level of general principles of right, but finds in it an ally in defending the specific nature of the legal sphere. Thus Kelsen's 'pure science of law' *(reine Rechtslehre)* has lent support to the anti-sociological trend of legal positivism.

In the United States, the group of *New Analytical Jurists*, whose programme is presented by R. S. Summers,[156] is related to the positivist movement. The questions raised by this movement started by jurists are of primordial interest to philosophers. In what context are expressions such as responsibility, crime, purchase, sale, debt, marriage, etc., valid as 'legal' concepts? The concept of responsibility is, in this regard, a good touchstone since the operations whereby an action is attributed to someone are perfectly accessible. Legal decisions whereby a judge rules that this is a valid contract, that this is a murder in the second degree and not a murder in the first degree, themselves constitute a body of distinct and paradigmatic propositions. In an important article entitled 'The ascription of responsibility and rights' (1948–1949),[157] H. L. A. Hart dwells on a remarkable characteristic of legal pronouncements: they may be contested either by denying the alleged facts, or by pleading circumstances that are capable of weakening, attenuating or even voiding a claim to a right or an accusation of a crime. Hart uses the verb 'to defeat' in speaking of this effect on claims and

154. Brimo, *Les grands courants de la philosophie du droit et de l'Etat* (1967).

155. Cf. Bergbohm, *Jurisprudenz und Rechtsphilosophie* (1892).

156. Summers, 'The New Analytical Jurists' (1966).

157. See also by Hart, *The Concept of Law* (1961); *Law, Liberty and Morality* (1963); *The Morality of the Criminal Law* (1965); *Punishment and Responsibility* (1968); and the already cited work by Hart & Honoré, *Causation in the Law* (1959). For Hart's work in general, see Montefiore, 'Law and morality: the work of H. L. A. Hart' (1971).

For a representative selection of articles in jurisprudence by academic lawyers see *Oxford Essays in Jurisprudence*, edited by A. G. Guest (1961); and for essays in the philosophy of law, *Essays in Legal Philosophy* (1968) and *More Essays in Legal Philosophy: General Assessments of Legal Philosophies* (1971), both edited by R. S. Summers. Reference should also be made to *The Concept of a Legal System* by Joseph Raz (1970), a work of particularly powerful analysis.

accusations, and he uses the word 'defeasible' in speaking of the character of the legal judgement whereby it is susceptible of this kind of contestation and confutation. He then goes on to say that actions capable of being 'ascribed' are also those which are capable of being 'defeated', invalidated and abrogated. At first sight it is surprising to see what originally appeared to be only a secondary characteristic of legal judgements turned into a criterion, but Hart attacks two prejudices which stand in the way of promoting invalidability to the rank of criterion.

(a) In the first place, there is a misunderstanding about the nature of legal reasoning itself. It is readily considered a mere subsuming of an individual case under a general rule which would require nothing more than recognition of the necessary and sufficient conditions for application of the general rule. A legal judgement is in point of fact an interpretation which attempts to ascertain which previous cases the new case most resembles, which establishes the bundle of precedents and identifies the '*ratio decidendi*' inherent in these similar cases. This is the interpretation to which an examination of nullity clauses leads. A legal concept is not defined solely by specifying the necessary and sufficient conditions, but also by considering the list of exceptions and negative cases. Thus in contracts, it is not enough to have two parties, an offer by one and acceptance by the other, etc.; all these propositions are valid 'unless' such or such circumstances obtain to invalidate the contract, and these circumstances can be enumerated. Thus invalidability is not a secondary characteristic but is the touchstone of legal reasoning and of the legal judgement itself. This is why Hart goes so far as to equate the possibility of 'ascribing' and the possibility of invalidating.

(b) Hart denounces a second prejudice: there is a tendency to think that the list of invalidation clauses, which is heterogeneous or even heteroclite, indicates in a negative way the absence of a central positive factor such as full information, express intention or firm will. The invalidation clause (defence) supposedly relies on the evidence of absence of a factor called 'true consent' in the case of contracts, and 'criminal intent' (*mens rea*) in the case of a person committing a crime (*reus*). If such were the case, the ascription of responsibility would be the same as the description of this mental factor. Hart shares with many positivist philosophers the conviction that this mental factor is a myth. Consequently, the criteria of invalidation cannot derive from evidence of the absence of such a factor. On the contrary, to speak of such a factor is an abbreviated way of saying that the exceptions do not apply to the case under consideration; but then this is no longer a description, but a decision – the decision that, in the case under consideration, the invalidation clauses are not operative. The ascription of responsibility is therefore based on the multiple criteria or arguments susceptible of invalidating the negation of responsibility. The criterion for the ascription of responsibility is the admission or non-admission of invalidity clauses. For these reasons, the judicial decision is indeed a decision: this is a valid contract, or that is a murder in the second degree – as such, a decision of this kind is neither true nor false, but good or bad.

These reflections, in point of fact, go beyond the limited framework as defined by the existing law, which does not extend further than the codified sphere of legal propositions. Ascriptive or prescriptive propositions cover the entire range of complex situations in which someone is said to have caused a wrong to or done a service for someone else, whether it be intentionally, by negligence or without being accusable of fault, or deserved of others responses which we call praise or blame, punishment or any other legal compensation. The theory of responsibility, to which Joel Feinberg, for example, has devoted his essays[158] in the wake of Hart's article, ranges back and forth across the boundaries between the Philosophy of Mind, ethics and law; and it is precisely for this reason that it constitutes a philosophical contribution to the theory of law. We shall consider below[159] works under the heading of 'action philosophy' which, on the basis of semantic factors, may contribute to a clarification of the concepts involved in legal reasoning: action and the agent of action; causes, motives and reasons; commission and accusation; intentions and excuses; imputability, punishment and compensation; right and duty; etc.

In a less abstract, less deductive and, in short, less *a priori* fashion, these are the very problems which the neo-Kantians tried to solve; but instead of seeking the meaning of legal propositions in elucidation of the question *'quid jus'*, prior to the question *'quid juris'*, the analysts start from the existence of a wealth of expressions, judgements and reasonings usually recognized or accepted as having a legal or juridical character, it being legitimate to try to ascertain the epistemological status of these expressions, judgements and reasonings in their functioning, that is, in their applicability to situations (protests, claims, conflicts) which result in court decisions. This method tackles problems raised by neo-Kantianism, but with an apparatus allied to legal positivism, while at the same time searching for criteria of a legal or juridical nature in a broader field than that of the existing law to which legal positivism confines itself.

(iv) With **sociological theories,** the specific nature itself of law is in question. Law is seen both as a product of social life and as a force making itself felt in society. According to the sociological conception, there is no unbridgeable epistemological barrier between what is (*Sein*) and what ought to be *(Sollen)*; as Professor Knapp says elsewhere in the present *Study*, 'each term reacts on the other, and knowledge of the first is a precondition of knowledge of the second'.[160] Here opposition to legal positivism is quite clear-cut: scientific knowledge of the law can never confine itself to the analysis and interpretation of legal norms; it is subordinate to knowl-

158. FEINBERG, *Doing and Deserving. Essays in the Theory of Responsibility* (1970).

159. Cf. below, Section V, *Man and Action*, sub-section A, § 3, 'Practical reasoning and the theory of argumentation, pp. 1387-1391.

160. For a general account of the sociological conception, see above, Chapter VI on *Legal Science* by Viktor Knapp, Section Two, II, 3,. pp. 960-965, and the bibliographical references given therein.

edge of the social conditions which determine law and the social milieu from which the law is born and in which it exists as a living law. Of course the sociological study of law also requires study of the influence exerted by law on society. But there again it is the imprint left on that society, that is to say the totality of the changes introduced into social dynamics by the law, which enhances understanding of the legal norm itself. In our view, a sociology of law does not eliminate, but points up, the question, by virtue of what can the subject-matter of sociological study aspiring to the title of legal sociology be termed legal. To consider law as social phenomenon is consequently 'to integrate law with the social sciences', as writers like Huntington Cairns and Jerome Hall propose. This is undoubtedly a legitimate and fruitful undertaking and one that cannot fail to have an effect on the definition of law: a phenomenon whose position in the great scheme of social interactions is better understood is viewed in a new light even by the specialists. In that respect, the isolation in which legal positivism kept legal phenomena is no longer admissible and turns out to be more impoverishing than fruitful. But can it not be said that a better understanding of the dialectic of social phenomena makes it still more necessary to delimit the phenomena under consideration?

This is indeed what has happened: a *functional* definition of law tends to replace a definition which might be called *essential*. To state what *forces* law *reinforces* is to state something about the law itself. To state what developments it retards or favours is to indicate the place that law occupies in the play of forces and forms in the overall social complex.

In this respect, Marxism may be regarded as akin to sociological trends in the philosophy of law, and that for two reasons: first, because it lumps together the science of State and the science of law, as relating to two aspects of one and the same social superstructure, and second it considers this social superstructure, of which State and law are elements, as ultimately determined, as Viktor Knapp points out, 'by the social basis and its development, that is to say, if we may have recourse to simple terms, by the material conditions (economic relationships) peculiar to the given society, which involve the fundamental relationships between the classes. The social factors – and, in the end, the social basis – which determine law are the motive power which promotes the historical development of law and its adaptation to the changes in the social factors by which it is determined. The law, in turn, as a component of the social superstructure, influences the society which has moulded it, and its historical evolution.'[161]

We shall deal separately with the research rends currently emerging in historical materialism after rounding off our study of trends in the philosophy of law with a study of trends in political philosophy, since political science and legal science are separate branches of study in the English-speaking countries and have recently become so in continental Europe.

161. Cf. above, Chapter VI, *Legal Science*, Section Two, III: 'Socialist countries (The Marxist-Leninist theory of law)' *ad* 5, pp. 967-968. See also KNAPP, 'La philosophie du droit dans les pays socialistes' (1971).

Let us confine ourselves for the moment to suggesting, in the context of the present discussion, that one of the main interests of the most recent research along Marxist-Leninist lines is precisely the specific nature of law within the general theory of the State. Viktor Knapp[162] speaks of a revival of interest in the system of law, and in particular in the question of the existence of an autonomous economic law independent of the civil law. Research done on the concept of 'socialist legality' follows the same trend: our colleague himself recognizes[163] that despite serious differences there is a 'common denominator' between this notion and those of '*primauté de la loi*' (in France), 'rule of law' (in Britain and the United States) and '*Rechtsstaat*' (in the legal science of the Federal Republic of Germany).

Thus it is the very development of contemporary society which forces jurists and philosophers to reflect on the question in what way and under what conditions can law protect the democratic system against the abuse of power and against violation of fundamental human rights. It is society itself which has become aware of the importance of this social rôle of law and formulates 'the categorical requirement to observe the law as a guarantor of democratic practice, and further to observe the procedures legally prescribed by legislation'.[164] The philosophy of law is thus entrusted with a question which it can neither water down nor avoid. It is all very well to look to society for the origin of law, and where it makes its impact, but it must also have some specific properties of its own even when it springs from society and returns thereto. Nor must it be identified with State coercion, but must be kept distinct from it, however close a relationship is posited between the legal phenomenon and the phenomenon of the State. In short, the question *quid jus*? rises ever anew, Phoenix-like, from its ashes, as an irreducible question, albeit it is to the question *quid juris*? that the non-legal disciplines increasingly devote themselves.

(b) *Political philosophy*

Political philosophy demands a treatment which is distinct but not divorced from that of the philosophy of law. Its subject-matter is in one sense more general than that of the philosophy of law, and in another sense more specific.

On the one hand, political philosophy is heir to the Aristotelian view of 'politics' as 'the most authoritative or architectonic science';[165] its end is 'the true good of mankind', man being essentially a political animal, a member of the 'polis' or community. From this standpoint, political philosophy embraces the philosophy of law as well, and the sciences which we now call sociology, jurisprudence and political science are simply individual fragments of that Science (in the Greek meaning of the word, which seems to have survived in the German *Wissenschaft,* at least in

162. Chapter VI, *Legal Science*, Section Two, III, *ad* a (ii), pp. 970-971.

163. *Ibid.*, Section Three, V: 'The law and democratic practice', *ad* 3, pp. 1007-1010.

164. V. Knapp, *loc. cit.*

165. *Nicomachean Ethics*, 1094 a 26 (Bk. 1, Chap. 2).

the time of German Idealism, and which perhaps requires a gloss such as 'rational knowledge'). For political philosophy in its widest sense is concerned with the quality of *good sense*, of *intelligibility*, of *satisfying* the *reason* (or however one wishes to put it) which human existences derive from their membership of the human community. Hence its basic task is to subsume under this all-embracing rationality the partial rationality of practical activities whose aims go to make up the aims of politics, and then to co-ordinate this rational characteristic of life in a community with all the other characteristics of good sense, intelligibility and reason displayed by man's other theoretical and practical pursuits.

On the other hand, political philosophy is the heir to a narrower conception of politics, one which prevailed in Hobbes and Machiavelli. Its central theme, according to this second tradition, is the *State,* i.e., the institution which in every age demands legal and political supremacy. What has always fascinated the philosopher, in his preoccupation with analysis, is the phenomenon of authority. He is therefore tempted to define politics solely in terms of the exercise of authority and to identify the State with whatever institution possesses organized power. Political philosophy is thus polarized between a sweeping view, which takes in the 'political animal' in all his aspects, and a narrow gaze, concentrated on the paramount will which bears the attributes of sovereignty. Rarely has it succeeded in holding the balance between these two extremes.

Perhaps the last occasion on which it did was with Hegel, in his *Philosophy of Right* – which, it must be remembered, is a development of the section in the *Encyclopaedia of the Philosophical Sciences* entitled 'Objective Mind (or Spirit)'. On the one hand, the subject-matter of political philosophy is 'the realization of freedom'; what makes the life of the State meaningful is the fact that life within an institution gives freedom a concrete and actual (realized) character, which Hegel opposes to the quality of mere requiredness (*Sollen*) possessed by subjective morality; life in the State is nothing unless it conforms to the very definition of right, as the realm of realized freedom. On the other hand, the State is the authority in which the will of the citizens is realized, so that the theory of the State rounds off a theory of the will; in this way the phenomenon of authority is brought to the centre of the stage. That this authority, however, be a force alien to all our wishes, a supreme arbiter, a hidden power, even that it may involve violence which would make it the very opposite of discourse, none of this is philosophically important; what is important is that this force offers to the individual satisfactions which are unknown in private life – a public existence in which he recognizes the deepest aspirations of his freedom. The two oldest demands of political philosophy are thus combined in the definition of the State as the actualized reality of freedom.[166] Unfortunately, this balance is only preserved at the cost of equivocation. The Hegelian State is neither a description of any of the

166. *Philosophy of Right*, § 260.

empirical States hitherto produced by the history of the world, nor merely an exemplary, Utopian ideal, which would mean consigning it to the much disdained realm of *Sollen*: it is the modern State seen from the point of view of whither it is tending, the leaven at work in the realities of modern politics, which makes the modern State a State, even when it is unaware of the fact and despite the forces that threaten to tear it apart. But then, if we are still to be able to equate the rational and the real, the real must be defined in terms which do not coincide with the empirical and the observable; or, to put it bluntly, in so far as the real is the rational, it is not the empirical and the observable.

This paradoxical character of the Hegelian 'real' has become even more open to question since the appearance of a *political science* which claims to be empirical and descriptive and therefore breaks with the concept of Science in Greek philosophy and German Idealism. The political science of today has more affinity with the sciences characterized above in terms of social sciences than it has with political philosophy, seen as a branch of speculative philosophy. For the philosopher, the question still remains whether and how this science can define its subject-matter.

The chapter of the present *Study* in which Professor W. J. M. Mackenzie examines the main trends of research in political science[167] is highly instructive in this respect.

The 'astonishing discrepancies'[168] which Professor Mackenzie finds in the definitions given by writers on political science merely reproduce, within the context of an empirical science, the oppositions found in more traditional political philosophy. At one extreme (Oakeshott), politics is the process of 'attending to the general arrangements of a society';[169] at the other (Maurice Duverger), what characterizes politics is 'in any society, organized power, the institutions of command and control'.[170] Ranging between these two extremes are conceptions that tend either to a broader definition of the overall social phenomenon, like that of Talcott Parsons, who, to quote Mackenzie, describes politics as 'the instrumental aspect of social organization (the deliberate creation of purposive social structures)'[171] or to definitions centred more on the phenomenon of State compulsion, such as that of David Easton, who defines politics as 'the authoritative allocation of values'.[172] We may note here that the Marxist exposition (to which we shall return in § 2 below) is twofold as well: on the one hand, it makes the exercise of compulsion the ultimate principle of political organization;

167. *Main Trends of Research in the Social and Human Sciences, Part I: Social Sciences* (1970), pp. 166-224.

168. *Ibid.*, p. 168. The subsequent examples are also taken from MACKENZIE's account.

169. Quoted by MACKENZIE, *loc. cit.*

170. 'En toute société, le pouvoir organisé, les institutions de commandement et de contrôle' – DUVERGER, *Introduction à la politique* (1964).

171. MACKENZIE, *loc. cit.*

172. EASTON, *The Political System. An Inquiry into the State of Political Science* (1953), quoted by MACKENZIE, *loc. cit.*

on the other it treats the political sector or aspect of social life as a relatively subordinate sphere or structure.

Hence political science also fluctuates between a broad conception centred on the 'political animal' and a narrow one conceived in terms of 'compulsion by the State'. Admittedly a science has less call to define its subject-matter than might be imagined, since it is not accountable for its intelligibility beyond the limits of the scientific treatment to which it subjects it; if need be, it can start not with a definition, but with a methodological supposition, namely 'that all human societies have an aspect which can be recognized as political in some sense'.[173] But a science which abjures definitions, to use Mackenzie's terms,[174] is faced with the question of its significance in a different form: what part does 'science of the State' play in political science? For although political institutions are geographically co-extensive with the territory over which the sovereignty of the State is exercised, the activity of groups and organizations which are not in themselves the State presents avowedly political aspects. Professor Mackenzie lists them briefly: first, there are political forms – particularly among new States – which have not acquired the form of a State, the theory of which originated in Aristotelian, Hobbesian and Hegelian political philosophy. Secondly, the most complex modern States have witnessed the emergence of public corporations, private organizations recognized as being in the public interest, an administration and a bureaucracy which have now become separate objects of political science, even though they are not, strictly speaking, State structures. Thirdly, nineteenth-century political theory – unlike Greek philosophy in this respect – centred almost exclusively on the juristic aspect of sovereignty at the expense of the economic and social forces operating in the infrastructure; political science, in breaking free from the juristic monopoly, claims to attend to all aspects of society, whether codified in juridical terms of political organization or not. Lastly, Professor Mackenzie draws attention to the need for specific study of the small groups in which decision-making is concentrated; on the political plane this micro-sociology does not coincide politically either with the theory of mass social dynamics which is the concern of politics or with the juridical theory of the organization of power.[175]

173. MACKENZIE, *op. cit.*, p. 169.

174. *Loc. cit.*

175. Studies by anthropologists (not only of the forms and bases of authority in traditional societies, but also – and this is even more important – of the present evolution of new national entities and their adaptation to the modern world) have done much to make the fundamental ideas of political theory less narrow, less rigid and less absolute; they have given rise to a process of critical reflection, often directed to essentials, that tends towards a reappraisal of the quality of universality somewhat rashly attributed to the habitual categories and principles of political science. A bibliography of these studies will be found in Section C, § 2 of an earlier chapter in the present work, entitled 'Social and cultural anthropology', by Maurice Freedman, pp. 39-42. Among the major recent works in this field reference may be made to the following: BAILEY, *Stratagems and Spoils. A Social Anthropology of Politics* (1969); and BALANDIER, *Anthropologie politique* (1967) (English translation,

Powerful forces are thus driving political science beyond the confines of a theory of the State, strictly speaking. But then, faced with the threat of being swallowed up by science in general, an underlying instinct of self-preservation diverts it towards the analysis of a quite different entity, the world system of States. Once again the comparative and differential study of States swings the pendulum in the direction of economics and ecology. And this oscillating movement itself demands a definition of the political in terms which are broader than those of State compulsion and more precise than those of organization or system. In this connection, the most interesting research, philosophically, relates to the concept of a 'system of norms': delimitation of the political then consists in identification of the norms which are peculiar to politics; some of these are legal (such as constitutional provisions) and others not (e.g., the norm of obedience to statutory authority), while some are technical (rules governing elections) and others general (the list of human rights).

Political philosophy and political science (in the empirical sense of the word 'science') are inextricably associated in the establishment and identification of these criteria and norms, since political scientists could scarcely determine criteria for what is political unless human beings could see those criteria as conditions for their own rational action and as means to the satisfaction which they pursue in their political life. Therefore the subject-matter of political science cannot be defined empirically without there being a theory to relate it to the meaning which men seek for their actions. In other words, the political can only be subject-matter for science if it is at the same time the meaning constituted by the very activity of men in their search for a rational life and genuine freedom. Political theory therefore necessarily looks two ways: towards empirical description and towards philosophical reflection. This Janus-aspect doubtless imposes a greater strain today, in so far as political science is struggling to be independent of both the legal and economic sciences and political philosophy itself; its fight to avoid being whittled down to a mere 'science of the State' seems a reaction, even, against political philosophy to the extent that the latter centred on questions of power, sovereignty and compulsion. Political science has therefore had to forgo something of political philosophy in order to develop and in order to assimilate the modern methods of the social sciences, such as statistics, cybernetics and simulation. This breaking free from the leading strings of philosophy is particularly marked in contemporary research on the science of administration and bureaucratic structures; in asserting its independence of the theory of the State this area of political

---

Political Anthropology, 1970) and *Sens et puissance. Les dynamiques sociales* (1971). The habitual structures of economic thinking are likewise being eroded; Maurice Godelier, for example, in the light of anthropology, has challenged their claim to rationality and universality. Cf. GODELIER, *Rationalité et irrationalité en économie* (1971), especially Pt. III, 'Rationalité des systèmes économiques'.

There is little need to dwell on the specifically political significance and scope of these conceptual revolutions.

science necessarily also asserts independence of traditional political philosophy.

However, as political science moves away from traditional philosophical speculation on concepts such as State, constitution, sovereignty, power and compulsion, there emerges the need for a political theory which will gather up the questions of meaning and intelligibility left behind by a science that is increasingly anxious to be recognized as empirical by the other social sciences. This is clear, on the borderline between science and philosophy, from the highly reflective character of fundamental political studies such as, for example in France, those of Bertrand de Jouvenel[176] and Raymond Aron,[177] in which ideas inherited from a long tradition of liberal thought are newly elaborated in the light of a detailed examination of the functioning of authority and group relationships in modern societies and economies.

It is actually empirical science which presents political theory, in the form of a fundamental residual question, with the most crucial problems concerning man's political existence. The most empirical of sciences ultimately leads to two kinds of questions which are inescapably philosophical in character: that of the minimal conditions of political existence without which a State would not be a State, and that of the optimum conditions for a State not merely to have a political existence but to be 'approved'. Both types of condition – of bare political existence and of approval – inevitably reintroduce the concepts of reason and intelligibility which bring into play political philosophy. Consider the following propositions, which manifestly apply to the 'approved' State: '[. . .] Government is necessary. [. . .] But a citizen in a constitutional State has rights against his government as well as duties towards it. [. . .] These rights are protected by law, and [require] regular procedure in all cases [before] law-courts.' These propositions are postulates, not hypotheses in the scientific meaning of the word.[178] Yet the same normative and philosophical character attaches to all political concepts if their function is to relate political science as a whole to the principle of 'political man'.

It is no accident, therefore, that a political science which is becoming

176. Jouvenel, *Du pouvoir. Histoire naturelle de sa croissance* (1945, 1972) (English translation, *On Power. Its Nature and the History of its Growth*, 1948, 1949); *De la souveraineté* (1955) (English translation, *Sovereignty. An Enquiry into the Political Good*, 1957); *The Pure Theory of Politics* (1963); *Du principat* (1972).

177. Of Raymond Aron's many works, the following in particular may be mentioned: *Dix-huit leçons sur la société industrielle* (1962) (English translation, *Eighteen Lectures on Industrial Society*, 1967); *Paix et guerre entre les nations* (1963) (English translation, *Peace and War. A Theory of International Relations*, 1967); *La lutte des classes. Nouvelles leçons sur les sociétés industrielles* (1964); *Démocratie et totalitarisme* (1965) (English translation, *Democracy and Totalitarianism*, 1968/1969); *Essai sur les libertés* (1965) (English translation, *An Essay on Freedom*, 1970); *Trois essais sur l'âge industriel* (1966) (English translation, *The Industrial Society. Three Essays on Ideology and Development*, 1968); *Etudes politiques* (collection of essays, 1972).

178. Mackenzie, *op. cit.*, p. 206 and n. 56.

increasingly scientific now finds a counterpart in works of political philosophy which are increasingly conceived in terms of philosophy.

The *Philosophie politique* of Eric Weil (1956) is expressly aimed at 'the comprehension of politics in its entirety and in its unity of structure, as being the comprehension of human action in history'.[179] The return to Aristotle is avowed explicitly: 'The term *politics* will be used in this book in its old Aristotelian meaning of *πολιτικη πραγματεια* the consideration of the common life of men according to its essential structures.'[180] Political categories are thus categories of *reasonable action*; politics itself is the *philosophical science of reasonable action*. As such, it is distinct from ethical action, and also from social action. Politics are distinguished from ethics by the fact that this reasonable action concerns not the individual, as the representative of all individuals, but the human race as a whole; as an individual, man can only concern himself with the negative universal of formal ethics, whereas politics is automatically concerned with the concrete universal. But politics are also distinguished from social action which takes its significance from the organization of modern society in a progressive struggle against alien nature; in modern society, the individual is fundamentally unsatisfied, and a conflict arises between the need for a living morality and the kind of rationality which is developed by the technological society. At this point, life in the State offers its own rationality. It is sufficient to quote, without further comment, the handful of propositions on which Weil proceeds to build up a political philosophy: 'The State is the organization of a historical community. Organized into a State, the community is capable of taking decisions.'[181] – 'The decisions of the community as the State basically apply for the duration of a specific (historical) community.'[182] 'The modern State is actualized in and by formal universal law.'[183] – 'For purposes of decision and implementation, the government of the modern State relies on *administration*.'[184] – These propositions are meaningless except in relation to a theory of rational action, which is simply an element of philosophy, itself conceived as *coherent discourse*. Consequently, the philosophical nature of these propositions, far from being disputed, is specifically asserted.[185]

179. '... une compréhension de la politique dans sa totalité et dans son unité structurée, comme compréhension de l'action humaine dans l'histoire' – WEIL, *Philosophie politique*, preface, p. 7.

180. 'Le terme *politique* sera pris dans ce livre en son acception antique, aristotéticienne, *πολιτικη πραγματεια*, considération de la vie en commun des hommes selon les structures essentielles de cette vie.' – *Ibid.*, p. 11.

181. 'L'état est l'organisation d'une communauté historique. Organisée en Etat, la communauté est capable de prendre des décisions.' – *Ibid.*, p. 131.

182. 'Les décisions de la communauté dans l'Etat visent essentiellement la durée d'une communauté particulière (historique).' – *Ibid.*, p. 132.

183. 'L'Etat moderne se réalise dans et par la loi formelle et universelle.' – *Ibid.*, p. 142.

184. 'Pour la délibération et l'exécution, le gouvernement de l'Etat moderne s'appuie sur l'*administration*.' – *Ibid.*, p. 148.

185. See also WEIL, *Essais et conférences*, 2 vols. (1970 and 1971), especially Vol. II, *Politique*.

Julien Freund's *Essence du politique* (1965) belongs to the same stream as Eric Weil's *Philosophie politique*, but with this difference, that Freund pays greater attention to the distinctive characteristics of political concepts in themselves, and in relation to non-political concepts, than to the possibility of including them in a comprehensive system of discourse. The task of political philosophy is to explain how the essence of politics differs from that of ethics, from that of religion and from that of philosophy, and to distinguish between the major categories of which the essence of politics is composed (public and private, friend and enemy, etc.); these major categories are set against a background of what is political rather than linked together to form a structure for rational action. Thus an outright form of essentialism, centred on the discrete elements which demarcate the field of politics, is opposed not only to scientific empiricism but also to theoretical deductivism.

In their different ways, each of these two styles of political philosophy testifies to the originality and irreducibility of political philosophy, concerned as it is with the analysis of questions of ground or foundations, vis-à-vis political science with its increasingly descriptive and empirical formulations.[186]

186. We may add that in the view of proponents of the most radical forms of social and political criticsm – those which can be described, to use what is now a household term, as 'anti-Establishment' (*contestation*) – the only reason political science that purports to be positive science claims to be neutral and objective is the better to exercise its true function of *ideological* production (in the pejorative sense), as a tool in the service of the established order. It must therefore be exposed as such, in its premisses and what they cover, and its functioning as a 'practice' and as a 'discourse' must be examined and laid bare. As for the conditions for a genuine science, they can be defined in markedly differing terms, whether it is that in a conflict situation theoretical validity is conceivable solely in terms of action (action in history), or that the scientific level can only be attained by virtue of what Louis Althusser calls an 'epistemological breaking off' (*coupure épistémologique*) – itself inseparable from the act of espousing a revolutionary cause.

Account must be taken of the prominent position occupied in the forefront of the intellectual scene in recent years by this other mould of political thought, armed with the full panoply of method developed by semiology, structuralism, etc., drawing its inspiration from a multitude of historical and contemporary sources, but mainly nourished by Marxist and Freudian ideas freely recast (and sometimes combined, along lines adumbrated at an earlier stage by Wilhelm Reich), and less concerned with defining satisfactory and viable forms of organization than with attacking the absurd, alienating, unjust and, in the opinion of some, neurotic character of what, in the shape of a variety of competing but collusive and mutually sustaining societies and régimes, is designated 'the system' – the established social, economic, political, moral and intellectual order, the implicit conventions in which scientific research and philosophical speculation imprison themselves, etc. The violent action which may be required to destroy or paralyse the 'system' is justified by the violence inherent in the system itself; it may even, in the opinion of certain revolutionary theorists (Frantz Fanon, for example), have a purifying quality. But non-violent forms of action, the mere refusal to participate, the direct search for other modes of existence and associations, also have their advocates.

Seen from this angle, no problem can be investigated non-politically; depoliticization can only lead to *mystification*. On the other hand, a genuine revolution in the

Coming somewhere in between these two and relating at each end of its own spectrum to them both, a fairly considerable body of work has been and is being produced by analytically or positivistically influenced philosophers of one sort or another.* In fact, there has been a very substantial revival of interest in political philosophy on the part of such philosophers since the days when T. D. Weldon's *The Vocabulary of Politics* (1955) was characteristic of the very little that was being produced. Since what may count as political philosophy tends naturally to have very close relations with philosophy of law, philosophy of the social sciences, moral philosophy and philosophy of history, it is not surprising that some of the most prominent writers in this field are also mentioned under these other headings; such writers include Hart, Berlin, Rawls, Barry and, although it would be stretching the sense of labels beyond all reasonable plausibility to regard him as any sort of analytic philosopher, Karl Popper. Reference to the principal writings of these philosophers is already made in these other sec-

---

political order is inseparable from a fundamental change in human life and relationships, from a metamorphosis of civilization; in the final analysis, the state of affairs to which it tends cannot be anticipated intellectually without the risk of reintroducing the currently dominating ideology. Once the yoke of relationships based on power, exploitation and indoctrination has been broken, the creative spontaneity of history will produce new forms. Hence the first task is, by counteracting *mystification* and stimulating the spirit of revolt, to liberate human beings at their very roots – in other words to liberate man's *desire* for total self-assertion.

These are some of the ideas which, compounded in widely differing proportions, seem to offer a broad definition of a particular way of thinking, a related group of intellectual approaches which are often stimulating but which it would be rash to judge in terms of the potential contribution they can make to constructive political philosophy.

For a cross-section of this line of thought, see for example: Marcuse, *Reason and Revolution. Hegel and the Rise of Social Theory* (1941), *Eros and Civilization. A Philosophical Inquiry into Freud* (1955), *One-Dimensional Man. Studies in the Ideology of Advanced Industrial Society* (1964), and *An Essay on Liberation* (1969); Sartre, *Critique de la raison dialectique* (1960); Fanon, *Les damnés de la terre* (1961) and Sartre's preface to this; Goldmann, *Marxisme et sciences humaines* (1970); Lefebvre, *Critique de la vie quotidienne* (1948, 1958), *Introduction à la modernité* (1962), *Métaphilosophie* (1965), *La vie quotidienne dans le monde moderne* (1968), and *La fin de l'histoire* (1970); Gabel, *La fausse conscience. Essai sur la réification* (1962); Glucksmann, *Le discours de la guerre* (1967); Baudrillard, *Le système des objets. La consommation des signes* (1968); Faye, *Théorie du discours. Introduction aux langages totalitaires* (1972); Foucault, *Folie et déraison. Histoire de la folie à l'âge classique* (1961, 1972); Deleuze & Guattari, *Capitalisme et schizophrénie. L'anti-Oedipe* (1972); Blackburn (ed.), *Ideology in Social Science. Readings in Critical Social Theory* (1972).

And, as examples of some adverse or critical reactions: Arendt, *On Violence* (1970); Aron, *Progress and Disillusion. The Dialectics of Modern Society* (1968) and *Marxismes imaginaires* (1969, 1970); Ellul, *De la révolution aux révoltes* (1972); MacIntyre, *Herbert Marcuse. An Exposition and a Polemic* (1970); Ruyer, *Eloge de la société de consommation* (1969) and *Les nuisances idéologiques* (1972).

* We are indebted for the following remarks and information to Alan Montefiore.

tions;[187] but special mention should here be added of John Rawls's recently published *A Theory of Justice* (1971/1972). This is an exceptionally wide-ranging essay in what may be called a neo-contractualist mode of thought, in which a systematic attempt is made to see what principles for the running of society might be agreed on by all rational agents were they placed behind a complete 'veil of ignorance' as to what their own positions in society were to be – an ignorance which is supposed to extend even to their own tastes and desires.

All the writers just mentioned may be regarded as continuing in one way or another the broad tradition of liberal political theory; and in this connection mention must certainly be made of a writer to whom Popper in particular recognizes an outstanding debt – F. A. Hayek, whose monumental work *The Constitution of Liberty* was published in 1960. It should not be thought, however, that this style of philosophical writing is indissolubly wedded to liberal political theory. R. P. Wolff, for example, in his *The Poverty of Liberalism* (1968), develops a critique of liberalism from within the individualist tradition; and C. B. Macpherson's earlier *The Political Theory of Possessive Individualism: Hobbes to Locke* (1962) shows a distinct Marxist influence.[188] Of course, not all of the political philosophy that must be regarded as in the forefront of British and American philosophical writing during this period has been of the analytic or positivist families at all. One most distinguished exception has been, for example, Hannah Arendt, whose work may be characterized as belonging to the grand traditional style of political, social and moral reflection. Among her works special mention may be made of *The Human Condition* (1958) and, more recently, *On Violence* (1970).

187. See, for example, in this same sub-section B, § 1, item (a) 'Philosophy of law', pp. 1278-1286; in sub-section A above, § 1, 'The logic of explanation in the social sciences', pp. 1225-1238, and, in § 4, 'Epistemology of historical sciences', item (a), 'English-speaking philosophers', pp. 1262-1267; finally, in Section V below, *Man and Action*, sub-section B, § 1, ' "Analytic" philosophy and the problems of human action', pp. 1392-1403.

188. During recent years there have been a number of major studies in the history of political thought, themselves constituting important contributions to the subject. For example: SABINE, *A History of Political Theory* (the first edition was published as far back as 1937, but the third edition of 1961 contains extensive rewriting and additional material); PLAMENATZ, *Man and Society* (1963); CUMMINGS, *Human Nature and Society* (1969); and WOLIN, *Politics and Vision. Continuity and Innovation in Western Political Thought* (1960).

It is possible to gain a very good idea of the way in which political philosophy as written by recent English-speaking writers has developed from an on the whole rather narrowly conceived and marginal exercise in linguistics and positivist philosophy to a much more richly varied and constructive type of conceptual enquiry from the three volumes of essays published successively in 1956, 1962 and 1967 under the title of *Philosophy, Politics and Society*; the first of these volumes was edited by Peter LASLETT, the second and third by LASLETT in conjunction with W. C. RUNCIMAN.

§ 2. *Historical materialism**

Historical materialism is a materialistic conception of the history of mankind – unlike pre-Marxist materialism. The latter assumes a scientific conception of the world as totality of physical, chemical and other like processes. Historical materialism is something different: in addition, it brings out the specific material basis of social life, which cannot be reduced to a natural process going on outside human existence. From the point of view of historical materialism, social being is material because it takes shape, exists and develops independently of social consciousness. Social life in its most general form is the union of social consciousness and social being. In this relation of opposites that have no separate existence, the determinative aspect is social being, or the material life of society, which is founded on material production. Material production in turn implies certain natural conditions which change in the course of its historical development. Social consciousness is essentially different from mental consciousness, as a property of highly organized matter, although it presupposes its existence and represents a higher (viz. social) level in its development. To put it differently, the possibility of the mental is conditioned, on the one hand, by the existence of the brain and, on the other, by the existence of the outer material world, whereas social consciousness presupposes in addition the presence of the social being of mankind, which is in a certain way related to a 'second nature' created by it.

In expounding this general thesis, Marxists complain that their critics impute to historical materialism a naturalistic understanding of the material and the spiritual, such as was found in pre-Marxist philosophical materialism.

(a) It is in fact **the theory of reflection** that distinguishes historical from pre-Marxist materialists, particularly those of eighteenth-century France. They understood the process whereby the outside world is reflected in men's consciousness in a simplified, non-dialectical way: for them, true human representations and judgements reflect the real, but ones that are incorrect, erroneous – for instance reilgious representations – do not. In this simplified understanding, the process of reflection received only a limited epistemological interpretation: truth differs from error in that it reflects reality. Actually, Feuerbach, in his doctrine of religion as reflection of alienated human essence, made a considerable step forward towards development of the reflection principle and its sociological application. However, he did not succeed in extending this principle to the entire content of consciousness; and for him the concept of reflection was synonymous with the concept of truth.

Historical materialism seeks the objective grounding of the content of consciousness, not in its relation to the truth or falsity of that content, but

* Based on an account written by T. I. Ojzerman as part of his general contribution to this work.

in its relation to the social being of men. A false consciousness is by no means devoid of content – witness the history of errors, the history of religion, the history of many ideologies. In consciousness the boundary between the real and the illusory is not given, at least not immediately: for instance, until Copernicus men were unaware that their representation of a moving sun and an unmoving earth was illusory. It is the gnoseological delimitation of essence and appearance that reveals the objective difference, which exists independently of consciousness, yet is not given in immediate sensible reflection of the real, in representation and in thought, independently of its practical or theoretical investigation in particular phenomena.

In this regard historical materialism is opposed to Husserl's phenomenology, which, it charges, took up in a garbled form this objective rule discovered by Marx and Engels and travestied it in its theory of intentional consciousness and transcendental reduction, which makes the difference between essential and inessential something merely parenthetical.[189]

The notion of reflection takes more than one form. Below the level of science, the most elementary act of the human consciousness – that is to say, the awareness that I myself am not an object I perceive – is the reflection of objective reality at the level of ordinary consciousness. At a second level, science, in so far as it attains to objective truth, is an adequate reflection of the real, conditional on the active, goal-directed and selective character of cognitive processes, and the development of special methods for research and for verifying the results obtained. However, alongside science there exist other forms of reflection of the real, the essential significance of which does not in any sense lie in a necessary relation to the objective truth they contain. The rôle of religion in the history of mankind is the best confirmation of this.

For Marxism, religion, at least in its contemporary forms, is an illusory reflection of the dominance exerted over men by the elemental forces of social development. Its fanciful images accordingly have a perfectly real objective content, of which, however, the believer is usually in total ignorance. The collective representations discussed by Durkheim and his followers are not a distinct spiritual, metaphysical reality: they are the result of men's acting in common and they exist in the heads of particular men, or in books, paintings, films or other objectifications of spiritual activities.

(b) Second topic: **the relation of individual to social consciousness.** Individual consciousness, like social consciousness, is a reality. Better, it is a uniting of opposites, a correlating of subjective and objective. This reality of the individual consciousness is not to be understood as something independent, that generates its own existence. The unity of individual and social consciousness does not efface the essential difference between these dialectical opposites, since individual consciousness partakes of the nature

189. The relation of historical materialism to Husserl's phenomenology has received special study in N. MOTROŠILOVA's monograph *Principy fenomenologii Ė. Gusserlja* (= The Principles of E. Husserl's Phenomenology) (1968).

of personality, is subjective in form and cannot contain within itself the multiplicity of social consciousness.

The unity of individual and social in the sphere of consciousness must also be distinguished from a unity of personal and social interests, which is far from being always realized in any social system. Within the underlying unity of individual and social consciousness, the consciousness of individuals is possessed of a relative degree of independence of social consciousness – though certainly not in the sense of containing anything that is in origin or character non-social. The antisocial, the asocial, also, are products of social development. Even the awareness the personality acquires of its particularity is conditioned by the specifically human form of existence and in a certain sense it, too, is social in character.

Thus the social character of the individual's consciousness does not debar individual peculiarities; conversely, talent, genius, and also individual sexual love, egoism, altruism must be considered products of social development, especially of the social division of labour – which of course does not in the least diminish the rôle of the personality in developing its own proper attitudes.

We must distinguish between developed individuality, as social phenomenon of an era in which social relations and needs are many and differentiated, and personality as characteristic attribute of man, marking him off from the animal. Even if his individuality be undeveloped on the cultural, occupational, moral and other planes, man is nonetheless a personality, that is, a being with a personal life, with personal needs, propensities, passions, interests, independently of the degree to which these propensities and interests are peculiarly and originally his. Individualistic contempt for the 'masses', historically bound up with thousands of years of rule by an exploiting minority over an exploited majority, often shows up in philosophy and sociology as anti-democratic denial of personality in 'simple' man, man who is insufficiently cultivated, having no share in the attainments of spiritual culture. Those who take this attitude lose sight of the fact that the tendency towards depersonalization of man in contemporary bourgeois mass society, 'consumer society', can only arise to the extent that human personalities exist: otherwise, a depersonalizing process would be quite simply impossible. Moreover it is human personality that provides the underlying stratum in the struggle of 'ordinary people' against all forms of social subjugation, exploitation, discrimination and war, and their efforts to bring about a rational reordering of social relations.

The basic forms of social consciousness are morality, art and ideology, which constitute qualitatively different types of reflection of social being. Not that the world of natural objects is not reflected in art or other forms of social consciousness; the point is that the procedure by which this reflection is realized is conditioned by social being. For example, in mediaeval art there is next to no landscape: and this is certainly to be explained by exclusively social causes. The domination of the elemental forces of nature over man which is reflected in primitive religion is not a natural but a social fact, depending on the very low level of development of the forces of production. The *relation of social being to natural being* is the central problem for historical materialism. According to this philosophy man differs from animal through the fact that consciously, deliberately, with the help of means of production he has created and perfected, he changes nature. Thereby, through a necessary though unconscious and spontaneous process, he also changes his own nature. This unity between the changes in external nature and in the nature of man, between subjective, conscious human actions and their objective result, is the basic starting-point of historical materialism, that which rules out all forms of essentialism and points the way towards a scientific understanding of social determination, excluding fatalism and voluntarism alike.

The social significance of men's productive activity, the significance of the social form of production, the rôle of production in the development of every aspect of social life – Marxists consider that theirs is the first philosophy to have posed all these problems in these terms. In their view, pre-Marxist and sociological studies envisaged production merely as external necessity. Now, the essential thing is not that without production human life is impossible but rather that production is also production of social relations, and therefore responsible not only for what has gone to make *homo sapiens* what he is but also for his entire further development. Because it has a social character, it is realized by means of definite relations which Marx and Engels called social relations governing production, or production relations. Production relations in their entirety form the economic structure of society, material basis of the political and legal structure and of the corresponding forms of social consciousness. Consequently, they are not merely forms of development of the forces of production but also the determinative basis of society's political and spiritual life, its transmutation and development. This complex significance of production relations is a very important characteristic of social life in its aspect as historical process. Ultimately, that is through a series of intermediate links, this historical process is governed by the development of productive forces.[190]

In challenging the notion that the necessary relation between material production and spiritual life is an external relation, Marx and Engels were opposing the idealist conception of history. For the founders of Marxism, material production not only creates objects of consumption but also engenders definite social relations, economic to begin with but also, though in a mediatized manner, political and ideological. Production of material and spiritual goods is by its very nature a social production. Production of such goods by an isolated individual ('Robinson Crusoeism', as Marx ironically put it) is an empty abstraction. That is why the first problem of the science of society is to investigate the social form of production.

190. On this subject, cf. APTHEKER, *History and Reality* (1955).

It is well known that the transition from feudal to capitalistic production relations contributed to the development of production and thereby to development of social relations as a whole. Marxism looks to socialist production relations to bring about a still greater acceleration in the development of productive forces and of all social progress. If, then, one pictures historical materialism as a doctrine confined to showing the determinative rôle of productive forces, one greatly restricts its content; worse, one falsifies its principal meaning. In all the multiplicity of social relations, it stresses the fundamental and primary rôle of production relations as determining, all told, every other relation of man. Research into the necessary link between production relations and others, both material and spiritual, is a major task of historical materialism. Since the development of production forces is accomplished in a historically determined social form, Marxist theory of social production is a theory of the unity of productive forces and production relations, giving rise to a mode of production which characterizes the level of historical development the society in question has reached.

In this overall approach the theory of socio-economic formations has played a key rôle by making it possible to elaborate a scientific theory of history conforming to historical materialism.[191] This theory consists in examining the phenomena of every historical period from the point of view of its specific structure, the elements of which form a system of social relations. A social system of this sort, covering long eras of history, constitutes a socio-economic form. Partial social phenomena like feudalism, the artisan system, the non-development of social division of labour, a barter economy, the predominance of country over town, the peasant community, the spiritual dictatorship of the Church, heresies, wars of religion and so on are scientifically explicable only as elements of a historically determined socio-economic system – the feudal system in this case – founded on a specific mode of production governing the property relations, the legal and ideological superstructure of the society, as well as the way in which all these social relations are reproduced within the framework of the system.

Thus only the concept of socio-historical formation provides a theoretical means for qualitative analysis of social phenomena in their structures and their regulated development.

An idealist theory of history is as a matter of principle incapable of explaining particular events in universal history as social revolutions. It necessarily becomes guilty of mystification, by transposing the content of the revolutionary transformation of social relations into an ideology which obscures its meaning. The task of historical materialism is to bring to light

191. On this subject, cf. the research done by the Bulgarian Marxist V. STEFANOV, *Teorija i metod obščestvoznanija* (= Theory and Method of the Social Sciences) (Russian translation, 1967) and the monograph by the Soviet philosopher B. A. GRUŠIN, *Očerki logiki istoričeskogo issledovanija* (= Essays on the Logic of Historical Research) (1961).

all the laws governing social revolutions, as the objectively necessary means for progressively transforming antagonistic production relations and their corresponding political and legal superstructures. For Marxism, the economic basis of social revolutions consists in conflict between the productive forces working for progressive development and antagonistic production relations whose transformation is constantly impeded by the ruling class which owns the basic means of production. Conflict between productive forces and production relations does not in itself suffice to generate a revolution; the oppressed classes and social groups in favour of the progressive transformation of production relations must also be made aware of the conflict. This is where the subjective factor comes in: mass awareness and organization of the masses, political preparation, determination, the readiness of a revolutionary class to seize the historic initiative.

The theory of the revolutionary situation was elaborated by Lenin.[192] He showed the mechanism of the transition from the spontaneous socio-historical process, during which the economic basis for revolution and the revolutionary situation are constituted, to the consciously realized transformation of social relations, to social creation on the part of the masses. In this conception of revolution the Marxist understanding of historical determination takes on its concrete character. Revolutions no longer appearing, now, as accidental, a violent rupture in the normal course of social development is seen as the highest form of the class struggle in an antagonistic society. This society itself results from the exploitation of men paralleling the exploitation of nature. Consequently, it is not surprising that the class struggle and the workers' liberation movement are a necessary expression of antagonistic production relations and the type of State to which they give rise – slave-owning, feudal, capitalist. The working masses are thus not only the chief productive force but also the chief political force.[193]

It is accordingly clear that mankind itself creates the objective conditions that determine its development. This position impugns not only supranaturalism, with its representation of transcendent forces directing the history of society, but also sociological naturalism, which would have it that history is determined by external nature and the nature of man himself. For Marxism, neither geographical conditions nor the other natural factors forming the *conditio sine qua non* of human life itself determine the character of the socio-historical process. Their changing – and in the course of history decreasing – influence on social life depends on the level of

192. Lenin, *Sočinenija* (= Complete Works), 5th edn. in Russian, Vol. 26, p. 219: 'Not every revolutionary situation gives rise to a revolution, but only a situation such that in addition to the objective changes listed above there is a subjective one, namely the fitness of a revolutionary class for mass revolutionary actions strong enough to break or crush the old government – which will never "fall", even in a time of crisis, unless it is "brought down".' (*translation*)

193. A. Gramsci did valuable work in developing the Marxist-Leninist theory of revolutionary activity, which he regards as one of the factors or, better, the criteria of social progress (Gramsci, *Opere*, Vol. II, 1953).

development of the society's productive forces, that is, of its means of production and also of its men, the most important productive force. The present level of the forces of production is partly inherited from the past and is partly the result of the activity of the human generations alive today. This correlation between living labour and accumulated labour, between the reified activity of living men and the reified activity of past generations, constitutes one of the most important economic factors in the objective determinations of history, wherein subjective and objective are mutually transformed one into the other – and it is this which qualitatively differentiates historical necessity from the necessary connection existing among natural phenomena.

In this way, historical materialism is opposed to abstract sociological theories in regard to 'irresistibly determinative historical trends' – a thesis so often ascribed to Marxists by its critics. The materialistic conception of history, Lenin explains, entails research into the real relation between objective trends in the socio-economic process and the activity of the classes and social groups that promote or oppose them. That is why historical materialism takes exception to philosophico-historical schematism, the idea of a pre-established plan for human history, the beginning of which would as it were contain, in a pre-formed state, its higher level. The contradictory unity of subjective and objective, the struggle of opposed social forces to determine which of various possibilities will be realized, precludes *a priori* the predetermination of historical periods, since the determining force of history is men themselves, the masses, the classes that create history under particular conditions which they partly inherit from the past but partly create through their own activity.

To superimpose necessity, regarded as an injunction from above, on the real historical activity of the masses is something quite foreign to the materialistic understanding of history. For instance it was possible for fascism to be viewed by Marxists as not accidental, since tied in with the anti-democratic tendencies of the ruling classes, and at the same time not inevitable. One must remember that men make their own history. On the other hand the socio-historical process is objectively conditioned: its course is independent of the consciousness of the men who actively participate in it.[194] The present has no power over the past – but also is only partially determined by it. New generations, in ever-increasing measure, create the productive forces they have at their disposal. Circumstances change men only to the degree that they are created or changed by men. Thus it is men, not gods, who create and develop the forces of production; and the historical necessity which is built up on this basis is the unity of the

194. The question of the relation of the classes and different social groups to the objective trends of the socio-historical process forms one of the most important facets in the Marxist solution of the problem of freedom and necessity. On this subject, cf. the work of the Bulgarian Marxist A. Popov, *Neobhodimieta i svobodata* (= Necessity and Freedom) (1957); cf. also Agosti, *Nacija i kul'tura* (= Nation and Culture) (1963).

living, objectifying and objectified activity of men.

It follows that the dialectic of subjective and objective in the socio-historical process precludes any one-sided determination of human activity, although this is a conception often attributed to Marxism by its critics. The unity of conscious human activity and objective necessity does not consist merely in men's submitting to that necessity once they are aware of it. Historical necessity itself represents the unity of the objective and the subjective. It is objective, as a process conditioned by the previous development of society; it is subjective as the need, interest, actual activity of particular classes, social groups, parties. That is why objective conditions can play the determinative rôle that is theirs in social life only to the extent that they are manifested, intervene and actualize themselves in human activity. This in its turn, to the degree that it is objectified, thereby creates objective conditions and itself plays a determinative rôle in the development of society. This dialectic of subjective and objective, which takes place not only in the sphere of production but in all other fields of activity, can be characterized on a broader philosophical plane as the unity of the processes of externalizing human activity and internalizing its objective conditions.

At a time when the scientific and technical revolution is radically transforming production and consumption, many philosophers and sociologists who are non-Marxist or even anti-Marxist are inclined to emphasize the decisive rôle of production in social development; and it even happens that on this score they may show a certain measure of agreement with historical materialism, while continuing to interpret it as a technological conception of the historical process. This superficial view of Marxist social theory is usually based on a one-sided interpretation of isolated formulas of Marx and Lenin, especially on a confusion between productive and technical forces. It is forgotten that the most important productive force is man himself, with his creative talents, which develop historically. To be sure these forces are independent of men's consciousness and will, but every individual realizes and develops his own potentialities himself, without being able to hand the responsibility over to others.[195]

The identification of productive forces with technology corresponds in the first place to an over-simplified understanding of the motive forces in the development of production, one which fails to take into account the complex interaction between human activity and existing technical means of production. In the second place it reflects a failure, typical of the idealist interpretation of history, to attach sufficient importance to the rôle of the popular masses in the development of production and society in general. Here Marxism is opposed to theorists who, weighing the likely consequences of the scientific and technical revolution, come to the conclusion that in future the production process will be carried on exclusively by the tech-

195. On this subject, cf. the work of certain German Marxists: KLAUS & SCHULZE, *Sinn, Gesetz und Fortschritt in der Geschichte* (1967); BOLLHAGEN, *Gesetzmässigkeit und Gesellschaft* (1967).

nical means of production themselves, that is to say, without the active participation of workers. In the third place, to equate productive forces with technology means divesting the production process of its specific social form (production relations) and thereby dodging the question of abolition of private ownership of the means of production; for example, the productive forces are geared to radically different ends in the capitalist and the socialist modes of production.

Finally, historical materialism poses the traditional philosophical problem of freedom and necessity in terms of concrete history, by characterizing freedom as control by man of nature, of himself and of his social relations. Hence there is a real connection between the advance of productive forces, the progressive transformation of social relations and the development of freedom. In an antagonistic society, the development of freedom has an antagonistic character – for example liberation in regard to the elemental forces of nature can co-exist with enslavement of the personality by the elementary forms of social development, as is the case in capitalist society. The socialist transformation of social relations, the building of communism, puts an end to the antagonistic development of freedom. It was in this sense that the founder of Marxism characterized the era of transition from capitalism to communism as a leap from the realm of necessity to the realm of freedom. But the basis of the realm of freedom – the expression is of course metaphorical – is the realm of necessity.[196] According to Marxist philosophers, the socialist transformation of social relations obeys qualitatively new objective rules characterized by the increased importance of the subjective factor,[197] though this does not mean that the rôle of objective conditions is diminished: on the contrary, it is through this that their influence necessarily and specifically makes itself felt.

As a contribution to elaborating the Marxist theory of scientific communism, historical materialism is today much concerned with the philosophical aspects of constructing communism: the relationship of social consciousness and social being, the dialectical transformation of necessity into freedom, the unity of personal and social interests and the rôle of the popular masses in developing socialist society.[198]

196. MARX, *Capital*, Vol. 3, in MARX-ENGELS, *Complete Works* (Russian-language edn.), Vol. 25, Pt. 2, p. 387: 'The true realm of freedom, which however can only attain to full development in the realm of necessity, as on its basis.' (*translation*)

197. Cf. the collective work of Soviet philosophers, *O dialektike razvitija sovetskogo socialističeskogo obščestva* (= The Dialectic of the Development of Soviet Socialist Society), edited by N. V. FILIPENKO (1968).

198. See also, for similarly oriented work on practical problems, the collective work by German and Japanese Marxists, *Historische Materialismus und Sozialforschung*, edited by H. SCHELER (1966), with essays by H. SCHELER, G. HEIDEN, H. ULLRICH, R. THIEL, W. FRIEDRICH, KAMBA Toshio, SHIBATA Shingo, TANAKA Seisuke, KAZAI Ioshishige, and MATZANARI Ioshie.

Cf. also the research done by certain Polish Marxists: SZCZEPAŃSKI, *Elementarne pojęcia socjologii* (= Elementary Notions of Sociology) (1963) (Russian translation, ŠČEPAN'SKIJ, *Èlementarnye ponjatija sociologii*, 1969); WIATR, *Szkice o materializmie*

§ 3. *Man and social reality in a philosophical anthropology*

We shall look, to end with, at a way of philosophizing on social reality which puts the emphasis on man himself. There is a close relationship between man's way of being and the way of being of social reality which has been brought out both in Marxist philosophy and in reflection on legal structures and models of inter-personal relations. In either context the relationship emerges as paradoxical: on the one hand, man makes his history, on the other he is made by it; he constitutes social reality and is constituted by it. Reflecting on this paradoxical relation is the task of ontology – an ontology we shall call philosophical anthropology. It draws sustenance from the Eastern tradition as well as from the Western. In India and Japan, the problem has its roots in the distant past, even when its formulation is modern.* In Western thought, political philosophy bears the imprint of a multiple, though more recent, heritage – Kantian, Hegelian, Marxist, Kierkegaardian, Nietzschean.

We shall arrange our treatment of the subject under three heads, borrowing in turn from these different traditions and heritages.

(a) *Freedom and institutions*

The dialectics of freedom between the individual and society has been seen and formulated in significant ways in Japanese and Indian thought. While Seizo Ohe emphasizes the dialectical relationship between nature and culture in the context of human freedom, Mutai Risaku does so in regard to what he calls the social and the existential dimensions of man. Tanabe, on the other hand, has brought out the significance of the dialectical relationship between mankind, the nation-state and the individual for the preservation and furtherance of human freedom.

Ohe argues that human behaviour at the voluntary level, where alone the issue of freedom arises, presupposes both relevant knowledge and emotional concern and that both of these require some sort of mediation and control mechanisms. He calls these the 'intellectual mediation' mechanism and the 'emotion control' mechanism. The relation of nature and culture in regard to both of these, according to him, is such that *culture frees man from nature and nature frees man from culture*. The first part of the dialectic is generally better known to most persons than the second. The emotional patterns first formed in the sociocultural process of human growth get stereotyped in the course of time, and can only be broken from their rigid moulds by unbridled passion or, in other words, by a return to the original bio-natural sensations of pleasure and pain. Similarly, the public body of knowledge which any established culture always has, tends to become rigid and

historycznym i socjologii (= Essay on Historical Materialism and Sociology) (2 vols., 1967).

* The references to Eastern philosophies in the various subdivisions of § 3 are due to Daya Krishna, who kindly wrote the relevant passages himself.

reject whatever is not in consonance with the established paradigm of knowledge, and this rigidity needs to be broken from time to time by creative genius which questions the fundamental assumptions of the framework itself. The freedom from nature, on the other hand, which culture bestows on man is too well known to be adumbrated here. Thus in both domains, of knowledge and emotions, there is a constant dialectical interplay between nature and culture which alone sustains and assures the freedom of the individual, who belongs to both realms simultaneously but, perhaps, dialectically. The recurrent demands for a 'return to nature' would thus be as essential to human freedom as the aspiration and demand for this return to be patterned into culture once again, without which it would be just a return to the instinctive world of the animal, which man could never be.

The dialectics which seems to give freedom to men, according to Risaku, consists in the interplay between the social and the existential dimensions of man, which, in a sense, constitute the total human reality.[199] The relation between the two, however, is not one of equality, for the existential presupposes the social for its existence. Yet it is the existential aspect which is subtler and responsible for the essential humanizing of man, and also for preventing history from stagnating and lessening the estrangement of man.

For Tanabe, on the other hand, the dialectics which ensures freedom is that between the individual, the nation-state and mankind.[200] The nation mediates between the individual on the one hand and humanity on the other. Yet, unless there were the larger perspective of mankind before the individual, it would be difficult for him not to be submerged by the nation. On the other hand, mankind is too amorphous and general a grouping to ensure the individuality of the individual unless there is such a concrete socio-cultural entity as a nation to which he may belong.

Unless the trans-social dimension of man is given a co-ordinate, if not superior, status to his social being, it is contended by many Indian thinkers that freedom of the human individual can neither be assured nor perhaps even fully understood. Vivekananda, Aurobindo, Tagore, Radhakrishnan and Bhagwan Das may be said to belong to this tradition.[201] Man's autonomy thus has to be safeguarded not only against nature but against society also. However, the relation has not been conceived in a dialectical manner, as appears to be the case with Japanese thinkers.

In Western thought, especially since Kant, the usual way of relating social to human reality is to show in the institution in general how freedom, in itself arbitrary, becomes *rational*. There are two advantages to this way of

199. RISAKU, 'Two conditions of human reality' (1959).

200. TANABE, 'The pressing need of the philosophy of politics' (1946).

201. VIVEKANANDA, *The Complete Works* (1955 edn.); AUROBINDO, *The Life Divine* (1949); *The Human Cycle* (1949); *The Ideal of Human Unity* (1950); RADHAKRISHNAN, *An Idealist View of Life* (1932); DAS, *Essential Unity of All Religions* (5th edn., 1955).

approaching the problem. On the one hand political reality, in the broadest sense of the word, is no longer mere subject-matter for a descriptive or explanatory science, as it was in our narrower treatment in § 1 above. It becomes a project of man himself, to the degree that it is involved in the concrete realization of freedom in history. On the other hand, and conversely, freedom no longer has the guise of an abstract postulate or illusory claim unless crowned by the actual winning of particular freedoms – political and personal, economic and social. There is, then, a mode of philosophical thought for which political reflection is not an optional detour but one of the main checkpoints on the itinerary. Indeed, abstract reflection on freedom – i.e., reflection which excludes from the process of reasoning any political considerations and even makes such exclusion an essential part of the metaphysical interpretation of freedom – is in danger of degenerating into ideology. What was at first a genuine reflective conquest turns into an instrument for concealment as regards the concrete, material conditions in which freedom is exercised and even into an instrument of repression as regards those conditions – such as exploitations and alienation – which hinder its effective realization. For reflection abstractly considered, freedom is merely the capability of saying yes or no. And one has direct access to this capability without a lengthy detour via the political and the social, simply by turning back on oneself. But this reflection dissembles its abstract character unless it faces up to the question as to the real, concrete status of reflection itself.

Kantian philosophy can hardly be exonerated on this count. Not that it was blind to the essential nature of the problems raised by right. In the *Metaphysics of Morals* we find the following definition: 'Right is the sum total of conditions making it possible to reconcile our free will with that of others in accordance with a universal law of freedom.' In Kant there is far more, even, than a philosophy of right – a philosophy of culture and a political philosophy which find their consummate expression in his *Zum Ewigen Frieden*. This analysis of the conditions of freedom even takes a very concrete turn: Kant was constantly preoccupied with the play of forces which in society oppose each other and contend until, eventually reconciled, they usher in cultural and political conditions that favour the reign of freedom based on morality. But, however highly one may rate Kant's writings on right, the State, war and peace, one cannot regard them as part and parcel of critical philosophy. And if they remain external to the three *Critiques* it is because the political realization of freedom forms no part of the definition of its concept – only of the external conditions favourable or unfavourable to its establishment. To have determined the essentially moral character of freedom is an achievement sufficient unto itself, its sole value lying in the provision of a foundation or ground. For this reason critical philosophy remains a philosophy of intention, more precisely moral intention, not of the actual realization of the reign of freedom in history.

This failure of the moral philosophy deriving from Kantianism can be overcome only if we try, simultaneously, to define freedom and social institutions respectively in such a way that the need for each can be deduced from the other. This is the course taken by a number of philosophers, in reviving certain aspects of Rousseau's and Hegel's thought.

Let us cite Rousseau's statement of the problem – which is, he said, 'to find a form of association which will defend and protect with the whole concerted strength the person and property of each associate, and through which each associate, in joining all the others, will yet obey only himself and remain as free as before.' This statement gives what at first sight looks to be an insoluble antinomy the form of a problem to be solved. The problem is to find a common foundation or ground for both individual freedom and the community: it is in fact to show the simultaneous emergence of sovereignty, enthroned above all, and of freedom in its civic form – in short, to make what gives rise to institutions coincide with what gives rise to freedom.

In recent years a great deal has been published on Rousseau's *Contrat social.*[202] Some writers charge that the social contract is a fiction without historical basis. Others see it as a foundational act implicit in every body politic, which it is the task of political philosophy to seek to make explicit. To generate at once freedom and sovereignty, this act necessarily takes the form of a *dispossessing*. It is an 'alienation' in the legal sense of the term – that is, a transfer to others – but an alienation the beneficiary of which cannot be another individual (for example the Prince, as in Hobbes) but the community as a whole. This alienation of each to all, thought Rousseau, cannot lead to coercion of one man by another: 'since each gives himself up to all, he gives himself up to no one'. Equality in relinquishment guarantees the citizens' mutual independence: 'since each gives himself entirely, conditions are equal for all; and conditions being equal for all, no one has anything to gain by making them onerous to others'.

But it has also been abundantly stressed that the provision of the social contract is not so much the solution to the problem as a simple repetition of it, the enigma merely being carried back to the celebrated 'general will' which is not a sum of particular wills.

That is why some philosophers are returning to Hegel: not of course with the impossible design of simply retracing his steps in the twentieth century but because they find in him the example of a dialectical construction in which the sort of surmounted ('sublated') contradiction which Rousseau verged on with his 'general will' is deployed in a series of consecutive stages instead of being crystallized in a single, all-embracing social compact.

202. Cf., for example, *Etudes sur le* CONTRAT SOCIAL *de Jean-Jacques Rousseau. Actes des journées d'études tenues à Dijon les 3, 4, 5 et 6 mai 1962* (1964); DERATHÉ, *Jean-Jacques Rousseau et la science politique de son temps* (1949, 1970); by the same author, *Le rationalisme de Jean-Jacques Rouseau* (1952); ALTHUSSER, 'Sur le *Contrat social* (Les décalages)' (1967); BERMAN, *The Politics of Authenticity* (1971); POLIN, *La politique de la solitude. Essai sur J.-J. Rousseau* (1971); etc.

What these 'Hegelizing' authors seek in Hegel is therefore twofold. First, the concept of freedom and will which had served as a benchmark for abstract reflection, from the Stoics to Kant, must be completely recast in such a way as to make room for the establishment of institutions. Secondly, the concept of the institution, retrieved from jurists, economists and students of government, must be recast in such a way as to delineate a quest for freedom.

Thus, on the one hand, this re-examination bears on the concept of freedom: freedom to do no matter what is not yet concrete freedom, but merely its first element, the simplest and poorest. Isolated, it is nothing but the 'freedom of the vacuum' that is expressed on the political level by the prior rejection of institutions. This first movement of indifference, indeterminateness, must be met with a second, counter movement – whereby we enter into particularity. We are not free so long as we abstain and hold ourselves completely aloof from all determination, for we have not yet done, or *made*, anything. To do – to make – is to positively will and thus take over as one's own the finished form of a work; in this way, concrete freedom is freedom which accepts for itself the law governing such work, which is itself the law governing what is accomplished. It is works in all their forms – technological, artistic, intellectual, etc. – which express the concrete dialectic of freedom.

Institutions, on the other hand, no longer take the guise of something given, an observable reality, spread out for inspection by a descriptive and explanatory science. They are works, in this same sense of Hegelian determinateness or particularity. Right or law as a whole, then, can be regarded as the sphere within which freedom is realized in the type of works we call institutions; and 'institution' will mean a set of rules relating to the acts of social life by virtue of which each person's freedom can be realized without impairing the freedom of others. This definition has the merit of mutually engendering institutions and freedom. It transfers institutions to an intelligible dimension which we can say, following Hegel, comprises the sum total of works in which freedom is realized. By laying down, right at the outset, the reciprocity of law and freedom, such a definition shatters two abstractions – the abstract conception of law which would contain no reference to freedom and would confine legal science to the study of positive law, and the abstract conception of freedom as something quite extraneous to the establishment of institutions.

It is at this point that the philosopher comes up against the political phenomenon *par excellence*, namely the State. The first thing that strikes him about it is this alien, disturbing, dangerous force which he calls power – the power of decision and of coercion. And this power is, indeed, at the heart of the political phenomenon of the State. Eric Weil, as we have seen,[203] defined this as follows: 'The State is the organization of a historical community. Organized into a State, the community is capable of taking

203. Cf. above, § 1, 'Philosophy of law and political philosophy', (b), 'Political philosophy', pp. 1286-1295.

decisions'.[204] By defining the State in these terms, the philosopher puts it squarely within the sphere of a philosophy of freedom. Is not decision-making just what reflective philosophy attributed to free will? And so philosophy is forced to relate the sovereign power of deciding, which is the State's, to the arbitrary power of choosing, which is in principle the property of every individual. In short, philosophy is forced to give an account of the relationship between our powers and power.

Resolution of this enigma is what gives political philosophy of Hegelian inspiration its aim and its meaning. For this philosophy, the organic set of institutions which constitutes the State would be profoundly meaningless did it not make restitution and diffuse a new quality of freedom to the individuals who are its members – a quality which is inseparable from the mere fact of living in the State and bears the stamp of the 'concrete universal'. What the State means is to bring about the convergence of the joint will and of subjective freedom, experienced in the individual consciousness as the pursuit of a particular goal.

But what does Hegel have in mind when he speaks of 'the State'? For some, he is the theorist of submission to a 'divine State'; for others, he simply described one empirical State, the Prussian; for others again, despite his virulent criticism of idealizations and abstractions he himself projected an ideal, a Utopia. Perhaps he did none of these things. For all that he is accused of deifying the State, Hegel never held political reality to be the highest reality: it is such only from the point of view of a philosophy of right, that is to say one limited division of Objective Spirit. Above and beyond the State there still reigns the tribunal of history, and above and beyond that the trilogy of art, religion and philosophy. The State, then, cannot possibly be the final instance of truth and reality. Nor ought we perhaps to apply too strictly to the Hegelian theory of the State the dichotomy between empirical and *a priori*: the State described by Hegel is rather an idea at work, a meaning in gestation. Not that there is a State anywhere in the world which fully realizes this meaning, but every modern State, *qua* State, exists only through this idea at work, in the manner of a seed. This idea of the State, then, is neither the description of an empirical State nor the projection of a Utopia. It is a teleological reality, of which signs are displaced by every State that is not a pure tyranny, and which can be deciphered through such signs. The final cause of every State then is to provide the locus where on the one hand a community accedes to decision – that is accedes, in a certain way, to collective freedom of choice – and where on the other hand individuals have a rôle and a task, carry out activities, pursue and achieve goals, which give them a certain satisfaction and which heighten their own consciousness. Thus the State is a very dense and pregnant reality – substantial, says Hegel – like the community which subtends it, and is maintained by its own strength. At the same time it is a reality willed by responsible citizens: it is in each indi-

204. WEIL, *Philosophie politique* (1956), p. 131. See also the same author's *Hegel et l'Etat* (1950).

vidual that the State becomes conscious of itself; it has its objectives, but they are effective only if recreated in thousands of individual purposes. In this sense it is, in Hegel's phrase, 'the actualized reality of concrete freedom'.

A political philosophy faithful to the Hegelian tradition will not therefore begin by stressing power and violence: the first thing to think of in the State is not conditioned coercion, but its constitution. In the constitution the community takes form, deliberate form; in it the community knows itself and is known. But in the constitution, also, freedoms are defined and interrelated, pass into the realm of the feelings and also into that of reality. Thus the constitution is perhaps the sole point in human experience at which freedom and necessity are united: freedom as expressed in freedoms, necessity as embodied in public authority. This idea of a State, germinating in actual States, is the highest pitch of thought in political philosophy. This actualized idea, if one may so express it, is not external to the philosophy of freedom but rather represents an advanced stage on the road towards the realization of freedom.

(b) *Power and violence*

The above analysis rested on the conviction that the real and the rational coincide in the meaning which the individual's life acquires within the State. The meeting of freedom and institutions is, indeed, the very exemplar of this coincidence of the rational and the real.

But for another kind of analysis, more existential and more tragic, the very locus in which this reconciliation comes about has an incurably irrational side. Whereas a political philosophy of Hegelian stamp sees the State from the standpoint of an institution, another analysis, more attentive to the play of underlying forces and the irruption into politics of emotions and subconscious drives, places the emphasis on a curious paradox, namely that it is at the highest level of rationality that we find the most intractable irrationality. The rationality displayed in life within the State has for its 'dark side' the violence which seems to be inseparable from the exercise of power.

This kind of consideration is familiar to the Oriental schools of thought. For Indian philosophers, the basic issue in relation to power and violence seems to relate to the question whether the violence and injustice of an authority entrenched in power can be met with without the perpetuation of violence and injustice on one's own part. The issue was raised long ago by Buddha. Since then, many thinkers have raised the problem off and on. But it is only in recent times that the theoretical and practical aspects of the issue have received a sustained and unparalleled attention in the work and life of Mahatma Gandhi which, in his own words, was a continuous experiment with truth. The voluminous writings of Gandhi[205] have hardly

205. His *Collected Works* have been published in 1960.

attracted the interest or attention of political or moral philosophers yet. Nor has a sustained attempt been made to reflect theoretically on his actual practice and the various techniques that he employed in his fight against injustice and tyranny. There has been some interesting work on the part of G. N. Dhawan, Joan Bondurant, Arne Naess, Gene Sharp, Frederic Streng and others.[206] Also, he has had tremendous influence as an example on many of the non-violent resistance movements after the Second World War, the most important of which undoubtedly was the one led by Martin Luther King, Jr., in the United States. The philosophical links that perhaps come closest to the ideas of Gandhi may be found in the philosophy of dialogue that is associated with the name of Martin Buber in contemporary philosophy.[207] But a real philosophical articulation of Gandhi's thought and practice still awaits the future.

The other, and perhaps even deeper, problem may be said to relate to the question of whether a non-violent polity is possible. Or, in other words, can one *rule* non-violently? Buddha formulated the problem when he pondered 'over the question as to whether it is possible to reign with dhamma, without killing or causing to kill, without conquering or causing to conquer, without grieving or causing to grieve'.[208] Buddha's answer seems ambiguous, though it perhaps leans more to the view that it is not possible. In fact, in the traditional story, the tempter appears at this point and 'asserting its possibility, prompts him to take up the life of a king'. The Buddha sees through the temptation and rejects the office.[209] Gandhi, on the other hand, never faced the problem directly, though his recommendation to the Congress Party to dissolve itself after the achievement of independence and his refusal to accept any office in the government may be taken as pointing in the direction of Buddha's answer. But if that is the answer, the relation of power to violence would seem intrinsic and necessary, and this makes any hope of a non-violent solution impossible in principle.

This situation of conflict emerges from the Hegelian analysis itself. Every institution is at once a victory over the arbitrary will of individuals and an occasion for exalting the arbitrary will of the man or men in power. This arbitrary character inherent in the concrete exercise of power resuscitates the arbitrary will of the individual and provokes a defensive and offensive rejoinder from the side of individual freedom. This exaltation of the arbitrary, manifest even in the victory of institutions over the untamed freedom of individuals, forms the paradox of politics. This is true even of

206. Dhawan, *The Political Philosophy of Mahatma Gandhi* (1946); Bondurant, *Conquest of Violence: The Gandhian Philosophy of Conflict* (rev. edn., 1965); Naess, *Gandhi and the Nuclear Age* (1965); Sharp, *The Politics of Nonviolent Action* (1971); Streng, 'The ethics of moral coercion: Gandhi and political revolution' (1973).

207. Buber, *Ich und Du* (1923) (English translation, *I and Thou*, 1970); *Das Problem des Menschen* (1948) (in English, *Knowledge of Man*, 1965).

208. Upadhyaya, *Early Buddhism and the Bhagvadgita* (1971), p. 484 (quoted from *Saṃyutta Nikāya*, I. 116).

209. *Ibid.*, p. 485 (*Saṁyutta Nikāya*, I. 117).

penal law, which Hegel saw as the first stratum of institutional life, hence the first stage of rational freedom. Punishment is rational to the extent that it annuls by violence the violence that was annulling the law; but this rationality by negation of negation still gives free rein to an untamed underlying surge of anger and revenge which continues to infect the justice of men. This ambiguous mixture of rationality and violence which we find in the penal code reappears in more dangerous forms at the political level proper, which forms the apex of the Hegelian philosophy of right. As sovereign power, the State displays, raised to the highest pitch, the admixture of irrationality that attaches to the very notion of sovereign power – first in its internal exercise, in the individual and subjective form pertaining to the taking of decisions; secondly in its external exercise, in the form of opposition between individual States or, rather, between State entities and, ultimately, in the form of war. Great weight has been laid on these two aspects by Julien Freund in his major work on the *Essence du politique* (1965).[210] Following Carl Schmitt,[211] he sees the friend-enemy relationship as one of the essential categories in politics. But the internal exercise of power is no less subject to arbitrariness: we know of no internal organization of powers that does not adjoin to the legislative power – which defines and establishes the universal, i.e., the law for all – and to the executive power – which subsumes particular spheres and cases under the general rule – the practice of a subjectivity that has the supreme power of decision in the State. The subjective element in power seems to be to the sphere of constitutional law what the residual element of revenge was to the sphere of penal law. The progress made by rationality towards bringing under its sway, in modern States, the legislative and the executive power has as its counterpart the irrationality of this third element which Hegel called the power of the prince; in this respect the category command-obey, essential to the internal exercise of power, seems to be as irreducible as the friend-enemy category, essential to its external exercise. It is perhaps a fact that we must recognize that it is the State, which represents the most momentous endeavour in the history of man to totalize the endless swarm of individual existences and experiences in a single coherent configuration, it is this rational totalization that gives rise to a specific evil, that of counter-rational totalitarianism.

Philosophers are divided when it comes to interpreting the violence of power. For Marxists, politics is a superstructure and the ills that affect it echo contradictions at the underlying socio-economic level. For others, who adopt a pessimistic view of politics, there is a special pathology of politics which calls for a special therapy: what is more, these authors are afraid

210. See also § 1 above, 'Philosophy of law and political philosophy', (b), 'Political philosophy', pp. 1286-1295.

211. See FREUND's preface to SCHMITT's *La notion de politique* followed by *Théorie du partisan* (1972), the French translations (in a single volume) of *Der Begriff des Politischen* (1927, 1963) and *Theorie des Partisanen* (1963).

that the attempt to suppress economic exploitation and reduce inequalities and injustices has made the political power still stronger, by giving it the monopoly of choice in all fields of economic, social and cultural life, and presented it with a cast-iron excuse for a recrudescence of tyranny. This is an important point at issue between Marxist and non-Marxist philosophers: will political compulsion be eliminated once the various forms of alienation resulting from individual appropriation of the means of production are done away with? or is it part and parcel of the exercise of power, from the moment any historical community organized as a State becomes capable of taking decisions? The solution of this question is of great consequence. If the political form of alienation, relatively to all other forms, is *sui generis* and autonomous, it is quite impossible to infer political progress from economic and social progress. By the same token political liberalism no longer goes hand in hand with economic liberalism. If we give the name 'political liberalism' to the various technical means employed to bring about the separation and mutual limitation of powers, and to achieve a balance between the private sphere of freedom and the public sphere of power, then the question is whether this highly pragmatic art of negotiation is necessarily bound up with the ruling class of the capitalist era or whether it expresses values that transcend the sphere of interest of this class and have a genuinely universal bearing. In the former case the concept of liberal government has no application outside the economic and social sphere in which it was conceived. In the latter case, it is not tied to the social theories and practices of capitalism, and can be transplanted to other soil; what it then denotes is the art of dealing with and resolving the disputes that continually arise from the encounter of two arbitrary wills – that of the individual and that of the State.

(c) *Violence and discourse*

For a philosophical anthropology, reflection on the violence associated with political power leads to reflection on the scope of the whole phenomenon of violence and on its peculiarly human significance.

The fact is that only man is violent. We are quite off the mark when we talk of the violence of nature – hurricanes, tidal waves, earthquakes. Nature is not violent; she only appears so when she puts an obstacle in the way of some human scheme for using physical forces. She is only violent when she breaks man's will to bend her to his designs, when she carries off, topples, destroys his towns and his goods, his works and his very life itself. It is also quite wrong to speak of the violence of our desires, pleasures or sufferings. To be sure it is no longer nature outside us but nature within us that is unleashed, but this nature, once again, only seems violent to us by contrast with our peculiarly human idea of taming the underlying energies of life, harnessing them to cultural ends, in short sublimating them.

If nature without and within us is violent only in relation to some human design for dominating physical and vital forces, the reason is that man alone is violent. And he is violent in his relationship to other men. Strictly speaking, violence begins with murder, that is, the act of deliberately taking another man's life. All other forms of violence are stages on the road to murder, or else variations on it. For one can kill a man in many ways, according to whether one breaks his body, his sense of human life or his self-respect and human dignity; by torture, deprivation of liberty, racial, political or religious discrimination, by stifling personality or by subjugating thought and preventing its expression.

But of all the forms of violence which man is given to employing against his fellow man, private violence is not the most baffling. Murder as the extreme form of private violence poses no very difficult problem. All civilized societies set out to eliminate it by substituting a civil order for the struggle of each against all, and by removing the practice of violence from the individual. Violence is truly a problem when (as in the foregoing pages) we find it linked with the creation of that very political order which is intended to replace the practice of violence by the individual.

It is when violence proves to be thus bound up with man's political existence that it takes on its tragically human meaning. To be convinced of this we need only think of Plato's meditations on the tyrant, true antithesis of the philosopher, and the prophets of Israel with their denunciations of the high and the mighty; we need only remember that it was the Greek city-state that killed Socrates the Wise and the Roman Empire that killed Jesus the Just. The most ineradicable violence actually seems to be connected with the very exercise of power. As the preceding analysis suggests, political evil seems to be commensurate with political greatness, which is irreducible to any other forms of human greatness, whether these result from work or property or exchange. Violence is the counterpart and, as we said before, the 'reverse side' of this greatness.

But the shocking thing about this violence is that it appears as identical with the established order. So we can speak of a *state of violence*, no longer simply of an *act of violence*. People tend too much, in fact, to equate violence with disorder and with the most spectacular forms of disorder, such as bloody riots. We forget that lasting deprivation of material goods, social advantages and cultural benefits is for whole groups of disinherited persons a veritable state of violence. Violence is not merely acts that explode in a situation of crisis and rupture; it is also a continuing condition endured without crisis or fanfare by an entire population.

It is because of this coincidence of violence and order that social change likewise takes the shape of violence. When this happens violence is called revolution. It is in some degree the counterpart of the state of violence that may coincide with the established order. Every modern State without exception is the result of some revolution in the remote or recent past. Revolution, like war, is a universal phenomenon of which no group, no nation can claim to be innocent. It reflects the fact that throughout history most States have been incapable of reforming themselves in step with the profound changes that take place in society. At a given moment power ceases to serve the interests of the whole community. For the very reason that it was born of the monopolization of legitimate physical coercion, it may join its lot indissolubly with that of a decaying social class or a governing group – be it merely the political class, leaders of the community or the 'Establishment' – whose interests no longer coincide in some way or other with the common good. In that event revolution becomes the sole resource of the outcasts and the dispossessed, of all those for whom the State in its present form symbolizes the state of violence they undergo, now that its capacity for reform and for adjustment to social change is exhausted.

But then – and this the philosopher has to repeat – resort to violence opens up terrifying possibilities by unleashing obscure forces which the revolutionaries will not necessarily be able to control. Revolution calls for a nicety of calculation exceeding that required for the exercise of power regulated by the law.

If we now ask the philosopher why he cannot undertake the apologia of violence, seeing that he recognizes its indelible mark on political history, if we ask him whether violence is always wrong, even when it is a lesser evil as in the case of wars of liberation or legitimate popular revolts, his answer is not in doubt: *violence is the opposite of discourse.* If we were not convinced of this, violence would not be a problem. A being who speaks and who by speaking pursues meaning, a being who has had experience of discussion, that is, of the search for agreement by means of language – it is for such a being that violence is a problem. Violence and discourse are the two opposite poles of human existence. This is what Eric Weil brings out at the beginning of his *magnum opus*, the *Logique de la philosophie* (1950).[212] Violence is always the interruption of discourse: discourse is always the interruption of violence. A violence that speaks is already a violence that is trying to be in the right, that is exposing itself to the gravitational pull of Reason and already beginning to renegue on its own character as violence. The prime example of this is that the 'tyrant' always tries to get discourse on his side. The tyrant, for Plato, is the opposite of the philosopher, the man of rational discourse. But in order to succeed tyranny has to seduce, persuade, flatter; it has never been the dumb exercise of brute force. Tyranny only puts itself across to the public by perverting language. The tyrant prefers the sophist's services to the executioner's; he needs the sophist to find words and phrases that stir up hatred and involve others ineluctably as accomplices in his crime. The aura cast by discourse is no less sought after by the violence that is war. As for revolutionary violence, it hardly needs saying that it can only succeed in harnessing people's energies if revolutionary thought, working through words, has already made them conscious of being subjected to a state of violence and has led to

212. 'Introduction. Philosophie et violence', pp. 1-86 in *Logique de la philosophie.* See also, by the same author, *Philosophie morale* (1961) and 'La morale de l'individu et la politique' (1965, 1970).

the formulation of a revolutionary plan attested by a number of key phrases that put hope into words.

This misuse of discourse by violence gives food for thought. It means that violence forces its way into history only by seducing its opposite, discourse. It is the task of the philosopher to denounce such miscegenation and restore the fundamental truth that violence and discourse are the most fundamental opposites in human existence. It is because we, as men, have chosen discourse – that is, discussion, seeking agreement by means of verbal confrontation – that the defence of violence for violence' sake is forever forbidden us. We have no other reason for considering that violence is an evil – even when we resort to it as to a lesser evil – than this fundamental opting for discourse which is what philosophy is.

Only then is it possible to recognize violence wherever it may be and resort to it when needs must. The violent man who knows that violence is the opposite of discourse, and goes on calling it by its right name, will be forever safeguarded from defending it as something good in itself, disguishing it or believing it transcended when in reality and truth it is not so.

CHAPTER VIII

Language, action and humanism

CONTENTS

SECTION FOUR

MAN AND LANGUAGE

SECTION FOUR: MAN AND LANGUAGE

It we were to attempt to cover all the work which testified to the interest philosophers have taken in language over the last fifty or sixty years, we should have to take into account almost the whole philosophical production of our time. For this interest in language is one of the dominant features of philosophy today. Of course, language has always had a place of honour in philosophy, in that man's understanding of himself and his world is articulated and expressed in language. This has been recognized ever since the time of Plato's *Cratylus* and Aristotle's *De Interpretatione*.

What distinguishes our age is that a conceptual knowledge of language as such should be held by many philosophers to be a necessary preliminary to solving the fundamental problems of philosophy, if by these we mean the problems handed down by tradition which bear, not on signs, but on things themselves: man, or the world. The idea that a theory of *signs* can and must be prior to a theory of *things* is characteristic of much of the philosophy of our time.

A. THE EPISTEMOLOGY OF LINGUISTICS

The philosophies of language of which we shall give an account in subsection B below are those of an age which has seen the extraordinary rise of linguistics into a leading position among the human sciences. The highly theoretical nature of this science makes it a necessary starting-point for any epistemological thinking, and there is then no difficulty in finding in linguistics descriptive and explanatory models for all the semiotic aspects of human reality. Thus the degree of precision and the generalizing power of linguistic theory force the modern philosopher, in his reflections on language, to study linguistic science before he can formulate a philosophy of language. Sub-section B will show quite clearly that such was not always the case, and that only recently has the philosophy of language taken into account the principles, methods and results of linguistics.

§ 1. *Structural and transformational linguistics*

Ferdinand de Saussure's *Cours de linguistique générale*, which was given in lecture form at the beginning of the century and published by his pupils in 1916, established the structural view of language which largely dominates modern linguistics, whatever disputes there may be between the different schools. We should recall that Saussure distinguished *langue* (language), which was a social system, from *parole* (speech), which was the individual's use of it; from *langue* he eliminated the acoustic and psychological 'substance', confining it to 'form', i.e., the purely differentiating, contrasting relations between units, whether these were phonemes or lexical meanings. To grasp these units' systematic interdependence and unity, which constituted *langue*, he then detached synchronic linguistics, which was to study the units entering into the system at any one time, from diachronic linguistics, which was to work on the changes leading from one state of the system to another. Most of these principles had been apprehended by his contemporaries and his predecessors, Baudoin de Courtenay and Kruszewski, but Saussure expressed them in a clearer, purer form; in this way the principles of the *Cours* became common ground in linguistic studies. However, Saussure also passed on to posterity many unsolved puzzles. The first of these concerns the sign itself: Saussure considered sign the 'identities' on which the system was based; in the end he adopted the Stoic view of the verbal sign as the intelligible *signatum*: this allowed him to eliminate the relationship to the thing, which is external to linguistics, and to retain only an internal difference within the sign itself. At the same time, he still gave a psychological interpretation of the *signans* and the *signatum*, viewing the former as an acoustic image and the latter as a concept, and considering *langue* itself as a treasure deposited unawares in each individual. As for the relation between *signans* and *signatum*, he was quite prepared to regard it as perfectly arbitrary, despite his remarks about 'relative motivation' which limited the arbitrary element, especially in word formation and derivation; he has been accused of confusing the relation between sign and thing with that between *signans* and *signatum*, the latter of which involves more 'iconic' factors than Saussure brought out. Further, Saussure asserted the linearity of the *signans* as forcefully as he did the arbitrary nature of the sign; but is it true that all is linear in language? Post-Saussurian linguistics, as can be seen in the works of Benveniste, stresses the hierarchic and architectonic qualities of language which impose overall shape on the component elements. When at the end of his *Cours* Saussure comes to deal with the 'mechanics of language', i.e., the way units function rather than their identification and classification, he follows Kruszewski in showing that the operations by which they combine presuppose two types of relationship: the first, based on the substitutability of equivalent terms not present in the actual message ('connexion *in absentia*)', was called 'associative' or 'paradigmatic'; the second, involving the combination of terms jointly present ('connexion *in praesentia*') was called 'syntagmatic'. Are paradigmatic series as arbitrary as Saussure would have it? And do not syntagmatic relations require a theory of the sentence which the *Cours* does not give, and which would reopen the whole question of the cut-and-dried distinction between *langue* and *parole*? But the most important puzzles which Saussure left unsolved are the dichotomies and antinomies which form the basis of the *Cours: langue* and *parole*, linguistic identity and system, synchronic and diachronic, *signans* and *signatum*, linearity and hierarchy, paradigm and syntagm. A feature of much post-Saussurian linguistics is the continual drive to reconcile these internal dichotomies.

First, however, it was left to the Prague and Copenhagen schools to consolidate the basic principles of structural linguistics.

The Prague Linguistic Circle was founded in 1926 on the initiative of V. Mathesius, with three Russian linguists participating: S. Karčevskij, R. Jakobson and N. S. Trubetzkoy. The Circle presented its theses to the first *International Congress of Linguists* at The Hague in 1928. The international success story of structural linguistics starts with the publication of proceedings of this conference, with language viewed as a functional system, primacy given to synchronic analysis and the hypothesis of converging development applied to diachronic studies. Above all, however, the Prague school is known for its structural treatment of phonology: this approach excludes objective physiological facts from linguistics, in favour of relations between phonemes, the latter being reduced to the rôle of differential entities.[1] Nevertheless, the privileged position of phonology should not blind us to the very broad literary and cultural interests which found expression in the eight-volume transactions of the Circle (1929-38): several contributions apply to literary and poetic language the methods of the Russian formalists who, following Propp,[2] had shown Russian fairy tales and popular legends to be simply built up from a limited repertoire of dramatic elements.

The school which took the main theses of the *Cours* to their full formal and abstract purity was that of Copenhagen, with Bröndal[3] (who in 1939 founded the *Acta Linguistica* of the Copenhagen Linguistic Circle) and above all Hjelmslev. The latter's *Prolegomena to a Theory of Language*,[4] published in Danish in 1943 and for the first time in English in 1953, defines the rigorous conditions to be applied to any theorizing in the linguistic field, subjecting such thinking to the requirements of logical empiricism: Hjelmslev states the principle that there is a system corresponding to every process; therefore it is possible to isolate the patterns of *langue* from the text of *parole*; in addition, both on the level of contents and on that of expression language applies a whole logical apparatus of combinations, a whole network of functions representing dependency and interdependency; so that it is possible for linguistics to be a theory in the fullest sense of the term. In this way Hjelmslev, eliminating any remnants of psychology or sociology, arrives at the conception of an algebra of language: of this algebra, however, it has sometimes been asked whether it can be used to describe a language without constant recourse to intuition.

Meantime American linguistics was growing away from European linguistics as it developed along its own particular lines: putting greater emphasis on describing the facts, paying attention to spoken languages with no

1. TRUBETZKOY, *Grundzüge der Phonologie*, published in 1939 in the Transactions of the Prague Linguistic Circle (English translation, *Principles of Phonology*, 1969).

2. PROPP, *Morfologija skazki* (1928, rev. 1969) (English translation from the 1st edn., *Morphology of the Folk Tale*, 1958, 1968; there is a French translation from the 2nd edn., *Morphologie du conte*, 1970).

3. Cf. BRÖNDAL, *Essais de linguistique générale* (1943).

4. HJELMSLEV, *Omkring sprogteoriens gundlæggelse*. See also, among HJELMSLEV's later works, *Essais linguistiques* (1959) and *Sproget* (1963) (English translation, *Language. An Introduction*, 1970).

written tradition, distrusting semantics for what it took to be that subject's mentalism and metaphysical outlook, and appealing to mathematical techniques of segmentation and the calculation of frequency distribution. While Sapir[5] had still managed to apply his genius for flexible synthesis to linguistic data in such a way as to take in all the many aspects of 'linguistic form' – the symbolic function of words, grammatical structure, formal configuration, reference to the world of concepts, etc. – Bloomfield[6] laid down a mechanistic and behaviouristic view of linguistic data, based on the famous stimulus-response pattern: SrsR (an external stimulus (S) acting on one speaker induces him to speak (r); this first, linguistic response constitutes a linguistic stimulus (s) in the second speaker, provoking a practical response (R) from him); the ideas, concepts, images and intentions to which the 'mentalist' appeals are no more than unsophisticated shorthand for extremely complex bodily movements; consequently, it must be possible to reduce the *signatum* of language to the sum total of practical events to which an utterance is linked. In this way, Bloomfield was trying to align linguistics on the type of discourse used in the natural sciences; as can be seen, what is involved is the very idea of scientific method applied to linguistics: does scientific necessarily imply mechanistic and behaviouristic? Whatever the answer, the philosopher will certainly note that Bloomfieldian tradition in linguistics is the only one, apart perhaps from that of Hjelmslev, of which it can be said that it is anti-philosophical, anti-mentalistic and anti-semantic: neither Saussure nor the founders of the Prague Linguistic Circle had drawn such a sharp line between linguistics on the one hand and psychology and phenomenology on the other. However, it must be said that the Bloomfieldian school in no way broke the unity of structural linguistics: it resembled European linguistics in that grammatical data were described in purely formal terms (bound and free forms, the component form and the complex, etc.). Thus during the forties American structuralism in the Bloomfieldian tradition concentrated on phenomena of 'distribution' within the phonetic stream, thus giving priority to the study of phrase structure. The automatic application of precise rules for segmentation and classification proved particularly productive in the description of little-known or unknown languages like the American Indian tongues; but productive though it was when applied to syntactic categories, this definition by distribution does not seem to have been so successful in semantics. Here Harris's handbook[7] gives the fullest and the most rigorous synthesis of Bloomfieldian linguistics as it was around the time of the fifties: everything other than formal or distributional differences between linguistic units is strictly excluded. With this work, taxonomic analysis reached maturity; and it was against this type of approach that transformational theory was to react.

These differences between European and American linguistics are secondary compared with the fundamental unity of linguistics during the past fifty

5. Sapir, *Language* (1921).
6. Bloomfield, *Language* (1933).
7. Harris, *Methods in Structural Linguistics* (1951).

years. In contrast to the great diversity of viewpoints which marked the preceding period, the key to such unity is no longer sought through the historical and comparative method; rather is it based on the primacy given, in description and explication, to 'inquiry into the verbal structure' and to recognition of the 'strictly relational and hierarchial nature of all its constituents'.[8]

Among these many currents two authors stand out: Roman Jakobson and André Martinet. By distinguishing different approaches to the act of linguistic communication, considering it from the point of view of speaker, hearer, message, context, contact and code, with which he correlates the conative, emotive, poetic, referential, phatic and metalinguistic functions of language, Jakobson offers philosophic thinking a model which is both exhaustive and solid. As early as 1928, in the same spirit of synthesis, he was going beyond the Saussurian dichotomy between synchronic and diachronic and the tendency to identify synchronic with static.[9] The same outlook inspires his criticism of the principle of the linearity of the *signans*: juxtaposing as he does operations of combination and operations of selection, he is led to his famous dichotomy between the metaphoric process (where selection within a sphere of similarity is the dominant factor) and the metonymic (where contiguity is dominant), a dichotomy which he claims is present in aphasic troubles as well as in the interplay of different types of literary expression.[10] Jakobson is equally famous for his rigorous application to phonology of the system of distinctive features, using a binary principle closely resembling the one on which information theory is based;[11] if it is true that the twelve binary distinctive-feature oppositions are universals, then Jakobson is right in thinking that linguistic theory is more concerned with why all languages are *language* than with why each language represents a different system from all other languages. And this is to raise the problem of the universality of linguistic categories, as formerly asserted by the theorists of universal grammar, and now taken up again by Chomsky, following a different line of approach.[12]

André Martinet differs from Jakobson in being more concerned with what distinguishes one language from another. Martinet's work interests the philosopher in several ways: first because of the actual idea of 'functional' linguistics, which distinguishes it from American structuralism; secondly because it introduces the principle of economy into diachronics,[13] which brings

8. Jakobson, 'Linguistics', p. 454 in *Main Trends of Research in the Social and Human Sciences, Part I: Social Sciences* (1970).

9. Theses presented to the First International Congress of Linguists, Nimeguen, under the title: 'Quelles sont les méthodes les mieux appropriées à un exposé complet et pratique de la grammaire d'une langue quelconque?' (cf. *Actes du Premier Congrès international des Linguistes*, pp. 33-36).

10. Cf. in particular Jakobson, 'Aphasia as a linguistic problem' (1955); 'Two aspects of language and two types of aphasic disturbances' (1956); 'Linguistics and poetics' (1960).

11. Cf. Jakobson & Halle, 'Phonology and phonetics' (1956).

12. Cf. Greenberg (ed.), *Universals of Language* (1963), in particular Jakobson's contribution, 'Implications of language universals for linguistics'.

13. Cf. Martinet, *Economie des changements phonétiques* (1955).

him close to Jakobson again; thirdly because of the clarity which the concept of the double articulation of the linguistic utterance brings to general linguistic studies: an articulation in monemes on the twin levels of expression and content, and an articulation in phonemes on the single level of expression. His most recent work extends his view of functional linguistics to syntax, constituting a fairly clean break with the American type of structuralism.[14]

Still in the Saussurian tradition, reference must be made to the broad synthesis represented by the work of one of the great masters of Indo-European comparative grammar, E. Benveniste. His *Problèmes de linguistique générale* (1966), a collection of his most important essays, drew attention to a distinction which had been too often overlooked: the distinction between semiology, i.e., those aspects of linguistic data governed by the items' place in the system of the language, and semantics, i.e., those aspects of the same data which derive from use of the items within the sentence, the sentence being considered as the first unit of discourse; while the system is timeless, 'the instance of discourse' is a fleeting event: it is in this 'instance of discourse' that language refers to the speaker, by means of certain 'indicators' (say *I, you* or *he*), and to reality, through the complex interplay of subject and predicate. This aspect of Benveniste's work invites comparison – and he himself touched on this subject – with Austin's analysis of the performative and more generally of the *speech-act*. Thus Benveniste's work contributes to overhauling the Saussurian principle of the contrast between *langue* and *parole*, even reinterpreted in terms of code and message, under the influence of information theory. A theory of discourse seems capable of accommodating the wide variety of codes and sub-codes which, according to circumstances, are involved in messages.

Where are we to place that most isolated of all linguists, Gustave Guillaume? In one sense he does not break the unity of structural linguistics in the broadest interpretation of the term. For what he is concerned with in his studies on the *tenses of the verb* and the *forms of the article*, and on grammatical categories in general, is once again the systematic organization of language: but in this case the organization is the outcome of an interplay of tendencies and tensions all of which can be described by psychology, but by a psychology radically different from behaviourism and mental atomism, viz. the psychology of rational operations. The last few years have seen the collection and publication of a large part of Guillaume's huge output, which is bristling with difficulties and remains unknown to most linguists. It foreshadows a possible reconciliation between the concept of structure and that of genesis: structural linguistics, at least at its beginnings, tended to set one against the other, because of the break with the comparative historical method which marked the founding of the school.[15]

14. Cf. MARTINET, *Eléments de linguistique générale* (1960) (English translation, *Elements of General Linguistics*, 1964) and *A Functional View of Language* (1962). See also, by the same author, *La linguistique synchronique* (1965).

15. GUILLAUME, *Langage et science du langage* (1964).

But, at least for the philosophy of language, the most important event in linguistics has been the appearance and rapid development of a new method of language analysis, that of the transformational grammar of Noam Chomsky and his followers. The use Chomsky makes of the ideas of the Port-Royal school, the Cartesians and Humboldt shows many strong links with linguistic philosophy in the widest sense; but the nucleus of his epistemology was formed as he worked on the solution of typically linguistic problems. For Chomsky, the main such problem is the 'competence' aspect of linguistic activity: to possess a language is to be able to produce a theoretically infinite number of sentences by using a finite number of elements making up the structure of the language. What has to be explained, then, is the production and recognition of new sentences. These sentences are grammatical, and intuitively perceived to be such by the user; but the nature of this grammaticality has still to be explained by linguistic theory. No theory explaining it can be arrived at by using even the most sophisticated models of statistical distribution or occurrence, which would remain on the level of the surface structure of the utterance; such a theory requires transformational operations on strings of utterances, bringing out the deep structure beneath the surface grammar; formalizing such transformations is the task of what can be called a generative grammar; in such a grammar, a finite number of operations is used to generate the whole system of rules whereby a specific structural description can be assigned to the infinite number of used sentences or new sentences.[16]

There are more ways than one in which generative grammar is relevant to philosophy. The distinction between 'competence' and 'performance' brings up again the question of the distinction between *langue* and *parole*, or that between system and process. In addition, the concept of competence involves the speaker and the intuition he has of the grammaticality of what he utters or understands; therefore it is not possible, as it was with Bloomfield, to carry out the analysis of utterances from a strictly external, describer's point of view, because grammatical interpretation is the pre-condition of phonological interpretation. Furthermore, transformational rules do not specify the distinctive individuality of a language, as did structure in the sense of the schools of Prague or Copenhagen; transformations are rooted in universal procedures which the grammarians of the eighteenth century explored before historical and structural linguistics brought to light the individual, autonomous and closed nature of the system characterizing each specific language. On a more basic level, perhaps, generative grammar sets up a new dichotomy between taxonomic and transformational descriptive models. It is then necessary to group together all the varieties of structural linguistics, based as they are on segmentation and classification procedures

16. See, in particular, CHOMSKY, *Syntactic Structures* (1957); 'On the notion "rule of grammar"' (1961); 'A transformational approach to syntax' (1962); *Current Issues in Linguistic Theory* (1964); *Aspects of the Theory of Syntax* (1965); *Cartesian Linguistics* (1966); *Language and Mind* (1968). And see RUWET, *Introduction à la grammaire générative* (1967).

which aim at drawing up finite inventories of invariants and result in a discrete hierarchical distribution of the different levels of invariants – phonemes, morphemes, etc. The other group contains the old grammars predating the two mutations which linguistics went through at the time of comparative historical studies and that of structural analysis, together with modern generative grammars. So it is not surprising that Chomsky links his concept of the generative rule to Humboldt's views on the subordination of structure to genesis.

The transformational approach has awakened great interest amongst psychologists, especially on the question of innate factors in language acquisition.[17] A discussion of the distinction between competence and performance will be found in Fodor and Garrett;[18] equally many difficulties have been raised by the theory of the innateness of syntactic universals.[19]

§ 2. *Linguistics and semiology*

It is not only for its epistemology as such that linguistics is of interest to the philosopher, but also for the fact that its descriptive and explicative models can be put to wider, more general use. Long before Saussure, Charles Sanders Peirce, following up a suggestion in Locke's *Essay concerning Human Understanding*, conceived the idea of a general theory of signs, or *semiotics*, of which linguistics would form part, being the study of one particular system of signs. Peirce was the first to attempt to classify signs as such using a variety of criteria; the best known of these is based on the variable proportions between the three main functions of the sign: index, icon and symbol. It is also possible to apply to the various semiotic systems a classification based on the functions referred to by Jakobson and mentioned above; systems would be classified in terms of the order and hierarchy of such functions. Saussure, too, who could not have known Peirce's writings because they were published only in posthumous editions, saw the science of verbal signs as included in a more general science of signs, which he called semiology: writing, signals, rites and customs are the main instances of non-linguistic systems. But both Peirce and Saussure, while subordinating linguistics to semiotics or semiology, perceived the peculiar nature of the relationship between them: for linguistics is not just any part of a general science of signs – it is also the model of that science. Language, indeed, is not only the most important semiological system, and linguistics the most advanced semiological science: all other systems of signs lead back in one way or another to language, although each has its individual characteristics.

Writing is the first problem raised: on the one hand, it becomes autono-

17. Cf. Lenneberg, *Biological Foundations of Language* (1967).

18. Cf. Fodor & Garrett, 'Some reflections on competence and performance' (1966).

19. Cf. Hook (ed.), *Language and Philosophy* (1969), in particular, in Part Two, the communications of W. v. O. Quine, N. Goodman, G. Harman, etc.

mous and develops its own characteristics only on the basis of a substitution relation with spoken language, a relationship on which Saussure places great emphasis; given this substitution, on the other hand, writing behaves like a system of specific signs. This being so, one may wonder whether the substitution does not require in the *signans* a structure which cannot be reduced to sound. Their written character gives signs not only a substance (equivalent to phonic substance, as Hjelmslev, who on this point differs from Saussure, had realized), but also formal properties of their own. Recording in writing, in itself, seems to involve a mode of spatial discontinuity which cannot be reduced to phonological distinctness.[20] Finally the writing-reading relationship has necessarily to be described in terms which cannot be reduced to those of dialogue, as we shall see below in the case of the theory of text, both in philosophic structuralism and in hermeneutics.

But, in the case of writing, the comparison is still with phonetic speech; in the case of systems of gesture, it is with that discursive character of language common to oral and written expression. The languages of gesture constitute semiotic systems distinct from discourse, though usually added on to discourse. In addition, as is shown in child psychology, the articulation of the language of gestures is itself dependent on the activity of language. This dependency bears witness to the primacy of language over other semiotic systems, a superiority which seems to be closely connected with the exceptional nature of the articulatory equipment used in speech. Unlike the equipment used for gesture, the former has no other purpose than the production of the message; moreover, its systematic articulation provides an unequalled distinctive apparatus; finally, it alone has the combinatory capacity to produce an infinite number of new messages on the basis of a finite repertoire.

A further semiotic comparison is provided by the distinction between natural and artificial languages; the latter are mainly formalized languages constructed for scientific or technical purposes, in accordance with the logical requirements of the mathematician. Three problems arise in relation to artificial languages, those of identifying, by means of a method of difference, the features of natural languages which remain unconvertible into formalized languages, of discovering what features of formalized languages remain dependent on the natural languages and of determining the scope and limits of collaboration between linguistics and mathematics in a science of mathematical linguistics.

On the first point, emphasis has been put on the large variety of modes of discourse other than the statement, and the fact that one cannot be reduced to the other: command, wish, question, etc. (we shall be meeting this problem again in connection with the philosophical theory of the speech-act). Stress has also been placed on the positive value of the multiple ambiguity of expressions in natural language, which has been related to the ability of such expressions to enter into an unlimited number of contexts, and consequently to the creativity of language; thus that same indeterminacy of meaning, that

20. Cf. Derrida, *De la grammatologie* (1965).

same sensitivity to context, which, from the point of view of symbolic logic, are serious faults, turn out to have a major functional rôle in a theory of ordinary language. The conclusion is that the description of natural languages cannot itself be derived from the theory of formal languages.

This answer to the first question contributes something towards the solution of the second: it is reasonable to think that formalized languages continue to draw on ordinary language for some capacity of imagination and invention. This accounts for the fact that mathematical symbols seem to remain rooted in the words of everyday language.

The third problem is that of collaboration between linguistics and mathematics. The idea of a mathematical linguistics in no way runs counter to recognition of the specific individuality of natural language in relation to formal; for a truly mathematical treatment of the structure of human language, implies no confusion between the two sorts of language, to the extent that mathematical theory is playing the part of a meta-language in relation to the language studied. Thus, Z. S. Harris, in his *Mathematical Structures of Language* (1968), does use a formalized language, that of set theory, to compare natural language and formalized systems.[21]

§ 3. *Linguistics and human sciences*

The relation between linguistics and a science of signs which includes linguistics points to a still higher-order relation: semiotics, or semiology, has itself always been considered a social science, or even a branch of sociology. As Jakobson states trenchantly, in the account which is used as a basis for this part of the report: 'If the cycle of semiotic disciplines is the nearest one to encompass linguistics, the next, wider concentric circle is the totality of communication disciplines'.[22] *Communication* then emerges as the broadest category, semiotics relating to the communication of messages of any sort, and linguistics being confined to the communication of verbal messages.

Communication concerns the psychologist as much as the sociologist. The psychologist considers it from the viewpoint of exchanges between individuals, the sociologist from that of supra-individual institutions and collective behaviour. Let us say that, to the psychologist, communication is interlocution, i.e., the exchange of messages seen from the point of view of the users. But it is common knowledge that the borderline between psychology and sociology is difficult to draw, especially if we think of the case of social psychology which either might claim for itself.

The relationship between linguistics and *psychology* is particularly deli-

21. We may add to this list of subjects the research currently being done on *machine translation*. A useful summary of the present state of the art will be found in BOOTH (ed.), *Machine Translation* (1967); for any discussion of the limitations of computerization in relation to the requirements of literary translation, the basic work is still BAR-HILLEL's *Language and Information* (1963).

22. 'Linguistics', p. 425 in *Main Trends . . ., Part I: Social Sciences* (1970).

cately balanced. Linguistics has in a sense taken a stand against any interpretation of linguistic units in terms of images and concepts, for while there are still traces of this in Saussure, in Hjelmslev there are none – though, conversely, Bloomfieldian linguistics owes its view of language in terms of stimulus and response to behaviourist psychology. What we today call 'psycho-linguistics' is indeed a result of linguistics having won its independence of psychology; it is an 'inter-disciplinary' science in the full sense of the word, since it requires both an excellent linguistic background on the part of the psychologist, and a keen sense of the specifically psychological nature of problems such as the reception and apprehension of messages, the perception of code, the acquisition of language, language disturbances (aphasia), and so forth.

One further correlation which can be made on this same level of individual psychology is that between linguistics and *psycho-analysis*; if it can be shown that the mechanisms of distortion, found for example in the workings of dreams, can be assimilated to processes like metaphor and metonymy, which Jakobson has shown to be universal in their nature, we are led to the hypothesis that the laws of the unconscious partake of the character, not so much of naturalistic or behaviouristic models, as of semiological ones. Conversely, the psycho-analyst's conception of the correlation between the *signans* and the *signatum* through the psycho-analytical relationship enriches the theory of discourse, which by definition is confined to the workings – the so-called normal workings – of ordinary language. This cross-fertilization between psycho-analysis and linguistics is the central theme in the work of Jacques Lacan.[23]

But the plane on which linguistics and human sciences meet most significantly is that of *sociology* itself, because of the inherently social nature of the linguistic code. Indeed, psycho-linguistics gradually merges into socio-linguistics; this is obvious when one considers the complicated interplay of codes and sub-codes within any linguistic community. Jakobson protests, as we have seen, against too strict a conception of the uniformity of the code for all speakers; each participates in many different communities, social milieux and situations which call for the development and use of a variety of codes adaptable to circumstance and milieu, and a speaker's language skill consists of regulating his 'performance' in the light of slight variations of codes on the level of 'competence', i.e., he shows metalinguistic perspicacity within the verbal exchange. Thus we see the possibility of a social dialectology, on the borderlines of linguistics, psychology and sociology.

In any comparison between codes themselves rather than between users of codes, the partnership is between linguistics and sociology proper. While 'linguistic sociology' is still concerned with the problem of linguistic variety and is therefore a mixed subject, a much more basic problem arises with the attempt to treat social phenomena themselves as coded phenomena, in the same way as linguistic phenomena are coded. Here we come upon the im-

23. LACAN, *Ecrits* (1965).

portant work of Claude Lévi-Strauss; for what, in the main, enables him to treat the social sciences themselves as a field in which the structural models of linguistics apply, is that he decides to treat the whole set of social functions in terms of communication – communication of messages, communication of goods, communication of women, etc. The extended field in which linguistic models are made to apply is the homogeneous field covered by a theory of communication, the various types of communication being differentiated only by their strategic levels. This gradual extension of his model can be traced through the work of Lévi-Strauss himself: taking as his starting-point the kinship systems which regulate the exchange of women in so-called primitive societies,[24] in *Structural Anthropology* [25] he proposes that all institutions should be viewed in terms of rules of exchange, and should consequently all be considered to belong to the same type of organization. This structural homology applies particularly well to mythology; [26] the identity of structure which it postulates enables the large units of more than one sentence with which the mythologist deals to be treated by the same model as are the shorter units of discourse which the linguist treats. A linguistic theory of myths conceived and worked out on this assumption refutes the old sociological idea of the 'primitive mind', and admits no differences between 'savage' and 'logical' thought other than those of use and strategy: any differences in logical structure are denied.

If we adopt Lévi-Strauss's suggestion that communication of messages is one of three forms of communication, the others being communication of women and goods, we are naturally led to consider the relationship between linguistics and *economics* from the point of view of modes of communication. The Saussurian concept of 'valeur' (the value or 'sense' of a message) – as opposed to its actual meaning – is explicitly based on a comparison between linguistic signs and monetary signs, in their dual relationship, firstly with the dissimilar things which such signs represent (merchandise), secondly with the similar things (other units in the same monetary system). And for their part, theories of wealth, merchandise and money have always gone hand in hand, as shown by Michel Foucault in *Les mots et les choses*,[27] with theories on the linguistic sign and on zoological classifications. It is therefore perfectly justifiable to follow Talcott Parsons in treating economic communication and the related monetary system as a transmission of messages governed by the appropriate code. Here too, we can look to language for more than a metaphorical basis for comparison; it can in fact provide us with a set of descriptive models and analytical methods.

24. LÉVI-STRAUSS, *Les structures élémentaires de la parenté* (1949) (English translation, *The Elementary Structures of Kinship*, 1969).

25. LÉVI-STRAUSS, *Anthropologie structurale* (1958) (English translation, *Structural Anthropology*, 1963).

26. See the four volumes of Cl. LÉVI-STRAUSS's *Mythologiques*, 1964, 1966, 1968 and 1971 (of which the first two, *Le cru et le cuit* and *Du miel aux cendres*, have been translated into English: *The Raw and the Cooked. Introduction to a Science of Mythology, I* (1969-1970) and *From Honey to Ashes. Introduction . . ., II*, 1973).

27. 1966 (translated into English under the title: *The Order of Things*, 1970).

In many ways, therefore, the idea is gaining ground that language and society have a common basis, or consist in phenomena which, in Claude Lévi-Strauss's words, fit into some sort of 'universal code, capable of expressing the properties common to the specific structures pertaining to each aspect'.[28]

§ 4. *Linguistics and natural sciences*

So far we have compared the verbal code with other systems of signs which presuppose language, accompany language and are translatable into the verbal register.

The comparisons which we shall now make as we turn to the natural sciences bring out instances of isomorphism which by no means presuppose that human language is prior.

This is the basic philosophical significance of the discovery of the *genetic code*.

Until this discovery was made, questions which traditionally arose from comparisons between the fields of linguistics and the natural sciences fell into three broad classes concerned respectively with: study of the biological preconditions of human language; its relationship to animal language; and the possible origins of human language in the patterns of pre-hominoid life.

The first head covers a vast variety of research in which new branches are being developed, and at the same time new groupings are forming between subjects; thus the physiology of speech brings together phonetics and biomechanics; neurobiology takes in cortical physiology and the pathology of speech; the study of language acquisition is the meeting point of the psychology of learning and the biology of adaptation; what is more, all arguments about the innate and the acquired assume full conversancy with the universal aspects of the phonology and syntax of natural languages.

The second head covers two types of problem which are at present under discussion: first the increasingly complex comparisons of animal and human language; with a better knowledge of the workings of animal messages, together with progress in general linguistics, we can pinpoint more precisely what constitutes the originality of human language, stressing the phenomenon of double articulation on which the lexicon-syntax dichotomy is based.[29] Immediately the second question, which is brought up by the general fact of evolution – the problem of the origin of signs – seems even more mysterious: how can one get from animal to human language? The anti-historical bias of

28. '[...] une sorte de code universel, capable d'exprimer les propriétés communes aux structures spécifiques relevant de chaque aspect' – Lévi-Strauss, p. 71 in *Anthropologie structurale* (1958) (quoted in French by Jakobson, 'Linguistics', p. 430 in *Main Trends . . ., Part I: Social Sciences*).

29. Cf., for example, Benveniste, 'Communication animale et langage humain' (1952, 1966) (English translation, 'Animal communication and human language. The language of the bees', 1953); Mounin, 'Communication linguistique humaine et communication non linguistique animale' (1960, 1970).

structural linguistics tends to 'place between brackets' the problem of the origin of signs, which was so important in the tradition of philosophic romanticism and in that of empiricism. However, progress has been made in another direction. The correlation between the introduction of signs, of tools and of the social law (mainly the prohibition of incest which draws a dividing line between permitted sexual partners and kin) is becoming increasingly clear; the work of Leroi-Gourhan, in this connection, is basic for a view of the simultaneous constitution of technology, language and sociality. As there is no palaeontology of language in the sense that there is a palaeontology of tools, it is in relation to the latter, and on the basis of a better understanding of the syntax of gesture, that progress on the problem of the origin of language is likely to be made.[30]

But the most important event of recent years is unquestionably the discovery of the 'genetic code'. This discovery is of the greatest possible importance for the philosophy of language. For the genetic code resembles the verbal code without in any way deriving from it. Moreover, the genetic code has one feature in common with the verbal code which is not shared by any other human semiotic system or any system of animal communication. For the genetic code is based on a finite number of discrete units (the letters of the 'alphabet'). These units, which like phonemes are devoid of meaning, in combination take on meanings comparable with those of 'words', the minimal meaningful units in our language; these words are arranged in 'sentences', the molecular messages which are written along the chromosomes; and these messages in turn show pattern and hierarchies comparable to those of a syntax. In addition, there is a sort of regulating interaction reminiscent of the phenomenon of interlocution. Finally, the genetic code recalls the predictive and programming functions of language; this enables the problem of teleology or teleonomy to be taken up again in terms of coding and decoding, without any reference to conscious intentions. Thus language constitutes, at both the biological and human levels, the very archetype of a teleological system.[31]

There then arises the question of the reason for this similarity of structure between the genetic code and the human verbal code. As R. Jakobson says in his contribution to the First Part of the present Study, 'One could venture the legitimate question whether the isomorphism exhibited by these two different codes, genetic and verbal, results from a mere convergence induced by similar needs, or whether, perhaps, the foundations of the overt linguistic patterns superimposed upon molecular communication have been modelled directly upon its structural principles'.[32]

30. Cf. LEROI-GOURHAN, *Le geste et la parole – Technique et langage* (1964) and *Le geste et la parole – La mémoire et les rythmes* (1965), but also the two earlier volumes: *Evolution et techniques – L'homme et la matière* (1943) and *Evolution et techniques – Milieu et techniques* (1945).

31. Cf. 'Vivre et parler', a television round-table discussion between JACOB, JAKOBSON, LÉVI-STRAUSS and L'HÉRITIER (1968).

32. 'Linguistics', p. 440 in *Main Trends . . ., Part I: Social Sciences.*

B. PHILOSOPHIES OF LANGUAGE

That the philosophy of language goes beyond the epistemology of linguistics is amply attested by the work of thinkers as different as Frege, Husserl, Russell, Wittgenstein, Carnap, Ryle, Austin and Quine. Not one of them holds the empirical study of language by linguistics to be the only approach to language; on the contrary, it might be thought surprising how little interest some of them have shown for the linguistics of linguists. The separate paths taken by philosophers, with their linguistic analysis, and linguists, with their empirical linguistics, are easily accounted for. If language is for the linguist a specific object, or even an autonomous system of purely internal dependences, to use Hjelmslev's term, a whole series of fundamental questions about language are thereby excluded from linguistics: firstly, the relation of language to the logical operations which cannot be reduced to a particular language structure or, more generally, the relation of linguistic structures to the operations of thought; secondly, the relation of linguistic communication to other phenomena of social communication, and to culture in general; finally, and most important of all, the relation of language to reality: the basic function of language is to *refer* to something other than itself; it is this vast problem which comes under the head of *reference*. With this problem, however, comes a paradox: the more linguistics is purified and reduced to the science of language, the more it expels from its field everything concerning the relationship of language to anything else but itself. This paradox is visible in Saussure: for the triadic relation (*signans* – *signatum* – thing) he substitutes the dyadic relationship of *signans* to *signatum*, which comes within the linguistic domain; but in doing so he eliminates the symbolic functon itself. As the French linguist E. Benveniste points out, '... language constitutes the highest form of a faculty which is inherent in the fact of being human, the faculty of *symbolizing*. By this we mean, very broadly, the faculty of representing the real with a sign and of understanding the "sign" as representing the real, thus establishing a "signifying" relationship between something and something else.' (*translation*) [33] The science which takes language as its subject, then, does not exhaust the question raised by language, the problem of its being the great mediator between man and the world and between one man and another: 'the existence of such a system of symbols reveals one of the essential facts, perhaps the most essential fact, of the human condition, namely that there is no natural, immediate and direct relationship between man and the world, or between one man and another; an intermediary is needed, this apparatus of symbols which has made thought and language possible. Outside the purely biological field, the ability to symbolize is man's most distinctive capability.' (*translation*) [34] We can thus appreciate the reasons which impel philosophies of language to examine the claims of language to represent reality.

We shall start with the movement of thought commonly called 'analytic'.

33. *Problèmes de linguistique générale* (1966), p. 26.
34. *Ibid.*, p. 29.

The postulate common to those philosophies which assume this name is that philosophy consists in explaining and clarifying the conceptual systems elaborated in the fields of science, art, ethics, religion, etc., explaining and clarifying them *in terms of that very language* in which the conceptual knowledge is expressed. In this way the *clarification* of language becomes the first, and finally the only, task of philosophy. All the important philosophical questions tend to be reduced to explaining and clarifying the grammar and syntax of natural language.

All the other philosophies of language are, in their various ways, attempts to go beyond the stage of *clarification*. The attempt may take two very different directions. Either the primacy of language is questioned, and the function of signs is reabsorbed in some vaster reality or activity in which the issue of language loses its special character and its exclusivity. It is in this subordinating, reintegrating fahion that, in their different ways, both phenomenology and Marxism proceed. Or else, recognizing the privileged position of language, the philosopher attempts to redefine reality itself in terms of language, being itself taking on the nature of language; a philosophy which considers being to be language becomes a philosophy of language in every sense of the word. This latter direction is that taken by two otherwise conflicting currents of thought, philosophical structuralism and hermeneutics.

§ 1. *The 'analytic' approach*

The 'analytic' conception of a philosophy of language is particularly well represented in the works of the period under consideration. *Logical empiricism* and the *philosophy of ordinary language* have continued to be equally influential, at least in the English-speaking world, while new tendencies have emerged. Both movements endow the task of clarification with a definitely therapeutic and preventative rôle with regard to metaphysical speculation, inasmuch as such speculation results from misuse of the natural languages through the uncritical and uncontrolled exercise of linguistic freedom. But there is still a considerable gap between the two movements. Logical empiricism undertakes this reductionist task by constructing artificial languages which eliminate such misuse, and for this purpose, works out conventions for the formation of sentences and their semantic interpretation which exclude metaphysical statements; the philosophies of ordinary language, on the other hand, remain within the confines of the natural languages in order to bring into the open the models which govern linguistic behaviour within the limited sphere of proper usage.

While the work of Carnap is for the most part earlier than the period under consideration, it continues to find an echo. His famous essay, '*Überwindung der Metaphysik durch logische Analyse der Sprache*' (1932), was republished by A. J. Ayer in *Logical Positivism* (1959) under the title 'The elimination of metaphysics through the logical analysis of language'. This essay's argument is remarkable in regard to the use of linguistics in philosophy. It states that what explains the inability of natural language to eliminate metaphysical utterances is the absence of certain conventions in such language. Therefore it is by taking 'the logical point of view' that we can criticize and distinguish what is syntactically adequate and inadequate; the philosophy of language consists in measuring the gap between logical syntax and grammatical syntax. In a language which satisfied the requirements of logical syntax, pseudo-statements could not be made. Here the most general presupposition is that any precise statement of a philosophic problem comes down to the logical analysis of language, that what is at stake in philosophical problems is a matter of language and not of the world, and that consequently such problems should be stated not in object-language but in metalanguage. The thesis that philosophical problems are purely linguistic ones endows the expression *linguistic philosophy* with a special meaning: it signifies that metaphysical statements about reality are pseudo-objective sentences which under a specious appearance of object sentences hide their real structure, which is that of syntactic sentences, showing for instance that particular syntactic properties go with particular words or expressions. The philosopher's job is therefore to re-translate sentences which bear on pseudo-objects from natural language into the syntactic sentences of an ideal language.[35]

But why distinguish genuine 'object'-sentences from pseudo-'object'-sentences? Is this not anti-metaphysical prejudice? For the possibility remains that they have reference in some sense other than that of the perception of object. In this case we need a semantic theory which is not itself a mere syntax; in his second major book, *Introduction to Semantics* (1942), Carnap had already abandoned all claim to reduce linguistic constructions to purely syntactic terms and brought in also semantics and pragmatics. Semantics adds to the syntactic rules other rules of designation which specify the things to which the expressions in the language refer, and rules of truth which pair off sentences and truth conditions. This semantic structure of language is the main theme in *Meaning and Necessity* (1947, republished in 1956). This brings out the fact that the relations holding in any language between the descriptive symbols and the derived expressions are based on semantic premises and rules analogous to the axioms and rules of inference in an axiomatic system.

The linguistic philosophy of Carnap and his followers raises a question of principle: by what right do they require that an artificial language should provide the linguistic conventions which are lacking in the natural languages? Do they base themselves on the prior conviction that metaphysical statements are meaningless? But if this conviction is not itself of linguistic origin, where does it come from? Is it the very structure of artificial language which re-

35. Cf. also WAISMANN, *The Principles of Linguistic Philosophy* (posth. publ., 1965), and also QUINE, *Word and Object* (1960) and DAVIDSON & HINTIKKA (eds.), *Words and Objections. Essays on the Work of W. v. Quine* (1970).

quires these restricting conventions? But has that structure a basis in necessity? [36]

These questions point the way to an examination of the natural languages, logical positivism dwelling on their shortcomings rather than on their deep semantic structure.

The philosophy of ordinary language shares with logical positivism the conviction that metaphysical statements are meaningless. But it pursues the same therapeutic end by applying a process of clarification to natural language; the idea is that the result of such work would show that natural language functions correctly as long as it remains within the limits of its proper use.

The foundations of this movement of thought were laid by Wittgenstein.* His *Philosophical Investigations* [37] were published in 1952; but subsequent years have seen continual progress in the understanding, interpretation and discussion of Wittgenstein's thought. Several of his miscellaneous writings have been published, and these enable us to plot his development and fill in large areas of silence in his major work. In addition, very valuable introductions and commentaries have been brought out; the most important of these are devoted to the *Tractatus* [38] and its view of the relationship between language and the world as distilled in the 'picture-theory'. These publications thus carry on the influence of the *Tractatus* beyond the date of publication of the *Investigations;* in this way, the 'logical' thinking of the *Tractatus* has by no means been eclipsed by the theory of language games and meaning as use which comes out of the *Philosophical Investigations.* [39]

What dominates the work of the Oxford school in the period under consideration is the tradition of the *Philosophical Investigations* rather than that of the *Tractatus.* Such work includes the writings of J. L. Austin, who died

36. Cf. SCHILPP (ed.), *The Philosophy of Rudolf Carnap* (1963), especially the contributions of K. R. Popper, A. J. Ayer, P. F. Strawson, etc.

* The rest of this analysis is based on a survey by L. Jonathan Cohen, from which we have reproduced a number of developments textually. We are also indebted to L. J. Cohen for some of the precisions given above (sub-section A) in connection with the epistemology of linguistics.

37. *Philosophische Untersuchungen/Philosophical Investigations.*

38. *Logisch-philosophische Abhandlung* (1921) / *Tractatus logico-philosophicus* (1922, 1961). Cf. ANSCOMBE, *An Introduction to Wittgenstein's* TRACTATUS (1959); STENIUS, *Wittgenstein's* TRACTATUS. *A Critical Exposition of Its Main Lines of Thought* (1960); BLACK, *A Companion to Wittgenstein's* TRACTATUS (1964).

39. The posthumous publication of Wittgenstein's post-*Tractatus* writings has continued with *Lectures and Conversation of Aesthetics, Psychology and Religious Belief,* edited by C. Barrett (1966) and *Zettel,* edited by G. E. M. Anscombe and G. H. von Wright (German text and English version, 1967). We may also mention *The Blue and Brown Books* (1958); 'Lecture on ethics' (1965); and the three volumes of *Schriften* published in German (1961-66). Several articles and books have been devoted to a discussion of Wittgenstein's work; the most recent: PITCHER (ed.), *Wittgenstein. The* PHILOSOPHICAL INVESTIGATIONS (1966/1968); WINCH (ed.) *Studies in the Philosophy of Wittgenstein* (1969). See also DE MAURO, *Ludwig Wittgenstein. His Place in the Development of Semantics* (1967).

in 1960, on 'Excuses' [40] and 'Performatives',[41] i.e., expressions in which the speaker uses language, without adding to the description of the world, in order to 'do' something. Such expressions are neither true or false; they may succeed or fail, may turn out well or ill (when I say: I take this man or woman to be my husband or wife, I actually do what I explicitly say; when I promise, I commit myself). It is noteworthy that Austin was trained in the classics, and knew more about Greek and Latin than about mathematics or the natural sciences; from this he derives an insistance on exactitude and scholarship which is very much that of the philologist. Thus it can be said that Austin's contribution to linguistic philosophy is to have given consistency to a philosopher's semantics parallel to the semantics of linguists, but independent of theirs. Austin's work has a second claim to fame: however limited it might seem from its close, one might even say extreme, attention to detail, equally it marks a broadening in comparison with the simplifications which logical positivism imposes on the theory of meaning; his care for words and sentences leads him to do justice to the extraordinary variety and complexity of uses in which they are meaningful. And certainly this honesty with regard to language is what allows him to loosen the straightjacket confining it to its assertoric and descriptive functions. Thus reduction and simplification must be tempered by differentiation. Where logical positivism strikes out a proposition as being a pseudo-assertion, and consigns it to the limbo of emotive meanings, Austin suggests that there are other relationships between linguistic utterances and the world than simply those of description. Austin brings yet another new contribution to linguistic philosophy when he suggests that the distinction between factual and performative statements is still too crude and works out a more general theory of the *speech-act*: the locutionary aspect (what the sentence means, its sense and reference), the illocutionary aspect (assertion, order, advice or plea: the attitude to which the speaker commits himself) and the perlocutionary aspect (the effect we actually produce with what we say) take their place in a general description of speech which must be put alongside Roman Jakobson's functions of communication.[42] Thus Austin shows the opposite prejudice to the logical positivists: if an expression has survived, that means that it has gathered from the usage of earlier generations an efficacity in producing distinctions and designating relationships which give it the right to be listened to before being corrected. This sort of 'linguistic phenomenology', in Austin's own phrase, represents a break with the suspicious, reductionist attitude of the logical positivists.

Alongside Austin's work we can put those theories of meaning which tend

40. 'A plea for excuses' (1956), republished in the *Philosophical Papers* (1961).

41. 'Performative utterances' (radio broadcast), republished in the *Philosophical Papers* (1961); *How to Do Things with Words*, edited by J. O. Urmson (1962); 'Performative-constative' (lecture given at the *Entretiens de Royaumont*), published in English in 1963.

42. Austin's ideas on illocutionary acts have been extended, modified or criticized by various writers: apart from *Speech-Acts* by SEARLE, which is mentioned below, see *Symposium on J. L. Austin*, edited by K. T. FANN (1969) and the important critical review by SEARLE, 'Austin on locutionary and illocutionary acts' (1970).

to bring back into the picture the *intention* of the speaker. While authors like Katz and Fodor, whom we mention below, and D. Davidson [43] define meaning in terms of formal characteristics of the proposition, other authors deem it impossible, without having recourse to the concept of intention, to account for the fact that certain utterances have significance, or that one means ('*veut dire*') something in uttering certain sounds or making certain marks. Thus, in the view of Paul Grice,[44] to say that a speaker means something by some utterance is the same as saying that he intends that utterance to produce a given effect on the listener through the latter's recognition of this intention. Thus the connection between meaning and intention is re-established and an essential feature of linguistic communication is restored. In *Speech-Acts. An Essay in the Philosophy of Language* (1969), John R. Searle attempts to push the 'speech-act' theory further than Austin and bring within its scope the analyses of Wittgenstein, Grice and Strawson. To speak a language, he says, is to *commit oneself to a type of rule-governed behaviour.* The speaker's command of such behaviour is reflexively understood by himself before any elaboration of 'criteria' which may verify the proposed characterizations of the units of the language. This type of action constituted by the act of speech is a complete act only when subjoined to a form of illocution (assertion, question, order or promise), and it involves, as a subordinate member, 'the propositional act', which in turn is broken down into 'predication' (what is said of the subject) and 'reference' (expressions which serve to identify a specific person or thing, e.g., proper names, nominal phrases preceded by the definite article, or pronouns). Thus the act of speech includes a *hierarchy of acts*, in accordance with *rules which constitute them* veritable forms of behaviour.

The appeal to ordinary language, which was common to most Oxford philosophers after 1945, was criticized, within the English-language tradition itself, by E. G. Gellner,[45] on the dual score of intellectual triviality and complacency; other critics go much further and challenge the movement's basic suppositions regarding the concept of meaning; [46] within the framework of a general discussion concerning the concept of meaning, L. J. Cohen attempts in *The Diversity of Meanings* (1962) to establish that Ryle and others often provided better support for their theories than their professed appeal to ordinary language suggested.

Whether as a result of these criticisms, or through a certain exhaustion of the élan of a movement which was critical and anti-metaphysical rather than constructive, the appeal to ordinary language is tending to disappear as a criterion of philosophical validity. Thus P. F. Strawson argues in *Individuals* (1959) for an ontology of relevant material bodies and persons as basic par-

43. Davidson, 'Truth and meaning' (1967).

44. Grice, 'Meaning' (1957) and 'Utterer's meaning and intentions' (1969).

45. Gellner, *Words and Things* (1959).

46. Cf. the collection of articles by H. H. Price and others edited by H. D. Lewis under the title: *Clarity Is Not Enough. Essays in Criticism of Linguistic Philosophy* (1963).

ticulars making possible identifying references in speech. It is noteworthy that Strawson presents his doctrine as 'descriptive metaphysics', not as linguistic philosophy; for what is involved is not so much the structure of language, even if this is assumed to be common to all languages, as a 'conceptual scheme', i.e., that scheme which is common to all human beings, in which material bodies and persons constitute the two basic particulars, while the other types of particulars are seen to be secondary in relation to these two 'categories'. Certainly particulars are referents of a language operation, the 'identification of particulars'; this operation consists basically in replying to a question of the type: who? what? which? But while identification belongs to language, the fact that the class of basic particulars to which our language is applied includes only bodies and persons is not a feature of language. Basic particulars typically bear proper names because they are basic particulars, rather than vice versa. Later, in *The Bounds of Sense* (1966), Strawson moved even further away from the 'ordinary language' style of philosophizing. In this book he attempts to distinguish those analytical and critical features of Kant's *Critique of Pure Reason* that he thinks valuable, and this leads on to a carefully weighted re-interpretation of the transcendental deduction of categories.

But while the philosophy of ordinary language has been tending to die away in Great Britain, it has begun to arise again in the United States in a new form. A number of philosophers, impressed by N. Chomsky's transformational generative grammar, have claimed it possible to solve certain long-standing philosophical problems within the framework of a type of general linguistics developed in accordance with Chomsky's ideas. This conjunction of 'linguistic analysis', in the philosophic sense of the word, and linguistic science did not occur in the philosophy of ordinary language. After the decline of this school we can see the moment coming when the logical requirements of logical empiricism, the descriptive attitude of the philosophy of ordinary language and the epistemology of linguistics will come together. It is too early as yet to say whether the theory of language derived from generative grammar will be the most suitable for achieving this far-reaching ambition. Some signs do, however, seem to point in this direction. J. J. Katz and J. A. Fodor, in their article 'The structure of a semantic theory' (1963),[47] worked out a theory of the semantic component which would fit into a Chomsky-type grammar; and Katz claimed, in *The Philosophy of Language* (1966), that this theory provided an adequate basis for the solution of several philosophical problems including those about analyticity. It may be questioned, however, whether Katz's claim is correct: for his theory cannot be used to determine whether or not mathematical propositions are analytic. Z. Vendler, in *Linguistics in Philosophy* (1967), argues that the methods of transformational generative grammar can be used to solve hitherto insoluble problems about the nature of singular terms, universal generalization, causes and effects, and so on. But at crucial points Vendler's argument tends to rely

47. Republished in FODOR & KATZ (eds.), *The Structure of Language* (1964).

on intuitions about the linguistic deviance, or non-deviance, of certain strings of words, where his opponents might be expected to have contrary intuitions. Does this appeal to the speaker's intuition constitute a serious objection? Would not an 'ordinary language' philosopher rather agree with a follower of Chomsky that no linguistic *criterion* can be deemed adequate except by virtue of the understanding we as speakers have of *our own* competence, i.e., *our* command of that remarkable, rule-governed skill called language?

All the above writers concern themselves with languages in the narrow sense. Nelson Goodman, in *Languages of Art* (1969), has contrived a general theory of linguistic representation which covers, in addition, maps, musical notations, paintings and so on. This book is perhaps the first of such importance and originality to deal with aesthetics in a rigorously analytical style.

§ 2. *The phenomenology of language*

How must reality be structured for signs to appear which represent and designate it? This question arose from the examination of language in earlier philosophies: it was the problem of reference and more precisely of the identifying reference which, with Strawson, pointed *outside* language and eventually led to the concept of the basic particulars, bodies and persons. This movement beyond and outside language towards its non-linguistic conditions is also that of phenomenology.

Husserlian phenomenology can be interpreted, after the rise of linguistics and in relation to analytic philosophy, as an attempt to solve the central paradox of language, namely that language does not on the one hand come first, nor is it even autonomous, but only the secondary expression of an apprehension of reality articulated at a lower level, yet on the other hand language is always the medium in which its own dependence on what is prior to it is expressed.

Husserl's work does not fall within the period under consideration; but like Wittgenstein, Husserl has been the subject of important editions and interpretations, several of which were brought together on the occasion of the centenary of his birth (1859-1959).[48] The originality of these works is evident if they are set against logical positivism or the philosophy of ordinary language; some of them engage the philosophy of the English-speaking countries in a dialogue which is bound to grow in volume during the next ten years. It is with this incipient dialogue in mind that we shall fish out from the great mass of contemporary phenomenological work a number of strands pertaining to the philosophy of language.

Looked at from this point of view, phenomenology is an attempt to relate language, taken as a whole, to the modes of apprehending reality which come to expression in discourse. This something, which is prior to discourse and which alone allows discourse to refer to anything, was already the sub-

48. Cf. Van Breda & Taminiaux (eds.), *Edmund Husserl, 1859-1959* (1959).

ject of Husserl's *Logische Untersuchungen*, the first of which, it will be recalled, is most relevantly entitled *Ausdruck und Bedeutung*. The whole purport of the book is to discern beneath the logical sense and its requirements of identity and univocity, the signifying function of language in general and, beneath this signifying function itself, an even more basic intentional function, which is common to all experience to the extent that this is 'consciousness-of' something. Husserl's later works were devoted to the signifying aspects of ante-predicative functions such as perception; the object perceived is already a presumed 'unity' of 'sense' ('sense-unity'), which may be validated or invalidated in the course of the object's subsequent appearances; thus there is sense before there is language. Language then emerges, not only as a mediator between man and the world, but rather as a means of switching between two requirements, the requirement that it should be logical, which endows it with a *telos*, and the requirement that it should be grounded in the ante-predicative, which gives it an *archê*. The symbolic function is to be understood in terms of these twin requirements.

In his last works [49] Husserl comes to a more exact awareness of the nature of the operation by which language is referred to the experience which precedes language. He calls this prior foundation the *Lebenswelt*. But the *Lebenswelt* is not purely and simply an immediate. It is itself subjected to an operation which occurs both within language and to language, consisting of a reflection, a reverse process, a recurrent questioning through which the whole of language perceives its own foundation in what is not language. What is entailed in this *Rückfrage* is that language, through a specific semantics, should itself state its own dependence on what, from the *Lebenswelt* side, makes language possible. This reverse process can be regarded as a kind of meta-language; but the meta-language is not confined to bringing out into the open the syntax of natural languages: it states the conditions under which the symbolic function as a whole becomes possible, and apprehends the relationship of sign to thing as a totality.

It is at this point that the dialogue with the philosophy of ordinary language becomes really interesting. The latter reminds phenomenology that the immediate is lost, that it is from the medium of language that language states its relation to something else which is not language; and phenomenology takes this point to the extent that it is wary of relapsing into a philosophy of the immediate, the experienced, and maintains the irreducible nature of the '*Rückfrage*'. In return, phenomenology can put the philosophy of ordinary language on its guard against the danger of losing itself in semantic exercises on the English phraseology; this danger is exorcised only if philosophy is capable of appealing over the head of language to those modes of apprehension which come to expression in language.

49. Cf. *Die Krisis der europäischen Wissenschaften und die transzendentale Phänomenologie* (partial publ., 1936; complete posth. edn., 1954), in particular, in the 1954 edition: Second Part, § 9 h ('Die Lebenswelt als vergessenes Sinnesfundament der Naturwissenschaft'), pp. 48-54; Third Part, A, §§ 28 sqq., pp. 105 sqq.; the 'Schlusswort', § 73, pp. 269-276; and the 'Beilagen', XVII, pp. 459-462, and XXV, pp. 491-501.

This problem of expression provides perhaps the best point from which to approach the last work of Merleau-Ponty; two important articles and one collection of articles appeared between 1956 and 1961, the year of his unexpected death, and an unfinished book was published posthumously two years later.[50]

These last works of his are not a mere repetition of the philosophy of signs presented in *Phenomenology of Perception*.[51] In the latter work, Merleau-Ponty was mainly concerned to assert, against behaviourism, the irreducibility of the signifying aspects of perception and behaviour and, against mentalism, the physical rooting of language in gesture. Thus the work of expression, the struggle to pass from the intention of meaning to the articulated utterance, was compared with what happens at the physical level when an existing pattern of behaviour is rearranged to serve a new purpose; thus a philosophy of the *parole parlante* (speaking speech), in which linguistic structures are subordinated to the effort of expression, was outlined. This enclosure of secondary speech as a form of deposit within 'parole parlante' allows us to take up anew the problem Saussure left unsolved, that of the relationship between *parole* and *language*, which he distinguished from each other but failed to co-ordinate. In his last works, Merleau-Ponty pays increasing attention to the problem of the transition from perceptual sense to language sense, according to a note from his unfinished manuscript published in 1963 under the title *Le visible et l'invisible*: 'The tacit *Cogito* must convey how language is not impossible, but cannot convey how it is possible; the problem remains of the transition from perceptual sense to language sense, from behaviour to thematization. Thematization itself, indeed, must be understood as higher-level behaviour – its relation to behaviour is a dialectic relationship: language, by breaking the silence, achieves what the silence willed and could not obtain.' (*translation*) [52] The idea of truth using language to cut its way forward is not a purely archelogical concept, but also a teleological; the higher level of truth attained by the untamed *logos* makes of expression a task: 'This is an act of creation called for and engendered by the *Lebenswelt* as operant, latent historicity, extending that *Lebenswelt* and bearing witness to it.' (*translation*) [53] And here Merleau-Ponty can point to

50. MERLEAU-PONTY, 'Le philosophe et son ombre' (1959); *Signes* (1960) (English translation, *Signs*, 1964); 'L'œil et l'esprit' (1961); *Le visible et l'invisible* (posth. publ., 1963) (English translation, *The Visible and the Invisible*, 1969).

51. *Phénoménologie de la perception* (1943) (English translation, *Phenomenology of Perception*, 1962).

52. 'Le *Cogito* tacite doit faire comprendre comment le langage n'est pas impossible, mais ne peut faire comprendre comment il est possible; reste le problème du passage du sens perceptif au sens langagier, du comportement à la thématisation. La thématisation elle-même doit être d'ailleurs comprise comme comportement de degré plus élevé – le rapport d'elle à lui est rapport dialectique: le langage réalise, en brisant le silence, ce que le silence voulait et n'obtenait pas.' – *Le visible et l'invisible*, pp. 229-230.

53. 'Il s'agit d'une création qui est appelée et engendrée par le *Lebenswelt* comme historicité opérante, latente, qui la prolonge et en témoigne.' – *Ibid.*, p. 228.

writers like Marcel Proust and painters like Cézanne, at the centre of whose work is the problem of creating a 'sense' which restores and manifests what went before. Thus confrontation with the arts can be seen to be just as essential as confrontation with linguistics and the human sciences; the silent reflection of the painter is at the dividing line between what can be seen and what can be said.

At the same time as broadening his field, Merleau-Ponty brings in several Heideggerian themes on being as an opening, on the *logos* that dominates the actions of speakers: language, according to *Le visible et l'invisible*, 'is a further step along the road to *articulation* which is the Being of all being' (*translation*);[54] 'It expresses an ontogenesis of which it itself forms part' (*translation*).[55] Being feels itself a spontaneous and inexhaustible source of transcendence towards the truth of meaning through the workings of expression. The phenomenology and ontology of expression are thus associated in a unique enterprise. Questions about the possibility of language are inseparable from questions about the possibility of reflection and its place in the architecture of being. Language is possible if something like a reflection on the unreflected is also possible. Thus at the most basic level the problem of language is not autonomous; before it comes the interlacing, the *chiasma*, which ensures the coming-together of my *flesh* with the *flesh* of the world and provides a 'formative environment' for any expressive effort. But this does not mean that Merleau-Ponty renounces the heritage of Husserl. His essay '*Le philosophe et son ombre*', written at the same time as the manuscript notes for *Le visible et l'invisible*, bears witness to his intention to remain faithful to the latent thought of Husserl, to what he calls his '*impensé*'. Certainly phenomenology cannot remain a reflection on states or actions of the conscious mind, for that would mean remaining a variety of idealism, some sort of neo-kantianism; its theme is the will to transcendance which imbues being this side of the intentional relation; but, he immediately adds, 'that which, within us, resists phenomonology – the natural being, the barbarian principle of which Schelling spoke – cannot remain outside phenomenology and must have a place within it' (*translation*).[56] The only possibility is an 'indirect' ontology, implied in phenomenological analysis. In this sense it is impossible to go beyond the phenomenological point of view. This rooting in phenomenology, that is, ultimately, in the reflection of the conscious mind on its actions, continues to ensure an equal rooting in the problems of language. For it is in the medium of articulation and discourse which is characteristic of language that the unreflected comes to reflection. Therefore the best tribute which could be paid to the work of Merleau-Ponty would be to link the philosophy of expression more clearly to some form of

54. '[...] continue un essai d'articulation qui est l'Etre de tout être.' – *Ibid.*, p. 168.

55. 'Il exprime une ontogenèse dont il fait partie.' – *Ibid.*, p. 139.

56. 'Ce qui résiste en nous à la phénoménologie – l'être naturel, le principe barbare dont parlait Schelling – ne peut pas demeurer hors de la phénoménologie, et doit avoir sa place en elle.' – 'Le philosophe et son ombre', p. 225 in *Signes.*

linguistics which is attentive to the creative aspects of language. In this connection, the controversy between phenomenology and structuralism, the disputes about the death of subject and the triumph of system, are in danger of obscuring this much more important prospect, that of an alliance between the phenomenology of expression and the linguistics of deep structures and those rule-governed transformations which are what make language a creative activity.

§ 3. *Marxism and the philosophical problems of language*

Marxism does not hold it necessary that a philosophy of language shall exist as a special discipline concerned with research into a substantial, self-sufficient reality determining various classes of phenomena. On the contrary, from the Marxist point of view, language can be understood correctly only in combination with other, non-linguistic phenomena which, to a greater or lesser degree, determine its fundamental characteristics. This view of the question, however, does not exclude research into the philosophical problems of the science of language, although of course it rules out the setting up of language as the main or sole subject of philosophy. It is this general characteristic which, in our opinion, places this approach alongside phenomenology.

The starting-point of Marxist philosophy is that the understanding of the essence of language presupposes appropriate philosophical notions concerning the relationship between matter and consciousness, the essence of man and the rôle of practice.* But by themselves these general notions, which do not relate only to language, are certainly insufficient: it is consequently necessary to undertake a philosophical analysis of the specific features of language.

The fact that language constitutes a subject of inquiry for linguistics, psychology, semantics, logic, sociology, information theory and cybernetics makes it a very complex phenomenon. Indeed it is because language has become the subject of many specialized sciences that the need for special philosophical research into the nature of language has clearly emerged. From this arise questions, for example, on the nature of meaning: is it an ideal essence, a material object or a psychic state which is a predisposition to action? An answer presupposes a definite philosophical position. Thus questions on the relationship of language to the external world, or to consciousness, on the part played by language in thought or knowledge, on the influence of linguistic form on the contents of knowledge – these problems are not exclusively linguistic problems, they are also philosophical problems.

A number of basic premisses concerning the nature of language were formulated by the founders of Marxism themselves. Thus, in Marx and

* The development which follows was worked out by V. S. Švyrev as part of the general survey prepared under T. I. Ojzerman's direction as a contribution to this work.

Engels's *German Ideology*, we read: 'Language is as old as consciousness; language *is* a real practical consciousness, existing for other men too, and it is only through that fact that it exists for me also; like consciousness, language proceeds only from need, the present necessity of relations with other men'.[57] For Marx, language is 'the immediate reality of thought' and of consciousness, in the sense that these exist and are manifested by means of objectivization, for example by means of various sounds linked to each other in a particular way, i.e., material processes, perceived by the senses and, in their specific linguistic form, 'engendered', arranged and controlled by man. Engels, in his study 'The rôle of work in the process of transformation from ape to man', starting off from the basic standpoints of historical materialism and drawing on the then scant scientific knowledge of the process of anthropogenesis, tried to sketch out the general lines of a solution to the question of the origin of discourse and language in connection with the development of the work activities of man. Although in the light of the facts available to modern science this problem appears much more complex, as can be noted in the work of Marxist philosophers,[58] the very attempt to link the origin and development of language to the development of production and forms of social communication was in itself of great methodological importance.

The problem of the gnoseological and social rôle of language was also raised by Lenin on many occasions. In his 'Philosophical Notebooks', he drew particular attention to the history of language as one of the fields of knowledge which must form, in his words, a starting-point for theory of knowledge and dialectic.[59]

Of course the Marxist approach to language does not rest solely on what the Marxist classics say on the subject, but on the principles of Marxist philosophy in their entirety, these forming the *ultima ratio* not only for Marxist philosophers, but also for Marxist psychologists, Marxist linguists and Marxist logicians. Thus one of the first Marxist attempts to enquire into the problems of language was the monograph by the psychologist L. S. Vygotsky, 'Thought and discourse' (1934),[60] in which the main distinguishing feature of human behaviour is seen in the fact that it is mediated through the man-created signs and symbols with which man dominates behaviour, whether his own or that of others. As the anthropological emergence of man means that the process of biological development changes into historical development, the development of working tools is directed towards the transformation of human behaviour, in the first place that of others, subsequently that of the subject himself. Thus discourse appears as a particular kind of 'social

57. *The German Ideology*, p. 29 in MARX-ENGELS, *Sočinenija* (= Complete Works), 2nd edn., Vol. 3.

58. See for example PORŠNEV, "O načale čelovečeskoj istorii' (= On the origins of human history) (1969).

59. LENIN, *Filosofskie tetradi* (= Philosophical Notebooks), p. 314 in *Sočinenija*, Vol. 29.

60. VYGOTSKIJ, *Myslenie i reč'* (= Thought and Discourse) (1934).

mechanism of behaviour', as a powerful means of controlling human activity by signs. Discourse, as a specific variety of sign-behaviour, forms the channel through which society acts on the individual and through which the behaviour of man in general is socially determined.

These ideas of L. S. Vygotsky's, which have something in common with those that were to be developed in France later by H. Wallon and his followers, have been further developed in recent years in the Soviet Union in the work of A. N. Leontiev and P. J. Galperin. Thus Vygotsky's idea of the rôle of language in the formation of the higher mental functions served as a methodological basis for one of the most interesting achievements of Soviet psychology, P. J. Galperin's theory of the formation of mental acts, the basic idea of which is the internalization of external acts involving objects on to the mental plane, through their transposition into the form of discourse.[61] Thus we see the crystallization, in the specialized work of a whole school of Marxist philosophers, of the Marxist philosophical concept of language as the immediate reality of thought and as a prerequisite for the working of the human consciousness, and of the link between language and thought and objective human reality.

Marxist teaching in regard to the social basis of language, its origin and development concomitant with socio-historical practice, have also provided a methodological framework for productive research by Marxist linguists. Indeed, some linguists, interpreting the notion of the social basis of language in an absolute sense, tended sometimes to underestimate the part played by the internal factors in language development, due to its relative autonomy (N. J. Marr and his followers). These errors and faults, however, should not blind us to the fact that such linguists were undoubtedly correct in seeking to relate language research to socio-cultural phenomena. Very important from this point of view is the analysis of language, or to be more precise of types of language, as specific 'codes', in which the inherent features of a given type of culture are reflected, especially those relating to the storage and transmission of social information. Language studied from this point of view becomes no less, but for specific purposes more, valuable evidence for the study of the social and cultural past of mankind than the monuments of material culture which archaeology takes into account.[62] In this, Marxism differs considerably from the common sociological viewpoint whereby language systems are some sort of mirroring of society, and can be handled in terms of simple linear relationships of dependency. The only thing which Marxism stresses is the fact that language is a necessary condition of sociality, and that the nature of this sociality, and the problems which arise from and are solved in it, must necessarily influence the character and nature of language. The concrete workings of this influence and the way the internal and external factors of language evolution interact – these are subjects for

61. GALPERIN, 'Razvitie issledovanij po formirovaniju umstvennyh dejstvij' (= Development of research on the formation of mental acts) (1959).

62. See the collection *Strukturno-tipologičeskie issledovanija* (= Structuro-typological Research) (1962).

special scientific inquiry and it is not the job of philosophy to anticipate the results of that inquiry.

The main Marxist research on the specific problem of language has been carried out in the last ten or fifteen years. Two factors explain this development: firstly, it reflects the rise of linguistics – semantics, structural and mathematical linguistics, mathematical logic; secondly, it is a reaction against those Western philosophical schools for whom the study of language is the main, if not the only concern of philosophy. These two facts explain why Marxist philosophers have recently been at pains to take up a clear position on the place and significance of the philosophical problems of language.[63]

Contemporary Marxist research on the problem takes as its starting-point the fact that, having emerged by historically complex processes, 'the linguistic picture of the world' acquires relative independence, and makes its own corrections to men's knowledge, consequently leaving some trace of its own on men's thought. At the same time that does not mean that there is a determining action of language on thought. Many indications which appear on the surface to be the determination of knowledge by language are really conditioned by extra-linguistic factors, and linked to socio-historical conditions. The complex dialectics of language and thought cannot be understood if language is considered only as the verbal expression of thought which exists as something prepared and given in advance, before and without language. Thought is not merely expressed in language, it is *realized* in language, and thus the reflection by thought of something real existing as something independent of language and thought is expressed through the medium of a linguistic system. However, what determines the nature and content of thought is not language, but the objective reality reflected in language.

Closely linked with the question of the relationship between language and thought is that of the nature of *meaning,* which is also being intensively studied at the present time in Marxist philosophical literature in the Soviet Union and elsewhere. In contrast with contemporary analytic philosophy, Marxism does not hold the problem of meaning to be the central theme of philosophy; rather, from the Marxist viewpoint, meaning is not a philosophico-gnoseological category, but a semiotic and linguistic category. However, the clarification of the nature of meaning is linked in the closest way to the solution of gnoseological problems. In Marxist literature, there are

63. For a Marxist criticism of viewpoints which make language the ultimate subject of all philosophical analysis, see especially BEGIAŠVILI, *Sovremennaja anglijskaja lingvističeskaja filosofija* (= Contemporary British Linguistic Philosophy) (1965); KORNFORT, *Marksizm i lingvističeskaja filosofija* (= Marxism and Linguistic Philosophy) (1968); SPASSOV, 'Philosophy of linguistics versus linguistic philosophy' (contribution to the XIVth International Congress of Philosophy, Vienna, 1968); collection *Jazyk i myslenie* (= Language and Thought) (1967).

Cf. also the following contributions to the XIVth International Congress of Philosophy (1968): ALBRECHT, 'Die Sprachphilosophie als Grunddisziplin der Philosophie'; BRUTIAN, 'On some aspects of language as an object of philosophical investigation'; AKHMANOVA & GINSBURG, 'Concerning the dialectics of language and thought'.

several different points of view on the nature of meaning.[64] Some see in meaning the fact of the object, phenomenon or relation being reflected in consciousness, a reflection which enters into the structure of the sign as its internal aspect. Others understand it as a social relation between men who understand each other in communication and are acting together (this viewpoint is developed by the Polish philosopher A. Schaff). Others see in the correlation of the sign with the particular phenomena of reality and their reflection in the consciousness the possibility of using the sign as a substitute for the corresponding phenomenon. Despite such divergences, we may agree with I. S. Narsky [65] when he says that, from the Marxist point of view, there are certain 'boundary conditions' for the solution of this problem: firstly, the sign must be considered as the dialectical unity of the meaning of the sign and its material representation within a particular system of signs: there is no meaning outside the sign, nor any material exponent of the sign besides meaning, but the material content of this exponent itself does not by any means disappear when it ceases to participate in a system of signs; secondly, there is no such thing as meaning in the sense of a specific 'essence' in the spirit of Husserlian idealism; thirdly, objective meaning (the meaning as the object designated), from the point of view of ontogenesis and phylogenesis, is at the root of all other sorts of meaning. For this reason we cannot handle meaning by isolating it from the processes of knowledge, but there is no ground for regarding objective meaning as the only sort of meaning there is. Meaning is not within the sign in the sense that it is not some essence vitalizing the materiality of the sign from within. On the other hand, meaning is in the sign in the sense that the sign does not exist alongside its meaning. Meaning is in the interpreter of the sign in the sense that it is precisely in the interpretation of the sign that its meaning is realized (in principle, a sign with no application is not a sign); however, meaning is not in the interpreter of the sign in the sense that it cannot exist outside this material component of the sign with which it constitutes a uniy.

The fact that Marxist philosophers take into account problems of semiotics, their clear awareness of the fruitfulness of the semiotic approach to language, the rejection of certain prejudices regarding the view of language as a system of signs, – all this shows considerable progress in Marxist research into language; the number of publications devoted to the philosophical problem of semiotics is on the increase in the Soviet Union and else-

64. Contemporary Marxist literature on the question of the nature of meaning is voluminous. We should note particularly the collection *Problema značenija v lingvistike i logike* (= The Problem of Meaning in Linguistics and Logic) (1963); Narsky (= Narskij), 'Das Problem der Bedeutung und die Dialektik'; Gorsky (= Gorskij), 'The problem of meaning (sense) of symbol expressions as the problem of their understanding' (contributions to the XIVth International Congress of Philosophy, 1968). See also below the bibliography on the question of semantics.

65. Narskij, 'Filosofskie problemy jazyka' (= Philosophical problems of language) (1968).

where.[66] While refusing to reduce knowledge to the status of a semiotic structure, by defending the theory of reflection, the Marxist philosophers nevertheless continually stress the immense part played by signs in knowledge. Without an analysis of the semiotic resources of knowledge, without an understanding of the specific features of the semiotic forms used in science and in art, without an understanding, finally, of the part played by signs as regulators of social and individual behaviour, no serious elaboration of philosophical problems is possible in our age.

The new stress placed on the semiotic approach is accompanied by increasing interest in the philosophical and methodological problems arising from the introduction of exact methods into linguistic research and from the development of structural and mathematical linguistics. The mathematization of knowledge, from the point of view of Marxist philosophy, is incontestably scientific progress. As regards structural linguistics, its concern, to use the definition of the Soviet linguist S. K. Shaumian,[67] is with constructing abstract codes to serve as formal models for the natural languages; to this end, structural linguistics no longer confines itself to the immediate search for sounds and meanings, as did traditional linguistics: rather, it studies the structures and relationships of which such sounds and meanings are the substrates and the components. This development has, fundamentally, moved linguistics from the empirical to the theoretical plane, or, in the terms of the materialist dialectic, from the search for the phenomenon to the search for the essence. For philosophy the problem then becomes one of studying the methodological and theoretical questions relating to the construction of formal models of language. In this connection, the Marxist philosophers refuse to make an absolute of the mere idea of the formalization of language; to them, the relativity of any formalization goes hand in hand with the primacy of an analysis of content, which inevitably goes beyond the framework of formalization.

Finally, the *logical* analysis of language has also become a special branch of language research, in close association with other socio-scientific and philosophical aspects. Although by language, within this framework, is meant specifically the language of science, nevertheless this analysis brings up a number of philosophical and logico-methodological questions which

66. See REZNIKOV, *Gnoseologičeskie voprosy semantiki* (= Gnoseological Questions in Semantics) (1964); ABRAMJAN, *Gnoseologičeskie problemy teorii znakov* (= Gnoseological Problems of the Theory of Signs) (1965); SCHAFF, *Wstep do semantyki* (1960) (Russian translation, ŠAFF, *Vvedenie v semantiku*, 1963; French translation, *Introduction à la sémantique*, 1968); by the same author, *Jezyk a poznanie* (1964) and *Skice z filozofii jezyka* (1967) (French translation, *Langage et connaissance*, followed by 'Six essais sur la philosophie du langage', 1969); KLAUS, *Sila slova* (= The Force of the Word) (1967); ALBRECHT, 'Die erkenntnis-theoretische Problematik des sprachlichen Zeichens' (1961); KLAUS, *Semiotik und Erkenntnistheorie* (1963).

67. ŠAUMJAN, *Filosofskie idei V. I. Lenina i razvitie sovremennogo jazykoznanija* (= The Philosophical Ideas of Lenin and the Development of Modern Linguistics) (1961).

have significance for the study of language in general; this field of the logical analysis of language is today being intensively explored by Marxist philosophers.[68]

Such then is the subtly-shaded attitude expressed in recent Marxist research into language. An earlier bias against the philosophically relevant achievements of linguistics has gone, but Marxism continues to reject any tendency to set up language as an absolute; in this it is resolutely opposed to logical positivism and any reduction of philosophy to 'linguistic analysis'. For Marxist philosophy, the genuinely philosophical problems are those of *Weltanschauung*; the relationship of matter and consciousness, and man's place in the world; in this it is closer to phenomenology, though opposed to the idealist interpretation of meaning as found in Husserl himself. Meanwhile, allowing for such reservations, Marxism too, under pressure from the sciences of language and the general philosophical situation, is coming to stress the function of language as a necessary condition of man's relationship to the world, and is beginning to give particular importance to those general philosophical problems which can be developed on the basis of questions concerning language.

§ 4. *Philosophical structuralism*

In the introduction to the present sub-section on the 'philosophies of language', we suggested dividing into two pairs of opposites those philosophical schools which seek to go beyond the stage of clarification. We have just looked at the first pair, formed by phenomenology and Marxism, with their attempt to put language back in the context of the non-linguistic activities which are its foundations. We shall now consider the other pair, which are diametrically opposed to each other, but have the common feature of taking language as a measure of being, and therefore of redefining reality itself in terms of language.

The first member of this new pair is philosophical structuralism.

We are distinguishing philosophical structuralism from structural linguistics, the principles of which we expounded in connection with epistemological thinking on linguistic science. Philosophical structuralism admittedly derives from such thinking, but it adds to it a tenet concerning reality which is a philosophical rather than a linguistic thesis. First we shall pinpoint what is derived from linguistics, before isolating the truly philosophical nucleus.

The principles that we shall bring forth as characterizing structural linguistics are the following: 1) Language or *langue*, in the Saussurian sense of

68. See especially Popovič, *O filosofskom analize jazyka nauki* (= On the Philosophical Analysis of the Language of Science) (1966) and 'Logika i metodologija nauki' (= Logic and methodology of science) (1967); *Logičeskaja struktura naučnogo znanija* (= The Logical Structure of Scientific Knowledge) (1965); Gorskij, 'Formal'naja logika i jazyk' (= Formal logic and language) (1962); *Logika i struktura jazyka* (= Logic and Structure of Language) (1965).

the term, consists in a system of differences without any absolute terms; what separates the phonemes or lexemes is the only reality of the language, which thus lacks 'substance', either physical or mental; 2) The code which governs the many superimposed systems does not derive from any speaking subject; rather, it is the categorical unconscious, making possible the use of speech by the speakers of the language; 3) The sign, which Saussure considered to be the fundamental entity of language, itself consists in a difference, that between a *signans* (signifier) and a *signatum* (signified); this difference is internal to the sign and, for this reason, comes within the universe of discourse; the sign requires no 'external' relationship, such as that between the *signum* and the *res* on which St. Augustine based his language theory. So, for the philosopher schooled in structural linguistics, language is a system without 'terms', a system without 'subject', and a system without 'things'.

Philosophical structuralism can be seen as a theory of reality in which this threefold character of language is taken as a model.

If language is *a system without 'terms'*, the difference which keeps units separate is more fundamental than the actual presence of the unit or 'thing' in sound or in the mind which we take to be the reality of language. The model of reality which language presents is thus radically opposed to that model of reality presented by naturalism or physicalism; the concept of a system of differences is suggestive, rather, of a constitution in which negativity is primary; to any doctrine of 'things', structuralism replies with an entirely 'de-reified', 'de-thinged' model. This generalization from the linguistic model is suggested by the linguist himself, who considers his discipline to be one province of a general science of signs or *semiology*, which C. S. Peirce had anticipated before F. de Saussure. What the structuralist philosopher has in mind is a *semiological* model of reality. He sees the first application of this in the view of social reality as a whole as a system of codified signs; if the various orders – economic, family, political and religious – can be held to be systems of communication governed by structure laws like those of language, then it is no longer possible to say that signs are of sociological origin; one must say, rather, that society is of semiological origin. This parallel between the various semiological systems can be used to interpret the relations between the various systems of codes in terms of intersignification, with the notion of structure mediating between infra- and supra-structures; disputes about the primacy of infra-structure, or the reflexive function of supra-structures, become meaningless as soon as such relations can no longer be expressed in terms of causation, only in those of structural homology. Such structural homology between social organization and ideological systems becomes manifest if it can be shown, as in the case of myths, that such systems are themselves governed by codes which, at the level of sequences of discourse, are comparable to those which govern units of language below the level of the sentence.

The work of Claude Lévi-Strauss is of capital importance in this connection; while it does not belong itself to philosophical structuralism, since it claims to be scientific anthropology, nevertheless it provides the school both with a methodological outline [69] and with a series of decoding exercises applied to specific mythical cycles.[70] These works suggest to the philosopher the possibility of boldly extrapolating from this semiological description of social reality and so arriving at a purely semiological definition of reality as a whole, whether that reality be man himself or the sum total of facts which he perceives, expresses and thus catches in the net of the *signans*. As it sets about this ambitious task of extrapolation, structuralist philosophy is fortified by an interpretation of Freudian psycho-analysis according to which 'the structure of the subconscious is similar to that of a language': to Jacques Lacan indeed [71] psycho-analysis is not a province of the natural sciences, despite the importance it attaches to drive (*Trieb*), but a branch of the semiological sciences, by virtue of the parallelism between the mechanisms of distortion in dream and neurosis and some rhetorical processes, such as metaphor and metonymy; in addition, the purely unconscious working of these mechanisms is evidence that sensory effects, at least those for which psychoanalytical theory accounts, do not proceed from the intentions of a subject, but resemble rather the workings of language, which has meaning independent of a speaking subject. It is at this point that the second principle comes into play.

If language is *a system without a 'subject'*, we can no longer take for granted the primacy of the subject which Western philosophy has asserted since Descartes, Kant and Fichte, and into which Husserlian phenomenology has breathed new life in the name of the intentional consciousness, of reduction and constitution. Far from constituting the sense, the subject himself is instituted by language. Like society, man is the product of language rather than its inventor. The question then arises whether the primacy of the subject, the justification of man as consciousness and ego, and all the moral and political 'humanism' which has grown up around this emphasis on subjectivity, do not simply denote a relatively recent ideology which is already decadent.[72] The intentions of speakers are merely surface phenomena compared with the deep interplay of *signans* and *signatum*, which takes place in the anonymous area of *langue*. This justification of the system at the expense of the subject portends a somewhat unexpected return to Spinoza; as with the author of the *Ethics*, the connection between ideas and things is not performed by some *cogito*; reflexivity is, rather, a secondary effect of this connection. Such anti-subjectivism alone, carrying on as it does the anti-mentalism of structural linguistics, is enough to distinguish structuralism from any idealism; or rather, the system imposes categories that transform

69. Lévi-Strauss, *Anthropologie structurale* (1958) (English translation, *Structural Anthropology*, 1963).

70. Lévi-Strauss, *Mythologiques: * Le cru et le cuit* (1964) (English translation, *The Raw and the Cooked. Introduction to a Science of Mythology: I*, 1969/1970); *** Du miel aux cendres* (1966) (English translation, *From Honey to Ashes. Introduction . . .: II*, 1973); *** *L'origine des manières de table* (1968); and **** *L'homme nu* (1971).

71. Lacan, *Ecrits* (1967).

72. Cf. below, section VI, sub-section § 3, pp. 1456 sqq.

the subject/object dualism which imbues Descartes, Kant and Husserl alike. This is the reason why L. Althusser has been able to put forward an interpretation of Marx in the spirit of structuralism which is parallel to Lacan's interpretation of Freud. In *Pour Marx* (1965) he dissociates mature Marxism from the philosophy of the young Marx, which is still a philosophical anthropology founded on the demand for the restoration of alienated subjectivity; Marxism proper derives from an epistemological break with all forms of 'humanism', and is based on a science of structures, forces and shapes which, like the semiology of subconscious mechanisms, requires no constituting consciousness, no subjectivity, no *cogito*. 'Structuralist' Marxism is diametrically opposed to the 'existential' Marxism of J.-P. Sartre in '*Questions de méthode*', the introduction to his *Critique de la raison dialectique* (1960).[73] The same anti-subjectivism also leads to a definite anti-historicism, witness the polemic against that same work of Sartre's at the end of Lévi-Strauss's *The Savage Mind*.[74] Whereas philosophical schools inspired by evolutionism held the ideas of genesis, development and historical derivation to be essential to the intelligibility of human reality, conversely historical understanding becomes suspect, and even problematical, to the structural intelligence, for which 'synchronic' patterns are the most easily intelligible. The history of culture seems more like a succession of epistemological configurations, each one having the coherence of a powerfully structured *epistēmē*, but the transitions between them remaining contingent, and consequently for the most part unintelligible. In *Les mots et les choses*,[75] Michel Foucault studies the organization of some of these *epistēmai* which have succeeded each other since the Renaissance, and which, at a sub-conscious, non-reflexive level, govern subsystems such as the theory of money, the theory of signs and zoological classification; each *epistēmē* is thus a major pattern governing partial systems, and the synchronic intelligibility of such patterns is prior to that of their diachronic development. In the organization of each of these *epistēmai* we can thus see disappearing, not only the constituting role of an empirical subjectivity, but also the part played by what naïve realism calls 'things' or 'reality'. And here the third principle becomes relevant.

For if, finally, language is *a system without 'things'*, if the universe of signs is made of signs which are entirely defined by some *internal* character of the sign, i.e., the difference between the *signans* and the *signatum*, it be-

73. SARTRE, 'Questions de méthode' (1957, republ. 1960, 1967) (English translation, *Search for a Method*, 1968).

74. LÉVI-STRAUSS, *La pensée sauvage* (1962) (English translation, *The Savage Mind*, 1966): cf. Chapter IX, 'Histoire et dialectique'. See also the series of pronouncements by Cl. LÉVI-STRAUSS on diachrony and history: 'Histoire et ethnologie' (1949; reproduced as the introductory chapter in *Anthropologie structurale*, 1958); *Race et histoire/Race and History* (1952); *Leçon inaugurale* (1960) (English translation, *The Scope of Anthropology*, 1967). One may refer, in the present work, to the chapter on *Social and Cultural Anthropology* by Maurice Freedman: see in particular Section *B*, § 6, pp. 29-31; Section *D*, § § 2, 3 and 4, pp. 76-87; etc.

75. FOUCAULT, *Les mots et les choses* (1966) (English translation, *The Order of Things*, 1970).

comes tempting to consider the interplay of the two to be the absolute itself, in the original sense of the word absolute: what is free of any bonds and is sufficient to itself. The universe of discourse elevated to the status of absolute, such is the philosophical aim of structuralism. A comparison with the analytic philosophy of language will show what this involves. Since Frege, Russell and Wittgenstein, the problem of reference has had pride of place in all analyses; the theory of proper names and definite descriptions, from which Strawson and Searle derive their theory of the identification of particulars, turns on that fundamental characteristic of language, that the truth of a proposition has to do not with its 'sense' (ideal), but with its 'reference', that is, in the last resort, its capacity to correspond to what is; however, the proposition is rooted in being by means of an act which consists in identifying one object, and one only; to comply with this condition, there must *exist* one object, and one only, to which the utterance applies and which is identified by means of the reference. Thus, no predication without reference, no reference without an existing referent. This stress on the reference and the referent is typical of a theory of language which concentrates on the speech-act and on the ordinary uses of language. In the structuralist view, the bias towards the referent does not take into account the linguistic revolution which enables the *signatum* to be distinguished from the things signified, and consequently from extra-linguistic reality. Furthermore, to a philosophy which derives its inspiration from the reduction of speech and subject, this concern with reference and referent tends to mask a basic potential inherent in language, i.e., that it functions for itself as an interplay of *signans* and *signatum*. This allows structuralism to escape from the positivistic fascination with fact, to throw off the mantle of ordinary language, where the same bias prevails, and to sit rather at the feet of language in its most 'de-reified' forms; 'literature', from Rimbaud and Joyce onwards, reveals what ordinary language hides: that language is a world to itself and that, while the referent is not essential to it, the *signans*, on the other hand, certainly is. Thus the main stress is coming to be placed, no longer on the identifying reference – such as the natural sciences and ordinary language would have us accept as necessary – but on the immanence and autonomy of language under the rule of the *signans*, such as is suggested by literature. But it is not just any literature which accords with the theory: the novel which conceives its 'fiction' as a quasi-reality, with quasi-referents, subjected to quasi-descriptions, does not fit in with it; nor does poetry conceived as a revelation of the hidden dimensions of reality; nor, in general, any discourse which might claim to have a 'message' and to say what is. No, the literature for which structuralism supplies the theoretical basis is an exercise which language performs on language, without reference or referent. Here we recognize the influence of Nietzsche and his fierce criticism of grammar and syntax, in which he perceives, in disguised form, the cult of the 'moral' and, in the distance, a sly revival of 'theology'. The death of God is the death of 'truthful' discourse and of the rules of 'correct expression'. Then, reference, farewell!

Logically, the structuralist will come to ask himself if the primacy of reference is not linked to the primacy of speech, the supreme quality of which is to empty itself in what it conjures up. But what if language is not exhausted by speech? Is not the differentiation, i.e., the gap between sign and sign, together with the dehiscence of the *signans* and the *signatum*, better preserved in writing than in speech, i.e., in a language operation less subject to interlocution and the immediacy of dialogue, involving also to a lesser degree the ostensive definition and designation of a referent, as denoting its actual existence? This question, with Jacques Derrida,[76] leads to the conclusion that the fate of structuralism depends not only on a generalized semiology, but on a 'grammatology', i.e., on a theory of *writing*. Though the theorist of ordinary language will hardly recognize in it what he calls language, a whole school of literature in which the writing is done in accordance with the criteria of structuralism and grammatology, is working to comply with a theory of language where differentiation takes the place of reference. No longer emptying itself in some world it purports to show forth, language proclaims itself the pure capability to differentiate, the act of differentiating.[77]

Thus structuralism is wavering between two aspirations, a 'Spinozan' aspiration towards expressing the *order* of the true without a subject, and a 'Nietzschean' aspiration towards expressing the *interplay* of the *signans* and the *signatum*, taking in on the way the death of God, man, the subject, standards, grammar and syntax.

§ 5. *The 'hermeneutic' current: the interpretation of language*

The theory of interpretation, like structuralism, tends to produce an image of reality which is very much determined by features borrowed from language. But whereas structuralism accepts the full consequences of the reduction of language merely to the form of language and the interplay of its internal oppositions, hermeneutics lays extreme stress on the *signatum* and what is beyond the *signatum*, i.e., the intention of the text and of its author. The same concept of writing and text is here used as the point of departure for a 'reconstruction' of that view of things, that *Weltanschauung*, which the text intends.

Recently there have been important developments in the theory of interpretation: in the first place, the intentions of its founders, Schleiermacher and Dilthey, have come to be better understood thanks to Kimmerle's publication, in 1959, of the former's early texts on hermeneutics,[78] in which 'grammatical' interpretation balances 'psychological' interpretation, thanks

76. Derrida, *De la grammatologie* (1967).

77. Derrida, *L'écriture et la différence* (1967). See also, by the same author, *La dissémination* (1972) and *Marges. De la philosophie* (1972).

78. Schleiermacher, *Hermeneutik* (1959, 1968), edited by H. Kimmerle. See Kimmerle's Introduction (1959) and his Postface (1968). See also Bultmann, 'Das Problem der Hermeneutik', pp. 214 sqq. in *Glauben und Verstehen*, Vol. II (1952).

also to Redeker's 1966 edition of the second part of Dilthey's *Leben Schleiermachers*.[79] These editions and the discussions which followed them are noteworthy for two characteristics. Firstly, the intention of founding a general hermeneutics to take in the various specialist types of hermeneutics (Biblical exegesis, classical philology, the understanding of history, and jurisprudence) emerges as the key to the interest taken in the notion of comprehension: what predominates is not some psychological interest in a mental operation, akin to *Einfühlung* or empathy, but the methodological inversion whereby the emphasis is transferred from individual texts to the conditions governing the interpretation of texts in general; this shift of emphasis is comparable to the Copernican revolution by virtue of which the question of objects gives way, with Kant, to the problem of the condition of objectivity in general. The second insight gained by the founders of hermeneutics is that any process of comprehension has a 'divining' aspect and a 'methodical' aspect; there are reasons of principle why a series of sentences does not constitute an immediately intelligible whole: the circular relationship between the comprehension of detail and the comprehension of the whole – a circularity which is the source of the famous 'hermeneutic circle' –, the cultural distance between the author's times and the age the interpreter lives in, the intentionally disguised or unintentionally distorted nature of the fundamental meaning, and the author's foreignness simply because he is another person, all these are reasons why interpretation must necessarily have an element of surmise and conjecture (divination), offset, though only partly, by methodical testing. If this is so, the act of putting oneself in another's psyche, which is stressed by Schleiermacher in his later writings, and then by Dilthey at the time of his famous article '*Die Entstehung der Hermeneutik*' (1900),[80] only makes up one part of interpretation, and is not its principle; rather, the principle is the interplay of conjecture and validation. Comprehension is the name given to the whole of this process, and not to the psychological, or even inter-psychological, stage of conjecture alone. That is why comprehension can be contrasted and compared with explanation; the interplay of conjecture and validation in interpretative disciplines is indeed wholly comparable to that of hypothesis and verification in the natural sciences, and is applicable to the field of *Geisteswissenschaften* as a whole.

Contemporary interpretation theory has moved further and further away from Dilthey's still psychologically inclined views: although he explicitly linked the notion of text and the transition from inter-subject understanding, such as is found in dialogue, to the 'interpretation of expressions of life committed to writing', Dilthey never explicitly connected the problems of comprehension with those of language. Hermeneutics has accordingly had to disengage itself from the psychological problems of transfer into another's life, and come to grips with the more ontological problems involved in comprehending 'being-in-the-world'. Here we see the influence of the early Heidegger, the Heidegger of *Sein und Zeit* (1927). Here comprehension is no longer

79. Dilthey, *Gesammelte Schriften*, Vol. XIV/2.
80. Cf. Dilthey, *Gesammelte Schriften*, Vol. V, pp. 317 sqq.

a psychological concept; it is dissociated from any idea of empathy, any perception of an alien consciousness; it is interpreted in ontological terms as one of the components of *Dasein*, which is no longer a consciousness but a being-in-the-world, a being which faces the question of being within the context of specific situations and plans, against a background of finiteness and mortality. Thus 'depsychologized', comprehension is brought nearer the primordial problem of language, for the understanding a human being may come to have of his own situation and plans can be explicated only, and therefore interpreted, through the articulatory medium of language. Thus *Auslegung*, or interpretation, emerges as the language stage of *Verstehen*, or comprehension. Interpretation is therefore not primarily a generalized offshoot of classical philology; it is rooted in the understanding of existence before being linked to written documents, to texts. For this reason, the well-known 'hermeneutic circle' which is met with particularly in Biblical exegesis (for belief there must be understanding, for understanding belief) is merely the literary facet of a more fundamental circle which can be traced at the level of the act of understanding, and consists in the being's arriving at self-understanding or the basis of patterns of conjecture. This notion of interpretation is so important in *Sein und Zeit* that the whole enterprise comprising existential analysis is itself a type of 'hermeneutic phenomenology', because of the forgotten, disguised, hidden nature of the fundamental structures of existence. For this reason too, any reading of the text of existence is a kind of philology writ large; all that is 'given' of the phenomenon of existence is prior to any interpretation, a mere sensory effect, a surface effect; to the extent that the immediate hides the essential, to that extent phenomenology *is* hermeneutic.

In Heidegger this revised view of the ontological 'locus' of language is accompanied by a revised view of the meaning of truth. A common assumption of those philosophies of language which we have so far considered was that truth consisted in a relationship of correlation, correspondence, conformity, 'adaequatio'. This is why those philosophies keep to the level of the proposition, and expound an *apophantic* theory of language. The step from truth as correspondence to truth as revelation entails going beyond the apophantic function to the hermeneutic function, in which 'fulfilled' (fully referential) language is that which 'shows', in the sense of 'letting be'. This step also means that the models must be changed: fulfilled language is neither the correct language which the logicians construct nor the ordinary language which linguistic analysis describes, rather it is the language of the fundamental poets and thinkers like the pre-Socratics. These are the witnesses of a dimension of language which Heidegger calls *Sagen* – 'telling' –, which is higher than *Sprechen* – the 'speaking' of ordinary or logicized language. Only telling is up to the task of revealing and showing, and it is this telling to which we must listen (*horchen*) and pay heed (*gehorchen*).

In the later works of Heidegger, primarily in *Unterwegs zur Sprache* (1959) [81], the ontological dimension of language is further emphasized; we

81. English translation, *On the Way to Language* (1971).

are faced, not with the *Dasein* which understands itself in the world, but with the need to 'move language as language towards language': *die Sprache spricht*, says the author in a deliberate truism; through this apparent tautology, the formula suggests the idea of a movement within one and the same medium, the idea of an interlacing of functions; 'on the way' (*unterwegs*), we learn to go beyond the determinations of speaking, in as much as it is a capacity, an activity, a possession of man – and that which has been explored by the philosophers of language, from Aristotle to Humboldt –, till we finally arrive at the power of *telling*, which expresses not our own power but that of the being-told, which indeed precedes and engenders our power over what we mean. Thus language is nothing, and its potential is nullified, if it is not internally bound by the very manifestation of what occurs when the thinker speaks and 'tells'. As can be seen from the vocabulary of Heidegger, the key words in this philosophy of language – let-be, let-belong, show, occur, etc. – belong neither to linguistics nor to linguistic analysis in the sense of the English philosophers, nor to phenomenology, nor even to the existential analysis developed in *Sein und Zeit*.

We shall say nothing of the influence which this type of hermeneutics had on Biblical exegetists like Rudolf Bultmann,[82] who still base themselves on an anthropological reading of *Sein und Zeit*, and then on the post-Bultmann exegetists, who had more feeling for the definitely anti-anthropological intentions of the later Heidegger. We should, on the other hand, say something about the work of someone like Gadamer, in which those conclusions of Heideggerian philosophy that are not only anti-psychological but also anti-methodological are developed. There must be a choice between truth and method: whence the title *Wahrheit und Methode*.[83] In this work, the author contrasts the power of truth which lies concealed in comprehension with any methodology, any technology which might be ascribed to the sciences of the mind. Comprehension is not some method which can be used to supplement the methods of the natural sciences, and in this connection Dilthey failed in his attempt to divide scientific method in two, and set the sciences of the mind alongside the natural sciences; Gadamer sees in the understanding which a man gains of his historical tradition and in his exercise of taste in aesthetic matters – sees in these the two great vehicles of such comprehension, which is irreducible to any methodology in the sense of modern science. The question 'what does understanding a language mean?' becomes a hermeneutic question only when the comprehension of language is thus related to historical and aesthetic comprehension. Then language emerges, not as a third area of comprehension alongside history and art, but as the universal medium in which any experience of meaning occurs. Situations such as

82. Cf. Bultmann, *Glauben und Verstehen* (4 vols., 1933, 1952, 1960, 1965); *Die Theologie des Neuen Testament* (1953); *Kerygma und Mythos* (3 vols., 4th edn., 1960); *Exegetica* (1967); etc. See below, Section VI, *Man and the Foundation of Humanism*, sub-section C, 'The philosophy of religion', § 5.

83. Gadamer, *Wahrheit und Methode. Grundzüge einer philosophischen Hermeneutik* (1960). Cf. also *Kleine Schriften*, I: *Philosophie, Hermeneutik* (1967).

agreement (agree with . . ., agree on . . .), listening, translating from one language into another, offer so many insights into the general problem of understanding, which is always a struggle against distance, against the alien character of what at first is not understood. Writing is pre-eminently the means of distancing; and to read it entails recovering the meaning. This is why writing is also pre-eminently the object of hermeneutics; and as every tradition, through writing, becomes a linguistic reality, so all understanding, be it historical or aesthetic, involves some measure of *Sprachlichkeit*. Taken in this way, language is that through which man 'has a world': a world, not merely an environment; a world which he keeps at a distance and holds in front of himself. This is the abiding truth of Humboldt's thesis that languages are visions of the world – with the proviso that every tradition is open to every other tradition, and that every world, since it is constituted by language, is itself open to every possible comprehension, and capable of infinite extension. So what makes a language is neither its grammar nor its lexicon but its power to articulate what is said in tradition. This conviction inevitably removes the debate from the linguistic plane and sets it down alongside Hegel and the Greek and Christian writers of antiquity.

This orientation of hermeneutics, firstly anti-psychological, secondly anti-methodological, ushered in a crisis within the hermeneutic movement; in correcting the 'psychologizing' tendency of Schleiermacher and Dilthey, ontological hermeneutics sacrificed the concern for validation which in the founders' world made up for the element of divination. Another current of thought took over the 'methodological' side of hermeneutics, and so initiated a discussion, a dialogue between interpretation theory and the sciences of man.

Some, opposing to the work of Gadamer that of E. Betti,[84] in whom they see the true heir of Dilthey, develop hermeneutics in the direction of a logic of proof; thus E. Hirsch,[85] basing himself firstly on Husserls' *Logische Untersuchungen*, stresses the objectivity of 'the verbal signification' of the text; this signification, which is identifiable and reproducible, must be distinguished from 'significance', which comprises all the non-intentional implications of the text and comes under the head, not of interpretation, but of criticism; then, basing himself on the logic of probability of Keynes, Reichenbach and K. Popper,[86] the author develops a theory of 'validation' of the probable meaning, corresponding in hermeneutics to the theory of 'verification' in the empirical sciences. In this way, contact is re-established between hermeneutics and logical positivism.

The contrast between interpretation and explanation is also restated by other authors, who compare hermeneutics with the structuralist models of

84. Betti, *Die Hermeneutik als gemeine Methodik der Geisteswissenschaften* (1952) and *Teoria generale dell' interpretazione* (1955).

85. Hirsch, *Validity in Interpretation* (1969).

86. See above, Section II, *Man and Natural Reality*, sub-section A, 'Epistemology of natural sciences', § 2, 'Probability', pp. 1154-1158.

linguistics and psychoanalysis. Thus P. Ricoeur[87] considers a semantics of the sentence, closely related to the theory of the speech-act, to afford a transition between the structural analysis of a text and its interpretation. Appropriation of the meaning by a subject, who by understanding the signs written down in writing, increases his comprehension of himself, is seen as the last act in comprehension; this final step is necessarily preceded by all the objective procedures of structural analysis and conceptual clarification. In this way explanation would no longer be opposed to comprehension, but would rather comprise all the objective intermediate steps leading up to appropriation of the meaning. Lastly, the appropriation of meaning is seen as inseparable from the process of adjudicating between conflicting interpretations, whether based on exercise of the critical faculty, as with Marx, Nietzsche or Freud, or on an attempt to recapture the authentic meaning, as in the ontological hermeneutics tradition of Heidegger and Gadamer.

'Methodical' hermeneutics has yet another side, that of the *critique of ideologies;* the part played by 'self-interest' in knowledge, and the interaction of practice and theory, bring out the social and political implications of any meaning attributed to cultural signs or to interpretation itself. The non-reducibility of human sciences to models derived from the natural sciences is once again attested by this signifying and teleological, self-interested and ideological nature of social phenomena; the result is to produce yet more links as well as contrasts between the natural and the human sciences. Most of the work of the 'Frankfurt philosophers' [88] spans the borderline between hermeneutics and the critique of ideologies; for this critique [89] is one type of hermeneutics, in the sense that the explanation of social phenomena is based on interpretation of the cultural meanings in which the interests underlying conscious activities are expressed. So we see that hermeneutics has common ground with the Marxist theory of ideologies and superstructures; in this connection, works like Adam Schaff's [90] are good examples of the stages by which it is possible to go from semantics to the critique of ideologies and interpretation theory.

In this way 'methodical' hermeneutics can enter into a dialogue with linguistic science, conceptual analysis, structuralism and Marxism, at the same time keeping up its contacts with 'ontological' hermeneutics.

87. RICOEUR, *De l'interprétation. Essai sur Freud* (1965) (English translation, *Freud and Philosophy. An Essay in Interpretation*, 1969); *Le conflit des interprétations* (1969).

88. See in particular: ADORNO, *Negative Dialektik* (1966); HABERMAS, *Theorie und Praxis* (1963) and *Zur Logik der Sozialwissenschaften* (1967); APPEL, 'Das Verstehen' (1968); HABERMAS (ed.), *Hermeneutik und Ideologiekritik* (1971).

89. See also in the preceding chapter, Section III, *Man and Social Reality*, subsection A, § 1, 'The logic of explanation in the social sciences', item c), 'Neutrality and value', p. 1234.

90. SCHAFF, *Wstep do semantyki* (1960) (French translation, *Introduction à la sémantique*, 1968). See § 3 above.

§ 6. *Language and ontology according to recent thinkers in the Orient**

Can language provide a clue to the nature of reality? And, even if it can, what is its own ontological status? Is it just a mirror in which we may see directly and clearly that which is reflected in it, but which essentially is outside it? Or, is it a net whose structure determines what types of reality we catch through it? Or, has it a creative dimension which brings certain types of reality into being, which could not possibly have been there without it? Is there, so to say, linguistic being? – if such a phrase be permitted.

To each of these questions, there have been some significant answers amongst recent thinkers in the Orient. K. C. Bhattacharyya has suggested a radical distinction between types of thought to which linguistic articulation is essential and those to which it is not. The latter he characterizes as empirical thought which, according to him, is 'the theoretic consciousness of a content involving reference to an object that is perceived or imagined to be perceived, such reference being part of the meaning of content'.[91] On the other hand, thought to which linguistic articulation is essential deals with a type of content which is not intelligible and, in fact, does not have even any being except as thus articulated. Philosophy, according to him, deals with such a type of thought. As he writes, 'pure thought (with which philosophy is concerned) is not thought of a content distinguishable from it', and just because of this reason it is 'sometimes regarded as a fiction, philosophy being rejected as a disease of speech'.[92] But to view philosophy in such a manner, that is, as fiction or a disease of speech, is to think that empirical thought is the only kind of thought we have. And this is what the positivists generally do. But that is to ignore the nature of philosophical thought itself.

A reflection on thought and its relation to language in terms of contingency or necessity reveals thus the nature of philosophy to a certain extent. The reflection on 'ego-centric particulars', on the descriptive and 'indicative' or, rather, imperative uses of language, and on the cognitive function of the paradoxical use of language are some of the other directions which thought about language has taken amongst these thinkers.

The 'ego-centric particular' which K. C. Bhattacharyya picks up for specific reflection and which, according to him, leads to an awareness of the unique character of subjectivity, is the word 'I'. If it be accepted that 'a meaning that is conveyed by a word must be intelligible to the hearer as what he would himself convey by the use of it', then the word 'I' obviously goes counter to the condition of meaningfulness. But this, in a sense, is true about all words which are concerned with what have been called 'ego-centric particulars'. The word 'I', on the other hand, has certain other features also. For example, the word 'I' conveys the communicating-act of the speaker which generally is never a part of the meaning of a word. Also, while most words mean either something individual or general, the 'I' is neither defi-

* Prepared by Daya Krishna as part of his general contribution to this work.

91. Bhattacharyya, K. C., *Studies in Philosophy*, Vol. II (1958), p. 102.

92. *Ibid.*, p. 103.

nitely singular nor general. As *used*, the term has a uniquely singular reference; but as *understood*, it is general. 'It is thus unlike a term meaning an objective content, which has the same reference – singular or general – to the speaker as to the hearer.' [93] Similarly, while all empirical objects may be pointed to without the use of a word – which is the basis for ostensive definitions –, that which is meant by 'I' cannot be so indicated. What one may ostensively point to is, rather, the body and not the subject which is intended by the word 'I'. The empiricist programme of translating all words referring to non-perceptual objects into those that refer to perceptual objects alone so that what they mean may ostensively be pointed out, and thus render all language superfluous, meets its essential limitation in words such as 'I', if what they intend cannot, in principle, be ostensively pointed out.

The distinction between the descriptive and the 'indicative', or what is usually called the imperative use of language, has been used by N. V. Banerjee [94] to point out the radical ontological distinction between time as 'before' and 'now', on the one hand, and time as 'after', on the other. Normally, the three are thought of as all of one piece. But the former two, that is, 'before' and 'now', have content, while the latter, in principle, can have none. There is thus a radical difference between the two, and it is 'after' which reveals more the essence of time than 'before' and 'now'. Similarly, man's involvement in temporality is essentially with respect to time as 'after', that is, the dimension of action which is conveyed by the 'indicative' or imperative use of language.

The rôle of paradox in language has been the subject of interesting reflection on the part of the Japanese philosopher, Tomonobu Imamichi. Paradoxes, it is well known, arise whenever we try to talk about totality, or the total whole. Mathematics tries to avoid them in an ad hoc manner – usually, through what has come to be known as *the theory of types*. However, that only leads to a postponement of the problem, rather than coming to grips with it. The alternative seems to remain, that either one does not talk about totality at all, or one talks about it in paradoxes. It is well known that mystics all the world over have used paradoxical language to convey their sense of reality. On the other hand, philosophers have generally felt paradoxes to be a sign of unreality, believing as they do in the law of contradiction as the basic law of thought. But if philosophy wants to comprehend the whole and is concerned with totality in a fundamental sense, then, Imamichi argues, it should cease to treat them as such. Rather, it should see that if totality is to be grasped linguistically and formulated at all, then it can only be done in the form of a paradox. The search for exact definitions as the goal of knowledge which, since Socrates, has been predominantly accepted in the West, gives us only the knowledge of determinate beings, or what in German is called *Seiende*. But if we seek the knowledge of *Sein*, and that is what metaphysics ultimately is, then the path of exact definition and grammatical or logical thought is the wrong path to take. Kant saw that antinomies were the

93. *Ibid.*, p. 20.
94. Banerjee, *Language, Meaning and Persons* (1963).

result of using understanding beyond its bounds. In other words, the moment we entered metaphysics, contradictory sentences seemed equally well supported by reason. But he took it as a sign of the misuse of understanding, instead of seeing it as the sign that now alone he was talking metaphysics proper. Imamichi, however, warns that though all paradox has an element of the self-contradictory, every self-contradictory statement is not a paradox. But he does not give any criterion for distinguishing between the two.[95]

Another interesting problem has been raised in connection with the interpretation of oral literature, which in many cultures may be the only thing to interpret. Most of Africa happens to be in such a situation. And, though the problem has been raised specifically by Paulin J. Hountondji [96] in connection with the attempts to explicate a philosophy which is supposed to be implicit in it, the issue obviously is of wider interest. Also, even if written texts or treatises exist, the problem of the interpretation of oral tradition and its collation, corroboration and confrontation with the written one would remain.

Another discussion which is of some interest in this connection relates to the question whether there are any universals of language and, if there are, are they just empirical generalizations from an examination of existing languages, subject to modification and falsification in the future, or just *a priori* conditions which any language has to fulfil in order to be a 'language'? Also, how far does the universal structure, if any, reflect the structure of the real or just the structure of the knowing process or even just that of communication itself? Further, would not the idea of a universal structure raise all the problems which Kant encountered, including that of the *synthetic a priori*, once more?

95. IMAMICHI, *Betrachtungen über das Eine* (1968).

96. Cf. HOUNTONDJI, 'Remarques sur la philosophie africaine contemporaine'/ 'Comments on contemporary African philosophy' (1970). See especially his comments on Kagame's work, p. 123 in *Diogenes*, No. 71. Also see in this connection Franz CRAHAY's article entitled 'Le décollage conceptuel: conditions d'une philosophie bantoue'/'Conceptual take-off: conditions for a Bantu philosophy' (1965), especially p. 58 in *Diogenes*, No. 52. Cf. KAGAME, *La philosophie bantu-rwandaise de l'être* (1956) and TEMPELS, *La philosophie bantoue* (French translation, 1949).

SECTION FIVE

MAN AND ACTION

SECTION FIVE: MAN AND ACTION

A. THE THEORY OF PRACTICAL DISCOURSE

We have already described what might be called the epistemology of action in connection with the epistemology of the social and human sciences; [1] the concepts of behaviour and action, as they are used in psychology and sociology, in fact provided the main thread running through this review of studies in epistemology. It is for that reason that we have kept the present sub-section of Section V for an examination of studies based on mathematics and logic dealing less with action itself, taken as something to be described, than with the discourse wherein action is expressed. The theory of practical discourse is accordingly taken to mean the systematic study of the notions, propositions and rational processes relating to human action.

We shall begin this review with decision theory and game theory, since these have been derived directly from the mathematics of probability as a forecasting technique applied to the contingent aspects of purposive behaviour.

We shall subsequently deal with attempts to constitute a logic of practical discourse similar to certain parts of the logic of theoretical discourse, in particular modal logic (the logic of necessity and possibility); these centre around deontic logic.

Lastly, we shall consider the theories of practical reasoning which include the actual decision as part of the reasoning process.

This order of presentation is in itself an expression of choice on the part of the rapporteur, who sees in this sequence a progressive recognition of the strictly *practical* nature of the discourse of action. Decision and game theory goes no further than a theoretical discourse on action. The logic of practical discourse does admittedly deal with the discourse of action, but is properly limited to the formal aspects (norms, imperatives and values) by virtue of which practical discourse is analogous to theoretical discourse. Only the theories of practical reasoning provide a theory of action in the sense of decision which is both free and rational.

§ 1. *Decision theory and game theory*

Decision and game theory is *formal* in the mathematical sense of the term, i.e., it constitutes a body of theorems deduced from a set of axioms, and therefore in no way represents an investigation into actual reasoning as employed in real action. It is only as an applied science that it can relate to specific economic, military or judicial situations which give rise to real conflicts and real decisions. It is even open to question whether the alternatives of real life do not comprise essential features which are specifically excluded

1. Cf. above, Section III, sub-section A, § 2, pp. 1238-1249.

by a formal structure. Thus one must not look to this theory to provide criteria for real decisions in all situations; nor was this the intention when it was devised. It is neither an account of actual reasoning processes nor a system of criteria for rational choice.

A distinction is drawn between *decision theory* and *game theory*.

Decision theory is the application of calculus to conjecture. The following passage from Jacques Bernoulli may be taken as defining its scope: 'For that which is certain and beyond doubt, we speak of knowledge and understanding; for all the rest, we say simply conjecture or opinion; to conjecture something is to measure its degree of probability; thus we define conjectural or stochastic knowledge as the knowledge of how to measure degrees of probability as accurately as possible, in order that in our decisions and actions we may always choose or accept what appears to us to be the best, the most satisfactory, the most sure, the most prudent – the sole purpose and aim of all the wisdom of the philosopher, all the foresight of the politician' (*translation*). The mathematical theory of decision is thus an attempt to base a methodology of rational conduct on the calculus of probabilities. In this respect game theory is merely the most striking form of decision theory, firstly because of the rôle of model provided by the strategy of certain games of chance, and also because of the place taken by this theory in the history of the problem since Pascal, Leibniz and the Bernoullis. It is for this reason that we begin by describing at length the problem of the mathematical theory of decision before going on to consider game theory. Viewed at this general level it differs from the logic of argumentation[2] in that it deals with quantitative determinations of probability. As such it is an art of conjecture which in addition employs the calculus of probabilities. This kind of art entails on the one hand the fullest possible enumeration of contingencies, i.e., of circumstances affecting the outcomes of the action, regardless of the volition of the actor, and the assignment of probabilities to the various contingencies; and, on the other hand, a classification of the various types of decision.

We shall leave aside situations of certainty, where it is theoretically possible to list in full the elements of the problem. Examples are operational research and linear programming (where the function to be maximized or minimized and the constraints are linear forms of the variables); the chief purpose of such mathematical models is to achieve the best use of available resources. Whenever the problem becomes one of establishing an order of preferences, we have situations of uncertainty, in which probabilities intervene: these include conflict situations, which lead us on directly to game theory.

The writings of Finetti, Ramsey, Savage and Jeffrey lay the basic foundations of a science of decisions which points the way to a veritable logic of human action.

Opinion occupies an essential place in the work of Finetti, who develops to the furthest possible limit the subjective element in probability. The betting

2. Cf. below, § 3, pp. 1387-1391.

method affords a means whereby probability may be measured quantitatively, even though the starting-point is a system of purely qualitative axioms concerning degrees of subjective probability; the logic of probability is thus seen to be an essential instrument in all reasoning involving uncertainty, a judgement as to what is a practical certainty or practical impossibility, in a word an evaluation of the probability of any given event.[3]

Ramsey studies belief, not on the basis of introspection, but at the level of behaviour whereby it can be measured. The measurement of degrees of belief is based on the determination of scale of preference as between the consequences in relation to a condition which is ethically neutral, i.e., a condition of belief graded ½.[4]

Savage, who continues the work of Finetti, sets out to describe the highly idealized behaviour of individuals in a situation of uncertainty; idealization makes it possible to keep the theory both normative and descriptive, and to define a rational individual capable of fully ordering his actions in accordance with his preferences for the consequences. In this writer's view, such consistent behaviour can be taught, since it remains a norm; in this respect it comes within the sphere not of psychology, but of logic.[5]

With Jeffrey the logical element is accentuated, leaving no place at all to psychology: his 'logic of decision' is a theory of preferences in which judgements of probability concern not events but propositions concerning events. Deliberation consists in assessing lines of action that are acceptable in terms of consequences which may depend on circumstances beyond the power of the actor to predict or control: the principle of deliberation is thus to perform an act which maximizes the expected utility, and the order of preferences is determined by the matrices of probability and utility.[6]

Formal decision theory thus deals with the question of 'how choices among alternatieve courses of action would be made if such choices were to satisfy certain criteria of "rationality" '.[7] Decision theory thus presupposes a rational actor, i.e., one whose choices of action can be completely defended on the basis of what the theory proves should be the preferred outcomes.

It will be for the philosophy of action, and not for decision theory, to

3. De Finetti, 'La décision et les probabilités' (1963), 'Probabilità composte a teoria delle decisioni' (1964), and 'Probability. The subjectivistic approach' (1968) (see in particular the references to the literature of the subject in this last article). On the subjective theory of probability, the reader may refer to the collection of fundamental essays edited by Kyburg & Smokler, *Studies in Subjective Probability* (1964) (with a critical introduction and a comprehensive bibliography).

4. Cf. F. P. Ramsey's article dating back to an earlier period: 'Truth and probability' (1931).

5. Savage, *Foundations of Statistics* (1954); Dubins & Savage, *How to Gamble if You Must* (1965). See also Ellsberg, 'Risk, ambiguity and the Savage axioms' (1961).

6. Jeffrey, *The Logic of Decision* (1965).

7. Rapoport, :The theory of games and its relevance to philosophy', p. 451 in Klibansky (ed.), *Contemporary Philosophy/La philosophie contemporaine*, Vol. II (1968). See also Rapoport, *Games, Fights and Debates* (1960) and 'The use and misuse of game theory' (1962).

clarify the intuitive notion of a rational decision to which the choice between alternative courses of action (the subject-matter of the theory) conforms. The question whether the action has outcomes, whether the actor makes a choice, whether he has the possibility of preferring one action to another depending on the outcomes, whether he regards rationality itself as a value – all these come within the sphere of philosophical psychology or ethics, not decision theory. The same is true of the meaning attached to the principle of maximum expected utility and to the notions of risk and gain; for the psychologist and the moralist these notions are actualized in concrete situations, for example those in which gain, and consequently also risk and utility, can be defined in monetary terms. For the theoretician, concrete situations in no way afford intuitive validation of these norms; at the most they serve as examples or illustrations. All that is required is that one should be able to define an event formally, i.e., to specify the conditions for an action which are independent of the actor and which are thus contingent; by 'strategy' is then meant a plan governing a series of actions, where the series is contingent on other eventualities, all possible eventualities being foreseen.[8]

The transition from decision theory to *game* theory is made by introducing the notion of two or more actors, together with the notion of conflict, i.e., a non-coincidence between the preferences of a number of actors, with two or more wills in a situation of conflict partly dependent on chance.

Game theory can thus be viewed either as a branch of formal decision theory (which is how it has been presented above) or as a formalized theory of conflict; it is in the latter sense that we shall now consider it.

In *games*, the other eventualities are either reduced to the choices made by the other players, as is the case with games of pure strategy (e.g., chess), or else they include a purely contingent initial situation which can be assimilated to a first choice made by a fictitious player, Chance, a choice which is itself not known to the other players (as in card games). By extension of this latter case, any situation of uncertainty may be assimilated to a struggle against nature viewed as a competing player.

The game model is thus a kind of idealized type proposed by the mathematician to the economist and the war scientist. Its fruitfulness is due to the fact that the element of chance is not natural but artificial, i.e., contribed by man, thus making it possible to study chance which can be controlled. Even in the second kind of game, for example bridge, the contingent initial situation, consisting of the dealing of cards, can produce only a finite, though very large, number of combinations. With the first kind of game, the solution is purely one of strategy; it involves only intelligence and the ability to perceive the situation as a whole, with its consequences. Since the outcome depends solely on a series of foreseeable actions, there exists a best strategy for each player, and game theory then consists in conceptualizing the pattern of pure strategy solutions. With the second kind of game, finesse and bluffing

8. CHURCHMAN, *Prediction and Optimal Decision. Philosophical Issues of a Science of Values* (1961); LUCE & SUPPES, 'Preference, utility and subjective probability' (1965).

are added to the logic of choice. In all cases, however, the actors are defined, without any reference to psychological traits, solely by their position as players, i.e., by being subject to the rules which govern the succession and the range of permissible choices, and which define both the manner of terminating the game and the set of situations resulting from the termination which constitute the outcome of the action as a whole. Game theory considers only the logico-mathematical structure of games: it is in no way concerned with the individual psychological peculiarities of the players.[9]

This inherent logic of games makes it possible to distinguish between games with two players and games with N players and, among the former, between *constant-sum* games and *non-constant-sum* games. All constant-sum games can be treated like *zero-sum* games, in which the winnings of one player equal the losses of the other (their algebraic sum = 0). The two-person zero-sum game is the simplest case, from which the game theory of von Neumann and Morgenstern is derived; it is the very epitome of a duel. Among non-constant-sum games we may distinguish those which permit negotiation, i.e., a co-ordination of the players' choices, involving communication and bargaining, with a view to achieving an outcome which both prefer. The notion of a negotiated (or co-operative) game is specially interesting in that it introduces the idea of collective rationality, where the advantage is increased for all players if one strategy is preferred to another. Further, negotiated strategies bring to light, in a pure theoretical form, notions which are frequently illustrated by actual conflict situations, such as security levels, equilibrium points, co-ordinated strategy or threat strategy, arbitration or calculated risk (the risk being that the threat will be carried out if agreement is not reached).

The theory of co-operation had already been propounded in the economic sphere by Pareto, who defined the notion of 'maximum collective ophelimity' as an equilibrium in which no further change could be made to the collective advantage without harming one of the members. Game theory has incorporated this notion by adding to the initial model of the two-person zero-sum game the more complicated model of the N-person non-constant-sum game, and by considering a variety of strategies ranging from conflict to coalition or partial co-operation.

The gap between this enlarged game theory and decision theory can be bridged in the following way.[10] The rules of decision can in fact be assimilated to a game, if nature is viewed as a will pitted against my will; statistics then becomes a game against nature. The problem of decision can then be stated in the following terms: to begin with, there is a set of possible decisions, and a certain amount of information on the relative probability of events: in conclusion, the value attached to the consequences of the decisions has been fixed; the problem consists in choosing a 'decision function', and

9. Von NEUMANN & MORGENSTERN, *Theory of Games and Economic Behavior* (1944).

10. Cf. WALD, *Statistical Decisions Functions* (1950, 1966). See also, on a related issue, APOSTEL, 'Game theory and the interpretation of deontic logic' (1960).

the solution lies in determining the 'risk function' pertaining to the 'decision function'. Thus the problem of decision is assimilated to a two-person zero-sum game, by virtue of the equivalence between decision function and strategy.

We may accordingly consider extending to partial or total uncertainty situations of different kinds the decision criteria which the theory proposes to represent the various forms of calculated risk; the criterion of 'mathematical expectation' (which is a number, i.e., the product of the gain, or pay off, and its probability); the criterion of 'utility'; the betting method (which takes account of the probability of the solutions envisaged and is thus an extension of Bayes' theory of subjective probability); the 'minimax' criterion (used where the future is entirely unknown, i.e., where there is no *a priori* probability distribution; this criterion is based on the possible losses attaching to the various outcomes; we give preference to the decision that minimizes the maximum loss which the state of the universe can expose us to: it is called the minimax criterion because one begins by defining what is, for the first player, the best of the worst positions, i.e., the greatest minimum gain or 'maximin'; the best choice for the second player then corresponds to the smallest of the maximum gains of the first player and is known as the 'minimax'; by definition, maximin and minimax coincide in team games); and the 'regret' criterion (when making his choice, the individual player thinks forward to the time of the consequences and attempts to minimize the regrets he will feel; regret, in the mathematical sense of the term, is the difference between the real loss sustained and the minimum loss which would have been sustained if one had known in advance that this state of the universe would come about, and if one had then made the decision corresponding to the minimum loss for that state of the universe; the criterion accordingly consists of choosing the decision that minimizes the greatest regret which the state of the universe can expose one to).

Critical remarks

Stress has been laid on the normative nature of decision theory and game theory. But what is meant by normative? Are we to think in terms of a *command*, thereby eliminating any kind of psychological consideration and even possibly any application to behaviour in real life? Or is what is meant a logical *tool* for use in real action, whether in the economic, political or legal spheres? Or is the theory expected to produce *models* to describe the real behaviour of individuals? All three interpretations are under study today.

The first is undoubtedly that which is implied by the theory on a formal, axiomatic view: as said above, its indifference to applications is basic to its very conception. It is for this reason that formal decision and game theory should never be applied directly or without due precautions to real behaviour. The 'rational man' whose conduct is governed by such a logic of decision is a fictitious being. This in no way constitutes an objection of principle to the theory, which in this respect remains true to the scientific method whereby over-complicated real phenomena are always replaced by simpler fictitious ones. In order that the theory may be applied to reality it is necessary for the real situations themselves to be reduced to schematic features approximating to the ideal theoretical type, with set alternatives and consequences, quantitative utility values, assignable probabilities and partial control of the outcome of the action by two or more equally rational actors. It is in fact chiefly in economics, war science and, to a lesser degree, in judicial logic, where deliberative procedures are sufficiently formalized, that the theory has the most striking success, particularly since its refinement by the introduction of N-person games and variable-sum games, as well as bargaining, negotiatory and reconciliatory strategies. In a word, it is when conflict situations can themselves be reduced to highly idealized and simplified patterns that the theory can be applied to them. We then come very close to the type of players presupposed by the theory, namely players who are rigorously logical and endowed with an unlimited memory, in short, players devoid of normal psychological characteristics, such as varying degrees of intelligence, a variety of temperaments and prejudices of all kinds.

Secondly, if by a 'normative science' is meant a logical *tool* for use in real action, we are led to consider the applications of the science of games, using 'application' in its pragmatic sense; we look to the science of games to provide an art of rational decision which can be taught. We are in fact justified in expecting the theory to provide the individual with the means whereby he can plan his decisions better by weighing the pros and cons, and thus practise an orderly art of considered choice; this project for training the individual to choose rationally by giving a normative science a pragmatic extension now receives wide support in the economic, political, military and judicial fields, i.e., wherever decision-making partakes of the nature of strategy. At the same time, one should never overlook the simplified schematic character of the models; as said above, the mathematician offers the economist and the war scientist ideal types, but the application of the theory of these fields appears fruitful only in systems which can be assimilated to a game of strategy where the players are restricted in number and act rationally in accordance with the principle of seeking the maximum utility. Here concerted strategies offer opportunities for generalization which are far from having been exhausted. Nevertheless the specialists themselves warn against the illusion that it is possible, by applying game theory directly, to devise a so-called 'winning' strategy. What is more, as one of them recalls, it is game theory itself, in its most highly developed form, i.e., the theory of non-constant-sum, N-person games, which has revealed 'a plethora of dilemmas inherent in the very concept of self-interest, and the need for a radical re-evaluation of this concept in the light of rigorous analysis of the logical structure of conflicts. Therein lies, perhaps, the most significant contribution of game theory of philosophy.' [11] The exercise of collective rationality by each actor, as in negotiated strategy, frequently produces more advantageous results than the

11. Rapoport, 'The theory of games and its relevance to philosophy', *op. cit.* (1968), pp. 471-472.

exercise of individual rationality. This principle has often been perceived intuitively by philosophers without being given rigorous form. The same is true for concepts formerly held to be simple, for example that of the 'greatest good for the greatest number'; these are in reality so many bundles of antinomies, and it is largely thanks to the theory that these antimonies can be bought into the open.

In a third sense of the word 'normative', the theory may be used as a descriptive model. Here we touch on the most disputed point: are the models of game theory an extent approximation to the real choices of human beings in situations other than a situation of contrived chance? The desire to extend normative theory in a descriptive sense doubtless attuned with the need for a deductive science of human behaviour, or praxeology, which was formulated independently of game theory and prior to von Neumann and Morgenstern's chief work (cf. von Mises [12] and the Polish school of praxeology [13]).

Much can be said in favour of extending mathematical models. Firstly, game theory provides more efficient models for describing real human behaviour than do strictly mechanistic models, which break down when confronted with the fortuitousness and unpredictability of behaviour; probabilistic models are more likely to resemble closely the human reality to be found in the no man's land between the necessary and the fortuitous, the intermediate zone of probability. Further, the very psychological nature of the choice presents a fundamental analogy with the abstract situations in a game; choosing is in effect 'betting on an outcome, and doing so on the basis of preferences, i.e., with a specific order in mind' (*translation*).[14] The order of the actions is largely governed by the order of preference in which the outcomes of the action are viewed; the outcomes have a characteristic called by economists 'utility' or 'ophelimity', by means of which, on the basis of an indifference curve comparison between several goods, the subjective value of a particular good can be determined and consequently the level of satisfaction of the subject. The analogy with a game situation is that, setting aside the element of risk, the individual acts by applying a sort of law of least effort, whereby he constantly seeks the maximum pleasure for the minimum effort. It can be shown by appropriate experiments that the way in which a real person lists goods in order of utility, with a view to building up subjective scales of value approaches to a spontaneous grasp of the axioms of the theory. If in addition we reintroduce the element of risk, i.e., of uncertainty as regards the consequences of the action, whether due to ignorance of the contingent factors or because the consequences will depend on the choice to be made by an adversary, it can be shown that by linking utility

12. Von Mises, L., *Nationalöconomie: Theorie des Handelns und Wirtschaftens* (1940) (English translation, *Human Action*, 1949).

13. See on this school the general survey by Kotarbiński, 'L'évolution de la praxéologie en Pologne' (1968) and the annexed bibliography. Cf. Kotarbiński's fundamental work, *Traktat a dobrej robocie* (1955) (English translation, *Praxeology. An Introduction to the Science of Efficient Action*, 1965).

14. '[. . .] parier sur un avenir, et c'est le faire en fonctions de préferences,c'est-à-dire d'un ordre' – Bresson, 'Les décisions' (1965).

with the probability assigned to the event's happening, we are virtually reconstituting a betting situation which confers a descriptive value on the normative model. Experimental situations have been set up for this purpose, making it possible to measure probable utility.

There is one further argument in favour of the descriptive value of the theory. Where real behaviour deviates from the norms of rational behaviour, a differential method can be used to isolate 'ordinary' psychological factors such as the rejection of axioms, the propensity of those who accept axioms in principle to make mistakes, opinions which are consistent but not in conformity with the pattern, and so on. This approximate representation of real choice processes by means of phenomena of deviation can itself assume a mathematical form, since simulated situations can be set up experimentally whereby significant differences can be shown between theoretical results obtained in contrived situations of rational choice and real choices experimentally observed. Simulation thus makes it possible, through differentiation, to identify more exactly the real factors in decision-making.

But is the gulf separating reality from the norm purely a phenomenon of deviation from axioms? This is also open to doubt, for several reasons. Firstly, the real-life decisions of human beings can be explained not only by aims or intentions, i.e., by anticipated outcomes and preferred future states; they can also be explained in terms of antecedent determinants and forces which 'push from behind'. Game theory provides tools only for the first kind of approach. Secondly, the theory permits of axiomatizing only a limited category of preferred future states. If the rational subjective probabilities of the theoretical fictitious man do not closely reflect the irrational subjective probabilities of real-life man, it is primarily because probabilities themselves are not usually quantitative; they are rather assessments of verisimilitude, not based directly on frequencies or symmetries. As for the actual orders of preference, they cannot all be reduced to a comparison of utility. Reference has been made above to the fact that different things have different kinds of utility, but utility itself is only one value among others, that which regulates the order in which various actions are preferred by their outcomes; the motivational field includes many other values – aesthetic, ethical and so on. We thus come up against the phenomenon of the heterogeneity of values, which will become a major problem for axiological logic, to be considered later.

We may then ask whether the descriptive scope of decision theory is not limited to situations which hang on a calculation of interest, and this is borne out by all the concepts involved in the above-mentioned criteria for decision-making: gain and loss, cost, risk and regret. Going even further, we may ask whether the theory does not have an irresistible affinity, which fundamentally limits its scope, with a profit-oriented economy, which is in fact the field of reality that lies nearest to the models elaborated by game theory. The range of activities to which the theory could be applied in practice or of which it could provide a descriptive approximation would thus be limited from the outset, as a result of the axiological field being reduced to a single value, self-interest, as a quantity which can be measured, preferably in terms of monetary gain. Successful application of the theory would then entail both as a prerequisite and as a counterpart, the 'putting in brackets' of the heterogeneous character of the axiological field in which real-life human action takes place.

§ 2. *Logic of practical discourse: deontic logic**

We are reserving this term, borrowed from G. H. Von Wright,[15] for logical research into certain uses of language other than descriptive. We repeat 'certain uses', since not all non-descriptive language is practical: for example that which expresses 'emotion' or 'sentiment' is not practical, practical language being solely that which expresses an action which can be performed by human actors. Similarly not all language comes within the purview of the logician, for example language which expresses the intention behind an act, or the manner of its accomplishment, but solely that which expresses the reasons motivating it. In addition, these reasons must be divided into those which can be *formalized* and those which cannot, namely those which can be assessed regardless of real actions and real actors and those which cannot be separated from the actual decision they supposedly help to justify. From this we deduce the existence of two types of logic of practical discourse, one formal and the other non-formal. If it is thought that the term 'formal logic' is pleonastic, the first group of theories may be given the name of 'the logic of practical discourse', and the second that of the 'theory of practical reasoning'. This is the terminology we have chosen.

The kind of practical discourse studied by deontic logic is the normative use of language. It is easy to understand why the notion of the norm is expressed in a series of operators known in ordinary languages and in the history of moral theory as obligation, prohibition, permission (and even indifference). These notions are obviously analogous to notions such as necessity, impossibility, possibility (– and contingence) which have for long been the subject of an entirely formalized logical theory, that of modal logic. It is on this analogy that the attempt to establish a deontic logic, or logic of obligation and related notions, rests, and it also explains how it is that deontic logic now constitutes the nucleus of the logic of practical discourse. Round this formalized nucleus are grouped other practical uses of language, which can be formalized to a greater or lesser degree, for example imperatives ('do this', 'do that') and value judgements (e.g., 'good', 'bad', 'better'): the logic of imperatives and axiological logic thus constitute more or less independent

* The working out of this paragraph has benefited from detailed notes provided by Kit Fine.

15. Cf. von Wright, 'The logic of practical discourse' (1968) and the annexed bibliography (at pp. 162-167 of Klibansky (ed.), *Contemporary Philosophy/La philosophie contemporaine*, Vol. I). See also, for an overall view, Kalinowski, 'Les thèmes actuels de la logique déontique' (1965), and Rescher, *The Logic of Commands* (1966).

branches of deontic logic within the framework of the logic of practical discourse.

Deontic logic is logic in the full sense of an entirely formalized theory. In the first place, the arrangement of the primitive notions listed above abstracts entirely from the obligatory things, that is to say from the 'states of affairs' expressed in language describing the action or its outcome or commanding what is to be done; above all, the logical relations between deontic notions are without respect of persons, i.e., of the agents of the action, their attitudes, intentions or decisions. This is the crucial point for subsequent discussion. In the second place, this arrangement has the effect of constructing systems which are comparable to the system of formal logic and subject to the same axiomatic requirements. The primary, most fundamental relations are those whereby the primitive notions are inter-defined: what is obligatory is not forbidden; what is not forbidden is permitted; and what is neither obligatory nor forbidden is indifferent (on the understanding that deontic indifference is also a deontic state, unlike deontic neutrality, which would denote a gap in the system: it remains to be seen whether a deontic system is in fact complete, and therefore closed, or incomplete and therefore open). The choice of axioms determines the form of the theory and the sequence of theorems. Finally, 'paradoxes' play the same part in deontic logic as in theoretical logic; much of the development of deontic logic consists in solving 'paradoxes' resulting from one or other choice of axioms.

The diversity of the systems of deontic logic is thus related to the nature of the *variables* which occur in the deontic formulae. We have just spoken in a vague way of obligatory, forbidden or permitted *things*. What are these things? For the logician, the inter-definition of the primitive notions means that all the propositions of deontic logic are derived from a single form OX, in which O represents the obligation and X what the obligation is. What exactly are O and X?

1. One type is where X (what the obligation is) can be replaced by the name of an action (or by a sentence describing an action);
2. A second type is where X represents *any* sentence (e.g., it is raining);
3. A third type is where X represents a sentence which is itself imperative;
4. A fourth type is where X represents a complex action, e.g., firstly this state of affairs, and then another state of affairs.

The interpretation of X governs that of O. In the first case O (the obligation) is read 'this is obligatory' or 'it is obligatory that you perform the act of . . .'. In the second case it is read 'it is obligatory to bring it about that . . .', in the third case 'it is obligatory to . . .' and in the fourth case 'it is obligatory to bring about the change that . . .'. These four types of formulation are in no way coincidental. Here the decisive test is that of iteration, i.e., the right to elaborate, in well-formed expressions, deontic propositions of a higher order. In the first case, OOX, i.e., an iteration of the deontic operator O (the obligation), is not permitted, since OX is neither a sentence describing an action nor the name of an action. In the second case, since OX is a sentence, iteration is possible. In the third case, iteration is permitted if OX is another imperative sentence, and forbidden if it is an indicative sentence. The iteration test thus provides a classification principle for the systems of deontic logic.

On the one hand there are non-iterative systems (the first case) and on the other there are iterative systems (the second, third and fourth cases).

(a) The former are the more ancient, and go back to the seminal work of von Wright, 'Deontic logic' (1951). In the system originated by von Wright the variables are names for types of acts (robbery), not individual acts (The Great Train Robbery). Since the variable is not a sentence and OX is not a name, iteration is forbidden. Within this narrow frame of reference von Wright nevertheless made the basic discovery of deontic logic, namely its analogy with one of the branches of theoretical logic, viz. modal logic. The modal concepts of necessity, impossibility, possibility and contingency are related by inter-definition in the same way as the concepts of obligation, prohibition, permission and (deontic) indifference. It is in fact this rigorous analogy, or rather this isomorphism with the formal theory of the modalities which gave rise to deontic logic and promoted its development, since both deontic logic and modal logic 'stand on the shoulders' of two-valued propositional logic in the sense that they accept its logical truths (the 'tautologies') and employ its rules of inference. Deontic logic derives its theoretical form from the axioms governing the propositional logic of its field of investigation. For example, a disjunctive act is permitted if, and, only if, one of the alternatives is permitted; either the performance of an act is permitted or its non-performance is permitted.

As with theoretical logic, the development of the theory originated in the discovery of paradoxes. Some of these paradoxes are well known in the literature of the subject, for example the Paradox of Derived Obligation or of Commitment, according to which to do something forbidden commits one to perform anything whatever! The most famous was called by Prior the Paradox of the Good Samaritan. Applying a theorem of von Wright (if something is forbidden, then its conjunction with any other thing is forbidden) the argument is that, if it is forbidden to rob someone, it is also forbidden to help someone who has been robbed: this is the case of the Good Samaritan who, by helping a robbed man, produced a state of affairs which must not be, since its existence entails that a robbery has taken place. It is the resolution of such paradoxes which makes it necessary to develop the system.[16]

(b) In iterative systems, where the obligation attaches to *any* sentence, including the sentence introduced by the obligation, the problems concern the actual meaning to be attached to iteration. The analogy with modal logic

16. The following articles mark the successive stages of this research: PRIOR, 'The paradoxes of derived obligation' (1954); von WRIGHT, 'A note on deontic logic and derived obligation' (1956); RESCHER, 'An axiom system for deontic logic' (1958); ANDERSON, 'On the logic of "commitment"' (1959); ÅQVIST, 'A note on commitment' (1963); CHISHOLM, 'Contrary-to-duty imperatives and deontic logic' (1963); von WRIGHT, 'Deontic logics' (1967); RESCHER, 'Semantic foundations for conditional permission' (1967).

provides a basis for several types of solution. Anderson's approach [17] deserves comment on account of its important philosophical implications. This author proposed that the analogy between deontic logic and modal logic should be interpreted as a rigorous reduction of the former to the latter. All that is required for a transition from obligation to necessity is to define obligation as the necessity that a sanction will be applied if, and only if, the action concerned does not happen: permission on the other hand will be defined as the possibility that, if the action is performed, the sanction may not come about. Thus it is sufficient to introduce a constant – the proposition expressing the sanction ('you will be unhappy', 'you will be punished', 'the world will be a worse place', etc.) – in order to replace the deontic operator (obligation, permission) by a modal operator (necessity, possibility). We may thus say that something is obligatory if, and only if, its non-performance or non-occurrence necessarily entails the sanction. This is what is done in penal codes, which do not attempt to state what is forbidden, but provide for a specific sanction to correspond to a specific delict ('unlawful behaviour'). The norm which stipulates the sanction is then the primary norm, and the norm which lays down what is forbidden the secondary norm; the latter is, in fact, contained by the former and its enunciation may thus be considered superfluous. If therefore obligation can be defined as the necessity that the non-occurrence of an act or an event entails a specific 'bad thing', it does indeed appear that deontic logic is reduced to modal logic.

(c) In the third case under consideration, the obligation attaches to an imperative,[18] namely the order that an indicative be brought about. One must then work out the operators for this conversion of imperatives into indicatives, and *vice-versa.*[19] This type of system raises the more general problem of the connection between imperative and obligational logic, to which we shall return in our critical remarks.

(d) The fourth type of system brings in the connection between deontic logic and the logic of action.[20] Here several kinds of interrelationships between states of affairs have to be substituted for the obligatory 'thing': instead of 'it is obligatory that...' we have 'it is obligatory, given that..., that...'. We then have a system of conditional or hypothetical obligation. New axioms must be introduced as a basis for this dyadic deontic logic (as opposed to the monadic system represented by the first case, which then appears as a limiting case of deontic logic, in which abstract or categorical obligation is a dyadic obligation relative to tautological conditions).

17. Cf. ANDERSON, *The Formal Analysis of Normative Systems* (1956, 1967). See also, by the same author, 'The logic of norms' (1958) and 'On the logic of "commitment"' (1959), already cited in the preceding footnote.

18. Cf. FISCHER, 'A system of deontic-alethic modal logic' (1962); CASTAÑEDA, 'Acts, the logic of obligation and deontic calculi' (1968).

19. The study of this question dates back to the articles by HOFSTADTER & MCKINSEY, 'On the logic of imperatives' (1939) and by BOHNERT, 'The semiotic status of commands' (1945).

20. Cf. von WRIGHT, *An Essay on Deontic Logic and the General Theory of Action* (1968), pp. 1394-1396.

Another way of connecting deontic logic with a logic of concrete action is, as Hintikka and Castañeda [21] have done, to introduce quantifiers over acts, persons, times or elements of action. It is then possible to get back to individual acts, which are excluded from the preceding systems since these take into account only generic acts (murder, robbery, etc.).

Comments on the limitations of deontic logic

We suggest comments on the following three lines:

(a) What is it that constitutes the deontic nature of this logic? The question may be raised taking as a starting-point the analogy between deontic and modal logic,[22] the latter being, as we know, a logic of truth and falsehood. We have said that deontic logic is logical by virtue of its isomorphism with modal logic. But what gives the deontic its specific character? We have seen that Anderson reduced deontic to modal logic by means of the sanction clause. This clause makes it possible to consider as obligatory that which, if it is not performed, necessarily entails the occurrence of a 'bad thing'. But what kind of necessity have we in mind? Is not the necessity which attaches a sanction to non-performance that which in everyday language we denote by using the word 'deserve'? A specific crime entails a specific sanction because a specific authority is legitimately empowered to impose it, legitimately on the ground that crime must be punished; must be, i.e., merits or deserves to be . . .

Thus if we are to avoid falling into the 'naturalistic fallacy' once denounced by G. E. Moore,[23] we must recognize that the necessity we denote by the word 'deserve' is in itself deontic.

This difficulty with regard to the connection between deontic and modal logic makes it possible to state the problem in more general terms: Can the logician employ the purely formal characteristics of discourse to give an account of what is irreducibly deontic in the normative uses of language? Is it not by a preliminary analysis of 'practical' concepts as such that he can hope to determine the strictly deontic nature of the primitive notions such as obligation, prohibition and permission? And does not such an analysis of practical concepts lead beyond the formal frame of reference which by definition limits deontic logic? The dilemma of deontic logic may be stated as follows: either deontic logic cannot be reduced to modal logic, in which case its analogy with modal logic is a superficial phenomenon resulting from the operation of abstraction whereby the obligation is detached from the actions

21. Cf. Hintikka, 'Quantifiers in deontic logic' (1957); Castañeda, 'The logic of obligation' (1959).

22. Cf. von Wright, *An Essay on Modal Logic* (1951); see also Lemmon's restatement of the question, 'Deontic logic and the logic of imperatives' (1965).

23. Cf. Lewy, C., 'G. E. Moore on the naturalistic fallacy', in Strawson (ed.), *Studies in the Philosophy of Thought and Action* (1968), and Moore, G. E. 'A reply to my critics', in Schilpp (ed.), *The Philosophy of G. E. Moore* (1942). On Moore's work and its influence, see below, sub-section B, § 1, point (b), 'Ethical philosophy proper'.

and agents, in a word from the area of decision-making in which this logic is rooted and which confers on it its practical nature; or else deontic logic can be entirely reduced to modal logic, with the axioms of the former becoming the theorems of the latter, in which case the deontic necessity which connects the non-performance of the obligation to the occurrence of the 'bad thing' is indistinguishable from natural necessity – and we succomb to the 'naturalistic fallacy'.

The choice of the first alternative leads us to interpret the notion of a logic of practical discourse non-formally, or less formally than is the case with deontic logic.

Many other considerations bear this out:

(b) The question of the limitations of deontic logic is raised by the existence of two other types of use of practical language similar to the uses of normative language, namely imperative concepts ('do this', 'don't do that') and axiological concepts ('good – bad', 'beautiful – ugly', etc.).

To begin with the latter, axiological concepts present many more difficulties than normative concepts: these difficulties arise much earlier than with normative discourse, which can be formalized without it being necessary to determine the ultimate meaning of these primitive notions. Here the difficulties arise as it were from the outset: Are the pairs of opposites even homogeneous? And are they denumerable? Do we not have to have recourse from the start to operations involving the subjects which perform the evaluations? If this is the case, can the operation of abstracting from actions and persons which is involved in the formalization of normative concepts be as easy, convincing and legitimate in the axiological system?

These difficulties explain why the specifically logical investigations into this field are less numerous and fruitful than those into deontic logic. Such investigations are based on a fundamental characteristic of the axiological order, namely the existence of comparative degrees of value (better – worse); these degrees make it possible to establish a logic of preference, in which the formalization processes are comparable to those of deontic logic.[24] Nevertheless, even supposing that this logic of preference can be developed to a considerable extent, the question of its relation to the logic of norms remains. And this question no longer appears to be a question of logic, but of ethics, since it can be formulated in the terms 'What is the relation between axio-

24. Cf. HALLDÉN, *On the Logic of 'Better'* (1957) and the criticism by CHISHOLM & SOSA, 'On the logic of "intrinsically better"' (1966); MONTEFIORE, 'Goodness and choice' (1963); von WRIGHT, *Norm and Action* (1963) and *The Logic of Preference* (1963); RESCHER (ed.), *The Logic of Decision and Action* (1967), in particular RESCHER's own contribution to this collection: 'Semantic foundations for the logic of preference'.

On the fringes of logic proper, we may mention, e.g., DAVIDSON, MCKINSEY & SUPPES, 'Outline of a formal theory of value, I' (1955), and HARTMAN, 'Value theory as a formal system' (1958-59), 'The logic of description and valuation' (1960), 'The logic of value' (1961), 'Discussion. Axiology as a science. A rejoinder note to Professor Neri Castañeda's review of *La estructura del valor. Fundamentos de la axiología científica*' (1962).

logical and normative concepts, between "better than" and "obligation"?' [25]

This question brings us back once again to that of the choice and the number of the primitive notions underlying practical discourse.

(c) The question of the limitations of deontic logic is also raised by the logic of imperatives ('do this', 'don't do that'). Imperative logic appears to lend itself to formalized treatment in the same way as the logic of norms,[26] but the imperative introduces at least one new feature, which is that it may or may not be 'satisfied' or obeyed. The pair of notions 'obedience – disobedience' brings in the idea of persons, namely those who receive the order, understand it, assent or dissent, accept or refuse, and those who give the order and are qualified to do so. A command has also non-accidental features such as appropriateness, relevance or otherwise to the situation, and consequently capability of fitting intto the structure of means and ends which direct human action.[27] Last but not least, a command is authoritative or has the force of law, or else it is invalid, depending on the position of the giver of the order within a real command hierarchy which confers or does not confer on him the right to have an order carried out. This position of the giver of an order is not irrelevant to the actual notion of imperative, and even appears to be inherent in one of its essential characteristics, validity.[28] That is why it is much more difficult to apply to imperatives a system of purely formal relations such as entailment, contradiction or equivalence.[29] By contrast a logic of imperatives can be successfully applied to the hierarchical relations between authorities, orders and norms.[30] The relationships between higher and lower orders recalls similar problems of iteration in modal and deontic logic. The notions of satisfying (obeying) and failing to satisfy (disobeying) an order have also a certain similarity with the notions of truth and falsehood, which are expressed in everyday language in closely related terms (satisfy, fulfil, realize) and used by the logician with a view to extending the scope of propositional logic by analogy. But to what point can we abstract from the position of the persons giving the orders, or of those asked to com-

25. Cf. CHISHOLM, 'Supererogation and offence. A conceptual scheme for ethics' (1963), and 'The ethics of requirement' (1964).

26. Cf. above, footnotes 18 and 19, p. 1383.

27. Cf. CASTAÑEDA, 'The logic of obligation' (1959), 'Imperative reasonings' (1960), 'Imperatives, decisions, and "oughts". A logico-metaphysical investigation' (1963).

28. Cf. LEMMON, 'Deontic logic and the logic of imperatives' (1965); Lemmon's position was criticized by SOSA, 'The semantic of imperatives' (1967).

29. This difficulty is often called 'Jørgensen's dilemma', from the name of one of the Scandinavian authors who initiated the study of these questions: cf. JØRGENSEN, 'Imperatives and logic' (1937-38); ROSS, A., 'Imperatives and logic' (1941); MORITZ, 'Der praktische Syllogismus und das juridische Denken' (1954).

Among recent contributions to this discussion, see: GEACH, 'Imperative inference' (1963); WILLIAMS, B. A. O., 'Imperative inference' (1963); RESCHER & ROBISON, 'Can one infer commands from commands?' (1964); GOMBAY, 'Imperative inference and disjunction' (1965); ÅQVIST, 'Choice-offering and alternative-presenting disjunctive commands' (1965).

30. Cf. von WRIGHT, *Norm and Action* (1963); HANSON, 'A logic of commands' (1966).

ply with the command, without in fact robbing this logic of its specifically practical character?

In this respect the status of the logic of legal reasoning provides an excellent touchstone for a discussion of the limits to formalization in the field of practical discourse. Does it merely express the application of deontic logic to a real-life field, that of the law? [31] Or does consideration of the agents form an integral part of the structure of practical reasoning? [32] On the reply to this question depends the right of deontic logic to cover the whole field of the logic of practical discourse. For those who consider that the logical study of the idea of law is an application of deontic logic, juristic logic is simply one aspect of applied deontic logic. For those who lay greater stress on its specific characteristics, the logical study of law requires a logical framework other than deontic logic (and even than the logic of imperatives), but designed on the same formal model.[33]

These objections in no way add up to a condemnation of any study of deontic logic, possibly supplemented by a logic of imperatives and an axiological logic; they are intended solely to bring out the limits to its validity and in so doing to show first, its close relationship with other forms of the logic of practical discourse and second, its subordination to another kind of investigation bearing on the specific characteristics of the notions of practical discourse. The first point will be dealt with in the following sub-section, and the second in sub-section B below.

§ 3. *Practical reasoning and the theory of argumentation*

The theory of argumentation is, with deontic logic, the second sector of the logic of practical discourse undergoing investigation today.[34] Here however the word 'logic' no longer retains the formal connotation it has in deontic logic; the question raised by the argumentation theorists is precisely whether

31. Cf. TAMMELO, 'Sketch for a symbolic juristic logic' (1955-56); KALINOWSKI, *Introduction à la logique juridique* (1965) and 'De la spécificité de la logique juridique' (1966); see the whole of the special issue of *Archives for Philosophy of Law and Social Philosophy* (1966) entitled *La logique du droit*, where this latter article appeared.

32. Cf. PERELMAN, 'Logique formelle, logique juridique' (1960) and 'Raisonnement juridique et logique juridique' (in *La logique du droit, op. cit.*, 1966).

33. Cf. KLUG, *Juristische Logik* (1951), and the critical review of this work by HOROWITZ, 'Ulrich Klug's *Legal Logic.* A critical account' (1966); PARADIES, *Die Technik des Rechts und ihre Logik* (1958); CONTE, *Saggio sulla completezza degli ordinamenti giuridici* (1962); KALINOWSKI, 'Logique déontique et logique juridique' (1965); ZIEMBIŃSKI, *Logiczne podstawy prawoznawstwa* (= The Logical Foundations of Jurisprudence) (1966).

See also COSSIO, *La teoría egológica del derecho y el concepto jurídico de libertad* (2nd edn., 1964); GARCÍA MAYNEZ, *Ensayos filosófico-jurídicos* (1959), and the more specialized work cited by von WRIGHT (1968).

34. Cf. PERELMAN, 'Le raisonnement pratique', with the annexed bibliography compiled by P. Gochet, and JOHNSTONE, 'Theory of argumentation', in KLIBANSKY (ed.), *Contemporary Philosophy/La philosophie contemporaine*, Vol. I (1968).

it is possible and whether it is right to abstract from actions and persons. According to them, practical reasoning is that which justifies a decision. But a decision depends on the person taking it, and consequently cannot be 'formal'.[35] We must therefore classify deontic logic with modal logic as being theoretical, and draw the borderline between the theoretical and the practical not, as formerly, between the operators of modal logic (necessity and possibility) and those of deontic logic (obligation and permission), but between demonstration and argumentation, the former of which abstracts from actions and persons, whereas the latter does not.

Here reasoning is a discussion technique for convincing or refuting an adversary, and for coming to an agreement with him on the legitimacy of a decision. But the reasons used, which purport to form a logical sequence, are interwoven with the power of decision, which is not annulled at any stage of the sequence, though it continues to be based and grounded on them. We have here a circular relation between the decision and the reasons for it, between freedom and order, which is absolutely specific. It is this original feature which places practical reasoning in between logical necessity, both modal and deontic, and the arbitrary. Similarly, the theory of practical reasoning has two boundaries: on one side, that which separates it from the logic of demonstration, and on the other, that which separates it from a theory of motivation, since reasons are motives which have undergone the test of practical reasoning, i.e., of argumentation. Argumentation is a logic to the extent that the motives for a decision are treated as reasons which validate and legitimize it, and which are therefore eminently communicable, comprehensible, capable of being recognized, assented to, adopted. This is why a theory of motivation is in no way dependent on a theory of practical reasoning.

Motivation may be viewed in terms of a behaviourist type of psychology, where it is reduced to natural causality, decision-making appearing solely as an extremely complex form of the stimulus-response pattern, extended so as to accommodate a large number of intermediate variables, both conscious and unconscious. This is the most current acceptation in academic psychology. Motivation may on the other hand also be seen in terms of a psychology

35. On this point meet PERELMAN and TOULMIN, whose work *The Uses of Argument* (1958) also gave birth to an important movement of thought in this field. Toulmin tends, however, to give more weight to the criterium of formal validity than Perelman does. For the most recent developments of TOULMIN's thought, see his *Human Understanding*, Vol. I (1972).

For a comparison between Perelman's and Toulmin's positions, cf. GIULIANA. *Il concetto di prova* (1961).

As illustrations of the trend of thought initiated by Toulmin, see in particular: CASTAÑEDA, 'On a proposed revolution in logic (Toulmin's *Uses of Argument*)' (1960); PIKE, 'Rules of inference in normal reasoning' (1961); BIRD, 'The re-discovery of the topics: Professor Toulmin's inference warrants' (1961); KERNER, 'Approvals, reasons and moral arguments' (1962); WATSON, 'Rules of inference in Stephen Toulmin's *The Place of Reason in Ethics*' (1963); and see HARE's views in *Freedom and Reason* (1963), and NIELSEN, 'Good reason in ethics. An examination of the Toulmin-Hare controversy' (1958).

of personality, dominated by concepts of integration and totality. Or again, it may be viewed from the standpoint of psychoanalysis, with the so-called reasons being treated as rationalizations, i.e., as one of the ways in which repression deforms unconscious representations and affects. Yet again, motivation theory may form part of the sociology of cognition, where apparent reasons put forward in discourse express, under various disguises, which have to be penetrated, conflicts between forces at the level of social reality. The theory of practical reasoning is distinct from all these variants of motivation theory in that it takes reasoning as it presents itself and at face value, i.e., at the actual level of argumentation it adopts, where it can be criticized and justified, making it possible for the decision itself to be blamed or approved.

Practical reasoning in fact regards these motives as so many reasons capable of justifying it, since it accords them the character of a norm or a value.[36] This is the point it has in common with deontic logic and axiological logic. The theory of argumentation does however part company with these two types of logic in that it takes into account the conditions under which agents in real life recognize and accept these norms as forming an integral part of them. All practical reasoning starts from norms recognized at its inception, and seeks, by means of a specific technique of argumentation, to ground a fresh decision on these recognized norms [37] – a decision which will in its turn set a precedent for future decisions, thus widening the normative field. It is accordingly impossible for practical reasoning to originate in a total axiological or normative vacuum. This does not mean however that such norms are universal and eternal; all that is required is that they be recognized by an audience which may vary in size but is always limited, the audience to which practical discourse is addressed, within the specific time of the argument. So far are we from presupposing that norms are eternal that the temporal dimension of argumentation is as essential as is timelessness for logical proof.[38] A discussion not only takes time, it is subject to time-limits; in addition it is based on related decisions which serve as precedents. The decision itself, once taken, has the force of a judgement at law; it in turn sets a precedent, and there are even cases where with the passage of time the legal process of prescription changes *de facto* situations into *de jure* ones.[39] These temporal characteristics reveal in marked degree the peculiar features of action, or rather its peculiar logical features, since the

36. Cf. BAIER, *The Moral Point of View: A Rational Basis for Ethics* (1958); SINGER, *Generalization in Ethics* (1961); MORRIS, 'Values, problematic and unproblematic, and science' (1961) and also *Signification and Significance* (1964); MONTEFIORE, 'Goodness and choice' (1961).

37. Cf. NIELSEN, 'The "good reasons approach" and "ontological justifications" of morality' (1959) and 'The good reasons approach revisited' (1965); PERELMAN, 'Jugements de valeur, justification et argumentation' (1961).

38. Cf. PERELMAN & OLBRECHTS-TYTECA, 'De la temporalité comme caractère de l'argumentation' (1958).

39. Cf. KELSEN, 'Der Begriff der Rechtsordnung' (1958); PERELMAN, *L'idéal de rationalité et la règle de justice* (1961).

theory of reasoning is not as such divorced from the field of logic, but rather finds an echo and support, firstly in the theory of orthopractic reasoning, which Castañeda [40] defines as reasoning concerning the appropriate thing to do in a context of ends and means and which he views as a kind of imperative logic; secondly in the *tense-logic* applied by A. N. Prior to indicative propositions involving a time copula (was, is, will be), which provides an intermediate link between the theory of modality (necessity, possibility) and the logic of imperatives.[41]

Practical reasoning thus excludes equally firmly the idea of creating values from nothing – an axiological vacuum – and the idea of a harmonious, perfect and intangible order. In the first place, practical reasoning is applicable to situations in which values and norms already obtain, since otherwise there would be no basis for reasoning; in the second place, it presupposes either a situation in which values are conflicting or the system of norms appears incomplete or a new situation with regard to which no decision has yet been taken and for which it is accordingly not known what rules should be applied – since otherwise there would be no point to it.

What models are relevant in the theory of argumentation? Two models are proposed, the juristic and the rhetorical, though it is sought to show that they converge. Legal reasoning, according to Chaïm Perelman, should lead to a fresh analysis of practical reasoning in the same way as mathematical reasoning has regenerated formal logic.[42] By legal reasoning is meant both the judge's decision in particular situations and the legislator's decision regarding principles: in any event the two forms of legal argumentation overlap on account of the application of case law and the development of law through legal precedents.[43] Legal reasoning is a model here in the sense of an exemplifying case, a major illustration. Rhetoric is a model in another sense, in that it provides the normative procedures and techniques of argumentation; in this respect the theory of argumentation is a '*new rhetoric*', based on an attempt to bring into a single system both Aristotelian rhetoric and all the techniques of persuasion developed since Aristotle.[44]

40. See the notes of § 2 above, pp. 1380-1387.

41. PRIOR, *Time and Modality* (1957), *Past, Present, and Future* (1967), and *Papers on Time and Tense* (1968).

42. Cf. PERELMAN, 'Ce qu'une réflexion sur le droit peut apporter au philisophe' (1962) (English translation, 'What the philosopher may learn from the study of law', 1966).

43. Cf. VIEHWEG, *Topik und Jurisprudenz. Ein Beitrag zur rechtswissenschaftlichen Grundlagenforschung* (1953; 3rd edn., 1965); FORIERS, 'Réflexions sur l'argumentation juridique' (1961); PERELMAN, 'La spécificité de la preuve juridique' (1963) (English translation, 'The specific nature of judicial proof', 1963); STONE, *Legal System and Lawyers' Reasonings* (1964); collection *La logique du droit* (*Archives for Philosophy of Law*, 1966).

44. Cf. PERELMAN & OLBRECHTS-TYTECA, *La nouvelle rhétorique. Traité de l'argumentation* (2 vols., 1958); OLBRECHTS-TYTECA, 'Rencontre avec la rhétorique' (1963).

For a genuine reduction of the theory of argumentation to rhetoric, see EHNINGER & BROCKRIEDE, *Decision by Debate* (1963) and also, by the same authors, 'Toulmin on argument. An interpretation and application' (1960).

One may nevertheless ask whether the persuasion techniques of rhetoric are not a special province of argumentation, to be distinguished from the type of validity required by legal reasoning. Thus juristic logic is disputed territory between the supporters of deontic logic (Kalinowski, von Wright) and those of the '*new rhetoric*' (Perelman).[45]

B. PHILOSOPHIES OF ACTION

Contemporary philosophies of action can be related to three main types of approach. According to the first, the rôle of philosophy can be seen as the elucidation of concepts, and since G. E. Moore this has been the principal trend of English-speaking philosophy. We shall deal with this school first, since it comes closest to a logic of action and can even be called a logic if by that term is meant, not only the theory of reasoning, but the analysis of statements and concepts. We shall give a few samples of this type of analysis as it is used in the realm of ethics. The second approach stems from Kant,

45. Cf. PERELMAN's articles 'Logique formelle, logique juridique' (1960) and 'Raisonnement juridique et logique juridique' (1966), and KALINOWSKI's article 'De la spécificité de la logique juridique' (1966), already cited in the preceding paragraph; von WRIGHT, 'The logic of practical discourse', § 9 (p. 115 in KLIBANSKY (ed.), *Contemporary Philosophy/La philosophie contemporaine*, Vol. II, 1968), and also, by the same author, 'Practical inference' (1963).

Among other contributions to this study (for additional references, see the bibliographies of PERELMAN's and JOHNSTONE's contributions, already cited, to KLIBANSKY (ed.), *Contemporary Philosophy/La philosophie contemporaine*, Vol. I, 1968): CRAWSHAY-WILLIAMS, *Methods and Criteria of Reasoning* (1958); DIESING, *Reason in Society: Five Types of Decisions and Social Conditions* (1962); EICHHORN, *Wie ist Ethik als Wissenschaft möglich?* (1964); GAUTHIER, *Practical Reasoning. The Structure and Foundations of Prudential and Moral Arguments and Their Exemplification in Discourse* (1963); JOHNSTONE, *Philosophy and Argument* (1959); KATTSOFF, *Making Moral Decisions* (1965); KENNY, 'Practical inference' (1966); NAESS, *Communication and Argument* (1966); PASSMORE, *Philosophical Reasoning* (1961); PERELMAN, 'Logique, langage et communication (1958) and *Justice et raison* (1963) (English translation, *The Idea of Justice and the Problem of Argument*, 1963); RESCHER, 'Practical reasoning and values' (1966); ROSS, A., 'On moral reasoning' (1964); collections: *L'Argumentation* (*Revue internationale de Philosophie*, 1961); *La Théorie de l'argumentation. Perspectives et applications* (*Logique et analyse*, 1963).

The reflections generated by the application of the theory of argumentation to philosophical controversy would deserve a development which would exceed the limits of the present survey. The reader may be referred to the following works: BEDNAROVSKI and TUCKER, 'Philosophical argument' (coupled articles, 1965); BLANCHÉ, 'L'argumentation philosophique vise-t-elle la vérité?' (1963); GRYNPAS, 'Remarques sur le problème de la critique philosophique en philosophie' (1961); GUÉROULT, 'Logique, architectonique et structures constitutives des systèmes philosophique' (1957) and 'Logique, argumentation et histoire de la philosophie' (1963); HALL, *Philosophical Systems* (1960); JOHNSTONE, *Philosophy and Argument* (1959); RORTY, 'Recent metaphilosophy' (1961); WAISMANN, 'How I see philosophy' (1956); collections: *Symposium sobre la argumentación filosófica* (XIIIth International Congress of Philosophy, 1963); *Philosophical Argument* (*The Monist*, 1964); NATANSON & JOHNSTONE (eds.), *Philosophy, Rhetoric and Argumentation* (1965).

Kierkegaard, Hegel and Nietzsche and continues down through Husserl's phenomenology and the existentialists; and the question it asks is what type of being human freedom must have in order for phenomena such as decision-making and responsibility to be possible. This second approach is more frequently to be found on the continent of Europe, though it has its followers in other parts of the world. The third approach, which stems from Marx, Engels and Lenin, applies the principles of dialectical materialism to human action and seeks to establish two things: in the first place, the primacy of practice, in both the social and physical contexts over abstract and speculative thought; and, secondly, the extreme complexity of the dialectical interplay between theory and practice which develops on the basis of the latter's primacy.

In discussing these three trends of thought we shall emphasize, above all, the heterogeneous character of the questions that can be asked. As we understand it, the elucidation of ethical concepts, the ontological investigation of 'being free' and the dialectic of theory and practice are not so much different ways of solving a single problem as different ways of dividing up the questions to be asked. And it is at the fundamental level of this inner structure of the field of inquiry that philosophical discussion should be initiated.

§ 1. *'Analytic' philosophy and the problems of human action*

Under this rubric may be classed a great variety of studies which are nevertheless methodologically related. They deal in the main with expressions found in everyday speech which describe, appraise or direct human action. These methodologically related works cover a variety of fields.[46]

(a) *Philosophical psychology*

We shall start with a consideration of those studies which, though related to ethics in several respects, belong more to the field of philosophical psychology than to moral philosophy. They concern under various heads the language of action. In her book *Intention* (1957), Miss Anscombe concentrates on *intention,* for which she seeks the criteria in expressions such as: I did this because . . . Intentional actions are thus seen to be those which form the object of statements replying to the double question: what did you do? and why? The reply to this question implies a certain meaning of the word 'because . . .' which excludes the notion of antecedent cause and denotes the 'reasons for' which action is being taken.

In his book *Thought and Action* (1960), Stuart Hampshire seeks to clarify certain well-known distinctions. He discusses the contrast between what is fixed and inevitable and what is contingent and changing in the structure of

46. For an overall view, cf. WARNOCK, M., *Ethics since 1900* (1960); KERNER, *The Revolution in Ethical Theory* (1966); WARNOCK, G. J., *Contemporary Moral Philosophy* (1967).

human thought; between human thought and its natural expression in speech and action; between the situation facing man and his response to it; between knowledge and decision; between abstract philosophical opinion and the diversity of concrete experience.[47] In the same way, an important discussion has revolved round the distinction to be made between motive and cause, and the relation between motive and reason for (doing something).[48]

We shall deal more especially with Anthony Kenny's *Action, Emotions and Will* (1963), a work which is particularly typical of the regeneration of philosophical psychology through linguistic analysis. Kenny takes the concept of action in its widest sense and derives from it a theory of volition, basing himself on the special logical properties of those transitive verbs used to describe action in response to a question of the type:'what did A do?'. The way this analysis develops is worth noting. It begins by establishing the affinity between transitive verbs (Brutus killed Caesar) and verbs of relation (Peter is taller than Paul). The next step is to differentiate them by drawing a distinction between the object of an action and the term of a relationship. Verbs are then further divided into various categories – state verbs, performance verbs and action verbs –, mainly on the basis of their different temporal behaviour. Here, the analysis borders on the idea of 'tense-logic' as elucidated by Prior, of which it extends the scope. Intention, as analysed by Miss Anscombe, is a property of performance verbs. All performance verbs possess the imperative form, and voluntary action is precisely that type of action which can be commanded; looking at it the other way round, only action which can be commanded can be the object of a decision or constitute the immediate object of an intention. The last step is to distinguish between different types of object, the 'intentional' object of a mental attitude and the 'non-intentional' object of an action expressed by the object of a transitive verb, any object capable of complementing an action verb being its formal object (fuel, in the case of the verb 'to burn'): the object is what makes the action specific by specifying a formal object. As can be seen, this type of analysis affords the opportunity for direct comparison with the broader notion of 'intentionality' as found in Husserlian and post-Husserlian phenomenology. A theory of volition or 'willing' can then be established on this basis, i.e., limited to a logical inquiry into the implications of the verb 'to will' when taken as expressing an attitude of approval with regard to any particular state of affairs. Willing is only the verbal noun formed from the verb, and from this point of view, to will, to desire and to command are closely related forms; the only difference between 'to will *p*' and 'to desire *p*' is that performance of what is willed depends in the first case on him who wills it. Willing or volition can therefore be regarded as a command addressed to oneself.

There is an obvious affinity between these analyses of the concepts of action, intention and volition and, on the one hand, the related analyses of

47. By Stuart Hampshire, see also *Freedom of the Individual* (1965).

48. On these controversies see Strawson (ed.), *Studies in the Philosophy of Thought and Action* (1968); White, A. R. (ed.), *The Philosophy of Action* (1968).

Aristotle – also linguistically based – concerning 'movement' (*kinesis*), 'activity' (*energeia*), 'creation' (*poiesis*) and 'action' (*praxis*), the categories of the *pros ti*, comparative verbal constructions and so on and, on the other, the distinctions which the schoolmen were so fond of drawing in regard to the *intentio* of mental acts and their respective objects.

At the same time, these analyses may be compared with those dealt with above in sub-section A, § 2, regarding the logic of practical discourse. At that point, we took logic to mean the deontic (prescriptive) equivalents of propositional logic, modal logic, and axiomatic logic. However, the term 'logic' can also be used, as it commonly is by English-speaking philosophers, to mean elucidation of the concept of action by means of a study of the fundamental properties of action verbs. In this respect, the two paradigms act-object and act-reason-for-the-act serve as a basis for a logic of ordinary speech, whose function is conceived to be one of conceptual clarification of the ways in which these concepts are used. This logic occupies a position, as it were, underlying the formal logic of imperatives and norms, and this in fact defines the field of action where it can be used.

(b) *Ethical philosophy proper* *

The origins of 'analytical' ethics go back to the beginning of the century with the work of G. E. Moore and his *Principia Ethica* (1903). It is common knowledge that this work is mainly devoted to the analysis of moral propositions. For Moore, ethics is characteristically concerned with the predicate 'good', and its principal question is to find out what 'good' means. The definition which Moore is interested in is not the verbal definition; so it would be pointless to propose synonyms for the term. What Moore is actually looking for is the property designated by the term 'good'. And his answer is well known: the property or quality designated by the term 'good' is a simple quality, and therefore not susceptible to analysis. In consequence, the term 'good' is not definable – in the analytic sense of 'definition' – inasmuch as only those properties which are complex can be defined, because the definition can then show the different parts of which the property in question is composed. As the property denoted by 'good' is not a complex property, it is impossible to define 'good', for there are no parts to be shown by the definition. The fact that something is good cannot be demonstrated, but only acknowledged as such by a simple apprehension. Any attempt to define 'good' is, in Moore's view, a fallacy, especially when the attempted definition is in terms of natural properties. He called this latter species of fallacy 'the naturalistic fallacy'. For Moore, all known forms of moral philosophy stem from the fallacy of identifying goodness with some other property which good things have. Thus good things are those approved of or those which produce happiness or those which contribute to the self-realization of the agent, and

* The developments under (b) and (c) have been worked out for the most part on the basis of a general survey by José Hierro; we owe to Daya Krishna the references to philosophic currents in the East.

so on. But goodness is not identical with any of these properties, and therefore 'good' cannot be defined in terms of any of them. To do so would be to commit a fallacy. Other moral notions, such as 'right', 'duty', etc., are, in Moore's view, definable, namely in terms of goodness. The right action will be that action which, in a given situation, will produce the greatest amount of good. In this way, moral rules can be determined by the foreseeable consequences of action, and by the goodness of these consequences. Moore did not much like the term 'intuition', but despite that, it does not seem unreasonable to call him an 'intuitionist'. After all, goodness is, in his view, known directly, and neither reasoning nor empirical observation play any part in the process. This last point concerns us here, in view of the place allotted to moral reasoning at the beginning of this review, more particularly to those categories of ethical discourse which can be most readily formalized, namely obligation and the other associated notions of deontic logic. For Moore's disciples, the rôle played by reasoning, by the search for reasons, is regarded as non-essential. To Prichard's way of thinking, the mistake that moral philosophers have traditionally made is to think that the question about what our duty is can be answered by giving reasons; the answer to this question can only be discovered by a direct consideration of the proposed action: duty is self-evident. Whatever the rôle of reasoning – in particular, that of deciding between two mutually contradictory courses of action – the abyss dividing moral from non-moral properties always remains.[49]

This question may be taken as the second question requiring consideration prior to the formulation of any logic of practical discourse, the first concerning the field of action itself, the second its ethical basis. It may be said that the two questions taken together distinguish the practical as such from the theoretical. The realm of the practical may be taken as defined by those propositions which answer two types of question: 'what did A do?' and 'is what A did good?'. Answers to the first type of question deal with the object of action-verbs, while answers to the second type deal with the goodness of the action.

However, if moral propositions are not reducible to statements of fact, that is to say to descriptive propositions, can it be held that they express knowledge and possess cognitive meaning? This question is important, for on the answer to it depends the possibility of developing a meaningful and coherent discourse on action, and consequently the possibility of constructing a logic of action (whether it be logic in the narrower meaning of deontic logic or in the broader meaning of the theory of practical reasoning). Now – and this is the point – the neo-positivist school has drawn from a consideration of G. E. Moore's intuitionism the conclusion that moral propositions possess no cognitive value. This idea is already to be found in Wittgenstein's

49. Cf. PRICHARD, *Moral Obligation. Essays and Lectures* (1949). See, however, W. D. Ross's less optimistic stressing of conflicts of duties: *The Right and the Good* (1930) and *The Foundations of Ethics* (1939).

Tractatus.[50] Wittgenstein writes there that the sense of the world must be outside the world, and therefore it is impossible for there to be ethical propositions; ethics cannot be expressed. And in a lecture, a few years later, he put forward the view that ethics was an attempt to go beyond the limit imposed by language, to express what it is impossible to express.[51]

Some neo-positivists have tried to determine what is the logical status of moral propositions, once it is clear that they have nothing to do with knowledge. Carnap, for instance, suggests that they are commands in disguise,[52] and Schlick tries to reduce ethics to the causal explanation of moral conduct, that is, to a part of psychology.[53] But the most influential has been Ayer's doctrine, foreshadowed by the work of Ogden and Richards. In Ayer's view,[54] moral judgements, not being statements or assertions of any kind, have only one function – that of expressing the speaker's feelings and of arousing in the hearer similar feelings and ultimately stimulating him to act. Moral judgements are not assertions, thus they cannot be either true or false, and therefore, Ayer concludes, they cannot contradict each other, and there is no place in ethics for rational argument. Emotivism, as that doctrine has been called, receives its most complete treatment from Stevenson.[55] He defines the emotive meaning of a phrase as a strong and persistent tendency, built up in the course of linguistic history, to give direct expression (quasi-interjectionally) to certain of the speaker's feelings, emotions or attitudes and to evoke (quasi-imperatively) corresponding feelings, emotions or attitudes in those to whom the remarks are addressed. Stevenson makes much use of the distinction between beliefs and attitudes, pointing out that it is characteristic of moral judgements that they express, not beliefs, but attitudes, and try to exert appreciable influence on the attitudes of other people. Emotivism has shown that there is a close relationship between moral judgements and conduct, but its interpretation of this relationship is open to all sorts of criticism. For instance, one can express one's feelings in various ways which have nothing to do with moral judgements. Also, the aim of influencing the conduct of others can be achieved by non-moral means, such as by threats, flattery or propaganda. There is a tendency here to reduce moral discourse to a matter of success in convincing others rather than justifying by good reasons.

(c) *The rôle of reason in ethics*

While neo-positivism was busy undermining the very basis of ethical discourse, consideration of the problem was continuing from an entirely different standpoint, namely the affinity between the logic of ethical discourse and

50. *Logisch-philosophische Abhandlung/Tractatus logico-philosophicus* (1921, 1922, 1961): cf. propositions 6.4 to 6.421 and 6.432.

51. Wittgenstein, 'Lecture on ethics' (1929 or 1930, publ. 1965).

52. Carnap, *Philosophy and Logical Syntax* (1935), p. 24.

53. Schlick, *Fragen der Ethik* (1930) (English translation, *Problems of Ethics*, 1939, 1962).

54. Ayer, *Language, Truth and Logic* (1936), in particular Chap. 6.

55. Stevenson, *Ethics and Language* (1944) and *Facts and Values* (1963).

the actual logic of theoretical discourse. At the beginning of this section, we set out this problem from the strictly logical point of view, and we now must return to it and examine it from the standpoint of its moral implications.

In this respect, Toulmin's *An Examination of the Place of Reason in Ethics* (1950) and Nowell-Smith's *Ethics* (1954) occupy a place of prime importance; they cover the same ground as Castañeda's influential logico-philosophical work concerning the imperative and prescriptive uses of language. We have referred to Castañeda's contribution in the context of logic, because he links it explicitly with the logic of imperatives, while distinguishing it from the logic of norms; however, it also provides a link with our present topic of discussion because of the part played in it by the notion of 'appropriate conduct': 'orthopractical' reasoning is a type of reasoning which relates to the appropriate course of action to take in a system of ends and means. Toulmin, who is more concerned with the similarity between science and ethics, puts forward an analysis of moral reasoning according to which a good moral reason is a reason which relates an action to an accepted social practice, whereas the moral justification of social practices depends on the comparative probable consequences of keeping or replacing those practices. Nowell-Smith, on the other hand, is more occupied with the everyday use of moral language and moral reasoning, of many different aspects of which he gives a very detailed analysis. His most fruitful suggestions include his concepts of 'contextual implication' and 'logical oddity', together with his work on semantics of ethical language, and in particular on ambivalent ('Janus principle') words such as 'good', so-called 'aptness-words' such as 'sublime' and gerundive words such as 'laudable'.

This illustrates the kind of gap that was developing between neo-positivist theory (which tended to consider any sort of significant and coherent moral discourse an impossibility) and the direct analysis of the rôle of reason in ethics.[56]

New methods of approach were opened up by the philosophy of ordinary language characteristic of post-war Oxford. This latter school differs both from intuitionism and from the neopositivism which preceded it. The neopositivists were able to deny any cognitive meaning to moral propositions, partly because their criterion for cognitive meaning was strictly assimilated to verifiability, and partly also because the intuitionists had implied that moral judgements are of a descriptive nature. There remained the possibility that ethical language, without being descriptive, might nonetheless convey some sort of meaning. This new concern with non-descriptive language is due in part to the conception of language formulated by Wittgenstein in *Philosophical Investigations* (1953). As is well known, language is here treated as an 'activity' the meaning of which is defined by its 'use'; however, what Wittgenstein concentrates on in his own analyses is exploring uses of

56. Among the partisans of a rationalist stand in ethics, one may also mention, e.g., BLANSHARD, *Reason and Goodness* (1961); FINDLAY, *Values and Intentions* (1961) and *Language, Mind and Value* (1963); GINSBERG, *Reason and Experience in Ethics* (1956).

language other than descriptive. In the same way as that of the later Wittgenstein, the work of J. L. Austin [57] has had a decisive influence on the philosophy of action and moral philosophy, to which he bequeathed in particular the distinction between descriptive expressions and performative expressions – intention, desire and command being pre-eminently performatives. In addition, Austin incorporated his notion of the performative within the framework of a more basic distinction which has provided all philosophers working on the logic of ordinary language with a valuable tool. This was the distinction he made between the act *of* saying something (locutionary act), the act done *in* saying something (illocutionary act), and the act done *by* saying something (perlocutionary act). In other words, he distinguished when something is said, that which is done in saying it, and the effects of saying it. For instance, that which is said may be the words: 'Shut the door', the act done in saying it may be an act of commanding (or perhaps one of requesting or of beseeching, and so on); and the effects achieved by saying it may be that you shut the door (or that you open it, or that you strike me, and so on). The most important fact to emerge from the analysis was undoubtedly the classification of illocutionary acts (commanding, asserting, evaluating, etc.) into families. This doctrine helped a good deal in clarifying the problem of the meaning of moral judgements. Another contribution of Austin to the analysis of moral language which calls for mention here is his discussion of sentences about free will. This problem has always been at the root of the keenest controversies. It was Moore who put the discussion on a new footing by bringing the dispute between determinists and libertarians down to an analysis of the meanings they respectively attach to the word 'could' or 'could have' in the sentence: 'I (or he) could have acted otherwise'. For Moore,[58] this has the meaning of: 'I would if I had chosen'. More recently, Nowell-Smith has taken up the problem; [59] in his view, we call free those actions which a person could have chosen not to perform, which in turn means that that person would have chosen not to perform it if certain conditions had been different. Austin's argument [60] is aimed at showing that neither Moore nor Nowell-Smith is right. He distinguishes the conditional and indicative uses denoted by 'he could have'. In the indicative it means 'he was able to', and that is independent of what could be said about what he would have done in particular circumstances. Moore and Nowell-Smith attempted to show that freedom of choice and universal causation are compatible, and therefore that the dispute between libertarians and determinists can be solved. Austin's analysis, by showing that 'I could' is not always analysable into 'I would if', has once more cast doubt on the possibility of reconciling determinism and freedom of choice. In the end, we would be left with the ques-

57. Cf. in particular *How to Do Things with Words* (1962). On Austin's work, cf. the preceding section, sub-section B, § 1, pp. 1340-1341.

58. Moore, G. E., *Ethics* (1912).

59. In *Ethics* (1954), already cited.

60. Austin, 'Ifs and cans' (1956, reprinted in Austin, *Philosophical Papers*, 1961).

tion whether it is true that I had, in a given situation, the possibility of acting otherwise, or not. The discussion has also dealt with the question whether these are simply different descriptions or explanations of human action which are assumed to run parallel one with the other, or whether there is one of them which is more entitled than the other to be considered true. Farrer, in *The Freedom of the Will* (1958), concludes that the former interpretation is too 'weak' to be accepted, but that if we turn to the second, the question it raises cannot at present be answered with certainty.

These discussions which stem from the neo-positivist critique of Moore, and then from the philosophy of ordinary language, form the backdrop to the work of Hare with its new approach to the problem of moral propositions.[61]

Hare's prescriptivism starts from the idea that it is equally wrong to say, in the intuitionist fashion, that moral judgements are descriptive and to adopt the emotivists' position by saying that they are fundamentally a means to influence the conduct of others. In Hare's view, moral judgements have as their typical function to guide conduct, in other words to prescribe. If, in addition, they influence others, that is accidental. What matters in a study of moral judgements or propositions is not what one can achieve with them, but what one does in using them. Hence the importance which Hare attaches to the problem of moral argument, with which neither intuitionists nor emotivists were much concerned. Hare classifies prescriptive language into two subclasses, namely imperatives and value judgements. Imperatives are in their turn distinguished as singular and universal, and value judgements are subdivided into moral and non-moral. Hare regards as typical value judgements *par excellence* those containing the terms 'good' and 'ought'. Judgements containing these words may be moral or non-moral. Moral language can thus be seen to be a kind of value judgement and consequently a subclass of prescriptive language. However, Hare admits that not all moral judgements are value judgements; thus, a part of moral language would be non-prescriptive. It must be acknowledged that Hare does not make this point entirely clear. At any rate, he thinks that non-evaluative moral judgements owe their meaning to the evaluative ones, and that were it not for the logical difficulties of the latter, the former could be analysed naturalistically (that is, in terms of 'facts').

According to Hare, to prescribe is either to answer a question of the form 'What shall I do?', or to say something which entails such an answer. This kind of question is properly answered by imperatives, and therefore if a value judgement is prescriptive, it will be so because it entails an imperative. Even if, for all this, Hare does not intend to reduce value judgements to the status of imperatives, the entailment he posits gives rise to serious problems in the case of deliberate wrongdoing, studied by Montefiore.[62]

61. Hare, *The Language of Morals* (1952) and *Freedom and Reason* (1963); cf. also *Descriptivism* (1963) and *Practical Inferences* (1972).

62. Montefiore, 'Deliberate wrongdoing' (1964).

In his treatment of the meaning of 'good', Hare's analysis is in terms of the function which that word has in commending objects or people: moral value judgements containing the term 'good' are judgements in which men are commended either directly or indirectly, and they are commended precisely as men, and not in a specific capacity.

Value judgements and imperatives are logically different in that the former, but not the latter, are universalizable.[63] That means that any value judgement made in any particular case must be equally applicable to all other similar cases. This is not true of imperatives. Uttering an imperative on a given occasion does not commit me to uttering it again, nor to assenting to it, on a similar occasion. The difference stems from the fact that, unlike imperatives, value judgements do not only have prescriptive meaning but also descriptive meaning, in so far as they relate to certain features of the cases to which they refer. In the last analysis, the difference comes from the fact that, unlike imperatives, value judgements invoke reasons in their support.

However, application of the principle of universalizability to moral arguments raises the question of the criteria of similarity between different cases. A wily disputant can always make out that *his* case is different. Hare does not seem to be able to get round this problem without accepting in advance value judgements of the utilitarian type. It may be asked whether an initial evaluatory framework is really necessary in order to determine what cases are similar and apply the rule of universalizability to moral disputes. But of course, in Hare's system, the very purpose of this rule is to provide a tool for solving moral disputes by purely logical means, i.e., without recourse to value judgements. Another essential point in Hare's doctrine is the impossibility of deriving value judgements from factual statements alone. This doctrine entails a strict separation between the meaning of evaluative terms (such as 'good') and the criteria for attributing goodness. As a result, agreement may be reached on the meaning of evaluative terms without there being a similar agreement as to what things, actions or persons are to be described as good. This has as an important consequence that everybody is finally free to make those ultimate value judgements he wants to. Whatever the facts and whatever the rules for a meaningful use of language, moral norms are in the end a question of decision for the individual.

Hare's theses have been subject to a good deal of discussion, some of which has centred on the relationship between evaluative expressions and the criteria for evaluation.[64] Hare's definition of the term 'good' in terms of its commendatory function has also received critical attention.[65] The idea that

63. Cf. Hare, *Freedom and Reason* (1963).

64. Cf. Foot, 'Moral arguments' (1958), 'Moral beliefs' (1958-59), and 'Goodness and choice' (dialogue with Montefiore, 1961); Geach, 'Good and evil' (1956, reprinted in Foot (ed.), *Theories of Ethics*, 1967).

65. Cf. Searle, 'Meaning and speech acts' (1962); Montefiore, 'The meaning of "good" and the act of commendation' (1967).

it is impossible to derive value judgements from statements of fact is attacked by Searle and Black,[66] while other criticisms deal with the rôle played by universalizability; [67] Castañeda also, whose influential work on value judgements and imperatives has already been mentioned, questions Hare's analysis on this point.[68] The same point receives attention in G. J. Warnock's criticism [69] which, in addition, questions Hare's view that value judgements depend on individual choice. Warnock himself suggests giving up the formalistic inquiry into evaluative or prescriptive language, and moving instead to an analysis of the contents of morals.

In general sense, Hare's prescriptivism has been responsible for bringing attention back to the problems of moral reasoning, which had been neglected by the emotivists; and several books similar in approach to Hare's work [70] have since been devoted to a study of these problems.

An interesting argument relating to the whole question of the relation between the descriptive and the evaluative has been advanced by the Indian economist A. K. Sen, who has also written some articles on moral philosophy.[71] He suggests that if the fundamentally basic character of a value judgement is defined with respect to the fact whether any *possible* factual circumstance could lead us to revise the judgement, then while the non-basicness of a judgement in some persons' value system could be conclusively established, the opposite would be impossible in principle. For, to accept any given value judgement as basic would, at best, be to give it the benefit of doubt. And, this for the simple reason that nobody can conclusively establish that no possible state of affairs could ever be imagined which might force us to give up the value judgement we thought we would hold on to in every conceivable factual circumstance. But if it is so, then it will be intrinsically impossible to determine the 'basic' character of any value judgement.

66. SEARLE, 'How to derive "ought" from "is"' (1964, reprinted in FOOT (ed.), *Theories of Ethics*, 1967); BLACK, 'The gap between "is" and "should"' (1964).

See HARE's replies to those various objections: 'The promising game' (1964); and FLEW's replies, 'On not deriving "ought" from "is"' (1964-65).

On the differences between HARE's and Mrs. FOOT's positions, cf. MONTEFIORE, 'Fact, value and ideology' (1966), and also 'Deliberate wrongdoing' (1964). See also, on the relations of fact and value, the discussion between ATKINSON and MONTEFIORE on the subject '"ought" and "is"' (1958), and the intermediate positions taken by BROAD (pp. 563-579 in SCHILLP (ed.), *The Philosophy of C. D. Broad*, 1959); EMMET, *Fact and Obligation* (1958) and *Rules, Roles and Relations* (1966); EWING, *The Definition of Good* (1947) and *Second Thoughts in Moral Philosophy* (1959); KEMP, *Reason, Action and Morality* (1964); NOWELL-SMITH, 'Contextual implication and ethical theory' (1962); von WRIGHT, *The Varieties of Goodness* (1963).

67. Cf. MONRO, *Empiricism and Ethics* (1967).

68. CASTAÑEDA, 'Imperatives, decisions and "oughts"' (1963) and 'Imperatives, oughts and moral oughts' (1966).

69. WARNOCK, G. J., *Contemporary Moral Philosophy* (1967).

70. BAIER, *The Moral Point of View* (1958); GAUTHIER, *Practical Reasoning* (1963); TAYLOR, P. W., *Normative Discourse* (1961); SINGER, M. G., *Generalization in Ethics* (1961).

71. SEN, 'The nature and classes of prescriptive judgments' (1967).

A very useful introduction to moral philosophy will be found in Montefiore's manual.[72, 73]

(d) *Analytic philosophy and the theory of games**

For historical reasons, the Anglo-Saxon school of analytic philosophy has been relatively little influenced by recent developments in decision theory and technology. The philosopher who first sought to apply game-theoretical techniques to philosophical analysis was R. B. Braithwaite, in his *Theory of Games as a Tool for the Moral Philosopher* (1955). Although in this he explicitly concerns himself with variable-sum games, the technique he applies is essentially in the spirit of the classical zero-sum game of von Neumann and Morgenstern. For this reason he, like John Rawls in 'Justice as fairness' (1958, 1962) [74] can be criticized as giving a 'rational reconstruction' not so much of justice or of fairness as of prudence, owing to the emphasis placed on the rational maxim inherent in all strategies. The main development of game theory has been due to T. C. Schilling's *Strategy of Conflict* (1961) and other works by this author, who has turned our attention from zero-sum to variable-sum games and concentrated more on an intuitive and realistic analysis than a formal and mathematical one. His work has been developed more philosophically by Anatol Rapoport, *Games, Fights and Debates* (1960), and is of very considerable importance in showing us the dialectic movement involved in all debate and the conditions under which men are capable of reaching agreement and rationally concerting their action.

Philosophers have been much more aware of cybernetics. As the technology of computers has improved, the question has been posed more and more insistently whether there is any essential difference between computers and the human brain.[75] H. C. Longuet-Higgins, of Edinburgh University, is work-

72. MONTEFIORE, *A Modern Introduction to Moral Philosophy* (1958).

73. The American naturalist tradition still has followers, such as, e.g.: BRANDT, *Ethical Theory* (1959); HOURANI, *Ethical Value* (1956); PEPPER, *The Sources of Value* (1958). But most American moralists express qualified views of this subject: cf., for instance, AIKEN, *Reason and Conduct* (1962); FRANKENA, *Ethics* (1963); HALL, *Our Knowledge of Fact and Value* (1961); MURPHY, *The Theory of Practical Reason* (1965); SESONSKE, *Value and Obligation* (1957).

English traditional utilitarianism, still championed by, e.g., SMART, J. J. C., *An Outline of a System of Utilitarian Ethics* (1961), and ZINK, *The Concepts of Ethics* (1962), is being widely discussed. It is clearly repudiated by, among others, LORING, *Two Kinds of Value* (1966); RAWLS, 'Justice as fairness' (1958, reprinted in LASLETT & RUNCIMAN (eds.), *Philosophy, Politics and Society*, 2nd ser., 1962); and STRAWSON, 'Social morality and individual ideal' (1961). See on these various points the general review by EWING, 'Ethics in English-speaking countries', in KLIBANSKY (ed.), *Contemporary Philosophy/La philosophie contemporaine*, Vol. IV (1971)

* Based on a general survey by John Lucas, whose developments we have followed closely.

74. RAWLS's thesis on justice is amplified in his recent work: *A Theory of Justice* (1971/72), which we have considered above (Section III, sub-section B, § 1).

75. On this same issue, approached from various angles by contemporary philosophy, cf. above, Section II, sub-section B, § 3, 'Levels of reality', point (b), 'The

ing to see how far it is possible to devise cybernetical systems to simulate the workings of the human mind. *Machine Intelligence*, edited by B. Meltzer and B. Mitchie (1969), presents a general survey. The main point at issue is whether or not there exist features of the human mind – consciousness, for example – which establish a difference in principle between man and machine. There has also been considerable controversy over the argument put forward by J. R. Lucas in 'Minds, machines, and Gödel' [76] and more fully in his *Freedom of the Will* (1970), which uses a theorem enunciated by Gödel, concerning the foundation of mathematics on undecidable propositions,[77] in order to show that even if a computer were designed to simulate the exact workings of a human mind, it would nevertheless be possible to find something else which the computer could not do, and the human mind could. A similar argument was adduced by J. Bronowski.[78] The argument has been criticized by J. J. C. Smart, by Paul Benacerraf, and by Hilary Putnam.[79] If it is agreed that these criticisms can be met, the philosophical implications for our understanding of human agency are very great.

Except in cybernetics, the development of technology has not had much impact on speculative philosophy, although for political and social philosophy the increasing pace of technical advance and man's difficulty in accommodating either his individual responses or his development of social institutions to it is a matter of profound moment.[80]

§ 2. *Freedom and the philosophy of 'values'*

(a) *A diverse heritage*

As we said in the introduction to this sub-section, the second method of approach originates with Kant, Hegel, Kierkegaard and Nietzsche. It is from Kant that modern philosophy has derived its idea of a 'critique of practical reason', a notion that takes freedom to be the *ratio essendi* of moral law, and considers law the *ratio cognoscendi* of freedom (I know I am free because I am aware of my obligations; conversely, I can do my duty because I am free).

problem of the specific character of consciousness . . .', pp. 1183-1192, and Section III, sub-section A, § 2, 'The logic of explanation in psychology', point (b), 'Physicalism old and new', pp. 1244-1247.

76. Originally published in 1961; reprinted in SAYRE & CROSSON (eds.), *The Modeling of Mind* (1963) and in ANDERSON (ed.), *Minds and Machines* (1964).

77. Cf. Gödel, 'Über formal unentscheidbare Sätze der Principia Mathematica und verwandter Systeme' (1931).

78. Cf. BRONOWSKI, *The Identity of Man* (1965), pp. 22-23, and 'The logic of mind' (1966).

79. SMART, J. J. C., *Philosophy and Scientific Realism* (1963) and *Between Science and Philosophy* (1968); BENACERRAF, 'God, the Devil, and Gödel' (1967); PUTNAM, 'The mental life of some machines' (1967).

80. Some of these problems have been posed by Henryk SKOLIMOWSKI in 'The structure of thinking in technology' (1966) and in 'Technology and philosophy' (contribution to KLIBANSKY (ed.), *Contemporary Philosophy/La philosophie contemporaine*, Vol. II (1968) – with bibliographical references).

It is these two grounds, the ontological and the gnoseological, which will provide us with a guiding thread for the purposes of the present sub-section, and it is for this reason that we have chosen to give it a heading that might serve to recall the two lines of approach to the ethical problem. There are, however, hardly any contemporary philosophers who adopt in its entirety the view that law can be reduced to the empty form of universality; while agreeing with Kant as to the actual relationship between freedom and obligation, most of them seek to ascribe to obligation a specific historical and institutional content – in a word, a 'material' content in the sense used by Max Scheler. It is at this point that the influence of Hegel comes into play From Hegel modern philosophy has inherited not only the critique of Kantian formalism, but also the desire to reinstate moral obligation in the general context of legal, economic, political and other practical activities, which at the same time give a specific content to the notion of freedom. If Kant, then, has shown modern philosophy the interrelationship between freedom and obligation, Hegel has taught it that the extent to which freedom can be achieved depends on the increasingly complex and increasingly concrete rational constitution of the historical and cultural orders of reality in which this freedom is embodied. In other words, what Hegel did was to replace Kantian formalism (which only acknowledged the universality of an empty law) by a dialectic of the concrete forms of social existence, arranged in order from the most abstract to the most concrete, and matched by a similar dialectic dealing with the stages in the realization of freedom. In so doing, Hegelian philosophy opened the way to a phenomenological approach to the problem. But man, as conceived by Hegel, is someone for whom all reality is in the last analysis rational, just as all rationality is, ultimately, the meaning of something actually achieved or made real. It is this with which Kierkegaard's individual takes issue in a protest whose echo even today reverberates through the very core of modern philosophy. Admittedly Kierkegaard does not totally abandon Hegel's central idea, namely that the full realization of freedom is a gradual process which evolves as the human order becomes increasingly richer. Indeed, Kierkegaard's 'stadia of life' – the aesthetic, ethical and religious 'stadia' or stages – are also stages of freedom, with a certain correlation prevailing at each stage between freedom and value. However, from one stage to the next there is no mediation but discontinuity. It is this irrational element contained in the 'leap' of freedom out of the realm of necessity which contemporary existentialism has taken over from Kierkegaard, much more than his notion of the contemporaneity of the Christ of the Gospels, which for him was the hidden meaning of this 'absurd' freedom. Finally, it was Nietzsche from whom contemporary philosophy inherited the formulation of the problem in terms of value. It is common knowledge that the word itself is borrowed from the founders of political economy, who understood by value the yardstick against which 'utilities' could be compared and exchanged. What Nietzsche did was to generalize this notion, using it as the correlative term for all valuations originating with the will. This formulation of the notion of value is not restricted to a simple change in the use of a word, since, by substituting for the Kantian imperative the power of valuation, Nietzsche places will in a quite different context from that of commands and obedience. This relationship between the law which commands and the will which obeys constituted for Kant the ultimate requisite for any critique of practical reason – so much so that he called it a *Faktum* of reason, which left its imprint on sensibility in the form of the feeling of respect. For Nietzsche, however, far from being merely receptive to the demands of law, will creates its own valuations, and indeed represents the very heart of this power, the power of the will to power. The mastery and dominance of will consist precisely in its ability to alter its order of values. From this point on the philosophy of freedom and values becomes a philosophy linking degrees of value to the differing degrees of strength and weakness of the will. To be more exact, a philosophy of will is a philosophy which applies a method of systematic doubt or conjecture to the systems of values stemming from traditional metaphysics, especially from the fusion of Platonism and Christianity in Western culture. This conjectural method, first developed in philology and medicine, consists of treating the value systems of all moral idealism as a sign and symptom of weakness. And 'genealogy' (in the biological and semiological sense of the word) is the specific method used to expose the weakness of will which assertions of so-called 'superior', 'transcendent' and 'spiritual' values serve to disguise. Consequently, it is in terms of 'genealogy' that Nietzsche's philosophy complements Kant's critique, Hegel's dialectic and Kierkegaard's philosophy of stadia. By reason of this, it not only accentuates the phenomenological character (in the descriptive sense of the word 'phenomenological') which philosophic method had inherited from Hegel, but furthermore paves the way for transforming this phenomenology into a hermeneutics, in other words into a theory of interpretation. The appeal of Kierkegaard's individual is answered by Nietzsche's systematic doubt.

Modern philosophy thus depends on two types or models of rationality – the critical and the speculative, the reflective and the productive, the analytic and the synthetic. To these two types of rationality it adds the corrosive action of two great 'exceptions', to quote the expression of Karl Jaspers: 'With our attention fixed on the exception, we, who are not the exception, philosophize.'

(b) *Contemporary phenomenological research*

On the basis of this diverse inheritance contemporary philosophy began by further investigating the phenomenological aspect of the problem of decision. At this level, the freedom-value problem can be formulated in the following terms: can choice be at one and the same time a determination of self by self and also a determination which results from the interplay of motives? Such an inquiry is not without antecedents: first of all in the Aristotelian description of human action, to which we owe the earliest analyses of the voluntary and the involuntary, of desire and preference, of *praxis* and *poiesis*, of deliberation and the practical syllogism; then in the Schoolmen's discriminating analyses of the various types of intentionality and the various objects – both formal and material – of human activity. It draws on Descartes' fourth *Meditation* and his *Treatise on the Passions of the Soul*, on Kant's *Fundamental Principles of the Metaphysics of Morals* and his tracts on the philosophy of history and more recently on Kierkegaard for the scattered threads of a phenomenology of the act of decision. However, it is Husserl's phenomenological method which yields principles by virtue of which this descriptive approach can claim to be considered a 'rigorous' science and to be put on the same footing as the 'exact' sciences derived from mathematics. Husserl himself, it should be pointed out, did not develop a phenomenology of decision nor, in general, of practical acts; his phenomenology is primarily concerned with the description of objectifying acts (such as perception, representation and judgement), in other words of 'theoretical reason'. However, although he limited himself to the realm of theory, his method of correlating *noesis* (act) and *noema* (correlate-object), together with his formulation of the concepts of intentionality and the intentional object, of meaning, of doxic and practical thesis and passive and active synthesis, themselves point beyond this context. The fruitful results which can be looked for from using the phenomenological method for description of the practical realm can be seen above all in the production of post-war existential phenomenology, with Jean-Paul Sartre, Simone de Beauvoir, Maurice Merleau-Ponty, Paul Ricoeur.[81] Here Husserl's call for scientific rigour is coupled

81. SARTRE, already in 'La transcendance de l'Ego' (1936, 1965) (English translation, *Transcendence of the Ego*, 1957); *Esquisse d'une théorie des émotions* (1939, 1965) (English translation, *Sketch for a Theory of the Emotions*, 1962); *L'imaginaire* (1940) (English translation, *Imagination*, 1962); *L'être et le néant* (1943) (English translation, *Being and Nothingness*, 1957); for the after-war years, see *Réflexions sur la question juive* (1946) (English translation, *Antisemite and Jew*, 1965); *Baudelaire* (1947) (English translation, 1964); *Saint Genet, comédien et martyr* (1952) (English translation, *Saint Genet, Actor and Martyr*, 1963); *Critique de la Raison dialectique*, I: *Théorie des ensembles pratiques* (1960), preceded by 'Questions de méthode' (1957) (English translation, *Search for a Method*, 1968), with which borrowings from Marxism become preponderant; *L'idiot de la famille* (study of Flaubert, 3 vols. to date, 1971-72); the essays collected in the series *Situations I, II, III*, etc. (1947-); but also the whole of the literary (fiction and drama) and political work. Cf. JEANSON, *Le problème moral et la pensée de Sartre* (1966); LAING & COOPER, *Reason and Violence* (1964). BEAUVOIR, *Pyrrhus et Cinéas* (1944) and *Pour une morale de l'ambiguïté* (1947). MERLEAU-PONTY, *Humanisme et terreur* (collected essays, 1947) (English translation, *Humanism and Terror*, 1969); *Sens et non-sens* (*id.*, 1948) (English translation, *Sense and Non-sense*, 1964); *Éloge de la philosophie* (inaugural lecture at the Collège de France, 1953) (English translation, *In Praise of Philosophy*, 1963); *Signes* (collected essays, 1960) (English translation, *Signs*, 1964); *Résumés de cours, Collège de France, 1952-1960* (posth. publ., 1968) (English translation, *Themes from the Lectures at the Collège de France, 1952-1960*, 1970). Cf. DE WAELHENS, *Une philosophie de l'ambiguïté: l'existentialisme de Maurice Merleau-Ponty* (1951). RICOEUR, *Philosophie de la volonté*: I. *Le volontaire et l'involontaire* (1950); II. *Finitude et culpabilité:* 1. *L'homme faillible*; 2. *La symbolique du mal* (1960) (English translations, *Freedom and Nature. The Voluntary and the Involuntary*, 1966; *Fallible Man*, 1965; *The Symbolism of Evil*, 1967); *Histoire et vérité* (1955, 1964) (English translation, *History and Truth*, 1965). Cf. VANSINA,

with influences emanating from the young Hegel, from Kierkegaard and Nietzsche, and indeed from the early work of Marx, but the main strand remains Husserlian intentionality, following on the lead which had already been given in that sense at the turn of the century by the phenomenologists of the Munich school, such as Pfänder.

For the phenomenologist, the fundamental thing is the intentional moment of decision. In the phrase 'I decide *something*', this something, which is the correlate of 'decide', presents itself as a something 'to be done by me'. This 'object' of volition is therefore dependent on me, to occur in the course of the world as my work; moreover, its occurrence within a specific time interval involves a temporal determination quite distinct from that involved in anticipating it, or even in desiring or fearing it. Thus the first determination of the decision is derived from its correlate, from its *noema*, i.e., the thing 'to be done by me'. Turning reflectively from the object to volition to the act of decision itself, we can describe the 'moment' of the determination of self by self: in deciding (determining to do) something, I determine *myself*, and it is this self-determination that constitutes the temporal emergence of the practical now; now I wish, I begin, I initiate, I place, I create. However, this determination of self by self must, in its turn, be such that it constitutes the final link in a motivational chain, and so affords a means of describing the complex interplay of motives and choice: I decide (determine to do) *something*, I determine *myself because* ... The fact that decision is thus buttressed by motivation is perhaps the crucial point presenting most difficulty in all the phenomenology of decision. On the one hand, motives determine choice; on the other hand, the direction in which my motives finally point is not itself determined until I make my choice. We have here a circular relationship which is incomprehensible if we view the link between motives and decision in terms of a model of causality, taken from the natural sciences; we cannot understand how motives can, in the words of Leibniz, 'influence without determining', unless we have a specific concept to designate a type of act which is at once determined and yet also self-determining. On the one hand, decision is responsive to its motives, listening to and following them; on the other hand, it asserts itself and begins something, not just in the world at large, but in the agent himself. This paradox – which in fact is only a paradox if considered in terms of natural causality – can be just as well expressed in temporal terms. The act of decision brings together the features of temporal discontinuity and continuity; the moment of choice is the moment in time when something begins, but the motivational chain which is concluded by the act of choosing is characterized by process and duration.

At this first, strictly phenomenological level, the philosophy of freedom and values lends itself to a fruitful comparison with the English-speaking school of linguistic analysis. On the one hand, for phenomenologists, the

---

'Esquisse, orientation et signification de l'entreprise philosophique de Paul Ricoeur' (1964). Cf. also, across the Atlantic, WILD, *Existence and the World of Freedom* (1963).

analysis of actual experience ('life') is inevitably the analysis of the various forms in which this experience is expressed; experience expresses itself and in so doing becomes knowable. Phenomenology is thus in fact a semantics of action and, for this reason, does not reduplicate behavioural psychology or a sociology of action; on the contrary it has much to learn from these disciplines, before reflecting on them, as indeed it has from the expressions and meanings of ordinary experience. Conversely, the phenomenologist might well ask if linguistic analysis is not in fact an unconscious form of phenomenology; indeed, are not the subtleties of ordinary language in some way the repository of an accumulated experience representing the latent meaning of the discourse of action? This convergence between linguistic analysis and the description of actual experience is especially noteworthy in regard to two crucial points already touched on in the preceding paragraph, namely the question of the object of volition and the question of motivation. On the first point, it is one and the same act-object relationship which is studied in the semantic analysis of action verbs and by the noesis-noema method of correlation. On the second point as well, the two schools of thought engage in comparable attempts to analyse the relation between motive and cause. One undertakes conceptual analysis of the notion of 'the reason why . . .'; the other sees in motivation the type of link which exists between the experiences of consciousness and between the 'meanings' of these experiences, once the relationship pertaining in natural attitudes is put in brackets. Such convergence may perhaps herald a period of cross-fertilization between the two methods.

(c) *Philosophy of 'values' proper*

At a second level – the ethical level proper – the philosophy of freedom and values attempts to offset the subjective aspect of the decision (which still imbues the concept of motivation with the objective aspect of the value.

The distinction between subjective will and objective will goes back to Kant and Hegel, who both distinguished the rational will (*Wille*) from the purely arbitrary will (*Willkür*). For the philosophy of values, which comes down from Max Scheler and Nicolaï Hartmann, moral behaviour is 'behaviour directed towards values'. This relationship, it is maintained, is intuitively apprehended in every situation which is not ethically neutral. The axiological character of every such situation is grasped immediately and absolutely: this is an indisputable, categorical fact of experience. For example, if I am brought face to face with a situation where someone's life is threatened, I feel that this life, in so far as it is 'to be saved', is valuable in itself; moreover, this feeling cannot be reduced to considerations of what is pleasant or what is useful, nor to a recognition of that course likely to induce happiness. Nor can the inherently valuable be considered as providing a psychological motive, even though our motives derive their moral character from it; in short, such a moral quality is irreducible and unconditioned. As can be seen, we are not very far here from Moore and his belief that the predicate 'good' is not susceptible to definition.

However, if what is valuable as such is not deducible, it does however comprise varying types of *relationship*. Understanding these relationships confers on the intuition of what is valuable an intellectual character which distinguishes it from mere emotion or from inarticulate and inexpressible sentiment.

In the first place, what is valuable is always an action, or the result of an action in a given situation; and this is what gives it its quality, the quality of being 'ethically relevant in such a situation'. In this sense, grasping a situation and understanding value go hand in hand. The fact that a category which possesses 'value' rather than 'reality' should be thus anchored in the reality of action expresses the first relativistic feature of what *prima facie* appears absolute.

Furthermore, the valuable comprises an inherent call for action, a 'to-be-done', thus bending back in the direction of reality a category which at first sight seems opposed to it, as what should be to what is. But it is of the very nature of the valuable that it should *be*. When we embrace a value, we are at one and the same time aware that we must realize it in the concrete situation which derives its moral character from it; thus the 'should be' aspect of a value involves its becoming real. This relationship between value and reality, expressed by the requirement that a value be realized, is the key to the following relationships.

This 'should be' aspect of value is in fact a command to act *addressed* to a moral subject. The 'objective' value addresses itself to a 'subjective' ability to respond. Such response to value constitutes properly speaking the *responsibility* of the individual; the individual is subjectively responsible only in so far as that which possesses objective value calls upon him to make such a response. Carrying on from here, however, human action is marked by characteristics which can be classed valuable, and pegged directly to behaviour, which we subsequently label good or bad; the objective value towards which behaviour is thus directed becomes attached first to the action, and then to the individual himself. This attachment to the individual introduces the third relativity qualifying the absolute value which, as a result, refers at one and the same time to a situation, to an action and to an agent. However, for the philosophy of value it is not an aspect of the situation, nor anything to do with performance of the action, nor is it even a process of valuation or some such attribute of the agent's will, which constitutes these various relativities; they all result from a movement originating with the absolute value itself. Ethics does not proceed from man, but from values, and it is they which confer on action its characteristics of response and responsibility.

The last, crucial relativity is that which introduces the notion of freedom as the correlate of the notion of value. Freedom is the capability of responding or not to the demands of value. In the sense that practical response presupposes a notion of value, then it may be said that the notion of freedom develops from the notion of value. In another sense, however, this cannot be said to be true, since nothing in the notion of the inherently valuable suggests that the individual may, or above all may *not*, respond to the value. It is indeed the capability of disobeying which shows that freedom, as the ability to adopt a position in regard to value, is, like it, absolute in character. Accordingly, we find ourselves back with the bipolarity of Kantian philosophy. On the one hand, we only know anything of freedom because we are responsible and it is solely conduct referred to value that can in any way be said to display freedom, so that without value and response to value we should never enter the realm of freedom; from this standpoint, then, value can be quite properly considered, to use Kant's expression, as the *ratio cognoscendi* of freedom. As against this, freedom for its part is the *ratio essendi* of value. Without freedom the philosophy of value would have said its last word with the conclusion that 'should be' is irreducible to 'be', all its force would have been spent in distinguishing value judgements from judgements of fact; but that the valuable should become real, that the gulf between what should be and what is should itself be eliminated in action, is not this what freedom demands? Without the bringing about, the 'making-to-be' which militant freedom bequeaths to the world, would value be anything more than an 'in itself', to be contemplated in a quasi-aesthetic awareness of the resplendence of Good? Would it be even a '*should be*', wherein the 'be' is as strongly emphasized as the 'should'?

Furthermore, this idea of the alternation of the two poles of value and freedom is required by a different type of analysis. Let us consider a specific value, for example the rule of justice – what does it assert? It holds that a second person is worth as much as me and that his point of view should be recognized, respected and harmonized with mine by way of actions which allow both for my interest and for his. Consequently, another's freedom, that is to say the free development of his being, emerges as the very content of the philosophy of value. Value does not merely give rise to the anti-pole of freedom (as what may or may not respond to its demands); through the intermediary of the other person, that is to say, the person himself, value shows itself to be the value of freedom, and not simply value *for* freedom. Value is what is postulated by the co-existence of individual freedoms. It is your value, but it is also my value, in so far as I myself, like all other individuals, am another person, a second person for everybody else.

This correlation between value and freedom (where emphasis is put in turn on what has value *in itself* and on the value *of* the individual) is the fundamental principle of a system of material ethics which purports to be the answer to the formal ethics of Kant, and also to be able to meet the same demand for specificity as did Hegel's philosophy of social institutions and realized freedom, without surrendering to the rigid systematism of Hegel's *Encyclopaedia*. This material ethics leads quite naturally to a sort of ethical personalism, which stresses now the dignity and development of the individual confronted with the anonymous forces and depersonalizing influences of technological civilization, now the crucial significance of dialogue for the actual definition of freedom as a finite reality, dependent on recognition by others. This apologia of dialogue implies, in the context of politics, an unremitting censure of tyranny and authoritarian régimes, and a plea for discussion as also for the free expression and unrestricted interplay of all shades of opinion. In its turn, this political liberalism sometimes aligns itself with (and at others runs counter to) the desire to bring into being economic and social structures which would ensure to each and all the material conditions necessary for realizing freedom. It is this active and vocal concern on behalf of the community which gives the philosophy of freedom and values a common frontier with Marxism and its economic and social philosophy.[82]

For many contemporary thinkers in the East,* freedom itself may be seen as closely related to the value-bearing and value-seeking consciousness of man. In fact, Devaraja, an Indian thinker, goes so far as to define philosophy as a study of sharable value-consciousness and/or the value-bearing consciousness of man, that is, of 'the modes of cultural activity embodied in art, scientific thinking, morality and philosophy with a view to discovering the criteria or principles by which the relatively higher, maturer, or more adequate forms of these activities or experience could be distinguished from the inferior, less mature or less adequate forms'. On the other hand, it may also be seen as the end which all human action seeks to realize. In a sense, freedom is the foundation of all seeking for values, for without it no value, including that of cognitive truth, could even be sought. But, and this is not

82. On the main trends of recent philosophic thinking in the field of value-theory and ethics, the reader may refer to the contributions of A. C. EWING, G. BASTIDE and A. GUZZO to KLIBANSKY (ed.), *Contemporary Philosophy/La philosophie contemporaine*, Vol. IV (1971) and the accompanying bibliographies.

Among significant works of recent years we may mention, in addition to those already cited in the preceding footnotes: ABBAGNANO, *Possibilità e libertà* (1956); ARANGUREN, *Ética* (1958) and *Ética y política* (1963); AXELOS, *Pour une éthique problématique* (1972); BASTIDE, G., *Traité de l'action morale* (1961), 'Nature, situation et condition humaine' (1962), and *Essai d'éthique fondamentale* (1972); BATTAGLIA, *Linee sommarie di dottrina morale* (1958); BOLLNOW, *Wesen und Wandel der Tugenden* (1958); GADAMER, 'Über die Möglichkeit einer philosophischen Ethik' (1963) (French translation, 'Sur la possibilité d'une éthique philosophique', 1971); GARAUDY, *Perspectives de l'homme* (1959) and *L'alternative* (1972); GUZZO, *La moralità* (1967); HARTMAN, *La estructura del valor* (1959) (English translation, *The Structure of Value*, 1967) and *El conocimiento del bien* (1966); JANKELEVITCH, *Traité des vertus* (1949; augm. edn., 1968-70-); LEPLEY, *The Language of Value* (1957); LOMBARDI, *Il concetto di libertà* (1966), *Problemi della libertà* (1966), and *Il mundo degli uomini* (2 vols., 1967); MARCEL, *La dignité humaine et ses assises existentielles* (1964); NABERT, *Essai sur le mal* (1955) and *Le désir de Dieu* (posth. publ., 1966); NAGEL, Th., *The Possibility of Altruism* (1970); PAREYSON, *Eisistenza e persona* (3rd edn., 1966); PERELMAN, *Justice et raison* (1963); PUCELLE, *Etudes sur les valeurs*: I. *La source des valeurs* (1957); II. *Le règne des fins* (1959); III. *Le contrepoint du temps* (1967); REINER, *Die philosophische Ethik* (1964); ROBINSON, R., *An Atheist's Values* (1964); SCHNEIDER, Herbert, *Three Dimensions of Public Morality* (1956) and *Morals for Mankind* (1960); SCIACCA, *La libertà e il tempo* (1965); SOURIAU, M., *Ethique transréelle* (1961); STRAWSON, *Freedom and Resentment* (1962); THÉVENAZ, *L'homme et sa raison* (1956); WEIL, *Philosophie morale* (1961); collection: *Sein und Ethos. Untersuchungen zur Grundlegung der Ethik* (1963).

* We owe the following remarks to Daya Krishna.

generally realized, it may also be seen as their end also. The pursuit of values can only be undertaken if man is free, but the end of all value-realization is ultimately the complete fulfilment of human personality in its self-conscious being, which is just another name for freedom.[83]

The relation of freedom to action may, however, be seen in a different perspective also. The values themselves may be seen as radically divided into those which are essentially active, and those which are contemplative in nature, that is, those which are concerned with the realization of a state of one's own being or consciousness. The former, which alone give rise to action proper, however, make one not merely dependent on others, but also involve one in a chain of causality from which there is little possibility of release. This *dependence* and *involvement* not only jeopardize one's freedom, but lead, slowly and subtly, to action which cannot but be considered as immoral in character. The reason lies in the fact that most of the ends one wants to achieve, require the help of *others* for their achievement. Now, if others co-operate in the facilitation of what one wants to achieve, then all is well and good. But if they obstruct – and this is the rule rather than the exception – one has either to withdraw from action or *coerce* and cajole others, that is, *force* them to do what they do not want to do. Besides this, most action occurs within an institutional context, where one has to take responsibility for what one may consider to be wrong and/or where one may not feel any responsibility because of the impersonal anonymity of decisions which are arrived at in a collective manner.[84]

There thus emerges a basic antagonism between action and freedom in this perspective. It links it to some extent with the classical debate between action and contemplation in relation to freedom.

The Japanese thinker, Imamichi, has suggested that the relation of freedom to action is, so to say, mediated through what may be called 'practical reasoning'. But there can be different models of 'practical reasoning', and we may have more or less points for the exercise of freedom, depending upon the model we choose. He has given two such models of practical reasoning. The classical paradigm may be presented in the following form:

I desire A. (decision with respect to aim)
P, g, r, etc. will realize this desired end. (enumeration of means)
For a certain reason, I select p. (selection of means)

The second model, on the other hand, may be formulated as follows:

We have the power P. (the means are given)
This power P will realize the aims a, b, c, d, etc. (enumeration of ends which can be achieved by the power)
For a certain reason, we select aim a. (selection of the aim)

In the first model, there are two points where freedom may be exercised, the

83. Cf. Devaraja, *The Philosophy of Culture* (1963).
84. Krishna, 'The active and the contemplative values' (1969).

choice of end and the choice of means. In the second, in an important sense, the ends are predetermined by the given means, and the only choice that one may exercise relates to the end that one will choose from amongst them.

What Professor Imamichi perhaps wants to point out is the increasing imbalance between means and ends, as the means accumulate, so that not only the primacy of end is lost, but the means begin subtly to determine the ends that we want.

(d) *Towards ontology*

The philosophy of freedom and values, as the foregoing comments on finite freedom would suggest, seeks to go beyond a philosophical psychology or phenomenology of decision, beyond even an ethico-political theory of responsibility; it attempts to place the very concept of freedom on a specifically ontological level; in so doing, it seeks to answer the question: how must reality as a whole be constituted in order that man may be an *agent*, that is, the author of his acts, in the twofold sense of having the psychological power of choice and being morally responsible? Such a question could not be mooted until a series of turning-points had been passed which testify to the epochal nature of philosophy.

The first step was to conceive of the will as infinite, which was not at all in line with Aristotle's views on decision; he believed choice could only be applied to that on which one could deliberate; one does not deliberate on ends but on means; and the question of the ends aimed at is made a matter of desire, which is not deliberated. The definition of virtue as the golden mean also, within its own limits, points to that quest of finite perfection which is the hallmark of the Greek world; to quote Hegel's famous comment in the *Philosophy of Right*, the transition from finite perfection to infinite reflection is the transition from the Greek to the Christian world.

The second requisite for the concept of freedom is the philosophical emphasis on subjectivity; this in turn could only impress itself on reflective thought in conjunction with a general conception of being according to which all 'being' is displayed like a 'picture' in front of a subject of representation. The age of subjectivity, according to one of Heidegger's penetrating insights, is in the first place the age of *world as picture*;[85] the reign of representation ushers in the reign of objectivity; this in turn raises the question of certainty of the object; and the question of certainty gives rise to the awareness of a subject for whom each thing is an object of representation; with Descartes the psychology of freedom is drawn into the gravitational field of the *Cogito*. Thereafter the model for freedom is free reflective thought.

The third requisite for the concept of freedom is the establishment of a concept of nature completely subordinate to the notion of absolute legality. Once this idea of nature and its legality is established, freedom can only be

85. HEIDEGGER, 'Die Zeit des Weltbildes' (1938, 1950) (English translation, 'The age of the world-view', 1951).

posited in paradoxical fashion. Kant's resolution of the antinomy between liberty and nature is, after establishment of the infinity of the will and its subjectivity, the third turning-point that had to be passed by modern thinking in regard to freedom. Freedom is that *'absolute spontaneity* of cause, which *of itself* originates a series of phenomena which proceeds according to natural laws'.[86] This free causality is the transcendental condition under which moral freedom may be conceived, in the form in which reflective thought extracts it from the everyday experience of the good will; it is this transcendental freedom which, opposed to the necessary legality under which I conceive of nature, gives rise to an antinomy. It is essential that reason, whenever it is applied to speculative thinking on freedom, must undergo the trial of antinomy. Neither a phenomenology of decision nor an ethico-political philosophy of responsibility *think* until they reach this critical moment.

But is the antinomy of freedom and nature thought's last word? Certainly Kant considered the antinomy a product of reason (*Vernunft*), not only of the understanding (*Verstand*): the understanding orders phenomena, proceeds from conditions to conditions; reason asks the radical question concerning the origin and calls for an unconditioned at the source of the series of conditions; in this sense it is indeed reason which asks the radical question regarding a beginning to causality. Yet in another sense one may ask oneself, with Hegel and in opposition to Kant, whether a thought which separates, which divides, is not still a thought pertaining to the understanding. The Kantian antinomy, it seems, calls for a type of thought other than the critical, a thought that is not content simply to affront free causality and natural causality, but which combines them in a single whole.

Where to look for this ontology of freedom today?

Neo-scholastic philosophy offers considerable resources; leaning on the Aristotelian notions of being (act) and becoming (entelechy), and above all on the Thomist affirmation of the *actus essendi*, this philosophy attempts to define degrees of *being*, according to the extent to which causal necessity or spontaneity predominates; between these degrees of being it discerns relationships of dependence in regard to the lower and emergence of the higher, which ensure both the interconnectedness and the hierarchical ordering of the levels of reality.[87]

Reference may also be made to Whitehead and to *Process Philosophy*, very much alive in the United States today, in connection with the vision of an ontological dynamism capable of accounting not only for such differences in degree, but also for the jumps and transitions without which the emergence of freedom would be simply incomprehensible.

86. '[...] eine *absolute Spontaneität* der Ursachen, eine Reihe von Erscheinungen, die nach Naturgesetzen läuft, *von selbst* anzufangen [...]' – KANT, *Critique of Pure Reason*, Proof of the Thesis, Third Conflict of the Transcendental Ideas, *in fine*.

87. Cf. LECLERCQ, *La philosophie morale de saint Thomas devant la pensée contemporaine* (1955); JOLIVET, *Les activités de l'homme et la Sagesse* (1963).

A definitely evolutionist philosophy, such as that of the French thinker Teilhard de Chardin,[88] informed by palaeontology, or that of the Indian thinker Sri Aurobindo,[89] meets a similar concern to account for the continuities and discontinuities in the genesis of forms and between modes of reality; this philosophy has gained a wide audience, mainly among those who seek a global vision able to withstand the fragmentation of scientific work and a conception of the world wherein the realm of the spiritual is both justified and well bedded in solid ontological realism. Lastly, this type of thought is in favour among those seeking for a philosophy which satisfies both mystic admiration and scientific observation.

Another school of contemporary ontology, on the other hand, roundly accepts the Kantian antinomy and expresses it in a vision of reality characterized by division, schism and tearing asunder. Inspired by Kierkegaard and his notion of 'leap', this sort of ontology dramatizes the dichotomy between modes of being in a frank dualism between the free being and the natural being. One strain of existentialism tends to this type of exaltation of that ungenerable negativity whereby freedom wrenches itself out of the inert being of nature. This was the position of Jean-Paul Sartre's philosophy before the *Critique de la raison dialectique* (1960). Another current, initiated by Gabriel Marcel, lays as much stress on the corporal, biological and natural origins of liberty as on its power of subversion. The 'tragic' sense which this philosophy attaches to the discovery of liberty stems less from speculative antinomy than from an existential conflict between fascination with objects, the barrier to 'openness' ('*disponibilité*') and the source of avarice and despair, and the will to live, the generator of joy and hope. In this type of philosophy, resolution of the antinomy does not lie in speculative thinking able to withstand the divisions of Kantian thought by making play with mediation and synthesis in the manner of Hegel. What we have is a free decision which posits its own freedom, recognizes its dependence and accepts its responsibility.

One may however ask if these philosophies possess the instruments of thought to enable them to express conceptually the experience of freedom. The attention which our contemporaries continue to pay to Hegelian philosophy is no doubt due to the dissatisfaction of the mind faced with philosophies which feel more than they think, and which have not recaptured

88. Teilhard de Chardin, *Œuvres* (posth. publ., 11 vols., 1955-73), in particular: Vol. I. *Le phénomène humain* (1955); Vol. V. *L'avenir de l'homme* (1959); Vol. VI. *L'énergie humaine* (1962); Vol. VIII. *La place de l'homme dans la nature* (1963; first published as *Le groupe zoologique humain*, 1956); and Vol. XI. *Les directions de l'avenir* (1973) (English translations: *The Phenomenon of Man*, 1959; *The Future of Man*, 1964; *Human Energy*, 1969; *Man's Place in Nature*, 1966).

See Barthélémy-Madaule, *La personne et le drame humain chez Teilhard de Chardin* (1967); Cuénot, 'La morale et l'homme selon Pierre Teilhard de Chardin' (1960) and *Pierre Teilhard de Chardin. Les grandes étapes de son évolution* (1967); Tresmontant, *Introduction à la pensée de Teilhard de Chardin* (1956); Francoeur, *The World of Teilhard de Chardin* (1961); Raven, *Theilhard de Chardin, Scientist and Seer* (1962).

89. Aurobindo, *The Life Divine* (1949); *The Human Cycle* (1949); *The Future Evolution of Man: The Divine Life upon Earth* (new edn., 1963).

the dialectical style that seems to be required by the problem of freedom. How, in fact, is it possible, without a dialectic, to think coherently of the emergence of a power of decision which nevertheless remains rooted in desire, or of the mutual genesis of the theoretical and the practical, or to pass from subjective will, as described in phenomenologies of decision, to objective will, which is the subject of ethico-political thought on responsibility? At all events freedom seems to be the fundamental experience which compels us to think of reality itself in a dialectical manner. Reality, thus conceived, is not only nature as regulated by laws which is confronted by freedom as its contrary, but is also nature as capable of progressively becoming the work of freedom. This type of nature, generated by freedom, would no longer be the 'other' of freedom; freedom would no longer oppose it to itself except in so far as it would achieve itself in it. This, it may well be, is the kind of reality to which further thinking has still to be directed.

§3. *Theory and practice in Marxist philosophy**

Whereas Anglo-Saxon philosophy is concerned with *clarifying* concepts of action and moral philosophy, Marxism sets out to *change* them. In this undertaking, it finds itself in conflict with the Western metaphysical and speculative tradition culminating in Hegelianism. What it discerns in this tradition is the opposition between philosophy and practice, philosophy and politics, philosophy and the activities of the 'common' man who is ignorant of philosophy, but to whose service the humanistic mission of philosophy must needs be dedicated. Marxism will have nothing to do with such opposition which, it contends, continues to characterize contemporary non-Marxist thought. At the same time, it insists on underlining the positive character of this theoretical activity, the alienated state of which is in fact largely due to its being divorced from practice.

(a) *Historical excursus*

This relationship of reciprocal alienation between theory and practice is not something accidental in itself: it has deep and, in that sense, unavoidable historical roots; not that this prevents it from wearing, in certain respects, a progressive historical aspect. But it is these historical roots which Marxism is determined to begin by laying bare. The chief value to be derived from this study is to show the complex and contradictory processes which have transformed what was first a one-way relationship between theory and practice into a two-way relationship. A brief account of this historical development will suffice here.

Theoretical activity, especially scientifico-theoretical activity, has not always existed: the advent of philosophy (about the sixth century B.C.) also

* Drawn up by T. I. Ojzerman within his overall contribution to the present work.

marks the historical beginning of mankind's theoretical activity – an activity which it is certainly necessary to distinguish from everyday knowledge acquired through customary experience, without any particular cognitive effort. The early philosophers – founders of theoretical knowledge – were learned men with a very wide range of interests partly directed towards practice; but their researches had no practical bearing, for science had to attain certain precise stages of development, experimental ones in particular, before being able to direct practical – e.g., productive – activity effectively. Men learned how to till the soil, grow cereals, train wild animals, cast metals, build houses, boats and so on, without the aid of theory. Systematic scientifico-theoretical research introduced something new; in the first place it brought a considerable improvement in technical efficiency. However, it is only in modern times, with the discovery of theoretical mechanics, that the systematic technological exploitation of science has moved beyond the field of its possible technical applications. Up to then, over numerous centuries, social production and other aspects of practical human activity developed chiefly on the basis of experience, habit, know-how, acquired in practice and handed down from generation to generation. Thus the utilization of scientific facts in production remained sporadic. As for the application of science to other spheres of practical activity, this general question had not yet been specifically raised and the reflections of a handful of distinguished thinkers on the subject struck their contemporaries as utopian. As long, therefore, as production and the other spheres of mankind's practical activity developed more or less independently of scientifico-theoretical research, the relationship between theory and practice (especially with regard to production) remained broadly speaking one-way: science reaped the benefit of practical data acquired in the process of production, but itself gave incomparably less *to* production than what it derived *from* production. For a long time, applications of theoretical discoveries assumed principally the guise of an explication or generalization of an existing practice, of technological processes developed spontaneously or as the result of discoveries made by persons without any theoretical training. A good example of the theoretical explication and generalization of a practical discovery is provided by Carnot's ideal steam-engine, which enabled production in this field to be placed for the first time on a scientific basis.

The transformation of the primarily one-way relationship between theory and practice into a two-way relationship, characterized by innumerable forms of reciprocal, interpenetrating actions and by mutual transformations of opposites, has resulted in a qualitative change in practical and theoretical activity. Some of these transformations are particularly striking: the contemporary scientifico-technical revolution has transformed nearly all spheres of practical human activity into subjects of theoretical research; the application to production of production methods based upon mathematics, physics, chemistry and biology, combined with traditional mechanical procedures, affords clear proof of the reciprocal transformation of theory and practice. This process, unprecedented in past history, is borne out by two particularly striking facts. On the one hand, many recent production techniques are literally based upon natural science: thus union of theory and practice has become a tangible fact, and this unity has been achieved, strengthened and developed without science having in any way surrendered its inherent theoretical and abstract character; on the contrary, all branches of science have become incomparably more theoretical than previously, as is shown, especially, by the way research methods have been mathematized; even applied scientific disciplines are becoming more and more theoretical.[90] The second significant fact is the link between the social sciences and social practice. Just as the development of natural science has become a factor of capital importance in the formation of a dynamic – or, strictly speaking, dialectical – unity between theory and production practice, so the development of Marxism has pointed the road to a perpetual transformation of social relations based upon a scientific theory of the laws governing the development of human society.

This brief historical digression reveals, by contrast, the meaning of the metaphysico-speculative opposition between theory and practice rejected by Marxism. Marxism saw in such conceptual opposition a reflection of genuine contradictions, further complicated by the opposition between intellectual and manual work, the exploitation of manual workers and other forms of opposition. In this way, Marxism, with its principle of the unity of theory and practice and its understanding of the relationship between philosophical theory and social practice, claims to present a theoretical balance-sheet of the process of mankind's development throughout the centuries.

(b) *A new understanding of practice*

However, although Marxist criticism of metaphysical and speculative tradition results in rejecting the opposition between theory and practice, the correct interpretation of their unity which it proposes also involves transcending the traditional view of practice as a purely applied activity. The ugliest – and certainly the pettiest – form this view can take is that whereby practice is seen as an activity carried out at the level of customary experience with a view to attaining strictly commonplace, generally egoistic ends. According to Marx, not even L. Feuerbach ever succeeded in freeing himself from such a view of practice.

In this respect, the work of Gramsci is of great significance. By calling dialectical and historical materialism the 'philosophy of praxis', Gramsci wished to mark the fact that a radically new understanding of practice was involved, that practice was not opposed to intellectual activity as to something alien to it but was, on the contrary, itself a practico-intellectual activity. That is why the Marxist 'philosophy of praxis' – that is to say, the inclusion, in philosophy, as one of its most important sectors of practically oriented re-

90. Cf. on this subject the collection of articles *Praktika – Kriterij istiny v nauke* (= Practice – The Criterium of Truth in Science) (1960), with contributions by I. D. ANDREEV and others.

search – has aimed at raising the scientific level of philosophy, by transcending at one and the same time both the empty esotericism which has lost all historical justification and the refusal to treat essential social problems which form the content of universal history. 'The philosophy of praxis', wrote Gramsci, 'does not seek to maintain the masses of "ordinary folk" at the primitive, common-sense philosophical level. [. . .] On the contrary, it strives to develop in them a more complete type of human being. This philosophy calls for a close association between scholars and "ordinary folk" not with the idea of reducing the scholars to the same level of consciousness as that of the masses at the present time, but for the purpose of creating a close intellectual bond which, with political support, will bring about intellectual progress throughout the masses, not merely within a limited group of intellectuals'.[91]

In order to ensure the unity of philosophical theory and non-philosophical practice, it will be necessary therefore: (i) to change completely the character of philosophical theory (this has already been referred to in the rubrics on dialectical and historical materialism) and (ii) to change not only the understanding but also the very nature of practical activity, this change in turn entailing a greater complexity of action itself, in respect of its content, its structure and the level at which it takes place. In this sense, present-day Marxists recognize that the founders of Marxism had, with their theory of economic and political practice and their understanding of the gnoseological implications of practice, anticipated with a stroke of genius its subsequent development.

But what is meant by practice? For it takes various forms. There is the routine type of practical activity corresponding to an everyday productive or political experience repeating itself year in year out. From this point of view, Lenin was able to criticize the practice of international opportunism in the working-class movement, stigmatizing it as trimming one's sails to the wind rather than using the wind to set off on a new, revolutionary tack. Marxism rejects the non-critical cult of praxis – to which it regards pragmatism as committed – in that it evaluates practice from the point of view of the past and future development of society and knowledge. But this does not mean that it regards practical success as an all-sufficient indication of the correctness of the practical line adopted.

Practice, said Lenin, is the foundation of knowledge, but it is at the same time the crown of knowledge, in the sense that it is above not only empirical knowledge but also theoretical knowledge.[92] Practical activity is here being

91. Gramsci, *Opere*, Vol. II (posth. publ., 1948), p. 21. In another part of the *Prison Notebooks*, Gramsci states that the 'philosophy of praxis', marking as it does a new era in the development of world thought and, especially, in the development of dialectical world thought, 'sublates the whole mainstream of preceding philosophy, both in its materialistic and idealistic currents' (*op cit.*, pp. 131-132).

92. In this paragraph we are not specially considering the question of the part played by practice in cognition, since an exposition of this question would involve that of the whole of Marxist gnoseology. On the other hand, the gnoseological im-

considered according to different structural levels; and the reference to its supremacy over theoretical knowledge relates to practice that has been theoretically elucidated, and is guided by scientific theory, which presupposes the transformation of theory into practice, practical action being based upon a theory which is itself verified, corrected and developed through practice. Marxism, after laying the theoretical foundations for the creative union between theory and practice and then effectively bringing about that union, is naturally not going to banish theory from the realm of practical activity; to do so was precisely the flaw in pre-Marxist idealistic thought. Practice, as man's conscious, purposive activity in its most developed form, is an exploratory, creative, cognitive and organizing activity: this, primarily, is what is meant by the Marxist principle of the unity of theory and practice.

It follows that the essence of the Marxist theory of practice consists in the scientific understanding of practice as a unique form of human activity. Practice embraces all man's conscious, purposive activity, as long as it is not restricted solely to perception, representation, thought, emotion, that is to say, to single psychic acts. However, mental activities themselves become specific practical activities as soon as, for instance, we apply a computer or adding machine to the solution of mathematical problems, though there is a difference of opinion among Marxist thinkers on this question, certain exponents of dialectical materialism being unwilling to agree that thought – or any other reflection of objective reality within man's consciousness in particular conditions – can constitute a practical activity.[93]

Practice, in whatever form, consists in the use of certain objects with a view to the apprehension or transformation of reality, the satisfaction of a host of material and spiritual requirements of individuals, on both the personal and the social plane, and the organization of joint action. In this respect, we should bear in mind the relative nature of the difference between the use of 'objects' and the process which results from such use. It should also be noted that the word 'object' is placed here in inverted commas so as to avoid any one-sided interpretation: objects with the aid of which practical activity is carried out may be tools or equipment, technical means of knowl-

---

plications of practice have been dealt with so fully in Marxist research (e.g., in the Bulgarian philosopher T. Pavlov's basic monograph, *Teorija otraženija* (= Theory of Reflection) (1949) that it is hardly necessary to expound them here. We should merely note that the gnoseological theory of practice radically transforms the traditional view of materialistic sensationalism according to which cognition always begins with the impact of objects from the outside world upon our sensory organs. From the point of view of Marxist gnoseology, cognition always begins with practice, with man's action upon his environment. It is precisely thanks to practice that man is a cognitive being, and cognition itself, at any rate in its developed forms, is a selective, directed activity, not merely the result of the impact of objects from the outside world upon man's organs of perception. Cf. in this connection Kosing, *Probleme der marxistischen Erkenntnistheorie* (1959).

93. Cf., for instance, Čarakiev, *Dejstvennostta no misleneto. Aktivnata rolja na misleneto v processa na poznanieto i v praktičeskata dejnost* (= Reality in Thought. The Active Rôle of Thought in the Process of Cognition and in Practical Activity) (1961).

edge, human hands or any other part of the human body. However, this does not mean that there are special types of means and of needs but merely denotes certain functional uses of objects, needs or human aptitudes, which, in another respect, are not practical.

The chief, decisive form of practice is material production which presents itself – as is clearly apparent from the above description – not only as a means of satisfying certain needs, but as an activity which has historically moulded mankind and created the conditions which determine the historical forms of man's development in the economic sphere and in the other spheres of social life. Intellectual work is, of course, a practical activity too, but only to the extent that it comprises the use of material means, whether natural or man-made. Such is, for instance, organizing activity, not only in the economic of cultural domain, but also in the political life of society. Political activity, as a specific vehicle for human organization, mutual co-operation and joint action for the purpose of carrying out particular social tasks, is one of the most important forms of practical social life. From this point of view, the class struggle must also be regarded as a special form of socio-political practice, and one that takes on exceptional importance in the conditions of an antagonistic society. The transformation of the proletariat's struggle for freedom from spontaneous, inadequately organized unrest into a conscious, politically organized mass movement can be seen as one of the most striking expressions of the progressive development of social practice.

The theory of revolutionary practice developed by Marxism-Leninism is a new and exceedingly important trend in philosophical and sociological research, one whose possibility even had, for the most part, never occurred to pre-Marxist philosophy and sociology.[94]

94. The starting-point of this philosophico-sociological research is Marx's concept of revolutionary practice, whereby a spontaneous social movement is transformed into an organized and controlled social process, which makes it possible, to a considerable extent, to prevent, eliminate or reduce to a minimum the discrepancy between the social objectives aimed at and the results obtained: 'The coincidence of a change of circumstances and a human activity can only be regarded and rationally understood as a revolutionary practice' (Marx-Engels, *Sočinenija* (= Complete Works), Vol. III, p. 2). Practice in so far as it includes within itself the scientific knowledge of natural and social laws (this, of course, does not apply to all practice), is the dialectical transformation of necessity into freedom – a transformation that does not eliminate necessity, but that realizes all its manifold inherent possibilities in such a way as to serve the interests of men, scientifically understood from the point of view of their essential character, their necessity and the possible results of their satisfaction. This refers especially to revolutionary practice, and above all to that form of it which benefits from the scientific interpretation of social reality. In this way the problem of freedom and necessity, which throughout numerous centuries proved the stumbling-block of philosophical thought, finds its rational solution in the Marxist theory of the unity of scientific theory and practice (cf. on this subject Besse, *Pratique sociale et théorie*, 1963). The concept of freedom as the union between theory and practice is the basis of the scientifico-philosophical solution of the problem of responsibility, the most important stages in responsibility – resolution and action – forming the component elements of the theory-practice synthesis.

Although the transformation of nature and the transformation of social reality are the two most important spheres and forms of practical activity, this must only be understood in the sense that these forms of practice govern the development of society. However, it must not be deduced from this that other forms of practical activity are less essential or are of secondary importance. Scientific research (which does not aim at changing the world, at least not directly), inter-personal relations, including personal relations between members of opposing sexes, practical studies in production or in an educational institution, sporting activities, creative activity – all these are fundamental aspects of social practice, notwithstanding the fact that they do not constitute the determining basis of social development.

Throughout this sub-section, we have adopted as our basis the division of mankind's manifold activity into theory and practice. This is not, of course, the only possible division. Human activity can equally well be divided into productive and non-productive, spontaneous and conscious, compulsory and free, mental and physical, socially useful and detrimental, and so on. In other words, there are a considerable number of ways of dividing up human activity dichotomously, which is no reason, of course, for regarding such divisions as artificial or subjective. From the Marxist point of view, the division of human activity into practical and theoretical is the most general, the most all-inclusive, since every human activity, in so far as it is conscious and purposive (one can exclude physiological processes which occur spontaneously) is either of a practical or theoretical nature or is at the same time both practical and theoretical, i.e., is a fusion of the two.[95]

Any theory which restricted itself merely to the service of current practice would inevitably restrict its own possibilities, lose its ascendancy over empirical scientific knowledge and, in the final resort, display its inadequacy, even from the point of view of current practice. In this case, creative practice would cease to be quickened by theory; it would no longer apprehend, with the aid of theory, the possibilities and perspectives which cannot be brought to light, as practical propositions, except on the basis of theory.

Since theory comes before practice, anticipating its problems and the means for their solution, it must naturally remain relatively independent of practice. The whole history of science confirms this. Geometry originated as a method for measuring the earth, but it only became a real science when it passed beyond the bounds of this restricted relationship to practice; the same happens to every science once it has attained a sufficiently high theoretical level. Science, although derived from practice, subsequently breaks with it and proceeds in search of possibilities that are non-existent in practice, even outside the ordinary range of thought, only to make contact again with practice at a later stage and sweep it along still unexplored paths. Recognition of the guiding rôle played by theory in regard to practice is one of the main features of the Marxist interpretation of the union between the two.[96]

95. Cf. Koch, *Gesetzmässigkeit und Praxis. Zur Dialektik von objektiver gesellschaftlicher Gesetzmässigkeit und subjektiver praktischer Tätigkeit* (1968).

96. Sanchez Vásquez, *Filosofía de la praxis* (1967).

This unity of theory and practice possesses a mediate character, which is exemplified even by the technological sciences that are closest to practical productive activity, although every technological discovery has to pass through a special process of elaboration (including theoretical elaboration) before it can enter into production. In the case of theoretical discoveries relating to natural science, and *a fortiori* to the human sciences, the road leading to practical application is always more complicated and mediate in character. This is in no sense a drawback but an advantage attached to theory, which is ever apt to run far ahead of practice.

The relation between philosophy and social practice is still more complex and mediate in character; for it is in philosophy that the socio-economic changes and perspectives of the time find theoretical expression. Engels called the eighteenth-century Age of Enlightenment in France a philosophical revolution immediately preceding the great French Revolution. It is in this same sense that he referred to German classical philosophy as the philosophy of the revolution in Germany. These philosophical revolutions, of course, only anticipated and prepared the ideological conditions for the coming social and economic upheavals because they were the intellectual expression of these same deep-seated economic, political and cultural currents out of which, historically, the bourgeois revolutions were to come. Thus we see how progressive philosophy played an important part in the shaping of revolutionary practice.[97]

Marxism rejects equally the utilitarian, pragmatic (instrumentalist) conception of practice and the aestheticizing interpretation of theory (of science, philosophy, knowledge in general) which would seek to confer upon it a sort of existence of its own, independent of any ulterior social goal. Tolstoy passionately condemned the romantic notion of art for art's sake because of its intolerable indifference to the destiny of mankind, and laid particular emphasis on the ultimate humanistic purpose of all artistic creation. In this respect, Tolstoy's ideas come close to Marxist interpretation not only of art, but also of science and philosophy. It is illuminating to consider the historical fate of 'pure science', used as it has been to produce the atomic bomb, chemical and bacteriological means of warfare and other weapons of mass destruction. In our day, scientists who are honest with themselves recognize their moral responsibility with regard to the practical application of their discoveries. Marxist thinkers consider that the cult of science for its own sake has provided sufficient evidence of its negative aspect, and this, in their opinion, amply bears out their interpretation of the relationship between science and practice.

(c) *Discussions relating to the Marxist conception of practice*

The dialectical materialist interpretation of the relationship between theory and practice has exerted a considerable influence on certain non-Marxist

97. Opitz, *Philosophie und Praxis. Eine Untersuchung zur Herausbildung des marxistischen Praxisbegriffs* (1967).

philosophical and sociological theories: J. Dewey, K. Mannheim, J.-P. Sartre and a number of other thinkers, influenced by some of Marx's, Engels's and Lenin's ideas concerning the essential nature and social rôle of practice, have incorporated them in philosophical structures which are regarded by orthodox Marxism as being to a large extent subjectivist and irrational. Orthodox Marxism considers the theory of practice as the basis for a dialectical interpretation of the process of knowledge and for a scientific interpretation of historical necessity; on the other hand, in these other theories (as orthodox Marxism understands them) the link between knowledge and practice is interpreted from the point of view of negation of the objective content of knowledge, and social practice as the negation of objective social laws. However, these interpretations, regarded as non-Marxist, of what were originally Marxist philosophical ideas have, in their turn, had some influence on certain philosophers who consider themselves Marxists, and especially on those grouped around the review *Praxis*, published in Zagreb.* These thinkers take as their motto: 'Back to Marx!', thus rejecting all the accretions to Marxist thought due to Lenin and his successors. Their ideas are often presented under the title of philosophy of praxis, philosophy of man, or humanized Marxism. An idea of their attitude can be gained from the following statement by G. P. Petrović: 'Does not practice consist precisely in the authentic "modus" of being by the aid of which we discover the true meaning and ultimate essence of being? In other words, is not practice the starting-point thanks to which we are able to discern also the essence of other "inferior" modes of being, as well as the meaning of being "in general"?' [98]

To Soviet Marxists, such a statement seems decidedly lacking in clarity; while recognizing that it reproduces in its own way the initial proposition of Marxist gnoseology, they are apprehensive lest the ontological interpretation of practice as the authentic mode of being should, if taken seriously, introduce confusion into the exposition of the basic truths of Marxism; in their opinion, such a wide interpretation of practice carries with it the danger of obscuring the difference between practice and being, the latter term expressing reality independent of practice and being, the latter term expressing reality independent of practice. The Soviet critics see their fears confirmed in an affirmation by V. Mikecin to the effect that dialectical materialism 'relegates man and his practice to the background, merges him in nature or in nothingness, and is for that reason a species of pantheism'. Hence the need to 'abolish the term historical and dialectical materialism and substitute for it that of philosophy of praxis or Marxist philosophy'.[99]

Soviet critics see in this not just a change of terminology but a substitution of content, and if not a direct continuation, at any rate a reformulation in comparable terms, of the ideas of the early nineteenth-century Social-

* For the developments which follow, T. I. Ojzerman acknowledges the co-operation he has enjoyed from S. Brajović, scientific collaborator of the Institute of Philosophy of the Academy of Sciences of the U.S.S.R.

98. PETROVIĆ, *Filozofija i marksizam* (1968), p. 227.

99. MIKECIN, *Marksisti i Marx* (1968).

Democrat Russian thinker A. Bogdanov (*Revoljucija i filosofija*, 1905). Bogdanov maintained that practice, or organized experience, is the sole reality; therefore, to posit a reality outside practice, or the objective truth of human knowledge, is to revert to idealism, mysticism, theology. In the same way we find one of the Zagreb school, M. Kangrga, writing as follows: 'Dialectic exists only in human history in the form of practice; apart from human practice, there *is* no dialectic, for outside this activity there exists an abstract principle or nothing, "Nature" which by itself is nothing'.[100]

The question, therefore, arises whether a proposition which asserts that, apart from practice, nature is first and last only an abstraction or nothing can still be considered Marxist. Soviet Marxists deny this, notwithstanding the fact that certain non-Marxists, like J. Y. Calvez and P. Bigo,[101] claim that Marx himself adopted a similar line of thought.

What is at stake, in the light of this argument on practice, is the rôle played in Marxist theory by objective truth, that is to say, by a content of knowledge independent of the knowing subject. Are we to regard the postulation of such an objective truth as theology or mysticism, as A. Krešic should contend? 'Whoever – writes Krešic – recognizes the existence of an objective truth totally independent of the knowing subject presupposes God, or in other words, someone in whom this truth is vested'.[102]

Soviet Marxists affirm that this line of thought distorts the basic concepts of Marxist philosophy. Marxists, they say, have never maintained (as opposed to Husserl, for instance) that truth can exist without cognition, independently of cognition. The gnoseology of dialectical materialism draws a distinction between the logical form of truth, which pertains to the knowing subject, and the content of truth, namely the part of reality existing outside of man which, in the present case, is known to man. This faithful reflection of the object of knowledge, in so far as it forms a content of knowledge, independent of knowledge, is precisely what is known as objective truth. Such a content of knowledge may be totally independent of the knowing subject (where we are dealing with phenomena existing without relation to man) but may also depend on man, if we have in mind an object created by man, for instance, a machine.

In short, Soviet Marxists refuse to regard the philosophical research carried out in the review *Praxis* as definitive. As far as they are concerned, it amounts to a re-formulation of views that were long ago discredited or, more damningly, declamatory exercises on philosophical themes. In their opinion, development of the Marxist conception of the relationship between theory and practice calls for a highly detailed study of the special features of contemporary science and contemporary social practice, with their continually changing forms of development.

100. Kangrga, *Etički problem u djelu K. Marksa* (= The Problem of Ethics in the Work of Karl Marx) (1963), p. 171.

101. Cf. Calvez, *La pensée de Karl Marx* (1956, 1970); Bigo, *Marxisme et humanisme* (1961).

102. Krešic, 'Marksistička i apstraktno-materijalistična filozofija' (1960).

The discussion is thus on two fronts, on the one hand in regard to the dialectical relationship between theory and 'praxis', on the other in regard to how this dialectic envisages objective truth.

It is from this dual standpoint that discussion might be fruitful, not only between schools of thought claiming adherence to Marxism, but also between Marxists and philosophers representing other streams of contemporary thought.

SECTION SIX

MAN AND THE FOUNDATION OF HUMANISM

SECTION SIX: MAN AND THE FOUNDATION OF HUMANISM

A. MAN ACCORDING TO EASTERN MODES OF THINKING *

The various conceptions about man are not just different conceptions, some of which may be true and the others false. Rather, each of them has a profound influence, not only on the way man thinks about himself, but also on the way he shapes himself. A concept, in a certain sense, always has more sharply defined boundaries than any actuality that we may ever meet with in experience. This gives rise, in the realm of nature, to the well-known discrepancy between the mathematical precision of the concept and the perceptual entity to which it is supposed to apply, a discrepancy which is bridged by increasing approximation for purposes of measurement and calculation. However, when the concept happens to relate to human reality, the discrepancy is seen in more or less moral terms, and the concept becomes some sort of an ideal shaping the behaviour of man in some degree or other.

The central point, then, would relate to the question as to how man conceives of himself in different traditions. And this obviously would relate to the way man conceives of the nature and rôle of consciousness which he enjoys and the relation between himself and the reality which is an object of his awareness and of which he feels himself, to some extent, to be a part also. In this context, the way the relation is conceived with respect to nature, or rather non-human reality, and the way in which it is conceived with respect to human or social reality, may be significantly different.

Contemporary non-Western thought may, then, be distinguished by the focal importance which it gives to consciousness and the consciously lived and enjoyed experience. This is that *in* which, *through* which and *for* which man lives and has his being. In the words of the Arabic philosopher Osman Amin,[1] 'Philosophy cannot go beyond human consciousness'. And though most of consciousness is made up of what has usually been called 'sense-experience', that is, experience derived from or occasioned by the senses, this is certainly not true of all of it. The positivists, in a certain sense, have seen this, but somehow they have always felt constrained to confine experience to sense-experience alone. No such limitation seems to have been felt by the phenomenologists and the existentialists, and it is in them that one finds subtle articulation of the various shades and nuances of experience as it is felt and lived through in consciousness. Amongst the thinkers oriented towards language, it is the ordinary language philosophers who display this tendency to some extent. But they do so only to the extent they find it reflected in language, and not directly, for they too are caught in what may be called the 'language-centric predicament'.

The centrality of consciousness and lived-through experience which is not confined just to sense-experience gives a certain character to non-Western

* Specially written by Daya Krishna for inclusion in this work.

1. In his contribution to the present study.

thought which may theoretically be characterized, in the words of the Japanese philosopher, Seizo Ohe, as the denial of clear-cut dichotomies, leading to the denial of the law of excluded middle which may be said to be based on it. But it is not only the law of excluded middle which presupposes for its *application* a world in which clear-cut dichotomies prevail. So does the law of contradiction also. But if one gives up both these laws, one gives up thought itself or, at least, one appears to do so. At any rate, thought tends to assume another shape which seems, to thinking based on clear-cut dichotomies, as almost no thinking at all. It is then that the logic of the indefinite takes over. The most fundamental principle on which logic could stand in that case would, in the words of K. C. Bhattacharyya, 'be none other than the bare dualism of the definite and indefinite',[2] it being understood 'that the line between the definite and the indefinite is itself indefinite'.[3]

The fact that there are no clear-cut dichotomies in experience may, however, take one in different directions. One of these, which was very much favoured in the classical times and which still is favoured to some extent by many thinkers, is to be suspicious of all differences and not grant them any ultimate reality in the ontological sense of the term. Or the differences could be seen as integrally related to that which transcended them and which, because of this essential transcendence, gave them their indefinite character. The difference between the two positions may itself be conceived of either in radical terms or just as a matter of degree. The former position is predominantly associated with one of the most significant and influential schools of Indian philosophy, that is, Advaita Vedanta, while the latter may be said to characterize most of Chinese and Japanese thought on this matter. But both the radical school of Indian philosophy and the Oriental philosophies of Japan and China would reject the thinking based on clear-cut dichotomies which may be said to predominantly characterize classical Western thought.

The awareness, then, that all differences are basically provisional in character, just because they are embedded in a unity which envelops and transcends them, may be said to be one of the results of taking the reality of consciousness and lived-through experience as the central reality from which all thought has to take its start. Yet, though no dichotomies in experience happen to be either clear-cut or absolute, certain dualities seem inherent in it, which, when reflected upon, give rise to alternatives which may be and, in fact, have been taken by thought. The first, and perhaps the best known, is the subject-object duality which all experience involves. The second may be characterized as the duality between the actual and the possible, or the actual and the ideal, or even the perceived and the imagined. However characterized, the duality, it should be remembered, is a feature of human experience itself. It, so to say, falls within it and not outside it. If these first two dualities lie primarily in the field of cognition and conation, the third may be said to lie in the realm of feeling. It may be characterized as the

2. Bhattacharyya, K. C., *Studies in Philosophy*, Vol. II (1958), p. 227.
3. *Ibid.*, p. 228.

fulfilment-frustration duality which haunts all lived-through experience each moment of one's life.

The self-conscious awareness of the diverse types of duality, where each infuses the other, gives rise to what K. C. Bhattacharyya has called the concept of alternative absolutes. The notion of the absolute involves the annulment of the duality in all its forms as revealed to self-conscious reflection on experience. But as the duality revealed is itself diverse in character, and as there is a basic indefiniteness characterizing the duality itself, the absolutes annulling the duality have to be diverse and alternative in character. The logic of the indefinite leads in K. C. Bhattacharyya's thought to the logic of alternation – a logic that has been worked out in greater detail by his son Kalidas Bhattacharyya, in his book *Alternative Standpoints in Philosophy* (1953). The duality between subject and object, or, as K. C. Bhattacharyya puts it, between consciousness and content, for example, may be overcome in diverse ways. 'The content may be freed from its reference to consciousness, i.e., from its contenthood. Or consciousness may be freed from its reference to content. Or the implicational relation itself may be freed from its terms as a definite self-subsistent unity'.[4] Thus different directions may be taken in the quest for the annulling of duality, which is the quest for the absolute. But there are *different* absolutes depending upon the direction one takes. Perhaps one should not even say that there are three absolutes. As K. C. Bhattacharyya writes, 'it is meaningless to cognitively assert that there are three absolutes or one absolute. The absolute has, however, to be formulated in this triple way. Each is absolute but what are here understood as *three* are only their verbal symbols, they themselves being understood together but not *as together*'.[5]

The reason why the absolutes involved in conscious experience are not only diverse but *alternative* in character has been elaborated by Kalidas Bhattacharyya in what he calls the logic of alternation. The absolutes apprehended as possibilities for realization when reflecting on self-conscious experience (which all philosophy ultimately is) are by their nature related in such a way that either the realization of one involves the negation of the others or the one that is realized is just indifferent to whether the others are realized or not.[6]

The absolutes, however, in this sense are the same as ideals, and the apprehension of them as involved in experience through philosophic reflection brings to light the duality of the ideal and the actual which also belongs to all experience. The duality, in fact, is apprehended at many specific levels, but it is only philosophic reflection which reveals the possibility of the absolute as annulling the duality altogether. The important point which the two Bhattacharyyas bring to light is that there are, so to say, *alternative* ideals

4. *Op. cit.*, p. 141.
5. *Ibid.*, pp. 141-142, author's italics.
6. See above, Section I, sub-section A, § 5, 'Logic and ontology in contemporary Indian and Japanese thought', pp. 1108-1109, for a somewhat fuller discussion of this point.

which are implicit in experience itself and which philosophic reflection brings into the open along with the fact that they are *alternative* in character. A very interesting view of philosophy is involved in this Bhattacharyya way of looking at things. It is seen as a theoretic reflection on the totality of experience which gives rise to the awareness of certain alternative existential possibilities implicit in the experience itself, but which may be actualized only by a *praxis* which is non-theoretic and non-reflective in nature.

The duality of the ideal and the actual seen in another way reveals the essential value-seeking character of man, which has been emphasized by a number of thinkers. As Devaraja has urged,[7] the moving force of human life may be seen in the awareness of values and their qualitative distinctions – an awareness which constantly urges man on to creative effort for their realization in progressively higher and nobler forms. Science itself, in this perspective, is seen as the seeking of an ideal value rather than as a system of theories more or less reflecting the nature of reality. And if this is so, the alleged conflict between science and values, it has been contended, should be seen as a conflict between different kinds of values rather than between values and something else which is not a value.

Yet if values are diverse and sometimes conflicting in nature, then the differences between men and at least some of the conflicts between them may be seen as due to differences in the values. The differences between societies may be seen as rooted in the diversity of cultures, and the diversity of cultures is merely another name for diversity in the values that men have apprehended and tried to pursue in their lives and embody in their institutions. Both man and society are thus seen in what may be called a value-centric perspective. Consciousness is a value-centric phenomenon, and to see the reality of man as essentially centred in the lived-through experience characterized by self-consciousness is to see it as centred in value-awareness. The human reality, therefore, is not one from which values have been banished and which has been disinfected and sterilized from all value considerations. Rather, values are at the very heart of human reality and constitute its essential core.

The awareness of values, however, involves a dichotomy between the 'is' and the 'ought' at the very centre of human consciousness. In regard to lived-through experience, this dichotomy appears as the duality which we have earlier characterized as 'frustration-fulfilment'. But as no dichotomies are clear-cut, the one between the 'is' and the 'ought' is not clear-cut either.

The duality of fulfilment and frustration that is found in lived-through experience is, in a sense, wider than the duality between the 'is' and the 'ought'. Frustration may be seen as primarily due not to what 'is', but to our attitude towards what 'is'. A transformation in the attitude then becomes the prime desideratum for overcoming the duality experienced as fulfilment-frustration. The predominant Japanese direction in this context may be said to be aesthetic in nature, and many of the contemporary Japanese thinkers

7. DEVARAJA, *The Philosophy of Culture* (1963).

seem to share this approach.[8] The Indian thinkers, on the other hand, seem to favour more the cultivation of a detached attitude resulting in non-involvement not merely with the objects that are experienced but rather, in a deeper sense, with respect to the fact of experiencing itself. The difference between this detachment and the distancing involved in the aesthetic attitude is perhaps difficult to articulate, but it is still there.

The transformation in consciousness, in whatever form and in whatever manner, is then the key to the change in the quality of lived-through experience which lies at the heart of the frustration-fulfilment duality involved in all experience. But as this duality is also the duality of bondage and freedom, or rather is merely another name for it, there has to be at least one limitation on the manner in which transformation may be brought about. Freedom may be achieved only if the transformation in consciousness is produced by an act of consciousness itself. The attempt to transform consciousness through any external means such as psychedelic drugs is, therefore, doomed from the very start as it can only lead to greater bondage rather than freedom of the psyche.

The notion of transformation of consciousness involves the idea that there are levels of consciousness other than the one predominantly enjoyed by man. Also, that reality is a correlate of consciousness, and that the problems of man *cannot* be solved by any external manipulation of circumstances, as they are structurally determined by the type of consciousness that he possesses. Sri Aurobindo [9] is the best exponent of this line of thought in modern India.

The emphasis on consciousness and lived-through experience tends to give an 'ego-centric' perspective to this kind of thought, even if the ego be conceived of as transcendental in character. Much of recent Indian and Japanese thought may be seen as trying to break out of this presumed or alleged 'ego-centricity', without giving up its essential truth and insights. The attempt to bring history, society and nature into this perspective and give it a dynamic transformational turn has been attempted, amongst others, by Tanabe, Risaku and Ohe in Japan (as we noted above),[10] and by Vivekananda,[11] Aurobindo and Gandhi [12] in India at both theoretical and applied levels. At a more technical-philosophical level, thinkers like Radhakrishnan,[13] N. V. Banerjee, N. K. Devaraja [14] and K. S. Murty have been trying to do the same. Amongst these, it is perhaps N. V. Banerjee who has gone farthest in argu-

8. This is an impression gathered from talks with certain leading Japanese thinkers in Kyoto, arranged through the courtesy of the Japanese National Commission for Unesco. But the impression might be mistaken.

9. AUROBINDO, *The Life Divine* (1949); *The Future Evolution of Man. The Divine Life upon Earth* (new edn., 1963).

10. Cf. *supra*, Section III, *Man and Social Reality*, sub-section B, § 1, (a), 'Philosophy of law', pp. 1278-1285.

11. VIVEKANANDA, *Complete Works* (1955 edn.).

12. GANDHI, *Collected Works* (1960 edn.).

13. RADHAKRISHNAN, *An Idealist View of Life* (1932).

14. DEVARAJA, *The Philosophy of Culture* (1963).

ing for the societal nature of the self even at the transcendental level. What knowledge and, perhaps, all experience presupposes for him is not the solitary *cogito* à la Descartes, but 'I *with* others', that is, *we*. It is this Realm of the Personal which is disclosed as the presupposition of experience by a transcendental reflection on it, and the basic cleavage that is revealed is between this presupposition and the day-to-day behaviour which reflects the centrality of the ego or the 'I' and its insulation from others.[15]

Another interesting Indian thinker in this connection is M. N. Roy, who developed the concept of Radical Humanism in the later part of his life. Starting as a Marxist, he was the organizer and leader of a political party. Yet, within the ambit of a materialistic metaphysics, he developed a humanistic philosophy which rejected all types of collectivistic creeds, and emphasized the reality and value of the individual human being. This he could do because he regarded human reason as a continuation of reason in nature, which he described as its 'law-governed character'. Rising against the background of law-governed nature, man himself was essentially rational, and Roy tried to derive his humanistic ethics from this essential rationality of man.

Both Banerjee and Roy reject the ideas of God or soul, as well as what usually go along with these ideas, that is, absolutism and mysticism. In this they are different from the other Eastern thinkers mentioned above, who all accept the idea of transcendence in some form or other and, in fact, argue that one could not have a viable Humanism without it.

B. MAN ACCORDING TO WESTERN MODES OF THINKING

§ 1. *The analytic conception of man and humanism, and the concept of mind**

Analytic philosophy contributes only indirectly to the debate on humanism. In general it contains little discussion of a strictly philosophical nature on the concept of humanism. On the other hand it does bring to this debate a preliminary clarification of what is meant by the concept of mind.

In fact, for reasons which no doubt go back to Descartes, the philosophy of man is known, in the Anglo-American tradition, as the philosophy of mind. Philosophy of mind stands between epistemology and ethics: the part of it which concerns the cognitive powers of the intellect and the senses overlaps with philosophy of knowledge, and the part of it which concerns the will, the emotions and the springs of action has close links with moral philosophy.

In what follows an attempt will be made to bring together the topics already analysed in earlier sections of this study and situate them in relation to this discussion of man and humanism.

15. BANERJEE, *Language, Meaning and Persons* (1963).
* Based very closely on the text of a synopsis by Anthony Kenny.

Epistemology has always been the foremost concern of philosophers in the British empiricist tradition, and there are still vigorous and influential thinkers who stand squarely in that tradition and devote their major attention to epistemology. The most typical representative of this school is A. J. Ayer, whose *Problem of Knowledge* was published in 1956. The best-known work of J. L. Austin, *Sense and Sensibilia* (1962), is a posthumously published set of lectures subjecting the bases of empiricist epistemology to radical criticism centring on the theory of sense-data.

However, epistemology has not been the principal concern of analytic philosophers of mind during the last three decades. Epistemological questions are not at all prominent in the two most influential works of the period, Wittgenstein's *Philosophical Investigations* (1953) and Ryle's *The Concept of Mind* (1949). This is not to say that these works do not contain material of interest to epistemologists; but they approach traditional topics from an angle unfamiliar to empiricist epistemology. Instead of asking how man acquires knowledge of the external world from private and subjective sense-data, these authors take statements of ordinary language about the world as basic, and inquire within this context how meaning can be given to the ascription of thoughts, sensations and mental states to human beings.

This emphasis on ordinary language is highly significant for the discussion of humanism. To take statements of ordinary language as basic is to accept the supposition that such language is the depository of the most appropriate expressions, as if, in the struggle among competing forms of discourse, only the best-formulated, best-adapted had survived; it also implies that it is in ordinary language that man's humanity is supremely evident. Herein lies the indirect contribution of analytic philosophy to the discussion of humanism.

Ryle and Wittgenstein are cases in point. Both were anti-Cartesian, but Ryle went further in the direction of behaviourism than Wittgenstein. While Ryle attempted, not always consistently, to reduce statements about mental events and states to statements about actual or hypothetical bodily performances, Wittgenstein adopted a subtler account of the relation between ascriptions of mental and bodily predicates. Mental phenomena, he believed, were neither reducible to their bodily expressions, nor connected with them in a merely contingent manner. Concepts of mental states could not be formed unless the connection between each such state and its possible expression in behaviour was understood: bodily behaviour of a certain kind was thus a criterion (i.e., direct and non-inductive evidence) for the occurrence of the appropriate mental state. The concept of pain, for instance, was tied to the normal causes and occasions of pain, and to the expression of pain both in involuntary symptoms and in motivated avoidance-behaviour. The heart of the *Philosophical Investigations* is the argument against private ostensive definition: the argument to show that a word like 'pain' cannot be given meaning merely by attention to one's own experience, but has significance only in the context of the public circumstances, in regard to cause and expression, in which it is in fact used. The interpretation of the *Philosophical*

Investigations continues to give rise to extensive debate.[16]

This also poses in more acute terms the problem of the criterion as to what is mental. In both the Cartesian and the empiricist traditions the line between mind and matter had been drawn in terms of private experience: the mental is whatever we are immediately aware of in ourselves. The effectiveness of this criterion had already been called in question by the theories of Freud; but Wittgenstein's rejection of any kind of epistemological privacy called for a complete restatement of the problem. There was thus a revival of interest in the notion of intentionality as a criterion for mentality; borrowed from Brentano and developed at length by the phenomenologist movement, this notion was given an analytic guise by R. M. Chisholm in his book *Perceiving* (1957). The study of intentionality is understood by analytic philosophers as the search for a purely logical criterion which will fit all psychological statements, and them only. One such suggested criterion is that of 'referential opacity' (to use a term of the logician W. v. O. Quine): if P is a sentence containing A, and Q is a sentence like P except that it contains B where P contains A, then P is intentional if P and A = B do not together imply Q. Thus the work on intentionality provides an overlap between the philosophy of mind and the philosophy of logic which we considered in Section I.

Another overlapping area between these two disciplines is epistemic logic, in which the most noteworthy contribution has been J. Hintikka's book *Knowledge and Belief* (1962). This treats the topics of its title not in the epistemological manner of inquiring about the justification of belief and the grounds of knowledge, but rather investigates the logical consequences of ascribing to a subject, a belief or claim to knowledge, whether such ascription is justified or not. A similar approach to the concept of belief is combined, in an informal manner, with traditional epistemological concerns in H. H. Price's *Belief* (1969).

But the philosophy of man (and indirectly, the discussion of humanism) possibly stands to gain more from the theory of action, dealt with in Section V of this study. Here we shall do no more than bring together the results of these investigations in so far as they concern the general theory of man. Here too the impetus derives initially from Ryle and Wittgenstein. In one of the most successful chapters of *The Concept of Mind* Ryle attacked the view that human actions are caused by indefinable but introspectible acts of the will. In his *Philosophical Investigations* Wittgenstein drew attention to the differences between the way in which human actions are explained by reasons and the way in which bodily movements are explained in terms of causes. J. L. Austin, in 'A plea for excuses' (1956), called on philosophers grappling with the problem of the freedom of human action to abandon the search for a characteristic peculiar to human actions, and to follow Aristotle in investigating such significant contrasts as those between freedom and duress, freedom and accident or freedom and mistake. The reader is referred

16. A full-length discussion, with a bibliography of over 200 titles, is given in SAUNDERS & HENZE, *The Private-Language Problem* (1967).

to the account we have given above [17] of Miss Anscombe's *Intention* (1957) and Stuart Hampshire's *Thought and Action* (1959) and *Freedom of the Individual* (1965). Reference should also be made in this connection to the exploration of the notions of responsibility and voluntariness in legal thinking; we shall cite only the work of H. L. A. Hart (already mentioned above),[18] several of whose papers on morals and law were collected in 1968 under the title *Punishment and Responsibility*.

A common feature of most of these analytic studies is their insistence on the special features involved in the explanation of human action, in more or less explicit reaction against behaviourist psychology, which tended to assimilate such explanation to the pattern afforded by the behaviour of inanimate objects. Here we shall do no more than refer the reader to the series of books entitled 'Studies in Philosophical Psychology', edited by R. F. Holland: using different arguments, R. S. Peters,[19] P. T. Geach,[20] A. I. Melden [21] and A. Kenny [22] highlight the differences between the methods used in interpersonal relationships for assigning reasons for actions and those used in scientific research into the causes of behaviour. These authors are thus led to criticize some of the presuppositions of current psychological research in the light of Wittgenstein's insights. Here Charles Taylor's *The Explanation of Behaviour* (1964) occupies a special position, in that it combines a defence of teleological explanation with a detailed critique of the attempts made to replace teleological concepts by concepts of the stimulus-response type.

The discussion of action theory has to some extent diverted attention from the kindred but more traditional problem of freedom in relation to determinism. In this connection Austin Farrer's *Freedom of the Will* (1968) was an original work, standing apart from the mainstream of analytic philosophy. Various signs can be seen of a reawakening of interest in the freedom of the will, M. R. Ayers' *The Refutation of Determinism* (1968) being highly symptomatic in this respect. In his *Four Essays on Liberty* (1969) Sir Isaiah Berlin traces the connections between freedom conceived as a political value and freedom as a prerequisite for genuinely human action. Freedom of the will is, indeed, a topic more familiar to moral philosophers than to philosophers of mind. This traditional problem has however been approached from a new angle, that of the theory of artificial intelligence, which is itself linked to the development of computers, and has already been referred to in an earlier part of this study.[23] For example, *Minds and Machines*, edited by A. R.

17. See above, Section V, *Man and Action*, sub-section B, § 1, 'Analytic philosophy and the problems of human action', pp. 1392-1403.

18. See above, Section III, *Man and Social Reality*, sub-section B, § 1 (a), 'The philosophy of law', pp. 1278-1285.

19. *The Concept of Motivation* (1958).

20. *Mental Acts* (1958).

21. *Free Action* (1961).

22. *Action, Emotion and Will* (1963).

23. See above, Section II, *Man and Natural Reality*, sub-section B, § 3, 'The

Anderson (1964), contains a number of papers by different authors which instil fresh life into the traditional problem of freedom and determinism.

Alongside freedom, immortality was in the past a traditional topic of the philosophy of man; in the current analytic tradition, if the theory of freedom has been to a greater or lesser extent submerged by the theory of action, immortality has almost completely disappeared as a subject of serious philosophical discussion. The decline of religious belief in England has been an extra-philosophical cause of this loss of interest; from the philosophical point of view, a pronounced distrust of almost all forms of dualism has had the same effect. Brief discussions of the topic are however found in A. Flew's *Body, Mind and Death* (1964) and P. T. Geach's *God and the Soul* (1969). But though immortality has found few defenders, very few philosophers have been prepared to go as far in the direction of materialism as a number of Australian philosophers, whose most vociferous spokesman is J. J. C. Smart. In his *Philosophy and Scientific Realism* (1963) Smart argued that mental events, when fully understood, would be seen to be identical with corresponding physiological processes in the brain; not, however, *necessarily* identical, since it is for science, not semantics, to reveal the identity, but *contingently* identical, in the same way as lightning is known to be identical with certain types of electrical discharge. This argument has been vigorously criticized and as vigorously defended.[24]

Many philosophers have come to the conclusion that it is unfruitful to investigate the relation between mind and body as a relation between two separate phenomena.

In *Individuals* (1959) P. F. Strawson argued that the primitive notion was rather the concept of a person, i.e., of an entity with both physical and mental characteristics, an entity to which both predicates ascribing corporeal characteristics and predicates ascribing mental characteristics were applicable. Among philosophers favouring this approach there has been discussion of the criterion of identity for an individual human person. Some have argued that memory provides this criterion, others, like S. Shoemaker in *Self-Knowledge and Self-Identity* (1963), have stressed the importance of the bodily criteria of spatio-temporal continuity. Recently David Wiggins, in *Identity and Spatio-Temporal Continuity* (1967), has argued that if the two criteria of memory and bodily continuity are properly understood they cannot conflict with each other.

One of the best-explored areas of philosophy of mind in recent years has been that of feeling, taken in its widest sense to include bodily sensations (such as those of heat and cold, hunger and thirst), emotions (such as fear, anger and surprise), desire and pleasure. Pleasure, in particular, a concept often applied but rarely examined in itself, at least in the empiricist and

levels of reality', pp. 1183-1192; and Section III, *Man and Social Reality*, subsection A, § 2, 'The logic of explanation in psychology', (b) 'Physicalism old and new', pp. 1244-1247.

24. A review of the early stages of the debate will be found in J. A. SHAFFER's paper 'Recent work on the mind-body problem' (1965).

utilitarian traditions, has been repeatedly analysed from the time of Ryle's 1954 paper entitled 'Pleasure' down to the publication of Gosling's book *Pleasure and Desire* in 1969. In this as in many other departments of the philosophy of mind there has been a revival of interest in the work of Aristotle, which had been left on one side during the period when the influence of Locke and Descartes was paramount.

However, if the main keynote of the work of the analytic school has been reaction against the Cartesian heritage, one of the least popular doctrines of Descartes, that of innate ideas, is now acquiring new respectability. Many philosophers have been interested in the work of the linguist Noam Chomsky who, in his *Cartesian Linguistics* (1966) and *Language and Mind* (1968), has explicitly defended the seventeenth-century rationalist tradition as the best framework for the understanding of man conceived as a language-user. Not all Chomsky's readers are convinced that the innate structures of mind postulated by his linguistic theory have more than the name in common with Descartes' innate ideas. None the less, philosophers who are commonly accorded the description 'linguistic' can hardly ignore the theory of mind presented by the most philosophical of linguists. It is too early to say what long-term influence Chomsky will have on the analytic philosophy of mind: but at least it is heartening to see a phenomenon which has been almost without precedent in the last decades, of a writer whose work is taken seriously by both analytic philosophers of mind and experimental psychologists working in the field.

§ 2. *Marxism and the discussion of humanism**

(a) Marxism claims to be a form of humanism. However, it parts company from the outset from any kind of philosophical anthropology which situates the unity of man or of all men above and beyond social differences. Thus each emergence of humanism – as a theory, a trend, an ideology – in the forefront of the historical scene is viewed by Marxists as linked to a crisis in the existing society. Christian humanism for instance emerged during the period when the institution of slavery was breaking up, and bourgeois humanism at the time of the crisis of feudalism; in the same way, the development of the contradictions of capitalist society and the awareness of its historically transient character contributed to the formation of Marxist or proletarian humanism.

The rejection of any anthropological foundation for social inequality is already expressed in the slogan 'Workers of the world unite', which Marx and Engels proposed to replace 'All men are brothers', the slogan of the 'Union of the Just', at that time influenced by Utopian socialism. The founders of Marxism did not thereby intend to revert to Hobbes's formula *homo homini lupus*, which is merely the inverse of this same anthropological viewpoint. What is expressed in the new slogan is the negation of the anthropological

* Text taken from T. I. Ojzerman's contribution to this study.

foundation to social inequality and the need to seek a non-anthropological, or specific social foundation; it follows that, however essential the anthropological study of man, it does not provide a key to the understanding of the history of mankind or the position of the individual in society.

It was for the same reason that Marx broke with Feuerbach, who placed the unity of all men above social differences. This anthropological materialism is deemed historically progressive in the struggle against feudal ideology (and also today, in the struggle against racialism, nationalism and other anti-humanist ideas), but has the basic flaw of ignoring the decisive significance of the social situation of men and their adherence to particular social classes and groups.

Thus the most characteristic feature of Marxist humanism is the introduction of the historical class viewpoint into the humanist conception of the world: in this way Marxist humanism regards itself as the dialectical negation of bourgeois humanism with its naturalistic, anthropological and naively republican illusions. At the same time it rejects only the bourgeois limitations to this humanism, not its humanist inspiration or its historically progressive content.

It is therefore necessary to stop talking about humanism as if it were a basically immutable conception linked to the meaning of the term 'humanist'. This essentialist analysis of humanism must be replaced by a typology of humanism – religious and atheistic, conservative and revolutionary, abstract and concrete, real and verbal. The differences between these types of humanism are at least as essential as the features they have in common. For contemporary Marxism, there are today three conflicting types of humanism – bourgeois, 'petit bourgeois' and Marxist. Thus ideas continue to conflict even within the mainstream of humanism.

The opposition to bourgeois humanism is obvious: Marxism sees it as no more than meliorist fine words claiming to put an end to exploitation within capitalism. The opposition to 'petit bourgeois' humanism is more complicated, since it too claims to represent a protest against the social institution of capitalism and an attack on the depersonalization of man; but since it sees in socialization of the means of production merely an extension, indeed the logical conclusion, of a basic trend of capitalism itself, its negative attitude to collectivism leads it to seek a third or intermediate path. For Marxists, the anti-capitalism of 'petit bourgeois' humanism is fictitious; the opposition between the personal and the social, the individual and the collective blinds it to the basic differences between various types of socialization. It is for this reason that in Marxist eyes the waverings of 'petit bourgeois' ideology make it a deliberate or involuntary accomplice of anti-communism.

In this dispute with non-Marxist humanism, Marxist humanism must deal with a basic objection, which is that Marxism overlooks the individual human being, and considers man unilaterally as identified with society and losing his unique nature, his unique significance to himself. According to Marxism, this objection rests on a premiss to which it is resolutely opposed, namely that the individual existence of man differs in principle from his social existence, which is always determined by external factors. In its contention such objections, based as they are on the assertion that individual existence is independent of social differences, and that man has a single, unique essence transcending accidental social differences, constitute an erroneous interpretation of the dialectical materialist view of the unity of the individual and society. Marxists readily agree that the emotional life of the individual cannot be reduced to a phenomenon of class-consciousness; the immediacy of feeling is not an economic, political or ideological fact; precisely because it is opposed to idealism, Marxism recognizes the existence of particular individuals differing from each other and with manifold individual destinies. But for Marxism the differences, oppositions and contradictions are much more essential than the natural kinship between individuals belonging to different classes. Briefly, what distinguishes Marxism from its opponents in this respect is the *class standpoint* of proletarian humanism.

In Marxist eyes this standpoint has nothing in common with that reduction of the individual to the social ascribed to it by its critics. To sociological reductionism Marxism opposes a genetic conception linked to a theory of development. Reductionism is in its view another form of essentialism: sociologism is indeed unable to explain why or how it comes about that a group of differing phenomena possessing common features reveal a particular essence. To answer these questions we must turn back from the essence to the individual phenomena, and seek to discover in the laws governing their development the degree of essentiality inherent in them. The essence of man, says Marx, is not a vague abstraction inherent in a particular individual, but an aggregate of historically determined social relationships. This very important definition of the essence of man must not be confused with the assertion that man is a social being; the latter assertion relates to all the manifestations of human life, whereas Marx's statement relates only to the essence of man, which does not exhaust all the manifold aspects of his activity as a living being. What is more, this definition does not even exhaust his essence which, like the essence of any concrete thing, is a unity of differing determinations, in principle irreducible to any single definition.

It is thus a caricature of the Marxist definition of the social essence of man to identify a given individual with the aggregate of social relationships which constitute him. Marx, trained on Hegelian dialectics, never equated the essence of man with individual human existence; it is an elementary dialectical truth that phenomena are richer than their essence. From the point of view of dialectical materialism, the concept of essence characterizes the bond, the mutual interaction between different phenomena in the same class, their identity, itself comprising essential differences. There is thus progressive enrichment when we add to the idea of essential humanity first, the idea of social essence, then that of the essentiality of the individual. In fact, the unity of mankind from the anthropological point of view refers as yet only to the unity of the anthropological essence of mankind. But the historical aggregate of social relationships characterizes in a more concrete manner the essence of man as the social essence of mankind. Lastly, human individuals, unlike trees and all other objects which cannot be described as having different essences, are characterized by the fact that their individuality constitutes one of the basic determinations of their essence. The essentiality of the individual as a human being is manifested not only by his relative independence, which is anthropologically, sociologically and historically conditioned, but also by the subjective content of his personal life, thanks to which things that are individual, unique, even inessential for society as a whole, may frequently be seen to play an essential rôle in a man's personal life. The whole sphere of a man's more or less intimate inner life is in part made up independently of his consciousness or volition (for example no one chooses his parents), in part results from his relative freedom of choice, relative inasmuch as restricted by particular circumstances and impulses. Reciprocally, personal life should also be considered from the historical point of view: what changes is not only the conditions, content, forms and direction, but also the level of personal life, its intensity, its standards, everything that goes to make its wealth. In other words, personal life, in all its inexhaustible subjectivity, is also social, but this in no way prevents it from remaining strictly personal.

(b) The theoretical discussion of humanism is complicated by a discussion of a more historical character regarding the unity of Marx's work and its classification by periods. It is in fact in his *Economic and Philosophical Manuscripts of 1844* that we find the following declaration: 'Man is a particular individual, and it is precisely this particularity that makes him, as an individual, a truly individual social being'.[25] Some authors, Marxist and non-Marxist alike, object that when the young Marx wrote these lines he was not yet a Marxist. Does Marx therefore have to be opposed to Marx? [26]

For those who argue in favour of Marx's progressive development, there

25. MARX-ENGELS, *Iz rannih proizvedenij* (= Early Works) (Moscow, 1956), p. 591.

26. A basic systematic study of the historical genesis of Marxism will be found in the four published volumes of A. CORNU's *Karl Marx et Friedrich Engels. Leur vie et leur œuvre* (1954-55 et seq.). The reader is also referred to T. Ojzerman's *Formirovanie filosofii marksizma* (= The Genesis of Marxist Philosophy) (1962) (German translation, *Die Entstehung der marxistischen Philosophie*, 1965). Ojzerman groups the works of the young Marx into four periods, the first comprising the works between 1839 and 1841, in which Marx shares the idealist views of the Young Hegelians, while distinguishing himself from them chiefly by a more firmly expressed political radicalism and atheism. In the second period, from 1842 to the beginning of 1843 (articles in the *Rheinische Zeitung*), there are already signs of the transition to materialism and communism, and the beginnings of a break with the Young Hegelians. The third period covers the second half of 1843 and the first half of 1844, during which Marx went over to dialectical and historical materialism and scientific communism, though retaining a number of propositions which he was soon to abandon. The fourth and last period begins with the *The Holy Family* and ends with the appearance of the first works of mature Marxism, *The Poverty of Philosophy* and the *Communist Manifesto*. In the works of this period it is possible to trace the theoretical reasoning behind the basic premisses of Marxist philosophy.

is no scientific sense in opposing his early works to those of his maturity, since what is most valuable, original and inspired in the early works consists in the foreshadowing of the major discoveries lying ahead. At the same time it is recognized that this foreshadowing can only be correctly appraised on the basis of the views advanced in *Das Kapital* and the other works of maturity. Without *Das Kapital*, Marxist parties and the institution of socialism, the *Economic and Philosophical Manuscripts of 1844* would have been of interest only to specialists. The ideological controversy provoked by the publication of these manuscripts from the 1930s onward is from this point of view revealing. Those who have taken up the cudgels in their defence have, in very many cases, been those holding anti-Marxist or what are regarded as revisionist views (S. Landshut and I. Maier, H. De Man, H. Marcuse).

Althusser's position deserves separate consideration: for him, the arguments in Marx's early writings are not so much the seeds of his later scientific beliefs as an obstacle to a scientific view of Marxism, in as far as they remain influenced by Hegel's essentialism and Feuerbach's abstract anthropology. In the scientific view, capitalism is no longer held to be a deformation of human nature, but a transitional historical stage in man's evolution. This scientific viewpoint excludes not only the philosophical humanism which we find in the young Marx associated with vestiges of Hegelianism and Feuerbachism, but theoretical humanism in general: 'From the strictly theoretical point of view, it is then possible and necessary to speak explicitly of Marx's *theoretical anti-humanism*, and to see in this the necessary condition for the absolute (negative) possibility of the (positive) knowledge of the world of humanity as such, and for changing it in practice.' (*translation*) [27] Althusser does not in fact deny the necessity of Marxist humanism: without such an ideology it is impossible for any society or social movement to exist. What he rejects is the possibility of a scientific ideology and with it that of a scientific humanist view of the world: 'Marx's rejection of theoretical humanism recognizes the need for humanism as an *ideology*, relating humanism to the conditions under which it exists.' (*translation*) [28] Theoretical humanism is a particular speculative, ideological conception; but in dealing a fatal blow to speculative philosophizing, Marx opposed to essentially non-scientific ideological views a sociological science in which there can be no room for theoretical humanism. Marxism is humanist only as a practical political and social movement, as a revolutionary *praxis*, renewing and transforming the world: it is in no way humanist as a theory.

T. Ojzerman and other Soviet Marxists strongly criticized this interpreta-

27. 'Sous le rapport strict de la théorie, on peut et on doit alors parler ouvertement d'un *anti-humanisme théorique de Marx*, et voir dans cet *anti-humanisme théorique* la condition de la possibilité absolue (négative) et de la connaissance (positive) du monde humain lui-même, et de sa transformation pratique'. – ALTHUSSER, 'Marxisme et humanisme', p. 10 in *La nouvelle Critique*, No. 164 (1965) (p. 236 in *Pour Marx*, 1965).

28. 'L'anti-humanisme théorique de Marx reconnaît, en le mettant en rapport avec ses conditions d'existence, une nécessité à l'humanisme comme *idéologie*.' – *Ibid.*, p. 11 (*Pour Marx*, p. 237).

tion of the relationship between socialist humanism and the scientific view of the world created by Marxism. What they object to is the breach between a strictly scientific system, leaving no room for theoretical humanism, and the ideology to which all forms of humanism are referred. It is true that Marx rejected as non-scientific and sentimental the criticism of capitalism on the ground that it was incompatible with humanity, and in this awareness of the immoral character of capitalism saw merely a subjective reflection of the fact that beneath the surface of society there gradually come into being the material conditions for the transition to a new social régime. Nevertheless, this criticism of abstract humanism did not prevent Marx from calling his theory a 'real humanism'.

For the critics of Althusser, Marxism is a scientific humanism, and scientific humanism is not to be understood as a particular division or segment of Marxist doctrine, as are, for example, the theories of surplus-value and the class struggle. Scientific humanism constitutes one of the basic aspects of Marxist doctrine, just like its revolutionary spirit, which obviously cannot be reduced to the fact that this doctrine comprises a theory of revolution. Inasmuch as Marxism is a scientific theory, Marxist humanism, based as it is on the whole content of Marxism, is naturally invested with a scientific character.[29]

In the language of this scientific humanism, Marxist theory appears as a life-affirming vision of the world, a historical optimism. But this optimism has nothing in common with a specious hypocritical pseudo-optimism taking solace in illusions and turning a blind eye to reality, nor with a humanism unconcerned with the sufferings and struggles of mankind. A historical optimism, it is based on a scientific study of the development of the objective conditions which alone can put an end to the exploitation of man and social inequality. It is a historical optimism in that the forces which create the history of man are human forces, even if they are not the forces of each of the generations taking part in the making of history. The objective reality which exists outside and independently of historical awareness thus appears as the boundless field for the conscious purposive activity of man. It is true that there exists no all-powerful nature, no all-powerful humanity, no omnipotence of any kind; but there is a potentially all-powerful humanity, as also a potentially all-powerful nature. And these potential infinities are never fully converted into reality. The road to the infinite remains infinite in each of its broken fragments. Not that there are unknowable 'things-in-themselves': the whole history of science and *praxis* shows on the contrary that 'things-in-themselves' are changed by man into 'things-for-us', thanks to which 'necessity-in-itself' becomes 'necessity-for-us'. This explains why, for Marxist hu-

29. Cf. the discussion of humanism, for and against Althusser's views, carried on in several numbers of *La nouvelle Critique* in 1965: articles by SEMPRUN and COHEN (No. 164), SIMON (No. 165), MACHEREY (No. 166), NAVARRI, BROSSARD and VERRET (No. 168), and GOLDBERG (No. 170). See also SÈVE, *Marxisme et théorie de la personnalité* (1969) and the postface to the 1972 edition.

manism, there are no limits to human free will, which is also the conscious purposive activity of man, based on the knowledge of objective necessity.

(c) *The question of alienation* provides an excellent touchstone for the entire materialistic critique of the cult of abstract man and the reassessment of humanist traditions by Marxism. For non-Marxists, alienation is a feature of essential human reality, and as such something given once and for all. Man is alienated in an alienated world, where he becomes a perpetual question-mark to himself. For Marxists, the topic verges on the rhetorical as soon as one ignores the qualitative differences between alienation in the different classes of bourgeois society. We read in *The Holy Family* that 'The propertied class and the proletariat represent the same human alienation. But the former feels at home in this alienation; it finds itself strengthened, recognizes in this self-alienation *its own power*, and possesses through it the *appearance* of human existence, whereas the latter feels itself annihilated, and sees in this alienation its powerlessness and the reality of an inhuman existence.' (*translation*)[30]

Thus the analysis of class differences within the alienation phenomenon puts an end to all speculation about alienation in general, and reveals its true content, which cannot be separated from basic socio-economic contradictions. To begin with, it is the concept of alienated labour which gives alienation a concrete historical content, as well as indicating how it can be finally eliminated. For Marx, the emergence of alienation is linked to the non-development of man's 'essential forces', that is to say an insufficient level of development of the forces of production, where the satisfaction of the needs of one part of the population is only possible at the cost of exploiting the other part, on the basis of the appropriation by private individuals of the means of production, and the constitution of a class society. It is thus economic alienation, as alienation of both the product and the actual activity of labour, which constitutes the basic alienation. This becomes the basis for the formation of alienated social relationships and the spiritual expression of alienation in religion, philosophy and ideology in general. The alienated consciousness then becomes a reflection of alienated being, and alienated political relations are a consequence of economic alienation. At the time of the *Economic and Philosophical Manuscripts of 1844* the concept of alienation formed a pivotal point on which Marx hung all manner of other oppositions, such as that between intellectual and physical labour, town and country, and so on – all the antagonistic contradictions of bourgeois society. But as the laws governing the appearance of these antagonistic contradictions became the main purpose of his study, the question of alienation lost its original significance for Marx; in this more limited form the concept of alienation became merely one of the concepts of historical materialism, alongside basic concepts such as the forces of production, production relations, the classes and the State. In *The German Ideology* Marx and Engels

30. MARX-ENGELS, *Die heilige Familie* (1845).

had already criticized the Young Hegelians and the pre-socialists for unduly extending the concept of alienation in such a way as to rob it of its real content and prevent a full investigation of social antagonisms. In their later works the founders of Marxism describe alienation as the transformation of the results of human activity into uncontrolled social forces dominating men, as the enslavement of man by social relationships formed at random. And in *Das Kapital* Marx identified as forms of the enslavement of the human personality the anarchy of capitalist production, crises of over-production, the subordination of personal needs to the law of value, and the transformation of workers into mere appendages of the machine.[31]

The Marxist concept of alienation gave rise to lively controversy both in and outside Marxism, particularly after the publication of the *Economic and Philosophical Manuscripts of 1844*. Previously, commentators on Hegel and Feuerbach had viewed this concept purely as a metaphorical description of the subject-object dialectic. But Marx's concept of alienation stood apart from the dialectical process in general, and appeared to be irreducible to Hegel's and Feuerbach's concepts of alienation. It was for this reason that all attempts to interpret Marxist theory as transposing the speculative conception of alienation into terms of political economy and sociology were bound to fail. It is not possible to consider Marx's doctrine as purely philosophical and speculative, or to interpret his economic and historical investigations simply as illustrations of a speculative conception, without repudiating the organic link between his work and the historical process of class struggle. In attacking the speculative interpretation of the Marxist conception of alienation, the aim of Marxist investigators has been to show that the specific social phenomenon of alienation is radically different from the objectivation of human activity – which is inseparable from production – and presupposes the existence of antagonistic social relationships or their survival under conditions of socialism.

The discussion of humanism and alienation is not confined to an exchange between Marxists and non-Marxists: it takes place within Marxism as well. Some investigators (A. Schaff, K. Kosik, N. Prouha, G. Petrovich) have put forward the view that alienation, unlike class conflicts, exploitation and the subjugation of personality, is a non-transient social phenomenon, resulting from the objective consequences of the conscious activity of men, to the extent that these consequences are independent of their volition or are to a certain degree of a spontaneous character. From this point of view, alienation is due not only to causes in respect of classes but also to certain non-economic, specifically anthropological and technological causes. In that case, it follows that alienation, assuming different historical forms, remains alienation to the extent that it basically represents a relationship which partakes of the nature of on the one hand a subject-object relationship and on the other

31. For a systematic study of the Marxist doctrine of alienation see *Marksistička teorija otudenja* (= The Marxist Theory of Alienation) by the Yugoslav philosopher D. Laković (1964); also *Trud i svoboda* (= Work and Freedom) by the Soviet philosopher Ju. Davydov (1966).

an interpersonal relationship – a relationship which thus, in the last analysis, expresses the contradictory character of practical activity, regardless of its historical forms. For orthodox Marxists, this conception does not clearly distinguish the concepts of alienation and of contradiction in social life, and ignores the difference between antagonistic and non-antagonistic contradiction; the Marxist interpretation of alienation is thus reduced to that of Hegel or Feuerbach, and loses both its specificity and its philosophical significance.

We can see what is at stake in the discussion: what is the relationship between alienation and the objectivation of human activity? Is alienation a historically transient phenomenon? For orthodox Marxists, while there may be room within socialism for carry-overs from the alienation peculiar to capitalism, there is no specific form of alienation peculiar to the socialist system.[32] Such vestiges of alienation differ from alienation proper in that they are bound to wither away as the socialist system develops and changes into a classless society.

But within socialist systems themselves there are practical tasks undertaken in the spirit of humanism (socialist legislation, the organization of work, social insurance, the health service, etc.) which entail fresh socio-anthropological correlations and these in turn have their impact on the philosophical discussion; in this connection some Marxists have found it necessary to elaborate the theory of a special philosophical discipline, philosophical anthropology, as a new constituent part of the philosophy of Marxism. According to the Romanian philosopher Gulian, there are to be found 'in all Marx's writings, starting with the early works preceding the *Communist Manifesto* and right up to *Das Kapital*, [...] in an undeveloped form, the premisses of a modern philosophical anthropology.' (*translation*) [33] To quote Gulian again: 'An objective enquiry into the nature and scope of the discipline now known as philosophical anthropology shows us that Marx's conception contains the fundamental premisses for an all-embracing conception of man, based primarily on a scientific foundation which enables it to achieve two objectives at once: 1) to subject man to a multi-dimensional dialectical analysis; 2) to represent in all its truth a turning-point in the history of the development of personality.' (*translation*) [34] But while the criticism of Feuerbach's anthropology, or any kind of anthropological basis to humanism, may be valid, attempts to create a Marxist philosophical anthropology must appear theoretically inconsistent, whatever the part played by an anthropological description of the human individual in the Marxist account of humanism. Such attempts can at most lay claim to some justification as re-

32. See the articles by GAYLORD, C. LEROY, T. APTHEKER, S. FINKELSTEIN, Howard L. PARSONS, Miklos ALMASI, T. I. OIZERMAN in the collection *Marxism and Alienation*, edited by H. APTHEKER, 1965. See also, in the Democratic Republic of Germany, the collective work *Entfremdung und Humanität* (1964) and the book by A. KURELLA, *Der Mensch als Schöpfer sich selbst* (1963).

33. GULIAN, 'Humanisme et anthropologie philosophique chez Marx', p. 39 in *Proceedings of the XIVth International Congress of Philosophy, Vienna, 1968*, Vol. II (1968) (Colloquium I, 'Marx and contemporary philosophy').

34. *Ibid.*, p. 34.

action against the failure of a number of relatively recent Marxist studies to attach due weight to the anthropological aspect. Fundamentally, however, philosophical anthropology, as a special philosophical discipline with its own particular subject of investigation, is, in the eyes of most Marxist authors, incompatible with a materialist view of history which explains the anthropological – that is to say the specifically human form of biological organization (in contrast to all other forms) – on the basis of its own premisses, which are in no way anthropological. The most extreme representatives of philosophical anthropology (e.g., P. Vranicki, of Yugoslavia) are for their part led to replace dialectical materialism with a 'philosophy of man', or a 'philosophy of praxis', the term used by the Czechoslovak philosopher M. Prouha; the critique of the 'scientistic' trend in Marxist philosophy leads to the statement that philosophy cannot be scientific, that its specific character consists in its lying beyond scientific and non-scientific conceptions of the world and that, like art, it elaborates its own world vision.[35] To most Marxists this can only seem like a return to philosophical individualism and, ultimately, to a mystical interpretation of man.

The traditional tenets of Marxist humanism have been called in question from yet another angle; the struggle for peace, and in particular the development of the policy of peaceful coexistence stemming from the danger of a thermonuclear holocaust, might lead to the view that the interests of mankind as a whole should now be placed before the demands of the class struggle.[36]

The Soviet philosophers P. N. Fedoseev and M. V. Mitin [37] reply that the thermonuclear threat only exists by virtue of the basic contradiction of the present time, namely that between imperialist and anti-imperialist forces. It is thus not possible to campaign for the prevention of atomic war and at the same time close one's eyes to the opposition between socialism and capital-

35. VRANICKI, 'Die Notwendigkeit verschiedener Varianten in der marxistischen Philosophie – Thesen', pp. 139-141 in *Proceedings of the XIVth International Congress of Philosophy*, Vol. II (1969) (Colloquium I, 'Marx and contemporary philosophy').

36. At the XIVth International Congress of Philosophy ('Marxism and contemporary humanism', pp. 116-121 in Vol. II (1968), Colloquium I, 'Marx and contemporary philosophy') the American Marxist John SOMERVILLE argued that 'From the point of view of man as man, all human mistakes are remediable save those that lead to thermonuclear war. This means that for the first time since Marx died there is in the world a problem more important to man and more important to Marxists than the transition from capitalism to communism, a problem the solution of which is an absolute precondition to the transition.' (*Loc. cit.*, pp. 120-121.) SOMERVILLE expounds his view in greater detail in his book *The Philosophy of Peace* (1949). Cf. also the paper by Howard PARSONS, *The Young Marx and the Young Generation* (1968), which contains a similar argument and a discussion of the problems raised by war and peace; and the same author's recent work *Humanism and Marx's Thought* (1971).

37. From a more general point of view, see MITIN's contribution to the XIVth International Congress of Philosophy, 'Marxist philosophy of real humanisms and its significance in our time', pp. 80-86 in Vol. II of the *Proceedings* (1968), Colloquium I, 'Marx and contemporary philosophy'.

ism. For this reason the majority of Marxist philosophers consider a dialogue possible between opposing ideologies, but refuse to base this dialogue on a return to abstract humanism. In their view, such a dialogue cannot possibly preclude ideological struggle; rather is it itself a specific form of such struggle.

§ 3. *The discussion of humanism in 'Continental' philosophy*

Over the last fifteen years or so – chiefly in France – there has been a dramatic change of course in 'Continental' philosophy that directly affects the concept of *humanism* on which we have focused this final Section of the present work. That concept has become the target for attacks which converge on it from different directions but are all stamped by *theoretical antihumanism.*

To understand this sudden change in the situation – bewildering not only to a British or American reader but to a Soviet Marxist imbued with the notion of scientific Marxist humanism – we have to remember to what type of philosophy this now ignominious label 'humanist' is attached. Therefore the first part of what follows will deal, not so much with the kind of humanism professed as with the kind denounced.

(a) The label 'humanist' then is attached to any philosophy of the subject deriving from the Cartesian *cogito* or from Kantian *transcendental* philosophy.

The term is applied, first, to the stream of neo-Kantianism which up to the First World War had often held pride of place in the philosophy faculties in France and Germany. To call this humanist is not wrong if by that we mean that the subject to whom the consideration of constituted forms of knowledge leads is the same as the responsible subject, on whom all ethical and political thought alike is based. There is nothing against using the word humanist, in a purely descriptive sense, for any philosophical viewpoint which has the carrier of language and knowledge, the moral agent and the citizen, coincide in the same subject. From this point of view neo-Kantianism had been the very epitome of humanism. On the one hand, it saw objectivity in the forms of knowledge as the work of a grounding subjectivity which, on the basis of its categorial structures, constitutes a world of objects and relates them one to another. On the other hand, it saw moral and political life as centred in the ethical individual, the acknowledged *fons et origo* of his own deeds and works just as the transcendental subject is the acknowledged source of the operations whereby the objective world is constituted. It is not mere chance that in Germany Cassirer, the author of *Philosophie der symbolischen Formen* (1923-29) [38] – therefore an investigator of the sym-

38. English translation, *Philosophy of Symbolic Forms* (1953-57); see also *Sprache und Mythos* (1925) (English translation, *Language and Myth*, 1946).

bolizing effect of the subject's action in language, myth and science – should also have written a book on *The Myth of the State* (1946), thrown himself eagerly into delineating Renaissance Man and Man of the Enlightenment, and produced *An Essay on Man* (published in 1945). The unity of Cassirer's work thus testifies to the unity of the subjective principle as it underlies both the symbolic function, on the theoretical plane, and the ethical function, on the practical and political plane. During the same period, in France the work of Lachelier's and Lagneau's disciples bore witness to this same 'transcendental' unity underlying theoretical subjectivity and the ethical, political individual. For the generation prior to the existentialist wave Léon Brunschvicg and Alain were the heralds of a philosophy which in the theoretical sphere appeals from things to mental operations and subordinates concepts to judgements, and therefore to the judging subject, whilst in the practical sphere it resuscitates the responsible subject, master of himself vis-à-vis customs, creeds, powers and authorities of every sort. The idea of the thinking subject thus leads on quite smoothly, with no discontinuity, to that of the 'citizen against the authorities'.[39]

Akin to transcendentalism, as a vehicle for humanism, stands the *reflective* philosophy of Jean Nabert.[40] For anyone not engaged in polemics, however, a great difference is evident. Where neo-Kantian philosophers are prompt to identify subjectivity and category, subjectivity and form, Nabert is intent on laying bare the acts that lie buried not only in the works of man but also in the objects of the external world, in bringing them to light from behind the signs through which they become known to other subjects as well as to one's own 'I'; and in order to recapture the act we must from the outset recognize that there is more than one 'focus of reflection' (notably, a theoretical and a practical focus) whereas transcendentalism lost no time about lumping them all together. Among the fields thus opened up for reflection, the field of human *action* and its structures is in the main what reflective philosophy puts first. In the view of a reflective philosopher, Kantian philosophy had not really reflected on the specific characteristics of the practical order, in its anxiety to find in that order the same distinctions between *a priori* and empirical, transcendental and varying, as prevail in the theoretical sphere. This attention to the structures peculiar to human action, as they can be observed directly, by examining action, without any cross-borrowing or transposition between the theoretical and the practical, gives Jean Nabert's philosophy its specifically ethical tone. Here ethical does not mean moral, exactly, if by moral philosophy we denote study of the norms that govern action, by analogy with the categories that govern knowledge. Ethics is reflection on the path actually followed by the consciousness in

39. Alain, *Le citoyen contre les pouvoirs* (1926).

40. See in particular Nabert, *Eléments pour une éthique* (1943, 1962) (English translation, *Elements for an Ethic*, 1969); *Essai sur le mal* (1955); *Le désir de Dieu* (publ. posth., 1966); and, for an earlier work, *L'expérience intérieure de la liberté* (1924).

search of the desire to be and the effort to exist which form its foundation. The starting-point of the journey is some poignant experience like wrongdoing, solitude, failure, in which the consciousness takes the measure of the distance between what it aims at and what it is in fact, between the originative affirmation which constituted it and the condition of fact which holds it captive. This recourse to the originative affirmation is one more mark of the distance between reflective philosophy and transcendental philosophy: this absolute affirmation of the absolute is much closer to the Plotinian One, Spinozan substance or, better, Fichte's 'thetic' act than to any empirical human consciousness. In this sense, man is not the measure of all things. However, the originative affirmation is not the final destination on this reflective journey. Rather is it the culminating point after which reflection turns and comes back down towards the values, works, institutions in which the consciousness attests the seriousness of its conversion. In this ascending-descending movement, all that reflective philosophy cares about is maintaining, as it comes to grips with the problems successively posed by existence, the difference in level and in tension between principle and actualization. In this respect, one fundamental experience is crucial: the experience of evil. Evil is not, as in Kant, the subversion of the norm in the maxim, the dereliction of duty. It expresses the unbridgeable gap separating oneself from oneself. It is another name for the 'impure causality' which is manifest both in the schism of individual consciousnesses – war, injustice, failure to recognize the other person – and in engulfment in the objectifications of everyday life, of technical operations, of social and political activities. Does this philosophy provide the foundation for a humanism? Yes, if we continue to apply the term 'humanism' to a philosophy which, while recognizing more than one focus of reflection, discerns their underlying unity within the 'consciousness'. No, if we say that humanism is a philosophy which places within man himself the origin of the affirmation that makes him man. Study of Jean Nabert's reflective philosophy gives rise to serious reservations as to the philosophical use of the word humanism. Either it is a vague rhetorical concept covering every attitude, theoretical or practical, that results in the 'defence of man', or it is a polemical concept defined by its adversary on the basis of positions he himself describes as anti-humanistic, or again it is a descriptive concept defined by the very persons who lay claim to it – thus in 1945 Sartre was saying 'Existentialism is a form of humanism'; before him Maritain had been presenting neo-Thomism as an 'integral humanism'; and Marxists, at least in the Soviet Union, accept the epithet humanist for their interpretation of Marxism. A reflective philosophy placing the origin of consciousness in an affirmation which transcends consciousness without being numerically distinct from it does not readily apply to itself the name of humanism. Rather it tends to stand aloof from any arrogant affirmation of man as 'measure of all things' and maintain that the self-assertion of man in his clear, sovereign consciousness is a secondary, derived power, which philosophy will always tend to subordinate to an affirmation that constitutes consciousness yet is independent of it.

Another type of philosophy which stands on the far side of the line drawn by anti-humanism is *phenomenology*. Less ethical in character than Nabert's philosophy, it is like neo-Kantian philosophy in some features, unlike it in others. What makes it akin to neo-Kantianism is, obviously, the transcendental theme and, even more basically, the central rôle of the *cogito* in grounding the forms of knowledge. What differentiates it is its recourse to the intuition of essences, which enables it to carry on its constitutive task in a descriptive style. But above all, when it claims to be a philosophy of the subject, in the case of phenomenology this means something entirely new. At the time of his *Logische Untersuchungen* (1900-1901),[41] the anti-psychologism of nascent phenomenology had led Husserl to make of his logical structures 'truths in themselves', lest they be reduced to the mental operations of an empirical subject. The 'objects of thought' then became the correlates of an intentional consciousness which 'intends' them *qua* units – 'unities' – of 'sense' in a multiplicity of acts. The intentional consciousness thus became the transcendental field in which all thinkable objects may be constituted. It is true that this consciousness was not necessarily an ego. But in the idealistic phase of phenomenology, from *Ideen* (1913)[42] to *Méditations cartésiennes* (1929, published 1931),[43] it seemed to Husserl that the relation between act and object – the intentional relation *par excellence* – required as source, as point of origin, as pole of unity and identity, a subject capable of gathering itself together reflectively as an ego. This 'egology'[44] was reinforced by the consideration of time, in the sense in which time is implied by constitution of the thematic unity of objects amidst their fluctuating outward appearances. In a manner differing from what we find in Kant, the unity of the 'I think' was restored to its basic position underlying all operations in which we can discern the constitution of a 'unity of sense' in a diversity of appearances. But Husserl's 'I think' is no longer the simple unifying function of an impersonal 'I', as seems to be the case in the Kantian critique, but the singular 'I' which the solipsist hypothesis isolates in a final movement of reflective ascesis. It is to such a subject that the grounding process (*mouvement fondationnel*) which moves from formal logic back to transcendental logic refers. For if judgements – not concepts – are the final instance to which the transcendental grounding of logic refers, then a judgement is still an act by someone, and subjectivity cannot be anonymous. As the fifth *Cartesian Meditation* shows, the problem of groundedness merges with that of solipsism. It is in so far as 'I' implies another 'I' that the world of the perceived and the thought can be, as Heraclitus said, the 'common world' and not the 'private world' of the dreamer or the madman. Thus it is that the transcendental grounding of objectivity requires not one subject – still less subjectivity in

41. English translation, *Logical Investigations* (1970).
42. English translation, *Ideas* (1931).
43. English translation, *Cartesian Meditations* (1960).
44. Cf. Broekman, *Phänomenologie und Egologie. Faktisches und transzendentales Ego bei Edmund Husserl* (1963).

general – but a network of subjects; in short a plural, inter-subjective, monadological foundation. Can this theory of grounding be called humanistic? Yes, if we apply the term 'humanism' to any resort to subjectivity in order to explain the conditions of objectivity; no, if we characterize humanism by the primacy of the human. For what phenomenology claims to do – and this is not the least of the obstacles to a proper understanding of what it aims at – is to attain to the transcendental point of view by a 'reduction' of the natural attitude, according to which everything that is there exists by its own force, and forces itself on us in a non-problematic manner. Now what in ordinary language and everyday life I call 'I, such and such a man' also comes in for this 'reduction'. In this sense, confusion between transcendental subjectivity and the empirical 'I' is debarred, even in the most 'egological' reading of the field within which objects are constituted. It is not denied that the parallelism of transcendental subject and empirical 'I' represents a very awkward difficulty and could be seen as a pointless reduplication or even as the confession of a disguised return to 'transcendental psychologism'. But, if we take phenomenology according to its avowed purpose, it does not propose to provide a philosophical justification for the claims of the 'human'. One can even imagine its being assailed for ignoring the 'human condition' or 'class viewpoint'! What it is specifically concerned with is a problem of foundation or ground. Its initial question is whether logical objects can have 'unity of sense', and its whole setting of the problem lies in the passage from the formal to the transcendental. No text of Husserl's shows this better than his *Formale und transzendentale Logik* (1929).[45] If phenomenology has humanistic implications, then, it is as something added – and perhaps against its grain, if we attach any importance to Husserl's distrust of 'philosophy of life' and questions of 'wisdom' [46] and if we continue to interpret phenomenology on the basis of his never denied concern for a 'scientific' approach.

But the phenomenological current is not to be identified with Husserl only. In Germany itself, at the beginning of the century, the Munich School had resisted the idealist interpretation of phenomenology and been less categorical than Husserl in opposing transcendental to psychological consciousness; thinkers of this school were even more anxious to regenerate psychology on the basis of the 'intentional' theme than to provide a foundation or ground for logic and science in general.

In this way we have seen a 'philosophical anthropology', that is a philosophy of man, drawing on phenomenology. With this anthropological stream we can associate Max Scheler, whose anti-formalist philosophy of values [47] culminates in the person, as the focus of ethical acts. But this does not mean that the person creates values, for all that it is the person which picks its

45. English translation, *Formal and Transcendental Logic* (1970).

46. Cf. the essay 'Die Philosophie als strenge Wissenschaft', first published in *Logos* in 1911 (English translation, 'Philosophy as rigorous science', 1965).

47. Cf. in particular *Der Formalismus in der Ethik* (1913, 1916); *Wesen und Formen der Sympathie* (1913, 1923) (English translation, *Nature of Sympathy*, 1970); *Die Stellung des Menschen im Kosmos* (1928) (English translation, *Man's Place in Nature*, 1963).

way among the different levels and orders of values, and integrates them. Their disposition in relation to one another is the work of an *ordo amoris* that is not dependent on individual discretion. Personalism is thus coupled with a moral illuminism which prevents man being made the measure of good and evil. Paul Landsberg,[48] emphasizing the personalist element in Max Scheler's work, enriched his philosophy and gave it a turn that brought it close to the personalism of Emmanuel Mounier, in France.[49] Mounier, who was first and foremost alive to the crisis of civilization, the exhaustion of bourgeois culture, the rise of varying forms of fascism, the danger of technology and technocracy, all the tragedy inherent in subjectivity deprived of hope, conceived of personalism as the philosophy of man's recurrent protest against being reduced to the level of ideas and things. So it is not exactly a reflective philosophy like Nabert's nor, like Husserl's, a philosophy specifically concerned with the problem of foundation or ground, but a 'fighting' philosophy that has never had time – save possibly in the *Traité du caractère* (1946) [50] – to discriminate between practice and theory. In this sense the changed emphasis of personalism, from Scheler to Landsberg and Mounier, can be interpreted as a move in the direction of 'humanism'. However, this is outweighed by the fact that for Landsberg and Mounier, the person is defined far less by the ability to posit oneself, by opposing oneself to oneself, than by receptiveness to other persons and openness to an order of values which is itself defined by its power to integrate persons in a community of persons. Moreover the sense of the hierarchy of values which these two thinkers have in common with Max Scheler, and which recalls Pascal's discussion of the three orders – of bodies, of minds, of charity – inclines them to see in the order of charity an order higher than the order of justice, in which free spirits mutually restrict each other in their presumptuous pride rather than participate in a being in common. That is why personalism hesitates to call itself humanism, except for saying that the end of social, political, community existence itself is the fulfilment of man in each individual person, not the reign of a collective entity ontologically distinct from persons and their destiny.

Phenomenology was to undergo an even more marked change of course, as a result of Husserl's descriptive method being fused with Heidegger's existential analytics and the existential reinterpretations of Kierkegaard, Nietzsche, the young Hegel, even the young Marx. Here, the works of Karl Jaspers [51] and Gabriel Marcel [52] form an essential link between the idealistic

48. Cf. 'L'acte philosophique de Scheler' (1936-37) and two posthumous works, *Essai sur l'expérience de la mort* (1951) and *Problèmes du personnalisme* (1952).

49. Cf. especially *La révolution personnaliste et communautaire* (1935); *Manifeste au service du personnalisme* (1936); *Qu'est-ce que le personnalisme?* (1946) (English translation, *Personalism*, 1970).

50. English translation, *Character of Man* (1956/1957).

51. Cf. especially *Philosophie* (1932) (English translation, *Philosophy*, 1970); *Vernunft und Existenz* (1935) (English translation, *Reason and Existence*, 1956); *Von der Wahrheit* (1948) (English translation, *Truth and Symbol*, 1962); *Vernunft und Widervernunft in unserer Zeit* (1950) (English translation, *Reason and Anti-*

interpretation of phenomenology by its founder and its existential reinterpretation in France. Karl Jaspers sharply differentiated 'the existing', which philosophy attempts to clarify (*Existenzerhellung*), from 'consciousness in general', on which the categorical structure of objectivity is built. He thus severed 'the exception' from 'generality' and destroyed the tranquil unity of the neo-Kantian subject. In the same way, by opposing existence to objectivity, Gabriel Marcel opposed the concrete person to the Cartesian and Kantian *cogito*, 'guardian of the threshhold of the valid'. By the same token, meditation on the person could no longer continue within the same conceptual horizons as the reflection on objective knowledge. The philosophy of existence required a new language in which notions like incarnation, having and being, despair and hope, faithfulness and overtness or 'availability', are intelligible at the level of a living experience which it is the function of the new language to awaken.

It was in this atmosphere, profoundly influenced from various directions, that French existentialism developed and flourished in the immediate post-war years. Sartre's *L'être et le néant* (1943) [53] epitomizes the fusion of phenomenology, philosophy of existence and humanism. Human freedom posits itself absolutely vis-à-vis things, vis-à-vis beings, the plenum of which is the first obstacle to its limitless power to nihilate. Today this philosophy, dramatically illustrated by plays and novels, is the chosen target for anti-humanism – as if it were the final, fully developed expression of a philosophy of the subject. But its renown cannot blind us to the radically different formulation of the problem of foundation or ground presented by Husserl in *Formale und transzendentale Logik* or, consequently, the polysemic, not to say homonymous character of the term 'philosophy of the subject'. Besides, Sartre's more recent works, the harsh judgement which he today pronounces on his erstwhile existentialism, now labelled ideology, the incorporation of his former point of view in a philosophy which frankly claims to be Marxist – this whole long forceful evolution of Sartre's thought must keep us from restricting the discussion of humanism to a consideration of what happened to the existentialism which came to birth in the immediate post-war period.

There is another good and sufficient reason for not making this unwarranted and purely polemical identification of humanism with existentialism. Alongside Sartre, the work of Merleau-Ponty indicated another way in which phenomenology could be developed in the direction of a philosophy of existence. And here the question we are interested in just now, consciousness and the subject, was put in different terms. While the questions of significance

---

Reason in Our Time, 1952); 'Über meine Philosophie' (1951, 1969).

52. Cf. especially *Journal métaphysique* (1927) (English translation, *Metaphysical Journal*, 1967; *Etre et avoir* (1935) (English translation, *Being and Having. An Existential Diary*, 1965); *Du refus à l'invocation* (1940) (English translation, *Creative Fidelity*, 1964); *Homo Viator* (1945) (English translation, *Homo Viator. Introduction to a Metaphysic of Hope*, 1962); *L'homme problématique* (1955) (English translation, *Problematic Man*, 1967); 'La dominante existentielle dans mon œuvre' (1969).

53. English translation, *Being and Nothingness* (1957).

and meaning – which are done away with by a behaviourist psychology and a positivist philosophy, but which are forever brought up anew by actual perceptual experience – are still referred to a consciousness, a subject for whom 'there is meaning', that consciousness is no longer defined by the ability to reflect oneself, to recover oneself in the transparency of presence to self. It is rather something accomplished, a form of work – as in Freud we have the 'work' of dreaming – that goes on in the anonymity of a prereflected existence. That is why Merleau-Ponty's last philosophy, renouncing the vocabulary of subjectivity and the primacy of the constitutive consciousness, attaches special importance to the more difficult language of the 'flesh' and the 'savage being', in order to say what is not freely constituted by any consciousness.

Thus the philosophy of existence, which sprang from the philosophy of constitutive subjectivity, turns against it the very themes it had helped elaborate: the themes of the unreflecting, of passive synthesis, of corporeity, of temporality, etc. What was so to say the obverse of a philosophy of consciousness becomes the main focus of attention of another kind of philosophy, one which is not yet sure of its new name, but for which consciousness, from having been a certainty, has become an enigma.

This development, with philosophy of the subject turning against itself, especially in Merleau-Ponty's *Le visible et l'invisible* (published posthumously in 1963) but even in Sartre's *Critique de la raison dialectique* (1960), has been seen by the next generation as the telltale symptom of something deeper, as betokening on the level of philosophical reflectiveness the ebb and flow of the humanist theme in the bottommost depths of our culture.

(b) If it is hard to speak of a coherent 'humanist' front – we have suggested that the term may be not so much polysemic as homonymous – it is equally risky to speak of an 'anti-humanist' one.

The anti-humanist is philosophically interesting only if, not content with denouncing moralistic platitudes and the hypocrisy of edifying speeches, he joins issue with attempts to 'found', in other words to establish the philosophical ground or foundations of humanism, these coinciding, at least in 'continental' philosophy, with philosophies of the subject. It is therefore in regard to this problem of 'founding' (*fondement*) that theoretical anti-humanism can be said to take shape.

But by this term *fondement* one can mean a good many different things. We shall look at four possible meanings, to see what implications they have for the quarrel over humanism. In each case, what is in question is the primacy of consciousness in an undertaking directed towards providing foundations or grounds.

(i) The first meaning of the term we shall consider is the one we described at the very beginning of this study under the heading 'Foundational problems'.[54] We are coming back to it now from the point of view of philosophi-

54. Cf. Section I, sub-section A, § 3, pp. 1100-1105, in the preceding chapter.

cal attempts to 'found' humanism. From this point of view, criticism is well represented, in France, by Jean Cavaillès, who opposes to the philosophy of judgement, as it appears in Kant, Brunschvicg and Husserl, the philosophy of *concept*. Cavaillès's basic objection to grounding truth on the subjective ability to make various representations cohere together is this: how can the extraordinary capacity for conceptual development which characterizes both the demonstrative system of logic and science itself, as knowledge based on demonstrations, be derived from a power identical to itself? Are we to turn to some intuitive matter outside this formal power of thought? But how is the empty form of thinking to be linked intelligibly with a matter which is by definition foreign to the consciousness, incongruous with order, and formless in itself? For Cavaillès, as for Brentano, the problem posed by an entirely demonstrative, that is, logical science is to 'apprehend this principle in its generating movement, to identify this structure not by description but apodictically, through its self-development and self-demonstration' (*translation*).[55] Now phenomenology is of no help here, for we have no means of referring logical laws to a transcendental consciousness defined by its deliberate acts, its intentionality. This subjectivity has in fact to be normated (*normée*). But in that case we have one of two things: either this norm must depend on a higher subjectivity, or it must belong to an absolute logic which derives its authority from itself alone and which therefore cannot be transcendental. It is the whole of reflective philosophy, the whole philosophy of consciousness, that Cavaillès calls into question in considering the production of logical beings. 'With the phenomenological method and point of view', he wrote, subjective logic 'is confined to analysing acts and intentions which constitute transcendental subjectivity, that is to unweaving tangled webs of subjective elementary motivations and actions, without any interrogation of the logical entity itself. Clearly that entity cannot be interrogated, since no consciousness is witness of the production of its content by an act, since phenomenological analysis can never do anything but move in the world of acts or, for the corresponding *noemata*, discern conceptual patterns, but in either case will come to a halt before the simple elements – that is, the realities of consciousness which do not refer back to anything else – and, what is more, will even have to use fundamental notions to describe them.' (*translation*)[56] In other

55. '[....] d'appréhender ce principe dans son mouvement générateur, de retrouver cette structure non par description, mais apodictiquement en tant qu'elle se déroule et se démontre elle-même.' – CAVAILLÈS, *Sur la logique et la théorie de la science* (published posthumously in 1947), p. 25.

56. 'Avec la méthode et le point de vue phénoménologique, [la logique subjective] se borne à analyser actes et intentions constitutifs de la subjectivité transcendantale, c'est-à-dire à décomposer les enchevêtrements de motivations et d'actions élémentaires subjectives sans que l'entité logique elle-même soit interrogée. Il est évident qu'elle ne peut l'être, puisqu'aucune conscience n'est témoin de la production de son contenu par un acte, puisque l'analyse phénoménologique ne pourra jamais que se mouvoir dans le monde des actes ou, pour les noèmes correspondants, dissocier des architectures de contenus, mais dans les deux cas s'arrêtera devant les éléments simples, c'est-à-dire les réalités de conscience qui ne renvoient à rien

words, phenomenology can at the utmost display the chain it has linked together and, confronted with some new invention, testify to the impossibility, in actual experience of judging otherwise. But if consciousness is not origin, must we not question the idea itself and seek, in its very progress, the reason for the idea's perpetual movement of exploring and expunging its own contents? Far more, must we not admit that the consciousness itself is furthered by the idea's working on itself? That is what is suggested by the aphorism with which the book concludes: 'It is not a philosophy of consciousness, but a philosophy of concept, that can give a doctrine of science. The generating necessity is not that of an activity but of a dialectic.' (*translation*) [57]

(ii) While Cavaillès opposed philosophy of the concept to philosophy of the judgement, dialectic process to constitutive activity, *Heidegger* posed the question of 'founding' in radically different, though polemically convergent, terms. A truly fundamental enquiry cannot accept as radical the subject-object relation, which was at the nub of the problem as posed by all theories of knowledge at the time of the *Methodenstreit*, and to which every subjective 'founding' of objectivity referred. But what is a fundamental enquiry? It is one which revives the now-forgotten question of the meaning of being. The question of the *cogito*, which Kant repeated four times over (what can I know? what must I do? what may I hope for? what is man?), is demoted from first rank by the question as to what being means. Of course this question is first perceived in the act of constituting the being that we are, viz., of constituting a being that already comprehends being before articulating its meaning conceptually. But this being that we are is not, at first, the subject conscious of its own experiences; it is not the *cogito* of Descartes, Kant and Husserl. That is why Heidegger calls it *Dasein* – 'being there' – and not *Ich* – 'I'; the level at which it belongs to being is lower than the level where the knowing subject confronts the object of thought. We have spoken elsewhere of the part played by the analysis of *Dasein* in 'founding' language [58] and reality [59]; here we shall confine ourselves to its implications for the discussion of humanism.

This question is first raised in the essay 'Die Zeit des Weltbildes', dating from 1938.[60] In that essay Heidegger said that the *Cogito sum* is not an extra-temporal or presuppositionless statement. It was made at a certain period of time, the period when science itself was emerging as a mode of intelligibility in which the *Seiende* – '*what-is*' – is made available to the representation that

---

d'autre, et, bien plus, pour les qualifier devra utiliser même les notions fondamentales.' – *Ibid.*, p. 75.

57. 'Ce n'est pas une philosophie de la conscience, mais une philosophie du concept, qui peut donner une doctrine de la science. La nécessité génératrice n'est pas celle d'une activité, mais d'une dialectique.' – *Ibid.*, p. 78.

58. Cf. above, Section IV, *Man and Language*, sub-section B, § 5, 'The "hermeneutic" current: the interpretation of language', pp. 1359-1364.

59. Cf. above, Section II, *Man and Natural Reality*, sub-section B, § 7, 'Phenomenology and ontology', pp. 1214-1222.

60. Reprinted in *Holzwege* (1950), pp. 69-104 (English translation, in a journal, 'The age of the world view', 1951).

we form. The first presupposition, then, is that of the process of objectivation and representation through which we claim to master reality in an experience of certitude. 'The acquisition of knowledge, in so far as it is research, calls *what-is* to account, as to how far it can be made available for representation. [...] The possibility of scientific knowledge arises only when the being of *what -is* is sought in such an objectivity.' (*translation*)[61] It is in this experience of the objective representation, as sure and certain, that we become a subject. However the 'subject' was not always ourselves; at first it was the 'substrate', that which lies beneath, that which gathers all things unto itself. With Descartes, man becomes the first and real subject, the foundation or ground – and at the same time the centre to which the existent – *what-is* – as such is referred. But this has only been possible because the world itself has become a *Bild*, a picture, an image, displayed before our eyes: 'The world's becoming an image is one and the same occurrence as man's becoming a subject in the midst of *what-is*.' (*translation*) [62] This being so, all contemporary questions about the primacy of the individual or the community rest upon an anthropological basis: 'It is only because, and in so far as, man has become in essence a subject that he comes up against the explicit question: How does he wish to be, how must he be – as the "I", surrendered to the limitations of its own arbitrary caprice, or as the *we* of society? [...] Only where man is already essentially a subject is there the possibility of slipping aside into the "dis-essence" of subjectivism in the sense of individualism.' (*translation*) [63] So for Heidegger two processes intertwine. The more the world becomes an image, the more man asserts himself as subject: 'The more extensively and thoroughly the world is available as something conquered, and the more objectively the object appears, then the more subjectively, that is assertively, the subject rises up, and the more irresistibly reflection on the world, theory of the world, turns into a theory of man, into anthropology. It is no wonder that humanism is in the ascendant only where the world becomes an image.' (*translation*) [64]

Even more directly, Heidegger's *Letter on Humanism (Brief über den*

61. 'Das Erkennen als Forschung zieht das Seiende zur Rechenschaft darüber, wie es und wie weit es dem Vorstellen verfügbar zu machen ist. [...] Zur Wissenschaft als Forschung kommt es erst, wenn das Sein des Seienden in solcher Gegenständlichkeit gesucht wird.' – *Holzwege*, p. 80.

62. 'Dass die Welt zum Bild wird, ist ein und derselbe Vorgang mit dem, dass der Mensch innerhalb des Seienden zum Subjectum wird'. – *Ibid.*, p. 85.

63. 'Nur weil und insofern der Mensch überhaupt und wesentlich zum Subjekt geworden ist, muss es in der Folge für ihn zu der ausdrücklichen Frage kommen, ob der Mensch als das auf seine Beliebigkeit beschränkte und in seine Willkür losgelassene Ich oder als das Wir der Gesellschaft [...] sein will und muss [...]. Nur wo der Mensch wesenhaft schon Subjekt ist, besteht die Möglichkeit des Ausgleitens in das Unwesen des Subjektivismus im Sinne des Individualismus.' – *Loc. cit.*

64. 'Je umfassender nämlich und durchgreifender die Welt als eroberte zur Verfügung steht, je objektiver das Objekt erscheint, umso subjektiver, d.h. vordringlicher, erhebt sich das Subjectum, umso unaufhaltsamer wandelt sich die Welt-Betrachtung und Welt-Lehre zu einer Lehre vom Menschen, zur Anthropologie. Kein Wunder ist, dass erst dort, wo die Welt zum Bild wird, der Humanismus heraufkommt.' – *Ibid.*, pp. 85-86.

'Humanismus', 1946, published in 1947, 1949) condemns as un-'founded' any philosophy that stops at the human entity – 'what is' – and does not carry the movement of revelation and dissimulation back into being itself. While it is true that the subject is negation, man only negates, as he only errs, on the basis of a nothingness that precedes him: 'The *Nicht* – "not" – can never derive from negation, from saying "no". Every "no" is merely an affirmation of "not". [. . .] It is within being itself that the essence of nihilation – of "not"-ting – stands revealed.' (*translation*) [65] In this sense man is not free, or does not set himself free, unless seized by the liberty that has, from all time, preceded him. The movement of this freedom is history, which before being the history of men is the history of being, the scene of its eras, its expansion and withdrawal, the stage where man's fate, especially Western man's, is played out. Just as nothingness precedes the negation we pronounce or promote, just as freedom takes hold of us before we entertain any idea of being free, so time comes over us before any temporalizing consciousness measures and orders the time of events. Therefore *logos* and *physis* express more adequately the antecedence of being than *subject, consciousness* or *man.*

Let us be clear about Heidegger's anti-humanism. What he 'dismantles' (*déconstruit*) is obviously not respect for man as the most valuable of beings, but the metaphysics of the subject on which some thinkers would have this ethics of respect hinge. That 'Who am I?' is an authentic question results from the very nature of the initial question: interrogation about being implicates and arouses the questioner that we are. The circular movement from questioned being to questioning being involves a repetition, a reformulation of the 'I am' which Descartes deduced from the 'I think'; in point of fact, the 'I am' lay hidden in the 'I think' set up as sovereign subject. But 'I am' is no longer a proposition; it remains a question for itself. It is because the meaning of the 'self' is hidden that the question 'Who am I?' is buried, at first, in the appearances of the indefinite pronoun 'one', in the pretentions of immediate self-knowledge, and even in the illusions of the reflective consciousness. And so the analysis of *Dasein* – 'being-there' – keeps on putting the question 'Who is it who is-there?' in a spirit of watchful mistrust: 'Does it seem *a priori* "evident" that access to *Dasein* must consist in a purely speculative reflection by the "I" as pole of the acts it poses? And what if it should turn out that in "giving itself" in this way *Dasein* was playing a trick on the existential analytic, what is more a trick rooted in the being of *Dasein*? [. . .] What if, in taking the "I" as given for its starting-point, the existential analytic were falling into a trap which *Dasein* itself has laid for it in the form of a falsely evident and falsely immediate interpretation of itself?' [66] This doubt

65. 'Das "Nicht" entspringt keinesfalls aus dem Nein-sagen der Negation. Alles Nein ist nur die Bejahung des Nicht. [. . . .] Das Nichten im Sein selbst west. . . .' – HEIDEGGER, *Brief über den 'Humanismus'* (pp. 152/153 in the bilingual German-French edition of 1957, entitled *Lettre sur l'humanisme*). For German reactions to Heidegger's presentation see *inter al.*: LÖWITH, *Heidegger, Denker in dürftiger Zeit* (1953); and W. MARX, *Heidegger und die Tradition* (1961).

66. *Sein und Zeit* (1927) (English translation, *Being and Time*, 1967), para. 25.

shows that the answer to the question: '*Who* is it that is-there?' can have no evidential value but only hermeneutic value which will itself depend on the interpretation of the relationship of the one who is-there to the world and to others.

In this way Heidegger dismantles humanism, not in order to destroy the basis on which ethics and politics rest but to 'found' them more deeply, more surely, in non-anthropological soil.

(iii) There is a third way in which the anthropological basis of humanism is assailed in contemporary philosophy – not, this time, by way of a theory of the foundations of science or a radical ontology, but as regards the *models of intelligibility* current in the human sciences.

At the beginning of the century, Dilthey still thought it possible to oppose to the explanatory models characteristic of the natural sciences a model proper to the human sciences, connoting understanding.[67] What we understand first and by principle, he thought, is the psychic life of other persons as expressed in outward signs. Consequently the *Verstehen* – 'understanding' – involved in history, sociology or philology can only be an extension of or a derivation from this first understanding, the effectiveness of which we all of us verify in ordinary everyday exchange of language. Thus the *Verstehen* was doubly subjective, since it went from a subject to a subject.

Now, with the brilliant developments in linguistics, psychoanalysis and structural anthropology, another model of intelligibility has come to the fore in the human sciences. For this model, *understanding* is no longer opposed to *explaining* and grasp of human facts no longer depends on consciousness of oneself or of others. This is the *semiological model* which is now so widely used by the various forms of philosophical structuralism.

The postulates of structural linguistics and their general application to all sign systems other than language have already been set forth above; [68] so here we shall do no more than note the influence of this model on the discussion of humanism. If the semiological model is a challenge to philosophy of the subject, it is to the extent that the problem of meaning or significance is seen from an entirely different point of view than that of a subject's intentional aims. Let us briefly recall the postulates of structural linguistics, to note their concatenation: First postulate, the dichotomy of language and speech. Second postulate, the subordination of the diachronic to the synchronic point of view. Third postulate, the reduction of the substantive aspects (the phonic and semantic substance) of language to the formal aspects. Language thus relieved of its contents is no more than a system of signs which are defined by their differentiae alone. What is involved in any structural hypothesis is then clear, and this it is that is enunciated in the fourth postulate: 'It is scientifically legitimate to describe language as being essen-

67. Cf. especially *Einleitung in die Geisteswissenschaften* (1883).

68. Cf. above, Section IV, *Man and Language*, especially sub-section A, § 1, 'Structural and transformational linguistics', pp. 1323-1330, and § 2, 'Linguistics and semiology', pp. 1330-1332, and sub-section B, § 4, 'Philosophical structuralism', pp. 1354-1359.

tially an autonomous entity of internal dependencies, in short a structure.' (*translation*) [69] This last postulate which may be called the postulate of the closed character of a system of signs, subsumes all the others. It is the one that offers the major challenge to phenomenology. For phenomenology, language is not an object but a mediation, that by and through which we direct ourselves towards reality; it consists in saying something about something; thereby it escapes towards what it is saying; it transcends itself and establishes itself in an intentional movement of reference. For structural linguistics, language is self-sufficient; all its differences are immanent in it; and it is a system which precedes the speaking subject. One can understand then how it is that the semiological model has set certain philosophers off in a deliberately anti-subjectivist, anti-humanist direction. Language appears as a self-sufficient self-contained system which has no object and opens neither on the world it would designate nor on the man who would infuse it with life and use it to say the world: self-reference and world-reference disappear simultaneously. Here literature has been an essential link between the operative structuralism of the linguists and the structuralism of the philosophers; the writer himself is fascinated by the process of literary production, of which he considers himself the scene rather than the agent: 'His whole work may be taken up with saying what the work ought to be, and what it cannot be, since in the last analysis it is only its own rough draft, or a construction "in abyss".' (*translation*) [70] With literature, theory and practice are exchanging rôles. The critic is inspiring the artist with the will for system; for example, the structures of the new novel are made to order for structural analysis, writer and reader reciprocally stimulating each other in a mirror relationship.

It is on this point that psychoanalysis here joins forces with linguistics. Its attack against philosophy of the subject is even more radical, in that it carries the assault to the very point where Descartes believed he had found firm ground of certainty. Freud burrows under the effects of meaning which make up the field of consciousness and lays bare the play of fantasies and illusions masking our desire. Indeed, the challenge to the primacy of the consciousness goes still further: for the so-called 'topical' psychoanalytic explanation consists in instituting a *topos*, a place, a *locus*, or rather a series of *loci*, without regard to the subject's internal perception. These *loci* – 'unconscious', 'preconscious', 'conscious' – are defined not at all by descriptive, phenomenological properties but rather as systems, that is as sets of representations and affects governed by specific laws and entering into mutual relations irreducible to any quality of consciousness, any determination of actual experience. Thus explanation begins with a general suspension of the

69. 'Il est scientifiquement légitime de décrire le langage comme étant essentiellement une entité autonome de dépendances internes, en un mot, une structure – Hjelmslev, 'Linguistique structurale' (1948), p. 28 of *Essais linguistiques* in the re-edition of 1970.

70. 'Il arrive que toute son œuvre s'emploie à dire ce que devrait être l'œuvre, et qu'elle ne peut être, puisqu'elle n'est enfin que son propre projet ou bien une construction en abîme.' – Mikel Dufrenne, *Pour l'homme* (1968), p. 71.

properties of consciousness. It is an anti-phenomenology which demands not reduction to the consciousness but reduction of the consciousness. When Freud went on to superimpose a second 'topic' – 'ego', 'id', 'superego' – on the first one he pushed the dispossession of the subject still further. It is not only the undermost part of the ego (the id) that is unconscious, but also the highest part (the superego). In other words, the unconscious is a characteristic not only of what we have repressed but of the very complex processes through which we interiorize imperatives and rules derived from social authority – chiefly parental authority, the primary source of prohibition during early childhood.

This convergence of linguistics and psychoanalysis goes far beyond a simple methodological parallelism in interpretations of psychoanalysis which might themselves be called linguistic. For Jacques Lacan, 'the subject's unconscious 'is the discourse of the other'. (*translation*)[71] 'The unconscious is that part of transindividual, i.e., concrete discourse which is lacking to the subject's mental make-up in order to restore the continuity of his conscious discourse.' (*translation*) [72] But this discourse is not an act that the subject could take back. All we have in it is 'the *signans* of a *signatum* which the subject's consciousness has repressed'. (*translation*) [73] The unconscious, then, is identified with the system or order of *signantia*. As to the desire underlying the discourse, it is itself 'not so much unalloyed passion in regard to the signatum as purely action on the part of the *signans*'. (*translation*) [74] Thus the science of the unconscious confirms the non-coincidence of discourse and self-consciousness. The subject is now reached only as 'dethroned, no longer centred on consciousness of self'. (*translation*) [75]

Alongside this conjunction of linguistics and psychoanalysis there is the conjunction of linguistics and structural ethnology or anthropology. We have shown, in its place, what structural anthropology means for the epistemology of the human sciences.[76] The philosophical implications of this methodological realignment are considerable. If cultural phenomena can be treated as systems of signs, actual individual experience becomes as irrelevant to them as the speaking subject of the phenomena of language: '[Anthropology] regards social life as a system of which all the aspects are organically connected. [. . .] When the anthropologist endeavours to create models, it is always with the underlying motive of discovering *a form that is common* to the

71. '[. . .] l'inconscient du sujet, [c'est] le discours de l'autre.' – 'Fonction et champ de la parole et du langage en psychanalyse' (1953/1956), p. 265 in *Ecrits* (1966).

72. 'L'inconscient est cette partie du discours concret en tant que transindividuel, qui fait défaut à la disposition du sujet pour rétablir la continuité de son discours conscient'. – *Ibid.*, p. 258.

73. '[. . .] le signifiant d'un signifié refoulé de la conscience du sujet.' – *Ibid.*, p. 280.

74. '[. . .] moins passion pure du signifié que pure action du signifiant' – 'La direction de la cure et les principes de son pouvoir' (1958/1961), p. 629 in *Ecrits*.

75. '[. . .] décentré de la conscience de soi' – 'Fonction et champ de la parole', p. 282 in *Ecrits*.

76. See above, Section IV, *Man and Language*, sub-section A, § 3, 'Linguistics and human sciences', pp. 1332-1335.

various manifestations of social life.' [77] Kinship systems present a paradigm in this respect. Like phonological systems, they are worked out by the mind at the level of unconscious thought: 'Linguistics [. . .] presents us with a dialectical and totalizing entity but one outside (or beneath) consciousness and will. Language, an unreflecting totalization, is human reason which has its reasons and of which man knows nothing. [. . .] Discourse never was [. . .] the result of a conscious totalization of linguistic laws'.[78] Consequently, the mind is not something peculiar to a psychological or transcendental subject. The mind at work in language, in kinship systems and in the various symbolic systems that make up society merges with what it produces; it is the institution itself, it is culture. This is the price at which objective knowledge is possible: thought is already in things, in social facts; and the thought of the wise man is only the place where that thought is reflected. We can go even further. If the mind is structure and if structure is in things, why not say that the mind is a thing? 'As the mind too is a thing, the functioning of this thing teaches us something about the nature of things'.[79] One understands how Lévi-Strauss could write that 'the ultimate goal of the human sciences [is] not to constitute, but to dissolve man'.[80]

Thus opposition to humanism on the theoretical plane has many facets: philosophies of the subject may be criticized from the point of view of the ground or foundations or from the point of view of models. In the former category we have discerned two main lines of attack, the one focusing on logic, the other on ontology, while as far as models are concerned we have noted at least three areas in which the idea of structure and system is at work – semiology, psychoanalysis and anthropology.

(iv) But these three different approaches – logical, ontological and in terms of models – are not the only possible approaches to the problem of ground or foundations. In these three cases the problem is viewed synchronically, i.e., by the abstraction of history. Now philosophies of the subject have very often been rooted in historical considerations; we have seen how Husserl and Merleau-Ponty assimilate constitutive consciousness and tem-

77. '[L'anthropologie] voit, dans la vie sociale, un système dont tous les aspects sont organiquement liés. [. . .] Quand l'anthropologue cherche à construire des modèles, c'est toujours en vue, et avec l'arrière-pensée, de découvrir *une forme commune* aux diverses manifestations de la vie sociale.' – Lévi-Strauss, *Anthropologie structurale* (1958), Chap. XVII, p. 399 (English translation, *Structural Anthropology*, 1963, p. 365).

78. 'La linguistique nous met en présence d'un être dialectique et totalisant, mais extérieur (ou inférieur) à la conscience et à la volonté. Totalisation non réflexive, la langue est une raison humaine qui a ses raisons, et que l'homme ne connaît pas. [. . . Le] discours n'a jamais résulté [. . .] d'une totalisation consciente des lois linguistiques.' – Lévi-Strauss, *La pensée sauvage* (1962), Chap. IX, 'Histoire et dialectique', p. 334 (English translation, *The Savage Mind*, 1966, p. 252).

79. 'Comme l'esprit aussi est une chose, le fonctionnement de cette chose nous instruit sur la nature des choses.' – *Ibid.*, p. 328, note (English translation, p. 248, note).

80. '[. . .] le but dernier des sciences humaines n'est pas de constituer l'homme, mais de le dissoudre.' – *Ibid.*, p. 326 (English translation, p. 247).

porality. And criticism of these philosophies too is also voiced in diachronic terms.

One way of devesting the subject of his constitutive ambitions is to show that he misses changes in systems as well as their inner, synchronic structure. And this is Michel Foucault's aim in establishing an 'archaeology of knowledge'.[81] Every epistemological field – or, as it is called, every *epistēmē*, construed as 'universe of knowledge' – has a coherent structure. Thus, the three positives investigated in *Les mots et les choses* – life, work, language – make a system at every era of knowledge; but from one era to another there is discontinuity and mutation, and 'these mutations are so sudden as to rule out any conception of continuity or progress in knowledge: our science is in no way the heir of earlier science, cannot recognize itself there; our modernity is brand new. Thus archaeology repudiates history – and at the same time that which makes for continuity of the historic: the permanence of a human nature structured by the *a priori*' (*translation*).[82] As for man himself, he is merely a vanishing figure in a temporary system of concepts: a finite being who only truly exists for as long as the system calls him forth, provides him with foundations and accords him a privileged place. By man, let us understand that which is promoted to epistemological existence by Cartesian philosophy and the human sciences. Here archaeological epistemology is very close to the philosophy of the concept postulated by Cavaillès as well as to Heidegger's philosophy of eras of being and the criticism of humanism which we were looking at earlier in his essay 'Die Zeit des Weltbildes'.

(c) We have set forth the record of the opposition to theoretical humanism in a deliberately antinomic style – philosophies of the subject on this side, philosophies of system on that. It is not the task of the rapporteur to propose a synthesis which, as the discussion now stands, could scarcely rise above the level of an eclectic harmonization. The most he can do is call attention to studies which in one way or another, and scattered over a very wide front, avoid this antinomy because they change or shift its terms.

The first place to look for a possible fresh rephrasing of man's relation to discourse is in a *criticism of the semiological models* that are borrowed from linguistics and extended to all the human sciences. Chiefly under the influence of Chomsky's transformational linguistics (mentioned, not for the first time, at the end of § 1 in the present sub-section), the structuralism of the Prague, Geneva and Copenhagen Schools is being integrated into models

81. Cf. especially FOUCAULT, *Les mots et les choses* (1966) (English translation, *The Order of Things*, 1970); *L'archéologie du savoir* (1969) (English translation, *The Archaeology of Knowledge*, 1972); and *L'ordre du discours* (1971).

82. '[. . .] ces mutations sont si soudaines qu'elles interdisent de concevoir une continuité ou un progrès du savoir: notre science n'est en rien héritière du savoir antérieur, elle ne peut se reconnaître en lui; notre modernité est toute neuve: ainsi l'archéologie récuse-t-elle l'histoire – en même temps que ce qui assure la continuité de l'historique: la permanence d'une nature humaine structurée par l'*a priori*.' – DUFRENNE, *Pour l'homme* (1968), pp. 40-41.

that are less taxonomic in character. According to this transformational model, the distinction between competence and performance follows very different lines from the distinction between language and speech or between system and process. Since Chomsky's *Aspects of the Theory of Syntax* (1965), a distinction has been made in the notion of competence itself between syntactic and semantic competence. And when the semantic dimension of language is taken into consideration we have a notion of discourse much richer than the Saussurian notion of speech, which tended to come down to contingent execution of the linguistic system by any and every speaker. Discourse is based on units – sentences – which are irreducible to the units of language – signs; it consists in an original operation, predication, irreducible to the relation of difference and opposition among a system's signs. It involves at once reference to something, a world, and reference to a speaker, a locutor, who designates himself as the one who is speaking by the use of personal pronouns. Finally, by referring to its own locutor, discourse refers to an interlocutor, an audience, whom the speaker is addressing.

But it is not only the notion of the speaking subject that is treated in this way. The notion of code, hypostatized by the philosophies of system, is itself tending to become diversified, as Roman Jakobson reminds us in his chapter on trends in linguistic research in *Part I* of this *Study*: 'Any verbal code is convertible and necessarily comprises a set of distinct subcodes or, in other words, functional varieties of language. Any speech community has at its disposal (a) more explicit and more elliptic patterns, with an orderly scale of transitions from a maximal explicitness to an extreme ellipsis, (b) a purposive alternation of more archaic and newfangled dictions, (c) a patent difference between rules of ceremonial, formal and informal, slovenly speech. The areally distinct and manifold sets of rules permitting, prescribing, or prohibiting talk and silence are destined to serve as a natural preface to any veritably generative grammar.' [83] What we call the verbal competence of speakers therefore implies their ability to discriminate among codes, to choose the right code for a given situation at the right moment. Thus speakers are aware of variations, distinctions and changes in the linguistic model, which tends to prove that 'metalanguage [is] a crucial intralinguistic factor'.[84]

Theory of discourse, therefore, seems to open up a new approach to the speaking subject. Towards it there converge from one side the strictly linguistic analyses of Benveniste, Jakobson and Chomsky, and from the other the strictly philosophical analyses of the speech-act which we discussed in an earlier passage.[85] So these converging analyses of discourse seem to demand a parallel and correlative revision of the notions of system and the speaking subject.

83. 'Linguistics', p. 429 in *Main Trends of Research in the Social and Human Sciences, Part I: Social Sciences* (1970).

84. *Loc. cit.*

85. See above, Section IV, *Man and Language*, sub-section B, § 1, 'The analytic approach', pp. 1338-1344.

The hermeneutic works we have also presented earlier, in the same sub-section,[86] point in the same direction. The idea that man understands himself only by interpreting the signs of his humanity hidden in literatures and cultures calls for no less radical a transformation of the concept of subject as of that of cultural text. For one thing, the indirect understanding of oneself implied by the hermeneutic act rejects the intuitionism of a philosophy founded on the *cogito,* with its claim to constitute itself in self-sufficiency and consistency, and attests the dependence of its meaning on the meaning of what it understands outside itself. For another thing, comprehension of a text does not end with discovering the codes that make up its structures but with revealing the image of the world, the mode of being, towards which it points. But this revelation is in turn only the counterpart of the dethroning of the subject who takes the roundabout route via the world of signs in order to understand himself. Thus the hermeneutic circle marks the simultaneous abandonment of the notions of system and of subject.

The need for connecting system and subjectivity in some new way is discernible not only in the philosophy of language (even in the broad sense in which we have used the term in this study) but also in *political philosophy.* It is true that there is no apparent relation between these two fields of theoretical investigation; but to bring them into relation is precisely what we are entitled to expect from any comprehensive examination of the theoretical ground or foundations of humanism. If anti-humanism is true, there is also no theoretical basis on which the legal subject can oppose the abuse of political authority. However, the concepts of 'rule of law' in the West and 'socialist legality' in the East require that the notion of a legal entity, possessing rights, be interpreted in a rationally consistent manner. Such postulates as 'Government is necessary; – A citizen in a constitutional State has rights against his government as well as duties towards it; – These rights are protected by law, and law depends on law-courts' [87] cry out to be linked with a conception of the citizen which itself entails a subjective, personal foundation for ethical and political thought. If we relegate them to the sphere of ideology, and cut ideology off from science, we are admitting that the whole edifice of the rights of man is arbitrary and un-'founded'.

This is an especially serious difficulty with a form of Marxism like Althusser's which insists on the 'epistemological cleavage' between science and ideology and relegates to the sphere of ideology forms like religion, morality, philosophy, which are recognized as structures essential to the historical life of societies and which even a communist society could not do without. The hiatus between ideology and science poses as many problems as it solves, seeing that ideology expresses men's experiential relation to their world and consequently covers revolutionary action itself. For one thing, inasmuch as they constitute an ideological relationship with reality the concrete actions of

86. See above, Section IV, sub-section B, § 5, 'The "hermeneutic" current: the interpretation of language', pp. 1359-1364.

87. Cf. MACKENZIE, 'Political Science', p. 206 in *Main Trends of Research in the Social and Human Sciences, Part I: Social Sciences* (1970).

men are 'vested in [an] imaginary relation: one that *expresses* a *will* (conservative, conformist, reforming, or revolutionary), or even a hope or a longing, rather than describes a reality' (*translation*).[88] The ideological concepts of socialist humanism – 'alienation, scission, fetishism, man whole and complete' [89] – are justified only by their practical function, namely that they give rise to new forms of organization in economic, political and ideological life, particularly for the phase in which the dictatorship of the proletariat withers away and is transcended. But between ideological concept, thus vested in practice, and theory, elevated to the formal purity of knowledge, the gulf remains as broad as ever.

There are in these circumstances two ways open for resolving the dilemma of humanism and theory. We can, with Mikel Dufrenne in *Pour l'homme*, try to return to Merleau-Ponty's standpoint and recapture the idea of a compact between man and nature, at a lower level than any theoretical or practical behaviour: 'A philosophy of man must recognize that it is man's privilege to be a correlate of the world; it must do justice to his intentional acts as to his *praxis*, and everywhere measure their constitutive power. But it must also, so to say, do justice to the world, to the weight of the real and the might of becoming. Because man is in the world, and also because he may be in truth, however truth is defined, it must be somehow related to reality.' (*translation*)[90] But here too, as in Merleau-Ponty's last philosophy, man is no longer the subject which constitutes all objectivity. Before ever he is constitutive of objects he is a prey to the world. Correspondingly, the universe of discourse is no longer self-sufficient. 'The garment of ideas that covers the world is a shirt of Nessus; it adheres to the world by all the fibres of truth, and man weaves it only because the world gives him its measurements.' (*translation*)[91] Thus system is no more sufficient unto itself than man is the giver of mean-

88. '[...] investie dans [un] rapport imaginaire: rapport qui *exprime* plus une *volonté* (conservatrice, conformiste, réformiste ou révolutionnaire), voire une espérance ou une nostalgie, qu'il ne décrit une réalité.' – ALTHUSSER, 'Marxisme et humanisme' (1965), p. 240 in *Pour Marx* (1965).

89. '... aliénation, scission, fétichisme, homme total ...' – *Ibid.*, p. 246. Cf. the references in § 2 above to Marxist thinkers professing scientific humanism; cf. also SCHAFF, *Marksizm i jednostka* (1965) (French translation, *Le marxisme et l'individu*, 1968). And see D'HONDT's discussion of 'La crise de l'humanisme dans le marxisme contemporain' in the special number of the *Revue internationale de Philosophie* (1968) entitled *La crise de l'humanisme.*

90. 'Une philosophie de l'homme doit reconnaître ce privilège qu'a l'homme d'être corrélat du monde, elle doit rendre justice à ses actes intentionnels comme à sa *praxis*, et mesurer partout leur vertu constituante. Mais elle doit aussi, si l'on peut dire, rendre justice au monde, à la pesanteur du réel et à la puissance du devenir. Parce que l'homme est au monde, et aussi parce qu'il peut être dans la vérité, de quelque façon qu'on la définisse, il ne se peut que la vérité n'ait quelque rapport à la réalité.' – DUFRENNE, *Pour l'homme*, p. 126; cf. also the same writer's 'L'anti-humanisme et le thème de la mort', in *La crise de l'humanisme* (Revue *internationale de Philosophie*, 1968).

91. 'Le vêtement d'idées qui couvre le monde, c'est une tunique de Nessus: elle adhère au monde par toutes les fibres du vrai, et l'homme ne la tisse que parce que le monde lui donne ses mesures.' – *Ibid.*, p. 127.

ing. To find the pristine correlation between subject and object, the pretensions of both must be exploded.

There is another way of accepting, *in toto*, the divorce between humanism and theory without thereby consigning humanism to the sphere of ideology. It is that of Emmanuel Levinas in *Totalité et Infini* (1961).[92] Levinas grants theory all it lays claim to, and above all the right to make a system, to connect in a 'same' and 'single' discourse all 'other' and scattered terms. Everything that can thus be reduced to unity in an order, Levinas denotes as 'being'. 'Being' is the realm of the 'same' – the realm of totality and therefore also of totalizing, 'totalitarian' power. So under 'being' we find, unexpected neighbours, Hegel and Heidegger, structuralism and all philosophies of subjectless system. Levinas sees them all as being more right than they think. The reign of totality is boundless and nothing that man says can lawfully escape the system. And yet something does elude this reign – something or rather someone. In the relationship with other persons there lies the possibility of *Otherness*, which, alone, does not come under the banner of the 'same'. The 'other' is what is always excluding or extruding itself from the system. The 'synchrony' of discourse would fain encompass what the other says, but is constantly foiled by the 'diachrony' of his saying. The 'other' 's speech does not enter into the system of the 'same' but, if it is true that everything is potentially contained in the system, is absolute externality: an externality which obtains only in the 'proximity' of one's neighbour but which guards this proximity as against any identification. Acts do not come under the category of *totality* but evince the opposing principle of *Infinity*. Whence the title: *Totalité et infini.*

Thus Levinas concedes everything to theory, system, being, only in order to withhold what is essential: 'the other', his face, his speech. But it is not the constitutive *cogito* of the philosophies of the subject that is thus put outside totality. It is not even 'I' but 'Thou'. And 'I' does not first arise as 'I' in the nominative but as 'me' in the accusative, propelled into responsibility by the proximity of the other. In that case the only possible form of metaphysics is ethics. But ethics lacks an ontological basis, since the subject-matter of ontology, being, comes under the principle of Totality. Ethics begins of itself from the moment I discover myself in the most vulnerable of situations – in hostage for the other.

This ethical passivity of the subject is somewhat reminiscent of the affective passivity which Michel Henry, in *L'essence de la manifestation* (1963), sees at the root of the movement of sublation or transcendence with which conquest of the object, domination of the world, and even the manipulation of men, begin.

Thus in many different ways, spurred on by criticism of philosophies of the subject, contemporary philosophy is attempting to bridge creatively the

92. See also, among articles by LEVINAS that are particularly germane to this approach to the problem: 'La trace de l'autre' (1963); 'Humanisme et anarchie' (published in *La crise de l'humanisme*, cited above, 1968); 'Le Dit et le Dire' (1971); and 'La proximité' (1971).

gulf between systems without a subject and a subject without truth. Or else, giving up the idea of reconciling the two in a higher synthesis at the cost of radically altering the opposing terms, it confines itself to reflecting on the insoluble character of the conflict. In this plight, humanism has no choice but to recognize that it is un-'founded' (*sans fondement*) – as a wager, or as a cry.

C. PHILOSOPHY OF RELIGION

Introduction

It is impossible to treat of the *foundations of humanism* without entering into the discussions in regard to the significance of the religious fact for philosophy. We have given our survey of these the title 'Philosophy of Religion', which simply means: How is the religious fact accounted for in philosophical discourse? The answers of atheistical humanism and of 'onto-theology' alike fall within the scope of this investigation. Their battleground is the complex field where man ponders the ultimate foundation and meaning of his existence.

To give a fair idea of the diversity of current trends in philosophy vis-à-vis religion, we shall begin (§ 1 and 2) by visualizing the two opposing poles which confer on this fundamental investigation its own peculiar tension. At one pole (§ 1) are those philosophies which, taking over into their discourse the religious affirmation of God or the absolute, endeavour to account for it rationally. At the other (§ 2) are philosophies for which religious statements have no meaning, or do not satisfy the logical or epistemological criteria that are valid for all scientific discourse. The first pole we have called the onto-theological pole, using an expression which Heidegger took over from Kant; the second may be called the analytical pole, in that it groups many – though not all – of the investigations of religious language and theological statements and arguments in the vein of British and American analytical philosophy.

In between these two poles we shall place writers who stand in a variety of relations to them and adopt a complex attitude to the religious fact. In § 3, we shall touch on works inspired by the human sciences, which quite often proffer much the same criticism of religious language as the analytical school, but which sometimes also point towards prospect of the renewal, not of language merely, but of the religious experience itself. The works examined in § 4 belong to the non-positivist current in the analysis of religious language and aim at identifying the peculiar characteristics of its grammar or logic, with no attempt either to reduce it or to justify it. Finally, in § 5, we shall bring together works that lie within the hermeneutic tradition studied earlier,[93] with their complex relationship to the ontological tradition, phe-

93. Cf. Section IV above, *Man and Language*, sub-section B, § 5: 'The "hermeneutic" current: the interpretation of language', pp. 1359-1364.

nomenology, analytical philosophy and the human sciences.

By way of conclusion we shall single out the principal intellectual attitudes in regard to the problem of the foundations of humanism. It is there that disagreement among the different types of philosophy is most plainly manifest.

§ 1. *Onto-theology: tradition and renewal**

The first '*locus*' of philosophy of religion falls in the category of the ontological tradition. Here thought about God or the absolute seeks to reflect in rational discourse the cultural heritage comprising not only what is originally present in religions but philosophico-theological elaboration of the religious fact in secular terms.

In many circles philosophers continue to investigate the question of being (onto-logy) as at once the most fundamental and within the range of human thought. They hold that it is of the nature of man to ask questions about his existence and that this leads him on to the question of the relation between being and 'what is', multiple and contingent. And they adduce various explanations of the fact that thought is capable of handling this question adequately; for in their theory of knowledge conformity between the thought of being and the truth of being is held to be certain because the very presence of the concept of being shows that it is there to be thought and because the question it raises points the mind to a fundamental, all-embracing field of knowledge. This field has a logic of its own, which the formal laws of a lower-level logic are powerless to contest.

In the sphere of ontological thought, we can see four main currents which are here arranged in the chronological order of the philosophical traditions they invoke.

(a) Those deriving from the Aristotelian-Thomist synthesis reflect Christian – and particularly Catholic – thinking.[94] They represent the tendency, first systematized by St. Thomas Aquinas, to link up – while distinguishing them in terms of their differing intelligibility – the God of Biblical monotheism and Greek-inspired metaphysical thought, with its investigation of the struc-

* Like § 3 below and the Conclusion of sub-section C, this is mainly based on a conspectus prepared, in the name of the International Academic Union, by Canon G. Verbeke in collaboration with A. Dondeyne, A. Vergote and other members of the teaching staff of the University of Louvain.

94. De Raeymaeker, *De metaphysiek van het zijn* (1947) and *The Philosophy of Being* (1964); Fabro, *Participation et causalité selon saint Thomas d'Aquin* (1961); Gilson, *God and Philosophy* (1941), *The Unity of Philosophical Experience* (1937) and *Being and Some Philosophers* (1952); Jolivet, *Le Dieu des philosophes et des savants* (1956); Lyttkens, *The Analogy between God and the World* (1952); Marc, *Dialectique de l'affirmation* (1952); Nink, *Ontologie* (1952); Smart, *Religious Belief and Philosophical Thought* (1963); Welte, *Auf der Spur des Ewigen* (1965) and 'Der philosophische Glaube bei Karl Jaspers und die Möglichkeit seiner Deutung durch die thomistische Philosophie' (1949).

tural relations between beings which are contingent (and consequently cannot of themselves render an account of their being) and the being which constitutes them through and through. In constitutive being, the final foundation of all beings, philosophers of this complexion recognize Being itself, the ultimate origin and cause, which their Biblical tradition has taught them to call God.

We should point out that this current of thought, while cast in an ancient tradition, is renewed and diversified on the one hand by the results of very searching historical inquiries into its origins and on the other hand by its systematic confrontation with recent trends of thought (metaphysical idealism, Blondel's philosophy of action, phenomenology, Heidegger).[95]

(b) Other thinkers on the absolute belong with Hegel, for whom reason and religion are united in philosophy as science.[96] One outstanding feature about the persistence and revival of metaphysics at the present time is undeniably the wealth of work that is being done on Hegelian philosophy, especially as regards reassessment of the *Phenomenology of Spirit* (or Mind). Three trends can be seen in this research. The first has to do with Hegel's conception of God as Absolute Mind, and consequently of the relation between religions (Christianity in particular) and philosophy as systematized knowledge. Next, many thinkers are working to clarify the relations between Hegel's philosophy, the philosophy of Kant, which he meant to transcend, and the philosophy of Heidegger, which restores the difference between being and beings and thereby claims to 'destroy' metaphysics *qua* thought of absolute subjectivity. The third field of research involves comparison of Hegelianism and existential philosophy, which derives from Kierkegaard and opposes to absolute knowing the radical individuality of responsibility and the contingency of man, a fundamentally historical being.

Special mention should be made of the persisting influence of Whitehead in philosophy of religion. His own work continues to attract commentaries and encourage a renewal of rational theology known as 'process theology'.[97]

95. BIRAULT, BOUILLARD *et al., L'existence de Dieu* (1961); BLONDEL, *La philosophie et l'esprit chrétien*, Vol. I (1944); COLLINS, *God in Modern Philosophy* (1960); DECHESNE *et al., Mens en God* (1963); DONDEYNE, *Foi chrétienne et pensée contemporaine* (1951); GIRARDI *et al., L'ateismo contemporaneo*, I (1967), pp. 165-196 (French translation, *L'athéisme dans la vie et la culture contemporaine*, 1967); JASPERS, *Der philosophische Glaube angesichts der Offenbarung* (1962) (English translation, *Philosophical Faith and Revelation*, 1967); LACHIÈZE-REY, *Le moi, le monde et Dieu* (1938, 1950); LE ROY, *Le problème de Dieu* (1930); LEWIS, *Philosophy of Religion* (1965); NINK, *Philosophische Gotteslehre* (1948); SCHULZ, *Der Gott der neuzeitlichen Metaphysik* (1957); SCIACCA, *Il problema de Dio e della religione nella filosofia attuale* (1964).

96. BRUAIRE, *L'affirmation de Dieu* (1964) and *Logique et religion chrétienne dans la philosophie de Hegel* (1964); CHAPELLE, *Hegel et la religion*: I. *La problématique* (1964); II. *La dialectique*: A. *Dieu et la Création* (1967) and Annexes (1967); TIELSCH, *Kierkegaards Glaube* (1964); VANDERKERKEN, *Inleiding tot de fundamentele filosofie* (1970).

97. CHRISTIAN, *An Interpretation of Whitehead's Metaphysics* (1959) and *Mean-*

Finally, the considerable body of work produced by Henry Duméry [98] has a place quite apart. In one sense it continues Blondel's dialectic of immanence, but places it on a more radical plane by the joint effect of Plotinian purification and Husserlian reduction; at the same time, religion is subjected to a distinctively Kantian philosophical critique.

(c) Husserl's phenomenology, while recognizing that the requirements of philosophy lead it 'to a teleology and a philosophical theology as a non-confessional road towards God', nevertheless makes a distinction between God and the absolute, by virtue of the method of phenomenological reduction, which begins by 'bracketing' naturalism and, with it, every form of *Weltanschauung*, in order to lead the reason back, by means of its self-elucidation, to the absolute ideality of the originative Logos. By doing this Husserl revives transcendental idealism, and the question of the relationship between reflection on absolute truth and an ontological thought resulting in the affirmation of God is kept on the horizon of philosophical speculation.

Husserl's phenomenology has opened up an area of questioning where many philosophers [99] try to bring transcendental reflection into relation with the metaphysic of Being or of the One by reviving Plotinianism, Aristotelian-Thomist ontology or a reflection on action, viz. the inner dynamism underlying both reflection and the purposive will. Sometimes these philosophers would situate the Absolute or the One above and beyond the opposition between theism and atheism, thus testifying to the difficulty of combining religious affirmation of God with a transcendental philosophy.

(d) Few thinkers have left so profound a mark on the philosophy of God as Heidegger, and that for four reasons. At the most superficial level, his analysis of existence (his 'anthropology') gave ontology a new point of departure by replacing theoretical thought about the world by the phenomenology of existence as being-in-the-world. In this way he introduced formal criteria in regard to the categories (such as cause or subject) by means of which it had sometimes been customary to think the being of God. By overturning the *a priori* nature of modern anthropocentricity, he restored to phi-

---

ing and Truth in Religion (1964); HARTSHORNE, *The Logic of Perfection and other Essays in Neo-classical Metaphysics* (1962); RICHMOND, *Theology and Metaphysics* (1971); OGDEN, *The Reality of God and other Essays* (1966/1967).

98. DUMÉRY, *Blondel et la religion* (1954); *Critique et religion* (1957); *Philosophie de la religion* (1957); *Le problème de Dieu en philosophie de la religion* (1957); *Phénoménologie et religion* (1958); *Raison et religion dans la Philosophie de l'Action* (1963).

99. BÖHM, *Vom Gesichtspunkt der Phänomenologie* (1968); HELD, *Lebendige Gegenwart. Die Frage nach der Seinsweise des transzendentalen Ich bei Edm. Husserl, entwickelt am Leitfaden der Zeitproblematik* (1966); HOHL, *Lebenswelt und Geschichte. Grundzüge des Spätphilosophie Edm. Husserls* (1962); LANDGREBE, *Phänomenologie und Metaphysik* (1949); LUYPEN, *Phenomenology and Atheism* (1964); MARCEL, *Le mystère de l'être* (1951) (English translation, *The Mystery of Being*, 1950-51); STRASSER, 'Das Gottesproblem in der Spätphilosophie Edmund Husserls' (1958).

losophers the dimension of the thought of Being. By his reflection on the difference between Being and 'what is' he dissociated the question of God from ontology; in doing so he opened up for some the possibility of locating God in the intervening gap – but on the basis of considerations other than those relevant in ontological thought (e.g., an analysis of the sacred). Finally, drawing on hermeneutics, he gave being an added dimension in the order of speech, which has widely influenced theological and philosophical thinking on the nature of the divine and on the religious relationship.[100]

§ 2. *The critique of religion in 'analytical philosophy'**

The second of the opposing poles which give the field we are considering its characteristic tension is constituted by the bulk of those works which depend on methods of formal or conceptual language-analysis. However, we shall not lump together here in § 2 all the works that might come under the heading 'Analytical philosophy of religion'. Within the so-called analytical school, in fact, there are very diverse trends.

We can make out a first group of thinkers for whom the preliminary question is: Do religious statements have a meaning? They usually reply to this question in the negative, on grounds of the linguistic or logical laws that govern human speech.

In a second group, thinkers apply themselves rather to the logical structure of religious arguments, particularly the classical proofs of the existence of God. While most of this group tend to conclude that theological discourse, *qua* demonstrative discourse, does not meet the logical and epistemological criteria that are valid for any discourse to qualify as scientific, we find a minority of writers who believe it possible to reformulate this or that traditional proof in a satisfactory way.

But there is a third group which can be associated with analytical philosophy in the broad sense but actually constitutes a quite separate trend, for which reason we shall deal with it separately.[101] Thinkers in this group, influenced by Wittgenstein II and by J. L. Austin, ponder the specific rules which hold together the grammar or logic of religious 'language games' or 'speech-acts'. Here we find a good many thinkers who, through the logic of religious discourse, are indirectly linked either with phenomenology or with hermeneutics, while retaining characteristically British and American modes of thought.

100. BULTMANN, *Glauben und Verstehen* (1933-52-60-65) (English translation, Vol. I. *Faith and Understanding*, 1969; Vol. II. *Essays: Philosophical and Theological*, 1955); LEVINAS, *Totalité et infini* (1961) and *En découvrant l'existence avec Husserl et Heidegger* (1967); RICOEUR, *Le conflit des interprétations* (1969); THEUNISSEN, *Der Andere* (1965).

* Based, like § 4, below, on a conspectus prepared by Jacques Poulain.

101. See § 4 below, pp. 1481-1484.

(a) *Has religious language a meaning?*

At no time has it been the practice of philosophers to leave the logical status of religious statements entirely to one side. St. Thomas Aquinas himself said that man cannot recognize God as He is in himself. But he held that by virtue of the analogical character of concepts man has the semantic ability to signify God. Several Thomist philosophers have gone more deeply into the analogical mode of our concepts, especially the concept of being. Their reflections on the 'names of God' (the qualities one can attribute to God through reason) have always stopped short of metaphysical idealism, with its belief in the ability to express the quintessence of the absolute. The idea of analogy is an attempt to discern the transcendent through human concepts, and to distinguish anthropomorphic images from true metaphysical concepts.

But it is above all in the setting of English-language analytical philosophy that the question of the sense or non-sense of religious statements has been treated with a rigour never before attained. The question relates specifically to the logical positivism of the Vienna Circle as carried on by their British and American disciples, especially A. J. Ayer,[102] and derives from the analyses of the relationship between the logical and the 'mystical' conducted by Bertrand Russell [103] and L. Wittgenstein [104] as contributions to the development of logical atomism. It is therefore inextricably bound up with the second question, which concerns the possibility of a rational theology, the applicability of mathematical logic to religious language and the consistency of certain types of rational theological discourse.

For Russell, the possibility of any rational theology depends on the truth of the propositions: 'God, the necessary being, exists' or 'God necessarily exists'. His theory of propositional functions prohibits the formation of such statements.[105] The term 'God' can only be a definite description the meaning of which does not imply the existence of the reality described. In the first case, we are driven to say that one individual and one only satisfies the propositional function 'to be *necessary*,' i.e., that this function is *possible*; in the other case, we forget that necessity can only be on the logical plane and can have no ontological significance. Wittgenstein characterizes religious 'propositions' as paradigmatically senseless: they try to say what can only be revealed – *das Mystische*, the mystical fact of the world's creation. Finally, God being neither a sensible phenomenon nor an observable, religious language appears meaningless to the logical positivists: we can no more establish experimental conditions which would let us ascertain the probability of the truth of its propositions than we can verify those propositions irrefutably. The only means of converting them into factual propositions is to eliminate

102. AYER, *Language, Truth and Logic* (1936).

103. RUSSELL, *Mysticism and Logic* (1910).

104. WITTGENSTEIN, *Logisch-philosophische Abhandlung / Tractatus logico-philosophicus* (1920; English translations, 1921 and 1961).

105. Cf. RUSSELL, 'The philosophy of logical atomism' (1918, reprinted in RUSSELL, *Logic and Knowledge. Essays 1901-1950*, 1956) and *Why I am not a Christian* (selected essays, 1957).

their specifically religious 'content'. They are therefore not analysable according to the logical patterns of *Principia Mathematica.*

The renewal of religious-language analysis is due to increased awareness of the difficulties inherent in logical empiricism and the dilemmas raised by the principle of verification. In his article 'Gods' (1944)[106] J. Wisdom re-opened the discussion: using the parable of the invisible gardener, he stressed a property common to theological and metaphysical language, namely their capacity for differentiating facts which cannot be differentiated without them and for showing the inner structure of the facts of the world. Their forms of expression give rise to *a priori* but non-deductive reasonings, use *a priori* but non-necessary logical connectives. This 'detranscendentalizing'[107] enabled A. Flew [108] to show how religious use of concepts from ordinary language ineluctably empties them of their meaning by making the propositions which contain them un-falsifiable. This criticism of religious discourse has made a potent contribution to the recent trend in religious thought which has been given the unifying label of 'death of God' theology. Often seeking support from sociological and psychoanalytical studies, as well as drawing on Biblical research, the 'death of God' thinkers [109] have proclaimed the idolatrousness or the vanity of religious philosophy. For discourse about God (theo-logy) they substitute a horizontal, ethical interpretation of the Christian message, although they take very divergent positions as to maintaining a reference to the God of Jesus Christ.[110]

(b) *Does theological discourse meet the logical and epistemological criteria that are valid in order for any discourse to qualify as scientific?*

Around this question range, as we have already indicated, works of widely differing trends. Indeed, in order to challenge the strictly empiricist and positivist interpretations of the principle of verification it was necessary for the discussion to be shifted to the plane of demonstration. But re-examination of the theological use of the concept of necessity led in its turn to posing the still more fundamental question whether religious language can be analysed or not, and what is the specific nature of the formal procedures it employs.

106. Reprinted in FLEW (ed.), *Logic and Language* (1951) and in WISDOM, *Philosophy and Psychoanalysis* (1953).

107. Cf. BAMBROUGH, *Reason, Truth and God* (1969).

108. Cf. FLEW, 'Theology and falsification (A)' (1950-51), reprinted in FLEW & MACINTYRE (eds.), *New Essays in Philosophical Theology* (1953).

For a general discussion of the epistemological status of theological statements, see FLEW, *God and Philosophy* (1966); DIAMOND & LITZENBORG (eds.), *Theology and Verification* (1972); RICHMOND, *Theology and Metaphysics* (1971); and also PEARS, *The Nature of Metaphysics* (1957, 1965).

109. Cf. VAHANIAN, *The Death of God. The Culture of Our Post-christian Era* (1961).

110. Cf. VAN BUREN, *The Secular Meaning of the Gospel based on an Analysis of Its Language* (1963, 1965).

J. N. Findlay [111] recalls Kant's criticism of the ontological argument and Russell's analysis of the necessary existence of God: if God is to be an adequate object of worship he cannot be something which does exist but might *not* exist; his existence must be identified with his essence. So this divine existence is meaningless or impossible, since logical necessity does not allow us to infer the necessity of a being's existence. For G. E. Hughes,[112] Findlay's argument suffers from a *petitio principii;* it presupposes that there can be no non-tautological necessary proposition. A. C. Rainer,[113] distinguishing between the property of necessary being attributed to God and the property of logical necessity, maintains that our assertion of the necessary existence of God remains contingent. We commit the converse sophism to Anselm's in saying that God cannot exist necessarily because we cannot assert his existence apodeictically. A. Kenny,[114] following Leibniz, defines the analytical proposition as a true proposition negation of which gives rise to a proposition that is self-contradictory. Furthermore, a necessary proposition is one which cannot change its truth-value. All analytical and *a priori* propositions are necessary, but it does not follow that all propositions which are necessary are analytical or *a priori.* The proposition 'God exists' is neither analytical nor *a priori.* But if God is eternal, the proposition 'God exists' is necessarily true since it cannot change its truth-value.

Though dissatisfied with Kantian criticism of the Anselmian proof, H. Scholz [115] reproaches St. Anselm for a psychologizing conception of logical laws that leads him to identify the necessary with that the non-existence of which is unthinkable. For B. L. Clarke [116] the ontological argument proves that if it is possible to formulate a definite description of God, then God necessarily exists. The description must be 'consistent' and formulated in terms of predicates with descriptive significance; it must be such that the individual described shall be necessary, that is, be an omnipresent individual. J. Bouveresse [117] and J. Vuillemin [118] find that the fragility of the speculative proofs of God's existence resides in the peculiar nature of the idea of God, which is shorn of all the elements that normally make it possible to demonstrate existence: no neutral yet adequate definite description of God is forth-

111. FINDLAY, 'Can God's existence be disproved?' (1948), reprinted in FLEW & MACINTYRE (eds.), *New Essays in Philosophical Theology* (1953).

112. HUGHES, 'Has God's existence been disproved?' (1949), reprinted in *New Essays in Philosophical Theology, op. cit.*

113. RAINER, 'Necessity and God' (1949), reprinted in *New Essays in Philosophical Theology, op. cit.*

114. KENNY, 'God and necessity', in WILLIAMS & MONTEFIORE (eds.), *British Analytical Philosophy* (1966).

115. SCHOLZ, 'Der anselmische Gottesbeweis' (1950-51), reprinted in *Mathesis Universalis* (1961).

116. CLARKE, *Language and Natural Theology* (1966).

117. BOUVERESSE, 'La théologie rationnelle et l'analyse logique du langage' (1968), reprinted in BOUVERESSE, *La parole malheureuse* (1971).

118. VUILLEMIN, *De la logique à la théologie. Cinq études sur Aristote* (1967).

coming. According to Vuillemin, rational theology 'is driven to assume in its premises an antinomy of properly mathematical type'.[119]

The first formalization of one of the Thomistic 'ways' (*ex motu*) was the work of J. Salamucha.[120] Taking into consideration the criticisms of Salamucha's work by B. L. Clarke, A. Kenny finds one formal structure that the five ways have in common: each of them uses a relational dyadic predicate to show that the relation indicated is non-reflexive and transitive and concludes therefrom either that there is an indeterminate series of things bound together in such manner or that there is something to which other beings can be related in that manner and which is not so related to any other being.[121] In distinguishing the propositions 'God exists' and 'God is' (with their respective contraries), P. Geach,[122] following St. Thomas, shows that 'is' is a true predicate which makes it possible to assemble all other predicates and safeguards a referential univocity. The possibility of attributing these descriptive predicates to God is based on analogy, which E. L. Mascall [123] sees as being essentially formal. Any analogical formulation points to an isomorphy between God and man: J. M. Bochenski [124] demonstrates that the relations which are then said to connect terms, beings and their properties unite them in the formal properties of non-reflexiveness, asymmetry and non-transitiveness.

Most of these ventures assume the analysability of religious language, the applicability of mathematical logic to descriptive religious propositions and the correctness of the operations by which religious discourse fulfils its referential function. That religious language can be analysed and logically structured is a legitimate presumption if it is shown we cannot demonstrate its unanalysability or that it is impossible that it should have meaning.[125] R. S. Heimbeck [126] establishes the falsifiability and verifiability of Christian theological statements in connection with the 'cognitive use' of theological propositions.

119. '[. . .] est conduite à supposer dans ses prémisses une antinomie de type proprement mathématique.' – VUILLEMIN, *Le Dieu d'Anselme et les apparences de la raison* (1971).

120. SALAMUCHA, 'Dovód "ex motu" na istienie Boda (Analiza logiczna argumentacji Sw. Thomasza z Akwinu)' (1934) (English translation, 'The proof "ex motu" for the existence of God', 1958).

121. KENNY, *The Five Ways* (1969).

122. GEACH, *God and the Soul* (1969).

123. MASCALL, *Existence and Analogy* (1949) and *Words and Images* (1957).

124. BOCHENSKI, *The Logic of Religion* (1965).

125. POULAIN, *Logique et Religion* (1973).

126. HEIMBECK, *Theology and Meaning. A Critique of Meta-theological Scepticism* (1969).

§ 3. *The impact of the human sciences on philosophy of religion**

Contemporary philosophy is largely a dialogue with the latest data of the sciences – of those sciences which study man from the socio-historical, psychological or linguistic points of view. Since philosophy of religion is the study of man in his concrete reality, in relation to what he considers to be his ultimate ground and meaning, it could not remain untouched by the enquiry these sciences have initiated into his manner of being, of understanding himself, of expressing himself. And it is not by chance that the emergence of these sciences coincides with what we are now given to calling the secularization of the world, if not of existence. It is by one and the same effort of will that modern man seeks to achieve technical mastery over nature, over society, over physical and mental phenomena, and embarks on scientific exploration of the laws governing their operation. This will to know and to control has led him to expel the sacred from large areas of existence, to dismiss religion as a key to understanding the things of this world, and even to criticize as illusory any discourse or attitude that refers to a world beyond. Consequently secularization sets the philosophy of religion new tasks which are busying many contemporary thinkers: either to interpret religious phenomena negatively on the basis of the critical implications of the human sciences, or to rethink religion by means of a critical confrontation with these negative interpretations. At all events, by virtue of the secularization that also implies making man the measure and reference of all things, it is religious man on whom all this research is focused; and reflection on the transcendent proceeds by way of philosophical anthropology, whichever of the human sciences it is based on.

(a) Today the *comparative history of religions* constitutes, for philosophy of religion, a vast storehouse of rites, symbols and myths, of beliefs and dogmas, which as it were document the religious fact. What these historical studies have brought to light is, first, the enormous variety of religions. By means of morphological systematizing as in Mircea Eliade,[127] categorial analysis in the manner of Kant as in Rudolf Otto [128] or a phenomenological reduction as in Max Scheler,[129] who revives Husserl's eidetic reduction, or in G. van der Leeuw,[130] who sees in this *epoche* the basis for a scientific understanding of the religious in the variety of its essential types, certain

* Based on the conspectus prepared, in the name of the International Academic Union, by G. Verbeke and other members of the teaching staff of the University of Louvain (cf. note * at the beginning of § 1 above).

127. ELIADE, *Traité d'histoire des religions* (1949) and *Das Heilige und das Profane* (1957) (English translation, *The Sacred and the Profane: the Nature of Religion*, 1968).

128. OTTO, *Das Heilige* (1917) (English translation, *The Idea of the Holy*, 1923).

129. SCHELER, *Vom Ewigen im Menschen* (1921, rev. edn. 1954) (English translation, *On the Eternal in Man*, 1960).

130. Van der LEEUW, *Phänomenologie der Religion* (1934; 3rd edn., 1970) (English translation, *Religion in Essence and Manifestation*, 2nd edn., 1964).

philosophers have tried to grasp the universal structures and permanent meaning of the sacred or the divine and the relationship of statements (myths) to symbolic behaviour (rites). They have been encouraged in this by the destruction the new historical research has wrought on nineteenth-century patterns of explanation, which presented a rather linear evolutionist conception of religious development. These philosophical reflections centre on a few fundamental themes: religious discourse (myth in its different aspects – originative and revelational language, its fanciful elaboration as related to the effects to which it is intended to give rise at the intersection of pre-philosophical and prescientific thought); religious symbolism as the representation of existence, one's body and the world in relation to a transcendent; and the sacred as the originative cultural element in mankind and, as certain thinkers hold, in each individual man.[131]

For some philosophers, what this research seems to yield is a realm above and beyond a metaphysical thought which they regard as cut off from its nourishing soil – the realm of sublime poetry, the only one in which the first and last questions in regard to existence can fruitfully be raised.[132] For others, the realm of myth remains the pre-scientific arena from which philosophy emerged once and for all victorious in classical Greece.[133]

(b) *The social sciences*, which are applied mainly to contemporary societies, tend rather to show how secularized the world has become by comparison with former cultures. In olden days, religions made it possible for culture to take root by combining the data of the world and of society into a comprehensive vision based on the sacred. The sacred conferred order on time and space and overcame looming chaos. The basis of legality which cultural institutions (family, tribe, city) needed they found in the divine order, whereas today it is conferred on them by purely human principles. In politics and learning, even in social ethics, religion is no longer directly functional.[134]

These observations, scientifically substantiated, oblige philosophers to examine the changing motivations of religion and its relations with ethics. And so we find philosophy of religion moving towards a fundamental reflection on the religious postulates of ethics or investigating the meaning of existence in a more radical manner which involves an increasingly critical differentiation between the functional motivations of the religious attitude and the fact that human destiny points in the direction of a transcendent existence, rooted in the inalienable freedom of man.

Other philosophers, following in the wake of Marxist thinking, try to account for religion as illusion or alienation by a dialectical reflection on man's

131. Ricoeur, *Philosophie de la volonté*: II. *Finitude et culpabilité* (2 vols., 1960) (English translation, *Fallible Man*, 1965, and *Symbolism of Evil*, 1967).

132. Toynbee, *An Historian's Approach to Religion* (based on his Gifford Lectures) (1956).

133. Vernant, *Mythe et pensée chez les Grecs* (1965).

134. Berger, *The Sacred Canopy. Elements of a Sociological Theory of Religion* (1967).

former relations with nature and society.[135] Many philosophical publications in the field of religion take up Feuerbach's and Marx's criticisms and systematize them or else strive to sublate them, passing beyond but at the same time taking into account this line of thought which inaugurated the philosophical justification of atheism.[136]

(c) *Psychology and, above all, psychoanalysis* have subjected religion to a very far-reaching critique, through study of the motives man can have for being religious.[137] Psychoanalytical interpretation of religion is not in itself in any way philosophical. But psychoanalysis has bequeathed to the philosopher who studies religion the task of making a fresh attack on two essential questions. First, the question of desire, which has been a fundamental theme in philosophy ever since Plato: psychoanalysis has shown what part the absolute, omnipotence and knowledge can play in the economy of desires. Debarred by the limits of its own intelligibility from the metaphysical realm, psychoanalysis was nevertheless intended by its founder to substitute metapsychology for metaphysics and offer a reductive interpretation of all the concepts that make up the philosophy of the absolute. So we find several philosophers, not content with a simple juxtaposition of the two types of understanding, trying to transcend from within a metapsychology which was originally meant to be exclusive. Their confrontation with psychoanalysis leads them to take man's desires, rather than the self-governing *cogito*, as their starting-point and to subject the concepts of omnipotence and absoluteness, the twin axes of religious thought, to a radical, though sometimes constructive, critique. Secondly, psychoanalysis obliges philosophers to take a fresh look at the great symbols and mythical representations which embody religious attitudes: the name of the Father, the Master-Servant relationship, the Fall as a figure of guilt and Paradise as typification of a state of innocence.

§ 4. *Religious 'language games' and 'speech-acts'**

In our account of analytical philosophy of religious language was left aside one trend which is very unlike the one going back to logical positivism. To it belong thinkers who, on the logical and epistemological plane, oppose all attempts at reframing ordinary language according to the logico-mathematical demands of 'well-formed languages' and ask that philosophy respect the

135. BLOCH, E., *Das Prinzip Hoffnung* (1954-55).

136. MURRAY, *The Problems of God. Yesterday and Today (The Thomas More Lectures*: I) (1964); ORTIGUES, *Le discours et le symbole* (1962).

137. Cf. BINSWANGER, *Grundformen und Erkenntnis menschlichen Daseins* (1942); RICOEUR, *De l'interprétation. Essai sur Freud* (1965) (English translation, *Freud and Philosophy. An Essay on Interpretation*, 1969); SUNDÉN, *Die Religion und die Rollen* (1966); SZONDI, *Ich-Analyse* (1956); VERGOTE, *Psychologie religieuse* (1966); von GEBSATTEL, *Prolegomena einer medizinischen Anthropologie* (1954).

* Based on a conspectus prepared by Jacques Poulain (cf. note * at the beginning of § 2 above).

infinite diversity of language games;[138] this approach to the problem originally stems from Wittgenstein II.[139] His influence has been reinforced by that of J. L. Austin with his distinction between performatives and constatives and, more generally, his description of the various 'illocutionary acts' which give discourse not only a meaning (propositional) but a force (the force of an assertion, order, promise, etc.).[140] This philosophy of language has brought forth a whole string of major works on the philosophy of religion which can be classed under the heading of analytical philosophy, but in a sense of the term different from, even opposed to, that in which it has been used above.

The meaningfulness of the terms used in religious language games is taken for granted by disciples of Wittgenstein II. Even so, the forms of their use have to be defined precisely. For I. M. Crombie,[141] theological use of words from ordinary language is legitimate if we define their field of reference, and what it is that theological statements explain. For R. L. Braithwaite[142] and R. Hepburn,[143] the significant force of religious parables lies in the 'agapeistic' behaviour pattern presupposed by the existential assertions they imply. The logical modifiers the believer employs to give concepts of current language a specific meaning may well produce paradoxes,[144] but, as is shown by I. T. Ramsey,[145] W Zuurdeeg[146] and D. M. High,[147] they can introduce the interlocutor to a personal dimension of the universe which inspires commitment; and according to T. Torrance[148] they establish the authority of the person who has revealed himself in them. The logic of authority which is inherent in religious language and is correlative with a logic of interpretation[149] is based on the believer's idiosyncratic vision of facts, which R. M. Hare[150]

138. See Section IV above, *Man and Language*, sub-section B, § 1, 'The "analytic" approach', pp. 1338-1344.

139. WITTGENSTEIN, *The Blue and Brown Books. Preliminary Studies for the Philosophical Investigations* (1958); *Philosophische Untersuchungen / Philosophical Investigations* (1953).

140. Cf. especially AUSTIN, *Philosophical Papers* (1961); *How to Do Things with Words* (1962).

The influence of G. E. MOORE (*Some Main Problems of Philosophy*, 1953; *Lectures of Philosophy*, 1966) is equally great; so is that of G. RYLE (*The Concept of Mind*, 1949; *Plato's Progress*, 1966).

141. CROMBIE, 'The possibility of theological statements' (1958).

142. BRAITHWAITE, *An Empiricist's View of the Nature of Religious Belief* (1955).

143. HEPBURN, *Christianity and Paradox* (1958).

144. WILLIAM, 'Tertullian's paradox' (1953).

145. RAMSEY, I. T., *Religious Language. An Empirical Placing of Theological Phrases* (1957), *Models and Mystery* (1964), *Christian Discourse. Some Logical Explorations* (1965) and *Myth and Symbol* (1966).

146. ZUURDEEG, *An Analytical Philosophy of Religion* (1958).

147. HIGH, *Language, Persons and Belief* (1967).

148. TORRANCE, T. F., *Theological Science* (1969).

149. MACQUARRIE, *God-Talk. An Examination of the Language and Logic of Theology* (1967).

150. HARE, 'Religion and morals' (1958).

calls '*blick*'. Even if – as D. Z. Phillips,[151] after L. Wittgenstein,[152] insists – the truth criteria for religious language must be found within a religious tradition, it has to be explained how the Christian ethical system is at once objective and founded on authority, and also how it commits the Christian to certain attitudes governed by teleological statements and enables him to act accordingly.[153] The paradigmatic Bible text showing that for the Christian acting is automatically involved in the act of speaking is, it is suggested, the story of the Creation [154]: the existence of the speaking subject and the form it will take are determined by what he says in it, and these statements express an acquiescence in their own truth, commit to a certain behaviour, and already carry out such commitment. By conceiving of creation on the model of action being entailed by an utterance relating to action, Biblical language brings the believer's religious language into correlation with the Word of God – likewise self-involving: by his Word God establishes his creatures in a dependent status, assigning them a rôle of their own, and enters into an engagement towards them by promising to maintain the order he has created. Thus, says J. Ladrière,[155] religious statements are the necessary condition for man's self-knowledge even if they do not in themselves suffice to enable him to obtain the object of the divine promise, salvation.

This research into language, its illocutionary and semantic capabilities, seems bound to constitute the main future thrust of philosophy of religion. Language is no longer regarded as the expression of an already constituted thought: we recognize that thought is constituted in and through language.

One consequently finds works in this group impinging on the field of phenomenologist ontology and even on that of classical ontology. Reflecting on the logic of the Plotinian schematism of procession and conversion, Stanislas Breton discovers in it the same pattern of relationships as in the operations of 'founding' identity, transitiveness and inversion which are characteristic of salvation.[156]

The frontiers with existential phenomenology and philosophy are no less interesting to explore, in so far as these make language the actual *locus* both of intersubjectivity and of manifestation of the being of things. Even psychoanalysis has links with this type of linguistic analysis, inasmuch as language and speech make up the entire arena in which it operates.

Finally, study of language games necessarily affects Biblical theology,[157] in so far as the religious discourse of the Bible consists of specific language games: stories, parables, hymns, doxologies, liturgical formulas and so on.

151. PHILLIPS, *Faith and Philosophical Inquiry* (1970).

152. WITTGENSTEIN, *Lectures and Conversations on Aesthetics, Psychology and Religious Belief* (1966).

153. WARD, *Ethics and Christianity* (1970).

154. EVANS, *The Logic of Self-Involvement* (1963).

155. LADRIÈRE, *L'articulation du sens* (1970).

156. BRETON, *Du principe* (1971) and *Foi et raison logique* (1971).

157. Cf. FERRÉ, *Language, Logic, and God* (1961); BEJERHOLM & HORNIG, *Wort und Handlung. Untersuchungen zur analytischen Religionsphilosophie* (1966); and *Religious Studies*, edited by H. D. LEWIS, Vol. I of which is dated 1965-1966.

The analysis of language games here comes close to the next type of approach we are going to discuss, namely the hermeneutic.

§ 5. *The hermeneutics of religious language*

Under this head we are grouping works which have the common aim of basing the renewal of theology – and more broadly the task of understanding religious discourse – on the interpretative disciplines discussed above in connection with the philosophy of language.

Applying hermeneutics to the religious fact seems appropriate enough, for several reasons. First, what we are here calling by the neutral name *religious fact* is never a naked mute experience, but one which is documented in texts – the Old Testament, the Gospels, the Koran, for the Western tradition – the meaning of which is forever being reinterpreted in a cultural tradition that differs from the one they were written in. Now, since Schleiermacher and Dilthey the notion of hermeneutics has been linked with that of interpreting 'expressions of life fixed by writing' (Dilthey). Moreover, these texts come down to us through a set of traditions which are themselves interpretations and which, to remain alive, demand a new interpretation. Finally, at the origin of the texts and of the traditions which convey their meaning there is a certain discourse – a 'message' – which purports to be an interpretation of ordinary human experience and is expressed in stories of confessional character, in parables, maxims, invocations, hymns.

Thus interpretation works at three levels which we may glance at in the following order: Interpreting religious texts means relating them to their intended referent (establishment of the people of Israel as the Chosen People, proclamation of the Kingdom of God, foundation of Islam as community of the faithful at both the social and the religious level, etc.). Interpreting traditions means linking their interpretations to the earliest interpretations and the earliest traditions. Finally, interpreting these earliest traditions means relating them to existence, of which they furnish a reading that goes below the surface and at the same time points beyond itself. The first two levels concern philosophy of religion only indirectly, only in so far as the exegesis and history of the kinds of theological discourse engaged in by religious or historical communities contribute to understanding of the fundamental interpretation the religious message brings to bear on human experience. It is the third level, then, which directly concerns philosophy of religion.

To begin with, we can situate the hermeneutics of religious language in relation to the other approaches discussed above as follows: First, the stories, parables, maxims, invocations, hymns that form the web of religious symbolism through which human experience is interpreted can be likened to the 'language games' studied by the (second type of) analytical school. Next, interpretation of existence through the medium of language comes very close to the phenomenology of religion as understood by van der Leeuw and Mircea Eliade, inasmuch as phenomenologists, too, see language as the stage on which the religious phenomenon takes specific shape. Lastly, this hermeneutics is continually crossing paths with a sociology of beliefs and ideologies and a psychoanalysis of symbols.

But the hermeneutic approach has one specific feature which sets it apart both from language analysis and from phenomenology and the human sciences, namely its recourse to the *correlation* method. It is this method, too, that marks its philosophical character as opposed to Karl Barth's assertion [158] that theological discourse is totally independent of all types of philosophical discourse, which are reduced to the mere rôle of contingent, instrumental expressions (even if Kantian references give more organic support to the theme of God's radical transcendence and subjectivity for himself).

Rudolph Bultmann [159] and Paul Tillich [160] are, in different ways, the originators of the correlation method. Bultmann's 'demythologizing' of Biblical expressions is the groundwork (required not merely by the modern conception of the world but by the Gospels' insistence on self-knowledge) for an existential interpretation in which the meaning of the 'new' existence announced by the word of God and brought out by exegesis is correlated with the profane reading of mundane existence brought out by Heidegger's analysis of *Dasein* at the time of *Sein und Zeit* (1927); hermeneutics thus consists in the mutual interpretation of languages one by another. Paul Tillich, basing himself on the dogmatic statements of the most abiding and widely accepted tradition in Christian theology rather than on exegesis of Biblical texts, effectuates a broad correlation between what he calls on the one hand 'situation' and on the other 'message'. He applies the method, first, at the level of statements about God, which he contrasts with statements about being; next at the level of statements about Christ as against anthropological considera-

158. Cf. Barth, *Der Römerbrief* (1919, 1963) (English translation, *The Epistle to the Romans*, 1933) and *Die kirchliche Dogmatik* (3 vols., 1945-46-51) (English translation, *Church Dogmatics*, 1936-). By Roger Mehl, who was influenced by Karl Barth, see *La condition du philosophe chrétien* (1947).

159. Cf. Bultmann, *Glauben und Verstehen* (4 vols., 1933-52-60-65) (English translation, *vide supra*, § 1, note 100), *Kerygma und Mythos* (3 vols., 1946-50-54) (English translation, *Kerygma and Myth*, 1948-55), *Theologie des N.T.* (1933) (English translation, *Theology of the N.T.*, 1952-55), and *Jesus* (1926) (English translation, *Jesus and the Word*, 1934). Bultmann's heritage on the exegetic, philosophical and theological planes is considerable. See in particular Ebeling, *Wort und Glaube* (1960); Fuchs, *Hermeneutik* (1950) and *Zum hermeneutischen Problem in der Theologie. Die existentiale Interpretation* (1959); Ott, *Geschichte und Heilgeschichte in der Theologie R. Bultmanns* (1955), *Denken und Sein. Der Weg Martin Heideggers und der Weg der Theologie* (1959) and *Dogmatik und Verkündigung* (1961). For an interpretation opposed to Bultmann's exegetic and theological approach, see Cullmann, *Christ et le temps. Temps et histoire dans le christianisme primitif* (1947) (English translation, *Christ and Time*, 1951) and *Heil als Geschichte* (1965) (French translation, *Le salut dans l'histoire*, 1966).

160. Cf. Tillich, *Systematic Theology* (3 vols., 1951-57-63). Tillich's disciples include, notably, Langdon Gilkey (*Naming the Whirlwind. The Renewal of God-Language*, 1969) and Nathan Scott (*The Wild Prayer of Longing: Poetry and the Sacred*, 1971).

tions, chiefly derived from existentialism and psychoanalysis; and lastly at the level of statements about Spirit, which he correlates with the problems raised by justice, power and love in the human community.

The correlation method has philosophical significance only to the extent that it rises above a lazy-minded eclecticism as a result of which correlation dissolves into mere juxtaposition and that it does not yield to the blandishments of a tranquil fideism for which the (profane) 'situation' would hold only questions and the (religious) 'message' only answers. Such a method is philosophically pertinent only when the question of the criteria for determining whether the two correlative terms form a mutually satisfactory and adequate fit is posed in a genuinely critical manner.[161]

It is this critical task that sets interpretation into play at the other two levels mentioned above. For one thing, to fit religious statements to the weightiest problems of human existence we must first match an appropriate interpretation of the major Scriptural 'motifs' to the confessional, doctrinal, symbolical and theological expressions of the different Jewish, Christian and Moslem traditions (to confine ourselves to the 'Abrahamic' religions).[162] For another thing, the correlation of religious symbolism with the main threads of common human experience has constantly to be reconsidered in the light of the evolution of the Scriptural sciences: the conflicts of method [163] between historical exegesis, structural exegesis and existential exegesis inevitably have repercussions at all other levels of interpretation; indeed, exegesis has hermeneutic or philosophical implications only if the interpreter avoids being ensnared by the analysis of inner structures and also eschews futile attempts at finding an 'author's intention' hidden behind the text, but rather exerts himself to make apparent the sort of being-in-the-world the text holds out in front of it as its transcendent reference, beyond its purely immanent sense. The problem of the textual referent in this way converges with the problem of the referent of all other forms of symbolism in which human experience is expressed.

In order to discharge its critical task hermeneutics thus also has to carry out a multiplicity of inter-level correlations, as between textual criticism, the critique of traditions and the critique of existence itself.

It is, however, the interpretation of common human experience in a religious sense that provides the acid test of the correlation method and its

161. What this involves is described in detail by Schubert M. OGDEN in 'The task of philosophical theology' (1971) and 'What is theology?' (1972); see also, by the same author, *The Reality of God and Other Essays* (1967).

162. On the concept of 'motif-research' see NYGREN, *Meaning and Method in Philosophy and Theology. Prolegomena to a Scientific Study of Religion* (1971), and on the concept of tradition MACKEY, *The Modern Theology of Tradition* (1962).

163. Cf. VAN HARVEY, *The Historian and the Believer* (1965); FUNK, *Language, Hermeneutic and the Word of God* (1966); ROBINSON, J. M., *A New Quest of the Historical Jesus* (1959); OGDEN, *Christ without Myth* (1961); VIA, *The Parables. Their Literary and Existential Dimension* (1967). On exegesis and literary criticism, see MARIN, *Sémiotique de la Passion* (1971); VIA, *The Parables . . ., op. cit.* (1967).

capacity for elaborating criteria as to what constitutes a mutually satisfactory and adequate fit.

A number of British and American philosophers are endeavouring to relate the phenomenology of human experience to the theories of 'meaning' and 'truth-value' of linguistic analysis.[164] Their work involves a sharp criticism of the concept of experience handed down by Hume and the British empiricists [165] and a corresponding rapprochement with existentialist phenomenologists, with their investigation of the pre-reflective, pre-conceptual, pre-thematic sphere of human experience. It is here that the hermeneutics of existence attempts to discern a religious dimension. For some writers, in the line of Heidegger's *An Introduction to Metaphysics*,[166] this point of negative entry is provided by dread. Others, with Tillich, emphasize the 'predicament' of modern man, torn between 'ultimate concern' and the variety of situations involving alienation. For others again, like Schubert Ogden,[167] it is on the contrary man's underlying confidence in the meaningfulness of life that implicitly connotes a religious dimension. Bernard Lonergan discerns the existence of a formally unconditioned factor, implied in scientific research and moral studies as well as in explicitly religious experience, and calls it 'being-in-love-without-qualification'.[168] In one way or another, all these analyses hark back, beyond the break made by the 'Wholly Other' theologians, to Schleiermacher's interpretation of religious feeling as a feeling of absolute dependence; [169] but in addition to picking up this thread they interweave with it the phenomenological concept of horizon, which serves as it were to bring transcendence into play at the very heart of ordinary human existence.[170]

Other philosophers, following in the line of von Hügel, William James or

164. Cf. HEIMBECK, *Theology and Meaning* (1969); MACQUARRIE, *God-Talk. An Examination of the Language and Logic of Theology* (1957); MARTIN, J. A., *The New Dialogue between Philosophy and Theology* (1966). A large amount of analytical material and relevant discussion will also be found in the following collections: MACINTYRE (ed.), *Metaphysical Beliefs* (1957); MITCHELL (ed.), *Faith and Logic* (1958); VIDLER (ed.), *Soundings. Essays concerning Christian Understanding* (1962); PHILLIPS (ed.), *Religion and Understanding* (1967); VESEY (ed.), *Talk of God* (1969); CASTELLI (ed.), *L'analyse du langage théologique. Le nom de Dieu* (1969); GEFFRÉ (ed.), *Le procès de l'objectivité de Dieu* (1969); DUBARLE (ed.), *La recherche en philosophie et en théologie* (1970). See also FLEW, *God and Philosophy* (1966).

165. Cf. SMITH, *Experience and God* (1958).

166. HEIDEGGER, *Einführung in die Metaphysik* (1953) (English translation, 1959). See the important commentary on this work by Jean WAHL, *Vers la fin de l'ontologie* (1956).

167. Cf. *Christ without Myth* (1961), already mentioned.

168. Bernard LONERGAN is the author of *Insight. A Study of Human Understanding* (1958); his most recent work is *Method in Theology* (1972). Among his disciples we may mention Michael NOVAK (*Ascent of the Mountain. Flight of the Dove. An Invitation to Religious Studies*, 1971) and John DUNNE (*A Search for God in Time and Memory*, 1969).

169. Cf. NIEBUHR, Richard R., *Experiental Religion. A Theology of Power and Suffering* (1972).

170. Cf. DUPRÉ, *The Other Dimension. A Search for the Meaning of Religious Attitudes* (1972).

Bergson, turn rather to the distilled and purified religious experiences of the great mystics. Distrustful of metaphysics, on epistemological or ontological grounds, they rely on what they regard as higher-level evidence concerning the meaning of existence and history and seek to integrate the truths of such experiences into an enlarged philosophy rooted not in natural, immediate experience of the world and of other persons but in the experience of morally exceptional men.[171]

In the work being done by French- and German-speaking authors, this absolute dimension is less frequently keyed to some lived-through, tested characteristic of human experience – still less to the exceptional experience of the mystics – than to human experience's ability to interpret itself through works of culture and the whole panoply of its historical traditions.[172]

Jean Nabert [173] sees the historical category of witness – witness to heroes of the spiritual life, and witness by those heroes to the absolute itself – as the bridge between the quasi-Fichtean idea of the absolute consciousness positing itself and the adulterated experiences of the empirical consciousness. In this way this 'hermeneutics of witness' brings out a special kind of correlation between reflective transcendental philosophy and interpretation of the historical and cultural signs of the spiritual life.

With the category of witness, the 'language character' of all human experience, as brought to light by the later Heidegger and by H. G. Gadamer's hermeneutic philosophy [174] again comes to the fore. Even the mystical experience of a St. John of the Cross [175] attests this mediative rôle played by a symbolic language at the very heart of the experience which is supposed to abolish discourse.

The works so far mentioned place their main emphasis on the religious implications of personal, even solitary, experience. Other works stress its historical, cultural, communal, ethical and political aspect and the necessary handing down of a collective tradition in any interpretation of human existence by itself.[176] Even with Gadamer, hermeneutics is a hermeneutics of his-

171. Cf. Lossky, *Vision de Dieu* (1962); Nédoncelle, 'Philosophie de la religion' (1958); Stein, *Endliches und Ewiges Sein* (1950).

172. Cf. the series of collections edited by Enrico Castelli: *Problema della demitizzazione* (1961); *Ermeneutica e tradizione* (1963) or *Herméneutique de l'herméneutique* (1963); *Démythisation et morale* (1965); *Mythe et foi* (1966); *Le mythe de la peine* (1967); *L'herméneutique de la liberté religieuse* (1968); *Analyse du langage théologique. Le nom de Dieu* (1969).

173. Nabert, *Le désir de Dieu* (1966).

174. Gadamer, *Wahrheit und Methode* (1960, 1965). Cf. Palmer, *Hermeneutics: Interpretation-Theory in Schleiermacher, Dilthey, Heidegger and Gadamer* (1969); Zahrnt (ed.), *Gespräch über Gott* (1968); Ott, *Denken und Sein. Der Weg Martin Heideggers und der Weg der Theologie* (1959) and *Dogmatik und Verkündigung* (1961); Robinson & Cobb (eds.), *New Frontiers in Theology Series*: I. *The Later Heidegger and Theology* (1963) and II. *The New Hermeneutic* (1964).

175. Cf. Morel, *Le sens de l'existence selon saint Jean de la Croix* (3 vols., 1960-61).

176. The influence of the German theologian Dietrich Bonhöffer, executed by the Nazis in 1944, is decisive in this respect. Cf. his posthumous *Ethik* (1958) (English translation, *Ethics*, 1955).

torical experience and tradition. More decisively, Jürgen Moltmann [177] links the theological strain of hope with a political reading of history. Metz [178] and his disciples, in their desire to remove the Christian interpretation of existence from the 'private' sphere, are founding a political theology which for some of them is frankly becoming a theology of revolution. Thus new correlations are being established between theological hermeneutics, philosophy of history [179] and criticism of ideologies in the style of either Marxist philosophers or the Frankfurt school.[180]

We shall make three remarks by way of conclusion. The diversity of trends in hermeneutics of the religious fact attests that there is no human experience that is not interpreted and that every interpretation involves a decision about the actual meaning of what is called modernity and about the 'contemporary relevance' of any discourse on man. This diversity further shows how far the correlation method has moved from the over-simplified conception of an apologetics whose task would merely be to key the 'answers' provided by the message to the 'questions' constituted by the situation; rather, the method brings to light increasingly complex cross-relations, between textual exegesis, the interpretation of historical traditions, and the deciphering of personal, communal and political existence. Lastly, this development of hermeneutic research has repercussions on theology properly speaking: these do not concern us here; suffice it to note that the age-old dispute between 'supernaturalism' and secular 'naturalism', and also between 'orthodoxy' and 'liberalism', is turned topsy-turvy by hermeneutic research. This shifting of positions in a field distinct from their own cannot fail to interest philosophers – including the atheistical philosopher, assuming he wants to know his adversary better and identify him more accurately.

Conclusion. Humanism in search of its foundation

Even though the exponents of secularization and Marxist, psychoanalytical and linguistic critics join hands in aiming at the demystification of religious thought, a number of thinkers remain convinced that humanism can only be grounded if thought and life are related to the divine ultimate and that it will end by destroying itself if it pushes its atheistic nihilism to the logical extreme. We shall mention only Aldous Huxley, Paul Tillich, Reinhold Niebuhr, Gabriel Marcel, Karl Jaspers, Arnold Toynbee. Their thought centres in situations and experiences that expose man to the presence of an ultimate or an all-encompassing which, if it cannot be named God, is yet in the nature of something at once absolute and divine.

177. Moltmann, *Theologie der Hoffnung* (1965), *Politische Theologie* (1969), and *Religion, Revolution and the Future* (1969).

178. Metz, *Zur Theologie der Welt* (1968).

179. Cf. Pannenberg, *Offenbarung als Geschichte* (1961) and *Was ist der Mensch? Die Anthropologie der Gegenwart im Lichte der Theologie* (1968).

180. Cf. Habermas (ed.), *Hermeneutik und Ideologiekritik. Theorie-Diskussion* (1971).

Over against these philosophers stand others who are convinced that all religious belief alienates man from the task of building his earthly humanity or at any rate keeps him captive to an illusion of thought and desire; only a radical affirmation of his finite nature will make him commit himself in all seriousness to the human adventure; thus the prerequisite for true humanism, whether founded on freedom, on the dialectical relations of man and nature or on self-governing reason, is the death of God. For thinkers in the first group mentioned, on the contrary, the death of God would mean the death of man.

Every trend in philosophy, every science, goes to bear out one or other humanist view; and those thinkers whose humanism is not bound up either with religious assent or with professed atheism are few and far between.

While contemporary philosophy of every school is at one in its ethical demands – as recorded in the Universal Declaration of Human Rights – it splinters into conflicting tendencies in its ultimate views of humanism. Philosophers do not agree among themselves when it comes to their convictions about the final liberation of man, and they may even oppose one another in seeking a ground or foundation for his humanity.

ANNEX: LIST OF WORKS CITED *

ABBAGNANO, Nicola, *Possibilità e libertà*, Turin, Taylor, 1956.

—, 'I compiti di una storiografia della filosofia', in ABBAGNANO, *Scritti scelti*, Turin, Taylor, 1967, pp. 191-198.

ABRAMJAN, L. A., *Gnoseologičeskie problemy teorii znakov* (= Gnoseological Problems of the Theory of Signs), Yerevan, 1965.

ACHINSTEIN, Peter, *Concepts of Science. A Philosophical Analysis*, Baltimore (Md.), Johns Hopkins Press, 1968.

ACKERMANN, Wilhelm, 'Begründung einer strengen Implikation', *The Journal of Symbolic Logic*, Vol. XXI, 1956, pp. 113-128.

Actes du Premier Congrès de Linguistes, 10-15 avril 1928, Leiden, 1928.

ADDISON, J. W., HENKIN, L. & TARSKI, A. (eds.), *The Theory of Models, Proceedings of the 1963 International Symposium at Berkeley*, Amsterdam, North Holland Publ. Co., 1965.

ADORNO, Theodor W., 'Zur Logik der Sozialwissenschaften' (discussion with POPPER), *Kölner Zeitschrift zur Soziologie und Sozialpsychologie*, 14 (233), 1962.

—, *Negative Dialektik*, Frankfurt am Main, Suhrkamp, 1966.

AGASSI, Joseph, 'Methodological individualism', *British Journal of Sociology* XI (3), Sept. 1960, pp. 244-270.

AGOSTI, E., *Nacija i kul'tura* (= Nation and Culture), Moscow, 1963.

AIKEN, Henry D., *Reason and Conduct. New Bearings in Moral Philosophy*, New York, Knopf, 1962.

AKČURIN, I. A., VEDENOV, M. F. & SAČKOV, Ju. V., *Poznavatel'naja rol' matematičeskogo modelirovanija* (= The Cognitive Role of Mathematical Modelling), Moscow, 1968.

AKHMANOVA, Olga & GINZBURG, Rozalia, 'Concerning the dialectics of language and thought' (summary), in "Sprachphilosophie' (Section III, XIVth International Congress of Philosophy, Vienna, 1968), *op. cit.* (1969), pp. 435-436.

Akten des XIV. Internationalen Kongresses für Philosophie, Wien, 2-9 September 1968 / Proceedings of the XIVth International Congress of Philosophy, Vienna, 2nd to 9 th September 1968 / Actes du XIVème Congrès international de Philosophie, Vienne, du 2 au 9 septembre 1968, Vienna, University of Vienna / Herder & Co., 1968-69.

[See also – 'Marx und die Philosophie der Gegenwart'; – 'Sprachphilosophie'.]

ALAIN, *Le citoyen contre les pouvoirs* (80 '*propos*'), Paris, Simon Kra, 1926.

ALBERT, Hans (ed.), *Theorie und Realität. Ausgewählte Aufsätze zur Wissenschaftslehre der Sozialwissenschaften*, Tübingen, J. C. B. Mohr, 1964.

ALBRECHT, Erhard, 'Die erkenntnis-theoretische Problematik des sprachlichen Zeichens', *Deutsche Zeitschrift für Philosophie*, 1961, No. 3.

—, 'Die Sprachphilosophie als Grunddisziplin der Philosophie' (summary), in 'Sprachphilosophie' (Section III, XIVth International Congress of Philosophy, Vienna, 1968), *op. cit.* (1969), pp. 430-431.

ALMOND, Gabriel A. & VERBA, Sidney, *The Civic Culture. Political Attitudes and Democracy in Five Nations*, Princeton (N.J.), Princeton University Press, 1963.

ALQUIÉ, Ferdinand, *Signification de la philosophie*, Paris, Hachette, 1971.

— *et al.*, *Structures logiques et structures mentales en histoire de la philosophie* (28 Feb. 1953 meeting of the French Society for Philosophy) = *Bulletin de la Société française de Philosophie*, 46th year, No. 3, June-Sept. 1953: communition by ALQUIÉ, pp. 89-107; discussion with statements by G. BACHELARD, BENICHOU, G. BERGER, P. BURGELIN, MESNAGE, SALZI, J.-M. SCHUHL, J. ULLMO, Jean WAHL and WOLFF, pp. 107-130; correspondence with G. BÉNÉZÉ, pp. 131-132.

* Of classics cited in the two preceding chapters, only those are listed here to which reference has been made in a particular modern edition or translation.

ALTHUSSER, Louis, 'Marxisme et humanisme', *Cahiers de l'Institut des Sciences économiques appliquées* (ISEA) (Paris), June 1964; reproduced in *La nouvelle Critique*, No. 164, Mar. 1965 (discussion on 'Marxisme et humanisme', *op. cit.*), pp. 2-21; and included in ALTHUSSER, *Pour Marx, op. cit.* (1965), pp. 224-249.
—, 'Note complémentaire sur l' "Humanisme réel" ', *La nouvelle Critique*, No. 164, Mar. 1965, pp. 32-37 (reply to the study by SEMPRUN in the discussion on 'Marxisme et humanisme', *op. cit.*); reproduced in ALTHUSSER, *Pour Marx, op. cit.* (1965), pp. 251-258.
—, 'Le marxisme n'est pas un historicisme', in ALTHUSSER *et al., Lire le* CAPITAL, *op. cit.* (1965), Vol. II, pp. 73-108.
—, *Pour Marx*, Paris, Maspero, 1965, 'Théorie' ser., No. 1 (English translation by Ben Brewster, *For Marx*, London, Allen Lane/New York, Pantheon Books, 1970; also in paperback edn., New York, Vintage Books, 1970).
—, 'Sur le *Contrat social* (Les Décalages)', in ALTHUSSER *et al., L'impensé de Jean-Jacques Rousseau = Cahiers pour l'Analyse. Travaux du Cercle d'Epistémologie de l'Ecole Normale Supérieure* (Paris), No. 8, 3rd quarter, 1967 (diffusion Editions du Seuil).
—, *Lénine et la philosophie* (communication presented to the French Society for Philosophy on 24 Feb. 1968, followed by a 'Note annexe'), Paris, Maspero, 1969, 'Théorie' ser.; reproduced, followed by 'Marx et Lénine devant Hegel', *ibid.*, 1972, 'Petite Collection Maspero' (English translation by Ben Brewster, *Lenin and Philosophy and Other Essays*, London, New Left Books, 1971/New York, Monthly Review, 1972).
—, 'Idéologie et appareils idéologiques d'Etat. Notes pour une recherche', *La Pensée* (Paris), No. 151, June 1970, pp. 3–38.
— *et al., Lire le Capital*, Paris, Maspero, 1965, 2 vols.; new edn., recast, in collaboration with Etienne BALIBAR, *ibid.*, 1968, 2 vols. (English translation by Ben Brewster, L. ALTHUSSER & E. BALIBAR, *Reading Capital*, London, New Left Books, 1970/New York, Pantheon Books, 1971).
AMBACHER, Michel, *Méthode de la philosophie de la nature*, Paris, Presses Universitaires de France, 1961.
ANDERSON, Alan Ross, *The Formal Analysis of Normative Systems*, New Haven (Conn.), Office of Naval Research, Group Psychology Branch, 1956 (Technical Report No. 2, Contract No. SAR/Nonr-609 (16)); reproduced in RESCHER (ed.), *The Logic of Decision and Action, op. cit.* (1967), pp. 147-213.
—, 'The logic of norms', *Logique et analyse*, 1958, No. 1, pp. 84-91.
—, 'On the logic of "commitment" ', *Philosophical Studies*, Vol. X, 1959, pp. 23-27.
— & BELNAP, Nuel D., 'The pure calculus of entailment', *The Journal of Symbolic Logic*, Vol. 27, 1962, pp. 19-52.
— & MOORE, Omar Khayyam, 'The formal analysis of normative concepts', *American Sociological Review* XXII (1), Feb. 1957, pp. 9-17.
[See also MOORE & ANDERSON.]
ANDERSON, Alan Ross (ed.), *Minds and Machines*, Englewood Cliffs (N.J.), Prentice-Hall, 1964.
ANDREEV, I. D.: see *Praktika – Kriterij istiny v nauke* (1960).
ANSCOMBE, G. E. M. (Elizabeth), *Intention*, Oxford, Basil Blackwell, 1957.
—, *An Introduction to Wittgenstein's Tractatus*, London, Hutchinson, 1959.
ANTONELLI, Maria-Teresa, *Filosofia e storia della filosofia. Idea di storia della filosofia*, Palermo, Manfredi, 1968, Biblioteca di 'Theorein', No. 3.
APOSTEL, Léo, 'Game theory and the interpretation of deontic logic', *Logique et analyse*, 1960, No. 10, pp. 70-90.
APPEL, K. O., 'Das Verstehen', in Joachim RITTER (ed.), *Historisches Wörterbuch der Philosophie*, Stuttgart, Schwabe & Co., 1971.
APTHEKER, Herbert, *History and Reality* (essays), New York, Cameron Associates, 1955; London, John Calder, 1956.
APTHEKER, Herbert (ed.), *Marxism and Alienation. A Symposium*, New York,

Humanities Press (published for the American Institute for Marxist Studies), 1965.

ÅQVIST, Lennart, 'Deontic logic based on a logic of "better"', *Acta Philosophica Fennica* (Helsinki), Vol. XVI, 1963, pp. 285-290.

—, 'A note on commitment', *Philosophical Studies* XIV (1/2), Jan.-Feb. 1963, pp. 22-25.

—, 'Choice-offering and alternative-presenting disjunctive commands', *Analysis* XXV (5), Apr. 1965, pp. 182-184.

ARANGUREN, José Luis, *Ética*, Madrid, Revista de Occidente, 1958.

—, *Ética y política*, Madrid, Guadarrama, 1963.

ARENDT, Hannah, *The Human Condition. A Study of the Central Dilemmas Facing Modern Man*, Chicago, The University of Chicago Press, 1958 (French translation by Georges Fradier, *Condition de l'homme moderne*, Paris, Calmann-Lévy, 1961, 'Liberté de l'esprit' ser.).

—, *On Violence*, New York, Harcourt, Brace & World / London, Allen Lane The Penguin Press, 1970 (French translation by G. Durand in ARENDT, *Du mensonge à la violence. Essais de politique contemporaine*, Paris, Calmann-Lévy, 1972, 'Liberté de l'esprit' ser.).

L'Argumentation = Revue internationale de Philosophie, No. 58, 1961, Fasc. 1, pp. 327-432.

ARMSTRONG, David M., *Berkeley's Theory of Vision*, Melbourne, Melbourne University Press, 1960.

—, *A Materialist Theory of the Mind*, London, Routledge, 1968.

ARON, Raymond, *Essai sur la théorie de l'histoire dans l'Allemagne contemporaine. La philosophie critique de l'histoire*, Paris, Vrin, 1938; new edn., *ibid.* 1950; republ. as *La philosophie critique de l'histoire. Essai sur une théorie allemande de l'histoire*, Paris, Vrin, 1969, 'Points' ser.

—, *Introduction à la philosophie de l'histoire. Essai sur les limites de l'objectivité historique*, Paris, Gallimard, 1938, Bibliothèque des Idées; new edn., rev., *ibid.*, 1948 (English translation by G. J. Irwin, *Introduction to the Philosophy of History. An Essay on the Limits of Historical Objectivity*, London, Weidenfeld / Boston, Beacon Press, 1961).

—, *L'opium des intellectuels*, Paris, Calmann-Lévy, 1955, 'Liberté de l'esprit' ser.; new edn., Paris, Gallimard, 1968, 'Idées' ser. (English translation by Terence Kilmartin, *The Opium of the Intellectuals*, New York, Norton, 1962).

—, *Dimensions de la conscience historique* (collection of essays), Paris, Plon, 1961; new edn., Paris, Union Générale d'Editions, 1965, '10/18' ser.

—, *Dix-huit leçons sur la société industrielle* (based on a series of lectures given at the Sorbonne during the academic year 1955-56), Paris, Gallimard, 1962, 'Idées' ser. (English translation by M. K. Bottomore, *Eighteen Lectures on Industrial Society*, London, Weidenfeld & Nicolson, 1967).

—, *Paix et guerre entre les nations*, Paris, Calmann-Lévy, 1963, 'Liberté de l'esprit' ser. (English translation by Richard Howard & Annette Baker Fox, *Peace and War. A Theory of International Relations*, London, Weidenfeld & Nicolson / New York, Doubleday, 1967).

—, *La lutte de classes. Nouvelles leçons sur les sociétés industrielles* (based on a series of lectures given at the Sorbonne during the academic year 1956-57), Paris, Gallimard, 1964, 'Idées' ser.

—, *Démocratie et totalitarisme* (based on a series of lectures given at the Sorbonne during the academic year 1957-58), Paris, Gallimard, 1965, 'Idées' ser. (English translation by Valence Ionescu, *Democracy and Totalitarianism*, London, Weidenfeld & Nicolson, 1968; New York, Praeger, 1969).

—, *Essai sur les libertés* (amplified version of three lectures delivered at the University of California in April 1963, at the invitation of the Committee of the *Jefferson Lectures*), Paris, Calmann-Lévy, 1965, 'Liberté de l'esprit' ser. (English translation by Helen Weaver, *An Essay on Freedom*, Cleveland (Ohio),

World Publ., 1970).
—, *Trois essais sur l'âge industriel* (1. 'Théorie du développement et idéologies de notre temps'; 2. 'Théorie du développement et philosophie évolutionniste'; 3. 'Fin des idéologies, renaissance des idées'), Paris, Plon, 1966, 'Preuves' ser. (English translation, *The Industrial Society. Three Essays on Ideology and Development*, New York, Praeger / London, Weidenfeld & Nicolson, 1967; republ. paperback, New York, Simon & Schuster, 1968).
—, *Progress and Disillusion. The Dialectics of Modern Society*, Chicago, Encyclopaedia Britannica / New York, Praeger, 1968, 'Britannica Perspectives' ser.; paperback edn., New York, The New American Library, 1969, Mentor Books.
—, *Les désillusions du progrès. Essai sur la dialectique de la modernité* (French version, adapted, of *Progress and Disillusion* (*op. cit.*, 1968), with the addition of a Postface), Paris, Calmann-Lévy, 1969, 'Liberté de l'esprit' ser.
—, *De la condition historique du sociologue* (Inaugural Lecture at the Collège de France, delivered on 1 Dec. 1970), Paris, Gallimard, 1971.
—, *Marxismes imaginaires. D'une sainte famille à l'autre*, Paris, Gallimard, 1970, 'Idées' ser. (includes, with three other essays dating back to 1946 and 1956, the substance of *D'une sainte famille à l'autre. Essai sur les marxismes imaginaires*, *ibid.*, 1969, 'Les Essais' ser., No. CXLVI).
—, 'Comment l'historien écrit l'épistémologie. A propos du livre de Paul Veyne [*Comment on écrit l'histoire*, 1971]', *Annales. Economies, Sociétés, Civilisations*, 26th Year, No. 6, Nov.-Dec. 1971, pp. 1319-1354.
—, *Etudes politiques* (collection of essays), Paris, Gallimard, 1972, Bibliothèque des Sciences humaines.
ARZAKANIN, C. G., 'K voprosy o stanovlenii istorii filosofii kak nauki', *Voprosy filosofii*, 1962, No. 6, pp. 95-107 (summary in English, ARZAKANYAN, Ts. G., 'The development of the history of philosophy into a science', *ibid.*, p. 186; integral English translation, ARZAKANIN, T. G., 'On the problem of the rise of a scientific conception of the history of philosophy', *Soviet Studies in Philosophy* (New York), I (3), Winter 1962-63, pp. 56-66).
ASHLEY, B. M., 'Does natural science attain nature or only the phenomena?', in Vincent Edward SMITH (ed.), *The Philosophy of Physics*, Jamaica (N.Y.), St. John's University Press, 1961, St. John's University Studies, Philosophical Ser., No. 2.
ATKINSON, R. F.: see *sub* '"Ought" and "Is"'.
AUBENQUE, Pierre, 'L'auto-interprétation de la philosophie', *Les Etudes philosophiques*, 1969, No. 1, pp. 57-68.
AUROBINDO, Sri, *The Human Cycle*, new edn., London, Luzac, 1949 (French translation through the care of The Mother, *Le cycle humain*, Paris, Buchet-Chastel, 1973).
—, *The Life Divine*, new edn., New York, Greystone Press, 1949 (French translation under the supervision of Jean Herbert, *La vie divine*, 4 vols.: First Section, Chaps. I to XII, Paris, Derain, 1958, 'Krishna' ser.; – II *La connaissance*, III and IV *La connaissance et l'ignorance* = Vols. V, VI and VII of *Œuvres complètes*, Paris, Albin Michel, 1956-58-59, 'Spiritualités vivantes' ser.), 'Hindouisme' sub-ser.).
—, *The Ideal of Human Unity*, new edn., London, Luzac, 1950 (French translation through the care of The Mother, *L'idéal de l'unité humaine*, Paris, Buchet-Chastel, 1972).
—, *The Future Evolution of Man. The Divine Life upon Earth*, edited by P. B. Saint-Hilaire, London, Allen & Unwin, 1963 (French translation, *L'évolution future de l'humanité. La vie divine sur la terre*, Paris, Presses Universitaires de France, 1962).
AUSTIN, J. L. (John Langshaw), 'Ifs and cans', *Proceedings of the British Academy*, 1956; reproduced in AUSTIN, *Philosophical Papers, op. cit.* (1961).
—, 'A plea for excuses' (The Presidential Address to the Aristotelian Society,

1956), *Proceedings of the Aristotelian Society*, Supp. Vol. LVII, 1956-57, pp. 1-30; reproduced in AUSTIN, *Philosophical Papers, op. cit.* (1961), pp. 123-152 (French translation by Robert Franck, 'Les excuses', *Revue de Métaphysique et de Morale*, 72th Year, No. 4, Oct.-Dec. 1967, pp. 414-445).

—, 'Performative utterances' (based on a B.B.C. broadcast, in AUSTIN, *Philosophical Papers, op. cit.* (1961), pp. 220-239.

—, *Philosophical Papers*, edited by J. P. Urmson & G. J. Warnock, Oxford, Clarendon Press, 1961.

—, *How to Do Things with Words* (The William James Lectures delivered at Harvard University), edited by J. O. Urmson, Oxford, Clarendon Press, 1962 (French translation by Gilles Lane, *Quand dire, c'est faire*, Paris, Seuil, 1970, 'L'ordre philosophique' ser.).

—, 'Performative-constative' (lecture delivered at the Entretiens de Royaumont), in C. E. CATON (ed.), *Philosophy and Ordinary Language*, Urbana, University of Illinois Press, 1963, pp. 22-54; French version, 'Performatif-constatif', in *La philosophie analytique*, Paris, Minuit, 1962, 'Cahiers de Royaumont', Philosophie, No. IV, pp. 271-304.

—, *Sense and Sensibilia* (*reconstructed from the manuscript notes by G. J. Warnock*), Oxford, Clarendon Press, 1962 (French translation by Paul Gochet, *Le langage de la perception* (*texte établi d'après les notes manuscrites de l'auteur par G. J. Warnock*), Armand Colin, Paris, 1971, 'U 2' ser.).

AXELOS, Kostas, *Pour une éthique problématique*, Paris, Minuit, 1972.

AYER, Alfred J., *Language, Truth and Logic*, London, Gollancz, 1936 (French translation by Joseph Ohana, *Langage, vérité et logique*, Paris, Flammarion, 1956, Bibliothèque de Philosophie scientifique).

—, *The Problem of Knowledge*, Harmondsworth (Middx.), Penguin Books, 1956.

—, 'Man as a subject for social science', in LASLETT & RUNCIMAN (eds.), *Philosophy, Politics and Society*, 3rd ser., *op. cit.* (1967).

—, *The Origins of Pragmatism. Studies in the Philosophy of Charles Sanders Peirce and William James*, London, Macmillan, 1968.

—, *Russell and Moore: The Analytical Heritage*, Cambridge (Mass.), Harvard University Press, 1971.

AYER, Alfred J. (ed.), *Logical Positivism*, London, Allen & Unwin / Glencoe (Ill.), The Free Press, 1959.

AYERS, M. R., *The Refutation of Determinism*, London, Methuen, 1968.

BACHELARD, Gaston, *Le nouvel esprit scientifique*, Paris, Alcan, 1934; republ., Presses Universitaires de France, 1963.

—, *La philosophie du non*, Paris, Presses Universitaires de France, 1940 (English translation by G. C. Waterston, *Philosophy of No. A Philosophy of the New Scientific Mind*, New York, Grossman, 1968).

—, *Le rationalisme appliqué*, Paris, Presses Universitaires de France, 1949.

—, *L'activité rationaliste de la physique contemporaine*, Paris, Presses Universitaires de France, 1951.

—, *Le matérialisme rationnel*, Paris, Presses Universitaires de France, 1953.

—, *L'engagement rationaliste*, collected essays (posth. publ.), Preface by Georges CANGUILHEM, Paris, Presses Universitaires de France, 1972.

Bachelard = L'Arc, No. 42, 3rd quarter, 1970 (articles by Catherine BACKÈS, Bernard PINGAUD *et al.*).

BACHELARD, Suzanne, *La logique de Husserl. Etude sur Logique formelle et logique transcendantale*, Paris, Presses Universitaires de France, 1957, 'Epiméthée' ser. (English translation by Lester E. Embree, *A study of Husserl's Formal and Transcendental Logic*, Evanston (Ill.), Northwestern University Press, 1968, Studies in Phenomenology and Existential Philosophy ser.).

BACHRACH, Peter & BARATZ, Norton S., 'Two faces of power', *American Political Science Review* LVI (4), Dec. 1962, pp. 947-952.

BAIER, Kurt, *The Moral Point of View. A Rational Basis of Ethics*, Ithaca (N.Y.), Cornell University Press, 1958.

BAILEY, F. G., *Stratagems and Spoils. A Social Anthropology of Politics*, Oxford, Basil Blackwell, 1969 (French translation by Jean Copans, *Les règles du jeu politique. Etude anthropologique*, Paris, Presses Universitaires de France, 1971).

BALADONI, Nicola, 'Filosofia, storia, e storia della filosofia nel marxismo', *Rivista critica di Storia della Filosofia* (Florence), Vol. XIX, 1964, pp. 62-86.

BALANDIER, Georges, *Anthropologie politique*, Paris, Presses Universitaires de France, 1967 (English translation by A. M. Sheridan Smith, *Political Anthropology*, London, Allen Lane The Penguin Press, 1970).

—, *Sens et puissance. Les dynamiques sociales*, Paris, Presses Universitaires de France, 1971.

BAMBROUGH, Renford, *Reason, Truth and God*, London, Methuen, 1969.

BANERJEE, N. V., *Language, Meaning and Persons*, London, Allen & Unwin, 1963.

BANU, Ion, *Introducere in storia filozofiei*, Bucarest, State Presses for Political Literature, 1957.

BARG, M. A., 'O nekotoryh predposylkah formalizacii istoričeskogo issledovanija' (= Some prerequisites for a formalization of historical research), in *Problemy vseobščej istorii* (= Problems of World History), collection, Kazan, 1967.

—, 'Učenie ob obščestvenno-èkonomičeskih formacijah i konkretnyj analiz istoričeskogo processa' (= Study of socio-economic formations and concrete analysis of the historical process), in *Očerki metodologii poznanija social'nyh javlenij* (= Essays on a Methodology for the Understanding of Social Phenomena), *op. cit.* (1970).

BAR-HILLEL, Yehoshua, *Language and Information. Selected Essays on their Theory and Application*, Reading (Mass.), Addison-Wesley / Jerusalem, Jerusalem Academic Press, 1964.

BARKER, S. F., *Induction and Hypothesis. A Study of the Logic of Confirmation*, Ithaca (N.Y.), Cornell University Press, 1957, Contemporary Philosophy.

BARRY, Brian M., *Political Argument*, London, Routledge, 1965.

BARTH, Karl, *Der Römerbrief* (1919), 8th edn., Bern, G. A. Bäschlin, 1947; new edn., Zurich, EVZ Verlag, 1963 (English translation from the 6th German edition by Edwyn C. Hoskyns, *The Epistle to the Romans*, 1933; new edn., London, Oxford University Press, 1968).

—, *Die kirchliche Dogmatik*, Zurich, EVZ Verlag, 1945-46-51, 3 vols. (English translation by G. W. Bromiley & T. F. Torrance, *Church Dogmatics*, London, T. & T. Clark, 1969).

BARTHÉLEMY-MADAULE, Madeleine, *La personne et le drame humain chez Teilhard de Chardin*, Paris, Seuil, 1967.

BARTOLONE, Filippo, *Struttura e significato nella storia della filosofia*, Bologna, R. Patron, 1964.

BASTIDE, Georges, *Traité de l'action morale*, Paris, Presses Universitaires de France, 1961, 2 vols.: I. *Analytique de l'action morale*; II. *Dynamique de l'action morale*, 'Logos' ser.

—, 'Nature, situation et condition humaines', in *Existence et nature*, Paris, Presses Universitaires de France, 1962.

—, 'Ethique et philosophie des valeurs', in KLIBANSKY (ed.), *Contemporary Philosophy / La philosophie contemporaine, op. cit.* (1971), Vol. IV, pp. 23-32.

—, *Essai d'éthique fondamentale*, Paris, Presses Universitaires de France, 1972.

BATKIN, L. M., 'Tip kul'tury kak istoričeskaja celostnost'. Metodologičeskie zametki v svjazi s ital'janskim Vozroždeniem' (= The cultural type as a historical entity. Methodological notes on the Italian Renaissance), *Voprosy filosofii*, 1969, No. 9.

BATTAGLIA, Felice, *Linee sommarie di dottrina morale*, Bologna, R. Patron, 1958.

BAUDRILLARD, Jean, *Le système des objets. La consommation des signes*, Paris, Gallimard, 1968; new edn., Paris, Denoël / Gonthier, 1972, Bibliothèque

'Médiations'.
BEAUVOIR, Simone de, *Pyrrhus et Cinéas*, Paris, Gallimard, 1944; new edn., with *Pour une morale de l'ambiguïté*, *ibid.*, 1969, 'Idées' ser.
—, *Pour une morale de l'ambiguïté*, Paris, Gallimard, 1947; new edn., followed by *Pyrrhus et Cinéas*, *ibid.*, 1969, 'Idées' ser.
BEDNAROWSKI, W., 'Philosophical argument', *Proceedings of the Aristotelian Society*, Supp. Vol. XXXIX, 1965, pp. 10-46.
BEGIAŠVILI, A., *Sovremennaja anglijskaja lingvističeskaja filosofija* (= Contemporary British Linguistic Philosophy), Tbilisi, 1965.
BEIERWALTES, Werner, 'Geschichtlichkeit als Element der Philosophie', *Tijdschrift voor filosofie* (Louvain-Utrecht), Vol. XXX, 1968, pp. 248-263.
BEJERHOLM, Lars & HORNIG, Gottfried, *Wort und Handlung. Untersuchungen zur analytischen Religionsphilosophie*, Gütersloh, G. Mohn, 1966.
BELAVAL, Yvon, *Leibniz critique de Descartes*, Paris, Gallimard, 1960.
— *et al.*, *L'histoire de la philosophie et son enseignement* (25 Nov. 1961 meeting of the French Society for Philosophy = *Bulletin de la Société française de Philosophie*, 56th Year, No. 2, Apr.-June 1962: communication by BELAVAL, pp. 41-58; discussion with statements by J. FILLIOZAT, M. P. DE GANDILLAC, H. GOUHIER, HEIDSIECK, J. MOREAU, Lucie PRENANT, R. QUENEAU, A. ROBINET and Jean WAHL, pp. 58-73; correspondence with É. SOURIAU, M. FRÉCHET, Th. RUYSSEN, G. BÉNÉZÉ, H. MAVIT, M. SOURIAU, Lucie PRENANT, pp. 74-85.
BELL, David: see MACINTYRE and BELL.
BENACERRAF, Paul, 'God, the Devil, and Gödel', *The Monist* (La Salle, Ill.), LI (1), Jan. 1967, pp. 9-37.
BENNETT, Jonathan, *Rationality. An Essay towards an Analysis*, London, Routledge / New York, Humanities Press, 1964.
—, *Kant's Analytic*, Cambridge, Cambridge University Press, 1966.
BENVENISTE, Emile, 'Communication animale et langage humain', *Diogène*, No. 1, Nov. 1952, pp. 1-8 (parallel publication of English translation, 'Animal communication and human language. The language of the bees', *Diogenes*, No. 1, 1953, pp. 1-7); reproduced in BENVENISTE, *Problèmes de linguistique générale*, *op. cit.* (1966), pp. 56-62.
—, *Problèmes de linguistique générale* (collection of essays), Paris, Gallimard, 1966, 'Bibliothèque des Sciences humaines' (English translation by Mary E. Meek, *Problems in General Linguistics*, Miami (Fla.), University of Miami Press, 1971).
BERGBOHM, K., *Jurisprudenz und Rechtsphilosophie*, Leipzig, 1892.
BERGER, Peter L., *The sacred canopy. Elements of a Sociological Theory of Religion*, New York, Doubleday, 1967.
— & LUCKMANN, Thomas, *The Social Construction of Reality. A Treatise in the Sociology of Knowledge*, New York, Doubleday, 1966 / Harmondsworth (Middx.), Penguin Books, 1967.
BERGMAN, Samuel Hugo, 'Philosophy and the history of philosophy', in *Further Studies in Philosophy*, Jerusalem, The Magnes Press, 1968, pp. 127-137.
BERGMANN, Gustav, 'Ideology', *Ethics* LXI (3), Apr. 1951, pp. 205-218.
BERGSON, Henri, *L'évolution créatrice*, Paris, Alcan, 1907; republ., Presses Universitaires de France (English translation by Arthur Mitchell, *Creative Evolution*, New York, 1911).
—, 'L'intuition philosophique' (lecture delivered at the Bologna Congress of Philosophy on 10 Apr. 1911), *Revue de Métaphysique et de Morale* XIX (6), Nov. 1911, pp. 809-827; reproduced in BERGSON, *La pensée et le mouvant. Essais et conférences*, Paris, Alcan, 1934, pp. 117-142; republ., Presses Universitaires de France (English translation by Mabelle Andison, 'Philosophical intuition', in BERGSON, *The Creative Mind*, New York, 1946).
BERLIN, Sir Isaiah, *Historical Inevitability*, London, Oxford University Press, 1954.
—, 'History and theory. The concept of scientific history', *History and Theory*

I (1), 1960, pp. 1-31.
—, *Four Essays on Liberty*, London, Oxford University Press, 1969.
BERMAN, Marshall, *The Politics of Authenticity*, London, Allen & Unwin, 1971.
BESSE, Guy, *Pratique sociale et théorie*, Paris, Editions sociales, 1963.
BETH, E. W., 'Semantics of physical theories', *Synthèse*, 12, 1960, pp. 172-175.
BETTI, Emilio, *Die Hermeneutik als gemeine Methodik der Geisteswissenschaften*, Tübingen, J. C. B. Mohr, 1952.
—, *Teoria generale della interpretazione*, Milan, Giuffré, 1955.
BHATTACHARYYA, Kalidas, *Alternative Standpoints in Philosophy*, Calcutta, Das Gupta & Co., 1953.
BHATTACHARYYA, Krishna Chandra, *Studies in Philosophy*, edited by Gopinath Bhattacharyya, Calcutta, Progressive Publishers, 1956-58, 2 vols.
BIBLER, V. S., *O sisteme kategorij dialektičeskoj logiki* (= On the System of Categories of Dialectical Logic), Dushambe, 1958.
BIGO, Pierre, *Marxisme et humanisme. Introduction à l'œuvre économique de Karl Marx*, Preface by Jean Marchal, Paris, Presses Universitaires de France, 1961, Bibliothèque de la Science économique.
BINSWANGER, Ludwig, *Grundformen und Erkenntnis menschlichen Daseins*, Munich, Ernst Reinhardt, 1942; 4th edn. 1964.
BIRAULT, Henri, BOUILLARD, Henri *et al.*, *L'existence de Dieu* (communications and discussions of the Septièmes Rencontres doctrinales held at La Sarte in April 1960), Tournai-Paris, Casterman, 1961, Cahiers de l'Actualité religieuse, No. 16.
BIRD, Graham H., *Kant's Theory of Knowledge. An Outline of One Central Argument in the Critique of Pure Reason*, London, Routledge / New York, Humanities Press, 1962.
BIRD, Otto, 'The re-discovery of the topics: Professor Toulmin's inference warrants', *Mind* LXX (230), Octo. 1961, pp. 534-539.
BIRKHOFF, Garret & NEUMANN, John von, 'The logic of quantum mechanics', *Annals of Mathematics*, 37, 1936, pp. 823 sqq.
BIRNBAUM, Norman, *The Sociological Study of Ideology (1940-1960). A Trend Report and Bibliography / L'étude sociologique de l'idéologie (1940-1960). Tendances actuelles de la recherche et bibliographie* (with a French summary of the exposition), *Prepared for the International Sociological Association with the Support of Unesco / Textes établis pour l'Association internationale de Sociologie avec le concours de l'Unesco = Current Sociology / La sociologie contemporaine* IX (2), 1960, issued in 1962 (Oxford, Basil Blackwell).
BLACK, Max, 'Notes on the meaning of "rule"', *Theoria* (Lund), Vol. XXIV, 1958, Pt. 2, pp. 107-126 and Pt. 3, p. 139-161.
—, 'Some questions about Parsons' theories', in BLACK (ed.), *The Social Theories of Talcott Parsons. A Critical Examination, op. cit.* (1961), pp. 268-288.
—, *Models and Metaphors. Studies in Language and Philosophy*, Ithaca (N.Y.), Cornell University Press, 1962.
—, *Philosophical Analysis*, Englewood Cliffs (N.J.), Prentice-Hall, 1963.
—, *A Companion to Wittgenstein's Tractatus*, Ithaca (N.Y.), Cornell University Press / Cambridge, Cambridge University Press, 1964.
—, 'The gap between "Is" and "Should"', *The Philosophical Review* 73 (2), Apr. 1964, pp. 165-181.
—, Article 'Induction', in EDWARDS (ed.), *The Encyclopedia of Philosophy, op. cit.* (1967), Vol. IV, pp. 169-181.
—, Article 'Probability', in EDWARDS (ed.), *The Encyclopedia of Philosophy, op. cit.* (1967), Vol. VI, pp. 464-479.
BLACK, Max (ed.), *The Social Theories of Talcott Parsons. A Critical Examination*, Englewood Cliffs (N.J.), Prentice-Hall, 1961.
BLACKBURN, Robin (ed.), *Ideology in Social Science. Readings in Critical Social Theory*, London, Fontana, 1972.

BLAKE, Christopher, 'Theories of history', *Mind*, N.S., LXIV (253), Jan. 1955, pp. 61-78.

BLALOCK, Hubert M., Jr., *Causal Inference in Nonexperimental Research*, Chapel Hill, University of North Carolina Press, 1964.

BLANCHÉ, Robert, 'L'argumentation philosophique vise-t-elle la vérité?', in *La Théorie de l'argumentation, op. cit.* (*Logique et analyse*, 1963), pp. 195-205.

BLANSHARD, Brand, *Reason and Goodness* (based on the Gifford Lectures at the University of St. Andrews and Nobel Lectures at Harvard University), New York, Macmillan / London, Allen & Unwin, 1961.

BLAUBERG, I. V., SADOVSKIJ, V. H. & JUDIN, E. G. (eds.), *Sistemnye issledovanija* (= Systems Research), Moscow, 1969.

BLOCH, Ernst, *Das Prinzip Hoffnung*, Berlin, Aufbau-Verlag, 1954-55, 2 vols.; republ., Frankfurt am Main, Suhrkamp, 1969.

BLOCH, Marc, *Apologie pour l'histoire, ou Métier d'historien* (posthumous publication of an essay written in 1941), Paris, Armand Colin, 1949; republ., 1961 (English translation by Peter Putnam, with an Introduction by Joseph R. Strayer, *Historian's Craft*, New York, Knopf / Manchester, Manchester University Press, 1954).

BLONDEL, Maurice, *La philosophie et l'esprit chrétien*, Vol. I: *Autonomie essentielle et connexion indéclinable*, Paris, Presses Universitaires de France, 1944.

BLOOMFIELD, Leonard, *Language* (new rev. edn. of *Introduction to the Study of Language* published in 1914), New York, Henry Holt, 1933; republ., 1961 (French translation by Janick Gazio, *Le Langage*, Paris, Payot, 1970).

BLUMENBERG, Hans, 'Paradigmen zu einer Metaphorologie', *Archiv für Begriffsgeschichte*, Vol. VI, 1960.

BOCHENSKI, I., *The Logic of Religion*, New York, New York University Press, 1965.

BOEHM, Rudolf, *Vom Gesichtspunkt der Phänomenologie. Husserl-Studien*, The Hague, Martinus Nijhoff, 1968, 'Phaenomenologica' ser., No. 26.

BOGDANOV, A., *Revoljucija i filosofija*, St. Petersburg, 1905.

BOGDANOV, B. W., 'Methodologische Probleme der Philosophiegeschichte', in *Symposium der Philosophie Historiker (Moskau, 1967), op. cit.* (1968), pp. 171-184.

BOHM, David, *Causality and Change in Modern Physics*, London, Routledge, 1957.

—, 'Hidden variables in the quantum theory', in David R. BATES (ed.), *Quantum Theory*, Vol. III, New York-London, Academic Press, 1962, pp. 345-387.

—, *Problems in the Basic Concepts of Physics*, lecture delivered at Birkbeck College on 13 Feb. 1963.

BOHNERT, Herbert Gaylord, 'The semiotic status of commands', *Philosophy of Science* XII (4), Oct. 1945, pp. 302-315.

BOHR, Niels, *Atomic Physics and Human Knowledge*, New York, Wiley / London, Chapman, 1958.

—, 'Quantenphysik und Philosophie', in *Atomphysik und menschliche Erkenntnis*, Moscow, 1961.

—, *Essays 1958 to 1962 on Atomic Physics and Human Knowledge*, New York, Random House, 1963, Vintage Books.

BOLLACK, Jean, *Empédocle*, Paris, Minuit, 4 vols., 'Le sens commun' ser.: I. *Introduction à l'ancienne physique*, 1965; II. *Les origines. Edition critique et traduction des fragments et des témoignages*, 1969; III. *Les origines. Commentaire des fragments et des témoignages*, 2 vols., 1969.

—, BOLLACK, Mayotte & WISMANN, Heinz, *La lettre d'Epicure*, Paris, Minuit, 1971, 'Le sens commun' ser.

— & WISMANN, Heinz, *Héraclite, ou la séparation*, Paris, Minuit, 1972, 'Le sens commun' ser.

BOLLHAGEN, P., *Gesetzmässigkeit und Gesellschaft* (= Legality and Society), Berlin, 1967.

BOLLNOW, Otto Friedrich, *Wesen und Wandel der Tugenden*, Frankfurt am Main. Ullstein Taschenbücher, 1958.

BONDURANT, Joan V., *Conquest of Violence: The Gandhian Philosophy of Conflict*, rev. edn., Berkeley, University of California Press, 1965.

BONHÖFFER, Dietrich, *Ethik* (posthumous publ.), edited by Eberhard Bethge, Munich, Chr. Kaisér Verlag, 1958; 7th edition, 1966 (English translation by Neville Horton Smith, *Ethics*, London / New York, 1955).

BOOTH, Andrew Donald (ed.), *Machine Translation*, Amsterdam, North Holland Publ. Co. / New York, Wiley, 1967.

BORN, Max, *Natural Philosophy of Cause and Change* (Waynflete Lectures, 1948), London, Oxford University Press, 1949.

—, 'Vorhersagbarkeit in der klassischen Mechanik', *Zeitschr. Phys.*, No. 153, 1958, pp. 372-388.

BOURDIEU, Pierre, CHAMBOREDON, Jean-Claude & PASSERON, Jean-Claude, *Le métier de sociologue*, Book I: *Préalables épistémologiques*, Paris, Mouton / Bordas, 1968, 'Les textes sociologiques' ser. (Book II: *La problématique*, and Book III: *Les outils*, to be published later).

BOURGEOIS-PICHAT, Jean, 'La démographie', in *Tendances principales de la recherche dans les sciences sociales et humaines, Partie I: Sciences sociales, op. cit.* (1970), pp. 427-503 (parallel publication of English version, 'Demography', in *Main Trends of Research in the Social and Human Sciences, Part I: Social Sciences, op. cit.* (1970), pp. 351-418).

BOURRICAUD, François, 'En marge de l'œuvre de Talcott Parsons: la sociologie et la théorie de l'action' = Introduction (pp. 1-107) to PARSONS, *Eléments pour une sociologie de l'action* (French translation), *op. cit.* (1955).

BOUVERESSE, Jacques, 'La théologie rationnelle et l'analyse logique du langage', *Critique* (Paris), No. 254, July 1968, pp. 678-714; reproduced in J. BOUVERESSE, *La parole malheureuse*, Paris, Minuit, 1971.

BRAITHWAITE, R. B., *Scientific Explanation*, Cambridge, Cambridge University Press, 1953.

—, *An Empiricist's View of the Nature of Religious Belief*, Cambridge, Cambridge University Press, 1955.

—, *The Theory of Games as a Tool for the Moral Philosopher*, Cambridge, Cambridge University Press, 1955.
[See also *sub* RAMSEY.]

BRANDT, Richard Booker, *Ethical Theory. The Problems of Normative and Critical Ethics*, Englewood Cliffs (N.J.), Prentice-Hall, 1959.

BRAUDEL, Fernand, *Ecrits sur l'histoire* (collection of previously published essays), Paris, Flammarion, 1969, 'Science' ser.

BRAYBROOKE, David (ed.), *Philosophical Problems of the Social Sciences*, New York, Macmillan, 1965.

BRÉHIER, Emile, 'Introduction', in BRÉHIER, *Histoire de la philosophie*, Vol. I. *L'Antiquité et le Moyen Age*, Fasc. 1, *Introduction. Période hellénique*, Paris, Alcan, 1926; republ., Presses Universitaires de France.

—, *La philosophie et son passé*, Paris, Alcan-Presses Universitaires de France, 1940, 'Nouvelle Encyclopédie philosophique' ser.

BRELAGE, Manfred, 'Die Geschichtlichkeit der Philosophie und die Philosophiegeschichte', *Zeitschrift für philosophische Forschung* (Meisenheim am Glan), Vol. XVI, 1962, pp. 375-405.

BRESSON, François, 'Les décisions', in Paul FRAISSE & Jean PIAGET (eds.), *Traité de psychologie expérimentale*, Paris, Presses Universitaires de France, 1965, Chap. XXIX.

BRETON, Stanislas, 'Monde et nature', in *Idée du monde et philosophie de la nature, op. cit.* (1966).

—, *Du principe*, Paris, BSR/Aubier-Montaigne/Editions du Cerf/Delachaux & Niestlé/Desclée de Brouwer, 1971.

—, *Foi et raison logique*, Paris, Seuil, 1971.

BRIDGMAN, P. W., 'P. W. Bridgman's *The Logic of Modern Physics* after thirty years' (a re-assessment, by the author himself, of his own work, *The Logic of Modern Physics*, published in 1928 by The Macmillan Company, New York), in *Daedalus* 88 (3), Summer 1959 (*Current Work and Controversies*), pp. 518-526.

BRILLOUIN, Léon, *Scientific Uncertainty and Information*, New York, Academic Press, 1964.

BRIMO, Albert, *Les grands courants de la philosophie du droit et de l'Etat*, Paris, Pedone, 1967.

BRITTON, K., *J. S. Mill*, Harmondsworth (Middx.)-London, Penguin Books, 1953.

BROAD, C. D.: see SCHILPP (ed.), *The Philosophy of C. D. Broad.*

BRODBECK, May (ed.), *Readings in the Philosophy of the Social Sciences*, New York, Macmillan, 1968.

BROEKMAN, Jan M., *Phänomenologie und Egologie. Faktisches und transzendentales Ego bei Edmund Husserl*, The Hague, Martinus Nijhoff, 1963, 'Phaenomenologica' ser.

BROGLIE, Louis de, *Certitudes et incertitudes de la science*, Paris, Albin Michel, 1966, 'Sciences d'aujourd'hui' ser.

BRÖNDAL, Viggo, *Essais de linguistique générale*, Copenhagen, Munksgaard, 1943.

BRONOWSKI, Jacob, *The Identity of Man*, New York, Natural History Press (Doubleday), 1965; republ., New York, Heinemann, 1966; rev. edn. (paperback), New York, Nat. Hist. Press, 1971.

—, 'The logic of mind', *American Scientist*, Vol. LIV, 1966, pp. 1-14.

BROWN, Robert R., *Explanation in Social Science*, London, Routledge, 1963.

BRUAIRE, Claude, *L'affirmation de Dieu, Essai sur la logique de l'existence*, Paris, Seuil, 1964.

—, *Logique et religion chrétienne dans la philosophie de Hegel*, Paris, Seuil, 1964.

BRUNNER, Fernand, 'Histoire de la philosophie et philosophie', in *Etudes sur l'histoire de la philosophie en hommage à Martial Guéroult, op. cit.* (1964), pp. 179-204.

BRUNSCHVICG, Léon, *Le progrès de la conscience dans la philosophie occidentale*, Paris, Alcan, 1927, 2 vols.; republ., Paris, Presses Universitaires de France, 1953.

BRUTIAN, George, 'On some aspects of language as an object of philosophical investigation' (summary), in 'Sprachphilosophie' (Section III, XIVth International Congress of Philosophy, Vienna, 1968), *op. cit.* (1969), pp. 431-433.

BUBER, Martin, *Ich und Du*, Leipzig, Insel Verlag, 1923; reproduced in BUBER, *Die Schriften über das dialogische Prinzip*, Heidelberg, Lambert Schneider Verlag, 1954 (English translation by Walter Kaufman, *I and Thou*, New York, Scribner, 1970).

—, *Das Problem des Menschen*, Heidelberg, Lambert Schneider Verlag, 1948 (English translation by Maurice Friedman & Ronald Gregor, *Knowledge of Man. Selected Essays*, London, Allen & Unwin, 1965; New York, Harper, 1966).

BUBNER, Rüdiger, 'Philosophie ist ihre Zeit, in Gedanken erfasst', in *Hermeneutik und Dialektik, op. cit.* (1970), Vol. I, p. 317 sqq.

BÜCHEL, Wolfgang, *Philosophische Probleme der modernen Physik. Vorlesungen über Naturphilosophie*, Pullach bei München, Verlag Berchmanskolleg, 1957.

BUHR, Manfred & IRRLITZ, G., *Der Anspruch der Vernunft*, Vol. I, Berlin, 1968.

BUHR, Manfred (ed.), *Zur Kritik der bürgerlichen Ideologie*, series of pamphlets, 11 issues published down to 1971, Berlin.

BULTMANN, Rudolf, *Jesus* (1st edition, Berlin, 1926), republ., Tübingen, J. C. B. Mohr, 1964 (English translation, *Jesus and the Word*, New York/London, 1934).

—, *Die Theologie des Neuen Testaments*, Tübingen, J. C. B. Mohr, 1933 (English translation, *Theology of the New Testament*, New York/London, 1952-55, 2 vols.; republ., New York, Scribner, 1970).

—, *Glauben und Verstehen*, Vols. I, II, III and IV, Tübingen, J. C. B. Mohr, 1933-52-60-65 (English translation by Louise P. Smith, edited and with an Introduction by Robert W. Fink, *Faith and Understanding*, Vol. I, New York, Harper & Row, 1969; Vol. II has been translated under the title: *Essays: Philosophical and Theological*, London, 1955).
—, *Kerygma und Mythos*, Vols. I, II and III, Hamburg, Reich & Heindrich Evans Verlag, 1946-50-54 (English translation by Hans W. Bartsch, *Kerygma and Myth. A Theological Debate*, New York, Harper & Row, paperback, undated).
—, 'Das Problem der Hermeneutik', in BULTMANN, *Glauben und Verstehen*, Vol. II, *op. cit.* (1952), pp. 214 sqq.
—, *History and Eschatology* (Gifford Lectures, 1955), Edinburgh, Edinburgh University Press, 1957; also published under the title: *Presence of Eternity. History and Eschatology*, New York, Harper, 1957; German version, *Geschichte und Eschatologie*, Tübingen, J. C. B. Mohr, 1958.
—, *Exegetica. Aufsätze zur Erforschung des Neuen Testaments*, Tübingen, J. C. B. Mohr, 1967.
BUNGE, Mario, *Causality*, Cambridge (Mass.), Harvard University Press, 1959.
—, *Foundations of Physics*, Berlin/New York, Springer, 1967, Springer Tracts in Natural Philosophy, Vol. 10.
BUNGE, Mario (ed.), *Quantum Theory and Reality* = Vol. II of BUNGE (ed.), *Scientific Research: Studies in the Foundations, Methodology, and Philosophy of Science*, *op. cit.* (1967).
—, *Scientfiic Research. Studies in the Foundations, Methodology, and Philosophy of Science*, Berlin-New York, Springer, 1967, 4 vols.

CALVEZ, Jean-Yves, *La pensée de Karl Marx*, Paris, Seuil, 1956; new edn., rev. and abr., *ibid.*, 1970, 'P. Politique' ser.
CANGUILHEM, Georges, 'Le concept et la vie', *Revue philosophique de Louvain*, Vol. LXIV, May 1966, pp. 193-223 (reproduced in CANGUILHEM, *Etudes d'histoire et de philosophie des sciences*, Paris, Vrin, 1968, pp. 335-364).
—, *Le normal et le pathologique*, Paris, Presses Universitaires de France, 1966.
—, 'Biologie et philosophie', in KLIBANSKY (ed.), *La philosophie contemporaine / Contemporary Philosophy*, Vol. II, *op. cit.* (1968), pp. 387-394.
[See also *sub* BACHELARD, G., *L'engagement rationaliste.*]
ČAPEK, Milič, 'Relativity and the state of space', *Review of Metaphysics*, IX, 1955.
—, *The Philosophical Impacts of Contemporary Physics*, New York, Van Nostrand Reinhold Cy., 1961.
ČARAKIEV, A., *Dejstvennostta no misleneto. Aktivnata rolja na misleneto v processa na poznanieto i praktičeskata dejnost* (= Reality in Thought. The Active Rôle of Thought in the Process of Cognition and in Practical Activity), Sofia, 1961.
CARNAP, Rudolf, 'Über die Abhängigkeit der Eigenschaften des Raums von denen der Zeit', *Kantstudien*, No. XXX, 1925, H. 3/4, pp. 331-345.
—, 'Psychology in physical language' (original publ. in *Erkenntnis*, Vol. II, 1931, pp. 432-465), reproduced in English in AYER (ed.), *Logical Positivism, op. cit.* (1969), pp. 165-198.
—, 'Überwindung der Metaphysik durch logische Analyse der Sprache', *Erkenntnis*, Vol. II, 1932; reproduced in English translation, 'The elimination of metaphysics through the logical analysis of language', in AYER (ed.), *Logical Positivism, op. cit.* (1959).
—, *Philosophy and Logical Syntax*, London, Routledge, 1935.
—, *Introduction to Semantics*, Cambridge (Mass.), Harvard University Press, 1942, Studies in Semantics, Vol. I.
—, *Meaning and Necessity. A Study in Semantics and Modal Logic*, Chicago, The University of Chicago Press, 1947, Phoenix Books, No. 30; 2nd edn., *ibid.*, 1956.
—, *Logical Foundations of Probability*, Chicago, The University of Chicago Press,

1950; London, Routledge, 1962.
—, *The Continuum of Inductive Methods*, Chicago, The University of Chicago Press, 1952.
—, 'The methodological character of theoretical concepts', in FEIGL & SCRIVEN (eds.), *The Foundations of Sciences* = Vol. I of *Minnesota Studies in the Philosophy of Science, op. cit.* (1956), pp. 38-76.
—, 'The aim of inductive logic', in NAGEL, SUPPES & TARSKI (eds.), *Logic, Methodology and Philosophy of Science, op. cit.* (1962), pp. 303-318.
[See also SCHILPP (ed.), *The Philosophy of Rudolf Carnap.*]
CARPIO, Adolfo P., 'El pasado filosófico', *Cuadernos filosóficos* (Rosario), No. 4, 1963, pp. 5-36.
CARR, E. H. (Edward Hallett), *What Is History?*, New York, Knopf, 1962.
CARRERAS Y ARTAU, Joaquím, *Balance, estado actual y perspectivas sobre la historia de la filosofía española*, Madrid, 1951.
CASSIRER, Ernst, *Philosophie der symbolischen Formen*, Berlin, D. Cassirer, 1923-29, 3 vols. (English translation by R. Manheim, *The Philosophy of Symbolic Forms*, Preface by Ch. W. HENDEL, New Haven (Conn.), Yale University Press, 1953-57, 3 vols.).
—, *Sprache und Mythos*, Berlin, 1925 (English translation by Susanne K. Langer, *Language and Myth*, New York, Dover, 1946).
—, *Individuum und Kosmos in der Philosophie der Renaissance*, Leipzig / Berlin, 1927; republ. by photographic reproduction, Darmstadt, Wissenschaftliche Buchgesellschaft, 1963.
—, *Die Philosophie der Aufklärung*, Tübingen, J. C. B. Mohr, 1932 (French translation by Pierre Quillet, *La philosophie des Lumières*, Paris, Fayard, 1966, 'L'histoire sens frontières' ser.).
—, *An Essay on Man. An Introduction to a Philosophy of Human Culture*, New Haven (Conn.), Yale University Press, 1945.
—, *The Myth of the State* (edited by Charles W. HENDEL), New Haven (Conn.), Yale University Press, 1946.
— *et al.*, *Renaissance Philosophy of Man* (prepared under the general editorship of Helmut KUHN), Chicago, University of Chicago Press, 1948.
CASTAÑEDA CALDERÓN, Hector Neri, 'The logic of obligation', *Philosophical Studies*, Vol. X, 1959, pp. 42-48.
—, 'On a proposed revolution in logic (Toulmin's *Uses of Argument*)', *Philosophy of Science* XXVII (3), July 1960, pp. 279-292.
—, 'Imperative reasonings', *Philosophy and Phenomenologcial Research* XXI (1), Sept. 1960, pp. 21-49.
—, 'Imperatives, decisions, and "oughts". A logico-metaphysical investigation', in H. N. CASTAÑEDA & Georges NAHNIKIAN (eds.), *Morality and the Language of Conduct*, Detroit, Wayne State University Press, 1963, pp. 219-299.
—, 'Imperatives, oughts and moral oughts', *Australian Journal of Philosophy* XLIV (3), Dec. 1966, pp. 277-300.
—, 'Acts, the logic of obligation, and deontic calculi', *Philosophical Studies* XIX (1/2), Jan.-Feb. 1968, pp. 13-26.
CASTAÑEDA CALDERÓN, Hector Neri (ed.), *Intentionality, Minds, and Perception, Discussions on Contemporary Philosophy. A Symposium*, Detroit, Wayne State University Press, 1967.
CASTELLI, Enrico *et al.*, *La filosofia della storia della filosofia* (symposium, Istituto di Studi filosofici, Università di Roma), Milan-Rome, 1954; French version, *La philosophie de l'histoire de la philosophie*, Paris, Vrin, 1956: papers by E. CASTELLI, F. DE CORTE, DEMPF (*op. cit.*), A. DEL NOCE, Eugenio O. GIGON, Henri GOUHIER, GUÉROULT (*op. cit.*), Edmund HUSSERL (unpublished papers on the theology of history, edited by P. Valori), E. LOMBARDI, P. VALORI and A. WAGNER.
—, *Herméneutique de l'herméneutique* (Proceedings of the international Sym-

posium 'Herméneutique et tradition'/'Ermeneutica e tradizione', Rome, 10-16 Jan. 1963), Paris, Vrin, 1963, Bibliothèque d'histoire de la philosophie (also published in Italian under the title: *Ermeneutica e tradizione*, Rome, 1963).

—, *Démythisation et morale* (Symposium organized by the Centre international d'Etudes humanistes and by the Rome Istituto di Studi filosofici), Paris, Aubier-Montaigne, 1965.

—, *Mythe et foi* (international symposium, Rome, 6-12 Jan. 1966), Paris, Aubier-Montaigne, 1966.

—, *Le mythe de la peine* (international symposium, Rome, Jan. 1967), Paris, Aubier-Montaigne, 1967.

—, *L'herméneutique de la liberté religieuse* (symposium, Rome, 7-12 Jan. 1968), Paris, Aubier-Montaigne, 1968.

—, GOUHIER, Henri, LEVINAS, Emmanuel, PANIKKAR, Raimundo, RICOEUR, Paul, HABACHI, René *et al.*, *Théologie de l'histoire*, Paris, Aubier, 1971, 2 vols.: Vol. I. *Herméneutique et eschatologie*; Vol. II. *Révélation et histoire.*

CASTELLI, Enrico (ed.), *Il problema della demitizzazione* (*Atti del Convegno indetto dall' Istituto di Studi filosofici*, Roma, 16-21 gennaio 1961), Rome, Istituto di Studi filosofici-Università, 1961.

—, *L'analyse du langage théologique. Le nom de Dieu*, Paris, Aubier-Montaigne, 1969.

CAVAILLÈS, Jean, *Sur la logique et la théorie de la science* (posthumous work), Paris, Presses Universitaires de France, 1947.

CENCILLO, Luis, 'Consecuencias de un criterio ecuménico en la historia de la filosofía', *Pensamiento* (Madrid), XXIV (93-94), Jan.-June 1968, pp. 57-69.

CERNEA, M., *Despre dialectica construirii socalismului* (= On the Dialectic of Constructing Socialism), Bucarest, 1967.

CERRONI, Umberto, *Kant e la fondazione della categoria giuridica*, Milan, A. Giuffrè, 1962.

CHAPELLE, Albert, *Hegel et la religion*, Paris, Editions Universitaires: Vol. I. *La problématique*, 1964; Vol. II. *La dialectique*: A. *Dieu et la création* (thesis, Institut Catholique, Paris), 1967; B. *Annexes – Les textes théologiques de Hegel*, 1967.

CHÂTELET, François (ed.), *Histoire de la philosophie*, Paris, Hachette, 1972-73, 8 vols.

CHIODI, Pietro, 'Filosofia, storia e realtà umana', *Rivista di filosofia* (Turin), Vol. LVI, 1965, pp. 123-137.

CHISHOLM, Roderick M., 'The contrary-to-fact conditional', *Mind*, Vol. LV, 1946, pp. 289-307; reproduced in FEIGL & SELLARS (eds.), *Readings in Philosophical Analysis, op. cit.* (1949).

—, 'Law statements and counterfactuals inference', *Analysis*, Vol. XV, 1955; reproduced in E. MADDEN, *The Structure of Scientific Thought. An Introduction to Philosophy of Science*, under the editorship of Lucius Garvin, Boston, Houghton Mifflin / London, Routledge, 1960.

—, *Perceiving. A Philosophical Study*, Ithaca (N.Y.), Cornell University Press, 1957; London, Oxford University Press, 1958.

—, 'Supererogation and offence: a conceptual scheme for ethics', *Ratio*, Vol. V, 1963, pp. 1-14.

—, 'Contrary-to-duty imperatives and deontic logic', *Analysis* XXIV (2), Dec. 1963, pp. 33-36.

—, 'The ethics of requirement', *American Philosophical Quarterly* I (2), Apr. 1964, pp. 147-153.

— & SOSA, Ernest, 'On the logic of "intrinsically better" ', *American Philosophical Quarterly* III (3), July 1966, pp. 244-249.

CHOMSKY, Noam, *Syntactic Structures*, The Hague, Mouton, 1957 (French translation by Michel Braudeau, *Structures syntaxiques*, Paris, Seuil, 1969).

—, 'A review of B. F. Skinner's *Verbal Behavior*', *Language* XXXV (1), 1959, pp. 26-58; reproduced in FODOR & KATZ (eds.), *The Structure of Language, op. cit.* (1964), pp. 547-578.

—, 'On the notion "rule of grammar"', in JAKOBSON (ed.), *The Structure of Language and Its Mathematical Aspects, op. cit.* (1961), pp. 6-24; reproduced in FODOR & KATZ (eds.), *The Structure of Language, op. cit.* (1964), pp. 119-136 (French translation, 'La notion de "règle de grammaire"', in RUWET (ed.), *La grammaire générative, op. cit.* (*Languages*, 1966), pp. 81-104).

—, 'A transformational approach to syntax', in A. A. HILL (ed.), *Proceedings of the Third Texas Conference on Problems of Linguistic Analysis in English*, Austin, The University of Texas, 1962, pp. 124-158; reproduced in FODOR & KATZ (eds.), *The Structure of Language, op. cit.* (1964), pp. 211-245 (French translation, 'Une conception transformationnelle de la syntaxe', in RUWET (ed.), *La grammaire générative, op. cit.* (*Langages*, 1966), pp. 39-80).

—, 'Formal properties of grammars', in LUCE, BUSH & GALANTER (eds.), *Handbook of Mathematical Psychology*, Vol. II, *op. cit.* (1963), pp. 323-418 (French translation, 'Propriétés formelles des grammaires', in CHOMSKY & MILLER, *L'analyse formelle des langues naturelles, op. cit.* (1968), pp. 59-168).

—, *Current Issues in Linguistic Theory*, The Hague, Mouton, 1964.

—, *Aspects of the Theory of Syntax* (M.I.T. Research Laboratory of Electronics, Special Technical Report No. 11), Cambridge (Mass.), The M.I.T. Press, 1965 (French translation by Jean-Claude Milner, *Aspects de la théorie syntaxique*, Paris, Seuil, 1971).

—, *Cartesian Linguistics. A Chapter in the History of Rationalist Thought*, New York, Harper & Row, 1966 (French translation by Nelcya Delanoë & Dan Sperber, *La linguistique cartésienne. Un chapitre de l'histoire de la pensée rationaliste*', followed by 'La nature formelle du langage', Paris, Seuil, 1969, 'L'ordre philosophique' ser.).

—, *Language and Mind*, New York, Harcourt, Brace & World, 1968 (French translation by Louis-Jean Calvet, *Le langage et la pensée*, Paris, Payot, 1970, Petite Bibliothèque Payot).

— & MILLER, George A., 'Introduction to the formal analysis of natural languages', in LUCE, BUSH & GALANTER (eds.), *Handbook of Mathematical Psychology*, Vol. II, *op. cit.* (1963), pp. 269-321 (French translation, 'Introduction à l'analyse formelle des langues naturelles', in CHOMSKY & MILLER, *L'analyse formelle des langues naturelles, op. cit.* (1968), pp. 1-58).

—, *L'analyse formelle des langues naturelles* (*Introduction to the Formal Analysis of Natural Languages*), French translation by Ph. Richard & N. Ruwet (comprises the 2 articles: CHOMSKY & MILLER, 'Introduction à l'analyse formelle des langues naturelles', *op. cit.*, and CHOMSKY, 'Propriétés formelles des grammaires', *op. cit.*, and the bibliography given with the original English text of these articles in LUCE, BUSH & GALANTER (eds.), *Handbook of Mathematical Psychology*, Vol. II, *op. cit.*, 1963), Paris, Gauthier-Villars Mouton, 1968, Ecole Pratique des Hautes Etudes (Sorbonne), Sixième Section: Sciences économiques et sociales.

CHRISTENSEN, Darrel E., 'Philosophy and its history', *The Review of Metaphysics* (New Haven, Conn.), XVIII (1/69), Sept. 1964, pp. 58-83.

CHRISTIAN, William A., *An Interpretation of Whitehead's Metaphysics*, New Haven (Conn.), Yale University Press, 1959.

—, *Meaning and Truth in Religion*, Princeton (N.J.), Princeton University Press, 1964.

CHURCH, Alonzo, 'Mathematics and logic', in NAGEL, SUPPES & TARSKI (eds.), *Logic, Methodology and Philosophy of Science, op. cit.* (1962), pp. 181-186; reproduced in KLIBANSKY (ed.), *Contemporary Philosophy / La philosophie contemporaine*, Vol. I, *op. cit.* (1968), pp. 295-307.

CHURCHMAN, Charles W., *Prediction and Optimal Design. Philosophical Issues of a Science of Values*, Englewood Cliffs (N.J.), Prentice-Hall, 1961.

CLARKE, Bowman L., *Language and Natural Theology*, The Hague, Mouton, 1966.
COHEN, L. Jonathan, *The Diversity of Meaning*, London, Methuen, 1962.
—, *The Implications of Induction*, London, Methuen, 1970.
COLLINS, James Daniel, *God in Modern Philosophy*, New York, Regnery, 1959; London, Routledge, 1960.
COLLINGWOOD, Robin G., *An Essay on Metaphysics*, London, Oxford University Press, 1940.
—, *The Idea of History*, London, Oxford University Press, 1946.
CONTE, A. G., *Saggio sulla completezza degli ordinamenti giuridici*, Turin, Università di Torino, 1962, Memorie dell'Istituto Giuridico, Serie II, Memoria CXI.
CORNFORTH, Maurice, *Marxism and the Linguistic Philosophy*, London, Lawrence & Wishart, 1955.
CORNU, Auguste, *Karl Marx et Friedrich Engels. Leur vie et leur œuvre*, Paris, Presses Universitaires de France, 4 vols. published: Vol. I. *Les années d'enfance et de jeunesse. La gauche hégélienne (1818/1820 – 1844)*, 1955; Vol. II. *Du libéralisme démocratique au communisme. La 'Gazette rhénane'. Les 'Annales franco-allemandes' (1842-1844)*, 1958; Vol. III. *Marx à Paris*, 1962; Vol. IV. *La formation du matérialisme historique (1845-1846)*, 1970. (German version, *Karl Marx und Friedrich Engels. Leben und Werke*, Berlin, 1954 sqq.).
COSSÍO, Carlos, *La teoría egológica del derecho y el concepto jurídico de libertad*, 2nd edn., Buenos Aires, 1964.
COSTA DE BEAUREGARD, Olivier, *La notion de temps. Equivalence avec l'espace*, Paris, Hermann, 1963.
—, *Le second principe de la science du temps: entropie, information, irréversibilité*, Paris, Seuil, 1963.
COWLING, Maurice, *The Nature and Limits of Political Science*, Cambridge, Cambridge University Press, 1963.
CRAHAY, Franz, 'Le "décollage" conceptual: conditions d'une philosophie bantoue' (based on a lecture delivered at the Goethe Institute at Léopoldville on 19 March 1965), *Diogène*, No. 52, Oct.-Dec. 1965, pp. 61-84 (parallel publication in English, translation by Victor A. Velen, 'Conceptual take-off: conditions for a Bantu philosophy', *Diogenes*, No. 52, Winter 1965, pp. 55-78).
CRANSTON, M., *Sartre*, London, Oliver & Boyd, 1962.
CRAWSHAY-WILLIAMS, Rupert, *Methods and Criteria of Reasoning. An Inquiry into the Structure of Controversy*, London, Routledge / New York, Humanities Press, 1958.
La crise de l'humanisme = Revue internationale de Philosophie, No. 85-86, 1968, Fasc. 3-4, pp. 261-402.
CROCE, Benedetto, *Logica*, 1909 (English translation, from the 3rd edn., by D. Ainslee, *Logic*, London, 1917).
—, *Teoria e storia della storiografia*, Bari, Laterza, 1917 (English translation by D. Ainslee, *History. Its Theory and Practice*, New York, 1921).
—, *La storia come pensiero e come azione*, Bari, Laterza, 1938 (English translation by Sylvia Sprigge, *History as the Story of Liberty*, New York, Norton / London, Allen & Unwin, 1941).
CROMBIE, I. M., 'The possibility of theological statements', in MITCHELL (ed.), *Faith and Logic, op. cit.* (1958).
CROSSLEY, John N. & DUMMETT, M. A. E. (eds.), *Formal Systems and Recursive Functions. Proceedings of the 8th Logic Colloquium, Oxford, 1963*, Amsterdam, North Holland Publ. Co., 'Studies in Logic and the Foundations of Mathematics'.
CUÉNOT, Claude, *Pierre Teilhard de Chardin. Les grandes étapes de son évolution*, Paris, Plon, 1958.
—, 'La morale et l'homme selon Pierre Teilhard de Chardin', in *Morale chrétienne et morale marxiste*, Paris, La Palatine, 1960, pp. 117-147.

CULLMANN, Oscar, *Christ et le temps. Temps et histoire dans le christianisme primitif*, Neuchâtel, Delachaux & Niestlé, 1947; 'Série théologique de l'actualité protestante'; 3rd edn., Zurich, EVZ Verlag, 1962 (English translation by Floyd V. Filson, *Christ and Time. The Primitive Christian Conception of Time and History*, London, SCM Press, 1951).

—, *Heil als Geschichte. Heilsgeschichtliche Existenz im Neuen Testament*, Tübingen, J. C. B. Mohr, 1965 (French translation by Marc Kohler, *Le salut dans l'histoire. L'existence chrétienne selon le Nouveau Testament*, Neuchâtel-Paris, Delachaux & Niestlé, 1966, 'Bibliothèque théologique').

CUMMINGS, Robert D., *Human Nature and History. A Study of the Development of Liberal Political Thought*, Chicago, University of Chicago Press, 1969, 2 vols.

CURRY, Haskell B., *Foundations of Mathematical Logic*, New York-San Francisco-Toronto-London, McGraw-Hill, 1963, 'McGraw-Hill Ser. in Higher Mathematics'.

—, 'Combinatory logic', in KLIBANSKY (ed.), *Contemporary Philosophy / La philosophie contemporaine*, Vol. I, *op. cit.* (1968), pp. 295-307.

— & FEYS, Robert, *Combinatory Logic*, Vol. I, Amsterdam, North Holland Publ. Co., 1958, 'Studies in Logic and the Foundations of Mathematics'.

— & HINDLEY, J. R., *Combinatory Logic*, Vol. II, Amsterdam, North Holland Publ. Co., 1973, 'Studies in Logic and the Foundations of Mathematics'.

DAHL, Robert Alan, *A Preface to Democratic Theory*, Chicago, University of Chicago Press / Cambridge, Cambridge University Press, 1956.

DAHRENDORF, Ralf, *Class and Class Conflict in Industrial Society*, Stanford (Calif.), Stanford University Press, 1959.

—, *Essays in the Theory of Society*, Stanford (Calif.), Stanford University Press, 1968.

DAL PRA, Mario, 'Storia e verità della filosofia', *Rivista critica di storia della filosofia* (Florence), Vol. XXVI, 1971, pp. 439-449.

DANILOV, A., 'K voprosu o metodologii istoričeskoj nauki' (= On the question of methodology in historical science), *Kommunist*, 1969, No. 5.

DANILOVA, L. V. (ed.), *Problemy istorii dokapitalističeskih obščestv* (= Problems in the History of Precapitalist Societies), Vol. 1, Moscow, Nauka, 1968.

DANTO, Arthur C., *Analytical Philosophy of History*, Cambridge, Cambridge University Press, 1965.

—, *Nietzsche as Philosopher*, New York, Macmillan, 1965.

DAS, Bhagwan, *Essential Unity of All Religions*, 5th edn., Adyar, Theosophical Publishing House, 1955.

DAVIDSON, Donald, 'Actions, reasons and causes', *The Journal of Philosophy* XL (23), 7 Nov. 1963, pp. 685-700; reproduced in Alan Richard WHITE (ed.), *The Philosophy of Action*, London, Oxford University Press, 1968, pp. 79-84.

—, 'Truth and meaning', *Synthese* 17 (3), Sept. 1967, pp. 304-323.

—, MCKINSEY, John C. & SUPPES, Patrick C., 'Outlines of a formal theory of value', I, *Philosophy of Science* XXII (2), Apr. 1955, pp. 140-160.
[See also HOFSTADTER & MCKINSEY.]

DAVIDSON, Donald & HINTIKKA, Jaakko (eds.), *Words and Objections. Essays on the Work of W. V. Quine*, Dordrecht, Reidel / New York, Humanities Press, 1970.

DAVIS, J. W., HOCKNEY, D. J. & WILSON, W. K. (eds.), *Philosophical Logic*, Dordrecht, Reidel, 1969, Synthese Library.

DAVIS, Kingsley, 'The myth of functional analysis as a special method in sociology and anthropology', *American Sociological Review* XXIV (6), Dec. 1959, pp. 757-772.

DAVYDOV, Ju., *Trud i svoboda* (= Work and Freedom), Moscow, 1966.

DAY, John P., *Inductive Probability*, London, Routledge / New York, Humanities Press, 1961.
—, 'Authority', *Political Studies* XI (3), Oct. 1963, pp. 257-271.
DE BROGLIE, Louis: see BROGLIE, Louis de.
DECHESNE, B. *et al.*, *Mens en God. Wijsgerige beschouwingen over het religieuze*, Utrecht, J. Bijleveld, 1963.
DE FINETTI, Bruno, *La probabilità e la statistica nei rapporti con l'induzione secondo i diversi punti di vista* (Corso CIME su Induzione e statistica a Varenna), Rome, 1959.
—, 'La décision et les probabilités', *Rev. Math. Acad. Rep. Pop. Roumaine*, Vol. VIII, 1963, pp. 405-413.
—, 'Probabilità composte e teoria delle decisioni', *Rend. di Mat. Roma*, Vol. XXIII, 1964, pp. 128-134.
—, 'Probability: the subjectivistic appròach', in KLIBANSKY (ed.), *Contemporary Philosophy / La philosophie contemporaine*, Vol. II, *op. cit.* (1968), pp. 45-53.
DE GRAAF, J.: see GRAAF, J. de.
DELEUZE, Gilles, 'Spinoza et la méthode générale de M. Guéroult', *Revue de Métaphysique et de Morale*, 74th Year, No. 4, Oct.-Dec. 1969, pp. 426-437.
— & GUATTARI, Félix, *Capitalisme et schizophrénie. L'anti-Oedipe*, Paris, Minuit, 1972, 'Critique' ser.
DELHOMME, Jeanne, 'Histoire, histoire de la philosophie, philosophie', in *L'Histoire, science humaine du présent* (Centre international de Synthèse, XXVth Semaine de Synthèse, 6-13 June 1963) = *Revue de Synthèse*, general series, Vol. LXXXVI, 3rd ser., No. 37-39, Jan.-Sept. 1965, pp. 305-316.
DELLA VOLPE, Galvano, *Chiave della dialettica storica*, Rome, Samonà & Savelli, 1968, 'Saggistica' ser., No. 2.
DE MAURO, Tullio, *Ludwig Wittgenstein. His Place in the Development of Semantics*, New York, Humanities Press, 1967.
DEMERATH, Nicholas Jay & PETERSON, Richard Austin (eds.), *System, Change and Conflict. A Reader on Contemporary Sociological Theory and the Debate over Functionalism*, New York, Free Press / London, Macmillan, 1967.
DEMPF, Alois, 'Philosophie de l'histoire de la philosophie', in CASTELLI *et al.*, *La philosophie de l'histoire de la philosophie, op. cit.* (1956), pp. 69-80 in the French version.
DE RAEYMAEKER, Louis, *De metaphysiek van het zijn*, 2nd edn., Antwerp, Standaardboekhandel / Nijmegen, Dekker & Van de Vegt, 1947.
—, *Philosophie de l'être. Essai de synthèse métaphysique*, 2nd edn., Louvain, Publications Universitaires de Louvain / Paris, Béatrice-Nauwelaerts, 1947 (English translation by Edmund H. Ziegelmeyer, *The Philosophy of Being*, New York-London, Herder, 1954).
DERATHÉ, Robert, *Jean-Jacques Rousseau et la science politique de son temps*, Paris, Presses Universitaires de France, 1949-50; 2nd edn., updated, Paris, Vrin, 1970.
—, *Le rationalisme de Jean-Jacques Rousseau*, Paris, Presses Universitaires de France, 1952.
DERRIDA, Jacques, *La voix et le phénomène. Introduction au problème du signe dans la phénoménologie de Husserl*, Paris, Presses Universitaires de France, 1962, 'Epiméthée' ser.
—, 'Violence et métaphysique. Essai sur la pensée d'Emmanuel Levinas', *Revue de Métaphysique et de Morale*, 69th Year, 1964, No. 3, July-Sept., pp. 322-354 and No. 4, Oct.-Dec., pp. 425-473; reproduced in DERRIDA, *L'écriture et la différence, op. cit.* (1967), pp. 117-228.
—, *De la grammatologie*, Paris, Minuit, 1967, 'Critique' ser.
—, *L'écriture et la différence*, Paris, Seuil, 1967, 'Tel Quel' ser.
—, *La dissémination*, Paris, Seuil, 1972, 'Tel Quel' ser.
—, *Marges. De la philosophie* (collection of essays), Paris, Minuit, 1972, 'Critique'

ser.

DESCHEPPER, Jean-Pierre, 'Le colloque de l'Institut des Hautes Etudes de Belgique (Les méthodes en histoire de la philosophie), Bruxelles, 6-7 mars 1972', *Revue philosophique de Louvain*, Vol. LXXII, 1972, pp. 277-278.

DESSAUER, Friedrich, *Naturwissenschaftliches Erkennen. Beiträge zur Naturphilosophie*, Frankfurt am Main, Knecht, 1958.

DESTOUCHES, Jean-Louis, 'Physique moderne et philosophie', in KLIBANSKY (ed.), *Philosophy in the Mid-Century / La philosophie au milieu du XXe siècle*, Vol. I, *op. cit.* (1958; 3rd edn., 1967), pp. 265-291.

DESTOUCHES-FÉVRIER, Paulette, *La structure des théories physiques*, Paris, Presses Universitaires de France, 1951, 'Philosophie de la Matière' ser.
[See also FÉVRIER, Paulette.]

DEVARAJA, N. K., *The Philosophy of Culture*, Allahabad, Kitab Mahal Private, 1963.

DE WAELHENS, Alphonse, *Une philosophie de l'ambiguïté: l'existentialisme de Maurice Merleau-Ponty*, Louvain, Publications universitaires de Louvain, 1951, Bibliothèque philosophique de Louvain, No. 9 (Institut supérieur de Philosophie).

DHAWAN, G. N., *The Political Philosophiy of Mahatma Gandhi*, Bombay, The Popular Book Depot, 1946.

D'HONDT, Jacques, 'La crise de l'humanisme dans le marxisme contemporain', in *La crise de l'humanisme, op. cit.* (*Revue internationale de Philosophie*, 1968), pp. 369-378.

DHONDT, Jan, 'Histoire et reconstruction du passé, in PERELMAN (ed.), *Raisonnement et démarches de l'historien, op. cit.* (1963) pp. 83-105.

—, 'L'histoire récurrente', *Diogène*, No. 75, July-Sept. 1971, pp. 26-59 (parallel publication of English version, translation by Simon Pleasance, 'Recurrent history', *Diogenes*, No. 75, Fall 1971, pp. 24-57).

Dialectica metodelor în cercetarea ştiinţifica (= The Dialectic of Methods in Scientific Research), collection, Bucarest, 1966.

DIAMOND, Malcolm & LITZENBURG, Thomas (eds.), *Theology and Verification*, Indianapolis (Ind.), Bobbs-Merrill, 1972.

DIESING, Paul, *Reason in Society. Five Types of Decisions and Their Social Conditions*, Urbana, University of Illinois Press, 1962.

DILTHEY, Wilhelm, *Einleitung in die Geisteswissenschaften* (1883) (French translation by Louis Sauzin, *Introduction à l'étude des sciences humaines. Essai sur le fondement qu'on pourrait donner à l'étude de la société et de l'histoire*, Paris, Presses Universitaires de France, 1942).

—, 'Die Entstehung der Hermeneutik' (= The Birth of Hermeneutics) (1900), republished in DILTHEY, *Gesammelte Schriften*, 5th edn., Vol. 5, Göttingen, Vandenhoek & Ruprecht / Stuttgart, Teubner, 1968, pp. 317 sqq.

—, *Weltanschauungslehre. Abhandlungen zur Philosophie der Philosophie = Gesammelte Werke*, 4th edn., Vol. VIII, Göttingen, Vandenhoek & Ruprecht / Stuttgart, Teubner, 1968.

—, *Leben Schleiermachers* (= Life of Schleiermacher), 2nd pt., edited by Redeker = *Gesammelte Schriften*, Vol. XIV/2, Göttingen, Vandenhoek & Ruprecht / Stuttgart, Teubner, 1966.

DONAGAN, Alan, 'Historical explanation. The Popper-Hempel model reconsidered', *History and Theory* IV (1), 1964, pp. 3-26.

DONDEYNE, Albert, *Foi chrétienne et pensée contemporaine. Les problèmes philosophiques soulevés dans l'Encyclique 'Humani Generis'*, Louvain, Publications Universitaires de Louvain / Paris, Béatrice-Nauwelaerts, 1951.

DRAY, William H., *Laws and Explanation in History*, London, Oxford University Press, 1957.

DRAY, William H. (ed.), *Philosophy of History*, Englewood Cliffs (N.J.), Prentice-Hall, 1964.

DREYFUS, Hubert L., 'Why computers must have bodies in order to be intelligent',

The Review of Metaphysics XXI (2/81), Sept. 1967, pp. 13-32.
—, *What Computers Can't Do. A Critique of Artificial Reason*, New York, Harper & Row, 1971.
DUBARLE, Dominique, 'Epistémologie et cosmologie', in *Idée du monde et philosophie de la nature, op. cit.* (1966).
—, 'Sur une formalisation de la logique hégélienne', *Epistémologie sociologique, Cahiers semestriels* (Editions Anthropos, Paris), No. 7, 1st quarter, 1969 = *Logique et sociologie*, pp. 49-61.
— & DOZ, André, *Logique et dialectique*, Paris, Larousse, 1972, 'Sciences humaines et sociales' ser.
DUBARLE, Dominique (ed.), *La recherche en philosophie et en théologie*, Paris, Editions du Cerf, 1970, 'Cogitatio Fidei' ser.
DUBINS, Lester E. & SAVAGE, L. J., *How to Gamble If You Must. Inequalities for Stochastic Processes*, New York, McGraw-Hill, 1965.
DUFRENNE, Mikel, 'L'anti-humanisme et le thème de la mort', in *La crise de l'humanisme, op. cit.* (*Revue internationale de Philosophie*, 1968), pp. 296-307.
—, *Pour l'homme. Essai*, Paris, Seuil, 1968.
DUMÉRY, Henry, *Blondel et la religion*, Paris, Presses Universitaires de France, 1954.
—, *Critique et religion. Problèmes de méthode en philosophie de la religion*, Paris, S.E.D.E.S., 1957.
—, *Philosophie de la religion. Essai sur la signification du christianisme*, Paris, Presses Universitaires de France, 1957, 2 vols.
—, *Le problème de Dieu en philosophie de la religion*, Paris, Desclée de Brouwer, 1957.
—, *Phénoménologie et religion. Structures de l'institution chrétienne*, Paris, Presses Universitaires de France, 1958, 'Initiation philosophique' ser.
—, *Raison et religion dans la Philosophie de l'Action*, Paris, Seuil, 1963.
—, 'Doctrine et structure', in *Etudes sur l'histoire de la philosophie en hommage à Martial Guéroult, op. cit.* (1964), pp. 155-176.
DUNCAN, Graeme & LUKES, Steven, 'The new democracy', *Political Studies* XI (2), June 1963, pp. 156-177.
DUNN, John, 'The identity of the history of ideas', *Philosophy* (London), Vol. XLIII, 1968, pp. 85-104.
DUNNE, John S., *A Search for God in Time and Memory*, New York, Macmillan, 1969.
DUPRÉ, Louis, *The Other Dimension. A Search for the Meaning of Religious Attitudes*, New York, Doubleday, 1972.
DUVERGER, Maurice, *Introduction à la politique*, Paris, Gallimard, 1964.

EASTON, David, *The Political System. An Inquiry into the State of Political Science*, New York, Knopf, 1953.
EBELING, Gerhard, *Wort und Glaube*, Tübingen, J. C. B. Mohr, 1960, 2 vols.
'Economics', in *Main Trends of Research in the Social and Human Sciences, Part I: Social Sciences, op. cit.* (1970), pp. 283-350 (parallel publication in French, 'La science économique', in *Tendances principales de la recherche dans les sciences sociales et humaines, Partie I: Sciences sociales, op. cit.* (1970), pp. 340-426).
EDWARDS, Paul (ed.), *The Encyclopedia of Philosophy*, New York, Macmillan, 1967, 8 vols.
EHNINGER, Douglas & BROCKRIEDE, Wayne E., *Decision by Debate*, New York, Dodd, 1963.
EHRHARDT, W. E., *Philosophiegeschichte und geschichtlicher Skeptizismus. Untersuchungen zur Frage: Wie ist Philosophiegeschichte möglich?*, Bern-Munich, Franke, 1967.
EHRLICH, Walter, *Philosophie der Geschichte der Philosophie*, Tübingen, Niemeyer, 1965.

—, 'Principles of a philosophy of the history of philosophy' (translated from German by Thomas Parry Freeman), in *Philosophy of the History of Philosophy, op. cit.* (*The Monist*, 1969), pp. 532-562.

EICHHORN, Wolfgang, *Wie ist Ethik als Wissenschaft möglich?*, Berlin, Deutscher Verl. d. Wissenschaften 1964.

ÈJDLIN, L., 'Idei i fakty. Neskol'ko voprosov po povodu idei kitajskogo Vozroždenija' (= Ideas and facts. Some questions with regard to the idea of a Chinese Renaissance), *Inostrannaja literatura* (= Foreign Literature), No. 8, 1970.

ELIADE, Mircea, *Traité d'histoire des religions*, Preface by Georges Dumézil, Paris, Payot, 1949, 'Bibliothèque scientifique'.

—, *Das Heilige und das Profane*, Hamburg, Rowohlt, 1957, Rowohlt Deutsche Enzyklopädie (English translation by Willard Trask, *The Sacred and the Profane. The Nature of Religion*, New York, Harcourt, Brace & World, 1968).

ELLSBERG, Daniel, 'Risk, ambiguity, and the Savage axioms', *Quarterly Journal of Economics* LXXV (4), Nov. 1961, pp. 643-669.

EMMET, Dorothy Mary, *Facts and Obligations*, London, Doctor Williams's Trust, 1958.

—, *Function, Purpose and Powers. Some Concepts in the Study of Individuals and Societies*, London, Macmillan, 1958.

—, *Rules, Roles and Relations*, London, Macmillan, 1966.
[See also MACINTYRE & EMMET.]

The Encyclopedia of Philosophy (1967): see EDWARDS (ed.).

Entfremdung und Humanität. Marx und seine klerikalen Kritiker (= Alienation and Humanity. Marx and his Clerical Critics), collective work edited by H. REINHARDT *et al.*, with contributions by Alfred ARNOLD *et al.*, Berlin, Dietz, 1964.

Etudes sur le Contrat social de Jean-Jacques Rousseau. Actes des journées d'études tenues à Dijon les 3, 4 5 et 6 mai 1962, Paris, 1964, Société Les Belles Lettres, Publications de l'Université de Dijon, XXX.

Etudes sur l'histoire de la philosophie en hommage à Martial Guéroult (L'histoire de la philosophie. Ses problèmes. Ses méthodes), Paris, Librairie Fischbacher, 1964: papers by Pierre-Maxime SCHUHL, GOLDSCHMIDT (*op. cit.*), VUILLEMIN (*op. cit.*), Yvon BELAVAL, Heinz HEIMSOETH, Jean HYPPOLITE, Henri GOUHIER, Leslie J. BECK, BRUNNER (*op. cit.*), PERELMAN (*op. cit.*), DUMÉRY (*op. cit.*), GRANGER (*op. cit.*), Jean-Louis BRUCH and Livio TEIXEIRA.

EVANS, Donald M., *The Logic of Self-Involvement. A Philosophical Study of Everyday Language, with special reference to the Christian Use of Language about God as Creator*, Toronto, Ryerson Press (The United Church Publ. House) / London, The SCM Press, 1963.

EVANS-PRITCHARD, E. E., *Nuer Religion*, London, Oxford University Press, 1956.

—, *Theories of Primitive Religion*, London, Oxford University Press, 1965 (French translation by M. Matignon, *La religion des primitifs à travers les théories des anthropologues*, Paris, Payot, 1971, 'Petite Bibliothèque Payot').

EWING, A. C., *The Definition of Good*, New York, Macmillan, 1947; London, Routledge, 1948.

—, *Second Thoughts in Moral Philosophy*, London, Routledge / New York, Macmillan, 1959.

—, 'Ethics in English-speaking countries', in KLIBANSKY (ed.), *Contemporary Philosophy / La philosophie contemporaine*, Vol. IV, *op. cit.* (1971), pp. 5-22.

FABRO, Cornelio, *Participation et causalité selon saint Thomas d'Aquin*, with a Preface by Louis de Raeymaeker, Louvain-Paris, Nauwelaerts, 1961, Université catholique de Louvain, Institut Supérieur de Philosophie.

FAHRENBACH, Helmut, 'Die logisch-hermeneutische Problemstellung in Wittgensteins *Tractatus*', in *Hermeneutik und Dialektik, op. cit.* (1970), Vol. II, pp. 25 sqq.

FAIN, Haskell, *Between Philosophy and History. The Resurrection of Speculative Philosophy of History within the Analytic Tradition*, Princeton (N.J.), Princeton University Press, 1970.
FANN, K. T. (Kuang Tih) *et al.* (eds.), *Symposium on J. L. Austin*, London, Routledge / New York, Humanities Press, 1969.
FANON, Frantz, *Les damnés de la terre*, with a Preface by Jean-Paul SARTRE (*op. cit.*), Paris, Maspero, 1961.
FARRELL, B. A., 'The status of psychoanalytic theory', *Inquiry* (Oslo), VII (1), Spring 1964, pp. 104-123.
FARRER, Austin, *The Freedom of the Will*, London, Black, 1958.
FAUROT, J. H., 'What is history of philosophy?', in *Philosophy of the History of Philosophy, op. cit.* (*The Monist*, 1969), pp. 642-655.
FAYE, Jean-Pierre, *Théorie du récit. Introduction aux langages totalitaires*, Paris, Hermann, 1972.
FEDOSEEV, P. N., *Dialektika sovremennoj èpohi* (= Dialectic of the Present Era), Moscow, 1966.
FEFERMAN, S., 'Systems of predicative analysis', in Jaakko HINTIKKA (ed.), *The Philosophy of Mathematics*, London, Oxford University Press, 1969, pp. 95-127.
FEIGL, Herbert, 'The mental and the physical', in FEIGL *et al.* (eds.), *Concepts, Theories and the Mind-Body Problem* = Vol. II of *Minnesota Studies in the Philosophy of Science, op. cit.* (1958), pp. 370-497.
FEIGL, Herbert & MAXWELL, Grover (eds.), *Current Issues in the Philosophy of Science (Proceedings of Section L of the American Association for the Advancement of Science, 1959)*, New York, Holt, Rinehart & Winston, 1961.
[See also *sub Minnesota Studies in the Philosophy of Science.*]
— & SELLARS, Wilfrid S. (eds.), *Readings in Philosophical Analysis*, New York, Appleton-Century-Crofts, 1949.
FEINBERG, Joel, *Doing and Deserving. Essays in the Theory of Responsibility*, Princeton (N.J.), Princeton University Press, 1970.
FERRÉ, Frederick Pond, *Language, Logic, and God*, New York, Harper & Row, 1961 (French translation by C. Besseyrias, *Le langage religieux a-t-il un sens? Logique moderne et foi*, Paris, Editions du Cerf, 1970, 'Cogitatio Fidei' ser., No. 47).
FÉVRIER, Paulette, *Déterminisme et indéterminisme*, with a Preface by Edouard LE ROY, Paris, Presses Universitaires de France, 1955, 'Philosophie de la Matière' ser.
FEYERABEND, Paul K., 'Naturphilosophie', in *Fischer-Lexicon. Enzyklopädie des Wissens*, Frankfurt am Main-Berlin, Fischer-Bücherei, 1958.
—, 'Problems of microphysics', in Robert G. COLODNY (ed.), *Frontiers of Science and Philosophy*, Pittsburgh, University of Pittsburgh Press, 1962, pp. 189-283.
FILIPENKO, N. V. (ed.) *O dialektike razvitija sovetskogo socialističeskogo obščestva* (= The Dialectic of the Development of Soviet Socialist Society), collection, Moscow, 1968.
Filosofskaja Ènciklopedija, Moscow: Vol. II, 1962; Vol. IV, 1967.
Filosofskie problemy istoričeskoj nauki (= Philosophical Problems in Historical Science), collection, Moscow, Academy of Sciences of the U.S.S.R., Institute of Philosophy, 1969.
Filosofskie voprosy estestvoznanija (= Philosophical Questions of the Natural Sciences), collection, Moscow, 1958 to 1960, Vols. 1-3.
FINDLAY, John N., 'Can God's existence be disproved?', *Mind* LVII (226), Apr. 1948, pp. 176-183; reproduced in FLEW & MACINTYRE (eds.), *New Essays in Philosophical Theology, op. cit.* (1953).
—, *Hegel. A Re-examination*, London, Allen & Unwin, 1958.
—, *Values and Intentions. A Study in Value-Theory and Philosophy of Mind*, London, Allen & Unwin / New York, Macmillan, 1961.
—, *Language, Mind and Value. Philosophical Essays*, London, Allen & Unwin /

New York, Humanities Press, 1963.
FINETTI, Bruno de: see DE FINETTI, Bruno.
FISCHER, Mark, 'A system for deontic-alethic modal logic', *Mind*, N.S. LXXI (282), Apr. 1962, pp. 231-236.
FLEW, A. G. N. (Anthony), 'Theology and falsification (A)', *University* (periodical, Oxford), 1950-51; reproduced in FLEW & MACINTYRE (eds.), *New Essays in Philosophical Theology, op. cit.* (1953).
—, *Hume's Theory of Belief*, London, Routledge, 1961.
—, *God and Philosophy*, London, Hutchinson / New York, Harcourt, Brace & World, 1966.
FLEW, A. G. N. (Anthony) (ed.), *Logic and Language*, Oxford, Basil Blackwell, 1951.
—, *Essays on Conceptual Analysis*, London, Macmillan, 1956.
—, *Body, Mind and Death. A Book of Readings with an Introduction*, New York, Macmillan, 1964.
— & MACINTYRE, Alasdair (eds.), *New Essays in Philosophical Theology*, London, SCM Publ., 1953.
FODOR, Jerry A., *Psychological Explanation*, New York, Random House, 1968.
— & GARRETT, M., 'Some reflections on competence and performance', in J. LYONS & R. J. WALES (eds.), *Psycholinguistic Papers. The Proceedings of the 1966 Edinburgh Conference*, Edinburgh, Edinburgh University Press, 1966, pp. 135-154 (followed by a discussion by N. S. SUTHERLAND, pp. 154-163, and by L. Jonathan COHEN, pp. 163-173, and by a general discussion, summarized on pp. 173-179).
FODOR, Jerry A. & KATZ, Jerrold J. (eds.), *The Structure of Language. Readings in the Philosophy of Language*, Englewood Cliffs (N.J.), Prentice-Hall, 1964.
FOGARASI, Béla, *Logik*, Berlin, Aufbau-Verlag, 1955.
FOOT, Philippa R., 'Moral arguments', *Mind* LXII (268), Oct. 1958, pp. 502-513.
—, 'Moral beliefs', *Proceedings of the Aristotelian Society*, Vol. LIX, 1958-59, pp. 83-104.
—, 'Goodness and choice', *Proceedings of the Aristotelian Society*, Supp. Vol. XXXV, 1961, pp. 45-60 (followed by an article by Alan MONTEFIORE on the same subject, pp. 61-80).
FORIERS, P., 'Réflexions sur l'argumentation juridique', in *L'Argumentation, op. cit.* (*Revue internationale de Philosophie*, 1961), pp. 400-409.
FOUCAULT, Michel, *Folie et déraison. Histoire de la folie à l'âge classique*, Paris, Plon, 1961, 'Civilisations d'hier et d'aujourd'hui' ser.; new edn., augm. Paris, Gallimard, 1972.
—, *Les mots et les choses. Une archéologie des sciences humaines*, Paris, Gallimard, 1966, Bibliothèque des Sciences humaines (English translation, *The Order of Things. An Archaeology of the Human Sciences*, New York, Pantheon Books, 1970).
—, *L'archéologie du savoir*, Paris, Gallimard, 1969, Bibliothèque des Sciences humaines (English translation by A. M. Sheridan Smith, *The Archaeology of Knowledge*, London, Tavistock Publ., 1972).
—, *L'ordre du discours* (Inaugural Lecture at the Collège de France, delivered on 2 Dec. 1970), Paris, Gallimard, 1971.
XIVth International Congress of Philosophy (Vienna, 2-9 Sept. 1968): see *Akten des XIV. Internationalen Kongresses für Philosophie.*
FRANCHINI, Raffaelo, *Esperienza del storicismo*, Naples, F. Giannini, 2nd edn., rev., 1960, 'Storia e pensiero' ser., No. 3.
FRANCOEUR, Robert T., *The World of Teilhard de Chardin*, Baltimore (Md.), Helicon Press, 1961.
FRANKENA, William K., *Ethics*, Englewood Cliffs (N.J.), Prentice-Hall, 1963.
FREGE, Gottlob, 'Über Sinn und Bedeutung', *Zeitschrift für Philosophie und philosophische Kritik*, 100, 1892, pp. 25-50 (English translation, 'On sense and refer-

ence', in FREGE, *Philosophical Writings*, Oxford, Basil Blackwell, 1960).
—, *Grundgesetze der Arithmetik, begriffsschriftlich abgeleitet* (= Basic Laws of Arithmetics, Derived by way of Concept-scripts), Jena, 1893-1903, 2 vols.; photographic reproduction in 1 vol., Hildesheim, Gg. Olms, 1962; in 2 vols., Darmstadt, Wissenschaftliche Buchgesellschaft, 1962.

FREUD, Sigmund, *Die Traumdeutung* (1900), 7th edn., Vienna, F. Deuticke, 1945; cf. *Gesammelte Werke*, Vols. II and II (English translation, *Interpretation of Dreams*, in Standard Edition, Vols. IV and V, London, Hogarth Press, pp. 1-621 and New York, Modern Library).

—, *Drei Abhandlungen zur Sexualtheorie*, Vienna, 1905; cf. *Gesammelte Werke*, Vol. V, pp. 29-145 (English translation by James Strachey, *Three Essays on the Theory of Sexuality*, in the *Standard Edition*, pp. 123-243, Vol. VII, and New York, Basic Books, 1962).

—, *Metapsychologie* (collection of 5 essays subsisting from among the 12 essays which Freud had first destined in 1915 for a collection to be entitled 'Zur Vorbereitung einer Metapsychologie', *viz.*: 'Triebe und Triebschicksale'; 'Die Verdrängung'; 'Das Unbewusste'; 'Metapsychologische Ergänzung zur Traumlehre'; and 'Trauer und Melancholie'), *Gesammelte Werke*, Vol. X (English translation by James Strachey, in *Standard Edition*, Vol. XIV).

—, *Aus den Anfängen der Psychoanalysis: Briefe an Wilhelm Fliess, Abhandlungen und Notizen aus den Jahren 1887-1902*, edited by Marie Bonaparte *et al.*, London, Imago Publishing, 1950 (English translation by Eric Mosbacher & James Strachey, Introduction by Ernst Kris, *The Origins of Psychoanalysis, Letters to Wilhelm Fliess. Drafts and Notes 1887-1902*. London, Imago / New York, Basic Books, 1954).

FREUND, Julien, *L'essence du politique*, Paris, Sirey, 1965, 'Philosophie politique' ser.

—, Preface (pp. 7-38) to SCHMITT, *La notion de politique*, followed by *Théorie du partisan* (French translation), *op. cit.* (1972).

FROLOV, I. T., *O pričinnosti i celesoobraznosti v živoj prirode* (= On Causality and Hormicity in Living Nature), Moscow, 1961.

—, *Genetika i dialektika* (= Genetics and Dialectic), Moscow, 1968.

FRUCHON, Pierre, 'Connaissance et interprétation du passé en histoire des idées', *Les Etudes philosophiques*, Vol. XXI, 1966, pp. 499-508.

—, 'Ressources et limites d'une herméneutique philosophique (H.-G. GADAMER, *Wahrheit und Methode*)', *Archives de Philosophie* XXX (3), July-Sept. 1967, pp. 411-438.

FUCHS, Ernst, *Hermeneutik*, Bad Cannstatt, R. Müllerschön, 1954; 4th edition, Tübingen, J. C. B. Mohr, 1970.

—, *Zum hermeneutischen Problem in der Theologie. Die existentiale Interpretation*, Tübingen, J. C. B. Mohr, 1959.

FUNK, Robert W., *Language, Hermeneutic and the Word of God*, New York, Harper & Row, 1966.

GABEL, Joseph, *La fausse conscience. Essai sur la réification*, Paris, Minuit, 1962, 'Arguments' ser., No. 11.

GADAMER, Hans-Georg, *Wahrheit und Methode. Grundzüge einer philosophischen Hermeneutik* (= Truth and Method. Elements of a Philosophic Hermeneutics), Tübingen, J. C. B. Mohr, 1960; 2nd edn., 1965.

—, 'Über die Möglichkeit einer philosophischen Ethik' (a lecture delivered at Walberberg in 1961), in *Sein und Ethos, op. cit.* (1963), pp. 11-24 (French translation by Pierre Fruchon, revised by Pierre Banges, 'Sur la possibilité d'une éthique philosophique', *Archives de Philosophie* XXXIV (3), July-Sept. 1971, pp. 393-408).

—, *Kleine Schriften, I. Philosophie, Hermeneutik*, Tübingen, J. C. B. Mohr, 1967.

—, 'Rhetorik, Hermeneutik und Ideologiekritik. Metakritische Erörterung zu

Wahrheit und Methode', in GADAMER, *Kleine Schriften, I. Philosophie, Hermeneutik, op. cit.* (1967), pp. 113-130 (French translation by Pierre Fruchon, revised by Pierre Bange, 'Rhétorique, herméneutique et critique de l'idéologie. Commentaires métacritiques de *Wahrheit und Methode*', *Archives de Philosophie* XXXIV (2), Apr.-June 1971, pp. 207-230).

—, 'Begriffsgeschichte als Philosophie', *Archiv für Begriffsgeschichte*, Vol. XIV, 1970.

GALLIE, Walter B., *Peirce and Pragmatism*, Harmondsworth (Middx.), Penguin Books, 1952.

—, *Philosophy and the Historical Understanding*, London, Chatto & Windus, 1964.

GALPERIN, P. Ja., 'Razvitie issledovanij po formirovaniju umstvennyh dejstvij' (= Development of research on the formation of mental acts), in *Psihologičeskaja nauka v SSSR* (= Psychology in the U.S.S.R.), Vol. I, Moscow, 1959.

GALTUNG, Johan, *Theory and Methods of Social Research*, Oslo, Universitetsforlaget, 1967.

GANDHI, M. K., *The Collected Works*, Ahmedabad, Navajivan Trust, 1960.

GARAUDY, Roger, *Perspectives de l'homme: Existentialisme, Pensée catholique, Marxisme (1929-1959)*, Paris, Presses Universitaires de France, 1959.

—, *L'alternative*, Paris, Robert Laffont, 1972, 'Libertés 2000' ser.

GARCÍA MÁYNEZ, Eduardo, *Ensayos filosófico-jurídicos*, Xalapa (Mexico), Biblioteca de la Facultad de Filosofía y Letras, 1959.

GARDINER, Patrick L., *The Nature of Historical Explanation*, London, Oxford University Press, 1952.

—, *Schopenhauer*, Harmondsworth (Middx.)-London, Penguin Books, 1963.

GARDINER, Patrick L. (ed.), *Theories of History. Readings from Classical and Contemporary Sources*, Glencoe (Ill.)-New York, Free Press, 1959; London, Allen & Unwin, 1960.

GARFINKEL, Harold, *Studies in Ethnomethodology*, Englewood Cliffs (N.J.), Prentice-Hall, 1967.

GARIN, Eugenio, *La filosofia come sapere storico*, Bari, Laterza, 1959 (important bibliography).

GAROTTI, Ricci, 'Filosofia e storicità', in *Studi in onore di Arturo Massolo*, edited by Livio SICHIROLLO = *Studi urbinati di storia, filosofia e letteratura* (Urbin), 16th Year, 1967, Fasc.. 41, N.S. b, No. 1-2.

GATES, John F., *Adventures in the History of Philosophy. An Introduction from a Christian Viewpont*, Grand Rapids (Mich.), Zondervan, 1961.

GAUTHIER, David P., *Practical Reasoning. The Structure and Foundations of Prudential and Moral Arguments and Their Exemplification in Discourse*, London, Oxford University Press, 1963.

GEACH, P. T. (Peter), 'Good and evil', *Analysis* XVII (2), 1956, pp. 33-42.

—, *Mental Acts*, London, Routledge, 1958.

—, 'Imperative inference', *Analysis*, Supp. XXIII, 1963, pp. 37-42.

—, *God and the Soul*, London, Routledge, 1969.

GEBSATTEL, Victor E. von, *Prolegomena einer medizinischen Anthropologie*, Berlin-Göttingen, Springer, 1954.

GEFFRÉ, Claude (ed.), *Procès de l'objectivité de Dieu. Les présupposés philosophiques de la crise de l'objectivité de Dieu* (symposium of the Dominican Faculties of the Saulchoir), Paris, Editions du Cerf, 1969, 'Cogitatio Fidei' ser., No. 41.

GEHLEN, Arnold, *Studien zur Anthropologie und Soziologie*, Berlin, Luchterhand, 1963.

GELDSETZER, Lutz, *Was heisst Philosophiegeschichte?*, Düsseldorf, Philosophia Verlag, 1968.

—, *Die Philosophie der Philosophiegeschichte im neunzehnten Jahrhundert. Zur Wissenschaftstheorie der Philosophiegeschichtsschreibung und -betrachtung*,

Meisenheim am Glan, Verlag Anton Hain, 1968.

GELLNER, Ernest, 'Explanations in history', *Proceedings of the Aristotelian Society, Supplementary Volume XXX*, 1956, pp. 157-176.

—, *Words and Things*, London, Gollancz, 1959; republ., Harmondsworth (Middx.), Penguin, 1968, Pelican Books.

—, 'Concepts and society', in *Transactions of the Fifth World Congress of Sociology, Washington, 1963*, I, 1962, pp. 153-183; reproduced in WILSON (ed.), *Rationality, op. cit.* (1970), pp. 18-49.

—, 'The new idealism', in I. LAKATOS & A. MUSGRAVE (eds.), *Problems in the Philosophy of Science (Proceedings of the International Colloquium on the Philosophy of Science, London, 1965)*, Vol. III. *Studies in Logic and the Foundations of Mathematics*, Amsterdam, North Holland Publ. Co., 1968, pp. 377-406; discussion, *ibid.*, pp. 407-432.

GENTILE, Giovanni, *La riforma della dialettica hegeliana*, Messina, Giuseppe Principato, 1913.

—, *Se e come è possibile la storia della filosofia*, 2nd edn., Padua, Liviana, 1966, 'Guide di cultura contemporanea'.

GENTZEN, Gerhard, 'Untersuchungen über das logische Schliessen', *Mathematische Zeitschrift*, Vol. XXXIX, 1934, pp. 176-210 and 405-431 (French translation with comments, by Robert Feys & Jean Ladrière, *Recherches sur la déduction logique*, Paris, Presses Universitaires de France, 1955, 'Philosophie de la matière' ser.).

Geschichte der Philosophie, published by the Academy of Sciences of the U.S.S.R., 2nd edn. in German, Vol. I, Berlin (East), 1962.

GIBSON, Quentin, *The Logic of Social Enquiry*, London, Routledge, 1960.

GILKEY, Langdon, *Naming the Whirlwind. The Renewal of God-Language*, New York, Bobbs-Merrill, 1969.

GILSON, Étienne, *The Unity of Philosophical Experience*, New York, Scribner, 1937; republ., 1955.
[For a review of this work, see GOUHIER, Henri.]

—, *God and Philosophy*, New Haven (Conn.), Yale University Press, 1941.

—, *L'être et l'essence*, Paris, Vrin, 1951.

—, *Being and Some Philosophers*, Toronto, Pontifical Institute of Medieaval Studies, Dept. of Publications, 2nd edn., rev. and augm., 1952.

GINSBERG, Morris, *Reason and Experience in Ethics* (Auguste Comte Memorial Lectures), London, Oxford University Press, 1956.

GIRARDI, Giulio *et al.*, *L'ateismo contemporaneo* (A cura della Facoltà filosofica della Pontificia Università salesiana di Roma), Vol. I, Turin, Società editrice internazionale, 1967 (French translation, *L'athéisme dans la vie et la culture contemporaines*, Paris, Desclée de Brouwer, 1967).

GIULIANI, Alessandro, *Il concetto di prova. Contributo alla logica giuridica*, Milan, Giuffré, 1961.

GLÉNISSON, Jean, 'L'historiographie française contemporaine: tendances et réalisations', in *La recherche historique en France de 1940 à 1965*, Paris, Editions du Centre national de la Recherche scientifique (C.N.R.S.), 1965, pp. IX-LXIV.

GLUCKSMANN, André, *Le discours de la guerre*, Paris, Editions de l'Herne, 1967, 'Théorie et stratégie' ser., No. 1.

GÖDEL, Kurt, 'Über formal unentscheidbare Sätze der Principia Mathematica und verwandter Systeme', Pt. 1, *Monatshefte für Mathematik und Physik*, Vol. XXXVIII, 1931, pp. 173-198.

GODELIER, Maurice, *Rationalité et irrationalité en économie*, Paris, Maspero, 1971, 2 vols., Petite Collection Maspero.

GOFFMAN, Erving, *The Presentation of Self in Everyday Life*, Garden City (N.Y.), Doubleday, 1959 (French translation by Alain Accardo, *La mise en scène de la vie quotidienne*, Paris, Minuit, 1973, 2 vols., 'Le sens commun' ser.: I. *La présentation de soi*; II. *Les relations en public*).

—, *Encounters. Two Studies in the Sociology of Interaction*, New York, Bobbs-Merrill, 1961.

GOLD, Thomas, 'Cosmic processes and the nature of time', in Robert G. COLODNY (ed.), *Mind and Cosmos*, Pittsburgh, University of Pittsburgh Press, 1966.

GOLDMAN, Alvin I., *A Theory of Human Action*, Englewood Cliffs (N.J.), Prentice-Hall, 1970.

GOLDMANN, Lucien, 'Matérialisme dialectique et histoire de la philosophie', *Revue philosophique de la France et de l'Etranger* LXXIII (2), June 1948; reproduced in GOLDMANN, *Recherches dialectiques, op. cit.* (1959), pp. 26-44.

—, *Sciences humaines et philosophie*, Paris, Presses Universitaires de France, 1952, 'Nouvelle Encyclopédie Philosophique' ser.; new edn., augm., Paris, Gonthier, 1966, Bibliothèque 'Médiations'.

—, Statements in the general discussion at the 'Colloque de Royaumont' on Descartes, in *Descartes*, Paris, Minuit, 1957, 'Cahiers de Royaumont', Philosophie, No. II, pp. 466-471 and 477.

—, *Recherches dialectiques* (collection of essays), Paris, Gallimard, 1959, Bibliothèque des Idées.

—, *Marxisme et sciences humaines* (collection of essays), Paris, Gallimard, 1970, 'Idées' ser.

GOLDSCHMIDT, Victor, 'Exégèse et axiomatique chez saint Augustin', in *Etudes sur l'histoire de la philosophie en hommage à Martial Guéroult, op. cit.* (1964), pp. 14-42.

—, *Platonisme et pensée contemporaine*, Paris, Aubier-Montaigne, 1970, 'Présence et pensée' ser.

—, 'Temps historique et temps logique dans l'interprétation des systèmes philosophiques', in GOLDSCHMIDT, *Questions platoniciennes*, Paris, Vrin, 1970, pp. 13-21.

GOLDSCHMIDT, Walter, *Comparative Functionalism. An Essay in Anthropological Theory*, Berkeley-Los Angeles, University of California Press, 1966.

GOMBAY, André, 'Imperative inference and disjunction', *Analysis* XXV (3), Jan. 1965, pp. 58-62.

GONSETH, Ferdinand, *Le problème du temps. Essai sur la méthodologie de la recherche*, Neuchâtel, Editions du Griffon / Paris, Dunod, 1964, 'Bibliothèque scientifique', No. 38.

GOOD, Irving J., *The Estimation of Probabilities. An Essay on Modern Bayesian Methods*, Cambridge (Mass.), The M.I.T. Press, 1965.

GOODMAN, Nelson, *Fact, Fiction and Forecast*, Cambridge (Mass.), Harvard University Press, 1955; 2nd edn., Indianapolis (Ind.), Liberal Arts Press, Bobbs-Merrill Co., 1965.

—, *Languages of Art. An Approach to a Theory of Symbols*, Indianapolis (Ind.) — New York, Bobbs-Merrill, 1968; London, Oxford University Press, 1969.

GORSKIJ, Dmitry P., 'Formal'naja logika i jazyk' (= Formal logic and language), in *Filosofskie voprosy sovremennoj formal'noj logiki* (= Philosophical Questions of Today's Formal Logic), Moscow, 1962.

—, *Problemy obščej metodologii nauk i dialektičeskoj logiki* (= Problems of General Scientific Methodology and Dialectical Logic), Moscow, 1966.

GORSKY (= GORSKIJ), Dmitry P., 'The problem of meaning (sense) of symbol expressions as the problem of their understanding', in 'Sprachphilosophie' (Section III, XIVth International Congress of Philosophy, Vienna, 1968), *op. cit.* (1969), pp. 464-468.

GORTARI, Elie de, *Introducción a la logica dialéctica*, Mexico-City/Buenos Aires, 1956.

GOSLING, J., *Pleasure and Desire*, Oxford, Clarendon Press, 1969.

GOUHIER, Henri, Review of GILSON, *The Unity of Philosophical Experience* (*op. cit.*, 1937), *La Vie intellectuelle*, 25 June 1938; reproduced in GOUHIER, *La philosophie et son histoire, op. cit.* (1944), pp. 127-134 (Appendix II).

—, *La philosophie et son histoire*, Paris, Vrin, 1944.
—, 'De l'histoire de la philosophie à la philosophie', in *Etienne Gilson, philosophe de la Chrétienté*, Paris, Editions du Cerf, 1949.
—, *L'histoire et sa philosophie*, Paris, Vrin, 1952, 'Problèmes et controverses' ser. [See also CASTELLI, GOUHIER, LEVINAS, PANIKKAR, RICOEUR, HABACHI *et al.*]
GOULIANE, C. I.: see GULIAN, Constantin.
GRAAF, J. de & BAKKER, R., *De bezinning over goed en kwaad in de geschiedenis van het menselijk denken. Geschiedenis der wijsgerige ethiek. De mondige mens tussen goed en kwaad*, Utrecht, J. Bijleveld, 2nd edn., 1967.
GRAMSCI, Antonio, *Opere*, Vol. II. *Il materialismo storico e la filosofia di Benedetto Croce* (texts from the 'Prison Notebooks' written from 1929 onwards), Turin, Einaudi, 1948 (posthumous publ.), 8th edn., 1966 (see the English translation, by Quintin Hoare & Geoffrey Nowell Smith, of the complete *Prison Notebooks*, London, Lawrence & Wishart, 1971).
GRANGER, Gilles-Gaston, *Pensée formelle et sciences de l'homme*, Paris, Aubier-Montaigne, 1960, 'Analyse et raisons' ser.; 2nd edn. (preceded by an 'Adresse au lecteur' on structuralism), *ibid.*, 1967.
—, 'Systèmes philosophiques et métastructures. L'argumentation du *Tractatus*', in *Etudes sur l'histoire de la philosophie en hommage à Martial Guéroult, op. cit.* (1964), pp. 139-154.
—, *Essai d'une philosophie du style*, Paris, Armand Colin, 1968, 'Philosophies pour l'âge de la science' ser.
—, 'L'explication dans les sciences sociales' (contribution to the symposium 'L'explication dans les sciences', organized in Geneva in Oct. 1970 by the Académie internationale de Philosophie des Sciences in collab. with the Centre international d'Epistémologie génétique), *Information sur les Sciences sociales / Social Science Information* X (2), Apr. 1971, pp. 31-44 (to appear also in the *Proceedings* of the symposium).
GREENBERG, Joseph H. (ed.), *Universals of Language*, Cambridge (Mass.), The M.I.T. Press, 1963; paperback edn., 1966.
GREIMAS, Algirdas Julien, *Sémantique structurale. Recherche de méthode*, Paris, Larousse, 1966, 'Langue et Langage' ser.
GRICE, H. P., 'Meaning', *Philosophical Review* LXVI (3), July 1957, pp. 377-388.
—, 'Utterer's meaning and intentions', *Philosophical Review* LXXVIII (2), Apr. 1969, pp. 147-177.
GROETHUYSEN, Bernard, 'Philosophie et histoire', *Médiations*, No. 2, 1961, pp. 19-28.
GROSS, Llewellyn (ed.), *Symposium on Sociological Theory*, Evanston (Ill.), Row, Peterson & Co., 1959.
GRÜNBAUM, Adolf, *Philosophical Problems of Space and Time*, New York, Knopf, 1963; London, Routledge, 1964.
GRÜNDER, Karlfried, 'Bericht [...] über das "Archiv für Begriffsgeschichte"', *Jahrbuch der Akademie der Wissenschaften und der Literatur* (Mainz), 1967.
—, 'Perspektiven für eine Theorie der Geschichtswissenschaft', *Saeculum* (Freiburg im Breisgau), Vol. XXII, 1971.
GRUNER, Rolf, 'Teleological and functional explanations', *Mind* LXXV (300), Oct. 1966, pp. 516-526.
GRUŠIN, B. A., *Očerki logiki istoričeskogo issledovanija: process razvitija i problemy ego naučnogo vosproizvedenija* (= Essays on the Logic of Historical Research: the Process of Development and Problems in Its Scientific Reconstruction), Moscow, Gosudarstvennoe izd. 'Vysšaja Škola', 1961.
GRYNPAS, Jerôme, 'Remarques sur le problème de la critique philosophique en philosophie', in *L'Argumentation, op. cit.* (*Revue internationale de Philosophie*, 1961), pp. 410-432.
GUÉROULT, Martial, *Leçon inaugurale* (delivered on 4 Dec. 1951), Paris, Collège de France, 1952.
—, 'La voie de l'objectivité esthétique', in *Mélanges d'esthétique et de science de*

l'art offerts à Etienne Souriau, Paris, Nizet, 1952, pp. 95-127.

—, *Descartes selon l'ordre des raisons*, Paris, Aubier, 1953, 2 vols.: I. *L'âme et Dieu*; II. *L'âme et le corps*.

—, 'Le problème de la légitimité de l'histoire de la philosophie', in CASTELLI *et al.*, *La philosophie de l'histoire de la philosophie, op. cit.* (1956), pp. 45-68.

—, Statements in the general discussion at the 'Colloque de Royaumont' in Descartes, in *Descartes*, Paris, Minuit, 1957, 'Cahiers de Royaumont', Philosophie, No. II, pp. 453-463 and 473-475.

—, 'Logique, architectonique et structures constitutives des systèmes philosophiques', in *L'Encyclopédie française*, Vol. XIX, Paris, 1957.

—, 'Bergson en face des philosophes', *Etudes bergsoniennes* (Paris), Vol. V, 1960.

—, 'Logique, argumentation et histoire de la philosophie chez Aristote', in *La Théorie de l'argumentation, op. cit.* (*Logique et analyse*, 1963), pp. 431-449.

—, 'The history of philosophy as a philosophical problem', in *Philosophy of the History of Philosophy, op. cit.* (*The Monist*, 1969), pp. 563-587).

— *et al.*, *Brunschvicg et l'histoire de la philosophie* (French Society for Philosophy, 30 Jan. 1954 meeting = *Bulletin de la Société française de Philosophie*, 48th Year, No. 1, 1954 (pp. 1-36): communication by GUÉROULT; discussion.

GUEST, A. G. (ed.), *Oxford Essays in Jurisprudence*, London, Oxford University Press, 1961.

GUILLAUME, Gustave, *Langage et science du langage*, Paris, Nizet / Quebec, Presses de l'Université Laval, 1964.

GULIAN, Constantin, *Problematica omului. Eseu de antropologie filozofica* (= The Problem of Man. An Essay in Philosophical Anthropology), Bucarest, Editura politica, 1966 (French translation by Jean Herdan, GOULIANE, C. I., *Le marxisme devant l'homme. Essai d'anthropologie philosophique*, Paris, Payot, 1968, Bibliothèque philosophique).

—, 'Humanisme et anthropologie philosophique chez Marx', in 'Marx und die Philosophie der Gegenwart' (Colloquium I, XIVth International Congress of Philosophy, Vienna, 1968), *op. cit.* (1968), pp. 34-40.

GULYGA, A. V., 'O haraktere istoričeskogo znanija' (= On the nature of historical knowledge), *Voprosy filosofii*, 1962, No. 9.

—, 'Istorija kak nauka' (= History as science), in *Filosofskie problemy istoričeskoj nauki, op. cit.* (1968).

GUREVIČ, A. Ja., 'Nekotorye aspekty izučenija social'noj istorii: obščestvenno-istoričeskaja psihologija' (= Some aspects of the study of social history: socio-historical psychology), *Voprosy istorii*, 1964, No. 10.

—, 'Social'naja psihologija i istorija. Istočnikovedčeskij aspekt' (= Social psychology and history: the 'study of source materials' aspect), in *Istočnikovodenie. Teoretičeskie i metodologičeskie problemy* (= The Study of Sources. Theoretical and Methodological Problems), *op. cit.* (1969).

—, 'Vremja kak problema istorii kul'tury' (= Time as a problem in the history of culture), *Voprosy filosofii*, 1969, No. 3.

—, 'Predstavlenie o vremeni v srednevekovoj Evrope' (= The conception of time in mediaeval Europe), in PORŠNEV & ANCIFEROVA (eds.), *Istorija i psihologija, op. cit.* (1970).

GUSDORF, Georges, *Introduction aux sciences humaines. Essai critique sur leurs origines et leur développement*, Paris, Les Belles Lettres, 1960, Publications de la Faculté des Lettres de l'Université de Strasbourg, Fasc. 140; new edn., Paris, Editions Ophrys, 1974.

—, *Les sciences humaines et la conscience occidentale*, Paris, Payot, 6 vols. published to date: I. *De l'histoire des sciences à l'histoire de la pensée*, 1966; II. *Les origines des sciences humaines*, 1967; III. *La révolution galiléenne*, 1969, 2 vols.; IV. *Les principes de la pensée au siècle des Lumières*, 1971; V. *Dieu, la nature, l'homme au siècle des Lumières*, 1972; VI. *L'avènement des sciences humaines au siècle des Lumières*, 1973; to appear next: VII. *La science de l'homme chez*

les idéologues et les romantiques.
—, *Les sciences de l'homme sont des sciences humaines*, Paris, Les Belles Lettres, 1967, Publications de la Faculté des Lettres de l'Université de Strasbourg.
Guzzo, Augusto, *L'uomo*, Vol. II. *La moralità*, 2nd edn., Turin, Edizioni di Filosofia, 1967.
—, 'Ethique: Publications italiennes', in Klibansky (ed.), *Contemporary Philosophy / La philosophie contemporaine*, Vol. IV, *op. cit.* (1971), pp. 33-47.

Habermas, Jürgen, *Theorie und Praxis. Sozialphilosophische Studien*, Neuwied am Rhein-Berlin, Luchterhand, 1963.
—, *Zur Logik der Sozialwissenschaften* = *Philosophische Rundschau* (Tübingen), Beiheft 5, 1967.
—, 'Erkenntnis und Interesse' (1965), in Habermas, *Technik und Wissenschaft als 'Ideologie'*, *op. cit.* (1968), pp. 146 sqq.
—, *Erkenntnis und Interesse*, Frankfurt am Main, Suhrkamp, 1968.
—, *Technik und Wissenschaft als 'Ideologie'*, Frankfurt am Main, Suhrkamp, 1968, 'Politica' ser., Vol. II (French translation by Jean-René Ladmiral, *La technique et la science comme 'idéologie'*, Paris, Gallimard, 1973, 'Les Essais' ser.).
Habermas, Jürgen (ed.), *Hermeneutik und Ideologiekritik. Theorie-Diskussion*, Frankfurt am Main, Suhrkamp, 1971.
Hacker, Peter, 'Are transcendental arguments a version of verificationism?', *American Philosophical Quarterly* IX (1), Jan. 1972, pp. 78-85.
Hacking, Jan, *Logic of Statistical Inference*, Cambridge, Cambridge University Press, 1965.
Haines, G. C., *The Revival of Natural Law Concepts*, Cambridge (Mass.), Harvard University Press, 1950.
Haldane, J. B. S., 'Animal ritual and human language', *Diogenes*, No. 4, Autumn 1953, pp. 61-73 (parallel publication of French translation, 'Rituel humain et communication animale', *Diogène*, No. 4, Oct. 1953, pp. 77-93).
Hall, Everett W., *Philosophical Systems. A Categorical Analysis*, Chicago, University of Chicago Press, 1960.
—, *Our Knowledge of Fact and Value*, Chapel Hill, University of North Carolina Press, 1961.
Halldén, Sören, *On the Logic of 'Better'*, Uppsala, 1957.
Halmos, Paul Richard, *Algebraic Logic*, New York, Chelsea Publishing Co., 1962.
Halpern, Ben, '"Myth" and "ideology" in modern usage', *History and Theory* I (2), 1961, pp. 129-149.
Hampshire, Stuart, *Spinoza*, Harmondsworth (Middx.)-London, Penguin Books, 1952.
—, *Thought and Action*, London, Chatto & Windus, 1959.
—, *Freedom of the Individual*, London, Chatto & Windus, 1965.
Handy, Rollo, *Methodology of the Behavioral Sciences. Problems and Controversies*, Springfield (Ill.), Charles C. Thomas, 1964.
Hanson, Norwood R., *Patterns of Discovery. An Inquiry into the Conceptual Foundations of Science*, Cambridge, Cambridge University Press, 1958.
—, 'Is there a logic of discovery?', in Feigl & Maxwell (eds.), *Current Issues in the Philosophy of Science*, *op. cit.* (1961), pp. 20-35.
—, 'The dematerialization of matter', in Ernan McMullin (ed.), *The Concept of Matter*, Notre Dame (Ill.), University of Notre Dame Press, 1963.
Hanson, W. H., 'A logic of commands', *Logique et analyse*, 1966, No. 9, pp. 329-348.
Hare, R. M. (Richard), *The Language of Morals*, London, Oxford University Press, 1952.
—, 'Religion and morals', in Mitchell (ed.), *Faith and Logic*, *op. cit.* (1958).
—, *Freedom and Reason*, London, Oxford University Press, 1963.
—, *Practical Inferences*, Berkeley, University of California Press, 1972.
Harré, Rom, *The Principles of Scienitfic Thinking*, London, Macmillan / Chicago,

University of Chicago Press, 1970.
— & Secord, P. F., *The Explanation of Social Behaviour*, Oxford, Basil Blackwell, 1972.
Harris, Zellig Sabbettai, *Mathematical Structures of Languages*, New York, Wiley, 1968, Tracts in Pure and Applied Mathematics, Vol. 21.
—, *Methods in Structural Linguistics*, Chicago, University of Chicago Press, 1951.
Hart, H. L. A. (Herbert), 'The ascription of responsibility and rights', *Proceedings of the Aristotelian Society*, Vol. XLIX, 1948-49, pp. 171-194.
—, *The Concept of Law*, Oxford, Clarendon Press, 1961.
—, *Law, Liberty and Morality*, London, Oxford University Press, 1963.
—, *The Morality of the Criminal Law*, London, Oxford University Press, 1965.
—, *Punishment and Responsibility* (collection of essays), Oxford, Clarendon Press, 1968.
— & Honoré, Antony M., *Causation in the Law*, London, Oxford University Press, 1959.
Hartman, Robert S., 'Value theory as a formal system', *Kant-Studien*, Vol. L, 1958-59, pp. 287-315.
—, *La estructura del valor. Fundamentos de axiología científica*, Mexico City, Fundo de Cultura Económica, 1959, serie de Diánoia (English translation, *The Structure of Value. Foundations of Scientific Axiology*, Carbondale, Southern Illinois University Press, 1967).
—, 'The logic of description and valuation', *The Review of Metaphysics* XIV (2), Dec. 1960, pp. 191-230.
—, 'The logic of value', *The Review of Metaphysics* XIV (3), Mar. 1961, pp. 389-432.
—, 'Axiology as a science. A rejoinder note to Professor Neri Castañeda's review of *La estructura del valor. Fundamentos de la axiología científica*', *Philosophy of Science* XXIX (4), Oct. 1962, pp. 412-433.
—, *El conocimiento del bien*, Mexico City-Buenos Aires, Fundo de Cultura Económica, 1966, serie de Diánoia.
Hartmann, Heinz, *Psychoanalysis and Moral Values*, New York, International Universities Press, 1960.
—, *Ego Psychology and the Problem of Management*, translation by D. Rapaport, New York, International Universities Press, 1964.
—, *Essays on Ego Psychology. Selected Problems in Psychoanalytic Theory*, New York, International Universities Press, 1965.
—, Kris, E. & Loewenstein, R., *Papers on Psychoanalytic Psychology*, New York, International Universities Press, 1964, Psychological Issues IV, Monograph No. 14.
Hartshorne, Charles, *The Logic of Perfection and Other Essays in Neoclassical Metaphysics*, LaSalle (Ill.), Open Court, 1962.
Havet, Jacques, Review of Perelman (ed.), *Raisonnement et démarches de l'historien* (*op. cit.*, 1963), *Revue de Métaphysique et de Morale*, 74th Year, No. 3, July-Sept. 1969, pp. 348-351.
Hayek, Friedrich A. von, 'Scientism and the study of society', 1st publ. in *Economica*, 1942-44; reproduced in *The Counter-Revolution of Science, op. cit.* (1952) (French translation by Raymond Barre, *Scientisme et sciences sociales. Essai sur le mauvais usage de la raison*, Paris, Plon, 1953, 'Recherches en sciences humaines', ser., No. 2).
—, *Individualism and Economic Order. A Critical Analysis of Socialist Economics and a Plea for the Preservation of True Individualism*, Chicago, University of Chicago Press, 1948; London, Routledge, 1949.
—, *The Counter-Revolution of Science. Studies on the Abuse of Reason*, Glencoe (Ill.), The Free Press / London, Allen & Unwin, 1952.
—, *The Constitution of Liberty*, London, Routledge, 1960.
Hegel, G. W. F., *Vorlesungen über die Geschichte der Philosophie, zuerst gehalten*

im Wintersemester 1805-06 in Jena, Einleitung, critical edition by Joh. Hoffmeister, in HEGEL, *Sämtliche Werke*, Vol. XV a, Leipzig, 1940; 3rd edn., abr., by Friedhelm Nicolin, 1959; reproduction, without change, *Einleitung in die Geschichte der Philosophie*, Hamburg, Felix Meiner, 1966.

—, *Vorlesungen über die Geschichte der Philosophie*, Vols. XVIII and XIX in *Jubiläumsausgabe*, edited by Hermann Glockner, Stuttgart, Frommann-Holzboog, 4th edn., 1961-68.

—, *Enzyklopädie der philosophischen Wissenschaften im Grundrisse* (1830), edited by Friedhelm Nicolin & Otto Pöggeler, Hamburg, Felix Meiner, 6th edn., 1959, Philosophische Bibliothek, No. 33.

HEIDEGGER, Martin, *Sein und Zeit. Erste Hälfte = Jahrbuch für Philosophie und phänomenologische Forschung*, Vol. VIII, 1927, pp. 1-438; further editions (without change) in book form, Halle an der Saale, Niemeyer, 1927, etc. (English translation by John Macquarrie & Edward Robinson, *Being and Time*, Oxford, Basil Blackwell, 1967).

—, *Über den Humanismus* (letter addressed to Jean Beaufret in December 1946), 1st publ. under the title 'Über den "Humanismus", Brief an Jean Beaufret', with *Platons Lehre von der Wahrheit*, Bern, Francke, 1947; separate publ. under the title *Über den Humanismus*, Frankfurt am Main, Klostermann, 1949 (French translation by Roger Munier, with German text on opposite pages, *Lettre sur l'humanisme*, Paris, Aubier-Montaigne, 1957; reproduced, without the German text, in HEIDEGGER, *Questions III*, Paris, Gallimard, 1966, 'Classiques de la Philosophie' ser.

—, *Holzwege*, Frankfurt am Main, Klostermann, 1950 (French translation by Wolfgang Brokmeier, edited by François Fédier, *Chemins qui ne mènent nulle part*, Paris, Gallimard, 1962, 'Classiques de la Philosophie' ser.).

—, 'Nietzsches Wort "Got ist tot"' (redraft of lectures delivered in 1943), in HEIDEGGER, *Holzwege, op. cit.* (1950), pp. 193-247 (French translation, 'Le mot de Nietzsche "Dieu est mort"', in HEIDEGGER, *Chemins qui ne mènent nulle part, op. cit.* (1962), pp. 173-219).

—, 'Der Ursprung des Kunstwerkes', in HEIDEGGER, *Holzwege, op. cit.* (1950), pp. 7-68; separate edition, Stuttgart, Reclams Universal Bibliothek, No. 8446/8447 (French translation, 'L'origine de l'œuvre d'art', in HEIDEGGER, *Chemins qui ne mènent nulle part, op. cit.* (1962), pp. 11-68).

—, 'Die Zeit des Weltbildes' (lecture delivered on 9 June 1938 under the title 'Die Begründung des nuzeitlichen Weltbildes durch die Metaphysik' [= The founding of the modern conception of the world through metaphysics]), included in HEIDEGGER, *Holzwege, op. cit.* (1950), pp. 69-104 (English translation by Marjorie Grene, 'The age of the world view', *Measure*, No. 2, 1951, pp. 269 sqq.; and French translation, 'L'époque des "conceptions du monde"', in HEIDEGGER, *Chemins qui ne mènent nulle part, op. cit.* (1962), pp. 69-100).

—, *Erläuterungen zu Hölderlins Dichtung*, 2nd edn., augm., Frankfurt am Main, Klostermann, 1951 (French translation by H. Corbin, M. Deguy, F. Fédier and J. Launay, *Approche de Hölderlin*, Paris, Gallimard, 1962).

—, *Einführung in die Metaphysik* (publication of a series of University lectures delivered in 1935), Tübingen, J. C. B. Mohr, 1953 (English translation by Ralph Manheim, *An Introduction to Metaphysics*, New Haven (Conn.). Yale University Press, 1959).

—, *Vorträge und Aufsätze*, Pfullingen, Neske, 1954 (French translation by André Préau, preface by Jean Beaufret, *Essais et conférences*, Paris, Gallimard, 1958).

—, *Zur Seinsfrage*, Frankfurt am Main, Klostermann, 1956 (English translation edited by W. Kluback & J. T. Wilde, *The Question of Being*, New York, 1958).

—, 'Hegel und die Griechen' (lecture delivered at a plenary meeting of the Academy of Sciences of Heidelberg on 26 July 1958), in *Die Gegenwart der Griechen im neueren Denken. Festschrift für Hans-Georg Gadamer zum 60. Geburtstag*, Tübingen, J. C. B. Mohr, 1960, pp. 43-57; republished separately,

Frankfurt am Main, Klostermann, 1967 (French translation by Jean Beaufret & Dominique Janicaud, 'Hegel et les Gregs', in HEIDEGGER, *Questions, II*, Paris, Gallimard, 1968, pp. 41-68) (French translation of a 1st version – lecture delivered by M. HEIDEGGER in the main lecture hall of the new Faculty at Aix-en-Provence on 20 Mar. 1958 – by Jean Beaufret & P. Sagave, *Cahiers du Sud* XLVII (349), Jan. 1959, pp. 355-368).

—, *Unterwegs zur Sprache*, Pfullingen, Neske, 1959 (English translation by Peter Hertz, edited by J. Glenn Gray & Fred Wieck, *On the Way to Language*, New York, Harper & Row, 1971).

—, *Nietzsche*, Pfullingen, Neske, 1961, 2 vols. (French translation by Pierre Klossowski, Paris, Gallimard, 1971, 2 vols., Bibliothèque de Philosophie).

HEIMBECK, Raeburne S., *Theology and Meaning. A Critique of Metatheological Scepticism*, Stanford (Calif.), Stanford University Press / London, Allen & Unwin, 1969.

HEIMSOETH, Helmut, 'Stand der philosophiegeschichtlichen Forschung', in WINDELBAND, *Lehrbuch der Geschichte der Philosophie*, edited by H. Heimsoeth, Tübingen, J. C. B. Mohr, 14th edn., 1950, pp. XI-XLVII.

HELD, Klaus, *Lebendige Gegenwart. Die Frage nach der Seinsweise des transzendentalen Ich bei Edm. Husserl, entwickelt am Leitfaden der Zeitproblematik*, The Hague, Martinus Nijhoff, 1966, 'Phaenomenologica' ser., No. 23.

HEMPEL, Carl Gustav, 'The function of general laws in history', *The Journal of Philosophy* XXXIX (2), 15 Jan. 1942, pp. 35-48.

—, 'The logic of functional analysis', in GROSS (ed.), *Symposium on Sociological Theory, op. cit.* (1959), pp. 271-307.

—, *Aspects of Scientific Explanation, and Other Essays in the Philosophy of Science*, New York, The Free Press / London, Collier-Macmillan, 1965.

— & OPPENHEIM, Paul, 'Studies in the logic of explanation', *Philosophy of Science*, Vol. XV, 1948, pp. 135-175; reproduced in HEMPEL, *Aspects of Scientific Explanation . . ., op. cit.* (1965), pp. 245-290.

HENRY, Michel, *L'essence de la manifestation*, Paris, Presses Universitaires de France, 1963, 2 vols., 'Epiméthée' ser.

—, *Philosophie et phénoménologie du corps. Essai sur l'ontologie biranienne*, Paris, Presses Universitaires de France, 1965, 'Epiméthée' ser.

HEPBURN, Ronald Wm., *Christianity and Paradox. Critical Studies in XXth-Century Theology*, London, Watts, 1958.

Hermeneutik und Dialektik. Hans-Georg Gadamer zum 70. Geburtstag, edited by Rüdiger BUBNER, Konrad CRAMER & Reiner WIEHL, Tübingen, J. C. B. Mohr, 1970, 2 vols.: I. *Methode und Wissenschaft, Lebenswelt und Geschichte*; II. *Sprache und Logik, Theorie des Auslegung und Probleme der Einzelwissenschaften.*

HESSE, Mary Brenda, *Models and Analogies in Science*, London-New York, Sheed & Ward, 1963.

HIGH, D. M., *Language, Persons and Belief. Studies in Wittgenstein's 'Philosophical Investigations' and Religious Uses of Language*, New York, Oxford University Press, 1967.

HINNERS, Richard, 'The ideological turn and its problem for the history of philosophy', in George F. MCLEAN (ed.), *Truth and Historicity of Man*, Washington (D.C.), The Catholic University of America Press, 1969.

HINTIKKA, Jaakko, 'Quantifiers in deontic logic', *Societas Scientiarum Fennica. Commentationes Humanarum Litterarum* (Helsinki), XXIII (4), 1957, pp. 2-23.

—, *Knowledge and Belief*, Ithaca (N.Y.), Cornell University Press, 1962.

—, 'Logic and philosophy', in KLIBANSKY (ed.), *Contemporary Philosophy. A Survey / La philosophie contemporaine. Chroniques*, Vol. I, *op. cit.* (1968), pp. 3-30.

—, 'Semantics for propositional attitudes, in DAVIS, HOCKNEY & WILSON (eds.), *Philosophical Logic, op. cit.* (1969), pp. 21-45.

HINTIKKA, Jaakko & SUPPES, Patrick (eds.), *Aspects of Inductive Logic*, Amsterdam, North Holland Publ. Co. / New York, Humanities Press, 1966, Studies in Logic and the Foundations of Mathematics.

—, *Information and Inference*, Dordrecht, Reidel, 1970, Synthese Library.

[See also DAVIDSON & HINTIKKA (eds.).]

HIRSCH, Eric D., *Validity in Interpretation*, New Haven (Conn.), Yale University Press, 1969.

HJELMSLEV, Louis, *Omkring sprogteoriens grundlæggelse*, Copenhagen, Akademisk Forlag, 1943 (English translation by Francis J. Whitfield, *Prolegomena to a Theory of Language*, 2nd edn., rev., Madison, University of Wisconsin Press, 1961).

—, *Essais linguistiques* (1st publ. under the title *Essais de linguistique générale*, Copenhagen, Nordisk Sprog- og Kulturforlag, 1959, Travaux du Cercle linguistique de Copenhague, No. XII); new edn., Paris, Minuit, 1971, 'Arguments' ser., No. 47.

—, *Sproget*, Copenhagen, Berlingske Forlag, 1963 (English translation by Francis Whitfield, *Language. An Introduction*, Madison, University of Wisconsin Press, 1970).

HOFSTADTER, Albert & MCKINSEY, J. C. C., 'On the logic of imperatives', *Philosophy of Science* VI (4), Oct. 1939, pp. 446-457.

[See also DAVIDSON, MCKINSEY & SUPPES.]

HOHL, Hubert, *Lebenswelt und Geschichte. Grundzüge der Spätphilosophie E. Husserls*, Freiburg-Munich, Karl Alber, 1962.

HOLLIS, Martin, 'The limits of irrationality', *Archives européennes de Sociologie / European Journal of Sociology* VIII (2), 1967 (entitled *Sympathy for Alien Concepts*), pp. 265-271; reproduced in WILSON (ed.), *Rationality, op. cit.* (1970), pp. 214-220.

HOMANS, George C., *Social Behavior. Its Elementary Forms*, New York, Harcourt, Brace & World, 1961.

—, *The Nature of Social Science*, New York, Harcourt, Brace & World, 1967.

HOOK, Sidney (ed.), *Psychoanalysis, Scientific Method and Philosophy*, New York, Grove Press, 1959.

—, *Dimensions of Mind. A Symposium*, Proceedings of the Annual Institute, 15-16 May 1959, of the New York University Institute of Philosophy, New York, New York University Press, 1960.

—, *Philosophy and History. A Symposium*, New York, New York University Press, 1963.

—, *Law and Philosophy. A Symposium* (New York University Institute of Philosophy), New York, New York University Press, 1964.

—, *Language and Philosophy*, New York, New York University Press, 1969 (in Pt. II, communications by W. v. O. QUINE, N. GOODMAN, G. HARMAN and others on the Chomskian theory of the innateness of syntactic universals).

HORKHEIMER, Max, *Eclipse of Reason*, London, Oxford University Press, 1947 (German translation and edition with complements by Alfred Schmidt, *Zur Kritik der instrumentellen Vernunft*, Frankfurt am Main, S. Fischer, 1967).

—, *Kritische Theorie: Eine Dokumentation* (collection of studies originally published for the most part in the periodical of the Institut für Sozialforschung in the thirties, with a preface and a letter from the author to the publisher – edited by Alfred Schmidt with a postface), Frankfurt am Main, S. Fischer, 1968.

HOROWITZ, J., 'Ulrich Klug's *Legal Logic*. A critical account', *Logique et analyse*, 1966, No. 9, pp. 78-144.

HORTON, R. & FINNEGAN, R. (eds.), *Modes of Thought. Essays on Thinking in Western and non-Western Societies*, London, Faber, 1973.

HOUNTONDJI, Paulin Jidenu, 'Remarques sur la philosophie africaine contemporaine', *Diogène*, No. 71, July-Sept. 1970, pp. 120-140 (parallel publication of English version, translation by Sally Bradshaw, 'Comments on contemporary

African philosophy', *Diogenes*, No. 71, Fall 1970, pp. 109-131).

HOURANI, George F., *Ethical Value*, Ann Arbor, University of Michigan Press, 1956.

HUACO, George A., 'The functionalist theory of stratification. Two decades of controversy', *Inquiry* (Oslo), IX (3), Autumn 1966, pp. 215-240.

HUGHES, George E., 'Has God's existence been disproved? - A reply to Professor J. N. Findlay', *Mind* LVIII (229), Jan. 1949, pp. 67-74; reproduced in FLEW & MACINTYRE (eds.), *New Essays in Philosophical Theology, op. cit.* (1953).

— & CRESSWELL, M. J., *An Introduction to Modal Logic*, London, Methuen / New York, Barnes & Noble, 1968.

HULL, Clark L., *Principles of Behavior. An Introduction to Behavior Theory*, New York, Appleton-Century-Crofts, 1943.

HUSSERL, Edmund, *Logische Untersuchungen*, Halle an der Saale, Niemeyer: I. *Prolegomena zur reinen Logik*, 1900; II. *Untersuchungen zur Phänomenologie und Theorie der Erkenntnis*, 1901; 2nd edn., rev., 1913 (*Prolegomena* and *Untersuchungen* I to V) and 1921 (VIth *Untersuchung: Elemente einer phänomenologischen Aufklärung der Erkenntnis*) (English translation from the 2nd German edn. by J. N. Findlay, *Logical Investigations*, London, Routledge / New York, Humanities Press, 1970, 2 vols.

—, 'Philosophie als strenge Wissenschaft', *Logos*, Vol. I (1910-11), pp. 289-341 (English translation by Quentin Lauer, 'Philosophy as rigorous science', in HUSSERL, *Phenomenology and the Crisis of Philosophy*, New York, Harper & Row, 1965).

—, *Ideen zu einer reinen Phänomenologie und phänomenologischen Philosophie*, Erstes Buch: *Allgemeine Einführung in die reine Phänomenologie = Jahrbuch für Philosophie und phänomenologische Forschung*, Vol. I, 1913; 2nd and 3rd edns. (without change), Halle an der Saale, Niemeyer, 1922 and 1928; new edn., edited by Walter Biemel, The Hague, Martinus Nijhoff, 1950, 'Husserliana' ser. (Husserl Archives in Louvain), Vol. III (English translation by W.-R. Boyce-Gibson, *Ideas*, London, Allen & Unwin / New York, Macmillan, 1931 – with, as a preface, an English translation of the 'Nachwort zu meinen *Ideen*', originally published in the *Jahrbuch*, XI, 1930).

—, *Erste Philosophie* (manuscripts, 1923-24), posthumous publ., edited by Rudolf Boehm, The Hague, Martinus Nijhoff, 1956-59, 2 vols., 'Husserliana' ser.: I. *Kritische Ideengeschichte*; II. *Theorie der phänomenologischen Reduktion* (French translation by Arion L. Kelkel, *Philosophie première*, Paris, Presses Universitaires de France, 1970-1972, 2 vols., 'Epiméthée ser.: I. *Histoire critique des idées*; II. *Théorie de la réduction phénoménologique*).

—, *Formale und transzendentale Logik. Versuch einer Kritik der logischen Vernunft = Jahrbuch für Philosophie und phänomenologische Forschung* (Halle an der Saale, Niemeyer), Vol. X, 1929, pp. 1-298 (English translation by Dorion Cairns, *Formal and Transcendental Logic*, New York, Humanities Press, 1969).

—, *Méditations cartésiennes. Introduction à la phénoméologie* (expanded and edited text of the lectures given at the Sorbonne on 23 and 25 Feb. 1929, translated into French by Gabrielle Peiffer & Emmanuel Levinas), Paris, A. Colin, 1931; republ., Paris, Vrin, 1947; original German text, *Cartesianische Meditationen und Pariser Vorträge*, edited by S. Strasser, The Hague, Martinus Nijhoff, 1950, 'Husserliana' ser., Vol. I (posthumous publ.) (English translations: *Cartesian Meditations. An Introduction to Phenomenology*, translated by Dorion Cairns, New York, Humanities Press, 1960; and *The Paris Lectures*, translated by Peter Koestenbaum with an introductory essay, The Hague, Martinus Nijhoff, 1965; New York, Humanities Press, 1966).

—, *Die Krisis der europäischen Wissenschaften und die transzendentale Phänomenologie. Eine Einleitung in die phänomenologische Philosophie*, 1st publ. (Pt. I only), *Philosophie* (Belgrade), Vol. I, 1936, pp. 77-176; posthumous publ., complete, with related texts and manuscripts, edited by Walter Biemel, The Hague, Martinus Nijhoff, 1954, 'Husserliana' ser., Vol. VI (English translations: *The Crisis of European Sciences and Transcendental Phenomenology. An Introduction to Phenomenological Philosophy*, translation by David Carr, Evanston (Ill.), Northwestern University Press, 1970; and 'Philosophy and the crisis of European man' (lecture delivered at the Vienna Kulturbund, 7 May 1935), translation with notes and an introduction by Quentin Lauer, in HUSSERL, *Phenomenology and the Crisis of Philosophy*, New York, Harper & Row, 1965).

—, Die Frage nach dem Ursprung der Geometrie als intentional-historisches Problem', posthumous publ., *Revue internationale de Philosophie* (Brussels), 1st Year, No. 2, 15 Jan. 1939 (issue devoted to Husserl), pp. 203-225 (French translation, with an introduction, by Jacques DERRIDA, *L'origine de la géométrie*, Paris, Presses Universitaires de France, 1962, 'Epiméthée' ser.).

Idée du monde et philosophie de la nature, Bruges-Paris, Desclée de Brouwer, 1966, 'Recherches de Philosophie' ser., No. VII.

IL'ENKOV, E. V., *Dialektika abstraktnogo i konkretnogo v 'Kapitale' Marksa* (= The Dialectic of Abstract and Concrete in Marx's *Das Kapital*), Moscow, 1960.

—, Article 'Ideal'noe' (= Ideal), in *Filosofskaja ènciklopedija*, Vol. II, Moscow, 1962.

IMAMICHI, Tomonobu A., *Betrachtungen über das Eine*, Tokyo, Institute of Aesthetics, Faculty of Philosophy, Tokyo University, 1968.

IOVČUK, M. T., 'O nekotoryh metodologičeskih problemah istorii filosofii', *Voprosy filosofii*, 1959, No. 11, pp. 63-78 (summary in English, JOVCHUK, M. T., 'About some methodological problems of the history of philosophy', *ibid.*, p. 187).

IRIARTE, Joaquím, 'Teoría de la historia de la filosofía', *Pensamiento* (Madrid), Vol. XVII, 1961, pp. 493-516.

—, 'Valor filosófico de la historia de la filosofía', *Razón y Fe* (Madrid), No. 170, 1964, pp. 171-188.

ISAJIW, Wsevolod W., *Causation and Functionalism in Sociology*, London, Routledge, 1968.

Istočnikovedenie. Teoretičeskie i metodologičeskie problemy (= The Study of Sources. Theoretical and Methodological Problems), collection, Moscow, 1969.

Istorija i psihologija: see PORŠNEV & ANCIFEROVA (eds.).

Istorija i sociologija (= History and Sociology), collection, Moscow, 1964.

JACOB, François, JAKOBSON, Roman, LÉVI-STRAUSS, Claude & L'HÉRITIER, Philippe, 'Vivre et parler' (T.V. round table), *Les Lettres françaises*, nos. 1221 and 1222, 14 and 21 Feb. 1968.

JAKOBSON, Roman, 'Aphasia as a linguistic problem', in H. WERNER (ed.), *On Expressive Language*, Worcester, Clark University Press, 1955, pp. 69-81 (French translation by Jean-Paul Boons & Radmila Zygouris, 'L'aphasie comme problème linguistique', in JAKOBSON, *Langage enfantin et aphasie*, Paris, Minuit, 1969, pp. 103-117).

—, 'Two aspects of language and two types of aphasic disturbances' = Pt. II (pp. 53-82) of JAKOBSON & HALLE, *Fundamentals of Language, op. cit.* (1956) (French translation by Nicolas Ruwet, 'Deux aspects du langage et deux types d'aphasie', in JAKOBSON, *Essais de linguistique générale, op. cit.* (1963), pp. 43-67).

—, 'Linguistics and poetics' ('Concluding statement: from the viewpoint of linguistics), in Thomas A. SEBEOK (ed.), *Style in Language* (outcome of a conference of linguists, anthropologists, psychologists and literary critics, held at the University of Indiana in the Spring of 1958), Cambridge (Mass.), The M.I.T. Press, 1960, pp. 350-377 (French translation by Nicolas Ruwet, 'Linguistique et poétique, in JAKOBSON, *Essais de linguistique générale, op. cit.* (1963), pp. 209-248).

—, *Essais de linguistique générale* (a selection of essays, translation into French by Nicolas Ruwet), Paris, Minuit, 1963.

—, 'Implications of language universals for linguistics', in GREENBERG (ed.), *Universals of Language, op. cit.* (1963), pp. 263-278.

—, 'Linguistics', in *Main Trends of Research in the Social and Human Sciences, Part I: Social Sciences, op. cit.* (1970), pp. 419-463 (parallel publication of French translation, 'La linguistique', in *Tendances principales de la recherche dans les sciences sociales et humaines, Partie I: Sciences sociales, op. cit.* (1970), pp. 504-556).

— & HALLE, Morris, 'Phonology and phonetics' = Pt. I (pp. 1-51) of JAKOBSON & HALLE, *Fundamentals of Language*, The Hague, Mouton, 1956, 'Janua linguarum', No. 1 (French translation by Nicolas Ruwet, 'Phonologie et phonétique', in JAKOBSON, *Essais de linguistique générale, op. cit.* (1963), pp. 103-149).

JAKOBSON, Roman (ed.), *The Structure of Language and Its Mathematical Aspects = Proceedings of the Twelfth Symposium in Applied Mathematics*, XII, Providence (R.I.), American Mathematical Society, 1961.

[See also JACOB, JAKOBSON, LÉVI-STRAUSS & L'HÉRITIER.]

JANKÉLÉVITCH, Vladimir, *Traité des vertus*, 1st edn., Paris, Bordas, 1949; new edn., recast and augmented, *ibid.*, 2 vols. published to date: I. *Le sérieux de l'intention*, 1968; II. *Les vertus et l'amour*, 1970.

JANNACCONE, Leonardo, *Il pensiero filosofico nella storia*, Florence, Giunti G. Barbera, 1967.

JARVIE, Ian C., *The Revolution in Anthropology*, London, Routledge, 1967.

—, *Concepts and Society*, London, Routledge, 1972, International Library of Sociology.

JASPERS, Karl, *Philosophie*, Heidelberg-Berlin, Springer, 1932, 3 vols. (English translation by E. B. Ashton, *Philosophy*, Chicago, University of Chicago Press, 1970, 3 vols.).

—, *Vernunft und Existenz*, Munich, Piper, 1935 (English translation by William Earle, *Reason and Existenz*, London, Routledge, 1956; paperback edn., New York, Farrar, Straus & Giroux).

—, *Von der Wahrheit* (Vol. I of *Philosophische Logik*), Munich, Piper, 1948 (English translation and Introduction by Jean T. Wilde *et al., Truth and Symbol.* New Haven (Conn.), College and University Press, 1962).

—, *Vom Ursprung und Ziel der Geschichte*, Munich, Piper / Zurich, Artemis, 1949 (English translation by Michael Bullock, *Origin and Goal of History*, New Haven (Conn.), Yale University Press, 1953).

—, *Vernunft und Widervernunft in unserer Zeit*, Munich, Piper, 1950 (English translation by Stanley Goodman, *Reason and Anti-Reason in Our Time*, New Haven (Conn.), Yale University Press, 1962; new edn., Camden (Conn.), Shoe String Press).

—, 'Über meine Philosophie', in JASPERS, *Rechenschaft und Ausblick. Reden und Aufsätze*, Munich, Piper, 1951; reproducd in KLIBANSKY (ed.), *Contemporary Philosophy / La philosophie contemporaine*, Vol. III, *op. cit.* (1969), pp. 177-205.

—, 'Essay on my philosophy', in KAUFMANN (ed.), *Existentialism from Dostoievsky to Sartre*, New York, 1956.

—, *Die grossen Philosophen*, Munich, Piper, 1957, 2 vols. (English translation by Ralph Manheim, edited by Hannah Arendt, *The Great Philosophers*, New York, Harcourt, Brace & World / London, Longmans, 1962-66, 2 vols.).

—, *Der philosophische Glaube angesichts der Offenbarung*, Munich, Piper, 1962 (English translation by E. B. Ashton, *Philosophical Faith and Revelation*, New York, Harper & Row / London, Collins, 1967).

[See also SCHILPP (ed.), *The Philosophy of Karl Jaspers.*]

Jazyk i myšlenie (= Language and Thought), collection, Moscow, 1967.

JEANSON, Francis, *Le problème moral et la pensée de Sartre*, followed by 'Un quidam nommé Sartre', with a Preface by Jean-Paul SARTRE, Paris, Seuil, 1966.

JEFFREY, Richard C., *The Logic of Decision*, New York, McGraw-Hill, 1965.
JOEL, Karl, *Wandlungen der Weltanschauung. Eine Philosophiegeschichte als Geschichtsphilosophie* (1928-34, 2 vols.), Tübingen, J. C. B. Mohr, 1965, photographic reproduction.
JOHNSTONE, Henry W., Jr., *Philosophy and Argument*, University Park, Pennsylvania State University Press, 1959.
—, 'Theory of argumentation', in KLIBANSKY (ed.), *Contemporary Philosophy / La philosophie contemporaine*, Vol. I, *op. cit.* (1968), pp. 177-184.
[See also NATANSON & JOHNSTONE (eds.).]
JOJA, Athanase, *Studii de logică* (= Studies in Logic), Bucarest, 1960-66, 2 vols.
JOLIVET, Régis, *Le Dieu des philosophes et des savants*, Paris, Fayard, 1956.
—, *Les activités de l'homme et la Sagesse (Cahiers du Centre d'Etudes de Carthage*, No. 2, Tunis), Lyon, Vitte, 1963 (distribution, Paris, Casterman).
JØRGENSEN, Jørgen, 'Imperatives and logic', *Erkenntnis* VII (4), 1937-38, pp. 288-296.
JOUVENEL, Bertrand de, *Du pouvoir. Histoire naturelle de sa croissance*, Geneva, A l'enseigne du cheval ailé, 1945; new edn., Paris, Hachette, 1972 (English translation by J. F. Huntington, *Power. The Natural History of Its Growth*, London, Hutchinson, 1948; another edition, *On Power. Its Nature and the History of Its Growth*, New York, Viking Press, 1949).
—, *De la souveraineté*, Paris, Librairie de Médicis M. T. Genin, 1955 (English translation by J. F. Huntington, *Sovereignty. An Enquiry into the Political Good*, Cambridge, Cambridge University Press / Chicago, University of Chicago Press, 1957).
—, *The Pure Theory of Politics*, Cambridge, Cambridge University Press/ New Haven (Conn.), Yale University Press, 1963 (French translation by Gabrielle Rolin, Guy Berger *et al.*, *De la politique pure*, Paris, Calmann-Lévy, 1963, 'Liberté de l'esprit' ser.).
—, *Du principat, et autres réflexions politiques* (collections of essays previously published, from 1956 to 1967), Paris, Hachette / Littérature, 1972.
Le Jugement moral = *Revue internationale de Philosophie*, No. 70, 1964, Fasc. 4, pp. 359-457.
JUHOS, Béla, 'Die Methode der fiktiven Prädikate', *Archiv für Philosophie* IX (1-4) and X (1-4), 1960.

KAGAME, Alexis, *La philosophie bantu-rwandaise de l'être*, Brussels, 1956.
KAHANE, Ernest, *La vie n'existe pas*, with a Preface by A. J. OPARINE, Paris, Les éditions rationalistes, 1962.
KALINOWSKI, Georges, *Introduction à la logique juridique*, Paris, Librairie générale de droit et de jurisprudence, 1965.
—, 'Logique déontique et logique juridique', *Les Etudes philosophiques*, Vol. XX, 1965, pp. 157-165.
—, 'Les thèmes actuels de la logique déontique', *Studia Logica* (Warsaw), Vol. XVII, 1965, pp. 75-107 (summaries in Polish, pp. 108-110, and in Russian, pp. 111-113).
—, 'De la spécificité de la logique juridique', in *La logique du droit, op. cit.* (*Archives de Philosophie du Droit*, 1966), pp. 7-23.
KAMENKA, Eugene, 'Marxism and the history of philosophy', *History and Theory*, Vol. V, 1965, pp. 83-104.
—, *Marxism and Ethics*, London, Macmillan, 1969.
KANE, William H., 'The extent of natural philosophy', *The New Scholasticism* (American Catholic Philosophical Association), XXXI (1), Jan. 1957, pp. 85-97.
KANGRGA, Milan, *Etički problem u djelu K. Marksa* (= The Problem of Ethics in the Work of Karl Marx), Zagreb, 1963.
KAPLAN, Abraham, *The Conduct of Inquiry. Methodology for Behavioral Science*, San Francisco, Chandler, 1964.
KATTSOFF, Louis O., *Physical Science and Physical Reality*, The Hague, Martinus

Nijhoff, 1957.
—, *Making Moral Decisions. An Existential Analysis*, The Hague, Martinus Nijhoff, 1965.
KATZ, Jerrold J., *The Problem of Induction and Its Solution*, Chicago, University of Chicago Press, 1962.
—, *The Philosophy of Language*, New York, Harper & Row, 1966 (French translation by Janick Gazio, *La philosophie du langage*, Paris, Payot, 1971).
— & FODOR, Jerry A., 'The structure of a semantic theory', *Language*, Vol. XXXIX, Apr.-June 1963, pp. 170-210; reproduced in FODAR & KATZ (eds.), *The Structure of Language, op. cit.* (1964), pp. 479-518.
— & POSTAL, Paul M., *An Integrated Theory of Linguistic Descriptions*, Cambridge (Mass.), The M.I.T. Press, 1964, Research Monograph No. 26.
KEDROV, B. M., *Razvitie ponjatija himičeskogo èlementa* (= The Development of the Concept of Chemical Element), Moscow, 1956.
—, *Edinstvo dialektiki, logiki i teorii poznanija*, Moscow, 1963 (French translation by Ivan Mignot, KEDROV, Boniface, *Dialectique, logique, gnoséologie: leur unité*, Moscow, Editions du Progrès, 1970).
KELSEN, Hans, 'Der Begriff der Rechtsordnung', *Logique et analyse*, 1958, No. 3-4, pp. 150-167.
—, *Aufsätze zur Ideologiekritik*, edited by Ernst Topitsch, Berlin, Luchterhand, 1964.
KEMP, John, *Reason, Action and Morality*, London, Routledge / New York, Humanities Press, 1964.
KENNY, Anthony J., *Action, Emotion and Will*, London, Routledge, 1963.
—, 'God and necessity', in WILLIAMS & MONTEFIORE (eds.), *British Analytical Philosophy, op. cit.* (1966).
—, 'Practical inference', *Analysis* XXVI (3), Jan. 1966, pp. 65-75.
—, *The Five Ways*, London, Routledge, 1969.
KERNER, George C., 'Approvals, reasons, and moral arguments', *Mind* LXXI (284), Oct. 1962, pp. 474-486.
—, *The Revolution in Ethical Theory*, London, Oxford University Press, 1966.
KEYNES, John Maynard, *A Treatise on Probability*, London, Macmillan, 1921; republ., New York, Harper, 1962.
KIMMERLE, Heinz, Introduction to SCHLEIERMACHER, *Hermeneutik, op. cit.*, 1959, and Postface to the 1968 Supplement to the same work.
KLAUS, G., *Semiotik und Erkenntnistheorie*, Berlin, 1963.
—, *Sila slova* (= The Force of the Word), Moscow, 1967.
— & SCHULZE, G., *Sinn, Gesetz und Fortschritt in der Geschichte* (= Meaning, Law and Progress in History), Berlin, 1967.
KLIBANSKY, Raymond (ed.), *Philosophy in the Mid-Century. A Survey / La philosophie au milieu du XXe siècle. Chroniques*, Florence, La Nuova Italia Editrice, 4 vols.: Vols. I-III, 1958; Vol. IV, 1959; 2nd edn., 1961-62; 3rd edn. of Vol. I, 1967.
—, *Contemporary Philosophy. A Survey / La philosophie contemporaine. Chroniques* (contributions in English, French or German), Florence, La Nuova Italia Editrice, 4 vols.: Vol. I. *Logic and Foundations of Mathematics / Logique et fondements des mathématiques*, 1968; Vol. II. *Philosophy of Science / Philosophie des sciences*, 1968; Vol. III. *Metaphysics; Phenomenology; Language and Structure / Métaphysique; phénoménologie; langage et structure*, 1969; Vol. IV. *Ethics; Aesthetics; Law; Religion; Politics; Historical and Dialectical Materialism; Philosophy in Eastern Europe, Asia and Latin America / Ethique; esthétique; droit; religion; politique; matérialisme historique et dialectique; la philosophie en Europe orientale, en Asie et en Amérique latine*, 1971.
KLUG, Ulrich, *Juristische Logik*, Berlin, Springer, 1951.
KNAPP, Viktor, 'La philosophie du droit dans les pays socialistes', in KLIBANSKY (ed.), *Contemporary Philosophy. A Survey / La philosophie contemporaine. Chro-*

niques, Vol. IV, *op. cit.* (1971), pp. 156-169.
KNEALE, William C., *Probability and Induction*, London, Oxford University Press, 1949.
KOCH, Gerhard, *Gesetzmässigkeit und Praxis. Zur Dialektik von objektiver gesellschaftlicher Gesetzmässigkeit und subjektiver praktischer Tätigkeit*, Berlin, Dietz, 1968.
KOCKELMANS, Joseph J., *Phenomenology and Physical Science. An Introduction to the Philosophy of Physical Science*, Pittsburgh, Duquesne University Press, 1966, Duquesne Studies, Philosophical ser. No. 21.
KON, Igor S., *Filosofskij idealizm i krizis buržuaznoj istoričeskoj mysli* (= Philosophical Idealism and the Crisis in Bourgeois Historical Thinking), Moscow, 1959; rev. and enl. version in German, authorized translation by Willi Hoepp, *Die Geschichtsphilosophie des 20. Jahrhunderts. Kritischer Abriss*, Berlin, Akademie-Verlag, 1964, 2 vols.; 2nd edn. in German, 1966.
—, Article 'Istorija', in *Filosofskaja ènciklopedija*, Vol. 2, *op. cit.* (1962).
—, *Der Positivismus in der Soziologie* (translated from the Russian by W. Hoepp), Berlin, Akademie-Verlag, 1968.
—, 'K sporam o logike istoričeskogo ob"jasnenija: shema Poppera-Gempelja i eë kritiki' (= Contribution to the discussion on the logic of historical explanation: the Popper-Hempel thesis and its critics), in *Filosofskie problemy istoričeskoj nauki, op. cit.* (1969).
—, 'Istorija i sociologija', *Voprosy filosofii*, 1970, No. 8 (French translation, 'Histoire et sociologie', *Social Science Information / Information sur les Sciences sociales* X (4), Aug. 1971, pp. 87-102).
KONRAD, M. A., *Zapad i Vostok* (= The West and the Orient), Moscow, 1966.
KOPNIN, P. V., *Dialektika kak logika* (= Dialectic as Logic), Kiev, 1961.
—, *Vvedenie v marksistskuju gnoseologiju* (= Introduction to Marxist Gnoseology), Kiev, 1966.
—, 'Zur Methode der philosophiegeschichtlichen Forschung' (translated from Russian by F. KUMPF), in *Symposium der Philosophie Historiker (Moskau, 1967), op. cit.* (1968), pp. 158-170.
—, *Filosofskie idei V. I. Lenina i logika* (= Logic and Lenin's Philosophical Ideas), Moscow, 1969.
KÖRNER, Stephan, *Kant*, Harmondsworth (Middx.), Penguin Books, 1955.
KORNFORT, M., *Marksizm i lingvističeskaja filosofija* (= Marxism and Linguistic Philosophy), Moscow, 1968.
KOSING, Alfred, *Probleme der marxistischen Erkenntnistheorie*, Berlin, 1959.
KOTARBIŃSKI, Tadeusz, *Traktat o dobrej robocie*, Warsaw, 1955 (English translation by Olgierd Wojtasiewicz, *Praxiology. An Introduction to the Science of Efficient Action*, London, Pergamon / New York, Macmillan, 1965).
—, L'évolution de la praxéologie en Pologne', in KLIBANSKY (ed.), *Contemporary Philosophy / La philosophie contemporaine*, Vol. II, *op. cit.* (1968), pp. 438-450.
KRAFT, Victor, 'Dreierlei Philosophiegeschichte', in *Philomathes*, The Hague, Martinus Nijhoff, 1971, pp. 293-305.
KREMJANSKIJ, V. I., *Strukturnye urovni živoj materii* (= Structural Levels of Living Matter), Moscow, 1969.
KREŠIC, Andrija, 'Marksistička i apstraktno-materijalistična filozofija', *Filozofija* (Belgrade), No. 2, 1960.
KRISHNA, Daya, 'Law of contradiction and empirical reality', *Mind*, Apr. 1957.
—, *Considerations towards a Theory of Social Change*, Bombay, Manaktalas, 1965.
—, 'The active and the contemplative values', *Philosophy and Phenomenological Research* XXIX (3), Mar. 1969, pp. 414-422.
—, *Social Philosophy – Past and Future*, Simla, Institute of Advanced Studies, 1969.
—, '"The self-fulfilling prophecy" and the nature of society', *American Sociological Review* XXXVI (6), Dec. 1971.

KRISHNA, Daya, MATHUR, D. C. & RAO, A. P. (eds.), *Modern Logic. Its Relevance to Philosophy*, Delhi, Impex India, 1969.

KRISTELLER, Paul O., 'History of philosophy and history of ideas', *Journal of the History of Philosophy* (Berkley-Los Angeles), II (1), Jan. 1964, pp. 1-14.

KUHN, Helmut, 'Ideologie als hermeneutischer Begriff', in *Hermeneutik und Dialektik, op. cit.* (1970), Vol. I, pp. 343 sqq.

KUHN, Thomas S., *The Structure of Scientific Revolutions*, Chicago-London, The University of Chicago Press, 1962, 'Foundations of the Unity of Science. Toward an International Encyclopedia of Unified Science' II (2); new edn., augm., paperback, 1970 (French translation, *La structure des révolutions scientifiques*, Paris, Flamarion, 1972).

—, 'Logic of discovery or psychology of research?' and 'Reflections on my critics', in LAKATOS & MUSGRAVE (eds.), *Criticism and the Growth of Knowledge, op. cit.* (1970), pp. 1-23 and pp. 231-278.

KUMPF, F., 'Übersicht' (general view of the proceedings), in *Symposium der Philosophie Historiker (Moskau, 1967), op. cit.* (1968), pp. 198-209.

KUNTZ, Paul G., 'The dialectic of historicism and anti-historicism', in *Philosophy of the History of Philosophy, op. cit.* (*The Monist*, 1969), pp. 656-669.

KURELLA, Alfred, *Der Mensch als Schöpfer sich selbst* (= Man as Creator of Himself), Berlin (East), 1963.

KUZNECOV, I. V., *Princip sootvetstvija v sovremennoj fizike i ego filosofskoe značenie* (= The Principle of Correspondence in Contemporary Physics and Its Philosophical Implications), Moscow, 1948.

KYBURG, Henry E., Jr., *Probability and the Logic of Rational Belief*, Middletown (Conn.), Wesleyan University Press, 1961.

KYBURG, Henry E., Jr. & SMOKLER, Howard E. (eds.), *Studies in Subjective Probability*, New York, Wiley, 1964.

LACAN, Jacques, 'Fonction et champ de la parole et du langage en psychanalyse' (Report of the Rome Congress held at the Istituto di Psicologia dell' Università di Roma on 26 and 27 Sept. 1953), *La Psychanalyse*, Vol. I, 1956; reproduced in LACAN, *Ecrits, op. cit.* (1966), pp. 237-322.

—, 'La direction de la cure et les principes de son pouvoir' (communication at an international symposium held at Royaumont from 10 to 13 July 1958), *La Psychanalyse*, Vol. VI, 1961, pp. 149-206; reproduced in LACAN, *Ecrits, op. cit.* (1966), pp. 585-645.

—, *Ecrits* (collected essays), Paris, Seuil, 1965.

LACHIÈZE-REY, Pierre, *Le moi, le monde et Dieu*, Paris, Boivin, 1938; new edn., rev. and augm., Paris, Aubier-Montaigne, 1950, 'Philosophie de l'Esprit' ser.

LACROIX, Jean, *Histoire et mystère*, Tournai (Belgium), Casterman, 1962, 'Cahiers de l'actualité religieuse' ser., No. 18.

LADRIÈRE, Jean, *Les limitations internes des formalismes. Etude sur la signification du théorème de Gödel et des théorèmes apparentés dans la théorie des fondements des mathématiques*, Louvain, E. Nauwelaerts / Béatrice-Nauwelaerts / Paris, Gauthier-Villars, 1957, 'Logique mathématique', ser. B, II.

—, 'Les limitations des formalismes et leur signification philosophique', *Dialectica* XIV (4), 15 Dec. 1960, pp. 279-320 (followed by a discussion, pp. 321-328).

—, 'Objectivité et réalité en mathématiques', *Dialectica* XX (2), 1966, pp. 215-241.

—, *L'articulation du sens. Discours scientifique et parole de la foi*, Paris, Editions BSR/Aubier-Montaigne/Cerf/Delachaux & Niestlé/Desclée de Brouwer, 1970.

LAING, Ronald D. & COOPER, David G., *Reason and Violence*, London, Tavistock Publ., 1964 (French translation by Jean-Pierre Cottereau, *Raison et violence. Dix ans de la philosophie de Sartre (1950-1960)*, Paris, Payot, 1972, 'Petite Bibliothèque Payot').

LAKATOS, Imre, 'Changes in the problem of inductive logic', in LAKATOS (ed.), *The*

Problem of Inductive Logic, op. cit. (1968), pp. 315-347.

—, 'Criticism and the methodology of scientific research programmes', *Proceedings of the Aristotelian Society*, Vol. LXIX, 1969, pp. 149-186.

LAKATOS, Imre (ed.), *Problems in the Philosophy of Mathematics*, Amsterdam, North Holland Publ. Co., 1967, Studies in Logic and the Foundations of Mathematics.

—, *The Problem of Inductive Logic (Proceedings of the International Colloquium in the Philosophy of Science, London, 1965, Volume 2)*, Amsterdam, North Holland Publ. Co., 1968, Studies in Logic and the Foundations of Mathematics.

— & MUSGRAVE, Alan (eds.), *Criticism and the Growth of Knowledge (Proceedings of the International Colloquium in the Philosophy of Science, London, 1965, Vol. 4)*, Cambridge, Cambridge University Press, 1970.

LAKOVIĆ, D., *Marksistička teorija otudenja* (= The Marxist Theory of Alienation), Belgrade, 1968.

LANDÉ, Alfred, *New Foundations of Quantum Mechanics*, Cambridge, Cambridge University Press, 1965.

LANDGREBE, Ludwig, *Phänomenologie und Metaphysik* (1st publ., 1949), new edn., Gütersloh, G. Mohn, 1968.

LANDOLT, Eduard, 'Cos' é storia della filosofia?', *Teoresi* (Catania), Vol. XXIV, 1969, pp. 153-164.

LANDSBERG, Paul L., 'L'acte philosophique de Scheler', *Recherches philosophiques*, VI, 1936-37, pp. 299-312; reproduced in LANDSBERG, *Problèmes du personnalisme, op. cit.* (1952).

—, *Essai sur l'expérience de la mort*, followed by *Le problème moral du suicide* (posthumous works), Paris, Seuil, 1951, 'Esprit' ser.

—, *Problèmes du personnalisme* (posthumous work), Paris, Seuil, 1952, 'Esprit' ser.

LASLETT, Peter (ed.), *Philosophy, Politics and Society*, 1st ser., Oxford, Basil Blackwell, 1956.

— & RUNCIMAN, Walter G. (eds.), *Philosophy, Politics and Society*, 2nd ser., Oxford, Basil Blackwell, 1962.

—, *Philosophy, Politics and Society*, 3rd ser., Oxford, Basil Blackwell, 1967.

LAZARSFELD, Paul F., *Philosophie des sciences sociales* (collection of basic writings presented by Raymond Boudon, translated into French by Pierre Birnbaum, François Chazel, Gérard Lagneau, Jacques Lautman and Bernard Lecuyer), Paris, Gallimard, 1970, Bibliothèque des Sciences humaines.

—, 'Sociology', in *Main Trends of Research in the Social and Human Sciences, Part I: Social Sciences, op. cit.* (1970), pp. 61-165 (parallel publication of French translation, 'La sociologie', in *Tendances principales de la recherche dans les sciences sociales et humaines, Partie I: Sciences sociales, op. cit.* (1970), pp. 69-197).

—, *The Language of Social Research*, New York, Free Press, 1971.

—, *Qualitative Analysis. Historical and Critical Essays*, Rockley (N.J.), Allyn & Bacon, 1971.

LAZARSFELD, Paul F. (ed.), *Mathematical Thinking in the Social Sciences*, Glencoe (Ill.), Free Press, 1954; New York, Russell & Russell, 1969.

LAZEROWITZ, Morris, *The Structure of Methaphysics*, London, Routledge, 1955.

LEBLANC, Hugues, *Statistical and Inductive Probabilities*, Englewood Cliffs (N.J.), Prentice-Hall, 1962.

LEBRUN, Pierre, 'Structure et quantification. Réflexions sur la science historique', in PERELMAN (ed.), *Raisonnement et démarches de l'historien, op. cit.* (1963), pp. 29-51.

LECLERCQ, Jacques, *La philosophie morale de saint Thomas devant la pensée contemporaine*, Louvain, Editions de l'Institut Supérieur de Philosophie, 1955.

LEE, Donald S., 'Hypothetic inference in systematic philosophy', *International Philosophical Quarterly* (New York / Louvain), Vol. IX, 1969, pp. 363-390.

LEEUW, Gerardus van der, *Phänomenologie der Religion* (1st publ., 1934), 3rd edn., Tübingen, J. C. B. Mohr, 1970 (English translation by J. E. Turner, *Religion in Essence and Manifestation*, 2nd edn. incorporating the additions of the 2nd German edition, New York, Harper Torch books, 1963; London, Allen & Unwin, 1964).

LEFEBVRE, Henri, *Critique de la vie quotidienne*, Paris, Grasset, 1948; new edn., Paris, L'Arche, 1958.

—, Statements made in the general discussion at the 'Colloque de Royaumont' on Descartes, in *Descartes*, Paris, Minuit, 1957, 'Cahiers de Royaumont', Philosophie, No. II, pp. 463-466 and 476.

—, *Introduction à la modernité*, Paris, Minuit, 1962, 'Arguments' ser., No. 9.

—, *Métaphilosophie*, Paris, Minuit, 1965, 'Arguments' ser., No. 26.

—, *La vie quotidienne dans le monde moderne*, Paris, Gallimard, 1968, 'Idées' ser.

—, *La fin de l'histoire*, Paris, Minuit, 1970.

LEINFELLNER, Werner, *Struktur und Aufbau wissenschaftlicher Theorien: Eine wissenschaftstheoretisch-philosophische Untersuchung*, Vienna-Würzburg, Physica Verlag, 1965.

LEMMON, E. J., 'Deontic logic and the logic of imperatives', *Logique et analyse*, 1965, No. 8, pp. 39-71.

LENIN, Vladimir Ilič, *Filosofskie tetradi* (= Philosophical Notebooks), in *Sočinenija* (= Complete Works), *op. cit.*, Vol. XXIX.

—, *Sočinenija* (= Complete Works), 5th edn., Moscow.

LENNEBERG, E. H., *Biological Foundations of Language*, New York, Wiley, 1967.

LEPLEY, Ray, *The Language of Value*, New York, Columbia University Press, 1957.

LEROI-GOURHAN, André, *Evolution et techniques. L'homme et la matière*, Paris, Albin Michel, 1943, 'Sciences d'aujourd'hui' ser.; 2nd edn., *ibid.*, 1971.

—, *Evolution et techniques. Milieu et techniques, ibid.*, 1945.

—, *Le geste et la parole. Technique et langage, ibid.*, 1964.

—, *Le geste et la parole. La mémoire et les rythmes, ibid.*, 1965.

LE ROY, Edouard, *Le problème de Dieu*, Paris, L'Artisan du livre, 1930.

LEŚNIEWSKI, S., 'Grundzüge eines neuen Systems der Grundlagen der Mathematik': §§ 1-11, *Fundamenta Mathematicae*, Vol. XIV, 1929, pp. 1-81; § 12, *Collectanea Logica*, Vol. I, 1938, pp. 61-144.

—, 'O podstawach matematyki' (= On the foundations of mathematics), *Przegląd Filozoficzny*, Vol. XXX, 1927, pp. 164-206; Vol. XXXI, 1928, pp. 261-291; Vol. XXXII, 1929, pp. 60-105; Vol. XXXIII, 1930, pp. 77-105; and Vol. XXXIV, 1931, pp. 142-170.

—, 'Über die Grundlagen der Ontologie', *Comptes rendus de la Société des Sciences et des Lettres de Varsovie*, Division III, Vol. XXIII, 1930, pp. 111-132.

LEVADA, Ju. A., 'Istoričeskoe soznanie i naučnyj metod' (= Historical consciousness and the scientific method), in *Filosofskie problemy istoričeskoj nauki, op. cit.* (1969).

LEVI, Isaac, *Gambling with Truth*, New York, Knopf / London, Routledge, 1967.

LEVINAS, Emmanuel, *Totalité et infini. Essai sur l'extériorité*, The Hague, Martinus Nijhoff, 1961, 'Phaenomenologica' ser., No. 8.

—, 'La trace de l'Autre', *Tijdschrift voor Filosofie*, Sept. 1963.

—, 'La signification et le sens', *Revue de Métaphysique et de Morale*, 69th Year, No. 2, Apr.-June 1964, pp. 125-156.

—, *En découvrant l'existence avec Husserl et Heidegger*, new edn., followed by further essays, Paris, Vrin, 1967, 'Bibliothèque d'Histoire de la Philosophie'.

—, 'Humanisme et an-archie', in *La crise de l'humanisme, op. cit.* (*Revue internationale de Philosophie*, 1968), pp. 323-337.

—, 'Le Dit et le Dire', *Le nouveau Commerce*, No. 18-19, Spring 1971, pp. 19-48.

—, 'La proximité', *Archives de Philosophie* XXXIV (3), July-Sept. 1971, pp. 373-391.

[See also CASTELLI, GOUHIER, LEVINAS, PANIKKAR, RICOEUR, HABACHI *et al.*]

LÉVI-STRAUSS, Claude, 'Histoire et ethnologie', *Revue de Métaphysique et de Morale*, 54th Year, No. 3-4, July-Dec. 1949, pp. 363-391; reproduced as Chap. I, 'Introduction. Histoire et ethnologie', in LÉVI-STRAUSS, *Anthropologie structurale, op. cit.* (1958), pp. 3-33 (English translation in *Structural Anthropology, op. cit.*, 1963).

—, *Les structures élémentaires de la parenté*, Paris, Presses Universitaires de France, 1949; 2nd edn., rev., with a new preface, Paris-The Hague, Mouton, 1967 (English translation by J. H. Bell, J. R. von Sturmer & R. Needham, *The Elementary Structures of Kinship*, London, Eyre & Spottiswoode, 1969).

—, *Race et histoire*, Paris, Unesco, 1952, 'La question raciale devant la science moderne' ser. (parallel publication of English version, *Race and History*, 'The Race Question in Modern Science'); original French text republished (with a study by Jean POUILLON), Paris, Gonthier, 1967, 'Médiations' ser., and in the collection *Le racisme devant la science*, Paris, Unesco, new edn., 1973, pp. 9-49 (English version, *Race, Science and Society*, to be published in 1975).

—, *Anthropologie structurale*, Paris, Plon, 1958 (English translation by Claire Jacobson & Brooke Grundfest Schoepf, *Structural Anthropology*, New York-London, Basic Books, 1963).

—, *Leçon inaugurale* (delivered on 5 Jan. 1960), Paris, Collège de France, 1960, Chaire d'Anthropologie sociale (English translation by Sherry Ortner Paul & Robert A. Paul, *The Scope of Anthropology*, London, Jonathan Cape, 1967).

—, *La pensée sauvage*, Paris, Plon, 1962 (English translation, *The Savage Mind*, London, Weidenfeld & Nicolson, 1966).

—, *Mythologiques*, Paris, Plon, 4 vols.: * *Le cru et le cuit*; ** *Du miel aux cendres*; *** *L'origine des manières de table*; **** *L'homme nu*, 1964, 1966, 1968, 1971 (English translation by John & Doreen Weightman, *The Raw and the Cooked. Introduction to a Science of Mythology*: I, New York, Harper / London, Jonathan Cape, 1969-70; and *From Honey to Ashes, ibid.*, 1973).
[See also JACOB, JAKOBSON, LÉVI-STRAUSS & L'HÉRITIER.]

LEWIS, Clarence Irving & LANGFORD, C. H., *Symbolic Logic*, New York, Dover, 1932; republ. 1950.

LEWIS, H. D., *Philosophy of Religion*, London, Barnes & Noble, 1965.

LEWIS, H. D. (ed.), *Contemporary British Philosophy. Personal Statements*, 3rd ser., London, Macmillan, 1956.

—, *Clarity Is Not Enough. Essays in Criticism of Linguistic Philosophy*, London, Allen & Unwin / New York, Humanities Press, 1963 (contributions by H. H. PRICE and others).
[See also *sub Religious Studies.*]

LEWIS, John, *Science, Faith and Scepticism*, London, Lawrence & Wishart, 1959.

LEWY, C., 'G. E. Moore on the naturalistic fallacy', in STRAWSON (ed.), *Studies in the Philosophy of Thought and Action, op. cit.* (1968), pp. 134-146.

LIBERTINI, Crescenzo, *Influsso della leggenda sulla storia della filosofia*, Naples, Ediz. academiche A.L.S.I., 1961.

LICHTHEIM, George, 'The concept of ideology', *History and Theory* IV (2), 1965, pp. 164-195; reproduced in LICHTHEIM, *The Concept of Ideology and Other Essays*, New York, Random House, 1967.

LINDNER, Herbert, *Der Entwicklungsgang des philosophischen Denkens*, Berlin, Deutscher Verlag der Wissenschaften, 1966.

LLEDÓ, Emilio, 'Lenguaje e historia de la filosofía', in *Hermeneutik und Dialektik, op. cit.* (1970), Vol. II, pp. 85 sqq.

LLERA, H. P. (ed.), *Idea de la historia de la filsosofía, de su eficacia didáctica y su importancia en el presente*, Habana, Sociedad cubana de Filosofía, Instituto de Filosofía, 1954.

LOEWENSTEIN, R., 'Rapport sur la psychologie psychanalytique de H. Hartmann, E. Kris et R. Loewenstein. Contribution à la théorie psychanalytique', *Revue française de Psychanalyse* XXX (5-6), Sept.-Dec. 1966 (*XXVIème Congrès des*

Psychanalystes de Langues romanes, Paris, 29-30-31 octobre et Ier novembre 1965), pp. 775-793; followed by statements made in the discussion, pp. 795-820. [See also Hartmann, Kris & Loewenstein.]

Logičeskaja struktura naučnogo znanija (= The Logical Structure of Scientific Knowledge), Moscow, 1965.

Logika i struktura jazyka (= Logic and Structure of Language), Moscow, 1965.

La Logique du droit, special issue of *Archiv für Rechts- und Sozialphilosophie/ Archives de philosophie du droit et de philosophie sociale/Archives for Philosophy of Law and Social Philosophy* (Wiesbaden), Vol. XI (1966), pp. 1-337.

Lombardi, Franco, *Il concetto della libertà*, Florence, Sansoni, 1966.

—, *Problemi della libertà*, Florence, Sansoni, 1966.

—, *Il mundo degli uomini*, Florence, Sansoni, 1967, 2 vols.

—, *Concetto e problemi della storia della filosofia* = Lombardi, *Opere*, Florence, Sansoni, Vol. IX, 3rd edn., 1970.

Lonergan, Bernard, *Insight. A Study of Human Understanding*, London, Longmans & Greene, 1958.

—, *Method in Theology*, New York, Harper, 1972.

Lorenzen, Paul, *Einführung in die operative Logik und Mathematik*, Berlin-Göttingen-Heidelberg, Springer, 1955, Die Grundlehren der mathematischen Wissenschaften, Vol. 78.

Loring, L. M., *Two Kinds of Values*, London, Routledge / New York, Humanities Press, 1966.

Lossky, Vladimir, *Vision de Dieu*, Neuchâtel-Paris, Delachaux & Niestlé, 1962, 'Bibliothèque orthodoxe'.

Louch, Alfred R., 'The very idea of a social science', *Inquiry* (Oslo), VI (4), Winter 1963, pp. 273-286.

—, *Explanation and Human Action*, Berkeley-Los Angeles, University of California Press, 1966; Oxford, Basil Blackwell, 1967.

Lourenço, Aranjo, P. J., 'História da história da filosofia', *Revista da Universidade de São Paulo* XXIII (40-41), 1961-62, pp. 50-74.

Löwith, Karl, *Von Hegel bis Nietzsche*, Zurich, 1941; republ. in rev. version as *Von Hegel zu Nietzsche*, 5th edn., Stuttgart, Kohlhammer, 1964 (French translation from the rev. text, by Rémi Laureillard, *De Hegel à Nietzsche*, Paris, Gallimard, 1969, Bibliothèque de Philosophie).

—, *Meaning in History. The Theological Implications of the Philosophy of History*, Chicago, University of Chicago Press, 1949; Cambridge, Cambridge University Press, 1950; republ. in German version as *Weltgeschichte und Heilsgeschehen. Die theologischen Voraussetzungen der Geschichtsphilosophie*, Stuttgart, Kohlhammer, 1953; 5th edn., *ibid.*, 1967.

—, *Heidegger, Denker in dürftiger Zeit* (= Heidegger, Thinker in a Needy Time), Frankfurt am Main, 1953; 2nd edn., Göttingen, Vandenhoeck & Ruprecht, 1960.

—, *Permanence and Change. Lectures on the Philosophy of History*, Capetown, Ham, 1969.

Lübbe, Hermann, 'Sprachspiele und "Geschichten"', *Kantstudien*, Vol. LII, 1960-61, pp. 220-243.

—, *Säkularisierung. Geschichte eines ideenpolitischen Begriffs*, Freiburg im Breisgau, Karl Alber, 1965.

Lucas, John R., 'Minds, machines, and Gödel', *Philosophy* XXXVI (137), April and July 1961; reproduced in Sayre & Crosson (eds.), *The Modeling of Mind, op. cit.* (1963), pp. 255-271 and in Anderson (ed.), *Minds and Machines, op. cit.* (1964), pp. 43-59.

—, *Freedom of the Will*, London, Oxford University Press, 1970.

Luce, Robert Duncan & Suppes, Patrick C., 'Preference, uitlity and subjective probability', in Luce *et al.* (eds.), *Handbook of Mathematical Psychology*, Vol. III, New York-London, Wiley, 1965, pp. 250-402.

Luce, Robert Duncan, Bush, Robert R. & Galanter, Eugene (eds.), *Handbook of*

Mathematical Psychology, Volume II (comprises, *inter alia*, the two articles: CHOMSKY & MILLER, 'Introduction to the formal analysis of natural languages', *op. cit.*, and CHOMSKY, 'Formal properties of grammars', *op. cit.*, followed by a bibliography), New York-London, Wiley, 1963.

LUKÁCS, Gyorgy [Georg], *Die Zerstörung der Vernunft*, Berlin, Aufbau-Verlag, 1954; republ. as Vol. IX of *Gesamtausgabe*, Berlin, Luchterhand, 1962 (French translation by Stanislas George, André Gisselbrecht & Edouard Pfrimmer, *La destruction de la raison*, Paris, L'Arche, 1958-59, 2 vols.).

—, *Schriften zur Ideologie und Politik*, edited by Peter Ludz, Neuwied-Berlin, Luchterhand, 1961.

LUKES, Steven, 'Alienation and anomie', in LASLETT & RUNCIMAN (eds.), *Philosophy, Politics and Society*, 3rd ser., *op. cit.* (1967).

—, 'Some problems about rationality', *Archives européennes de Sociologie / European Journal of Sociology* VIII (2), 1967 (entitled *Sympathy for Alien Concepts*), pp. 247-264; reproduced in WILSON (ed.), *Rationality, op. cit.* (1970), pp. 194-213.

—, 'Methodological individualism reconsidered', *British Journal of Sociology* XIX (2), June 1968, pp. 119-139.
[See also DUNCAN & LUKES.]

LUYPEN (= LUIJPEN), Wilhelmus A. M., *Phenomenology and Atheism*, translated by Walter van de Putte, Pittsburgh, Duquesne University Press / Louvain, Nauwelaerts, 1964, Duquesne University. Duquesne Studies.

LYONS, John, *Structural Semantics. An Analysis of Part of the Vocabulary of Plato*, Oxford, Basil Blackwell 1969, Publications of the Philosophical Society, No. XX.

LYTTKENS, Hampus, *The Analogy between God and the World. An Interpretation of Its Background and Interpretation of Its Use by Thomas of Aquino*, Uppsala, Uppsala Universitet Årsskrift, 1952.

McCOY, Charles & PLAYFORD, John (eds.), *Apolitical Politics. A Critique of Behavioralism*, New York, Thomas Crowell, 1967.

MACINTYRE, Alasdair, *The Unconscious. A Conceptual Analysis*, London, Routledge, 1958.

—, 'Is understanding religion compatible with believing?', in John Harwood HICK (ed.), *Faith and the Philosophers*, New York, St. Martin's Press / London, Macmillan, 1964; reproduced in WILSON (ed.), *Rationality, op. cit.* (1970), pp. 62-77.

—, *A Short History of Ethics*, New York, Macmillan, 1966.

—, *Herbert Marcuse. An Exposition and a Polemic*, London, Collins, 1970, 'Fontana Modern Masters' (French translation by Nouchka Pathé, *Marcuse*, Paris, Seghers, 1970, 'Maîtres modernes' ser.).

—, *Against the Self-Images of the Age. Essays on Ideology and Philosophy*, London, Duckworth / New York, Schocken, 1971.

— and BELL, D. (= David) R., two essays forming a discussion, under the title 'The idea of a social science', *Proceedings of the Aristotelian Society*, Supp. Vol. XLI, 1967, respectively pp. 95-114 and 115-132; MACINTYRE's essay, 'The idea of a social science', is reproduced in WILSON (ed.), *Rationality, op. cit.* (1970), 112-130.

— & EMMET, Dorothy, *Sociological Theory and Philosophical Analysis*, New York, Macmillan, 1970.

MACINTYRE, Alasdair (ed.), *Metaphysical Beliefs*, London, Allenson A.R./S.C.M. Press, 1957.
[See also FLEW & MACINTYRE (eds.).]

MACKENZIE, W. J. M., 'Political science, in *Main Trends of Research in the Social and Human Sciences, Part I: Social Sciences, op. cit.* (1970), pp. 166-224 (parallel publication of French translation, 'La science politique', in *Tendances principales de la recherche dans les sciences sociales et humaines, Partie I: Sciences sociales, op. cit.* (1970), pp. 198-273).

MACKEY, James P., *The Modern Theology of Tradition*, London, Darton, 1962; New York, Herder & Herder, 1963.

MACPHERSON, C. B., *The Political Theory of Possessive Individualism: from Hobbes to Locke*, Oxford, Clarendon Press, 1962 (French translation by Michel Fuchs, *La théorie politique de l'individualisme possessif: de Hobbes à Locke*, Paris, Gallimard, 1971, Bibliothèque des Idées).

MACQUARRIE, John, *God-Talk. An Examination of the Language and Logic of Theology*, New York, Harper & Row / London, S.C.M. Press, 1957.

MAGEE, Bryan (ed.), *Modern British Philosophy* (collection of radio broadcasts), London, Secker & Warburg, 1971.

Main Trends of Research in the Social and Human Sciences, Part I: Social Sciences, Paris-The Hague, Mouton / Unesco, 1970 (parallel publication in French, *Tendances principales de la recherche dans les sciences sociales et humaines, Partie I: Sciences sociales, ibid.*).

MALCOLM, Norman, 'Behaviorism as a philosophy of psychology', in Trenton William WANN (ed.), *Behaviorism and Phenomenology*, Chicago, The University of Chicago Press, 1964.

—, 'The conceivability of mechanism', *Philosophical Review* LXXVII (1), Jan. 1968, pp. 45-72.

MAMARDAŠVILI, M. K., 'Nekotorye voprosy issledovanija istorii filosofii kak istorii poznanija', *Voprosy filosofii*, 1959, No. 12, pp. 59-71 (summary in English, MAMARDASHVILI, M. K., 'Some problems of the investigation of the history of philosophy as the history of cognition', *ibid.*, p. 178).

—, 'K probleme metoda istorii filosofii', *Voprosy filosofii*, 1965, No. 6, pp. 93-103 (summary in English, MAMARDASHVILI, M. K., 'Concerning the problem of method in the history of philosophy', *ibid.*, p. 185).

MANDELBAUM, Maurice H., 'Societal facts', *British Journal of Sociology* VI (4), Dec. 1955, pp. 305-317.

—, 'Societal laws', *British Journal for the Philosophy of Science* VIII (31), Nov. 1957, pp. 211-224.

—, 'Historical explanation. The problem of covering laws', *History and Theory* I (3), 1961, pp. 229-260.

MANNHEIM, Karl, *Ideologie und Utopie*, Bonn, Verlag Schulte-Bulmke, 1929 (English translation by Louis Wirth & Edward Shils, *Ideology and Utopia. An Introduction to the Sociology of Knowledge*, New York, Harcourt, Brace & Co. / London, Routledge, 1936; 6th edn., 1952).

—, *Mensch und Gesellschaft im Zeitalter des Umbaus*, Leiden, Sijthoff, 1935 (English translation by Edward Shils, *Man and Society in an Age of Reconstruction. Studies in Modern Social Structure, with a Bibliographical Guide to the Study of Modern Society*, New York, Harcourt, Brace & Co. / London, Routledge, 1940).

—, *Essays on the Sociology of Knowledge*, edited by Paul Kecskemeti, London, Oxford University Press, 1952.

—, *Essays on the Sociology of Culture*, edited by Ernest Manheim & Paul Kecskemeti, London, Oxford University Press, 1956.

MANSER, A. R., *Sartre. A Philosophic Study*, London, Athlone Press, 1966.

MANSON, R. N., 'On the symmetry between explanation and prediction', *Phil. Rev.*, No. 68, 1950.

MARC, André, *Dialectique de l'affirmation. Essai de métaphysique réflexive*, Paris, Desclée de Brouwer / Brussels, L'édition universelle, 1952.

MARCEL, Gabriel, 'Existence et objectivité', originally published as an article in the *Revue de Métaphysique et de Morale*, April-June 1925, and made available as an offprint; reproduced as Appendix to the *Journal métaphysique, op. cit.* (1927).

—, *Journal métaphysique*, Paris, Gallimard, 1927 (English translation by Bernard Wall, *Metaphysical Journal*, Chicago, Regnery, Henry, Company, 1967).

—, *Etre et avoir* (collection of essays), Paris, Aubier-Montaigne, 1935; new edn., *ibid.*, 1968, 'Foi vivante' ser. (English translation, *Being and Having: An Existential Diary*, New York, Harper & Row / London, Collins, 1965).

—, *Du refus à l'invocation*, Paris, Gallimard, 1940 (English translation, with an Introduction, by Robert Rosthal, *Creative Fidelity*, New York, Farrar, Strauss & Co., 1964).

—, *Homo Viator*, Paris, Aubier-Montaigne, 1945 (English translation by Emma Craufurd, *Homo Viator: Introduction to a Metaphysic of Hope*, New York, Harper, 1962, 'Torchbooks').

—, *Le mystère de l'être*, Paris, Aubier-Montaigne, 1951, 2 vols., 'Philosophie de l'Esprit' ser.: I. *Réflexion et mystère*; II. *Foi et réalité* (English translation by G. S. Fraser & René Hauge, *The Mystery of Being*, Chicago, Harvill, 1950-51, 2 vols.).

—, *L'homme problématique*, Paris, Aubier-Montaigne, 1955 (English translation by Brian Thompson, *Problematic Man*, New York, Herder & Herder, 1967).

—, *La dignité humaine et ses assises existentielles* (Harvard lectures), Paris, Aubier-Montaigne, 1964, 'Présence et pensée' ser.

—, 'La dominante existentielle dans mon œuvre', in KLIBANSKY (ed.), *Contemporary Philosophy / La philosophie contemporaine*, Vol. III, *op. cit.* (1969), pp. 171-176.

MARCUS, Solomon, *Lingvistică matematică*, Bucarest, Editura didactica şi pedagogică, 1963 (French translation, recast and adapted by the author, *Introduction mathématique à la linguistique structurale*, Paris, Dunod, 1967, Monographies de linguistique mathématique, I).

—, *Algebraic Linguistics. Analytical Models*, New York-London, Academic Press, 1967, Mathematics in Science and Engineering, Vol. 29.

MARCUSE, Herbert, *Reason and Revolution. Hegel and the Rise of Social Theory*, London, 1941 (French translation by Robert Castel & P. Henri Gonthier, *Raison et révolution. Hegel et la naissance de la théorie sociale*, Paris, Minuit, 1968, 'Le sens commun' ser.).

—, *Eros and Civilization. A Philosophical Inquiry into Freud*, Boston, Beacon Press, 1955 (French translation by Jean-Guy Nény & Boris Fraenkel, revised by the author, *Eros et civilisation. Contribution à Freud*, Paris, Minuit, 1963, 'Arguments' ser., No. 18).

—, *One-dimensional Man. Studies in the Ideology of Advanced Industrial Society*, Boston, Beacon Press, 1964 (French translation by Monique Wittig, revised by the author, *L'homme unidimensionnel. Essai sur l'idéologie de la société industrielle avancée*, Paris, Minuit, 1968, 'Arguments' ser., No. 34).

—, *An Essay on Liberation*, Boston, Beacon Press, 1969 (French translation by Jean-Baptiste Grasset, *Vers la libération. Au-delà de l'homme unidimensionnel*, Paris, Minuit, 1969).

MARGENAU, Henry, *The Nature of Physical Reality*, New York, McGraw-Hill, 1951.

—, 'Naturphilosophie', in *Die Philosophie im XX. Jahrhundert*, 1959.

—, 'Does physical knowledge require *a priori* or undemonstrable presuppositions?', in P. W. BRIDGMAN, *The Nature of Physical Knowledge*, edited by Lawrence W. Friedrich, Milwaukee (Wis.), Marquette University Press, 1960.

MARIN, Louis, *Sémiotique de la Passion. Topiques et figures*, Paris, Aubier-Montaigne/Editions du Cerf/Delachaux & Niestlé/Desclée de Brouwer, 1971, 'Bibliothèque des Sciences religieuses'.

MARITAIN, Jacques, *Humanisme intégral. Problèmes temporels et spirituels d'une nouvelle chrétienté*, Paris, Aubier-Montaigne, 1947.

—, *On the Philosophy of History*, New York, Scribner, 1957 (French translation by Mgr. Charles Journet, *Pour une philosophie de l'histoire*, Paris, Seuil, 1959).

MARKOV, A. A., *Teorija algorifmov = Trudy Mat. Inst. Steklov*, Vol. 42, Academy of Sciences of the U.S.S.R., Moscow (English translation by Jacques H. Schorr and the staff of the P.S.T., *Theory of Algorithms*, published for the National Science Foundation, Washington, D.C. and the Department of Commerce, U.S.A., by the Israel Program for Scientific Translations, 1961).

MARROU, Henri-Irénée, *De la connaissance historique*, Paris, Seuil, 1954; 4th edn., rev. and augm., 1962.

—, *Théologie de l'histoire*, Paris, Seuil, 1968.

MARTIN, James A., *The New Dialogue between Philosophy and Theology*, New York, Seaburg, 1966.

MARTIN, Richard M., *Toward a Systematic Pragmatics*, Amsterdam, North Holland Publ. Co., 1959, Studies in Logic and the Foundations of Mathematics.

MARTINDALE, Don A. (ed.), *Functionalism in the Social Sciences*, Philadelphia, American Academy of Political and Social Science, Feb. 1965.

MARTINET, André, *Economie des changements phonétiques. Traité de phonologie diachronique*, Bern, Francke, 1955.

—, *Eléments de linguistique générale*, A. Colin, Paris 1960; republished in 'U' ser., 1967 (English translation by Elisabeth Palmer, *Elements of General Linguistics*, London, Faber & Faber, 1964).

—, *A Functional View of Language*, Oxford, Clarendon Press, 1962 (French translation by Henriette & Gérard Walter, *Langue et fonction*, Paris, Gonthier / Denoël, 1969, Bibliothèque 'Médiations').

—, *La linguistique synchronique. Etudes et recherches*, Paris, Presses Universitaires de France, 1965, 'Le linguiste' ser.

MARTY, François, 'Le rapport de la philosophie à son histoire. L'accès à la vérité pour un être dans l'histoire', in *La Vérité. Actes du XIIème Congrès des Sociétés de Philosophie de langue française* (Bruxelles), Louvain, Nauwelaerts, 1964, Vol. I, pp. 263-267.

MARX, Karl, *Economic and Philosophic Manuscripts of 1844*, English translation by Martin Milligan, edited by Dirk J. Struik, London, Lawrence & Wishart / New York, International Publ. Co., 1970.

—, *Misère de la philosophie. Réponse à la Philosophie de la Misère de M. Proudhon* (Paris and Brussels, 1847), new edn. with the French translation of F. ENGEL's Prefaces to the 1st and 2nd German edns. (1884 and 1892), Paris, Editions sociales, 1946.

MARX, Karl – ENGELS, Friedrich, *Iz rannih proizvedenij* (= Early Works), Moscow, 1956.

—, *Sočinenija* (= Complete Works), 2nd edn., Moscow, Vol. III.

MARX, Karl & ENGELS, Friedrich, *Die heilige Famile* (Francfort, 1845), in MARX-ENGELS, *Werke*, Vol. II, Berlin, 1959 (French translation by Erna Cogniot, presented and annotated by Nicole Meunier & Gilbert Badia, *La Sainte Famille, ou Critique de la critique critique. Contre Bruno Bauer et consorts*, Paris, Editions sociales, 1969).

—, *Manifest der Kommunistischen Partei* (1848), in MARX-ENGELS, *Gesamtausgabe*, Abt. I, Bd. 6, Moscow-Leningrad, 1933 (English translation, edited by H. J. Laski, *Communist Manifesto*, London, Lawrence & Wishart, 1948).

—, *Die deutsche Ideologie* (French translation, integral, by Henri Auger, Gilbert Badia, Jean Baudrillard & Renée Cartelle, presented and annotated by Gilbert Badia, *L'Idéologie allemande*, Paris, Editions sociales, 1968; English translation of Pt. I and selections from Pts. II and III, edited by C. J. Arthur, *The German Ideology*, London, Lawrence & Wishart / New York, International Publ. Co., 1970).

MARX, Werner, *Heidegger und die Tradition. Eine problemgeschichtliche Einführung in die Grundbestimmen der Seins*, Stuttgart, Kohlhammer, 1961.

'Marx und die Philosophie der Gegenwart' / 'Marx et la philosophie contemporaine' / 'Marx and contemporary philosophy' (Kolloquium I), in *Akten des*

XIV. Internationalen Kongresses für Philosophie, Wien, 2-9 September 1968, op. cit., Vol. II (1968), pp. 1-147.

'Marxisme et humanisme', discussion published in several issues of 1965 of *La nouvelle Critique* (Paris): No. 164, Mar. (ALTHUSSER, 'Marxisme et humanisme', *op. cit.*; study by Jorge SEMPRUN, pp. 22-31, previously published in *Clarté*, No. 58; reply by ALTHUSSER, 'Note complémentaire sur l' "Humanisme réel" ', *op. cit.*; statement by Francis COHEN, pp. 38-44); No. 165, Apr. (statement by Michel SIMON, pp. 96-132); No. 166, May (statement by Pierre MACHEREY, pp. 131-141); No. 168, Aug. (statements by Geneviève NAVARRI, by M. BROSSARD, and by Michel VERRET, pp. 68-121); and No. 170, Nov. ('Thèses sur l'humanisme', by Jacques GOLDBERG, pp. 129-134).

MASCALL, Eric L., *Existence and Analogy*, London, Longmans, Green & Co., 1949.

—, *Words and Images*, London, Longmans, Green & Co., 1957.

MASSOLO, Arturo, *La storia della filosofia come problema*, Florence, Vallechi, new edn., 1967.

MASSON-OURSEL, Paul, *La philosophie comparée*, Paris, 1923.

MEHL, Roger, *La condition du philosophie chrétien*, Neuchâtel-Paris, Delachaux & Niestlé, 1947, 'Série théologique de l'actualité protestante'.

Materialismul dialectic şi ştiinţele moderne (= Dialectical Materialism and Modern Science), collection, Bucarest, 1967.

MEILAND, Jack W., *Scepticism and Historical Knowledge*, New York, Random House, 1965.

MELDEN, Abraham I., *Free Action*, London, Routledge / New York, Humanities Press, 1961.

MELSEN, Andreas G. M. van, *The Philosophy of Nature*, Pittsburgh, Duquesne University Press / Louvain, Nauwelaerts, 1961, Duquesne Studies, Philosophical ser., No. 2.

—, *Evolution and Philosophy*, Pittsburgh, Duquesne University Press / Louvain, Nauwelaerts, 1965, Duquesne Studies, Philosophical ser., No. 19.

—, 'The philosophy of nature', in KLIBANSKY (ed.), *Contemporary Philosophy / La philosophie contemporaine*, Vol. II, *op. cit.* (1968), pp. 151-160.

MELTZER, Bernard M. & MITCHIE, Donald (eds.), *Machine Intelligence*, Vol. I, Edinburgh, Edinburgh University Press, 1969.

MERLEAU-PONTY, Maurice, *Phénoménologie de la perception*, Paris, Gallimard, 1943, Bibliothèque des Idées (English translation by Colin Smith, *Phenomenology of Perception*, London, Routledge / New York, Humanities Press, 1962).

—, 'Le doute de Cézanne', first published in *Fontaine*, No. 47, Dec. 1945, pp. 80-100; reproduced in MERLEAU-PONTY, *Sens et non-sens, op. cit.* (1948), pp. 15-49.

—, *Humanisme et terreur*, Paris, Gallimard, 1947 (English translation, with notes, by John O'Neill, *Humanism and Terror. An Essay on the Communist Problem*, Boston, Beacon Press, 1969).

—, *Sens et non-sens* (collection of essays), Paris, Nagel, 1948, 'Pensées' ser. (English translation by Hubert L. & Patrice Dreyfus, *Sense and Non-Sense*, Evanston (Ill.), Northwestern University Press, 1964).

—, *Eloge de la philosophie* (Inaugural Lecture given at the Collège de France, on Thursday, 15 Jan. 1953), Paris, Gallimard, 1953; republished with other essays, *ibid.*, 1965, 'Idées' ser., No. 75 (English translation by James Edie & John Wild, *In Praise of Philosophy*, Evanston (Ill.), Northwestern University Press, 1963).

—, *Les aventures de la dialectique*, Paris, Gallimard, 1955.

—, 'Le philosophe et son ombre', in *Edmund Husserl, 1859-1959, op. cit.* (1959), pp. 195-220; reproduced in MERLEAU-PONTY, *Signes, op. cit.* (1960), pp. 201-228 (English translation in *Signs*, 1964).

—, 'L'œil et l'esprit' (dated July-Aug. 1960), 1st publ. in the yearly *Art de France* (Paris), No. 1, 1961, pp. 187-208; reproduced in *Les Temps modernes* (Paris),

XVII (184-185), 1961, pp. 193-227; posthumous republication in booklet form, *L'œil et l'esprit*, Paris, Gallimard, 1964.

—, *Signes* (collection of essays), Paris, Gallimard, 1960 (English translation by Richard C. McCleary, *Signs*, Evanston (Ill.), Northwestern University Press, 1964).

—, *Le visible et l'invisible*, followed by 'Notes de travail' (materials for the work which the author was preparing when he died in 1961, edited by Claude Lefort), Paris, Gallimard, 1963, Bibliothèque des Idées (English translation by Alphonso Lingis, *The Visible and the Invisible*, Evanston (Ill.), Northwestern University Press, 1969).

—, *Résumés de cours. Collège de France, 1952-1960* (posthumous publication edited by Claude Lefort), Paris, Gallimard, 1968 (English translation by John O'Neill, *Themes from the Lectures at the Collège de France, 1952-1960*, Evanston (Ill.), Northwestern University Press, 1970).

MERTON, Robert K., *Social Theory and Social Structure*, rev. edn., New York, The Free Press, 1957 (French translation, adapted, of part of the work, by Henri Mendras, *Eléments de théorie et de méthode sociologiques*, Paris, Plon, 1965).

Metodologičeskie i istoriografičeskie voprosy istoričeskoj nauki (= Methodological and Historiographical Questions in Historical Science), collection, Tomsk, 1963 to 1966, Publications of Tomsk State University, Nos. 1-5.

METZ, J. B., 'La historicidad de la filosofía', *Convivium*, No. 22, 1966, pp. 29-38.

—, *Zur Theologie der Welt*, Mainz, Matthias Grünewald Verlag, 1968.

MICHELI, Gianni, 'Storia della scienza e storia della filosofia: problemi di metodo', *Rivista critica di Storia della Filosofia* (Florence), Vol. XXII, 1967, pp. 303-318.

MIKECIN, Vjekoslav, *Marksisti i Marx*, Zagreb, 1968.

MILLS, Charles Wright, *Sociological Imagination*, London-New York, Oxford University Press, 1959 (French translation by Pierre Clinquart, *L'imagination sociologique*, Paris, Maspera, 1967).

MINK, Louis O., 'Philosophical analysis and historical understanding', *The Review of Metaphysics* XXI (4/84), June 1968, pp. 667-698.

Minnesota Studies in the Philosophy of Science (Minnesota University Center for the Philosophy of Science), Minneapolis, University of Minnesota Press / London, Oxford University Press, 3 vols.: Vol. I. FEIGL, Herbert & SCRIVEN, Michael (eds.), *The Foundations of Science and the Concepts of Psychology and Psychoanalysis*, 1956; Vol. II. FEIGL, Herbert *et al.* (eds.), *Concepts, Theories and the Mind-Body Problem*, 1958; Vol. III. FEIGL, Herbert & MAXWELL, Grover E. (eds.), *Scientific Explanation, Space and Time*, 1962.

MISES, Ludwig von, *Nationalöconomie. Theorie des Handelns und Wirtschaftens*, Geneva, 1940 (English translation, *Human Action. A Treatise on Economics*, New Haven (Conn.), Yale University Press / London, Wm. Hodge & Co., 1949; rev. edn., New Haven (Conn.), Yale University Press, 1959).

MISES, Richard von, *Probability, Statistics and Truth* (rev. edn. in English prepared by Hilda Geiringer), New York, Macmillan / London, Allen & Unwin, 1957; republ., New York, Humanities Press, 1961.

—, *Mathematical Theory of Probability and Statistics* (edited by Hilda Geiringer), New York, Academic Press, 1958; 2nd edn., rev., 1964.

MITCHELL, Basil G. (ed.), *Faith and Logic*, Boston, Beacon Press, 1957; London, Allen & Unwin, 1958, 'Oxford Essays in Philosophical Theology'.

MITIN, Mark Borisowitsch (= Mark Borisovič), 'Marxist philosophy of real humanism and its significance in our time', in 'Marx und die Philosophie der Gegenwart' (Colloquium I, XIVth International Congress of Philosophy, Vienna, 1968), *op. cit.* (1968), pp. 80-86.

MOLTMANN, Jürgen, *Theologie der Hoffnung*, Munich, C. Kaiser, 1965.

—, *Politische Theologie*, Regensburg, Ärztliche, 1969.

—, *Religion, Revolution and the Future* (translated from the German by M. Douglas Meeks), New York, Scribner, 1969.

MONDOLFO, Rodolfo, *Problemi i metodi di ricerca nelle storia della filosofia*, Florence, La Nuova Italia Editrice, 1969.
MONOD, Jacques, *Le hasard et la nécessité. Essai sur la philosophie naturelle de la biologie moderne*, Paris, Seuil, 1970 (English translation, *Chance and Necescity*, New York, Knopf, 1971).
MONRO, David Hector, *Empiricism and Ethics*, Cambridge, Cambridge University Press, 1967.
MONTAGUE, Richard, 'Logical necessity, physical necessity, ethics, and quantifiers', *Inquiry*, Vol. IV, 1960, pp. 259-269.
—, 'Syntactical treatments of modality, with corollaries on reflexion principles and finite axiomatizability', in *Proceedings of a Colloquium on Modal and Many-valued Logics, Helsinki, 23-26 August, 1962 = Acta Philosophica Fennica*, Vol. XVI, 1963, pp. 153-169.
—, 'Pragmatics', in KLIBANSKY (ed.), *Contemporary Philosophy. A Survey / La philosophie contemporaine. Chroniques*, Vol. I, *op. cit.* (1968), pp. 102-122.
MONTEFIORE, Alan, *A Modern Introduction to Moral Philosophy*, London, Routledge, 1958.
—, 'Goodness and choice', *Proceedings of the Aristotelian Society*, Supp. Vol. XXXV, 1961, pp. 45-61.
—, 'Deliberate wrongdoing', in *Le Jugement moral, op. cit.* (*Revue internationale de Philosophie*, 1964), pp. 413-431.
—, 'Fact, value and ideology', in WILLIAMS & MONTEFIORE (eds.), *British Analytical Philosophy, op. cit.* (1966).
—, 'The meaning of "good" and the act of commendation', *The Philosophical Quarterly* XVII (67), Apr. 1967, pp. 115-129.
—, 'Law and morality. The work of H. L. A. Hart', in KLIBANSKY (ed.), *Contemporary Philosophy. A Survey / La philosophie contemporaine. Chroniques*, Vol. IV, *op. cit.* (1971), pp. 144-150.
[See also '"Ought" and "Is"', WILLIAMS & MONTEFIORE (eds.).]
MOORE, E. C. & ROBIN, R. S. (eds.), *Studies in the Philosophy of Charles Sanders Peirce*, 2nd ser., Amherst, University of Massachusetts Press, 1964.
MOORE, G. E. (George Edward), *Principia Ethica*, Cambridge, Cambridge University Press, 1903.
—, *Ethics*, London, Oxford University Press, 1912, Home University Library.
—, 'A reply to my critics', in P. A. SCHILPP (ed.), *The Philosophy of G. E. Moore*, Evanston-Chicago, Northwestern University Press, 1942.
—, *Some Main Problems of Philosophy*, New York, Macmillan / London, Allen & Unwin, 1953, 'Muirhead Library of Philosophy'; 2nd edn., 1966.
—, *Lectures on Philosophy*, edited by Casimir Lowy, London, Allen & Unwin, 1966.
[See also *sub* RAMSEY.]
MOORE, Omar Khayyam & ANDERSON, Alan Ross, 'Some puzzling aspects of social interaction', *The Review of Metaphysics* XV (3), Mar. 1962, pp. 409-433.
—, 'The structure of personality', *The Review of Metaphysics* XVI (2), Dec. 1962, pp. 212-236.
[See also ANDERSON & MOORE, O. K.]
MORAZÉ, Charles, *La logique de l'histoire*, Paris, Gallimard, 1967, 'Les Essais' ser., No. CXXXIX.
—, 'Logique et histoire dans l'événement et l'expression (un exercice)', *L'homme* VIII (2), June 1968, pp. 19-35.
MOREAU, Joseph, *La construction de l'idéalisme platonicien*, Paris, Boivin, 1939.
MOREL, Georges, *Le sens de l'existence selon saint Jean de la Croix*, Paris, Aubier-Montaigne, 1960-61, 3 vols.
MORGENBESSER, Sidney (ed.), *Philosophy of Science Today*, New York, Basic Books, 1967.
MORITZ, M., 'Der praktische Syllogismus und das juridische Denken', *Theoria*,

Vol. XX, 1954, pp. 78-127.
MORRA, G., 'Genesi e dissoluzione della categoria di progresso nella storiografia della filosofia', *Ethica* (Forlì), Vol. VII, 1968, pp. 45-58.
MORRIS, Charles W., *Foundations of the Theory of Signs*, Chicago, University of Chicago Press, 1938, International Encyclopedia of Unified Science, Vols. I-II, Foundations of the Unity of Science I (2).
—, *Signs, Language and Behavior*, Englewood Cliffs (N.J.), Prentice-Hall, 1946.
—, 'Values, problematic and unproblematic, and science', *Journal of Communication* XI (4), Dec. 1961, pp. 205-210 (French translation by Edgar André, 'Les valeurs problématiques ou non problématiques et la science', *Revue universitaire de Science morale* (Havré-lez-Mons, Belgium), Vol. IV, 1966, pp. 33-38).
—, *Signification and Significance. A Study of the Relations of Signs and Value*, Cambridge (Mass.), The M.I.T. Press, 1964.
MOSTOWSKI, Andrzej, 'Recent results in set theory', in LAKATOS (ed.), *Problems in the Philosophy of Mathematics, op. cit.* (1967), pp. 83-96.
MOTROŠILOVA, S., *Principy fenomenologii È. Gusserlja* (= The Principles of E. Husserl's Phenomenology), Moscow, 1968.
MOUNIER, Emmanuel, *La révolution personnaliste et communautaire*, Paris, Aubier-Montaigne, 1935.
—, *Manifeste au service du personnalisme*, Paris, Aubier-Montaigne, 1936.
—, *Qu'est-ce que le personnalisme?*, Paris, Seuil, 1946 (English translation by Philip Mairet, *Personalism*, London, Routledge / New York, Grove Press, 1952; 2nd edn., Notre Dame (Ill.), University of Notre Dame Press, 1970).
—, *Traité du caractère*, Paris, Seuil, 1946 (English translation by Cynthia Rowland, *Character of Man*, London, Rockliff, 1956; New York, Harper, 1957).
MOUNIN, Georges, 'Communication linguistique humaine et communication non linguistique animale', Les *Temps modernes*, Apr.-May 1960; reproduced in MOUNIN, *Introduction à la sémiologie*, Paris, Minuit, 1970, 'Le sens commun' ser., pp. 41-56.
MUKERJI, K. P., *Implications of the Ideology Concept*, Bombay, Popular Book Depot, 1966.
MURDOCH, Iris, *Sartre: Romantic Irrationalist*, Cambridge, Bowes & Bowes, 1953.
MURPHY, Arthur Edward, *The Theory of Practical Reason* (The Paul Carus Lectures, ser. 10, 1955), edited by A. I. Melden, La Salle (Ill.), Open Court, 1965.
MURRAY, John Courtney, *The Problem of God. Yesterday and Today* (The Thomas More Lectures, No. 1), New Haven (Conn.), Yale University Press, 1964.
MYRDAL, Gunnar, *Value in Social Theory. A Selection of Essays on Methodology*, London, Routledge / New York, Harper & Row, 1958.

NABERT, Jean, *L'expérience intérieure de la liberté*, Paris, Presses Universitaires de France, 1924.
—, *Eléments pour une éthique*, Paris, Presses Universitaires de France, 1943; new edn., with a Preface by Paul RICOEUR, Paris, Aubier-Montaigne, 1962, 'Philosophie de l'Esprit' ser. (English translation by William J. Petrek, with P. Ricoeur's Preface, *Elements for an Ethic*, Evanston (Ill.), Northwestern University Press, 1969).
—, *Essai sur le mal*, Paris, Presses Universitaires de France, 1955, 'Epiméthée' ser.
—, *Le désir de Dieu* (posthumous work), with a Preface by Paul RICOEUR, Paris. Aubier-Montaigne, 1966, 'Philosophie de l'Esprit' ser.
NADEL, Siegfrid F., *The Foundations of Social Anthropology*, New York, The Free Press, 1951.
NAESS, Arne, *Gandhi and the Nuclear Age*, Totowa (N.J.), Bedminster Press, 1965.
—, *Communication and Argument. Elements of Applied Semantics*, translated from the original Norwegian by Alastair Hannay, London, Allen & Unwin. 1966.
NAGEL, Ernest, 'The logic of historical analysis', 1st publ., *Scientific Monthly*, 1952;

reproduced in GARDINER (ed.), *Theories of History, op. cit.* (1959-60).
—, 'A formalization of functionalism', in Ernest NAGEL, *Logic without Metaphysics, and Other Essays in the Philosophy of Science*, Glencoe (Ill.), The Free Press, 1957.
—, *The Structure of Science. Problems in the Logic of Scientific Explanation*, New York, Harcourt, Brace & World / London, Routledge, 1961.
NAGEL, Ernest, SUPPES, Patrick C. & TARSKI, Alfred (eds.), *Logic, Methodology and Philosophy of Science. Proceedings of the 1960 International Congress*, Stanford (Calif.), Stanford University Press, 1962.
NAGEL, Thomas, *The Possibility of Altruism*, Oxford, Clarendon Press, 1970.
NARSKIJ, Igor S., 'Filosofskie problemy jazyka' (= Philosophical problems of language), *Filosofskie nauki* (= Philosophical Sciences), 1968, No. 4.
NARSKI (= NARSKIJ), Igor S., 'Das Problem der Bedeutung und die Dialektik (Die Bedeutung eines Wortzeichens ist sein Gebrauch in der Sprache)' (= The Problem of Meaning and Dialectics (The Meaning of a Verbal Sign is Its Use in Language)), in 'Sprachphilosophie' (Section III, XIVth International Congress of Philosophy, Vienna, 1968), *op. cit.*, Vol. III (1969), pp. 393-400.
NATANSON, Maurice A. (ed.), *Philosophy of the Social Sciences. A Reader*, New York, Random House, 1963.
NATANSON, Maurice A. & JOHNSTONE, Henry W., Jr. (eds.), *Philosophy, Rhetoric and Argumentation*, University Park, Pennsylvania State University Press, 1965.
NÉDONCELLE, Maurice, 'Philosophie de la religion', in KLIBANSKY (ed.), *Philosophy in the Mid-Century / La philosophie au milieu du XXe siècle*, Vol. III, *op. cit.* (1958), pp. 189-222.
NEGRI, Antimo, *Storia della filosofia e attività storiografica*, Roma, A. Armando, 1972 (important bibliography).
NEUMANN, John von, *Mathematical Foundations of Quantum Mechanics*, Princeton (N.J.), Princeton University Press, 1955, 'Investigations in Physics', Vol. II.
— & MORGENSTERN, Oskar, *Theory of Games and Economic Behavior*, Princeton (N.J.), Princeton University Press, 1944; republ. in paperback, New York, Wiley, Science Editions, 1964.
[See also BIRKHOFF & von NEUMANN.]
NIEBUHR, Reinhold, *The Nature and Destiny of Man. A Christian Interpretation*, New York, Scribner / London, Nisbet, 1941, 2 vols.
—, *Faith and History. A Comparison of Christian and Modern Views of History*, New York, Scribner / London, Nisbet, 1949.
NIEBUHR, Richard R., *Experimental Religion. A Theology of Power and Suffering*, New York, Harper & Row, 1972.
NIELSEN, Kai, 'Good reasons in ethics. An examination of the Toulmin-Hare controversy', *Theoria*, Vol. XXIV, 1958, pp. 9-28.
—, 'The "good reasons approach" and "ontological justifications" of morality', *The Philosophical Quarterly* IX (35), Apr. 1959, pp. 116-130.
—, 'The good reasons approach revisited', *Archiv für Rechts- und Sozialphilosophie / Archives de Philosophie du Droit et de Philosophie sociale / Archives for Philosophy of Law and Social Philosophy*, Vol. IV (1965), pp. 455-483.
NINK, Caspar, *Philosophische Gotteslehre*, Munich, Kösel, 1948.
NISHITANI, Keiji, 'Science and Zen', *The Eastern Buddhist*, N.S., I (1), Sept. 1965.
NOBLE, Denis, 'Charles Taylor on teleological explanations', *Analysis* 27 (3), Jan. 1967, pp. 96-103.
NØRLUND, Ib, *Kybernetik og marxisme. Menneskets nye muligheder* (= Cybernetics and Marxism. New Possibilities for Mankind), Copenhagen, Tiden, 1964.
NOTA, J. H., 'Geschiedschrijving en wijsbegeerte', *Algemeen Nederlands Tijdschrift voor Wijsbegeerte en Psychologie* (Naarden), Vol. LVIII, 1966, pp. 77-84.
NOVAK, Michael, *Ascent of the Mountain, Flight of the Dove. An Invitation to Religious Studies*, New York, Harper & Row, 1971.

NOVIKOV, A. I., 'Predmet i zadači istoriografii filosofii', *Voprosy filosofii*, 1964, No. 3, pp. 129-139 (English summary, 'The subject matter and tasks of philosophical historiography', *ibid.*, p. 186; integral English translation, 'Historiography of philosophy: subject-matter and aims', *Soviet Studies in Philosophy* (New York), III (2), Fall 1964, pp. 24-34).

NOWELL-SMITH, Patrick H., *Ethics*, Harmondsworth (Middx.), Penguin Books, 1954.

—, 'Are historical events unique?', *Proceedings of the Aristotelian Society*, N.S., Vol. LVII, Containing the papers read before the Society during the 78th Session, 1956-57 (London, Harrison & Sons), 1957, pp. 107-160.

—, 'Contextual implication and ethical theory', *Proceedings of the Aristotelian Society*, Supp. Vol. XXXVI, 1962, pp. 1-18.

NYGREN, Anders, *Meaning and Method in Philosophy and Theology*, edited by P. S. Watson, Philadelphia, The Fortress Press, 1971.

OAKESHOTT, Michael, *Rationalism in Politics and Other Essays*, London, Methuen, 1952.

'Ob"ektivacija' (= Objectivation), 'Opredmečivanie' (= id.), 'Oveščestvlenie' (= Hypostatization): articles in the *Filosofskaja ènciklopedija*, Vol. IV, *op. cit.* (1967).

Obščee i osobennoe v istoričeskom razvitii stran Vostoka. Materialy diskussi ob obščestvennyh formacijah na Vostoke (Aziatskij sposob proizvodstva) (= General and particular aspects of the historical development of the Eastern countries. Material for discussion on social formations in the East (the 'Asiatic mode of production')), collection, Moscow, 1966.

Očerki metodologii poznanija social'nyh javlenij (= Essays on a Methodology for the Understanding of Social Phenomena), collection, Moscow, 1970.

O'CONNOR, D. J., *John Locke*, Harmondsworth (Middx.)-London, Penguin Books, 1952.

O'CONNOR, D. J. (ed.), *A Critical History of Western Philosophy*, New York, The Free Press of Glencoe, 1964.

OEHLER, Klaus, 'Der Entwicklungsgedanke als heuristisches Prinzip der Philosophietheorie', *Zeitschrift für philosophische Forschung* (Meisenheim am Glan), Vol. XVII, 1963, pp. 604-613.

OGDEN, C. K. & RICHARDS, I. A., *The Meaning of Meaning*, London, Kegan Paul, 1923.

OGDEN, Schubert M., *Christ without Myth*, New York, Harper & Row, 1961.

—, *The Reality of God and Other Essays*, New York, Harper & Row, 1966; London, S.C.M. Publ. House, 1967.

—, 'The task of philosophical theology', in Robert A. EVANS (ed.), *The Future of Philosophical Theology*, Philadelphia, Westminster Press, 1971.

—, 'What is theology?', *Journal of Religion* LII (1), 1972.

OHE, Seizo, 'The multiple structure of our external knowledge', in *Philosophical Studies of Japan*, Vol. V, Tokyo, Japanese Society for the Promotion of Science, 1964.

—, 'Nature and culture in human freedom', *Ethics* LXXVII (4), July 1967.

—, 'Flexibility in human cognitive behaviour towards nature', *The Annals of the Japan Association for Philosophy of Science* III (3), Mar. 1968.

OJZERMAN, Teodor Ilič, *Formirovanie filosofii marksizma* (= The Genesis of Marxist Philosophy), Moscow, 1962 (German translation, *Die Entstehung der marxistischen Philosophie*, Berlin, 1965).

OLBRECHTS-TYTECA, L., 'Rencontre avec la rhétorique', in *La Théorie de l'argumentation. Perspectives et applications, op. cit.* (*Logique et analyse*, 1963), pp. 3-18. [See also PERELMAN & OLBRECHTS-TYTECA.]

OPARIN, A. I. (Aleksandr Ivanovič), *Life. Its Nature, Origin and Development*, translated from the Russian by Ann Synge, New York, Academic Press, 1961.

OPARIN, A. I. (Aleksandr Ivanovič) *et al.* (eds.), *The Origin of Life on Earth* (Pro-

ceedings of the International Symposium on the Origins of Life on the Earth, Moscow, 1957, edited for the International Union of Biochemistry by F. Clark & R. L. M. Synge), Oxford, Pergamon, 1959, International Union of Biochemistry Symposium ser., Vol. 1.
[See also *sub* KAHANE.]

OPITZ, Heinrich, *Philosophie und Praxis. Eine Untersuchung zur Herausbildung des marxistischen Praxisbegriffs*, Berlin, Dietz, 1967.

ORTEGA Y GASSET, José, *Ideas para una historia de la filosofía*, Madrid, 1942.

—, 'Dos Prologos' = Introduction to the Spanish translation, by Ortega y Gasset, of BRÉHIER, *Histoire de la philosophie* (*op. cit.*, 1926-), Madrid, Revista de Occidente, 1944.

ORTIGUÈS, Edmond, *Le discours et le symbole*, Paris, Aubier-Montaigne, 1962, 'Philosophie de l'Esprit' ser. (originally published as a thesis in 1959 under the title: *Langage et symboles*).

OTT, Heinrich, *Geschichte und Heilgeschichte in der Theologie R. Bultmanns*, Tübingen, J. C. B. Mohr, 1955.

—, *Denken und Sein. Der Weg Martin Heideggers und der Weg der Theologie*, Zürich, EVZ Verlag, 1959.

—, *Dogmatik und Verkündigung*, Zürich, EVZ Verlag, 1961.

OTTO, Rudolf, *Das Heilige. Über das Irrationale in der Idee des Göttlichen und sein Verhältnis zum Rationalen* (1917), 18th edn., Gotha, Klotz, 1929 (English translation by John W. Harvey, *The Idea of the Holy*, London, Oxford University Press, 1923; 4th edn., 1926, etc.).

' "Ought" and "Is" - A discussion between R. F. Atkinson and A. C. Montefiore', *Philosophy* XXXIII (124), Jan. 1958, pp. 29-49.

OVČINNIKOV, N. F., *Ponjatie massy i ènergii v ih istoričeskom razvitii i filosofskom značenii* (= The Concepts of Mass and Energy in Their Historical Development and Philosophical Significance), Moscow, 1957.

PALMADE, Guy, *L'unité des sciences humaines*, Paris, Dunod, 1961, 'Organisation et sciences humaines' ser.

PALMER, Richard, *Hermeneutics: Interpretation-Theory in Schleiermacher, Dilthey, Heidegger and Gadamer*, Evanston (Ill.), Northwestern University Press, 1969.

PANNENBERG, Wolfhart, *Offenbarung als Geschichte*, Göttingen, Vandenhoek & Ruprecht, 1961.

—, *Was ist der Mensch? Die Anthropologie der Gegenwart im Lichte der Theologie*, Göttingen, Vandenhoek & Ruprecht, 1968.

PARADIES, F., *Die Technik des Rechts und ihre Logik*, Amsterdam, 1958.

PAREYSON, Luigi, *Esistenza e persona*, 3rd edn., Turin, Taylor, 1966.

—, 'Originarietà dell' interpretazione', in *Hermeneutik und Dialektik, op. cit.* (1970), Vol. II, pp. 353 sqq.

PARRY, Geraint, *Political Elites*, London, Allen & Unwin, 1969.

PARSONS, Howard L., 'The young Marx and the young generation', *Horizons. The Marxist Quarterly* (Toronto), No. 26, Summer 1968; issued also in booklet form (reprint), *The Young Marx and the Young Generation*, Toronto, Progress Books, 1968.

—, *Humanism and Marx's Thought*, Springfield (Ill.), Charles C. Thomas, 1971.

PARSONS, Talcott, *Essays in Sociology Pure and Applied*, Glencoe (Ill.), The Free Press, 1949; rev. edn. under the title *Essays in Sociological Theory, ibid.*, 1954.

—, *The Social System*, Glencoe (Ill.), The Free Press, 1951.

—, *Social Structure and Personality*, Glencoe (Ill.), The Free Press, 1954.

—, *Structure and Process in Modern Societies*, Glencoe (Ill.), The Free Press, 1960.

—, *Eléments pour une sociologie de l'action* [being a selection of basic writings comprising various chapters from the *Essays in Sociology* (*op. cit.*, 1949, 1954) and from *The Social System* (*op. cit.*, 1951)], French translation preceded by

an introduction ('En marge de l'œuvre de Talcott Parsons: la sociologie et la théorie de l'action', *op. cit.*) by François BOURRICAUD, Paris, Plon, 1955, 'Recherches en sciences humaines' ser., No. 6.
—, *Societies. Evolutionary and Comparative Perspectives*, Englewood Cliffs (N.J.), Prentice-Hall, 1966.
—, *Sociological Theory and Modern Society*, Glencoe (Ill.), The Free Press, 1967.
—, *Politics and Social Structure*, Glencoe (Ill.), The Free Press, 1969.
— & SHILS, Edward A., *Toward a General Theory of Action*, Cambridge (Mass.), Harvard University Press, 1951.
PARSONS, Talcott, SHILS, Edward A., NAGELS, K. & PITTS, J. R. (eds.), *Theories of Society: Foundations of Modern Sociological Theory*, Glencoe (Ill.), The Free Press, 1961.
PARTRIDGE, P. H., 'Politics, philosophy, ideology', *Political Studies* IX (3), Oct. 1961, pp. 217-235.
—, 'Political philosophy and political sociology', *Australian and New Zealand Journal of Sociology*, I, 1965, pp. 3-20.
PASSMORE, John A., *Hume's Intentions*, Cambridge, Cambridge University Press, 1953.
—, 'The objectivity of history', *Philosophy* XXXIII (125), Apr. 1958, pp. 97-111.
—, *Philosophical Reasoning*, London, Duckworth, 1961.
—, *A Hundred Years of Philosophy*, London, Duckworth, 1957; republ., Harmondsworth (Middx.), Penguin Books, 1968, Pelican ser.
PAVLOV, Todor, *Teorija otraženija* (= Theory of Reflection), Moscow, 1949.
PAVLOV, Todor (ed.), *Leninskaja teorija otraženija i sovremennost'* (= The Leninist Theory of Reflection and the Present Day), Moscow / Sofia, 1969.
PEARS, D. F., *Bertrand Russell and the British Tradition in Philosophy*, London, Fontana, 1967.
PEARS, D. F. (ed.), *The Nature of Metaphysics* (essays by H. P. GRICE *et al.*), London, Macmillan, 1957; republ., New York, St-Martin's Press / London, Macmillan, 1965.
PEIRCE, Charles Sanders, *Collected Papers*, edited by C. Hartshorn, P. Weiss & A. Burks, Cambridge (Mass.), Harvard University Press, 1931 to 1935 and 1958.
PEPPER, Stephen C., *The Sources of Value*, Berkeley-Los Angeles, University of California Press, 1958.
PERELMAN, Chaïm, 'Logique, langage et communication', in *Atti del XIIo Congresso Internazionale di Filosofia*, I, 1958, pp. 123-135.
—, 'Logique formelle, logique juridique', *Logique et analyse*, 1960, No. 3, pp. 226-230.
—, *L'idéal de rationalité et la règle de justice = Bulletin de la Société française de Philosophie*, No. 55, 1961.
—, 'Jugements de valeur, justification et argumentation', in *L'Argumentation, op. cit.* (*Revue internationale de Philosophie*, 1961), pp. 325-335.
—, 'Ce qu'une réflexion sur le droit peut apporter au philosophe', *Archives de Philosophie du Droit*, Vol. VII, 1962, pp. 35-43; reproduced in PERELMAN, *Justice et raison, op. cit.* (1963), pp. 244-255 (English translation, 'What the philosopher may learn from the study of law', *Natural Law Forum*, Vol. XI (1966), pp. 1-12, and in PERELMAN, *The Idea of Justice . . ., op. cit.*, 1963).
—, *Justice et raison* (collection of previously published studies), Brussels, Presses Universitaires de Bruxelles, 1963, Université libre de Bruxelles, Travaux de la Faculté de Philosophie et Lettres (English translation by John Petrie, with an Introduction by H. L. A. HART, *The Idea of Justice and the Problem of Argument*, London, Routledge, 1963).
—, 'La spécificité de la preuve juridique', in PERELMAN, *Justice et Raison, op. cit.* (1963), pp. 206-217 (English translation, 'The specific nature of judicial proof', in PERELMAN, *The Idea of Justice and the Problem of Argument, op. cit.*, 1963).
—, 'Le réel commun et le réel philosophique', in *Etudes sur l'histoire de la philo-*

sophie en hommage à Martial Guéroult, op. cit. (1964), pp. 127-138.

—, 'Raisonnement juridique et logique juridique', in *La logique du droit, op. cit.* (*Archives de Philosophie du Droit*, 1966), pp. 1-6.

—, *Eléments d'une théorie de l'argumentation*, Brussels, Presses Universitaires de Bruxelles, 1968.

—, 'Le raisonnement pratique', in KLIBANSKY (ed.), *Contemporary Philosophy / La philosophie contemporaine*, Vol. I, *op. cit.* (1968), pp. 168-173 (with the collaboration of P. GOCHET for the bibliography, pp. 174-176).

— & OLBRECHTS-TYTECA, L., *La nouvelle rhétorique. Traité de l'argumentation*, Paris, Presses Universitaires de France, 1958, 2 vols., 'Logos' ser.; 2nd edn., Brussels, Editions de l'Institut de Sociologie de l'Université libre de Bruxelles, 1970.

—, 'De la temporalité comme caractère de l'argumentation', *Archivio di Filosofia*, 1958, No. 1, pp. 115-133.

PERELMAN, Chaïm (ed.), *Raisonnement et démarches de l'historien*, Brussels, Editions de l'Institut de Sociologie, Université libre de Bruxelles, 1963, Travaux du Centre national de Recherches de Logique (extract from *Revue de l'Institut de Sociologie*, 1963, No. 4).

PETERS, R. S. (Richard), *Hobbes*, Harmondsworth (Middx.)-London, Penguin Books, 1956.

—, *The Concept of Motivation*, London, Routledge, 1958.

PETROVIĆ, Gajo, *Filozofija i marksizam* (= Philosophy and Marxism), Zagreb, 1968.

PEURSEN, C. A. van, 'Fasen in de ontwikkeling van het menselijk denken', *Wijsgerig Perspectief of Maatschappij en Wetenschap* (Amsterdam), Vol. IV, 1963-64, pp. 105-115.

PHILLIPS, Dewi Z., *Faith and Philosophical Inquiry*, London, Routledge, 1970.

PHILLIPS, Dewi Z. (ed.), *Religion and Understanding*, Oxford, Basil Blackwell, 1967.

Philosophical Argument = The Monist, Vol. XLVIII, 1964, pp. 467-601.

'La philosophie et son histoire', in *L'homme et l'histoire. Actes du VIème Congrès des Sociétés de Philosophie de langue française (Strasbourg, 1952)*, Paris, 1952, pp. 337-403 (Section IV).

Philosophy of the History of Philosophy = The Monist (La Salle, Ill.), LIII (4), Oct. 1969; comprises: 'Introduction and Bibliography', by Lewis White BECK (pp. 523-531); papers by EHRLICH (*op. cit.*), GUÉROULT (*op. cit.*), John E. SMITH (pp. 583-605), Hayden V. WHITE (pp. 606-630), J. J. MULHERN (pp. 631-641), FAUROT (*op. cit.*), and KUNTZ (*op. cit.*).

PIAGET, Jean, 'La psychologie', in *Tendances principales de la recherche dans les sciences sociales et humaines, Partie I: Sciences sociales, op. cit.* (1970), pp. 274-339 (parallel publication of English translation, 'Psychology', in *Main Trends of Research in the Social and Human Sciences, Part I: Social Sciences, op. cit.* (1970), pp. 225-282).

PIKE, Nelson, 'Rules of inference in moral reasoning', *Mind* LXX (79), July 1961, pp. 391-399.

PIOVANI, Pietro, *Filosofia e storia delle idee*, Bari, Laterza, 1965, Biblioteca di Cultura moderna, No. 616.

PITCHER, George (ed.), *Wittgenstein. The Philosophical Investigations*, New York, Doubleday, 1966, 'Modern Studies in Philosophy'; another edition, London, Macmillan / Notre Dame (Ill.), University of Notre Dame Press, 1968.

PLACE, U. T., 'Is consciousness a brain process?', *The British Journal of Psychology*, Vol. XLVII, Pt. 1, Feb. 1956, pp. 44-50.

PLAMENATZ, John P., *Man and Society*, Vols. I-II, London, Longmans, 1963.

POLANYI, Michael, *Personal Knowledge*, London, Routledge, 1958; republ., New York, Harper, 1964, Harper Torchbooks.

—, *The Tacit Dimension*, New York, Doubleday, 1966.

POLIN, Raymond, *La politique de la solitude. Essai sur J.-J. Rousseau*, Paris, Sirey, 1971, 'Philosophie politique' ser.

POPOV, A., *Neobhodimieta i svobodata* (= Necessity and Freedom), Sofia, 1957.

POPOVIČ, M. V., *O filosofskoi analize jazyka nauki* (= On the Philosophical Analysis of the Language of Science), Kiev, 1966.

—, 'Logika i metodologija nauki' (= Logic and methodology of science), Moscow, 1967.

POPPER, Karl R., *The Poverty of Historicism* (1st publ., *Economica*, XI, 1944, pp. 86-103 and 119-137, and XII, 1945, pp. 69-89), London, Routledge / Boston, Beacon Press, 1957 (French translation by Hervé Rousseau, *Misère de l'historicisme*, Paris, Plon, 1956, 'Recherches en sciences humaines' ser., No. 8).

—, *The Open Society and Its Enemies*, London, George Routledge & Sons, 1945, 2 vols.; republ., Princeton (N.J.), Princeton University Press, 1950, 1 vol.; rev. re-edns., 1952, 1957, 1962, etc.

—, 'A note on natural laws and so-called "contrary-to-fact" conditionals', *Mind*, N.S., LVIII (1), Jan. 1949, pp. 62-66.

—, *The Logic of Scientific Discovery* (rev. and augm. English version, by the author himself, of his *Logik der Forschung*, Vienna, Springer, 1934), London, Hutchinson, 1959; several re-editions (French translation, *La logique de la découverte scientifique*, Paris, Payot, 1973).

—, 'Die Logik der Sozialwissenschaften', *Kölner Zeitschrift für Soziologie und Sozialpsychologie* 14 (233), 1962 (discussion with ADORNO).

PORŠNEV, B. F., *Social'naja psihologija i istorija* (= Social Psychology and History), Moscow, 1966.

—, 'O načale čelovečeskoj istorii' (= On the origins of human history), in *Filosofskie problemy istoričeskoj nauki* (= Philosophical Problems of Historical Science), Moscow, 1969.

PORŠNEV, B. F. & ANCIFEROVA, L. I. (eds.), *Istorija i psihologija* (= History and Psychology), collection published jointly by the Institute of World History and the Institute of Philosophy of the Academy of Sciences of the U.S.S.R., Moscow, 1970.

POSER, Hans, 'Philosophiegeschichte und rational Rekonstruktion. Wert und Grenze einer Methode', *Studia Leibnitziana*, Vol. III, 1971, pp. 67-76.

POSESCU, Al., 'De l'explication dans l'historiographie de la philosophie', *Revue roumaine des Sciences sociales*, série *Philosophie et Logique*, Vol. X, 1966, pp. 317-324.

POULAIN, Jacques, *Logique et religion*, The Hague-Paris, Mouton, 1973.

Praktika – Kriterij istiny v nauke (= Practice – Criterion of Truth in Science), collection, Moscow, Institute of Philosophy of the Academy of Sciences of the U.S.S.R., 1960 (contributions by I. D. ANDREEV and others).

PRICE, H. H., *Belief (The Gifford Lectures Delivered at the University of Aberdeen in 1960)*, London, Allen & Unwin, 1969.
[See also *sub* LEWIS, H. D. (ed.), *Clarity Is Not Enough*.]

PRICHARD, Harold A., *Moral Obligation. Essays and Lectures*, London, Oxford University Press, 1949.

PRINI, Pietro, *Introduzione critica alla storia della filosofia*, Rome, A. Armando, 1967; 2nd edn., rev., 1969.

PRIOR, A. N., 'The paradoxes of derived obligation', *Mind*, N.S., LXIII (249), Jan. 1954, pp. 64-65.

—, *Time and Modality* (John Locke Lectures, 1955-56), London, Oxford University Press, 1957.

—, *Past, Present and Future*, London, Oxford University Press, 1967.

—, *Papers on Time and Tense*, London, Oxford University Press, 1968.

PRO, Diego, 'Problemas de la historiografía de las ideas', in *Temas de Filosofía contemporanea (IIo Congreso nacional de Filosofía)*, Buenos Aires, Editorial Sudamericana, 1971, pp. 151-171.

Problema pričinnosti v sovremennoj fizike (= The Problems of Causality in Contemporary Physics), collective work, Moscow, 1960.

Problema značenija v lingvistike i logike (= The Problem of Meaning in Linguistics and Logic), collection, Moscow, Press of the State University, 1963.

Problemy formal'nogo analiza sistem (= Problems of the Formal Analysis of Systems), collective work, Moscow, 1968.

Proceedings of a Colloquium on Modal and Many-valued Logics (held in Helsinki, 23-26 August, 1962), Helsinki, Akateeminen Kirjakauppa, 1963, 'Acta Philosophica Fennica', Fasc. XVI.

Proceedings of the XIVth International Congress of Philosophy, Vienna, 2nd to 9th September 1968: see *Akten des XIV. Internationalen Kongressen für Philosophie.*

PROPP, Vladimir Ja., *Morfologija skazki*, Leningrad, Akademija, 1928; 2nd edn., rev. and augm., Leningrad, Nauka, 1969 (English translation, from the 1st edn., by Laurence Scott, *Morphology of the Folk Tale*, Bloomington, Indiana University Research Center in Anthropology, Folklore and Linguistics, 1958; 2nd edn., Austin-London, University of Texas Press, 1968) (French translations [from the 1st edn. and] from the 2nd edn., by Marguerite Derrida, *Morphologie du conte*, Paris, Seuil, 1970, 'Points' ser.).

PUCELLE, Jean, *Etudes sur la valeur*: Vol. I. *La source des valeurs. Les relations intersubjectives*, Lyons-Paris, Vitte, 1957, 'Problèmes et doctrines' ser.; Vol. II. *Le règne des fins. L'essence de la civilisation*, with a Preface by Jean NABERT, *ibid.*, 1959; Vol. III. *Le contrepoint du temps*, Louvain-Paris, Nauwelaerts, 1967.

PUTNAM, Hilary, 'An examination of Grünbaum's philosophy of geometry', in *Philosophy of Science, 1961-62 to 1962-63. Delaware University Seminar in the Philosophy of Science*, Vol. 2 (1962-63), New York-London-Sydney, Interscience Publishers, 1963, pp. 205-255.

—, 'Minds and machines', in HOOK (ed.), *Dimensions of Mind, op. cit.* (1960); reproduced in ANDERSON (ed.), *Minds and Machines, op. cit.* (1964), pp. 72-97.

—, 'The mental life of some machines', in CASTAÑEDA (ed.), *Intentionality, Minds, and Perception, op. cit.* (1967), pp. 177-200.

QUINE, Willard van Orman, 'New foundations for mathematical logic', *Amer. Math. Monthly*, Vol. XLIV, 1937, pp. 70-80.

—, *From a Logical Point of View. Nine Logic-philosophical Essays*, Cambridge (Mass.), Harvard University Press, 1953; London, Oxford University Press, 1954; 2nd edn., rev., New York, Harper & Row, 1963.

—, 'On what there is', in QUINE, *From a Logical Point of View, op. cit.* (1953), pp. 1-19 in the 2nd rev. edn. (1963).

—, *Word and Object*, New York, Wiley / Cambridge (Mass.), The M.I.T. Press, 1960.

—, *Philosophy of Logic*, Englewood Cliffs (N.J.), Prentice-Hall, 1970, Prentice-Hall Foundations of Philosophy Ser.

RÁBADE ROMERO, G., 'Hacia una revisión de concepto histórico de filosofía moderna', *Estudios* (Madrid), Vol. XVI, 1960, pp. 241-252.

RADHAKRISHNAN, Sarvepalli, *An Idealist View of Life*, London, Allen & Unwin, 1932.

RAINER, A. C., 'Necessity and God – A reply to Professor Findlay', *Mind* LVIII (229), Jan. 1949, pp. 75-77; reproduced in FLEW & MACINTYRE (eds.), *New Essays in Philosophical Theology, op. cit.* (1953).

RAMSEY, Frank P., 'Truth and probability', in F. P. RAMSEY, *The Foundations of Mathematics and Other Logical Essays*, edited by R. B. BRAITHWAITE with a Preface by G. E. MOORE, London, Routledge / New York, Harcourt, Brace & Co., 1931.

RAMSEY, Ian Thomas, *Religious Language. An Empirical Placing of Theological Phrases*, London, 1957; republ., New York, Macmillan, 1963; London, S.C.M. Press, 1967.

—, *Models and Mystery* (The Whidden Lectures, 1963), London, Oxford University Press, 1964.

—, *Christian Discourse. Some Logical Explorations* (Durham University Riddell Memorial Lectures, No. 35, delivered at the University of Newcastle-upon-Tyne, 5-7 Nov. 1963), London, Oxford University Press, 1965.

—, *Myth and Symbol*, edited by F. W. Dillistone, London, SPCK (Society for Promoting Christian Knowledge), 1966, 'Theological Collections', No. 5.

RANDALL, John H., *How Philosophy Uses Its Past* (Matchette Foundation Lectures, 1961), New York-London, Columbia University Press, 1964.

RAPAPORT, David, *Structure of Psychoanalytic Theory. A Systematizing Attempt*, New York, International Universities Press, 1967.

RAPOPORT, Anatol, *Games, Fights and Debates*, Ann Arbor, The University of Michigan Press, 1960 (French translation by Josette de la Thébeaudière, with Notes and Preface by J. Peirani, *Combats, débats et jeux*, Paris, Dunod, 1967, 'Organisation et sciences humaines' ser., No. 7).

—, 'The use and misuse of game theory', *Scientific American* CCVII (6), Dec. 1962, pp. 108-118 (French translation by Dominique Raoul-Duval, 'Bon et mauvais usage de la théorie des jeux', *Les Temps modernes*, 19th Year, No. 209, Oct. 1963, pp. 681-706).

—, 'The theory of games and its relevance to philosophy', in KLIBANSKY (ed.), *Contemporary Philosophy / La philosophie contemporaine*, Vol. II, *op. cit.* (1968), pp. 451-473.

RAVEN, Ch. E., *Teilhard de Chardin, Scientist and Seer*, London, Collins, 1962.

RAWLS, John, 'Two concepts of rules', *Philosophical Review* LXIV (1), Jan. 1955, pp. 3-32.

—, 'Justice as fairness', *The Philosophical Review* LXVII (2), Apr. 1958, pp. 164-194; reproduced in LASLETT & RUNCIMAN (eds.), *Philosophy, Politics and Society*, 2nd ser., *op. cit.* (1962), pp. 132-157.

—, *A Theory of Justice*, Cambridge (Mass.), Harvard University Press, 1971; Oxford, Clarendon Press, 1972.

RAZ, Joseph, *The Concept of a Legal System*, London, Oxford University Press, 1970.

REICHENBACH, Hans, *Philosophic Foundations of Quantum Mechanics*, Berkeley-Los Angeles, University of California Press, 1944; republ., *ibid.*, 1965.

—, *Theory of Probability. An Enquiry into the Logical and Mathematical Foundations of the Calculus of Probability* (English version established by Ernest H. Hutten & Maria Reichenbach), Berkeley-Los Angeles, University of California Press, 1949.

—, *The Direction of Time* (posthumous work edited by Maria Reichenbach), Berkeley-Los Angeles, University of California Press, 1956.

—, *The Philosophy of Space and Time*, English translation by Maria Reichenbach, New York, Dover, 1957.

REINER, Hans, *Die philosophische Ethik. Ihre Fragen und Lehren in Geschichte und Gegenwart*, Heidelberg, Quelle & Meyer, 1964.

Religious Studies, I, 1965-66 *et seq.*, edited by H. D. LEWIS, Cambridge.

RESCHER, Nicholas, 'On prediction and explanation', *Brit. Journ. Phil. of Sci.*, 1950, No. 8, pp. 281 sqq.

—, 'An axiom system for deontic logic', *Philosophical Studies*, Vol. IX, 1958, pp. 24-30.

—, *The Logic of Commands*, New York, Dover / London, Routledge, 1966.

—, 'Practical reasoning and values', *The Philosophical Quarterly* XVI (63), Apr. 1966, pp. 121-136.

—, 'Semantic foundations for conditional permission', *Philosophical Studies* XVIII

(4), June 1967, pp. 56-61.
—, 'Semantic foundations for the logic of preference', in RESCHER (ed.), *The Logic of Decision and Action, op. cit.* (1967).
—, 'Recent developments in philosophical logic', in KLIBANSKY (ed.), *Contemporary Philosophy / La philosophie contemporaine*: Vol. I, *op. cit.* (1968), pp. 31-40 (with bibliography).
—, *Topics in Philosophical Logic*, Dordrecht, Reidel, 1968, Synthese Library.
— & ROBISON, John, 'Can one infer commands from commands?', *Analysis* XXIV (5), Apr. 1964, pp. 176-179.
RESCHER, Nicholas (ed.), *The Logic of Decision and Action*, Pittsburgh, University of Pittsburgh Press, 1967.
REZNIKOV, L. O., *Gnoseologičeskie voprosy semantiki* (= Gnoseological Questions in Semantics), Leningrad, 1964.
RICHMOND, James, *Theology and Metaphysics*, New York, Schocken, 1971.
RICOEUR, Paul, *Philosophie de la volonté, I. Le volontaire et l'involontaire*, Paris, Aubier-Montaigne, 1950 (English translation by E. V. Kohak, *Freedom and Nature. The Voluntary and the Involuntary*, Evanston (Ill.), Northwestern University Press, 1966, 'Studies in Phenomenology and Existential Philosophy').
—, 'L'histoire de la philosophie et l'unité du vrai', *Revue internationale de Philosophie*, No. 29, 1954 (first published in German version in *Offener Horizont*, Festschrift in honour of Karl Jaspers, Munich, Piper, 1953); reproduced in RICOEUR, *Histoire et vérité, op. cit.* (1955; 2nd edn., augm., 1964), pp. 45-60.
—, *Histoire et vérité* (collection of essays), Paris, Seuil, 1955; 2nd edn., augm., *ibid.*, 1964 (English translation by C. A. Kelbley, *History and Truth*, Evanston (Ill.), Northwestern University Press, 1965).
—, *Philosophie de la volonté, II. Finitude et culpabilité*, Paris, Aubier-Montaigne, 1960, 2 vols.: 1. *L'homme faillible*; 2. *La symbolique du mal* (English translations: by Charles Kelbley, *Fallible Man*, Chicago, Regnery, 1965; and by Emerson Buchanan, *Symbolism of Evil*, New York, Harper, 1967).
—, *De l'interprétation. Essai sur Freud*, Paris, Seuil, 1965 (English translation by Denis Savage, *Freud and Philosophy. An Essay on Interpretation*, New Haven (Conn.), Yale University Press, 1969, Terry Lectures ser.).
—, *Le conflit des interprétations*, Paris, Seuil, 1969.
[See also CASTELLI, GOUHIER, LEVINAS, PANIKKAR, RICOEUR, HABACHI *et al.*]
RISAKU, Mutai, 'Two conditions of human reality', in *Philosophical Studies of Japan*, Vol. I, Tokyo, Japanese Society for the Promotion of Science, 1959.
ROBIN, Léon *et al.*, *Sur la notion d'histoire de la philosophie* (25 April 1936 meeting of the French Society for Philosophy = *Bulletin de la Société française de Philosophie*, 36th Year, 1936, pp. 103-140: communication by ROBIN; statements by Jean BARUZI, Léon BRUNSCHVICG, Henri BERR, P. DUCASSÉ, ÉTARD, Alexandre KOYRÉ, J. LÉVY, Dominique PARODI, Paul SCHRECKER and Jean WAHL.
ROBINET, André, 'De la vérité en histoire de la philosophie', in *La Vérité. Actes du XIIème Congrès des Sociétés de Philosophie de langue française (Bruxelles)*, Louvain, Nauwelaerts, 1964, Vol. I, pp. 272-275.
—, 'La communication philosophique à l'ère des ordinateurs', *Revue internationale de Philosophie* (Brussels), 23rd Year, Fasc. 4, No. 90, 1969.
—, 'Dialectique et histoire de la philosophie', in *La Dialectique. Actes du XIVe Congrès des Sociétés de Philosophie de langue française (Nice)*, Paris, Presses Universitaires de France, 1969, Vol. I, pp. 103-107.
—, 'Hypothèse et confirmation en histoire de la philosophie', *Revue internationale de Philosophie* (Brussels), 25th Year, 1971, Fasc. 1-2, No. 95-96 (special issue: *Hypothèse et confirmation*), pp. 119-146.
ROBINSON, Abraham, *Introduction to Model Theory and to the Metamathematics of Algebra*, Amsterdam, North Holland Publ. Co. / New York, Humanities Press, 1963, 'Studies in Logic and the Foundations of Mathematics'.
—, *Non-Standard Analysis*, Amsterdam, North Holland Publ. Co. / New York,

Humanities Press, 1966, 'Studies in Logic and the Foundations of Mathematics'.
ROBINSON, James M., *A New Quest of the Historical Jesus*, Naperville (Ill.), Alec R. Allenson, 1959.
ROBINSON, James M. & COBB, John B., Jr. (eds.), *New Frontiers in Theology Series*, New York, Harper: I. *The Later Heidegger and Theology*, 1963; II. *The New Hermeneutic*, 1964.
ROBINSON, Richard, *An Atheist's Values*, London, Oxford University Press, 1964.
ROKKAN, Stein, 'Cross-cultural, cross-societal and cross-national research', in *Main Trends of Research in the Social and Human Sciences, Part I: Social Sciences, op. cit.* (1970), pp. 645-689 (parallel publication of French version, 'Recherche trans-culturelle, trans-sociétale et trans-nationale', in *Tendances principales de la recherche dans les sciences sociales et humaines, Partie I: Sciences sociales, op. cit.* (1970), pp. 765-821).
ROGERS, Hartley, *Theory of Recursive Functions and Effective Computability*, New York, McGraw-Hill, 1967.
ROMERO, Francisco, *La estructura de la historia de la filosofía, y otros ensayos*, Buenos Aires, Losada, 1968.
ROMMEN, H., *Die ewige Wiederkehr des Naturrechts* (= The Eternal Return of Natural Law), Munich, 1947.
RORTY, Richard M., 'Recent metaphilosophy', *The Review of Metaphysics* XV (2/58), Dec. 1961, pp. 299-318.
—, 'Mind-body identity, privacy, and categories', *The Review of Metaphysics* XIX (1/73), Sept. 1965, pp. 24-54.
ROSS, Alf, 'Imperatives and logic', *Theoria*, Vol. VII, 1941, pp. 53-71; reproduced, *Philosophy of Science*, Vol. XI, 1944, pp. 30-46.
—, 'On moral reasoning', *Danish Yearbook of Philosophy*, I, 1964, pp. 120-132.
ROSS, W. D., *The Right and the Good*, London, Oxford University Press, 1930.
—, *The Foundations of Ethics*, London, Oxford University Press, 1939.
ROSSI, Paolo, 'Sulla storicità della filosofia e della scienza', *Rivista di Filosofia* (Turin), Vol. LV, 1964.
—, *Storia e filosofia*, Turin, Einaudi, 1970.
ROZENTAL', M. M., *Principy dialektičeskoj logiki* (= Principles of Dialectic Logic), Moscow, 1960.
—, *Lenin i dialektika* (= Lenin and Dialectic), Moscow, 1963.
—, *Voprosy dialektiki v 'Kapitale' Marksa* (= Questions of Dialectic in Marx's *Das Kapital*), Moscow, Gospolitizdat, 1955; 2nd edn., 1967.
RUBINŠTEJN, S. L., *Bytie i soznanie* (= Being and Consciousness), Moscow, Izd. Akad. Nauk, 1957.
RUDNER, Richard S., *Philosophy of Social Science*, Englewood Cliffs (N.J.), Prentice-Hall, 1966.
RUNCIMAN, Walter G., 'Sociological evidence and political theory', in LASLETT & RUNCIMAN (eds.), *Philosophy, Politics and Society*, 2nd ser., *op. cit.* (1962).
—, *Social Science and Political Theory*, Cambridge, Cambridge University Press, 1963.
[See also LASLETT & RUNCIMAN (eds.)]
RUSSELL, Bertrand, Lord, 'On denoting', 1st publ., *Mind*, N.S., Vol. XIV, 1905, pp. 479-493; reproduced in B. RUSSELL, *Logic and Knowledge. Essays 1901-1950, op. cit.* (1956), pp. 41-56 (French translation by Philippe Devaux, 'De la dénotation', *L'âge de la science* (Paris, Dunod), Vol. III, July-Sept. 1970, pp. 171-186).
—, *The Principles of Mathematics*, Cambridge, Cambridge University Press, 1903.
—, 'Mathematical logic as based on the theory of types', 1st publ., *American Journal of Mathematics*, 30, 1908, pp. 222-262; reproduced in B. RUSSELL, *Logic and Knowledge. Essays 1901-1950, op. cit.* (1956), pp. 59-102.
—, *Mysticism and Logic*, London, Longmans-Green, 1910 (French translation by P. de Menasce, *Le mysticisme et la logique, suivi d'autres essais*, Paris, Payot,

1922).
—, *The Problems of Philosophy* (1912), republ., London, Oxford University Press, 1959 (French translation by S. M. Guillemin, *Problèmes de philosophie*, Paris, Payot, 1968, Petite Bibliothèque Payot).
—, *Our Knowledge of the External World as a Field for Scientific Method in Philosophy*, New York, Open Court, 1914; 2nd edn., London, Allen & Unwin, 1916; re-edn., rev., *ibid.*, 1926 (French translation by Philippe Devaux, *La méthode scientifique en philosophie. Notre connaissance du monde extérieur*, Paris, Vrin, 1929; republ., Paris, Payot, 1971, Petite Bibliothèque Payot).
—, 'The philosophy of logical atomism', 1st publ., *The Monist*, 28, 1918, reproduced in B. RUSSELL, *Logic and Knowledge. Essays 1901-1950, op. cit.* (1956), pp. 177-281.
—, *The Analysis of Mind*, London, Allen & Unwin, 1921 (French translation, *L'analyse de l'esprit*, Paris, Payot, 1927).
—, *Logic and Knowledge. Essays 1901-1950*, edited by Robert C. Marsh, London, Allen & Unwin / New York, Macmillan, 1956.
—, *Why I Am Not a Christian* (selected essays, edited by Paul Edwards), London, Allen & Unwin, 1957 (French translation by Guy Le Clec'h, *Pourquoi je ne suis pas chrétien*, with a Foreword by Louis Rougier, Paris, J. J. Pauvert, 1964, 'Libertés nouvelles' ser., No. 11).
— & WHITEHEAD, Alfred North, *Principia Mathematica*, Cambridge, Cambridge University Press, 3 vols.: Vol. I. 1st edn., 1910; 2nd edn., 1925; Vol. II. 1st edn., 1912; 2nd edn., 1927; Vol. III. 1st edn., 1913; 2nd edn., 1927.
[See also *sub* WITTGENSTEIN, *Logisch-philosophische Abhandlung / Tractatus logico-philosophicus*.]
RUWET, Nicolas, *Introduction à la grammaire générative*, Paris, Plon, 1967, 'Recherches en sciences humaines' ser.
RUWET, Nicolas (ed.), *La grammaire générative = Langages* (Paris), No. 4, Dec. 1966.
RUYER, Raymond, *La genèse des formes vivantes*, Paris, Flammarion, 1956.
—, 'La philosophie de la nature et le mythe', *Revue internationale de Philosophie*, No. 36, 1956, No. 2 (special issue: *Philosophie de la nature*), pp. 166-173.
—, *L'animal, l'homme, la fonction symbolique*, Paris, Gallimard, 1964, 'L'avenir de la science' ser., No. 41.
—, *Paradoxes de la conscience et limites de l'automatisme*, Paris, Albin Michel, 1966, 'Les savants et le monde' ser.
—, *Eloge de la société de consommation*, Paris, Calmann-Lévy, 1969, 'Liberté de l'esprit' ser.
—, *Les nuisances idéologiques*, Paris, Calmann-Lévy, 1972, 'Liberté de l'esprit' ser.
RYLE, Gilbert, *The Concept of Mind*, London, Hutchinson, 1949.
—, *Dilemmas*, Cambridge, Cambridge University Press, 1954.
—, 'Pleasure', in RYLE, *Dilemmas, op. cit.* (1954).
—, *Plato's Progress*, Cambridge, Cambridge University Press, 1966.

SABINE, George H., *A History of Political Theory*, New York-London, Holt, Rinehart & Winston, 1957; 3rd edn., rev. and augm., *ibid.*, 1961.
SALAMUCHA, J., 'Dovod "ex motu" na istienie Boga (Analiza logiczna argumentacji Sw. Thomasza Akwinu)', *Collectanea theologica*, XV, 1934 (English translation, 'The proof "Ex Motu" for the existence of God', *The New Scholasticism*, Vol. XXXII, 1968).
SALERNO, Luigi, 'Opposizione relativa, progresso e storia del pensiero', *Sapienza* (Naples), Vol. XXIV, 1971, pp. 146-162.
SÁNCHEZ VÁSQUEZ, A., *Filosofía de la praxis*, Mexico City, Grijalbo, 1967.
SANDOR, Paul, 'Die Entwicklungsgesetze der Geschichte der Philosophie', *Hegel Jahrbuch*, 1968-69, pp. 264-269.
ŠANIN, M. A., *O nekotoryh logičeskih problemah aritmetiki* (= On Some Logical

Problems of Arithmetic), Moscow, Academy of Sciences of the U.S.S.R., 1955, Trudy Mat. Inst. Steklov, No. 43.

SAPIR, Edward, *Language*, New York, 1921 (French translation by S. M. Guillemin, *Le langage. Introduction à l'étude de la parole*, Paris, Payot, 1953).

SARTRE, Jean-Paul, 'La transcendance de l'Ego. Esquisse d'une description phénoménologique', *Recherches philosophiques* (Paris), Vol. VI, 1936-37, pp. 85-123; republ. in book form, Paris, Vrin, 1965 (English translation, annotated and with an Introduction, by Forrest Williams & Robert Kirkpatrick, *Transcendence of the Ego. An Existentialist Theory of Consciousness*, New York, The Noonday Press, 1957).

—, *Esquisse d'une théorie phénoménologique des émotions*, Paris, Hermann, 1939, 'Actualités scientifiques et industrielles', No. 838; republ., *ibid.*, 1965, 'L'esprit et la main' ser. (English translation by Philip Mairet, with a Preface by Mary Warnock, *Sketch for a Theory of the Emotions*, London, Methuen, 1962).

—, *L'imaginaire*, Paris, Gallimard, 1942, Bibliothèque des Idées (English translation, with an Introduction, by Forrest Williams, *Imagination. A Psychological Critique*, Ann Arbor, University of Michigan Press, 1962).

—, *L'être et le néant. Essai d'ontologie phénoménologique*, Paris, Gallimard, 1943, Bibliothèque des Idées (English translation by Hazel E. Barnes, *Being and Nothingness*, London, Methuen, 1957).

—, *L'existentialisme est un humanisme* (dialogue with Pierre Naville), Paris, Nagel, 1946.

—, *Réflexions sur la question juive*, Paris, Morihien, 1946; republ., Paris, Gallimard, 1954; republ. as paperback, *ibid.*, 1962, 'Idées' ser. (English translation by George J. Becker, *Antisemite and Jew*, New York, Schocken, 1965).

—, *Situations* (collected essays), 9 vols. published to date, Paris, Gallimard, 1947-72; comprising: *Situations, I*, 1947; *Situations, II. Qu'est-ce que la littérature?*, 1948; *Situations, III*, 1949; *Situations, IV. Portraits*, 1964; *Situations, V. Colonialisme et néo-colonialisme*, 1964; *Situations, VI* and *VII. Problèmes du marxisme, 1* and *2*, 1964-65; *Situations, VIII. Autour de 1968*, 1972; *Situations, IX. Mélanges*, 1972 (among the existing partial translations in English, in particular: *What Is Literature?*, translated by Bernard Frechtman, New York, Philosophical Library, 1949; London, Methuen, 1950; republ., with a new Introduction by Wallace Fowlie, New York, Harper & Row, 1965).

—, *Baudelaire*, Paris, Gallimard, 1947; republ. in paperback; *ibid.*, 1963, 'Idées' ser. (English translation by Martin Turnell, London, Hamish Hamilton, 1964; republ. as paperback, New York, New Directions, undated).

—, *Saint Genet, comédien et martyr*, Paris, Gallimard, 1952 (English translation by Bernard Frechtman, *Saint Genet, Actor and Martyr*, New York, Braziller, 1963).

—, 'Questions de méthode', 1st publ., *Les Temps modernes*, Nos. 139 and 140, 1957; constitutes the first part of *Critique de la raison dialectique*, *op. cit.* (1960); published separately in paperback edn., *Questions de méthode*, Paris, Gallimard, 1967, 'Idées' ser. (English translation by Hazel E. Barnes, *Search for a Method*, New York, Random House, 1968, Vintage Books).

—, *Critique de la raison dialectique: I. Théorie des ensembles pratiques* (preceded by 'Questions de méthode', *op. cit.*, 1957), Paris, Gallimard, 1960, Bibliothèque des Idées.

—, Preface to FANON, *Les damnés de la terre*, *op. cit.* (1961); reproduced in SARTRE, *Situations, V. Colonialisme et néo-colonialisme*, *op. cit.* (1964), pp. 167-193.

—, *L'idiot de la famille* (study on Flaubert), 3 vols. published to date, Gallimard, 1971, 1972, Bibliothèque de Philosophie.

SASS, Hans Martin, 'Philosophische Positionen in der Philosophiegeschichtsschreibung', *Vierteljahrsschrift für Literaturwissenschaft und Geistesgeschichte*, 46th Year, 1972, No. 3.

SASSO, G., *Passato e presente nella storia della filosofia*, Bari, Laterza, 1967.

ŠAUMJAN, S. K., *Filosofskie idei V. I. Lenina i razvitie sovremennogo jazykoznanija* (= The Philosophical Ideas of Lenin and the Development of Modern Linguistics), Moscow, Institute of Slavonic Studies, 1961, Short Communication No. 31.

SAUNDERS, John T. & HENZE, D. F., *The Private-Language Poblem. A Philosophical Dialogue*, New York, Random House, 1967.

SAUSSURE, Ferdinand de, *Cours de linguistique générale* (given from 1906 to 1911), edited by Ch. Bally & A. Sechehaye (1916), 5th edn., Paris, Payot, 1960; critical edition prepared by Tullio de Mauro, *ibid.*, 1972, 'Payothèque' ser. (English translation by Wade Baskin, *Course in General Linguistics*, New York, Philosophical Library, 3rd edn., 1959).

SAVAGE, Leonard J., *Foundations of Statistics*, New York, Wiley, 1954; republ., New York, Dover, 1972.

—, *La probabilità soggettiva nei problemi della statistica* (Corso CIME su Induzione e Statistica a Varenna), Rome, 1959.

SAYRE, Kenneth M., *Recognition. A Study in the Philosophy of Artificial Intelligence*, Notre Dame (Ill.), University of Notre Dame Press, 1965.

—, *Consciousness. A Philosophic Study of Minds and Machines*, New York, Random House College Division, 1969.

SAYRE, Kenneth M. & CROSSON, Frederick J. (eds.), *The Modeling of Mind: Computers and Intelligence*, Notre Dame (Ill.), University of Notre Dame Press, 1963.

—, *Philosophy and Cybernetics*, Notre Dame (Ill.), University of Notre Dame Press, 1967.

ŠČEPAN'SKIJ, Ja.: see SZCZEPAŃSKI, Jan.

SCHAAR, John H. & WOLIN, Sheldon S., 'Review essay. Essays on the scientific study of politics: a critique' (review of STORING, Herbert J. (ed.), *Essays on the Scientific Study of Politics*, New York, Holt, Rinehart & Winston, 1962), *American Political Science Review* LVII (1), Mar. 1963, pp. 125-150 (followed by Replies by Herbert J. STORING, Leo STRAUSS and others, *ibid.*, pp. 151-160).

SCHAFF, Adam, *Wstep do semantyki*, Warsaw, Państwowe Wydawnictwo Naukowe, 1960 (Russian translation, ŠAFF, A., *Vvedenie v semantiku*, Moscow, 1963; French translation by Georges Lisowski, *Introduction à la sémantique*, Paris, Editions Anthropos, 1968).

—, *Jezyk a posnanie*, Warsaw, Państwowe Wydawnictwo Naukowe, 1964 (French translation by Claire Brendel, *Langage et connaissance*, followed by 'Six essais sur la philosophie du langage', Paris, Editions Anthropos, 1969).

—, *Szkice z filozofii jezyka*, Warsaw, Ksiazka i Wiedza, 1967 (French translation by Claire Brendel, 'Six essais sur la philosophie du langage', with *Langage et connaissance, op. cit.* 1969).

—, *Marksizm i jednostka*, Warsaw, 1965 (French translation, *Le marxisme et l'individu. Contribution à la philosophie marxiste de l'homme*, Paris, A. Colin, 1968, 'Sciences politiques' ser.).

—, *Histoire et vérité. Essai sur l'objectivité de la connaissance historique*, translated from Polish by Anna Kaminska & Claire Brendel (original title: *Historia i prawda*), Paris, Editions Anthropos, 1971.

SCHELER, H. (ed.), *Historische Materialismus und Sozialforschung* (essays by H. SCHELER, G. HEIDEN, H. ULLRICH, R. THIEL, W. FRIEDRICH, KAMBA Toshio, SHIBATA Shingo, TANAKA Seisuke, KAZAI Iohishige, and MATZANARI Ioshie, Berlin, 1966.

SCHELER, Max, 'Über *Ressentiment* und moralische Werturteile', 1st publ., *Zeitschrift für Psychopathologie*, 1912; republ. in a rev. and expanded version under the title: 'Vom *Ressentiment* im Aufbau der Moralen' in SCHELER, *Abhandlungen und Aufsätze*, I, Leipzig, 1915, a collection reprinted under the title: *Vom Umsturz der Werte*, Leipzig, Der neue Geist Verlag, 1919, 2 vols. (English translation by William W. Holdheim, *Ressentiment*, edited with an

Introduction by Lewis A. Coser, New York, The Free Press of Glencoe, 1961).
—, *Der Formalismus in der Ethik und die materiale Wertethik*, 1st publ., *Jahrbuch für Philosophie und phänomenologische Forschung*, 1913, 1916 (French translation by Maurice de Gandillac, *Le formalisme en éthique et l'éthique matériale des valeurs. Essai nouveau pour fonder un personnalisme éthique*, Paris, Gallimard, 1955, 'Bibliothèque de Philosophie').
—, *Wesen und Formen der Sympathie*, 1923 (1st publ. under the title *Zur Phänomenologie und Theorie der Sympathiegefühle*, 1913) (English translation by Peter Heath, *Nature of Sympathy*, Hamden (Conn.), Shoe String Press, 1970).
—, *Die Stellung des Menschen im Kosmos* (1928) (English translation by Hans Meyerhoff, *Man's Place in Nature*, Boston, Beacon Press, 1961; republ., New York, Farrar, Straus & Giroux, 1963).
—, *Vom Ewigen im Menschen* (*Gesammelte Werke*, edited by Maria Scheler, Vol. V), Bern, Francke, 1954 (English translation by Bernard Noble, *On the Eternal in Man*, London, SCM Press, 1960; New York, Harper, 1961).
SCHELLING, Thomas C., *Strategy of Conflict*, Cambridge (Mass.), Harvard University Press, 1961.
SCHILPP, Paul A. (ed.), *The Philosophy of Karl Jaspers*, Cambridge, Cambridge University Press, 1957; New York, Tudor, 1958, 'The Library of Living Philosophers'.
—, *The Philosophy of C. D. Broad*, New York, Tudor / Cambridge, Cambridge University Press, 1959, 'The Library of Living Philosophers'.
—, *The Philosophy of Rudolf Carnap*, La Salle (Ill.), Open Court, 1963, 'The Library of Living Philosophers'.
[See also MOORE, G. E.]
SCHLEIERMACHER, (Daniel) Friedrich Ernst, *Hermeneutik*, edited by Heinz Kimmerle, Heidelberg, Carl Winter, 1959, Abhandlungen der Heidelberger Akademie der Wissenschaften, Philosophisch-historische Klasse, 1959, No. 2; Supplement, with a Postface by the editor, *ibid.*, 1968.
SCHLICK, Moritz, *Fragen der Ethik*, Vienna, 1930 (English translation, with an Introduction, by David Rynin, *Problems of Ethics*, New York, Dover, 1939; republ., *ibid.*, 1962).
SCHMITT, Carl, *Der Begriff des Politischen* (1st publ., *Archiv für Sozialwissenschaft und Sozialpolitik*, Vol. 58, 1927; reproduced in *Probleme der Demokratie*, Berlin, Politische Wissenschaft, 1928; republ. in book form, 1932, 1933, etc.), followed by *Theorie des Partisanen: Zwischenbemerkung zum Begriff des Politischen*, Berlin, Duncker & Humblot, 1963 (French translation by Marie-Louise Steinhauser, *La notion de politique*, followed by *Théorie du partisan*, Preface by Julien FREUND, *op. cit.*, Paris, Calmann-Lévy, 1972, 'Liberté de l'esprit' ser.).
SCHNEIDER, Helmut, *Das Verhältnis von System und Geschichte der Philosophie als Methodenproblem*, Witterschlick bei Bonn, Dissertationsdruckerei R. Schwarzfeld, 1968.
SCHNEIDER, Herbert W., *Three Dimensions of Public Morality*, Bloomington, Indiana University Press, 1956.
—, *Morals for Mankind* (The Paul Anthony Brick Lectures), Columbia, University of Missouri Press, 1960.
SCHOLZ, H., 'Der anselmische Gottesbeweis', 1950-51; reproduced in *Mathesis Universalis*, Basel, Schwabe Verlag, 1961.
SCHULZ, Walter, *Der Gott der neuzeitlichen Metaphysik*, Pfullingen, Neske, 1957.
SCHUTZ, Alfred, *Der sinnhafte Aufbau der sozialen Welt: eine Einleitung in die verstehende Soziologie* (1932), new edn., Vienna, 1960 (English translation by George Walsh & Frederick Lehnert, *The Phenomenology of the Social World*, Evanston (Ill.), Northwestern University Press, 1967, 'Studies in Phenomenology and Existential Philosophy').
—, *Collected Papers*, The Hague, Martinus Nijhoff, 3 vols., 'Phaenomenologica'

ser., Nos. 11, 15 and 22: Vol. I. *The Problem of Social Reality*, edited by Maurice Natanson, 1962; Vol. II. *Studies in Social Reality*, edited by Arvid Brodersen, 1964; Vol. III. *Studies in Phenomenological Philosophy*, edited by I. Schutz, 1970.

SCHWARZ, Baldwin, *Über das innere Prinzip der Periodisierung der Philosophiegeschichte*, Salzburg-Munich, Pustet, 1966, Salzburger Universitätsreden, No. 7.

SCIACCA, Michele Frederico, *Il problema de Dio e della religione nella filosofia attuale*, Milan, Marzorati, 1964; 5th edn., *ibid.*, 1968.

—, *La libertà e il tempo*, 2nd edn., Milan, Marzorati, 1965.

SCOTT, Marvin B. & LYMAN, Stanford M., 'Accounts', *American Sociological Review* XXXIII (1), Feb. 1968, pp. 46-62; reproduced in Stanford M. LYMAN & Marvin B. SCOTT, *Sociology of the Absurd*, New York, Appleton-Century-Crofts, 1970.

SCOTT, Nathan, *The Wild Prayer of Longing: Poetry and the Sacred*, New Haven (Conn.), Yale University Press, 1971.

SCRIVEN, Michael, 'Explanation and prediction in evolutionary theory', *Science*, No. 130, 1959, pp. 477 sqq.

—, 'Truisms as the grounds for historical explanation', in GARDINER (ed.), *Theories of History, op. cit.* (1959-60).

—, 'The key property of physical laws – inaccuracy', in FEIGL & MAXWELL (eds.), *Current Issues in the Philosophy of Science, op. cit.* (1961), pp. 91 sqq.
[See also *sub Minnesota Studies in the Philosophy of Science.*]

SEARLE, John R., 'Meaning and speech-acts', *The Philosophical Review* LXXI (4), Oct. 1962, pp. 423-432.

—, 'How to derive "Ought" from "Is"', *The Philosophical Review* LXXIII (1), Jan. 1964, pp. 43-58.

—, 'Austin on locutionary and illocutionary acts', *The Philosophical Review* LXXVII (4), Oct. 1968, pp. 405-424.

—, *Speech Acts. An Essay in the Philosophy of Language*, Cambridge, Cambridge University Press, 1969 (French translation by Hélène Pauchard, Preface by Oswald Ducrot, *Les actes de langage. Essai de philosophie du langage*, Paris, Hermann, 1972, 'Savoir' ser.).

SÉBAG, Lucien, *Marxisme et structuralisme*, Paris, Payot, 1964.

SEBBA, Gregor, 'What is "history of philosophy"?', *Journal of the History of Philosophy* (Berkeley-Los Angeles), Vol. VIII, 1970, pp. 251-262.

Sein und Ethos. Untersuchungen zur Grundlegung der Ethik (collection of essays), Mainz, Matthias Grünewald Verlag, 1963, Walberberger Studien, Vol. I.

SELLARS, Wilfrid S., 'Time and the world order', in FEIGL & MAXWELL (eds.), *Scientific Explanation, Space and Time* = Vol. III of *Minnesota Studies in the Philosophy of Science, op. cit.* (1962).

—, *Science, Perception and Reality*, London, Routledge / New York, Humanities Press, 1963.

—, 'The identity approach to the mind-body problem', *The Review of Metaphysics* XVIII (3/71), Mar. 1965, pp. 430-451.
[See also FEIGL & SELLARS (eds.).]

SEN, Amartya K., 'The nature and classes of prescriptive judgments', *The Philosophical Quarterly* XVII (66), Jan. 1967, pp. 46-62.

SESONSKE, Alexander, *Value and Obligation. The Foundations of an Empiricist Ethical Theory*, Berkeley-Los Angeles, University of California Press, 1957; republ., New York, Oxford University Press, 1964.

SÈVE, Lucien, *Marxisme et théorie de la personnalité*, Paris, Editions sociales, 1969; new edn., with addition of a Postface, *ibid.*, 1972.

SHAFFER, J., 'Recent work on the mind-body problem', *American Philosophical Quarterly*, 1965, No. 2, pp. 81-104.

SHARP, Gene, *The Politics of Nonviolent Action*, Philadelphia, Pilgrim Press, 1971.

SHOEMAKER, Sidney, *Self-Knowledge and Self-Identity*, Ithaca (N.Y.), Cornell University Press, 1963.

SIKORA, Joseph John, *The Scientific Knowledge of Physical Nature. An Essay on the Distinction between the Philosophy of Nature and Physical Science*, Tournai (Belgium), Desclée, 1967, 'Essais pour notre temps', 2, Section de Philosophie, No. 2.

SINGER, Marcus George, *Generalization in Ethics*, New York, Knopf, 1961.

Sistemnye issledovanija (= Systems Resarch), collective work, Moscow, 1969.

SKINNER, Burrhus Frederic, *Science and Human Behavior*, New York, Macmillan, 1953.

—, 'Critic of psychoanalytic concepts and theories', in FEIGL & SCRIVEN (eds.), *The Foundations of Science and the Concepts of Psychology and Psychoanalysis* = Vol. I of *Minnesota Studies in the Philosophy of Science, op. cit.* (1956), pp. 77-87.

—, *Verbal Behavior*, New York, Appleton-Century-Crofts, 1957.

SKLAR, Lawrence, 'Types of inter-theoretical reduction', *British Journal for the Philosophy of Science* XVIII (2), 1967-68, pp. 109-124.

ŠKLOVSKIJ, I. S., *Vselennaja, žizn', razum* (= The Universe, Life, and Reason), Moscow, 1962.

SKOLIMOWSKI, Henryk, 'The structure of thinking in technology', *Technology and Culture*, Vol. VII (1966), pp. 371-383.

—, 'Technology and philosophy', in KLIBANSKY (ed.), *Contemporary Philosophy / La philosophie contemporaine*: Vol. II, *op. cit.* (1968), pp. 426-437.

SMART, Harold R., *Philosophy and Its History*, La Salle (Ill.), Open Court, 1963.

SMART, John J. C., *An Outline of a System of Utilitarian Ethics*, Melbourne, Melbourne University Press, 1961; another edn., Cambridge, Cambridge University Press (on behalf of the University of Adelaide), 1961.

—, *Philosophy and Scientific Realism*, London, Routledge / New York, Humanities Press, 1963.

—, Article 'Space' in EDWARDS (ed.), *The Encyclopedia of Philosophy, op. cit.* (1967), Vol. 7.

—, *Between Science and Philosophy*, New York, Random House, 1968.

SMART, John J. C. (ed.), *Problems of Space and Time*, London, Collier / New York, Macmillan, 1964.

SMITH, John E., *Experience and God*, London, Oxford University Press, 1958.

SOMERVILLE, John, *The Philosophy of Peace*, New York, Gaer, 1949.

—, 'Marxism and contemporary humanism', in 'Marx und die Philosophie der Gegenwart' (Colloquium I, XIVth International Congress of Philosophy, Vienna, 1968), *op. cit.* (1968), pp. 116-121.

SOSA, Ernest, 'The semantic of imperatives', *American Philosophical Quarterly* IV (1), Jan. 1967, pp. 57-64.

SOURIAU, Étienne, *L'instauration philosophique*, Paris, Alcan, 1939, Bibliothèque de Philosophie contemporaine.

SOURIAU, Michel, *Ethique transréelle*, Paris, Minuit, 1961.

SPASSOV, Dobrin, 'Philosophy of linguistics *versus* linguistic philosophy' (summary), in 'Sprachphilosophie' (Section III, XIVth International Congress of Philosophy, Vienna, 1968), *op. cit.* (1969), pp. 401-402.

'Sprachphilosophie' / 'Philosophy of Language' / 'Philosophie du langage' = Sektion III, in *Akten des XIV. Internationalen Kongresses für Philosophie, Wien, 2.-9. September 1968, op. cit.*, Vol. III, 1969, pp. 391-504.

ŠTAERMAN, E. M., 'K probleme strukturnogo analiza v istorii' (= On the problem of structural analysis in history), *Voprosy istorii*, 1968, No. 6.

STEFANOV, V., *Teorija i metod obščestvoznanija* (= Theory and Method of the Social Sciences) (translated from the Bulgarian), Moscow, 1967.

STEIN, Edith, *Endliches und Ewiges Sein. Versuch eines Aufstieges zum Sinn des Seins* = *Werke*, edited by Lucy Gelber & Romaeus Leuven, Vol. II, Louvain, Nauwelaerts / Freiburg im Breisgau, Herder, 1950.

STENIUS, Erik, *Wittgenstein's Tractatus. A Critical Exposition of Its Main Lines of*

Thought, Oxford, Basil Blackwell / Ithaca (N.Y.), Cornell University Press, 1960.

STEVENSON, Charles L., *Ethics and Languages*, New Haven (Conn.), Yale University Press, 1944.

—, *Facts and Values. Studies in Ethical Analysis*, New Haven (Conn.), Yale University Press, 1963.

STONE, Julius, *Legal System and Lawyers' Reasonings*, Stanford (Calif.), Stanford University Press / London, Oxford University Press, 1964.

STOVER, Robert, *The Nature of Historical Thinking*, Chapel Hill, University of North Carolina Press, 1967.

STRASSER, Stephan, *Phänomenologie und Erfahrungswissenschaft vom Menschen. Grundgedanken zu einem neuen Ideal der Wissenschaftlichkeit* (translated by the author from the Dutch original: *Fenomenologie en empirische menskunde*), Berlin, de Gruyter, 1964 (French translation by Arion L. Kelkel, with a Preface by Paul RICOEUR, *Phénoménologie et sciences de l'homme. Vers un nouvel esprit scientifique*, Louvain, Publications Universitaires de Louvain / Paris, Editions Béatrice-Nauwelaerts, 1967, Bibliothèque philosophique de Louvain No. 21 (Institut supérieur de Philosophie)).

—, 'Das Gottesproblem in der Spätphilosophie Edmund Husserls', *Philosophisches Jahrbuch*, 1958, pp. 130-142.

STRAWSON, P. F., 'On referring', *Mind* LIX (235), July 1950, pp. 320-344; reproduced in FLEW (ed.), *Essays on Conceptual Analysis, op. cit.* (1956).

—, *Individuals. An Essay in Descriptive Metaphysics*, London, Methuen, 1959; republ., New York, Doubleday, 1963; London, Methuen, 1964 (French translation by A. Shalom & P. Drong, *Les individus. Essai de métaphysique descriptive*, Paris, Seuil, 1973, 'L'ordre philosophique' ser.).

—, 'Social morality and individual ideal', *Philosophy* XXXVI (1), 1961, pp. 1-17.

—, *Freedom and Resentment* (Hertz Trust Lectures), London, British Academy, 1962.

—, *The Bounds of Sense*, London, Methuen, 1966.

—, 'Introduction', in STRAWSON (ed.), *Philosophical Logic, op. cit.* (1967), pp. 1-16.

STRAWSON, P. F. (ed.), *Philosophical Logic*, London, Oxford University Press, 1967, Oxford Readings in Philosophy.

—, *Studies in the Philosophy of Thought and Action*, London, Oxford University Press, 1968.

STRENG, Frederick J., 'The ethics of moral coercion: Gandhi and political revolution', *Philosophy East and West* XXIII (3), July 1973, pp. 283-290.

STRETTON, Hugh, *The Political Sciences*, London, Routledge, 1969.

STROUD, Barry, 'Transcendental arguments', *The Journal of Philosophy* LXV (9), 2 May 1968, pp. 241-256.

Structuralisme et marxisme = *La Pensée* (Paris), No. 135, Sept.-Oct. 1967.

Struktura i forma materii (= The Structure and Forms of Matter), collective work, Moscow, 1967.

Strukturno-tipologičeskie issledovanija (= Structuro-typological Research), collection, Moscow, 1962.

SUMMERS, Robert S., 'The New Analytical Jurists', *New York University Review*, Vol. XLI, 1966, pp. 861 sqq.

SUMMERS, Robert S. (ed.), *Essays in Legal Philosophy*, Oxford, Basil Blackwell, 1968.

—, *More Essays in Legal Philosophy: General Assessments of Legal Philosophies*, Berkeley-Los Angeles, University of California Press, 1971.

SUNDÉN, Hjalmar, *Die Religion und die Rollen. Eine psychologische Untersuchung der Frömmigkeit* (German translation by Hermann Müller & Suzanne Öhman, approved by the author), Berlin, Töpelmann, 1966 (Swedish original, *Religionen och rollerna*).

SUPPES, Patrick C.: see DAVIDSON, MCKINSEY & SUPPES; HINTIKKA & SUPPES (eds.);

LUCE & SUPPES; NAGEL, SUPPES & TARSKI (eds.).
SVEČNIKOV, G. A., *Kategorija pričinnosti v fisike* (= The Category of Causality in Physics), Moscow, 1961.
Symposium der Philosophie Historiker (Moskau, 1967), Berlin, Sowjetw. Gesellschaftw. Beiträge, 1968.
'Symposium sobre la argumentación filosófica', in *XIIIo Congreso Internacional de Filosofía*, Mexico City, 1963.
SZCZEPAŃSKI, Jan, *Elementarne pojęcia socjologii* (= Elementary Notions of Sociology), Warsaw, Państwowe Wydawnictwo Naukowe, 1963 (Russian translation, ŠČEPAN'SKIJ, Ja., *Èlementarnye ponjatija sociologii*, Moscow, 1969).
SZONDI, Leopold, *Triebpathologie*, Vol. II: *Ich-Analyse. Die Grundlage zur Vereinigung der Tiefenpsychologie*, Stuttgart, Hans Huber, 1956.

TAMMELO, Ilmar, 'Sketch for a symbolic juristic logic', *Journal of Legal Education* VIII (3), 1955-65, pp. 277-306.
TANABE, Hajime, 'The pressing need of the philosophy of politics', Tokyo, 1946.
TARSKI, Alfred, 'Der Wahrheitsbegriff in den formalisierten Sprachen', *Studia Philosophica* I, 1935, pp. 261-405 (Polish original published in 1933).
—, *Logic, Semantics, Metamathematics. Papers from 1923 to 1938*, translated by J. H. Woodger, Oxford, Clarendon Press, 1969.
[See also ADDISON, HENKIN & TARSKI (eds.); NAGEL, SUPPES & TARSKI (eds.).]
TAVANEC, P. V. (ed.), *Problemy logiki naučnogo poznanija* (= Problems of the Logic of Scientific Knowledge), Moscow, Nauka 1964, Institute of Philosophy of the Academy of Sciences of the U.S.S.R.
TAYLOR, Charles, *The Explanation of Behaviour*, London, Routledge / New York, Humanities Press, 1964.
—, 'Marxism and empiricism', in WILLIAMS & MONTEFIORE (eds.), *British Analytical Philosophy, op. cit.* (1966).
—, 'Neutrality in political science', in LASLETT & RUNCIMAN (eds.), *Philosophy, Politics and Society*, 3rd ser., *op. cit.* (1967).
TAYLOR, Paul Warren, *Normative Discourse*, Englewood Cliffs (N.J.), Prentice-Hall, 1961.
TAYLOR, Richard, *Action and Purpose*, Englewood Cliffs (N.J.), Prentice-Hall, 1966.
TEILHARD DE CHARDIN, Pierre, *Œuvres* (posthumous publication), Paris, Seuil, 1955-73, 11 vols., in particular: Vol. I. *Le phénomène humain*, 1955; Vol. V. *L'avenir de l'homme*, 1959; Vol. VI. *L'énergie humaine*, 1962; Vol. VIII. *La place de l'homme dans la nature. Le groupe zoologique humain*, 1963 (published previously under the title *Le groupe zoologique humain*, Paris, Albin Michel, 1956, 'Les savants et le monde' ser.); and Vol. XI. *Les directions de l'avenir*, 1973 (English translations: *The Phenomenon of Man*, with an Introduction by Julian HUXLEY, London, Collins / New York, Harper, 1959; *The Future of Man, ibid.*, 1964; *Human Energy*, London-New York, Collins, Harcourt, Brace & World, 1969; *Man's Place in Nature*, New York, Harper, 1966; etc.).
TEIXEIRA, Livio, 'Filosofia e história da filosofia', *Facultade de Letras* (Lisbon), 1962, No. 6, pp. 50-67.
TEMPELS, Placide, *La philosophie bantoue*, translated from the Dutch, Paris, Présence africaine, 1949.
La Théorie de l'argumentation. Perspectives et applications = Logique et analyse (Louvain), 1963, No. 6, pp. 1-614.
THEOBALD, David W., 'Models and methods', *Philosophy*, No. 39, 1964, pp. 260-267.
THEUNISSEN, Michael, *Der Andere. Studien zur Sozialontologie der Gegenwart*, Berlin, de Gruyter, 1965.
THÉVENAZ, Pierre, *L'homme et sa raison* (I. *Raison et conscience de soi*; II. *Raison et histoire*), with a Preface by Paul RICOEUR, Neuchâtel, La Baconnière, 1956,

2 vols., 'Etre et penser' ser.

THIEME, H., 'L'apport du droit naturel au droit positif', in *Rapports généraux au VIIe Congrès international de Droit comparé, Uppsala, 1966*, Stockholm, 1968, pp. 74 sqq.

TIELSCH, Elfriede, *Kierkegaards Glaube. Die materiale und formale Glaubenslehre Sören Kierkegaards im Verhältnis zur klassischen Wertlehre*, Göttingen, Vandenhoeck & Ruprecht, 1964.

TILLICH, Paul, *Systematic Theology*, Chicago, University of Chicago Press, 1951-57-63, 3 vols.

TJUHTIN, V. S., *O prirode obraza* (= On the Nature of the Image), Moscow, 1963.

TOKAREV, S. A., 'Načalo frejdistskogo napravlenija v ètnografii i istorii religii' (= The beginnings of the Freudian orientation in ethnography and the history of religions), in PORŠNEV & ANCIFEROVA (eds.), *Istorija i psihologija, op. cit.* (1970).

TONELLI, Giorgio, 'Qu'est-ce que l'histoire de la philosophie?', *Revue philosophique de la France et de l'Etranger*, Vol. LXXXVII, 1962, pp. 290-306.

TONQUÉDEC, Joseph de, *La philosophie de la nature*, Paris, Lethielleux, 1956-59, 3 vols., 'Les principes de la philosophie thomiste' ser.

TOPITSCH, Ernst, *Vom Ursprung und Ende der Metaphysik. Eine Studie zur Weltanschauungskritik*, Vienna, Springer, 1958.

——, *Sozialphilosophie zwischen Ideologie und Wissenschaft*, Neuwied, Luchterhand, 1961; 2nd edn., *ibid.*, 1966.

——, *Mythos – Philosophie – Politik. Zur Naturgeschichte der Illusion*, 2nd edn., Freiburg in Breisgau, Rombach, 1969.

TOPITSCH, Ernst (ed.), *Logik der Sozialwissenschaften*, Cologne-Berlin, Kiepenheuer & Witsch, 1965.

[See also *sub* KELSEN, *Aufsätze zur Ideologiekritik*.]

TORRANCE, John, 'Rationality and the structural analysis of myth', *Archives européennes de Sociologie / European Journal of Sociology* VIII (2), 1967 (entitled: *Sympathy for Alien Concepts*), pp. 272-281.

TORRANCE, Thomas F., *Theological Science*, London, Oxford University Press, 1969.

TOULMIN, Stephen E., *An Examination of the Place of Reason in Ethics*, Cambridge, Cambridge University Presss, 1950.

——, *Philosophy of Science*, London, Hutchinson, 1953.

——, *The Uses of Argument*, Cambridge, Cambridge University Press, 1958; republ., *ibid.*, 1964.

——, *Human Understanding*: Vol. I. *General Introduction and Part One: Concepts. Their Collective Use and Evolution*, Oxford, Clarendon Press / Princeton (N.J.), Princeton University Press, 1972 (2 more volumes announced for publication).

TOURAINE, Alain, *Sociologie de l'action*, Paris, Seuil, 1965.

——, *Production de la société*, Paris, Seuil, 1973.

TOYNBEE, Arnold, *An Historian's Approach to Religion* (based on Gifford Lectures), London-New York-Toronto, Oxford University Press, 1956.

TRESMONTANT, Claude, *Introduction à la pensée de Teilhard de Chardin*, Paris, Seuil, 1956.

TRUBSTZKOY, Nicolas S., *Grundzüge der Phonologie*, Prague, 1939, Travaux du Cercle linguistique de Prague, No. VII; 3rd edn., Göttingen, 1962 (English translation by Christiane A. Baltaxe, *Principles of Phonology*, Berkeley, University of California Press, 1969).

UËMOV, A. I., *Vešči, svojstva i otnošenija* (= Things, Properties and Relations), Moscow, 1963.

UPADHYAYA, K. N., *Early Buddhism and the Bhagvadgita*, Delhi, Motilal Banarsidass, 1971.

URMSON, J. O., *Philosophical Analysis. Its Development between the Wars*, London, Oxford University Press, 1956.
[See also *sub* AUSTIN, *Philosophical Papers and How to Do Things with Words*.]

VAHANIAN, Gabriel, *The Death of God. The Culture of Our Post-Christian Era*, New York, Braziller, 1961.

VAN BREDA, Hermann Leo & TAMINIAUX, J. (eds.), *Edmund Husserl, 1859-1959* (commemorative collection published on the occasion of the hundredth anniversary of the philosopher's birth), The Hague, Martinus Nijhoff, 1959, 'Phaenomenologica' ser., No. 4.

VAN BUREN, Paul Matthews, *The Secular Meaning of the Gospel. Based on an Analysis of Its Language*, New York, Macmillan / London, SCM Press, 1963; republ., London, SCM Press, 1965.

VANDERKERKEN, Libert, *Inleiding tot de fundamentele filosofie*, Antwerp, Antw. Uitgeverij 'De Nederlandsche Boekh.', 1970, 'Filosofie en kultuur' ser., No. 9.

VAN DER LEEUW, Gerardus: see LEEUW, Gerardus van der.

VAN HARVEY, Austin, *The Historian and the Believer. The Morality of Historical Knowledge and Christian Belief*, New York, Macmillan, 1965.

VAN MELSEN, Andreas G. M.: see MELSEN, Andreas G. M. van.

VAN ORMAN QUINE, Willard: see QUINE, Willard van Orman.

VAN PEURSEN, C. A.: see PEURSEN, C. A. van.

VANSINA, Dirck F., 'Esquisse, orientation et signification de l'entreprise philosophique de Paul Ricoeur', *Revue de Métaphysique et de Morale*, 69th Year, No. 2, 1964, Apr.-June, pp. 179-208 and No. 3, July-Sept., pp. 305-321.

VENDLER, Zeno, *Linguistics in Philosophy*, Ithaca (N.Y.), Cornell University Press, 1967.

VENN, John, *The Logic of Chance. An Essay on the Foundations and Province of the Theory of Probability, with special reference to Its Application to Moral and Social Science*, London / Cambridge, 1866; 2nd edn., *ibid.*, 1876; 3rd edn., London-Cambrige, Macmillan, 1888.

VERDE, Felice M., 'Un debattito intorno a "filosofia" e storia della filosofia', *Sapienza* (Naples), Vol. XIII, 1969, pp. 406-426.

VERGOTE, Antoine, *Psychologie religieuse*, Brussels, Dessart, 1966, 'Psychologie et sciences humaines' ser.

VERNANT, Jean-Pierre, *Mythe et pensée chez les Grecs*, Paris, Maspero, 1965, 'Les textes à l'appui', ser., No. 13; 2nd edn., *ibid.*, 1969.

VESEY, G. N. A. (ed.), *Talk of God*, London, Macmillan, 1969, 'Royal Institute of Philosophy Lectures'.

VEYNE, Paul, *Comment on écrit l'histoire. Essai d'épistémologie*, Paris, Seuil, 1971, 'L'univers historique' ser.

—, 'Contestation de la sociologie', *Diogène*, No. 75, July-Sept. 1971, pp. 3-25 (parallel publication of English version, translated by Brenda Porster Amato, 'A contestation of sociology', *Diogenes*, No. 75, Fall 1971, pp. 1-23).

VIA, Dan Otto, Jr., *The Parables. Their Literary and Existential Dimension*, Philadelphia, The Fortress Press, 1967.

VIANO, Carlo Augusto, 'Storia della filosofia e sociologia', *Rivista di Filosofia* (Turin), Vol. LVII, 1966, pp. 251-283.

VIDAL MUÑOZ, Santiago, 'O método comparativo na investigação da história das idéias' (translated from the Spanish by L. W. Vita), *Revista brasileira de filosofia*, Vol. XIV, 1964, pp. 348-355.

VIDLER, Alec R. (ed.), *Soundings. Essays concerning Christian Understanding*, Cambridge, Cambridge University Press, 1962; republ., *ibid.*, 1967.

VIEHWEG, Th., *Topik und Jurisprudenz. Ein Beitrag zur rechtswissenschaftlichen Grundlagenforschung*, Munich, Beck, 1953; 3rd edn., *ibid.*, 1965.

VIET, Jean, *Les méthodes structuralistes dans les sciences sociales*, Paris-The Hague, Mouton, 1965, Maison des Sciences de l'Homme, Paris, Service d'Echanges

scientifiques, Publications, Série D, Méthodes et Techniques, I.
VILLEY, Michel, *La formation de la pensée juridique moderne (Cours d'histoire de la philosophie du Droit, 1961-1966)*, Paris, Montchrestien, 1968.
—, *Seize essais de philosophie du droit, dont un sur la crise universitaire*, Paris, Dalloz, 1969.
VIVEKANANDA, Swami, *The Complete Works*, Calcutta, Advaita Ashram, 1955.
VOLLENHOVEN (D. H.), Th., 'De consequent probleemhistorische methode', *Philosophia reformata* (Amsterdam), Vol. XXVI, 1961, pp. 1-34.
VON MISES, Richard: see MISES, Richard von.
VON NEUMANN, John: see NEUMANN, John von.
VON WEIZSÄCKER, Karl Friedrich: see WEIZSÄCKER, Karl Friedrich von.
VON WEIZSÄCKER, Viktor, see WEIZSÄCKER, Viktor von.
VON WRIGHT, Georg Henrik: see WRIGHT, Georg Henrik von.
Voprosy metodologii istoričeskoj nauki (= Questions of Methodology in Historical Science), collection, Moscow, Moscow State Institute of Historical Archives, 1967, Vol. 25.
VRANICKI, Pedrag, 'Die Notwendigkeit verschiedener Varianten in der marxistischen Philosophie – Thesen' (= The necessity of diverse variants in Marxist philosophy – Theses), in 'Marx und die Philosophie der Gegenwart' (Colloquium, XIVth International Congress of Philosophy, Vienna, 1968), *op. cit.* (1968), pp. 139-141.
VUILLEMIN, Jules, 'Sur les propriétés formelles et matérielles de l'ordre cartésien des raisons', in *Etudes sur l'histoire de la philosophie en hommage à Martial Guéroult, op. cit.* (1964), pp. 43-58.
—, *De la logique à la théologie. Cinq études sur Aristote*, Paris, Flammarion, 1967.
—, *Le Dieu d'Anselme et les apparences de la raison*, Paris, Aubier-Montaigne, 1971.
VYGOTSKIJ, L. S., *Myšlenie i reč'* (= Thought and Discourse), Moscow, 1934.

WAHL, Jean, *Existence humaine et transcendance*, Neuchâtel, La Baconnière, 1944.
—, *Poésie, pensée, perception* (collection of essays), Paris, Calmann-Lévy, 1948.
—, *Traité de métaphysique*, Paris, Payot, 1953.
—, *Vers la fin de l'ontologie. Etude sur l'Introduction dans la métaphysique par Heidegger*, Paris, S.E.D.E.S., 1956.
—, *L'expérience métaphysique*, Paris, Flammarion, 1965, Nouvelle Bibliothèque scientifique.
WAISMANN, Friedrich, 'How I see philosophy', in H. D. LEWIS (ed.), *Contemporary British Philosophy*, 3rd ser., *op. cit.* (1956).
—, 'The decline and fall of causality', in Alistair Cameron CROMBIE (ed.), *Turning Points in Physics*, New York, Harper, 1961 (also in German version, 'Niedergang und Sturz der Kausalität', in Dirk Ter HAAR & A. C. CROMBIE (eds.), *Wendepunkte in der Physik. Eine Vorlesungsreihe der Universität Oxford*, Braunschweig, Vieweg, 1963, pp. 67 sqq.).
—, *The Principles of Linguistic Philosophy*, edited by Rom Harré, London, Macmillan, 1965.
WALD, Abraham, *Statistical Decisions Functions*, New York, Wiley, 1950, 'Publications in Statistics' ser.; 6th edn., *ibid.*, 1966.
WALSH, William H., *An Introduction to Philosophy of History*, London, Hutchinson, 1951; republished as paperback under the title *The Philosophy of History*, New York, Harper, 1960.
—, 'The limits of scientific history', in James HOGAN (ed.), *Historical Studies*, III, New York, Hillary House, 1961; reproduced in DRAY (ed.), *Philosophy of History, op. cit.* (1964).
—, *Metaphysics*, London, Hutchinson, 1963.
—, *Hegelian Ethics*, London, Macmillan, 1969.
WARD, Keith, *Ethics and Christianity*, London, Allen & Unwin, 1970, 'Muirhead

Library of Philosophy.
WARNOCK, G. J. (Geoffrey), *Berkeley*, Harmondsworth (Middx.)-London, Penguin Books, 1953.
—, *English Philosophy since 1900*, London, Oxford University Press, 1958.
—, *Contemporary Moral Philosophy*, London, Macmillan, 1967.
[See also *sub* AUSTIN, *Philosophical Papers* and *Sense and Sensibilia.*]
WARNOCK, Mary, *Ethics since 1900*, London, Oxford University Presss, 1960, 'The Home University Library'.
—, *The Philosophy of Sartre*, London, Hutchinson, 1965.
[See also *sub* SARTRE, *Esquisse d'une théorie phénoménologique des émotions.*]
WATANABE, Satosi, *Knowing and Guessing. A Quantitative Study of Inference and Information*, New York-London-Sidney-Toronto, Wiley, 1969.
WATKINS, J. W. N., 'Ideal types and historical explanation', *The British Journal for the Philosophy of Science* III (9), Feb. 1952.
—, 'Historical explanation in the social sciences', *The British Journal for the Philosophy of Science* VIII (30), Aug. 1957, pp. 104-117.
WATSON, R., 'Rules of inference in Stephen Toulmin's *The Place of Reason in Ethics*', *Theoria*, Vol. XXIX, 1963, pp. 312-315.
WEIL, Éric, *Hegel et l'Etat*, Paris, Vrin, 1950.
—, *Logique de la philosophie*, Paris, Vrin, 1950.
—, *Philosophie politique*, Paris, Vrin, 1956.
—, *Philosophie morale*, Paris, Vrin, 1961.
—, 'La morale de l'individu et la politique', *Tijdschrift voor Filosofie*, 27th Year, No. 3, Sept. 1965, pp. 476-490; reproduced in WEIL, *Essais et conférences*, Vol. I, *Philosophie, op. cit.* (1970), Chap. VII, pp. 159-174.
—, *Essais et conférences* (collection of previously published papers), Paris, Plon, 2 vols., 'Recherches en sciences humaines' ser., Nos. 33 and 34: Vol. I. *Philosophie*, 1970; Vol. II. *Politique*, 1971.
WEIZSÄCKER, Karl-Friedrich von, *Zum Weltbild der Physik*, Stuttgart, Hirzel, 1958.
WEIZSÄCKER, Viktor von, *Am Anfang schuf Gott Himmel und Erde: Grundfragen der Naturphilosophie*, 6th edn., Göttingen, Vandenhoeck & Ruprecht, 1963.
WELDON, Thomas D., *The Vocabulary of Politics*, Harmondsworth (Middx.), Penguin Books, 1953; Johnson Reprint, 1971.
WELTE, Bernhard, 'Der philosophische Glaube bei Karl Jaspers und die Möglichkeit seiner Deutung durch die thomistische Philosophie', *Symposion. Jahrbuch für Philosophie* (Freiburg im Breisgau), Vol. II, 1949.
—, *Auf der Spur des Ewigen*, Freiburg im Breisgau, Herder, 1965.
WEYL, Hermann, *Philosophie der Mathematik und Naturwissenschaft* (= Fascicules 4 and 5 of A. BÄUMLER & M. SCHRÖTER (eds.), *Handbuch der Philosophie*): Pt. I, pp. 1-64; Pt. II, pp. 65-162; Munich-Berlin, Oldenbourg, 1926 (English version, amplified, *Philosophy of Mathematics and Natural Science*, rev. and augm. edn. based on a translation by Olaf Helmer, Princeton (N.J.), Princeton University Press, 1949.)
WHITE, Alan R. (ed.), *The Philosophy of Action*, London, Oxford University Press, 1968, 'Readings in Philosophy'.
WHITE, Morton G., *Foundations of Historical Knowledge*, New York, Harper & Row, 1965.
WHITEHEAD, Alfred North, *Process and Reality. An Essay in Cosmology* (Gifford Lectures delivered at the University of Edinburgh during the session 1927-28), New York, The Macmillan Co., 1929.
[See also RUSSELL & WHITEHEAD.]
WHITROW, G. J., *The Natural Philosophy of Time*, London, Nelson, 1961.
WHORF, Benjamin Lee, *Language, Thought and Reality*, Cambridge (Mass.), The M.I.T. Press, 1956 (French translation by Claude Carme, *Linguistique et anthropologie*, Paris, Denoël-Gonthier, 1971, Bibliothèque 'Médiations').
WIATR, Jerzy J., *Szkice o materializmie historycznym i socjologii* (= Essay on His-

torical Materialism and Sociology), Warsaw, Ksiazka i Wiedza, 1967, 2 vols.

WIEACKER, Franz, *Privatrechtsgeschichte der Neuzeit* (= History of Private Law in Modern Times), 2nd edn., Göttingen, Vandenhoeck & Ruprecht, 1967.

WIGGINS, D., *Identity and Spatio-temporal Continuity*, Oxford, Basil Blackwell, 1967.

WILD, John, *Existence and the World of Freedom*, Englewood Cliffs (N.J.), Prentice-Hall, 1963.

WILLIAMS, Bernard [B. A. O.], 'Tertullian's paradox', in FLEW & MACINTYRE (eds.), *New Essays in Philosophical Theology, op. cit.* (1953).

—, 'Imperative inference', *Analysis*, Supp. XXIII, 1963, pp. 30-36.

WILLIAMS, Bernard & MONTEFIORE, Alan (eds.), *British Analytical Philosophy*, London, Routledge / New York, Humanities Press, 1966.

WILSON, Bryan R. (ed.), *Rationality*, Oxford, Basil Blackwell / New York, Harper & Row, 1970; republished in paperback edn., New York, Harper & Row, 1971, Torchbooks.

WINCH, Peter, *The Idea of a Social Science and Its Relations to Philosophy*, London, Routledge / New York, Humanities Press, 1958; extracts republished in abridged form as 'The idea of a social science' in WILSON (ed.), *Rationality, op. cit.* (1970), pp. 1-17.

—, 'Understanding a primitive society', *American Philosophical Quarterly* I (4), Oct. 1964, pp. 307-324; reproduced in WILSON (ed.), *Rationality, op. cit.* (1970), pp. 78-111.

WINCH, Peter (ed.), *Studies in the Philosophy of Wittgenstein*, London, Routledge, 1969.

WISDOM, John, 'Gods', *Proceedings of the Aristotelian Society*, 1944; reproduced in FLEW (ed.), *Logic and Language, op. cit.* (1951) and in J. WISDOM, *Philosophy and Psychoanalysis*, Oxford, Basil Blackwell, 1953.

WITTGENSTEIN, Ludwig, 'Logisch-philosophische Abhandlung – mit Vorwort von B. RUSSELL', *Annalen der Naturphilosophie* (Leipzig), Vol. XIV, 1921, pp. 185-262; reproduced with an English translation by C. K. Ogden & F. P. Ramsey, *Tractatus logico-philosophicus* (with an Introduction by Bertrand RUSSELL), London, Kegan Paul, Trench, Trubner & Co., 1922, International Library of Psychology, Philosophy and Scientific Method; new edn., *Tractatus logico-philosophicus* (The German Text of Ludwig WITTGENSTEIN's *Logisch-philosophische Abhandlung*, with a new Translation by D. F. Pears & B. F. McGuinness and with the Introduction by Bertrand RUSSELL), London, Routledge / New York, Humanities Press, 1961, International Library of Philosophy and Scientific Method (French translation from the German original by Pierre Klossowski, with Bertrand RUSSELL's Introduction, *Tractatus logico-philosophicus*, followed by *Investigations philosophiques*, Paris, Gallimard, 1961, Bibliothèque des Idées).

—, *Philosophische Untersuchungen / Philosophical Investigations* (written between 1936 and 1949), German text and English translation by G. E. M. Anscombe, Oxford, Basil Blackwell / New York, Macmillan, 1952-53; 2nd edn., rev., Oxford, Basil Blackwell, 1958 (French translation from the German original by Pierre Klossowski, *Investigations philosophiques*, with the *Tractatus logico-philosophicus*, Paris, Gallimard, 1961).

—, *Remarks on the Foundations of Mathematics*, English translation by G. E. M. Anscombe, Oxford, Basil Blackwell, 1956.

—, *The Blue and Brown Books. Preliminary Studies for the Philosophical Investigations*, Oxford, Basil Blackwell, 1958; 2nd edn., 1964 (French translation by Guy Durand, with a Preface by Jean Wahl, *Le cahier bleu et le cahier brun*, followed by 'Ludwig Wittgenstein' by Norman Malcolm, Paris, Gallimard, 1965, 'Les Essais' ser.).

—, *Schriften*, Vols. I-III, Frankfurt am Main, Suhrkamp, 1961-66.

—, 'Lecture on Ethics' (delivered at Cambridge in 1929 or 1930), *The Philosophical*

Review LXXIV (1), Jan. 1965 (French translation by J. Fauve, 'Conférence sur l'éthique', with *Leçons et conversations*, Paris, Gallimard, 1971).

—, *Lectures and Conversations on Aesthetics, Psychology and Religious Belief*, edited by C. Barrett, Oxford, Basil Blackwell, 1966 (French translation by J. Fauve, *Leçons et conversations*, followed by 'Conférence sur l'éthique', Paris, Gallimard, 1971, 'Les Essais' ser.).

—, *Zettel*, German text and English translation, edited by G. E. M. Anscombe & G. H. von Wright, Oxford, Basil Blackwell, 1967 (French translation by J. Fauve, *Fiches*, Paris, Gallimard, 1971, Bibliothèque des Idées).

WOLFF, Robert Paul, *The Poverty of Liberalism*, Boston, Beacon Press, 1968.

WOLIN, Sheldon S., *Politics and Vision. Continuity and Innovation in Western Political Thought*, Boston, Little, Brown and Co., 1960.
[See also SCHAAR & WOLIN.]

WOLLHEIM, Richard, *F. H. Bradley*, Harmondsworth (Middx.)-London, Penguin Books, 1959.

WRIGHT, Georg Henrik von, 'Deontic logic', *Mind*, N.S., LX (237), Jan. 1951, pp. 1-15; reproduced in VON WRIGHT, *Logical Studies*, London, Routledge, 1957, pp. 58-74.

—, *An Essay on Modal Logic*, Amsterdam, North Holland Publ. Co., 1951, 'Studies in Logic and the Foundations of Mathematics'.

—, 'A note on deontic logic and derived obligation', *Mind*, N.S., LXV (260), Oct. 1956, pp. 507-509.

—, *The Logic of Preference. An Essay*, Edinburgh, Edinburgh University Press / Chicago, Aldine, 1963.

—, *Norm and Action*, London, Routledge, 1963.

—, 'Practical inference', *The Philosophical Review* LXXII (2), Apr. 1963, pp. 159-179.

—, *The Varieties of Goodness*, London, Routledge / New York, Humanities Press, 1963.

—, 'Deontic logics', *American Philosophical Quarterly* IV (2), Apr. 1967, pp. 136-143.

—, *An Essay in Deontic Logic and the General Theory of Action, with a Bibliography of Deontic and Imperative Logic*, Helsinki, Academic Bookstore / Amsterdam, North Holland Publ. Co., 1968, Acta Philosophica Fennica, Fasc. XXI.

—, 'The logic of practical discourse', in KLIBANSKY (ed.), *Contemporary Philosophy / La philosophie contemporaine*, Vol. I, *op. cit.* (1968), pp. 141-167.

—, *Explanation and Understanding*, edited by Max Black, Ithaca (N.Y.), Cornell University Press / London, Routledge, 1971.

XIRAU, Ramon, *Introducción a la historia de la filosofía*, Mexico City, UNAM, Facultad de Filosofía y Letras.

YOLTON, J. W., *John Locke and the Way of Ideas*, London, Oxford University Press, 1956.

ZAHRNT, Heinz (ed.), *Gespräch über Gott. Die protestantische Theologie im 20. Jahrhundert. Ein Textbuch*, Munich, Piper, 1968.

ZIEMBIŃSKI, Z., *Logiczne podstawy prawoznawstwa* (= The Logical Foundations of Jurisprudence), Warsaw, 1966.

ZINK, Sidney, *The Concepts of Ethics*, London, Macmillan / New York, St Martin's Press, 1962.

ZUURDEEG, Willem Frederik, *An Analytical Philosophy of Religion*, New York, Abingdon Press, 1958.

Subject index*

* Established by Monique Couratier (Paris) on the basis of indications from the general rapporteur.